INTEGRATING BUSINESS WITH TECHNOLOGY

By completing the projects in this text, students will be able to demonstrate business knowledge, application software proficiency, and Internet skills. These projects can be used by instructors as learning assessment tools and by students as demonstrations of business, software, and problem-solving skills to future employers. Here are some of the skills and competencies students using this text will be able to demonstrate:

Business Application skills: Use of both business and software skills in real-world business applications. Demonstrates both business knowledge and proficiency in spreadsheet, database, and Web page/blog creation tools.

Internet skills: Ability to use Internet tools to access information, conduct research, or perform online calculations and analysis.

Analytical, writing and presentation skills: Ability to research a specific topic, analyze a problem, think creatively, suggest a solution, and prepare a clear written or oral presentation of the solution, working either individually or with others in a group.

Business Application Skills

BUSINESS SKILLS	SOFTWARE SKILLS	CHAPTER
Finance and Accounting		
Financial statement analysis	Spreadsheet charts	Chapter 2*
	Spreadsheet formulas	Chapter 10
	Spreadsheet downloading and formatting	
Pricing hardware and software	Spreadsheet formulas	Chapter 5
Technology rent vs. buy decision Total Cost of Ownership (TCO) analysis	Spreadsheet formulas	Chapter 5*
Analyzing telecommunications services and costs	Spreadsheet formulas	Chapter 7
Risk assessment	Spreadsheet charts and formulas	Chapter 8
Retirement planning	Spreadsheet formulas and logical functions	Chapter 11
Capital budgeting	Spreadsheet formulas	Chapter 14
		Chapter 14*
Human Resources		
Employee training and skills tracking	Database design Database querying and reporting	Chapter 13*
Job posting database and Web page	Database design Web page design and creation	Chapter 15
Manufacturing and Production		
Analyzing supplier performance and pricing	Spreadsheet date functions Database functions Data filtering	Chapter 2
Inventory management	Importing data into a database Database querying and reporting	Chapter 6
Bill of materials cost sensitivity analysis	Spreadsheet data tables Spreadsheet formulas	Chapter 12*
Sales and Marketing		
Sales trend analysis	Database querying and reporting	Chapter 1

Customer reservation system	Database querying and reporting	Chapter 3
Improving marketing decisions	Spreadsheet pivot tables	Chapter 12
Customer profiling	Database design Database querying and reporting	Chapter 6*
Customer service analysis	Database design Database querying and reporting	Chapter 9
Sales lead and customer analysis	Database design Database querying and reporting	Chapter 13
Blog creation and design	Blog creation tool	Chapter 4

Internet Skills

Using online software tools to calculate shipping costs	Chapter 1
Using online interactive mapping software to plan efficient transportation routes	Chapter 2
Researching product information and evaluating Web sites for auto sales	Chapter 3
Using Internet newsgroups for marketing	Chapter 4
Researching travel costs using online travel sites	Chapter 5
Searching online databases for products and services	Chapter 6
Using Web search engines for business research	Chapter 7
Researching and evaluating business outsourcing services	Chapter 8
Researching and evaluating supply chain management services	Chapter 9
Evaluating e-commerce hosting services	Chapter 10
Using shopping bots to compare product price, features, and availability	Chapter 11
Using online software tools for retirement planning	Chapter 12
Redesigning business processes for Web procurement	Chapter 13
Researching real estate prices	Chapter 14
Researching international markets and pricing	Chapter 15

Analytical, Writing and Presentation Skills*

BUSINESS PROBLEM	CHAPTER
Management analysis of a business	Chapter 1
Value chain and competitive forces analysis Business strategy formulation	Chapter 3
Formulating a corporate privacy policy	Chapter 4
Employee productivity analysis	Chapter 7
Disaster recovery planning	Chapter 8
Locating and evaluating suppliers	Chapter 9
Developing an e-commerce strategy	Chapter 10
Identifying knowledge management opportunities	Chapter 11
Identifying international markets	Chapter 15

*Dirt Bikes Running Case on MyMISLab

Management Information Systems

MANAGING THE DIGITAL FIRM

TWELFTH EDITION

Kenneth C. Laudon
New York University

Jane P. Laudon
Azimuth Information Systems

Prentice Hall

Boston   Columbus   Indianapolis   New York   San Francisco   Upper Saddle River
Amsterdam   Cape Town   Dubai   London   Madrid   Milan   Munich   Paris   Montreal   Toronto
Delhi   Mexico City   Sao Paulo   Sydney   Hong Kong   Seoul   Singapore   Taipei   Tokyo

Library of Congress Cataloging-in-Publication Information is available.

Executive Editor: Bob Horan
Editorial Director: Sally Yagan
Editor in Chief: Eric Svendsen
Product Development Manager: Ashley Santora
Editorial Assistant: Jason Calcano
Editorial Project Manager: Kelly Loftus
Senior Marketing Manager: Anne Fahlgren
Senior Managing Editor: Judy Leale
Senior Production Project Manager: Karalyn Holland
Senior Operations Specialist: Arnold Vila
Operations Specialist: Cathleen Petersen
Senior Art Director: Janet Slowik
Cover Designer: Jill Lehan
Cover Illustration/Photo: Merve Poray/Shutterstock Images
Manager, Rights and Permissions: Hessa Albader
Media Editor: Denise Vaughn
Media Project Manager: Lisa Rinaldi
Composition: Azimuth Interactive, Inc.
Full-Service Project Management: Azimuth Interactive, Inc.
Printer/Binder: RR Donnelley
Typeface: 10.5/13 ITC Veljovic Std Book

Credits and acknowledgments borrowed from other sources and reproduced, with permission, in this textbook appear on the appropriate page within the text (or on page P1).

Microsoft® and Windows® are registered trademarks of the Microsoft Corporation in the U.S.A. and other countries. Screen shots and icons reprinted with permission from the Microsoft Corporation. This book is not sponsored or endorsed by or affiliated with the Microsoft Corporation.

Pearson Education Ltd., London
Pearson Education Singapore, Pte. Ltd
Pearson Education, Canada, Ltd
Pearson Education-Japan
Pearson Education Australia PTY, Limited

Pearson Education North Asia, Ltd., Hong Kong
Pearson Educación de Mexico, S.A. de C.V.
Pearson Education Malaysia, Pte. Ltd.
Pearson Education Upper Saddle River, New Jersey

Prentice Hall
is an imprint of

10 9 8 7 6 5 4 3
ISBN-13: 978-0-13-214285-4
ISBN-10: 0-13-214285-6

www.pearsonhighered.com

About the Authors

Kenneth C. Laudon is a Professor of Information Systems at New York University's Stern School of Business. He holds a B.A. in Economics from Stanford and a Ph.D. from Columbia University. He has authored twelve books dealing with electronic commerce, information systems, organizations, and society. Professor Laudon has also written over forty articles concerned with the social, organizational, and management impacts of information systems, privacy, ethics, and multimedia technology.

Professor Laudon's current research is on the planning and management of large-scale information systems and multimedia information technology. He has received grants from the National Science Foundation to study the evolution of national information systems at the Social Security Administration, the IRS, and the FBI. Ken's research focuses on enterprise system implementation, computer-related organizational and occupational changes in large organizations, changes in management ideology, changes in public policy, and understanding productivity change in the knowledge sector.

Ken Laudon has testified as an expert before the United States Congress. He has been a researcher and consultant to the Office of Technology Assessment (United States Congress), Department of Homeland Security, and to the Office of the President, several executive branch agencies, and Congressional Committees. Professor Laudon also acts as an in-house educator for several consulting firms and as a consultant on systems planning and strategy to several Fortune 500 firms.

At NYU's Stern School of Business, Ken Laudon teaches courses on Managing the Digital Firm, Information Technology and Corporate Strategy, Professional Responsibility (Ethics), and Electronic Commerce and Digital Markets. Ken Laudon's hobby is sailing.

Jane Price Laudon is a management consultant in the information systems area and the author of seven books. Her special interests include systems analysis, data management, MIS auditing, software evaluation, and teaching business professionals how to design and use information systems.

Jane received her Ph.D. from Columbia University, her M.A. from Harvard University, and her B.A. from Barnard College. She has taught at Columbia University and the New York University Graduate School of Business. She maintains a lifelong interest in Oriental languages and civilizations.

The Laudons have two daughters, Erica and Elisabeth, to whom this book is dedicated.

Brief Contents

Complete Contents

Chapter 4   Ethical and Social Issues in Information Systems 120

Chapter 7   Telecommunications, the Internet, and Wireless Technology 244

Chapter 12   Enhancing Decision Making 452

Part Four   Building and Managing Systems 475

Chapter 15   Managing Global Systems 558

(available on the Web at www.pearsonhighered.com/laudon)

BUSINESS CASES AND INTERACTIVE SESSIONS

Here are some of the business firms you will find described in the cases and Interactive Sessions of this book:

Chapter 1: Information Systems in Global Business Today
The New Yankee Stadium Looks to the Future
MIS in Your Pocket
UPS Competes Globally with Information Technology
What's the Buzz on Smart Grids?

Chapter 2: Global E-Business and Collaboration
America's Cup 2010: USA Wins with Information Technology
Domino's Sizzles with Pizza Tracker
Virtual Meetings: Smart Management
Collaboration and Innovation at Procter & Gamble

Chapter 3: Information Systems, Organizations, and Strategy
Verizon or AT&T—Which Company Has the Best Digital Strategy?
How Much Do Credit Card Companies Know About You?
Is the iPad a Disruptive Technology?
Will TV Succumb to the Internet?

Chapter 4: Ethical and Social Issues in Information Systems
Behavioral Targeting And Your Privacy: You're the Target
The Perils of Texting
Too Much Technology
When Radiation Therapy Kills

Chapter 5: IT Infrastructure and Emerging Technologies
BART Speeds Up with a New IT Infrastructure
New to the Touch
Is Green Computing Good for Business?
Salesforce.com: Cloud Services Go Mainstream

Chapter 6: Foundations of Business Intelligence: Databases and Information Management
RR Donnelley Tries to Master Its Data
What Can Businesses Learn from Text Mining?
Credit Bureau Errors—Big People Problems
The Terror Watch List Database's Troubles Continue

Chapter 7: Telecommunications, the Internet and Wireless Technology
Hyundai Heavy Industries Creates a Wireless Shipyard
The Battle Over Net Neutrality
Monitoring Employees on Networks: Unethical or Good Business?
Google, Apple, and Microsoft Struggle for Your Internet Experience

Preface

We wrote this book for business school students who want an in-depth look at how today's business firms use information technologies and systems to achieve corporate objectives. Information systems are one of the major tools available to business managers for achieving operational excellence, developing new products and services, improving decision making, and achieving competitive advantage. Students will find here the most up-to-date and comprehensive overview of information systems used by business firms today.

When interviewing potential employees, business firms often look for new hires who know how to use information systems and technologies for achieving bottom-line business results. Regardless of whether a student is an accounting, finance, management, operations management, marketing, or information systems major, the knowledge and information found in this book will be valuable throughout a business career.

WHAT'S NEW IN THIS EDITION

CURRENCY

The 12th edition features all new opening, closing, and Interactive Session cases. The text, figures, tables, and cases have been updated through November 2010 with the latest sources from industry and MIS research.

NEW FEATURES

- Thirty video case studies (2 per chapter) and 15 instructional videos are available online.
- Additional discussion questions are provided in each chapter.
- Management checklists are found throughout the book; they are designed to help future managers make better decisions.

NEW TOPICS

- Expanded coverage of business intelligence and business analytics
- Collaboration systems and tools
- Cloud computing
- Cloud-based software services and tools
- Windows 7 and mobile operating systems
- Emerging mobile digital platform
- Office 2010 and Google Apps
- Green computing
- 4G networks
- Network neutrality
- Identity management

- Augmented reality
- Search engine optimization (SEO)
- Freemium pricing models in e-commerce
- Crowdsourcing and the wisdom of crowds
- E-commerce revenue models
- Building an e-commerce Web site
- Business process management
- Security issues for cloud and mobile platforms

WHAT'S NEW IN MIS

Plenty. A continuing stream of information technology innovations is transforming the traditional business world. What makes the MIS field the most exciting area of study in schools of business is this continuous change in technology, management, and business processes. (Chapter 1 describes these changes in more detail.)

Examples of transforming technologies include the emergence of cloud computing, the growth of a mobile digital business platform based on smartphones, netbook computers, and, not least, the use of social networks by managers to achieve business objectives. Most of these changes have occurred in the last few years. These innovations enable entrepreneurs and innovative traditional firms to create new products and services, develop new business models, and transform the day-to-day conduct of business. In the process, some old businesses, even entire industries, are being destroyed while new businesses are springing up.

For instance, the emergence of online music stores—driven by millions of consumers who prefer iPods and MP3 players—has forever changed the older business model of distributing music on physical devices, such as records and CDs, and then selling them in retail stores. Say goodbye to your local music store! Streaming Hollywood movies from Netflix is transforming the old model of distributing films through theaters and then through DVD rentals at physical stores. Say goodbye to Blockbuster! The growth of cloud computing, and huge data centers, along with high-speed broadband connections to the home support these business model changes.

E-commerce is back, generating over $255 billion in revenue in 2010 and estimated to grow to over $354 billion by 2014. Amazon's revenue grew 39 percent in the 12-month period ending June 30, 2010, despite the recession, while offline retail grew by 5 percent. E-commerce is changing how firms design, produce, and deliver their products and services. E-commerce has reinvented itself again, disrupting the traditional marketing and advertising industry and putting major media and content firms in jeopardy. Facebook and other social networking sites such as YouTube, Twitter, and Second Life exemplify the new face of e-commerce in the twenty-first century. They sell services. When we think of e-commerce, we tend to think of selling physical products. While this iconic vision of e-commerce is still very powerful and the fastest growing form of retail in the U.S., cropping up alongside is a whole new value stream based on selling services, not goods. Information systems and technologies are the foundation of this new services-based e-commerce.

Likewise, the management of business firms has changed: With new mobile smartphones, high-speed Wi-Fi networks, and wireless laptop computers, remote salespeople on the road are only seconds away from their managers' questions and oversight. Managers on the move are in direct, continuous contact with their employees. The growth of enterprise-wide information systems with extraordinarily rich data means that managers no longer operate in a fog of confusion, but instead have online, nearly instant access to the important information they need for accurate and timely decisions. In addition to their public uses on the Web, wikis and blogs are becoming important corporate tools for communication, collaboration, and information sharing.

THE 12TH EDITION: THE COMPREHENSIVE SOLUTION FOR THE MIS CURRICULUM

Since its inception, this text has helped to define the MIS course around the globe. This edition continues to be authoritative, but is also more customizable, flexible, and geared to meeting the needs of different colleges, universities, and individual instructors. This book is now part of a complete learning package that includes the core text and an extensive offering of supplemental materials on the Web.

The core text consists of 15 chapters with hands-on projects covering essential topics in MIS. An important part of the core text is the Video Case Study and Instructional Video package: 30 video case studies (2 video cases per chapter) plus 15 instructional videos that illustrate business uses of information systems, explain new technologies, and explore concepts. Video cases are keyed to the topics of each chapter.

In addition, for students and instructors who want to go deeper into selected topics, there are over 40 online Learning Tracks that cover a variety of MIS topics in greater depth.

myMISlab provides more in-depth coverage of chapter topics, career resources, additional case studies, supplementary chapter material, and data files for hands-on projects.

THE CORE TEXT

The core text provides an overview of fundamental MIS concepts using an integrated framework for describing and analyzing information systems. This framework shows information systems composed of management, organization, and technology elements and is reinforced in student projects and case studies.

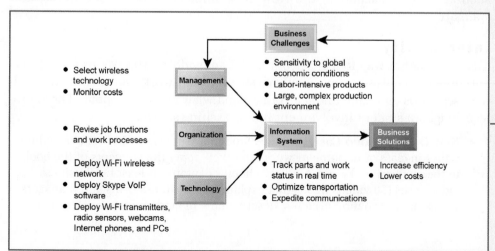

A diagram accompanying each chapter-opening case graphically illustrates how management, organization, and technology elements work together to create an information system solution to the business challenges discussed in the case.

Chapter Organization

Each chapter contains the following elements:
- A chapter-opening case describing a real-world organization to establish the theme and importance of the chapter
- A diagram analyzing the opening case in terms of the management, organization, and technology model used throughout the text
- A series of learning objectives
- Two Interactive Sessions with case study questions and MIS in Action projects
- A Hands-on MIS Projects section featuring two management decision problems, a hands-on application software project, and a project to develop Internet skills
- A Learning Tracks section identifying supplementary material on myMISlab
- A Review Summary section keyed to the learning objectives
- A list of key terms that students can use to review concepts
- Review questions for students to test their comprehension of chapter material
- Discussion questions raised by the broader themes of the chapter
- A pointer to downloadable video cases
- A Collaboration and Teamwork project to develop teamwork and presentation skills, with options for using open source collaboration tools
- A chapter-ending case study for students to apply chapter concepts

KEY FEATURES

We have enhanced the text to make it more interactive, leading-edge, and appealing to both students and instructors. The features and learning tools are described in the following sections.

Business-Driven with Business Cases and Examples

The text helps students see the direct connection between information systems and business performance. It describes the main business objectives driving the use of information systems and technologies in corporations all over the world: operational excellence, new products and services, customer and supplier intimacy, improved decision making, competitive advantage, and survival. In-text examples and case studies show students how specific companies use information systems to achieve these objectives.

We use only current (2010) examples from business and public organizations throughout the text to illustrate the important concepts in each chapter. All the case studies describe companies or organizations that are familiar to students, such as Google, Facebook, the New York Yankees, Procter & Gamble, and Walmart.

Interactivity

There's no better way to learn about MIS than by doing MIS. We provide different kinds of hands-on projects where students can work with real-world business scenarios and data, and learn first hand what MIS is all about. These projects heighten student involvement in this exciting subject.

- **New Online Video Case Package.** Students' can watch short videos online, either in-class or at home or work, and then apply the concepts of the book to the analysis of the video. Every chapter contains at least two business video cases (30 videos in all) that explain how business firms and managers are using information systems, describe new management practices, and explore concepts discussed in the chapter. Each video case consists of a video about a real-world company, a background text case, and case study questions. These video cases enhance students' understanding of MIS topics and the relevance of MIS to the business world. In addition, there are 15 instructional videos that describe developments and concepts in MIS keyed to respective chapters.

- **Management Decision Problems.** Each chapter contains two management decision problems that teach students how to apply chapter concepts to real-world business scenarios requiring analysis and decision making.

Management Decision Problems

1. Applebee's is the largest casual dining chain in the world, with 1,970 locations throughout the United States and nearly 20 other countries worldwide. The menu features beef, chicken, and pork items, as well as burgers, pasta, and seafood. The Applebee's CEO wants to make the restaurant more profitable by developing menus that are tastier and contain more items that customers want and are willing to pay for despite rising costs for gasoline and agricultural products. How might information systems help management implement this strategy? What pieces of data would Applebee's need to collect? What kinds of reports would be useful to help management make decisions on how to improve menus and profitability?

Two real-world business scenarios per chapter provide opportunities for students to apply chapter concepts and practice management decision making.

- **Collaboration and Teamwork Projects.** Each chapter features a collaborative project that encourages students working in teams to use Google sites, Google Docs, and other open-source collaboration tools. The first team project in Chapter 1 asks students to build a collaborative Google site.

- **Hands-on MIS Projects.** Every chapter concludes with a Hands-on MIS Projects section containing three types of projects: two management decision problems; a hands-on application software exercise using Microsoft Excel Access, or Web page and blog-creation tools; and a project that develops Internet business skills. A Dirt Bikes USA running case in myMISlab provides additional hands-on projects for each chapter.

Part One Organizations, Management, and the Networked Enterprise

ID	Store No	Sales Region	Item No	Item Description	Unit Price	Units Sold	Week Ending
1	1	South	2005	17" Monitor	$229.00	28	10/27/2010
2	1	South	2005	17" Monitor	$229.00	30	11/24/2010
3	1	South	2005	17" Monitor	$229.00	9	12/29/2010
4	1	South	3006	101 Keyboard	$19.95	30	10/27/2010
5	1	South	3006	101 Keyboard	$19.95	35	11/24/2010
6	1	South	3006	101 Keyboard	$19.95	39	12/29/2010
7	1	South	6050	PC Mouse	$8.95	28	10/27/2010
8	1	South	6050	PC Mouse	$8.95	3	11/24/2010
9	1	South	6050	PC Mouse	$8.95	38	12/29/2010
10	1	South	8500	Desktop CPU	$849.95	25	10/27/2010
11	1	South	8500	Desktop CPU	$849.95	27	11/24/2010
12	1	South	8500	Desktop CPU	$849.95	33	12/29/2010
13	2	South	2005	17" Monitor	$229.00	8	10/27/2010
14	2	South	2005	17" Monitor	$229.00	8	11/24/2010
15	2	South	2005	17" Monitor	$229.00	10	12/29/2010

Store & Region Sales Database

Record 1 of 95   No Filter   Search

Students practice using software in real-world settings for achieving operational excellence and enhancing decision making.

should be offered at full price, and which times of the year products should be discounted. Modify the database table, if necessary, to provide all of the information you require. Print your reports and results of queries.

Improving Decision Making: Using Intelligent Agents for Comparison Shopping

Software skills: Web browser and shopping bot software
Business skills: Product evaluation and selection

This project will give you experience using shopping bots to search online for products, find product information, and find the best prices and vendors.

You have decided to purchase a new digital camera. Select a digital camera you might want to purchase, such as the Canon PowerShot S95 or the Olympus Stylus 7040. To purchase the camera as inexpensively as possible, try several of the shopping bot sites, which do the price comparisons for you. Visit My Simon (www.mysimon.com), BizRate.com (www.bizrate.com), and Google Product Search. Compare these shopping sites in terms of their ease of use, number of offerings, speed in obtaining information, thoroughness of information offered about the product and seller, and price selection. Which site or sites would you use and why? Which camera would you select and why? How helpful were these sites for making your decision?

Each chapter features a project to develop Internet skills for accessing information, conducting research, and performing online calculations and analysis.

- **Interactive Sessions.** Two short cases in each chapter have been redesigned as Interactive Sessions to be used in the classroom (or on Internet discussion boards) to stimulate student interest and active learning. Each case concludes with two types of activities: case study questions and MIS in Action. The case study questions provide topics for class discussion, Internet discussion, or written assignments. MIS in Action features hands-on Web activities for exploring issues discussed in the case more deeply.

INTERACTIVE SESSION: ORGANIZATIONS

CREDIT BUREAU ERRORS—BIG PEOPLE PROBLEMS

You've found the car of your dreams. You have a good job and enough money for a down payment. All you need is an auto loan for $14,000. You have a few credit card bills, which you diligently pay off each month. But when you apply for the loan you're turned down. When you ask why, you're told you have an overdue loan from a bank you've never heard of. You've just become one of the millions of people who have been victimized by inaccurate or outdated data in credit bureaus' information systems.

Most data on U.S. consumers' credit histories are collected and maintained by three national credit reporting agencies: Experian, Equifax, and TransUnion. These organizations collect data from various sources to create a detailed dossier of an individual's borrowing and bill paying habits. This information helps lenders assess a person's credit worthiness, the ability to pay back a loan, and can affect the interest rate and other terms of a loan, including whether a loan will be granted in the first place. It can even affect the chances of finding or keeping a job: At least one-third of employers check credit reports when making hiring, firing, or promotion decisions.

U.S. credit bureaus collect personal information and financial data from a variety of sources, including creditors, lenders, utilities, debt collection agencies, and the courts. These data are aggregated and stored in massive databases maintained by the credit bureaus. The credit bureaus then sell this information to other companies to use for credit assessment.

The credit bureaus claim they know which credit cards are in each consumer's wallet, how much is due on the mortgage, and whether the electric bill is paid on time. But if the wrong information gets into their systems, whether through identity theft or errors transmitted by creditors, watch out! Untangling the mess can be almost impossible.

The bureaus understand the importance of providing accurate information to both lenders and consumers. But they also recognize that their own

The sheer volume of information being transmitted from creditors to credit bureaus increases the likelihood of mistakes. Experian, for example, updates 30 million credit reports each day and roughly 2 billion credit reports each month. It matches the identifying personal information in a credit application or credit account with the identifying personal information in a consumer credit file. Identifying personal information includes items such as name (first name, last name and middle initial), full current address and ZIP code, full previous address and ZIP code, and social security number. The new credit information goes into the consumer credit file that it best matches.

The credit bureaus rarely receive information that matches in all the fields in credit files, so they have to determine how much variation to allow and still call it a match. Imperfect data lead to imperfect matches. A consumer might provide incomplete or inaccurate information on a credit application. A creditor might submit incomplete or inaccurate information to the credit bureaus. If the wrong person matches better than anyone else, the data could unfortunately go into the wrong account.

Perhaps the consumer didn't write clearly on the account application. Name variations on different credit accounts can also result in less-than-perfect matches. Take the name Edward Jeffrey Johnson. One account may say Edward Johnson. Another may say Ed Johnson. Another might say Edward J. Johnson. Suppose the last two digits of Edward's social security number get transposed—more chance for mismatches.

If the name or social security number on another person's account partially matches the data in your file, the computer might attach that person's data to your record. Your record might likewise be corrupted if workers in companies supplying tax and bankruptcy data from court and government records accidentally transpose a digit or misread a document.

The credit bureaus claim it is impossible for

Each chapter contains two Interactive Sessions focused on management, organizations, or technology using real-world companies to illustrate chapter concepts and issues.

CASE STUDY QUESTIONS   MIS IN ACTION

1. Assess the business impact of credit bureaus' data quality problems for the credit bureaus, for lenders, for individuals.

2. Are any ethical issues raised by credit bureaus' data quality problems? Explain your answer.

3. Analyze the management, organization, and technology factors responsible for credit bureaus' data quality problems.

4. What can be done to solve these problems?

Go to the Experian Web site (www.experian.com) and explore the site, with special attention to its services for businesses and small businesses. Then answer the following questions:

1. List and describe five services for businesses and explain how each uses consumer data. Describe the kinds of businesses that would use these services.

2. Explain how each of these services is affected by inaccurate consumer data.

> Case study questions and MIS in Action projects encourage students to learn more about the companies and issues discussed in the case studies.

Assessment and AACSB Assessment Guidelines

The Association to Advance Collegiate Schools of Business (AACSB) is a not-for-profit corporation of educational institutions, corporations, and other organizations that seeks to improve business education primarily by accrediting university business programs. As a part of its accreditation activities, the AACSB has developed an Assurance of Learning program designed to ensure that schools teach students what the schools promise. Schools are required to state a clear mission, develop a coherent business program, identify student learning objectives, and then prove that students achieve the objectives.

We have attempted in this book to support AACSB efforts to encourage assessment-based education. The front end papers of this edition identify student learning objectives and anticipated outcomes for our Hands-on MIS projects. In the Instructor Resource Center and myMISlab is a more inclusive and detailed assessment matrix that identifies the learning objectives of each chapter and points to all the available assessment tools that ensure students achieve the learning objectives. Because each school is different and may have different missions and learning objectives, no single document can satisfy all situations. Therefore, the authors will provide custom advice to instructors on how to use this text in their respective colleges. Instructors should e-mail the authors or contact their local Pearson Prentice Hall representative for contact information.

For more information on the AACSB Assurance of Learning program and how this text supports assessment-based learning, visit the Instructor Resource Center and myMISlab.

Customization and Flexibility: New Learning Track Modules

Our Learning Tracks feature gives instructors the flexibility to provide in-depth coverage of the topics they choose. There are over 40 Learning Tracks available to instructors and students. A Learning Tracks section at the end of each chapter directs students to short essays or additional chapters in myMISlab. This supplementary content takes students deeper into MIS topics, concepts, and debates; reviews basic technology concepts in hardware, software, database design, telecommunications, and other areas; and provides additional hands-on software instruction. The 12th edition includes new Learning Tracks on cloud computing, managing knowledge and collaboration, creating a pivot table with Microsoft Excel PowerPivot, the mobile digital platform, and business process management.

AUTHOR-CERTIFIED TEST BANK AND SUPPLEMENTS

- **Author-Certified Test Bank.** The authors have worked closely with skilled test item writers to ensure that higher level cognitive skills are tested. The test bank includes multiple-choice questions on content, but also includes many questions that require analysis, synthesis, and evaluation skills.
- **New Annotated Interactive PowerPoint Lecture Slides.** The authors have prepared a comprehensive collection of 500 PowerPoint slides to be used in lectures. Ken Laudon uses many of these slides in his MIS classes and executive education presentations. Each of the slides is annotated with teaching suggestions for asking students questions, developing in-class lists that illustrate key concepts, and recommending other firms as examples in addition to those provided in the text. The annotations are like an instructor's manual built into the slides and make it easier to teach the course effectively.

STUDENT LEARNING-FOCUSED

Student learning objectives are organized around a set of study questions to focus student attention. Each chapter concludes with a review summary and review questions organized around these study questions.

MYMISLAB

MyMISlab is a Web-based assessment and tutorial tool that provides practice and testing while personalizing course content and providing student and class assessment and reporting. Your course is not the same as the course taught down the hall. Now, all the resources that instructors and students need for course success are in one place—flexible and easily organized and adapted for an individual course experience. Visit www.mymislab.com to see how you can teach, learn, and experience MIS.

CAREER RESOURCES

MyMISlab also provides extensive career resources, including job-hunting guides and instructions on how to build a digital portfolio demonstrating the business knowledge, application software proficiency, and Internet skills acquired from using the text. Students can use the portfolio in a resume or job application; instructors can use it as a learning assessment tool.

INSTRUCTIONAL SUPPORT MATERIALS

Instructor Resource Center

Most of the support materials described in the following sections are conveniently available for adopters on the online Instructor Resource Center (IRC). The IRC includes the Image Library (a very helpful lecture tool), Instructor's Manual, Lecture Notes, Test Item File and TestGen, and PowerPoint slides.

Image Library

The Image Library is an impressive resource to help instructors create vibrant lecture presentations. Almost every figure and photo in the text is provided and organized by chapter for convenience. These images and lecture notes can be imported easily into PowerPoint to create new presentations or to add to existing ones.

Instructor's Manual

The Instructor's Manual features not only answers to review, discussion, case study, and group project questions, but also in-depth lecture outlines, teaching objectives, key terms, teaching suggestions, and Internet resources.

Test Item File

The Test Item File is a comprehensive collection of true-false, multiple-choice, fill-in-the-blank, and essay questions. The questions are rated by difficulty level and the answers are referenced by section. The Test Item File also contains questions tagged to the AACSB learning standards. An electronic version of the Test Item File is available in TestGen, and TestGen conversions are available for BlackBoard or WebCT course management systems. All TestGen files are available for download at the IRC.

Annotated PowerPoint Slides

Electronic color slides created by the authors are available in PowerPoint. The slides illuminate and build on key concepts in the text.

Video Cases and Instructional Videos

Instructors can download step-by-step instructions for accessing the video cases from the Instructor Resources page at www.pearsonhighered.com/laudon. The following page contains a list of video cases and instructional videos.

Video Cases and Instructional Videos

Chapter	Video
Chapter 1: Information Systems In Global Business Today	Case 1: UPS Global Operations with the DIAD IV Case 2: IBM, Cisco, Google: Global Warming by Computer
Chapter 2: Global E-business and Collaboration	Case 1: How FedEx Works: Enterprise Systems Case 2: Oracle's Austin Data Center Instructional Video 1: FedEx Improves Customer Experience with Integrated Mapping, Location Data
Chapter 3: Information Systems, Organizations, and Strategy	Case 1: National Basketball Association: Competing on Global Delivery with Akamai OS Streaming Case 2: Customer Relationship Management for San Francisco's City Government
Chapter 4: Ethical and Social Issues in Information Systems	Case 1: Net Neutrality: Neutral Networks Work Case 2: Data Mining for Terrorists and Innocents Instructional Video 1: Big Brother Is Copying Everything on the Internet Instructional Video 2: Delete: The Virtue of Forgetting in a Digital Age
Chapter 5: IT Infrastructure: and Emerging Technologies	Case 1: Hudson's Bay Company and IBM: Virtual Blade Platform Case 2: Salesforce.com: SFA on the iPhone and iPod Touch Instructional Video 1: Google and IBM Produce Cloud Computing Instructional Video 2: IBM Blue Cloud Is Ready-to-Use Computing Instructional Video 3: What the Hell Is Cloud Computing? Instructional Video 4: What Is Ajax and How Does It Work? Instructional Video 5: Yahoo's FireEagle Geolocation Service
Chapter 6: Foundations of Business Intelligence: Databases and Information Management	Case 1: Maruti Suzuki Business Intelligence and Enterprise Databases Case 2: Data Warehousing at REI: Understanding the Customer
Chapter 7: Telecommunications, the Internet, and Wireless Technology	Case 1: Cisco Telepresence: Meeting Without Traveling Case 2: Unified Communications Systems with Virtual Collaboration: IBM and Forterra Instructional Video 1: AT&T Launches Managed Cisco Telepresence Solution Instructional Video 2: CNN Telepresence
Chapter 8: Securing Information Systems	Case 1: IBM Zone Trusted Information Channel (ZTIC) Case 2: Open ID and Web Security Instructional Video 1: The Quest for Identity 2.0 Instructional Video 2: Identity 2.0
Chapter 9: Achieving Operational Excellence and Customer Intimacy: Enterprise Applications	Case 1: Sinosteel Strengthens Business Management with ERP Applications Case 2: Ingram Micro and H&R Block Get Close to Their Customers Instructional Video 1: Zara's: Wearing Today's Fashions with Supply Chain Management
Chapter 10: E-commerce: Digital Markets, Digital Goods	Case 1: M-commerce: The Past, Present, and Future Case 2: Ford AutoXchange B2B Marketplace
Chapter 11: Managing Knowledge	Case 1: L'Oréal: Knowledge Management Using Microsoft SharePoint Case 2: IdeaScale Crowdsourcing: Where Ideas Come to Life
Chapter 12: Enhancing Decision Making	Case 1: Antivia: Community-based Collaborative Business Intelligence Case 2: IBM and Cognos: Business Intelligence and Analytics for Improved Decision Making
Chapter 13: Building Information Systems	Case 1: IBM: Business Process Management in a Service-Oriented Architecture Case 2: Rapid Application Development With Appcelerator Instructional Video 1: Salesforce and Google: Developing Sales Support Systems with Online Apps
Chapter 14: Managing Projects	Case 1: Mastering the Hype Cycle: How to Adopt the Right Innovation at the Right Time Case 2: NASA: Project Management Challenges Instructional Video 1: Software Project Management in 15 Minutes
Chapter 15: Managing Global Systems	Case 1: Daum Runs Oracle Apps on Linux Case 2: Monsanto Uses Cisco and Microsoft to Manage Globally

ACKNOWLEDGEMENTS

The production of any book involves valued contributions from a number of persons. We would like to thank all of our editors for encouragement, insight, and strong support for many years. We thank Bob Horan for guiding the development of this edition and Kelly Loftus for her role in managing the project. We also praise Karalyn Holland for overseeing production for this project.

Our special thanks go to our supplement authors for their work. We are indebted to William Anderson for his assistance in the writing and production of the text and to Megan Miller for her help during production. We thank Diana R. Craig for her assistance with database and software topics.

Special thanks to colleagues at the Stern School of Business at New York University; to Professor Edward Stohr of Stevens Institute of Technology; to Professors Al Croker and Michael Palley of Baruch College and New York University; to Professor Lawrence Andrew of Western Illinois University; to Professor Detlef Schoder of the University of Cologne; to Professor Walter Brenner of the University of St. Gallen; to Professor Lutz Kolbe of the University of Gottingen; to Professor Donald Marchand of the International Institute for Management Development; and to Professor Daniel Botha of Stellenbosch University who provided additional suggestions for improvement. Thank you to Professor Ken Kraemer, University of California at Irvine, and Professor John King, University of Michigan, for more than a decade's long discussion of information systems and organizations. And a special remembrance and dedication to Professor Rob Kling, University of Indiana, for being my friend and colleague over so many years.

We also want to especially thank all our reviewers whose suggestions helped improve our texts. Reviewers for this edition include the following:

Edward J. Cherian, *George Washington University*
Sherry L. Fowler, *North Carolina State University*
Richard Grenci, *John Carroll University*
Dorest Harvey, *University of Nebraska Omaha*
Shohreh Hashemi, *University of Houston—Downtown*
Duke Hutchings, *Elon University*
Ingyu Lee, *Troy University*
Jeffrey Livermore, *Walsh College*
Sue McDaniel, *Bellevue University*
Michelle Parker, *Indiana University—Purdue University Fort Wayne*
Peter A. Rosen, *University of Evansville*
Donna M. Schaeffer, *Marymount University*
Werner Schenk, *University of Rochester*
Jon C. Tomlinson, *University of Northwestern Ohio*
Marie A. Wright, *Western Connecticut State University*
James H. Yu, *Santa Clara University*
Fan Zhao, *Florida Gulf Coast University*

K.C.L.
J.P.L.

PART ONE

Organizations, Management, and the Networked Enterprise

Part One introduces the major themes of this book, raising a series of important questions: What is an information system and what are its management, organization, and technology dimensions? Why are information systems so essential in businesses today? Why are systems for collaboration and teamwork so important? How can information systems help businesses become more competitive? What broader ethical and social issues are raised by widespread use of information systems?

Chapter 1

Information Systems in Global Business Today

THE NEW YANKEE STADIUM LOOKS TO THE FUTURE

Although baseball is a sport, it's also big business, requiring revenue from tickets to games, television broadcasts, and other sources to pay for teams. Salaries for top players have ballooned, as have ticket prices. Many fans now watch games on television rather than attending them in person or choose other forms of entertainment, such as electronic games. One way to keep stadiums full of fans, and to keep fans at home happy as well, is to enrich the fan experience by offering more video and services based on technology. When the New York Yankees built the new Yankee Stadium, they did just that.

The new Yankee Stadium, which opened on April 2, 2009, isn't just another ballpark: It's the stadium of the future. It is the most wired, connected, and video-enabled stadium in all of baseball. Although the new stadium is similar in design to the original Yankee Stadium, built in 1923, the interior has more space and amenities, including more intensive use of video and computer technology. Baseball fans love video. According to Ron Ricci, co-chairman of Cisco Systems' sports and entertainment division, "It's what fans want to see, to see more angles and do it on their terms." Cisco Systems supplied the computer and networking technology for the new stadium.

Throughout the stadium, including the Great Hall, the Yankees Museum, and in-stadium restaurants and concession areas, 1,200 flat-panel high-definition HDTV monitors display live game coverage, up-to-date sports scores, archival and highlight video, promotional messages, news, weather, and traffic updates. There is also a huge monitor in center field that is 101 feet wide and 59 feet high. At the conclusion of games, the monitors provide up-to-the moment traffic information and directions to the nearest stadium exits.

The monitors are designed to surround fans visually from the moment they enter the stadium, especially when they stray from a direct view of the ball field. The pervasiveness of this technology ensures that while fans are buying a hamburger or a soda, they will never miss a play. The Yankees team controls all the monitors centrally and is able to offer different content on each one. Monitors are located at concession stands, around restaurants and bars, in restrooms, and inside 59 luxury and party suites. If a Yankee player wants to review a game to see how he played, monitors in the team's video room will display what he did from any angle. Each Yankee player also has a computer at his locker.

The luxury suites have special touch-screen phones for well-heeled fans to use when ordering food and merchandise. At the stadium business center, Cisco interactive videoconferencing technology will link to a library in the Bronx and to other New York City locations, such as hospitals. Players

and executives will be able to videoconference and talk to fans before or after the games. Eventually data and video from the stadium will be delivered to fans' home televisions and mobile devices. Inside the stadium, fans in each seat will be able to use their mobile phones to order from the concessions or view instant replays. If they have an iPhone, an application called Venuing lets them communicate with other fans at the game, find nearby facilities, obtain reviews of concessions, play pub-style trivia games, and check for news updates.

The Yankees also have their own Web site, Yankees.com, where fans can watch in-market Yankees games live online, check game scores, find out more about their favorite players, purchase tickets to games, and shop for caps, baseball cards and memorabilia. The site also features fantasy baseball games, where fans compete with each other by managing "fantasy teams" based on real players' statistics.

Sources: www.mlb.com, accessed May 5, 2010; Rena Bhattacharyya, Courtney Munroe, and Melanie Posey, "Yankee Stadium Implements State-of-the-Art Technology from AT&T," www.forbescustom.com, April 13, 2010; "Venuing: An iPhone App Tailor-Made for Yankee Stadium Insiders," NYY Stadium Insider, March 30, 2010; Dean Meminger, "Yankees' New Stadium Is More than a Ballpark," NY1.com, April 2, 2009.

The challenges facing the New York Yankees and other baseball teams show why information systems are so essential today. Major league baseball is a business as well as a sport, and teams such as the Yankees need to take in revenue from games in order to stay in business. Ticket prices have risen, stadium attendance is dwindling for some teams, and the sport must also compete with other forms of entertainment, including electronic games and the Internet.

The chapter-opening diagram calls attention to important points raised by this case and this chapter. To increase stadium attendance and revenue, the New York Yankees chose to modernize Yankee Stadium and rely on information technology to provide new interactive services to fans inside and outside the stadium. These services include high-density television monitors displaying live game coverage; up-to-date sports scores, video, promotional messages, news, weather, and traffic information; touch screens for ordering food and merchandise; interactive videoconferencing technology for connecting to fans and the community; mobile social networking applications; and, eventually, data and video broadcast to fans' home television sets and mobile handhelds. The Yankees' Web site provides a new channel for interacting with fans, selling tickets to games, and selling other team-related products.

It is also important to note that these technologies changed the way the Yankees run their business. Yankee Stadium's systems for delivering game coverage, information, and interactive services changed the flow of work for ticketing, seating, crowd management, and ordering food and other items from concessions. These changes had to be carefully planned to make sure they enhanced service, efficiency, and profitability.

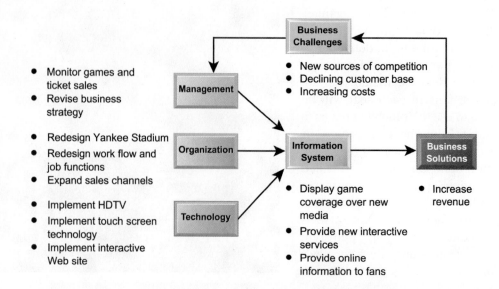

1.1 THE ROLE OF INFORMATION SYSTEMS IN BUSINESS TODAY

I t's not business as usual in America anymore, or the rest of the global economy. In 2010, American businesses will spend over $562 billion on information systems hardware, software, and telecommunications equipment. In addition, they will spend another $800 billion on business and management consulting and services—much of which involves redesigning firms' business operations to take advantage of these new technologies. Figure 1-1 shows that between 1980 and 2009, private business investment in information technology consisting of hardware, software, and communications equipment grew from 32 percent to 52 percent of all invested capital.

As managers, most of you will work for firms that are intensively using information systems and making large investments in information technology. You will certainly want to know how to invest this money wisely. If you make wise choices, your firm can outperform competitors. If you make poor choices, you will be wasting valuable capital. This book is dedicated to helping you make wise decisions about information technology and information systems.

HOW INFORMATION SYSTEMS ARE TRANSFORMING BUSINESS

You can see the results of this massive spending around you every day by observing how people conduct business. More wireless cell phone accounts were opened in 2009 than telephone land lines installed. Cell phones, BlackBerrys, iPhones, e-mail, and online conferencing over the Internet have all become essential tools of business. Eighty-nine million people in the United States access the Internet using mobile devices in 2010, nearly half the total

FIGURE 1-1   INFORMATION TECHNOLOGY CAPITAL INVESTMENT

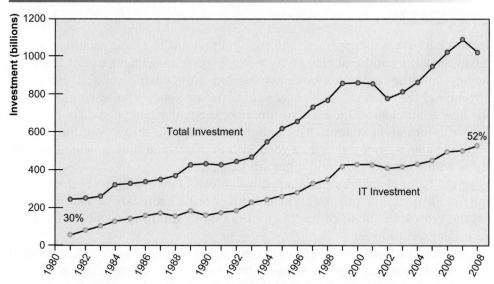

Information technology capital investment, defined as hardware, software, and communications equipment, grew from 30 percent to 52 percent of all invested capital between 1980 and 2009.

Source: Based on data in U.S. Department of Commerce, Bureau of Economic Analysis, *National Income and Product Accounts,* 2009.

Internet user population (eMarketer, 2010). There are 285 million cell phone subscribers in the United States, and nearly 5 billion worldwide (Dataxis, 2010).

By June 2010, more than 99 million businesses worldwide had dot-com Internet sites registered (Verisign, 2010). Today, 162 million Americans shop online, and 133 million have purchased online. Every day about 41 million Americans go online to research a product or service.

In 2009, FedEx moved over 3.4 million packages daily in the United States, mostly overnight, and the United Parcel Service (UPS) moved over 15 million packages daily worldwide. Businesses sought to sense and respond to rapidly changing customer demand, reduce inventories to the lowest possible levels, and achieve higher levels of operational efficiency. Supply chains have become more fast-paced, with companies of all sizes depending on just-in-time inventory to reduce their overhead costs and get to market faster.

As newspaper readership continues to decline, more than 78 million people receive their news online. About 39 million people watch a video online everyday, 66 million read a blog, and 16 million post to blogs, creating an explosion of new writers and new forms of customer feedback that did not exist five years ago (Pew, 2010). Social networking site Facebook attracted 134 million monthly visitors in 2010 in the United States, and over 500 million worldwide. Businesses are starting to use social networking tools to connect their employees, customers, and managers worldwide. Many Fortune 500 companies now have Facebook pages.

Despite the recession, e-commerce and Internet advertising continue to expand. Google's online ad revenues surpassed $25 billion in 2009, and Internet advertising continues to grow at more than 10 percent a year, reaching more than $25 billion in revenues in 2010.

New federal security and accounting laws, requiring many businesses to keep e-mail messages for five years, coupled with existing occupational and health laws requiring firms to store employee chemical exposure data for up to 60 years, are spurring the growth of digital information at the estimated rate of 5 exabytes annually, equivalent to 37,000 new Libraries of Congress.

WHAT'S NEW IN MANAGEMENT INFORMATION SYSTEMS?

Lots! What makes management information systems the most exciting topic in business is the continual change in technology, management use of the technology, and the impact on business success. New businesses and industries appear, old ones decline, and successful firms are those who learn how to use the new technologies. Table 1-1 summarizes the major new themes in business uses of information systems. These themes will appear throughout the book in all the chapters, so it might be a good idea to take some time now and discuss these with your professor and other students.

In the technology area there are three interrelated changes: (1) the emerging mobile digital platform, (2) the growth of online software as a service, and (3) the growth in "cloud computing" where more and more business software runs over the Internet.

IPhones, iPads, BlackBerrys, and Web-surfing netbooks are not just gadgets or entertainment outlets. They represent new emerging computing platforms based on an array of new hardware and software technologies. More and more business computing is moving from PCs and desktop machines to these mobile devices. Managers are increasingly using these devices to coordinate

TABLE 1-1 WHAT'S NEW IN MIS

CHANGE	BUSINESS IMPACT
TECHNOLOGY	
Cloud computing platform emerges as a major business area of innovation	A flexible collection of computers on the Internet begins to perform tasks traditionally performed on corporate computers.
Growth in software as a service (SaaS)	Major business applications are now delivered online as an Internet service rather than as boxed software or custom systems.
A mobile digital platform emerges to compete with the PC as a business system	Apple opens its iPhone software to developers, and then opens an Applications Store on iTunes where business users can download hundreds of applications to support collaboration, location-based services, and communication with colleagues. Small portable lightweight, low-cost, net-centric subnotebook computers are a major segment of the laptop marketplace. The iPad is the first successful tablet-sized computing device with tools for both entertainment and business productivity.
MANAGEMENT	
Managers adopt online collaboration and social networking software to improve coordination, collaboration, and knowledge sharing	Google Apps, Google Sites, Microsoft's Windows SharePoint Services, and IBM's Lotus Connections are used by over 100 million business professionals worldwide to support blogs, project management, online meetings, personal profiles, social bookmarks, and online communities.
Business intelligence applications accelerate	More powerful data analytics and interactive dashboards provide real-time performance information to managers to enhance decision making.
Virtual meetings proliferate	Managers adopt telepresence video conferencing and Web conferencing technologies to reduce travel time, and cost, while improving collaboration and decision making.
ORGANIZATIONS	
Web 2.0 applications are widely adopted by firms	Web-based services enable employees to interact as online communities using blogs, wikis, e-mail, and instant messaging services. Facebook and MySpace create new opportunities for business to collaborate with customers and vendors.
Telework gains momentum in the workplace	The Internet, netbooks, iPads, iPhones, and BlackBerrys make it possible for growing numbers of people to work away from the traditional office; 55 percent of U.S. businesses have some form of remote work program.
Co-creation of business value	Sources of business value shift from products to solutions and experiences and from internal sources to networks of suppliers and collaboration with customers. Supply chains and product development are more global and collaborative than in the past; customers help firms define new products and services.

work, communicate with employees, and provide information for decision making. We call these developments the "emerging mobile digital platform."

Managers routinely use so-called "Web 2.0" technologies like social networking, collaboration tools, and wikis in order to make better, faster decisions. As management behavior changes, how work gets organized, coordinated, and measured also changes. By connecting employees working on teams and projects, the social network is where works gets done, where plans are executed, and where managers manage. Collaboration spaces are where employees meet one another—even when they are separated by continents and time zones.

The strength of cloud computing and the growth of the mobile digital platform allow organizations to rely more on telework, remote work, and distributed decision making. This same platform means firms can outsource more work, and rely on markets (rather than employees) to build value. It also means that firms can collaborate with suppliers and customers to create new products, or make existing products more efficiently.

You can see some of these trends at work in the Interactive Session on Management. Millions of managers rely heavily on the mobile digital platform to coordinate suppliers and shipments, satisfy customers, and manage their employees. A business day without these mobile devices or Internet access would be unthinkable. As you read this case, note how the emerging mobile platform greatly enhances the accuracy, speed, and richness of decision making.

GLOBALIZATION CHALLENGES AND OPPORTUNITIES: A FLATTENED WORLD

In 1492, Columbus reaffirmed what astronomers were long saying: the world was round and the seas could be safely sailed. As it turned out, the world was populated by peoples and languages living in isolation from one another, with great disparities in economic and scientific development. The world trade that ensued after Columbus's voyages has brought these peoples and cultures closer. The "industrial revolution" was really a world-wide phenomenon energized by expansion of trade among nations.

In 2005, journalist Thomas Friedman wrote an influential book declaring the world was now "flat," by which he meant that the Internet and global communications had greatly reduced the economic and cultural advantages of developed countries. Friedman argued that the U.S. and European countries were in a fight for their economic lives, competing for jobs, markets, resources, and even ideas with highly educated, motivated populations in low-wage areas in the less developed world (Friedman, 2007). This "globalization" presents both challenges and opportunities for business firms

A growing percentage of the economy of the United States and other advanced industrial countries in Europe and Asia depends on imports and exports. In 2010, more than 33 percent of the U.S. economy resulted from foreign trade, both imports and exports. In Europe and Asia, the number exceeded 50 percent. Many Fortune 500 U.S. firms derive half their revenues from foreign operations. For instance, more than half of Intel's revenues in 2010 came from overseas sales of its microprocessors. Eighty percent of the toys sold in the U.S. are manufactured in China, while about 90 percent of the PCs manufactured in China use American-made Intel or Advanced Micro Design (AMD) chips.

It's not just goods that move across borders. So too do jobs, some of them high-level jobs that pay well and require a college degree. In the past decade, the United States lost several million manufacturing jobs to offshore, low-wage producers. But manufacturing is now a very small part of U.S. employment (less than 12 percent and declining). In a normal year, about 300,000 service jobs move offshore to lower wage countries, many of them in less-skilled information system occupations, but also including "tradable service" jobs in architecture, financial services, customer call centers, consulting, engineering, and even radiology.

INTERACTIVE SESSION: MANAGEMENT

MIS IN YOUR POCKET

Can you run your company out of your pocket? Perhaps not entirely, but there are many functions today that can be performed using an iPhone, BlackBerry, or other mobile handheld device. The smartphone has been called the "Swiss Army knife of the digital age." A flick of the finger turns it into a Web browser, a telephone, a camera, a music or video player, an e-mail and messaging machine, and for some, a gateway into corporate systems. New software applications for social networking and salesforce management (CRM) make these devices even more versatile business tools.

The BlackBerry has been the favored mobile handheld for business because it was optimized for e-mail and messaging, with strong security and tools for accessing internal corporate systems. Now that's changing. Companies large and small are starting to deploy Apple's iPhone to conduct more of their work. For some, these handhelds have become necessities.

Doylestown Hospital, a community medical center near Philadelphia, has a mobile workforce of 360 independent physicians treating thousands of patients. The physicians use the iPhone 3G to stay connected around the clock to hospital staff, colleagues, and patient information. Doylestown doctors use iPhone features such as e-mail, calendar, and contacts from Microsoft Exchange ActiveSync. The iPhone allows them to receive time-sensitive e-mail alerts from the hospital. Voice communication is important as well, and the iPhone allows the doctors to be on call wherever they are.

Doylestown Hospital customized the iPhone to provide doctors with secure mobile access from any location in the world to the hospital's MEDITECH electronic medical records system. MEDITECH delivers information on vital signs, medications, lab results, allergies, nurses' notes, therapy results, and even patient diets to the iPhone screen. "Every radiographic image a patient has had, every dictated report from a specialist is available on the iPhone," notes Dr. Scott Levy, Doylestown Hospital's vice president and chief medical officer. Doylestown doctors also use the iPhone at the patient's bedside to access medical reference applications such as Epocrates Essentials to help them interpret lab results and obtain medication information.

Doylestown's information systems department was able to establish the same high level of security for authenticating users of the system and tracking user activity as it maintains with all the hospital's Web-based medical records applications. Information is stored securely on the hospital's own server computer.

D.W. Morgan, headquartered in Pleasanton, California, serves as a supply chain consultant and transportation and logistics service provider to companies such as AT&T, Apple Computer, Johnson & Johnson, Lockheed Martin, and Chevron. It has operations in more than 85 countries on four continents, moving critical inventory to factories that use a just-in-time (JIT) strategy. In JIT, retailers and manufacturers maintain almost no excess on-hand inventory, relying upon suppliers to deliver raw materials, components, or products shortly before they are needed.

In this type of production environment, it's absolutely critical to know the exact moment when delivery trucks will arrive. In the past, it took many phone calls and a great deal of manual effort to provide customers with such precise up-to-the-minute information. The company was able to develop an application called ChainLinq Mobile for its 30 drivers that updates shipment information, collects signatures, and provides global positioning system (GPS) tracking on each box it delivers.

As Morgan's drivers make their shipments, they use ChainLinq to record pickups and status updates. When they reach their destination, they collect a signature on the iPhone screen. Data collected at each point along the way, including a date- and time-stamped GPS location pinpointed on a Google map, are uploaded to the company's servers. The servers make the data available to customers on the company's Web site. Morgan's competitors take about 20 minutes to half a day to provide proof of delivery; Morgan can do it immediately.

TCHO is a start-up that uses custom-developed machinery to create unique chocolate flavors. Owner Timothy Childs developed an iPhone app that enables him to remotely log into each chocolate-making machine, control time and temperature, turn the machines on and off, and receive alerts about when to make temperature changes. The iPhone app also enables him to remotely view several video cameras that show how the TCHO

FlavorLab is doing. TCHO employees also use the iPhone to exchange photos, e-mail, and text messages.

The Apple iPad is also emerging as a business tool for Web-based note-taking, file sharing, word processing, and number-crunching. Hundreds of business productivity applications are being developed, including tools for Web conferencing, word processing, spreadsheets, and electronic presenta-

tions. Properly configured, the iPad is able to connect to corporate networks to obtain e-mail messages, calendar events, and contacts securely over the air.

Sources: "Apple iPhone in Business Profiles, www.apple.com, accessed May 10, 2010; Steve Lohr, Cisco Cheng, "The Ipad Has Business Potential," *PC World*, April 26, 2010; and "Smartphone Rises Fast from Gadget to Necessity," *The New York Times*, June 10, 2009.

CASE STUDY QUESTIONS

1. What kinds of applications are described here? What business functions do they support? How do they improve operational efficiency and decision making?

2. Identify the problems that businesses in this case study solved by using mobile digital devices.

3. What kinds of businesses are most likely to benefit from equipping their employees with mobile digital devices such as iPhones, iPads, and BlackBerrys?

4. D.W. Morgan's CEO has stated, "The iPhone is not a game changer, it's an industry changer. It changes the way that you can interact with your customers and with your suppliers." Discuss the implications of this statement.

MIS IN ACTION

Explore the Web site for the Apple iPhone, the Apple iPad, the BlackBerry, and the Motorola Droid, then answer the following questions:

1. List and describe the capabilities of each of these devices and give examples of how they could be used by businesses.

2. List and describe three downloadable business applications for each device and describe their business benefits.

iPhone and iPad Applications Used in Business:

1. Salesforce.com

2. FedEx Mobile

3. iTimeSheet

4. QuickOffice Connect

5. Documents to Go

6. GoodReader

7. Evernote

8. WebEx

Whether it's attending an online meeting, checking orders, working with files and documents, or obtaining business intelligence, Apple's iPhone and iPad offer unlimited possibilities for business users. Both devices have stunning multitouch display, full Internet browsing, capabilities for messaging, video and audio transmission, and document management. These features make each an all-purpose platform for mobile computing.

On the plus side, in a normal, non-recessionary year, the U.S. economy creates over 3.5 million new jobs. Employment in information systems and the other service occupations is expanding, and wages are stable. Outsourcing has actually accelerated the development of new systems in the United States and worldwide.

The challenge for you as a business student is to develop high-level skills through education and on-the-job experience that cannot be outsourced. The challenge for your business is to avoid markets for goods and services that can be produced offshore much less expensively. The opportunities are equally immense. You will find throughout this book examples of companies and individuals who either failed or succeeded in using information systems to adapt to this new global environment.

What does globalization have to do with management information systems? That's simple: everything. The emergence of the Internet into a full-blown international communications system has drastically reduced the costs of operating and transacting on a global scale. Communication between a factory floor in Shanghai and a distribution center in Rapid Falls, South Dakota, is now instant and virtually free. Customers now can shop in a worldwide marketplace, obtaining price and quality information reliably 24 hours a day. Firms producing goods and services on a global scale achieve extraordinary cost reductions by finding low-cost suppliers and managing production facilities in other countries. Internet service firms, such as Google and eBay, are able to replicate their business models and services in multiple countries without having to redesign their expensive fixed-cost information systems infrastructure. Half of the revenue of eBay (as well as General Motors) in 2011 will originate outside the United States. Briefly, information systems enable globalization.

THE EMERGING DIGITAL FIRM

All of the changes we have just described, coupled with equally significant organizational redesign, have created the conditions for a fully digital firm. A digital firm can be defined along several dimensions. A **digital firm** is one in which nearly all of the organization's *significant business relationships* with customers, suppliers, and employees are digitally enabled and mediated. *Core business processes* are accomplished through digital networks spanning the entire organization or linking multiple organizations.

Business processes refer to the set of logically related tasks and behaviors that organizations develop over time to produce specific business results and the unique manner in which these activities are organized and coordinated. Developing a new product, generating and fulfilling an order, creating a marketing plan, and hiring an employee are examples of business processes, and the ways organizations accomplish their business processes can be a source of competitive strength. (A detailed discussion of business processes can be found in Chapter 2.)

Key corporate assets—intellectual property, core competencies, and financial and human assets—are managed through digital means. In a digital firm, any piece of information required to support key business decisions is available at any time and anywhere in the firm.

Digital firms sense and respond to their environments far more rapidly than traditional firms, giving them more flexibility to survive in turbulent times. Digital firms offer extraordinary opportunities for more flexible global organization and management. In digital firms, both time shifting and space shifting are the norm. *Time shifting* refers to business being conducted continuously, 24/7, rather than in narrow "work day" time bands of 9 A.M. to 5 P.M. *Space shifting* means that work takes place in a global workshop, as well as within national boundaries. Work is accomplished physically wherever in the world it is best accomplished.

Many firms, such as Cisco Systems. 3M, and IBM, are close to becoming digital firms, using the Internet to drive every aspect of their business. Most other companies are not fully digital, but they are moving toward close digital integration with suppliers, customers, and employees. Many firms, for example, are replacing traditional face-to-face meetings with "virtual" meetings using video-conferencing and Web conferencing technology. (See Chapter 2.)

STRATEGIC BUSINESS OBJECTIVES OF INFORMATION SYSTEMS

What makes information systems so essential today? Why are businesses investing so much in information systems and technologies? In the United States, more than 23 million managers and 113 million workers in the labor force rely on information systems to conduct business. Information systems are essential for conducting day-to-day business in the United States and most other advanced countries, as well as achieving strategic business objectives.

Entire sectors of the economy are nearly inconceivable without substantial investments in information systems. E-commerce firms such as Amazon, eBay, Google, and E*Trade simply would not exist. Today's service industries—finance, insurance, and real estate, as well as personal services such as travel, medicine, and education—could not operate without information systems. Similarly, retail firms such as Walmart and Sears and manufacturing firms such as General Motors and General Electric require information systems to survive and prosper. Just as offices, telephones, filing cabinets, and efficient tall buildings with elevators were once the foundations of business in the twentieth century, information technology is a foundation for business in the twenty-first century.

There is a growing interdependence between a firm's ability to use information technology and its ability to implement corporate strategies and achieve corporate goals (see Figure 1-2). What a business would like to do in five years often depends on what its systems will be able to do. Increasing market share, becoming the high-quality or low-cost producer, developing new products, and increasing employee productivity depend more and more on the kinds and quality of information systems in the organization. The more you understand about this relationship, the more valuable you will be as a manager.

Specifically, business firms invest heavily in information systems to achieve six strategic business objectives: operational excellence; new products, services, and business models; customer and supplier intimacy; improved decision making; competitive advantage; and survival.

Operational Excellence

Businesses continuously seek to improve the efficiency of their operations in order to achieve higher profitability. Information systems and technologies are some of the most important tools available to managers for achieving higher levels of efficiency and productivity in business operations, especially when coupled with changes in business practices and management behavior.

Walmart, the largest retailer on earth, exemplifies the power of information systems coupled with brilliant business practices and supportive management

**FIGURE 1-2   THE INTERDEPENDENCE BETWEEN ORGANIZATIONS AND
INFORMATION SYSTEMS**

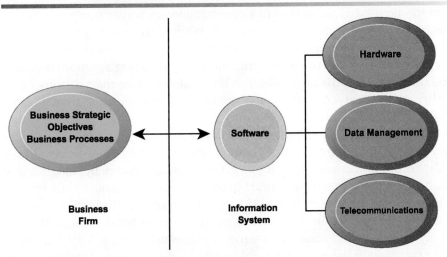

In contemporary systems there is a growing interdependence between a firm's information systems and its business capabilities. Changes in strategy, rules, and business processes increasingly require changes in hardware, software, databases, and telecommunications. Often, what the organization would like to do depends on what its systems will permit it to do.

to achieve world-class operational efficiency. In fiscal year 2010, Walmart achieved $408 billion in sales—nearly one-tenth of retail sales in the United States—in large part because of its Retail Link system, which digitally links its suppliers to every one of Walmart's stores. As soon as a customer purchases an item, the supplier monitoring the item knows to ship a replacement to the shelf. Walmart is the most efficient retail store in the industry, achieving sales of more than $28 per square foot, compared to its closest competitor, Target, at $23 a square foot, with other retail firms producing less than $12 a square foot.

New Products, Services, and Business Models

Information systems and technologies are a major enabling tool for firms to create new products and services, as well as entirely new business models. A **business model** describes how a company produces, delivers, and sells a product or service to create wealth.

Today's music industry is vastly different from the industry a decade ago. Apple Inc. transformed an old business model of music distribution based on vinyl records, tapes, and CDs into an online, legal distribution model based on its own iPod technology platform. Apple has prospered from a continuing stream of iPod innovations, including the iPod, the iTunes music service, the iPad, and the iPhone.

Customer and Supplier Intimacy

When a business really knows its customers, and serves them well, the customers generally respond by returning and purchasing more. This raises revenues and profits. Likewise with suppliers: the more a business engages its suppliers, the better the suppliers can provide vital inputs. This lowers costs. How to really know your customers, or suppliers, is a central problem for businesses with millions of offline and online customers.

The Mandarin Oriental in Manhattan and other high-end hotels exemplify the use of information systems and technologies to achieve customer intimacy. These hotels use computers to keep track of guests' preferences, such as their preferred room temperature, check-in time, frequently dialed telephone numbers, and television programs., and store these data in a large data repository. Individual rooms in the hotels are networked to a central network server computer so that they can be remotely monitored or controlled. When a customer arrives at one of these hotels, the system automatically changes the room conditions, such as dimming the lights, setting the room temperature, or selecting appropriate music, based on the customer's digital profile. The hotels also analyze their customer data to identify their best customers and to develop individualized marketing campaigns based on customers' preferences.

JCPenney exemplifies the benefits of information systems-enabled supplier intimacy. Every time a dress shirt is bought at a JCPenney store in the United States, the record of the sale appears immediately on computers in Hong Kong at the TAL Apparel Ltd. supplier, a contract manufacturer that produces one in eight dress shirts sold in the United States. TAL runs the numbers through a computer model it developed and then decides how many replacement shirts to make, and in what styles, colors, and sizes. TAL then sends the shirts to each JCPenney store, bypassing completely the retailer's warehouses. In other words, JCPenney's shirt inventory is near zero, as is the cost of storing it.

Improved Decision Making

Many business managers operate in an information fog bank, never really having the right information at the right time to make an informed decision. Instead, managers rely on forecasts, best guesses, and luck. The result is over- or underproduction of goods and services, misallocation of resources, and poor response times. These poor outcomes raise costs and lose customers. In the past decade, information systems and technologies have made it possible for managers to use real-time data from the marketplace when making decisions.

For instance, Verizon Corporation, one of the largest telecommunication companies in the United States, uses a Web-based digital dashboard to provide managers with precise real-time information on customer complaints, network performance for each locality served, and line outages or storm-damaged lines. Using this information, managers can immediately allocate repair resources to affected areas, inform consumers of repair efforts, and restore service fast.

Competitive Advantage

When firms achieve one or more of these business objectives—operational excellence; new products, services, and business models; customer/supplier intimacy; and improved decision making—chances are they have already achieved a competitive advantage. Doing things better than your competitors, charging less for superior products, and responding to customers and suppliers in real time all add up to higher sales and higher profits that your competitors cannot match. Apple Inc., Walmart, and UPS, described later in this chapter, are industry leaders because they know how to use information systems for this purpose.

Survival

Business firms also invest in information systems and technologies because they are necessities of doing business. Sometimes these "necessities" are driven by industry-level changes. For instance, after Citibank introduced the first automated teller machines (ATMs) in the New York region in 1977 to attract customers through higher service levels, its competitors rushed to provide ATMs to their customers to keep up with Citibank. Today, virtually all banks in the United States have regional ATMs and link to national and international ATM networks, such as CIRRUS. Providing ATM services to retail banking customers is simply a requirement of being in and surviving in the retail banking business.

There are many federal and state statutes and regulations that create a legal duty for companies and their employees to retain records, including digital records. For instance, the Toxic Substances Control Act (1976), which regulates the exposure of U.S. workers to more than 75,000 toxic chemicals, requires firms to retain records on employee exposure for 30 years. The Sarbanes–Oxley Act (2002), which was intended to improve the accountability of public firms and their auditors, requires certified public accounting firms that audit public companies to retain audit working papers and records, including all e-mails, for five years. Many other pieces of federal and state legislation in health care, financial services, education, and privacy protection impose significant information retention and reporting requirements on U.S. businesses. Firms turn to information systems and technologies to provide the capability to respond to these challenges.

1.2   PERSPECTIVES ON INFORMATION SYSTEMS

So far we've used *information systems* and *technologies* informally without defining the terms. **Information technology (IT)** consists of all the hardware and software that a firm needs to use in order to achieve its business objectives. This includes not only computer machines, storage devices, and handheld mobile devices, but also software, such as the Windows or Linux operating systems, the Microsoft Office desktop productivity suite, and the many thousands of computer programs that can be found in a typical large firm. "Information systems" are more complex and can be best be understood by looking at them from both a technology and a business perspective.

WHAT IS AN INFORMATION SYSTEM?

An **information system** can be defined technically as a set of interrelated components that collect (or retrieve), process, store, and distribute information to support decision making and control in an organization. In addition to supporting decision making, coordination, and control, information systems may also help managers and workers analyze problems, visualize complex subjects, and create new products.

Information systems contain information about significant people, places, and things within the organization or in the environment surrounding it. By **information** we mean data that have been shaped into a form that is meaningful and useful to human beings. **Data**, in contrast, are streams of raw facts representing events occurring in organizations or the physical environment before they have been organized and arranged into a form that people can understand and use.

A brief example contrasting information and data may prove useful. Supermarket checkout counters scan millions of pieces of data from bar codes, which describe each product. Such pieces of data can be totaled and analyzed to provide meaningful information, such as the total number of bottles of dish detergent sold at a particular store, which brands of dish detergent were selling the most rapidly at that store or sales territory, or the total amount spent on that brand of dish detergent at that store or sales region (see Figure 1-3).

FIGURE 1-3   **DATA AND INFORMATION**

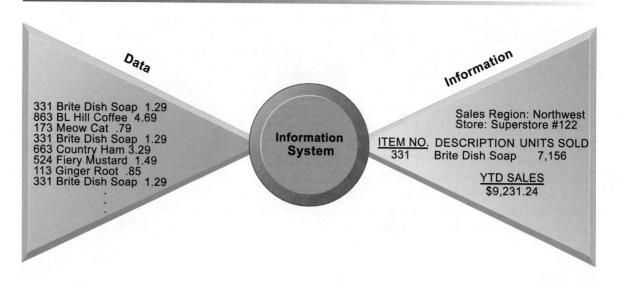

Raw data from a supermarket checkout counter can be processed and organized to produce meaningful information, such as the total unit sales of dish detergent or the total sales revenue from dish detergent for a specific store or sales territory.

Three activities in an information system produce the information that organizations need to make decisions, control operations, analyze problems, and create new products or services. These activities are input, processing, and output (see Figure 1-4). **Input** captures or collects raw data from within the organization or from its external environment. **Processing** converts this raw input into a meaningful form. **Output** transfers the processed information to the people who will use it or to the activities for which it will be used. Information systems also require **feedback**, which is output that is returned to appropriate members of the organization to help them evaluate or correct the input stage.

In the Yankees' system for selling tickets through its Web site, the raw input consists of order data for tickets, such as the purchaser's name, address, credit card number, number of tickets ordered, and the date of the game for which the ticket is being purchased. Computers store these data and process them to calculate order totals, to track ticket purchases, and to send requests for payment to credit card companies. The output consists of tickets to print out, receipts for orders, and reports on online ticket orders. The system provides meaningful information, such as the number of tickets sold for a particular game, the total number of tickets sold each year, and frequent customers.

Although computer-based information systems use computer technology to process raw data into meaningful information, there is a sharp distinction between a computer and a computer program on the one hand, and an information system on the other. Electronic computers and related software programs are the technical foundation, the tools and materials, of modern information systems. Computers provide the equipment for storing and processing information. Computer programs, or software, are sets of operating instructions that direct and control computer processing. Knowing how computers and computer programs work is important in designing solutions to organizational problems, but computers are only part of an information system.

FIGURE 1-4   **FUNCTIONS OF AN INFORMATION SYSTEM**

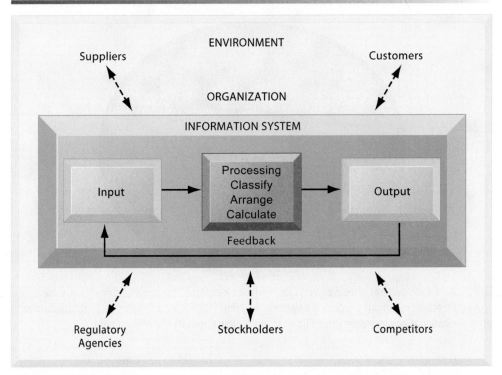

An information system contains information about an organization and its surrounding environment. Three basic activities—input, processing, and output—produce the information organizations need. Feedback is output returned to appropriate people or activities in the organization to evaluate and refine the input. Environmental actors, such as customers, suppliers, competitors, stockholders, and regulatory agencies, interact with the organization and its information systems.

A house is an appropriate analogy. Houses are built with hammers, nails, and wood, but these do not make a house. The architecture, design, setting, landscaping, and all of the decisions that lead to the creation of these features are part of the house and are crucial for solving the problem of putting a roof over one's head. Computers and programs are the hammers, nails, and lumber of computer-based information systems, but alone they cannot produce the information a particular organization needs. To understand information systems, you must understand the problems they are designed to solve, their architectural and design elements, and the organizational processes that lead to these solutions.

DIMENSIONS OF INFORMATION SYSTEMS

To fully understand information systems, you must understand the broader organization, management, and information technology dimensions of systems (see Figure 1-5) and their power to provide solutions to challenges and problems in the business environment. We refer to this broader understanding of information systems, which encompasses an understanding of the management and organizational dimensions of systems as well as the technical dimensions of systems, as **information systems literacy**. **Computer literacy**, in contrast, focuses primarily on knowledge of information technology.

The field of **management information systems (MIS)** tries to achieve this broader information systems literacy. MIS deals with behavioral issues as well

FIGURE 1-5   **INFORMATION SYSTEMS ARE MORE THAN COMPUTERS**

Using information systems effectively requires an understanding of the organization, management, and information technology shaping the systems. An information system creates value for the firm as an organizational and management solution to challenges posed by the environment.

as technical issues surrounding the development, use, and impact of information systems used by managers and employees in the firm.

Let's examine each of the dimensions of information systems—organizations, management, and information technology.

Organizations

Information systems are an integral part of organizations. Indeed, for some companies, such as credit reporting firms, there would be no business without an information system. The key elements of an organization are its people, structure, business processes, politics, and culture. We introduce these components of organizations here and describe them in greater detail in Chapters 2 and 3.

Organizations have a structure that is composed of different levels and specialties. Their structures reveal a clear-cut division of labor. Authority and responsibility in a business firm are organized as a hierarchy, or a pyramid structure. The upper levels of the hierarchy consist of managerial, professional, and technical employees, whereas the lower levels consist of operational personnel.

Senior management makes long-range strategic decisions about products and services as well as ensures financial performance of the firm. **Middle management** carries out the programs and plans of senior management and **operational management** is responsible for monitoring the daily activities of the business. **Knowledge workers**, such as engineers, scientists, or architects, design products or services and create new knowledge for the firm, whereas **data workers**, such as secretaries or clerks, assist with scheduling and communications at all levels of the firm. **Production or service workers** actually produce the product and deliver the service (see Figure 1-6).

Experts are employed and trained for different business functions. The major **business functions**, or specialized tasks performed by business organizations, consist of sales and marketing, manufacturing and production,

FIGURE 1-6　　**LEVELS IN A FIRM**

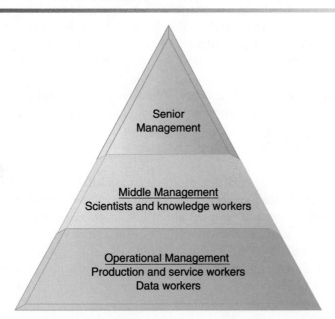

Senior
Management

Middle Management
Scientists and knowledge workers

Operational Management
Production and service workers
Data workers

Business organizations are hierarchies consisting of three principal levels: senior management, middle management, and operational management. Information systems serve each of these levels. Scientists and knowledge workers often work with middle management.

finance and accounting, and human resources (see Table 1-2). Chapter 2 provides more detail on these business functions and the ways in which they are supported by information systems.

An organization coordinates work through its hierarchy and through its business processes, which are logically related tasks and behaviors for accomplishing work. Developing a new product, fulfilling an order, and hiring a new employee are examples of business processes.

Most organizations' business processes include formal rules that have been developed over a long time for accomplishing tasks. These rules guide employees in a variety of procedures, from writing an invoice to responding to customer complaints. Some of these business processes have been written down, but others are informal work practices, such as a requirement to return telephone calls from co-workers or customers, that are not formally documented. Information systems automate many business processes. For instance, how a customer receives credit or how a customer is billed is often determined by an information system that incorporates a set of formal business processes.

TABLE 1-2 MAJOR BUSINESS FUNCTIONS

FUNCTION	PURPOSE
Sales and marketing	Selling the organization's products and services
Manufacturing and production	Producing and delivering products and services
Finance and accounting	Managing the organization's financial assets and maintaining the organization's financial records
Human resources	Attracting, developing, and maintaining the organization's labor force; maintaining employee records

Each organization has a unique **culture**, or fundamental set of assumptions, values, and ways of doing things, that has been accepted by most of its members. You can see organizational culture at work by looking around your university or college. Some bedrock assumptions of university life are that professors know more than students, the reasons students attend college is to learn, and that classes follow a regular schedule.

Parts of an organization's culture can always be found embedded in its information systems. For instance, UPS's concern with placing service to the customer first is an aspect of its organizational culture that can be found in the company's package tracking systems, which we describe later in this section.

Different levels and specialties in an organization create different interests and points of view. These views often conflict over how the company should be run and how resources and rewards should be distributed. Conflict is the basis for organizational politics. Information systems come out of this cauldron of differing perspectives, conflicts, compromises, and agreements that are a natural part of all organizations. In Chapter 3, we examine these features of organizations and their role in the development of information systems in greater detail.

Management

Management's job is to make sense out of the many situations faced by organizations, make decisions, and formulate action plans to solve organizational problems. Managers perceive business challenges in the environment; they set the organizational strategy for responding to those challenges; and they allocate the human and financial resources to coordinate the work and achieve success. Throughout, they must exercise responsible leadership. The business information systems described in this book reflect the hopes, dreams, and realities of real-world managers.

But managers must do more than manage what already exists. They must also create new products and services and even re-create the organization from time to time. A substantial part of management responsibility is creative work driven by new knowledge and information. Information technology can play a powerful role in helping managers design and deliver new products and services and redirecting and redesigning their organizations. Chapter 12 treats management decision making in detail.

Information Technology

Information technology is one of many tools managers use to cope with change. **Computer hardware** is the physical equipment used for input, processing, and output activities in an information system. It consists of the following: computers of various sizes and shapes (including mobile handheld devices); various input, output, and storage devices; and telecommunications devices that link computers together.

Computer software consists of the detailed, preprogrammed instructions that control and coordinate the computer hardware components in an information system. Chapter 5 describes the contemporary software and hardware platforms used by firms today in greater detail.

Data management technology consists of the software governing the organization of data on physical storage media. More detail on data organization and access methods can be found in Chapter 6.

Networking and telecommunications technology, consisting of both physical devices and software, links the various pieces of hardware and transfers data from one physical location to another. Computers and communications equipment can be connected in networks for sharing voice, data, images, sound, and video. A **network** links two or more computers to share data or resources, such as a printer.

The world's largest and most widely used network is the **Internet**. The Internet is a global "network of networks" that uses universal standards (described in Chapter 7) to connect millions of different networks with more than 1.4 billion users in over 230 countries around the world.

The Internet has created a new "universal" technology platform on which to build new products, services, strategies, and business models. This same technology platform has internal uses, providing the connectivity to link different systems and networks within the firm. Internal corporate networks based on Internet technology are called **intranets**. Private intranets extended to authorized users outside the organization are called **extranets**, and firms use such networks to coordinate their activities with other firms for making purchases, collaborating on design, and other interorganizational work. For most business firms today, using Internet technology is both a business necessity and a competitive advantage.

The **World Wide Web** is a service provided by the Internet that uses universally accepted standards for storing, retrieving, formatting, and displaying information in a page format on the Internet. Web pages contain text, graphics, animations, sound, and video and are linked to other Web pages. By clicking on highlighted words or buttons on a Web page, you can link to related pages to find additional information and links to other locations on the Web. The Web can serve as the foundation for new kinds of information systems such as UPS's Web-based package tracking system described in the following Interactive Session.

All of these technologies, along with the people required to run and manage them, represent resources that can be shared throughout the organization and constitute the firm's **information technology (IT) infrastructure**. The IT infrastructure provides the foundation, or *platform*, on which the firm can build its specific information systems. Each organization must carefully design and manage its IT infrastructure so that it has the set of technology services it needs for the work it wants to accomplish with information systems. Chapters 5 through 8 of this book examine each major technology component of information technology infrastructure and show how they all work together to create the technology platform for the organization.

The Interactive Session on Technology describes some of the typical technologies used in computer-based information systems today. UPS invests heavily in information systems technology to make its business more efficient and customer oriented. It uses an array of information technologies including bar code scanning systems, wireless networks, large mainframe computers, handheld computers, the Internet, and many different pieces of software for tracking packages, calculating fees, maintaining customer accounts, and managing logistics.

Let's identify the organization, management, and technology elements in the UPS package tracking system we have just described. The organization element anchors the package tracking system in UPS's sales and production functions (the main product of UPS is a service—package delivery). It specifies the required procedures for identifying packages with both sender and recipient information, taking inventory, tracking the packages en route, and providing package status reports for UPS customers and customer service representatives.

INTERACTIVE SESSION: TECHNOLOGY

UPS COMPETES GLOBALLY WITH INFORMATION TECHNOLOGY

United Parcel Service (UPS) started out in 1907 in a closet-sized basement office. Jim Casey and Claude Ryan—two teenagers from Seattle with two bicycles and one phone—promised the "best service and lowest rates." UPS has used this formula successfully for more than 100 years to become the world's largest ground and air package delivery company. It's a global enterprise with over 408,000 employees, 96,000 vehicles, and the world's ninth largest airline.

Today, UPS delivers more than 15 million packages and documents each day in the United States and more than 200 other countries and territories. The firm has been able to maintain leadership in small-package delivery services despite stiff competition from FedEx and Airborne Express by investing heavily in advanced information technology. UPS spends more than $1 billion each year to maintain a high level of customer service while keeping costs low and streamlining its overall operations.

It all starts with the scannable bar-coded label attached to a package, which contains detailed information about the sender, the destination, and when the package should arrive. Customers can download and print their own labels using special software provided by UPS or by accessing the UPS Web site. Before the package is even picked up, information from the "smart" label is transmitted to one of UPS's computer centers in Mahwah, New Jersey, or Alpharetta, Georgia, and sent to the distribution center nearest its final destination. Dispatchers at this center download the label data and use special software to create the most efficient delivery route for each driver that considers traffic, weather conditions, and the location of each stop. UPS estimates its delivery trucks save 28 million miles and burn 3 million fewer gallons of fuel each year as a result of using this technology. To further increase cost savings and safety, drivers are trained to use "340 Methods" developed by industrial engineers to optimize the performance of every task from lifting and loading boxes to selecting a package from a shelf in the truck.

The first thing a UPS driver picks up each day is a handheld computer called a Delivery Information Acquisition Device (DIAD), which can access one of the wireless networks cell phones rely on. As soon as the driver logs on, his or her day's route is downloaded onto the handheld. The DIAD also automati-

cally captures customers' signatures along with pickup and delivery information. Package tracking information is then transmitted to UPS's computer network for storage and processing. From there, the information can be accessed worldwide to provide proof of delivery to customers or to respond to customer queries. It usually takes less than 60 seconds from the time a driver presses "complete" on a DIAD for the new information to be available on the Web.

Through its automated package tracking system, UPS can monitor and even re-route packages throughout the delivery process. At various points along the route from sender to receiver, bar code devices scan shipping information on the package label and feed data about the progress of the package into the central computer. Customer service representatives are able to check the status of any package from desktop computers linked to the central computers and respond immediately to inquiries from customers. UPS customers can also access this information from the company's Web site using their own computers or mobile phones.

Anyone with a package to ship can access the UPS Web site to check delivery routes, calculate shipping rates, determine time in transit, print labels, schedule a pickup, and track packages. The data collected at the UPS Web site are transmitted to the UPS central computer and then back to the customer after processing. UPS also provides tools that enable customers, such Cisco Systems, to embed UPS functions, such as tracking and cost calculations, into their own Web sites so that they can track shipments without visiting the UPS site.

In June 2009, UPS launched a new Web-based Post-Sales Order Management System (OMS) that manages global service orders and inventory for critical parts fulfillment. The system enables high-tech electronics, aerospace, medical equipment, and other companies anywhere in the world that ship critical parts to quickly assess their critical parts inventory, determine the most optimal routing strategy to meet customer needs, place orders online, and track parts from the warehouse to the end user. An automated e-mail or fax feature keeps customers informed of each shipping milestone and can provide notification of any changes to flight schedules for commercial airlines carrying their parts. Once orders are complete, companies can print documents such as labels and bills of lading in multiple languages.

UPS is now leveraging its decades of expertise managing its own global delivery network to manage logistics and supply chain activities for other companies. It created a UPS Supply Chain Solutions division that provides a complete bundle of standardized services to subscribing companies at a fraction of what it would cost to build their own systems and infrastructure. These services include supply chain design and management, freight forwarding, customs brokerage, mail services, multimodal transportation, and financial services, in addition to logistics services.

Servalite, an East Moline, Illinois, manufacturer of fasteners, sells 40,000 different products to hardware stores and larger home improvement stores. The company had used multiple warehouses to provide two-day delivery nationwide. UPS created a new logistics plan for the company that helped it reduce freight time in transit and consolidate inventory. Thanks to these improvements, Servalite has been able to keep its two-day delivery guarantee while lowering warehousing and inventory costs.

Sources: Jennifer Levitz, "UPS Thinks Out of the Box on Driver Training," *The Wall Street Journal*, April 6, 2010; United Parcel Service, "In a Tighter Economy, a Manufacturer Fastens Down Its Logistics," *UPS Compass*, accessed May 5, 2010; Agam Shah, "UPS Invests $1 Billion in Technology to Cut Costs," *Bloomberg Businessweek*, March 25, 2010; UPS, "UPS Delivers New App for Google's Android," April 12, 2010; Chris Murphy, "In for the Long Haul," *Information Week*, January 19, 2009; United Parcel Service, " UPS Unveils Global Technology for Critical Parts Fulfillment," June 16, 2009; and www.ups.com, accessed May 5, 2010.

CASE STUDY QUESTIONS

1. What are the inputs, processing, and outputs of UPS's package tracking system?
2. What technologies are used by UPS? How are these technologies related to UPS's business strategy?
3. What strategic business objectives do UPS's information systems address?
4. What would happen if UPS's information systems were not available?

MIS IN ACTION

Explore the UPS Web site (www.ups.com) and answer the following questions:

1. What kind of information and services does the Web site provide for individuals, small businesses, and large businesses? List these services.
2. Go to the Business Solutions portion of the UPS Web site. Browse the UPS Business Solutions by category (such as shipment delivery, returns, or international trade) and write a description of all the services UPS provides for one of these categories. Explain how a business would benefit from these services.
3. Explain how the Web site helps UPS achieve some or all of the strategic business objectives we described earlier in this chapter. What would be the impact on UPS's business if this Web site were not available?

The system must also provide information to satisfy the needs of managers and workers. UPS drivers need to be trained in both package pickup and delivery procedures and in how to use the package tracking system so that they can work efficiently and effectively. UPS customers may need some training to use UPS in-house package tracking software or the UPS Web site.

UPS's management is responsible for monitoring service levels and costs and for promoting the company's strategy of combining low cost and superior service. Management decided to use computer systems to increase the ease of sending a package using UPS and of checking its delivery status, thereby reducing delivery costs and increasing sales revenues.

The technology supporting this system consists of handheld computers, bar code scanners, wired and wireless communications networks, desktop computers, UPS's data center, storage technology for the package delivery data, UPS in-house package tracking software, and software to access the World Wide Web. The result is an information system solution to the business challenge of providing a high level of service with low prices in the face of mounting competition.

IT ISN'T JUST TECHNOLOGY: A BUSINESS PERSPECTIVE ON INFORMATION SYSTEMS

Managers and business firms invest in information technology and systems because they provide real economic value to the business. The decision to build or maintain an information system assumes that the returns on this investment will be superior to other investments in buildings, machines, or other assets. These superior returns will be expressed as increases in productivity, as increases in revenues (which will increase the firm's stock market value), or perhaps as superior long-term strategic positioning of the firm in certain markets (which produce superior revenues in the future).

We can see that from a business perspective, an information system is an important instrument for creating value for the firm. Information systems enable the firm to increase its revenue or decrease its costs by providing information that helps managers make better decisions or that improves the execution of business processes. For example, the information system for analyzing supermarket checkout data illustrated in Figure 1-3 can increase firm profitability by helping managers make better decisions on which products to stock and promote in retail supermarkets.

Every business has an information value chain, illustrated in Figure 1-7, in which raw information is systematically acquired and then transformed through various stages that add value to that information. The value of an information system to a business, as well as the decision to invest in any new information system, is, in large part, determined by the extent to which the system will lead to better management decisions, more efficient business

Using a handheld computer called a Delivery Information Acquisition Device (DIAD), UPS drivers automatically capture customers' signatures along with pickup, delivery, and time card information. UPS information systems use these data to track packages while they are being transported.

FIGURE 1-7   THE BUSINESS INFORMATION VALUE CHAIN

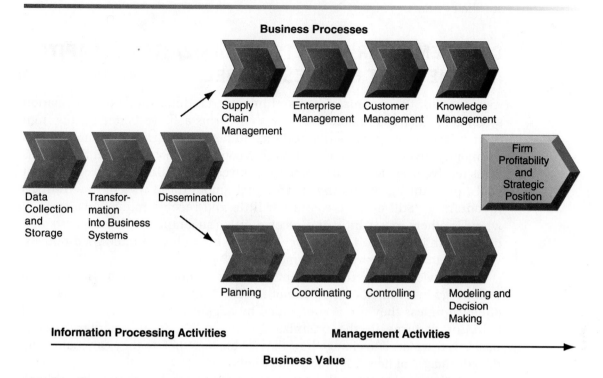

From a business perspective, information systems are part of a series of value-adding activities for acquiring, transforming, and distributing information that managers can use to improve decision making, enhance organizational performance, and, ultimately, increase firm profitability.

processes, and higher firm profitability. Although there are other reasons why systems are built, their primary purpose is to contribute to corporate value.

From a business perspective, information systems are part of a series of value-adding activities for acquiring, transforming, and distributing information that managers can use to improve decision making, enhance organizational performance, and, ultimately, increase firm profitability.

The business perspective calls attention to the organizational and managerial nature of information systems. An information system represents an organizational and management solution, based on information technology, to a challenge or problem posed by the environment. Every chapter in this book begins with a short case study that illustrates this concept. A diagram at the beginning of each chapter illustrates the relationship between a business challenge and resulting management and organizational decisions to use IT as a solution to challenges generated by the business environment. You can use this diagram as a starting point for analyzing any information system or information system problem you encounter.

Review the diagram at the beginning of this chapter. The diagram shows how the Yankees' systems solved the business problem presented by declining interest in baseball games and competition from television and other media. These systems provide a solution that takes advantage of new interactive digital technology and opportunities created by the Internet. They opened up new channels for selling tickets and interacting with customers that improved business performance. The diagram also illustrates how management, technology, and organizational elements work together to create the systems.

COMPLEMENTARY ASSETS: ORGANIZATIONAL CAPITAL AND THE RIGHT BUSINESS MODEL

Awareness of the organizational and managerial dimensions of information systems can help us understand why some firms achieve better results from their information systems than others. Studies of returns from information technology investments show that there is considerable variation in the returns firms receive (see Figure 1-8). Some firms invest a great deal and receive a great deal (quadrant 2); others invest an equal amount and receive few returns (quadrant 4). Still other firms invest little and receive much (quadrant 1), whereas others invest little and receive little (quadrant 3). This suggests that investing in information technology does not by itself guarantee good returns. What accounts for this variation among firms?

The answer lies in the concept of complementary assets. Information technology investments alone cannot make organizations and managers more effective unless they are accompanied by supportive values, structures, and behavior patterns in the organization and other complementary assets. Business firms need to change how they do business before they can really reap the advantages of new information technologies.

Some firms fail to adopt the right business model that suits the new technology, or seek to preserve an old business model that is doomed by new technology. For instance, recording label companies refused to change their old business model, which was based on physical music stores for distribution rather than adopt a new online distribution model. As a result, online legal

FIGURE 1-8 **VARIATION IN RETURNS ON INFORMATION TECHNOLOGY INVESTMENT**

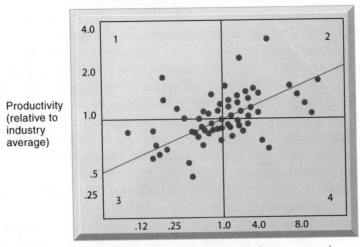

IT Capital Stock (relative to industry average)

Although, on average, investments in information technology produce returns far above those returned by other investments, there is considerable variation across firms.

Source: Based on Brynjolfsson and Hitt (2000).

music sales are dominated not by record companies but by a technology company called Apple Computer.

Complementary assets are those assets required to derive value from a primary investment (Teece, 1988). For instance, to realize value from automobiles requires substantial complementary investments in highways, roads, gasoline stations, repair facilities, and a legal regulatory structure to set standards and control drivers.

Research on business information technology investment indicates that firms that support their technology investments with investments in complementary assets, such as new business models, new business processes, management behavior, organizational culture, or training, receive superior returns, whereas those firms failing to make these complementary investments receive less or no returns on their information technology investments (Brynjolfsson, 2003; Brynjolfsson and Hitt, 2000; Davern and Kauffman, 2000; Laudon, 1974). These investments in organization and management are also known as **organizational and management capital**.

Table 1-3 lists the major complementary investments that firms need to make to realize value from their information technology investments. Some of this investment involves tangible assets, such as buildings, machinery, and tools. However, the value of investments in information technology depends to a large extent on complementary investments in management and organization.

Key organizational complementary investments are a supportive business culture that values efficiency and effectiveness, an appropriate business model, efficient business processes, decentralization of authority, highly distributed decision rights, and a strong information system (IS) development team.

Important managerial complementary assets are strong senior management support for change, incentive systems that monitor and reward individual innovation, an emphasis on teamwork and collaboration, training programs, and a management culture that values flexibility and knowledge.

TABLE 1-3 COMPLEMENTARY SOCIAL, MANAGERIAL, AND ORGANIZATIONAL ASSETS REQUIRED TO OPTIMIZE RETURNS FROM INFORMATION TECHNOLOGY INVESTMENTS

Organizational assets	Supportive organizational culture that values efficiency and effectiveness
	Appropriate business model
	Efficient business processes
	Decentralized authority
	Distributed decision-making rights
	Strong IS development team
Managerial assets	Strong senior management support for technology investment and change
	Incentives for management innovation
	Teamwork and collaborative work environments
	Training programs to enhance management decision skills
	Management culture that values flexibility and knowledge-based decision making.
Social assets	The Internet and telecommunications infrastructure
	IT-enriched educational programs raising labor force computer literacy
	Standards (both government and private sector)
	Laws and regulations creating fair, stable market environments
	Technology and service firms in adjacent markets to assist implementation

Important social investments (not made by the firm but by the society at large, other firms, governments, and other key market actors) are the Internet and the supporting Internet culture, educational systems, network and computing standards, regulations and laws, and the presence of technology and service firms.

Throughout the book we emphasize a framework of analysis that considers technology, management, and organizational assets and their interactions. Perhaps the single most important theme in the book, reflected in case studies and exercises, is that managers need to consider the broader organization and management dimensions of information systems to understand current problems as well as to derive substantial above-average returns from their information technology investments. As you will see throughout the text, firms that can address these related dimensions of the IT investment are, on average, richly rewarded.

1.3   CONTEMPORARY APPROACHES TO INFORMATION SYSTEMS

The study of information systems is a multidisciplinary field. No single theory or perspective dominates. Figure 1-9 illustrates the major disciplines that contribute problems, issues, and solutions in the study of information systems. In general, the field can be divided into technical and behavioral approaches. Information systems are sociotechnical systems. Though they are composed of machines, devices, and "hard" physical technology, they require substantial social, organizational, and intellectual investments to make them work properly.

FIGURE 1-9   CONTEMPORARY APPROACHES TO INFORMATION SYSTEMS

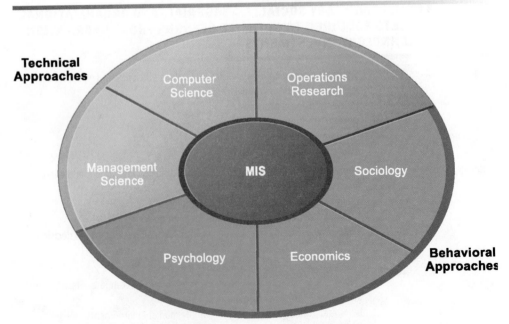

The study of information systems deals with issues and insights contributed from technical and behavioral disciplines.

TECHNICAL APPROACH

The technical approach to information systems emphasizes mathematically based models to study information systems, as well as the physical technology and formal capabilities of these systems. The disciplines that contribute to the technical approach are computer science, management science, and operations research.

Computer science is concerned with establishing theories of computability, methods of computation, and methods of efficient data storage and access. Management science emphasizes the development of models for decision-making and management practices. Operations research focuses on mathematical techniques for optimizing selected parameters of organizations, such as transportation, inventory control, and transaction costs.

BEHAVIORAL APPROACH

An important part of the information systems field is concerned with behavioral issues that arise in the development and long-term maintenance of information systems. Issues such as strategic business integration, design, implementation, utilization, and management cannot be explored usefully with the models used in the technical approach. Other behavioral disciplines contribute important concepts and methods.

For instance, sociologists study information systems with an eye toward how groups and organizations shape the development of systems and also how systems affect individuals, groups, and organizations. Psychologists study information systems with an interest in how human decision makers perceive and use formal information. Economists study information systems with an interest in understanding the production of digital goods, the dynamics of digital markets, and how new information systems change the control and cost structures within the firm.

The behavioral approach does not ignore technology. Indeed, information systems technology is often the stimulus for a behavioral problem or issue. But the focus of this approach is generally not on technical solutions. Instead, it concentrates on changes in attitudes, management and organizational policy, and behavior.

APPROACH OF THIS TEXT: SOCIOTECHNICAL SYSTEMS

Throughout this book you will find a rich story with four main actors: suppliers of hardware and software (the technologists); business firms making investments and seeking to obtain value from the technology; managers and employees seeking to achieve business value (and other goals); and the contemporary legal, social, and cultural context (the firm's environment). Together these actors produce what we call *management information systems*.

The study of management information systems (MIS) arose to focus on the use of computer-based information systems in business firms and government agencies. MIS combines the work of computer science, management science, and operations research with a practical orientation toward developing system solutions to real-world problems and managing information technology resources. It is also concerned with behavioral issues surrounding the development, use, and impact of information systems, which are typically discussed in the fields of sociology, economics, and psychology.

Our experience as academics and practitioners leads us to believe that no single approach effectively captures the reality of information systems. The successes and failures of information are rarely all technical or all behavioral. Our best advice to students is to understand the perspectives of many disciplines. Indeed, the challenge and excitement of the information systems field is that it requires an appreciation and tolerance of many different approaches.

The view we adopt in this book is best characterized as the **sociotechnical view** of systems. In this view, optimal organizational performance is achieved by jointly optimizing both the social and technical systems used in production.

Adopting a sociotechnical systems perspective helps to avoid a purely technological approach to information systems. For instance, the fact that information technology is rapidly declining in cost and growing in power does not necessarily or easily translate into productivity enhancement or bottom-line profits. The fact that a firm has recently installed an enterprise-wide financial reporting system does not necessarily mean that it will be used, or used effectively. Likewise, the fact that a firm has recently introduced new business procedures and processes does not necessarily mean employees will be more productive in the absence of investments in new information systems to enable those processes.

In this book, we stress the need to optimize the firm's performance as a whole. Both the technical and behavioral components need attention. This means that technology must be changed and designed in such a way as to fit organizational and individual needs. Sometimes, the technology may have to be "de-optimized" to accomplish this fit. For instance, mobile phone users adapt this technology to their personal needs, and as a result manufacturers quickly seek to adjust the technology to conform with user expectations. Organizations and individuals must also be changed through training, learning, and planned organizational change to allow the technology to operate and prosper. Figure 1-10 illustrates this process of mutual adjustment in a sociotechnical system.

FIGURE 1-10   A SOCIOTECHNICAL PERSPECTIVE ON INFORMATION SYSTEMS

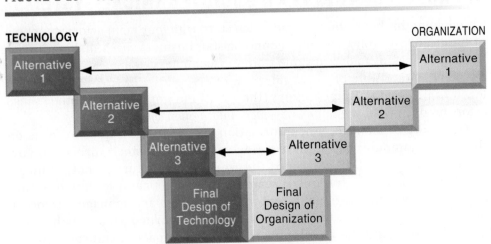

In a sociotechnical perspective, the performance of a system is optimized when both the technology and the organization mutually adjust to one another until a satisfactory fit is obtained.

1.4 HANDS-ON MIS PROJECTS

The projects in this section give you hands-on experience in analyzing financial reporting and inventory management problems, using data management software to improve management decision making about increasing sales, and using Internet software for developing shipping budgets.

Management Decision Problems

1. Snyders of Hanover, which sells more than 78 million bags of pretzels, snack chips, and organic snack items each year, had its financial department use spreadsheets and manual processes for much of its data gathering and reporting. Hanover's financial analyst would spend the entire final week of every month collecting spreadsheets from the heads of more than 50 departments worldwide. She would then consolidate and re-enter all the data into another spreadsheet, which would serve as the company's monthly profit-and-loss statement. If a department needed to update its data after submitting the spreadsheet to the main office, the analyst had to return the original spreadsheet and wait for the department to re-submit its data before finally submitting the updated data in the consolidated document. Assess the impact of this situation on business performance and management decision making.

2. Dollar General Corporation operates deep discount stores offering house-wares, cleaning supplies, clothing, health and beauty aids, and packaged food, with most items selling for $1. Its business model calls for keeping costs as low as possible. Although the company uses information systems (such as a point-of-sale system to track sales at the register), it deploys them very sparingly to keep expenditures to the minimum. The company has no automated method for keeping track of inventory at each store. Managers know approximately how many cases of a particular product the store is supposed to receive when a delivery truck arrives, but the stores lack technology for scanning the cases or verifying the item count inside the cases. Merchandise losses from theft or other mishaps have been rising and now represent over 3 percent of total sales. What decisions have to be made before investing in an information system solution?

Improving Decision Making: Using Databases to Analyze Sales Trends

Software skills: Database querying and reporting
Business skills: Sales trend analysis

Effective information systems transform data into meaningful information for decisions that improve business performance. In MyMISLab, you can find a Store and Regional Sales Database with raw data on weekly store sales of computer equipment in various sales regions. A sample is shown below, but MyMISLab may have a more recent version of this database for this exercise. The database includes fields for store identification number, sales region number, item number, item description, unit price, units sold, and the weekly sales period when the sales were made. Develop some reports and queries to make this information more useful for running the business. Try to use the information in the database to support decisions on which products to restock, which stores and sales regions would benefit from additional marketing and promotional campaigns, which times of the year products

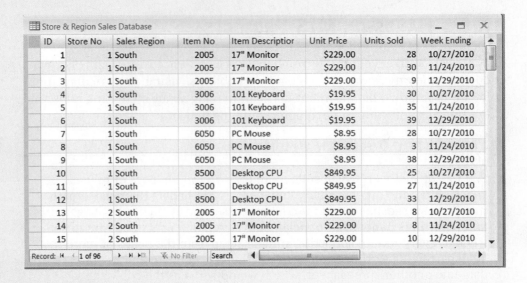

should be offered at full price, and which times of the year products should be discounted. Modify the database table, if necessary, to provide all of the information you require. Print your reports and results of queries.

Improving Decision Making: Using the Internet to Locate Jobs Requiring Information Systems Knowledge

Software skills: Internet-based software
Business skills: Job searching

Visit job-posting Web sites such as Monster.com or CareerBuilder.com. Spend some time at the sites examining jobs for accounting, finance, sales, marketing, and human resources. Find two or three descriptions of jobs that require some information systems knowledge. What information systems knowledge do these jobs require? What do you need to do to prepare for these jobs? Write a one- to two-page report summarizing your findings.

LEARNING TRACK MODULES

The following Learning Tracks provide content relevant to topics covered in this chapter:

1. How Much Does IT Matter?
2. Information Systems and Your Career
3. The Emerging Mobile Digital Platform

Review Summary

1. *How are information systems transforming business and what is their relationship to globalization?*

   E-mail, online conferencing, and cell phones have become essential tools for conducting business. Information systems are the foundation of fast-paced supply chains. The Internet allows many businesses to buy, sell, advertise, and solicit customer feedback online. Organizations are trying to become more competitive and efficient by digitally enabling their core business processes and evolving into digital firms. The Internet has stimulated globalization by dramatically reducing the costs of producing, buying, and selling goods on a global scale. New information system trends include the emerging mobile digital platform, online software as a service, and cloud computing.

2. *Why are information systems so essential for running and managing a business today?*

   Information systems are a foundation for conducting business today. In many industries, survival and the ability to achieve strategic business goals are difficult without extensive use of information technology. Businesses today use information systems to achieve six major objectives: operational excellence; new products, services, and business models; customer/supplier intimacy; improved decision making; competitive advantage; and day-to-day survival.

3. *What exactly is an information system? How does it work? What are its management, organization, and technology components?*

   From a technical perspective, an information system collects, stores, and disseminates information from an organization's environment and internal operations to support organizational functions and decision making, communication, coordination, control, analysis, and visualization. Information systems transform raw data into useful information through three basic activities: input, processing, and output.

   From a business perspective, an information system provides a solution to a problem or challenge facing a firm and represents a combination of management, organization, and technology elements. The management dimension of information systems involves issues such as leadership, strategy, and management behavior. The technology dimension consists of computer hardware, software, data management technology, and networking/telecommunications technology (including the Internet). The organization dimension of information systems involves issues such as the organization's hierarchy, functional specialties, business processes, culture, and political interest groups.

4. *What are complementary assets? Why are complementary assets essential for ensuring that information systems provide genuine value for an organization?*

   In order to obtain meaningful value from information systems, organizations must support their technology investments with appropriate complementary investments in organizations and management. These complementary assets include new business models and business processes, supportive organizational culture and management behavior, appropriate technology standards, regulations, and laws. New information technology investments are unlikely to produce high returns unless businesses make the appropriate managerial and organizational changes to support the technology.

5. *What academic disciplines are used to study information systems? How does each contribute to an understanding of information systems? What is a sociotechnical systems perspective?*

   The study of information systems deals with issues and insights contributed from technical and behavioral disciplines. The disciplines that contribute to the technical approach focusing on formal models and capabilities of systems are computer science, management science, and operations research. The disciplines contributing to the behavioral approach focusing on the design, implementation, management, and business impact of systems are psychology, sociology, and economics. A sociotechnical view of systems considers both technical and social features of systems and solutions that represent the best fit between them.

Key Terms

Review Questions

1. How are information systems transforming business and what is their relationship to globalization?
   - Describe how information systems have changed the way businesses operate and their products and services.
   - Identify three major new information system trends.
   - Describe the characteristics of a digital firm.
   - Describe the challenges and opportunities of globalization in a "flattened" world.

2. Why are information systems so essential for running and managing a business today?
   - List and describe six reasons why information systems are so important for business today.

3. What exactly is an information system? How does it work? What are its management, organization, and technology components?
   - Define an information system and describe the activities it performs.
   - List and describe the organizational, management, and technology dimensions of information systems.
   - Distinguish between data and information and between information systems literacy and computer literacy.

   - Explain how the Internet and the World Wide Web are related to the other technology components of information systems.

4. What are complementary assets? Why are complementary assets essential for ensuring that information systems provide genuine value for an organization?
   - Define complementary assets and describe their relationship to information technology.
   - Describe the complementary social, managerial, and organizational assets required to optimize returns from information technology investments.

5. What academic disciplines are used to study information systems? How does each contribute to an understanding of information systems? What is a sociotechnical systems perspective?
   - List and describe each discipline that contributes to a technical approach to information systems.
   - List and describe each discipline that contributes to a behavioral approach to information systems.
   - Describe the sociotechnical perspective on information systems.

Discussion Questions

1. Information systems are too important to be left to computer specialists. Do you agree? Why or why not?

2. If you were setting up the Web site for another Major League Baseball team, what management, organization, and technology issues might you encounter?

3. What are some of the organizational, managerial, and social complementary assets that help make UPS's information systems so successful?

Video Cases

Video Cases and Instructional Videos illustrating some of the concepts in this chapter are available. Contact your instructor to access these videos.

Collaboration and Teamwork: Creating a Web Site for Team Collaboration

Form a team with three or four classmates. Then use the tools at Google Sites to create a Web site for your team. You will need to a create a Google account for the site and specify the collaborators (your team members) who are allowed to access the site and make contributions. Specify your professor as the viewer of the site so that person can evaluate your work. Assign a name to the site. Select a theme for the site and make any changes you wish to colors and fonts. Add features for project announcements and a repository for team documents, source materials, illustrations, electronic presentations, and Web pages of interest. You can add other features if you wish. Use Google to create a calendar for your team. After you complete this exercise, you can use this Web site and calendar for your other team projects.

What's the Buzz on Smart Grids?
CASE STUDY

The existing electricity infrastructure in the United States is outdated and inefficient. Energy companies provide power to consumers, but the grid provides no information about how the consumers are using that energy, making it difficult to develop more efficient approaches to distribution. Also, the current electricity grid offers few ways to handle power provided by alternative energy sources, which are critical components of most efforts to go "green." Enter the smart grid.

A smart grid delivers electricity from suppliers to consumers using digital technology to save energy, reduce costs, and increase reliability and transparency. The smart grid enables information to flow back and forth between electric power providers and individual households to allow both consumers and energy companies to make more intelligent decisions regarding energy consumption and production. Information from smart grids would show utilities when to raise prices when demand is high and lower them when demand lessens. Smart grids would also help consumers program high-use electrical appliances like heating and air conditioning systems to reduce consumption during times of peak usage. If implemented nationwide, proponents believe, smart grids would lead to a 5 to 15 percent decrease in energy consumption. Electricity grids are sized to meet the maximum electricity need, so a drop in peak demand would enable utilities to operate with fewer expensive power plants, thereby lowering costs and pollution.

Another advantage of smart grids is their ability to detect sources of power outages more quickly and precisely at the individual household level. With such precise information, utilities will be able to respond to service problems more rapidly and efficiently.

Managing the information flowing in these smart grids requires technology: networks and switches for power management; sensor and monitoring devices to track energy usage and distribution trends; systems to provide energy suppliers and consumers with usage data; communications systems to relay data along the entire energy supply system; and systems linked to programmable appliances to run them when energy is least costly.

If consumers had in-home displays showing how much energy they are consuming at any moment and the price of that energy, they are more likely to curb their consumption to cut costs. Home thermostats and appliances could adjust on their own automatically, depending on the cost of power, and even obtain that power from nontraditional sources, such as a neighbor's rooftop solar panel. Instead of power flowing from a small number of power plants, the smart grid will make it possible to have a distributed energy system. Electricity will flow from homes and businesses into the grid, and they will use power from local and faraway sources. Besides increasing energy efficiency, converting to smart grids along with other related energy initiatives could create up to 370,000 jobs.

That's why pioneering smart grid projects such as SmartGridCity in Boulder, Colorado, are attracting attention. SmartGridCity represents a collaboration by Xcel Energy Inc. and residents of Boulder to test the viability of smart grids on a smaller scale. Participants can check their power consumption levels and costs online, and will soon be able to program home appliances over the Web. Customers access this information and set goals and guidelines for their home's energy usage through a Web portal. They also have the option of allowing Xcel to remotely adjust their thermostats during periods of high demand.

SmartGridCity is also attempting to turn homes into "miniature power plants" using solar-powered battery packs that "TiVo electricity," or stash it away to use at a later time. This serves as backup power for homes using the packs, but Xcel can also tap into that power during times of peak energy consumption to lessen the overall energy load. Xcel will be able to remotely adjust thermostats and water heaters and will have much better information about the power consumption of their consumers.

Bud Peterson, chancellor of the University of Colorado at Boulder, and his wife Val have worked with Xcel to turn their home into the prototype residence for the SmartGridCity project. Their house was supplied with a six-kilowatt photovoltaic system on two roofs, four thermostats controlled via the Web, a plug-in hybrid electric vehicle (PHEV) Ford Escape, and other high-tech, smart grid-compatible features. Xcel employees are able to monitor periods of high power consumption and how much energy the Petersons' Escape is using on the road.

A digital dashboard in the Petersons' house displays power usage information in dozens of different ways—live household consumption and production, stored backup power, and carbon emission reductions translated into gallons of gasoline and acres of trees saved each year. The dashboard also allows the Petersons to program their home thermostats to adjust the temperature by room, time of day, and season. Since the project began in the spring of 2008, the Petersons have been able to reduce their electricity use by one-third.

Xcel is not alone. Hundreds of technology companies and almost every major electric utility company see smart grids as the wave of the future. Heightening interest is $3.4 billion in federal economic recovery money for smart grid technology.

Duke Energy spent $35 million on smart grid initiatives, installing 80,000 smart meters as part of a pilot project in Charlotte, North Carolina, to provide business and residential customers with up-to-the-minute information on their energy use, as well as data on how much their appliances cost to operate. This helps them save money by curbing usage during peak times when rates are high or by replacing inefficient appliances. Duke now plans to spend $1 billion on sensors, intelligent meters, and other upgrades for a smart grid serving 700,000 customers in Cincinnati.

Florida Power and Light is budgeting $200 million for smart meters covering 1 million homes and businesses in the Miami area over the next two years. Center Point Energy, which services 2.2 million customers in the metropolitan Houston area, is planning to spend $1 billion over the next five years on a smart grid. Although residential customers' monthly electric bills will be $3.24 higher, the company says this amount will be more than offset by energy savings. Pacific Gas & Electric, which distributes power to Northern and Central California, is in the process of installing 10 million smart meters by mid-2012.

Google has developed a free Web service called PowerMeter for tracking energy use online in houses or businesses as power is consumed. It expects other companies to build the devices that will supply data to PowerMeter.

There are a number of challenges facing the efforts to implement smart grids. Changing the infrastructure of our electricity grids is a daunting task. Two-way meters that allow information to flow both to and from homes need to be installed at any home or building that uses electric power–in other words, essentially everywhere. Another challenge is creating an intuitive end-user interface. Some SmartGridCity participants reported that the dashboard they used to manage their appliances was too confusing and high-tech. Even Val Peterson admitted that, at first, managing the information about her power usage supplied through the Xcel Web portal was an intimidating process.

The smart grid won't be cheap, with estimated costs running as high as $75 billion. Meters run $250 to $500 each when they are accompanied by new utility billing systems. Who is going to pay the bill? Is the average consumer willing to pay the upfront costs for a smart grid system and then respond appropriately to price signals? Will consumers and utility companies get the promised payback if they buy into smart grid technology? Might "smart meters" be too intrusive? Would consumers really want to entrust energy companies with regulating the energy usage inside their homes? Would a highly computerized grid increase the risk of cyberattacks?

Jack Oliphant, a retiree living north of Houston in Spring, Texas, believes that the $444 he will pay Center Point for a smart meter won't justify the expense. "There's no mystery about how you save energy," he says. "You turn down the air conditioner and shut off some lights. I don't need an expensive meter to do that." Others have pointed out other less-expensive methods of reducing energy consumption. Marcel Hawiger, an attorney for The Utility Reform Network, a San Francisco consumer advocacy group, favors expanding existing air conditioner-cycling programs, where utilities are able to control air conditioners so they take turns coming on and off, thereby reducing demands on the electric system. He believes air conditioner controllers, which control temperature settings and compressors to reduce overall energy costs, provide much of the benefit of smart meters at a fraction of their cost.

Consumer advocates have vowed to fight smart grids if they boost rates for customers who are unable or unwilling to use Web portals and allow energy companies to control aspects of their appliances. Advocates also argue that smart grids represent an Orwellian intrusion of people's right to use their appliances as they see fit without disclosing facts about their usage to others. A proposal by officials in California to require all new homes to have remotely adjustable thermostats was soundly defeated after critics worried about the privacy implications.

Energy companies stand to lose money as individuals conserve more electricity, creating a disincentive for them to cooperate with conservation efforts like smart grids. Patience will be critical as energy companies and local communities work to set up new technologies and pricing plans.

Sources: Rebecca Smith, "What Utilities Have Learned from Smart-Meter Tests," *The Wall Street Journal*, February 22, 2010; "Smart Grid: & Reasons Why IT Matters," *CIO Insight*, March 24, 2010; Yuliya Chernova, "Getting Smart About Smart Meters," *The Wall Street Journal*, May 10, 2010; Bob Evans, "IT's Dark-Side Potential Seenin SmartGridCity Project," *Information Week*, March 24, 2009; Bob Violino, "No More Grid-Lock," *Information Week*, November 16, 2009; K.C. Jones, "Smart Grids to Get Jolt from IT," *Information Week*, March 23, 2009; Rebecca Smith, "Smart Meter, Dumb Idea?" *The Wall Street Journal*, April 27, 2009; Stephanie Simon, "The More Your Know..." *The Wall Street Journal*, February 9, 2009; and Matthew Wald and Miguel Helft, "Google Taking a Step into Power Metering," *The New York Times*, February 10, 2009.

CASE STUDY QUESTIONS

1. How do smart grids differ from the current electricity infrastructure in the United States?

2. What management, organization, and technology issues should be considered when developing a smart grid?

3. What challenge to the development of smart grids do you think is most likely to hamper their development?

4. What other areas of our infrastructure could benefit from "smart" technologies? Describe one example not listed in the case.

5. Would you like your home and your community to be part of a smart grid? Why or why not? Explain.

Chapter 2

Global E-business and Collaboration

LEARNING OBJECTIVES

After reading this chapter, you will be able to answer the following questions:

1. What are business processes? How are they related to information systems?

2. How do systems serve the different management groups in a business?

3. How do systems that link the enterprise improve organizational performance?

4. Why are systems for collaboration and teamwork so important and what technologies do they use?

5. What is the role of the information systems function in a business?

Interactive Sessions:

Domino's Sizzles with Pizza Tracker

Virtual Meetings: Smart Management

CHAPTER OUTLINE

2.1 **BUSINESS PROCESSES AND INFORMATION SYSTEMS**
Business Processes
How Information Technology Improves Business Processes

2.2 **TYPES OF INFORMATION SYSTEMS**
Systems for Different Management Groups
Systems for Linking the Enterprise
E-business, E-commerce, and E-government

2.3 **SYSTEMS FOR COLLABORATION AND TEAMWORK**
What Is Collaboration?
Business Benefits of Collaboration and Teamwork
Building a Collaborative Culture and Business Processes
Tools and Technologies for Collaboration and Teamwork

2.4 **THE INFORMATION SYSTEMS FUNCTION IN BUSINESS**
The Information Systems Department
Organizing the Information Systems Function

2.5 **HANDS-ON MIS PROJECTS**
Management Decision Problems
Improving Decision Making: Using a Spreadsheet to Select Suppliers
Achieving Operational Excellence: Using Internet Software to Plan Efficient Transportation Routes

LEARNING TRACK MODULES
Systems from a Functional Perspective
IT Enables Collaboration and Teamwork
Challenges of Using Business Information Systems
Organizing the Information Systems Function

AMERICA'S CUP 2010: USA WINS WITH INFORMATION TECHNOLOGY

The BMW Oracle Racing organization won the 33rd America's Cup yacht race in Valencia, Spain on February 18, 2010. The BMW Oracle boat USA, backed by software billionaire Larry Ellison, beat Alinghi, the Swiss boat backed by Ernesto Bertarelli, a Swiss billionaire. It's always a spectacle when two billionaires go head to head for the prize. Lots and lots of money, world-class talent, and in this case, the best technologies and information systems in the world. In the end, the 114-foot USA won handily the first two races of a best-of-three series, reaching speeds over 35 miles an hour, three times faster than the wind. As far as experts can figure, USA is the fastest sailboat in history.

So what kind of technology can you get for a $300 million sailboat? Start with the physical structure: a three hulled trimaran, 114 feet long, fashioned from carbon fiber shaped into a form descended from Polynesian outrigger boats over a thousand years old. The hull is so light it only extends six inches into the water. Forget about a traditional mast (that's the pole that holds up the sails) and forget about sails too. Think about a 233-foot airplane wing also made from carbon fiber that sticks up from the boat deck 20 stories high. Instead of cloth sails, think about a stretchy aeronautical fabric over a carbon fiber frame that is hydraulically controlled to assume any shape you want, sort of like a stretchy garment hugs the body's bones. The result is a wing, not a sail, whose shape can be changed from pretty near flat to quite curved just like an aircraft wing.

Controlling this wickedly sleek sailboat requires a lightning-fast collection of massive amounts of data, powerful data management, rapid real-time data analysis, quick decision making, and immediate measurement of the results. In short, all the information technologies needed by a modern business firm. When you can perform all these tasks thousands of times in an hour, you can incrementally improve your performance and have an overwhelming advantage over less IT-savvy opponents on race day.

For USA, this meant using 250 sensors on the wing, hull, and rudder to gather real-time data on pressure, angles, loads, and strains to monitor the effectiveness of each adjustment. The sensors track 4,000 variables, 10 times a second, producing 90 million data points an hour.

Managing all these data is Oracle Database 11g data management software. The data are wirelessly transferred to a tender ship running Oracle 11g for near real-time analysis using a family of formulas (called velocity prediction formulas) geared to understanding what makes the boat go fast. Oracle's Application Express presentation graphics summarize the millions of data points and present the boat managers with charts that make sense of the information. The data are also sent to Oracle's Austin data center for more in- depth analysis. Using powerful data analysis tools, USA managers were able to find relationships they had never thought about before. Over several years of practice, from day one to the day before the race, the crew of USA could chart a steady improvement in performance.

All this meant "sailing" had changed, perhaps been trans-

formed by IT. Each crew member wore a small mobile handheld computer on his wrist to display data on the key performance variables customized for that person's responsibilities, such as the load balance on a specific rope or the current aerodynamic performance of the wing sail. Rather than stare at the sails or the sea, the crew had to be trained to sail like pilots looking at instruments. The helmsman turned into a pilot looking at data displayed on his sunglasses with an occasional glance at the deck crew, sea state, and competitors.

Professional and amateur sailors across the world wondered if the technology had transformed sailing into something else. The billionaire winner Larry Ellison sets the rules for the next race, and the blogs are speculating that he will seek a return to simpler more traditional boats that need to be sailed, not flown like airplanes. Yet few really believe Ellison will give up a key IT advantage in data collection, analysis, presentation, and performance-based decision making.

Sources: Jeff Erickson, "Sailing Home with the Prize," *Oracle Magazine,* May/June 2010; www.america's cup.com, accessed May 21,2010; and www.bmworacleracing.com, accessed May 21, 2010.

The experience of BMW Oracle's USA in the 2010 America's Cup competition illustrates how much organizations today, even those in traditional sports such as sailing, rely on information systems to improve their performance and remain competitive. It also shows how much information systems make a difference in an organization's ability to innovate, execute, and in the case of business firms, grow profits.

The chapter-opening diagram calls attention to important points raised by this case and this chapter. The America's Cup contenders were confronted with both a challenge and opportunity. Both were locked in the world's most competitive sailing race. They staffed their crews with the best sailors in the world. But sailing ability was not enough. There were opportunities for improving sailing performance by changing and refining the design of the competing vessels using information systems intensively for this purpose.

Because Oracle is one of the world's leading information technology providers, the company was a natural for using the most advanced information technology to continually improve USA's design and performance. But information technology alone would not have produced a winning boat. The Oracle team had to revise many of the processes and procedures used in sailing to take advantage of the technology, including training experienced sailors to work more like pilots with high-tech instruments and sensors. Oracle won the America's Cup because it had learned how to apply new technology to improve the processes of designing and sailing a competitive sailboat.

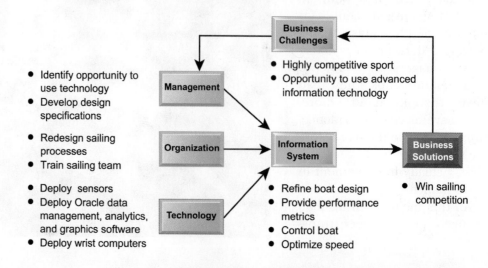

2.1 BUSINESS PROCESSES AND INFORMATION SYSTEMS

In order to operate, businesses must deal with many different pieces of information about suppliers, customers, employees, invoices and payments, and of course their products and services. They must organize work activities that use this information to operate efficiently and enhance the overall performance of the firm. Information systems make it possible for firms to manage all their information, make better decisions, and improve the execution of their business processes.

BUSINESS PROCESSES

Business processes, which we introduced in Chapter 1, refer to the manner in which work is organized, coordinated, and focused to produce a valuable product or service. Business processes are the collection of activities required to produce a product or service. These activities are supported by flows of material, information, and knowledge among the participants in business processes. Business processes also refer to the unique ways in which organizations coordinate work, information, and knowledge, and the ways in which management chooses to coordinate work.

To a large extent, the performance of a business firm depends on how well its business processes are designed and coordinated. A company's business processes can be a source of competitive strength if they enable the company to innovate or to execute better than its rivals. Business processes can also be liabilities if they are based on outdated ways of working that impede organizational responsiveness and efficiency. The chapter-opening case describing the processes used to sail the 2010 winning America's Cup boat clearly illustrates these points, as do many of the other cases in this text.

Every business can be seen as a collection of business processes, some of which are part of larger encompassing processes. For instance, designing a new sailboat model, manufacturing components, assembling the finished boat, and revising the design and construction are all part of the overall production process. Many business processes are tied to a specific functional area. For example, the sales and marketing function is responsible for identifying customers, and the human resources function is responsible for hiring employees. Table 2-1 describes some typical business processes for each of the functional areas of business.

TABLE 2-1 EXAMPLES OF FUNCTIONAL BUSINESS PROCESSES

FUNCTIONAL AREA	BUSINESS PROCESS
Manufacturing and production	Assembling the product
	Checking for quality
	Producing bills of materials
Sales and marketing	Identifying customers
	Making customers aware of the product
	Selling the product
Finance and accounting	Paying creditors
	Creating financial statements
	Managing cash accounts
Human resources	Hiring employees
	Evaluating employees' job performance
	Enrolling employees in benefits plans

Other business processes cross many different functional areas and require coordination across departments. For instance, consider the seemingly simple business process of fulfilling a customer order (see Figure 2-1). Initially, the sales department receives a sales order. The order passes first to accounting to ensure the customer can pay for the order either by a credit verification or request for immediate payment prior to shipping. Once the customer credit is established, the production department pulls the product from inventory or produces the product. Then the product is shipped (and this may require working with a logistics firm, such as UPS or FedEx). A bill or invoice is generated by the accounting department, and a notice is sent to the customer indicating that the product has shipped. The sales department is notified of the shipment and prepares to support the customer by answering calls or fulfilling warranty claims.

What at first appears to be a simple process, fulfilling an order, turns out to be a very complicated series of business processes that require the close coordination of major functional groups in a firm. Moreover, to efficiently perform all these steps in the order fulfillment process requires a great deal of information. The required information must flow rapidly both within the firm from one decision maker to another; with business partners, such as delivery firms; and with the customer. Computer-based information systems make this possible.

HOW INFORMATION TECHNOLOGY IMPROVES BUSINESS PROCESSES

Exactly how do information systems improve business processes? Information systems automate many steps in business processes that were formerly performed manually, such as checking a client's credit, or generating an invoice and shipping order. But today, information technology can do much more. New technology can actually change the flow of information, making it possible for many more people to access and share information, replacing sequential steps

FIGURE 2-1   THE ORDER FULFILLMENT PROCESS

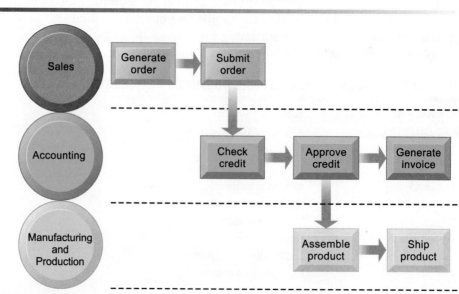

Fulfilling a customer order involves a complex set of steps that requires the close coordination of the sales, accounting, and manufacturing functions.

with tasks that can be performed simultaneously, and eliminating delays in decision making. New information technology frequently changes the way a business works and supports entirely new business models. Downloading a Kindle e-book from Amazon, buying a computer online at Best Buy, and downloading a music track from iTunes are entirely new business processes based on new business models that would be inconceivable without today's information technology.

That's why it's so important to pay close attention to business processes, both in your information systems course and in your future career. By analyzing business processes, you can achieve a very clear understanding of how a business actually works. Moreover, by conducting a business process analysis, you will also begin to understand how to change the business by improving its processes to make it more efficient or effective. Throughout this book, we examine business processes with a view to understanding how they might be improved by using information technology to achieve greater efficiency, innovation, and customer service.

2.2 TYPES OF INFORMATION SYSTEMS

Now that you understand business processes, it is time to look more closely at how information systems support the business processes of a firm. Because there are different interests, specialties, and levels in an organization, there are different kinds of systems. No single system can provide all the information an organization needs.

A typical business organization has systems supporting processes for each of the major business functions—systems for sales and marketing, manufacturing and production, finance and accounting, and human resources. You can find examples of systems for each of these business functions in the Learning Tracks for this chapter. Functional systems that operate independently of each other are becoming a thing of the past because they cannot easily share information to support cross-functional business processes. Many have been replaced with large-scale cross-functional systems that integrate the activities of related business processes and organizational units. We describe these integrated cross-functional applications later in this section.

A typical firm also has different systems supporting the decision-making needs of each of the main management groups we described in Chapter 1. Operational management, middle management, and senior management each use systems to support the decisions they must make to run the company. Let's look at these systems and the types of decisions they support.

SYSTEMS FOR DIFFERENT MANAGEMENT GROUPS

A business firm has systems to support different groups or levels of management. These systems include transaction processing systems, management information systems, decision-support systems, and systems for business intelligence.

Transaction Processing Systems

Operational managers need systems that keep track of the elementary activities and transactions of the organization, such as sales, receipts, cash deposits, payroll, credit decisions, and the flow of materials in a factory. **Transaction**

processing systems (TPS) provide this kind of information. A transaction processing system is a computerized system that performs and records the daily routine transactions necessary to conduct business, such as sales order entry, hotel reservations, payroll, employee record keeping, and shipping.

The principal purpose of systems at this level is to answer routine questions and to track the flow of transactions through the organization. How many parts are in inventory? What happened to Mr. Smith's payment? To answer these kinds of questions, information generally must be easily available, current, and accurate.

At the operational level, tasks, resources, and goals are predefined and highly structured. The decision to grant credit to a customer, for instance, is made by a lower-level supervisor according to predefined criteria. All that must be determined is whether the customer meets the criteria.

Figure 2-2 illustrates a TPS for payroll processing. A payroll system keeps track of money paid to employees. An employee time sheet with the employee's name, social security number, and number of hours worked per week represents a single transaction for this system. Once this transaction is input into the system, it updates the system's master file (or database—see Chapter 6) that permanently maintains employee information for the organization. The data in the system are combined in different ways to create reports of interest to management and government agencies and to send paychecks to employees.

Managers need TPS to monitor the status of internal operations and the firm's relations with the external environment. TPS are also major producers of information for the other systems and business functions. For example, the payroll system illustrated in Figure 2-2, along with other accounting TPS,

FIGURE 2-2   A PAYROLL TPS

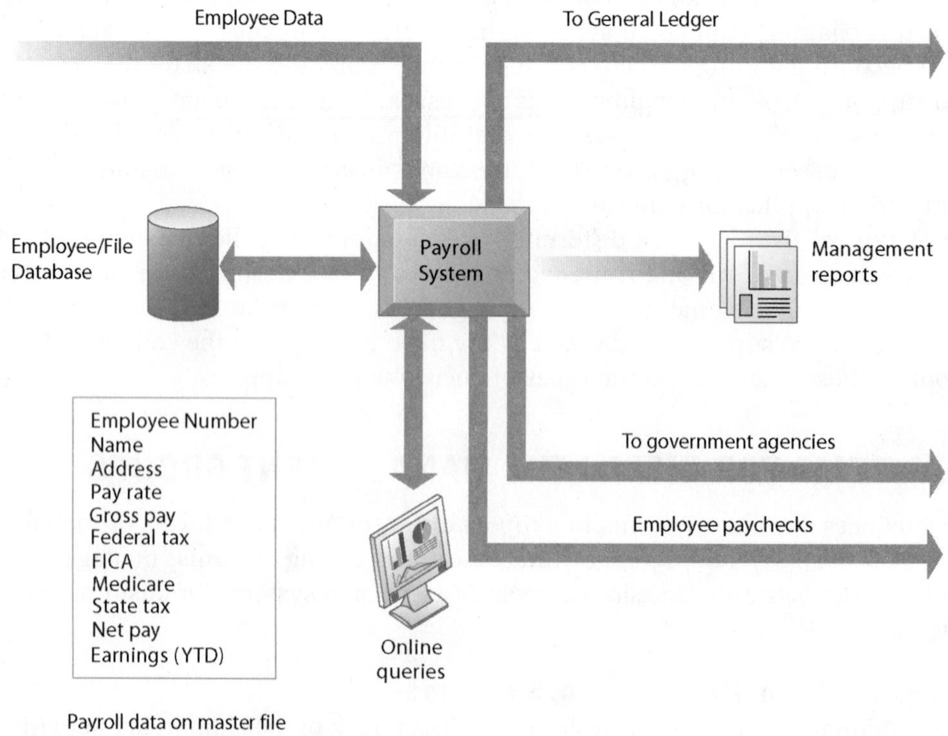

A TPS for payroll processing captures employee payment transaction data (such as a time card). System outputs include online and hard-copy reports for management and employee paychecks.

supplies data to the company's general ledger system, which is responsible for maintaining records of the firm's income and expenses and for producing reports such as income statements and balance sheets. It also supplies employee payment history data for insurance, pension, and other benefits calculations to the firm's human resources function and employee payment data to government agencies such as the U.S. Internal Revenue Service and Social Security Administration.

Transaction processing systems are often so central to a business that TPS failure for a few hours can lead to a firm's demise and perhaps that of other firms linked to it. Imagine what would happen to UPS if its package tracking system were not working! What would the airlines do without their computerized reservation systems?

Business Intelligence Systems for Decision Support

Middle management needs systems to help with monitoring, controlling, decision-making, and administrative activities. The principal question addressed by such systems is this: Are things working well?

In Chapter 1, we define management information systems as the study of information systems in business and management. The term **management information systems (MIS)** also designates a specific category of information systems serving middle management. MIS provide middle managers with reports on the organization's current performance. This information is used to monitor and control the business and predict future performance.

MIS summarize and report on the company's basic operations using data supplied by transaction processing systems. The basic transaction data from TPS are compressed and usually presented in reports that are produced on a regular schedule. Today, many of these reports are delivered online. Figure 2-3 shows how a typical MIS transforms transaction-level data from order process-

FIGURE 2-3   HOW MANAGEMENT INFORMATION SYSTEMS OBTAIN THEIR DATA FROM THE ORGANIZATION'S TPS

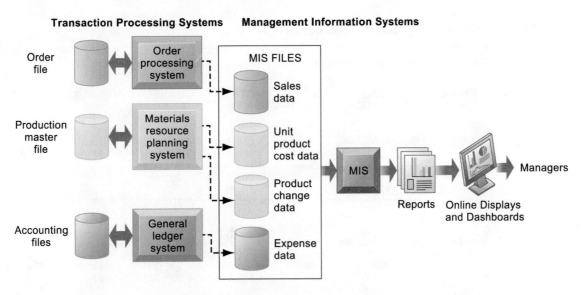

In the system illustrated by this diagram, three TPS supply summarized transaction data to the MIS reporting system at the end of the time period. Managers gain access to the organizational data through the MIS, which provides them with the appropriate reports.

ing, production, and accounting into MIS files that are used to provide managers with reports. Figure 2-4 shows a sample report from this system.

MIS serve managers primarily interested in weekly, monthly, and yearly results. These systems typically provide answers to routine questions that have been specified in advance and have a predefined procedure for answering them. For instance, MIS reports might list the total pounds of lettuce used this quarter by a fast-food chain or, as illustrated in Figure 2-4, compare total annual sales figures for specific products to planned targets. These systems generally are not flexible and have little analytical capability. Most MIS use simple routines, such as summaries and comparisons, as opposed to sophisticated mathematical models or statistical techniques.

In contrast, **decision-support systems (DSS)** support more non-routine decision making. They focus on problems that are unique and rapidly changing, for which the procedure for arriving at a solution may not be fully predefined in advance. They try to answer questions such as these: What would be the impact on production schedules if we were to double sales in the month of December? What would happen to our return on investment if a factory schedule were delayed for six months?

Although DSS use internal information from TPS and MIS, they often bring in information from external sources, such as current stock prices or product prices of competitors. These systems use a variety of models to analyze the data and are designed so that users can work with them directly.

An interesting, small, but powerful, DSS is the voyage-estimating system of a subsidiary of a large American metals company that exists primarily to carry bulk cargoes of coal, oil, ores, and finished products for its parent company. The firm owns some vessels, charters others, and bids for shipping contracts in the open market to carry general cargo. A voyage-estimating system calculates financial and technical voyage details. Financial calculations include ship/time costs (fuel, labor, capital), freight rates for various types of cargo, and port expenses. Technical details include a myriad of factors, such as ship cargo capacity, speed, port distances, fuel and water consumption, and loading patterns (location of cargo for different ports).

FIGURE 2-4 **SAMPLE MIS REPORT**

Consolidated Consumer Products Corporation Sales by Product and Sales Region: 2011

PRODUCT CODE	PRODUCT DESCRIPTION	SALES REGION	ACTUAL SALES	PLANNED	ACTUAL versus PLANNED
4469	Carpet Cleaner	Northeast	4,066,700	4,800,000	0.85
		South	3,778,112	3,750,000	1.01
		Midwest	4,867,001	4,600,000	1.06
		West	4,003,440	4,400,000	0.91
	TOTAL		16,715,253	17,550,000	0.95
5674	Room Freshener	Northeast	3,676,700	3,900,000	0.94
		South	5,608,112	4,700,000	1.19
		Midwest	4,711,001	4,200,000	1.12
		West	4,563,440	4,900,000	0.93
	TOTAL		18,559,253	17,700,000	1.05

This report, showing summarized annual sales data, was produced by the MIS in Figure 2-3.

The system can answer questions such as the following: Given a customer delivery schedule and an offered freight rate, which vessel should be assigned at what rate to maximize profits? What is the optimal speed at which a particular vessel can maximize its profit and still meet its delivery schedule? What is the optimal loading pattern for a ship bound for the U.S. West Coast from Malaysia? Figure 2-5 illustrates the DSS built for this company. The system operates on a desktop personal computer, providing a system of menus that makes it easy for users to enter data or obtain information.

The voyage-estimating DSS we have just described draws heavily on models. Other systems supporting non-routine decision making are more data-driven, focusing instead on extracting useful information from large quantities of data. For example, Intrawest—the largest ski operator in North America—collects and stores large amounts of customer data from its Web site, call center, lodging reservations, ski schools, and ski equipment rental stores. It uses special software to analyze these data to determine the value, revenue potential, and loyalty of each customer so managers can make better decisions on how to target their marketing programs. The system segments customers into seven categories based on needs, attitudes, and behaviors, ranging from "passionate experts" to "value-minded family vacationers." The company then e-mails video clips that would appeal to each segment to encourage more visits to its resorts.

All of the management systems we have just described are systems for business intelligence. **Business intelligence** is a contemporary term for data and software tools for organizing, analyzing, and providing access to data to help managers and other enterprise users make more informed decisions. You'll learn more about business intelligence in Chapters 6 and 12.

Business intelligence applications are not limited to middle managers, and can be found at all levels of the organization, including systems for senior management. Senior managers need systems that address strategic issues and long-term trends, both in the firm and in the external environment. They are

FIGURE 2-5　VOYAGE-ESTIMATING DECISION-SUPPORT SYSTEM

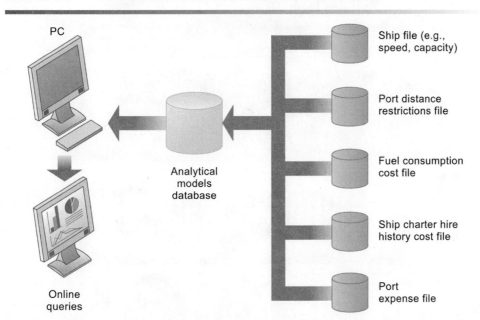

This DSS operates on a powerful PC. It is used daily by managers who must develop bids on shipping contracts.

concerned with questions such as these: What will employment levels be in five years? What are the long-term industry cost trends, and where does our firm fit in? What products should we be making in five years? What new acquisitions would protect us from cyclical business swings?

Executive support systems (ESS) help senior management make these decisions. They address non-routine decisions requiring judgment, evaluation, and insight because there is no agreed-on procedure for arriving at a solution. ESS present graphs and data from many sources through an interface that is easy for senior managers to use. Often the information is delivered to senior executives through a **portal**, which uses a Web interface to present integrated personalized business content. You will learn more about other applications of portals in Chapter 11.

ESS are designed to incorporate data about external events, such as new tax laws or competitors, but they also draw summarized information from internal MIS and DSS. They filter, compress, and track critical data, displaying the data of greatest importance to senior managers. Increasingly, such systems include business intelligence analytics for analyzing trends, forecasting, and "drilling down" to data at greater levels of detail.

For example, the CEO of Leiner Health Products, one of the largest manufacturers of private-label vitamins and supplements in the United States, has an ESS that provides on his desktop a minute-to-minute view of the firm's financial performance as measured by working capital, accounts receivable, accounts payable, cash flow, and inventory. The information is presented in the form of a **digital dashboard**, which displays on a single screen graphs and charts of key performance indicators for managing a company. Digital dashboards are becoming an increasingly popular tool for management decision makers.

Dundas Data Visualization's digital dashboard delivers comprehensive and accurate information for decision making. The graphical overview of key performance indicators helps managers quickly spot areas that need attention.

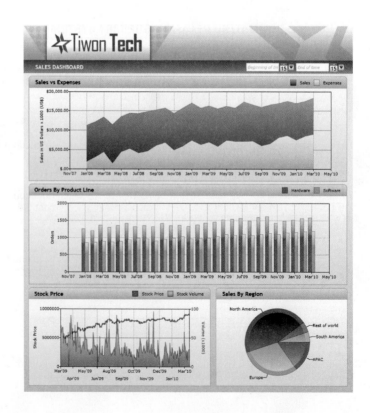

The Interactive Session on Organizations describes real-world examples of several types of systems we have just described that are used by a successful fast-food chain. Note the types of systems illustrated in this case and the role they play in improving business performance and competitiveness.

SYSTEMS FOR LINKING THE ENTERPRISE

Reviewing all the different types of systems we have just described, you might wonder how a business can manage all the information in these different systems. You might also wonder how costly it is to maintain so many different systems. And you might wonder how all these different systems can share information and how managers and employees are able to coordinate their work. In fact, these are all important questions for businesses today.

Enterprise Applications

Getting all the different kinds of systems in a company to work together has proven a major challenge. Typically, corporations are put together both through normal "organic" growth and through acquisition of smaller firms. Over a period of time, corporations end up with a collection of systems, most of them older, and face the challenge of getting them all to "talk" with one another and work together as one corporate system. There are several solutions to this problem.

One solution is to implement **enterprise applications**, which are systems that span functional areas, focus on executing business processes across the business firm, and include all levels of management. Enterprise applications help businesses become more flexible and productive by coordinating their business processes more closely and integrating groups of processes so they focus on efficient management of resources and customer service.

There are four major enterprise applications: enterprise systems, supply chain management systems, customer relationship management systems, and knowledge management systems. Each of these enterprise applications integrates a related set of functions and business processes to enhance the performance of the organization as a whole. Figure 2-6 shows that the architecture for these enterprise applications encompasses processes spanning the entire organization and, in some cases, extending beyond the organization to customers, suppliers, and other key business partners.

Enterprise Systems Firms use **enterprise systems**, also known as enterprise resource planning (ERP) systems, to integrate business processes in manufacturing and production, finance and accounting, sales and marketing, and human resources into a single software system. Information that was previously fragmented in many different systems is stored in a single comprehensive data repository where it can be used by many different parts of the business.

For example, when a customer places an order, the order data flow automatically to other parts of the company that are affected by them. The order transaction triggers the warehouse to pick the ordered products and schedule shipment. The warehouse informs the factory to replenish whatever has been depleted. The accounting department is notified to send the customer an invoice. Customer service representatives track the progress of the order through every step to inform customers about the status of their orders. Managers are able to use firm-wide information to make more precise and timely decisions about daily operations and longer-term planning.

INTERACTIVE SESSION: ORGANIZATIONS

DOMINO'S SIZZLES WITH PIZZA TRACKER

When it comes to pizza, everyone has an opinion. Some of us think that our current pizza is just fine the way it is. Others have a favorite pizza joint that makes it like no one else. And many pizza lovers in America agreed up until recently that Domino's home-delivered pizza was among the worst. The home-delivery market for pizza chains in the United States is approximately $15 billion per year. Domino's, which owns the largest home-delivery market share of any U.S. pizza chain, is finding ways to innovate by overhauling its in-store transaction processing systems and by providing other useful services to customers, such as its Pizza Tracker. And more important, Domino's is trying very hard to overcome its reputation for poor quality by radically improving ingredients and freshness. Critics believe the company significantly improved the quality of its pizza and customer service in 2010.

Domino's was founded in 1960 by Tom Monaghan and his brother James when they purchased a single pizza store in Ypsilanti, Michigan. The company slowly began to grow, and by 1978, Domino's had 200 stores. Today, the company is headquartered in Ann Arbor, Michigan, and operates almost 9,000 stores located in all 50 U.S. states and across the world in 60 international markets. In 2009, Domino's had $1.5 billion in sales and earned $80 million in profit.

Domino's is part of a heated battle among prominent pizza chains, including Pizza Hut, Papa John's, and Little Caesar. Pizza Hut is the only chain larger than Domino's in the U.S., but each of the four has significant market share. Domino's also competes with local pizza stores throughout the U.S. To gain a competitive advantage Domino's needs to deliver excellent customer service, and most importantly, good pizza. But it also benefits from highly effective information systems.

Domino's proprietary point-of-sale system, Pulse, is an important asset in maintaining consistent and efficient management functions in each of its restaurants. A point-of-sale system captures purchase and payment data at a physical location where goods or services are bought and sold using computers, automated cash registers, scanners, or other digital devices.

In 2003, Domino's implemented Pulse in a large portion of its stores, and those stores reported improved customer service, reduced mistakes, and shorter training times. Since then, Pulse has become a staple of all Domino's franchises. Some of the functions Pulse performs at Domino's franchises are taking and customizing orders using a touch-screen interface, maintaining sales figures, and compiling customer information. Domino's prefers not to disclose the specific dollar amounts that it has saved from Pulse, but it's clear from industry analysts that the technology is working to cut costs and increase customer satisfaction.

More recently, Domino's released a new hardware and software platform called Pulse Evolution, which is now in use in a majority of Domino's more than 5,000 U.S. branches. Pulse Evolution improves on the older technology in several ways. First, the older software used a 'thick-client' model, which required all machines using the software to be fully equipped personal computers running Windows. Pulse Evolution, on the other hand, uses 'thin-client' architecture in which networked workstations with little independent processing power collect data and send them over the Internet to powerful Lenovo PCs for processing. These workstations lack hard drives, fans, and other moving parts, making them less expensive and easier to maintain. Also, Pulse Evolution is easier to update and more secure, since there's only one machine in the store which needs to be updated.

Along with Pulse Evolution, Domino's rolled out its state-of-the-art online ordering system, which includes Pizza Tracker. The system allows customers to watch a simulated photographic version of their pizza as they customize its size, sauces, and toppings. The image changes with each change a customer makes. Then, once customers place an order, they are able to view its progress online with Pizza Tracker. Pizza Tracker displays a horizontal bar that tracks an order's progress graphically. As a Domino's store completes each step of the order fulfillment process, a section of the bar becomes red. Even customers that place their orders via telephone can monitor their progress on the Web using Pizza Tracker at stores using Pulse Evolution. In 2010, Domino's introduced an online polling system to continuously upload information from local stores.

As with most instances of organizational change of this magnitude, Domino's experienced some resistance. Domino's originally wanted its franchises to select Pulse to comply with its requirements for data security, but some franchises have resisted switching to Pulse and sought alternative systems. After Domino's tried to compel those franchises to use Pulse, the U.S. District Court for Minnesota sided with franchisees who claimed that Domino's could not force them to use this system. Now, Domino's continues to make improvements to Pulse in an effort to make it overwhelmingly appealing to all franchisees.

Pizza Hut and Papa John's also have online ordering capability, but lack the Pizza Tracker and the simulated pizza features that Domino's has successfully implemented. Today, online orders account for almost 20 percent of all of Domino's orders, which is up from less than 15 percent in 2008. But the battle to sell pizza with technology rages on. Pizza Hut customers can now use their iPhones to place orders, and Papa John's customers can place orders by texting. With many billions of dollars at stake, all the large national pizza chains will be developing innovative new ways of ordering pizza and participating in its creation.

Sources: PRN Newswire, "Servant Systems Releases Domino's Store Polling Software," PRN Newswire, April 14, 2010; Julie Jargon, "Domino's IT Staff Delivers Slick Site, Ordering System," *The Wall Street Journal*, November 24, 2009; www.dominosbiz.com, accessed May 17, 2010; Paul McDougall, "Interop: Domino's Eyes Microsoft Cloud," *Information Week*, April 26, 2010; "Domino's Builds New Foundation Under Proprietary Store Tech," *Nation's Restaurant News*, February 25, 2009; "and "Inside Domino's 'Pizza Tracker.' What It Does, Why, and How," *Nation's Restaurant News*, February 27, 2008.

CASE STUDY QUESTIONS

1. What kinds of systems are described in this case? Identify and describe the business processes each supports. Describe the inputs, processes, and outputs of these systems.

2. How do these systems help Domino's improve its business performance?

3. How did the online pizza ordering system improve the process of ordering a Domino's pizza?

4. How effective are these systems in giving Domino's a competitive edge? Explain your answer.

MIS IN ACTION

Visit Domino's Web site and examine the order placement and Pizza Tracker features. Then answer the following questions:

1. What steps does Pizza Tracker display for the user? How does the Pizza Tracker improve the customer experience?

2. Would the Pizza Tracker service influence you to order pizza from Domino's instead of a competing chain? Why or why not?

3. What improvements would you make to the order placement feature?

Supply Chain Management Systems Firms use **supply chain management (SCM) systems** to help manage relationships with their suppliers. These systems help suppliers, purchasing firms, distributors, and logistics companies share information about orders, production, inventory levels, and delivery of products and services so that they can source, produce, and deliver goods and services efficiently. The ultimate objective is to get the right amount of their products from their source to their point of consumption in the least amount of time and at the lowest cost. These systems increase firm profitability by lowering the costs of moving and making products and by enabling managers to make better decisions about how to organize and schedule sourcing, production, and distribution.

Supply chain management systems are one type of **interorganizational system** because they automate the flow of information across organizational boundaries. You will find examples of other types of interorganizational information systems throughout this text because such systems make it possible for firms to link electronically to customers and to outsource their work to other companies.

Customer Relationship Management Systems Firms use **customer relationship management (CRM) systems** to help manage their relationships with their customers. CRM systems provide information to coordinate all of the business processes that deal with customers in sales, marketing, and service to optimize revenue, customer satisfaction, and customer retention. This information helps firms identify, attract, and retain the most profitable customers; provide better service to existing customers; and increase sales.

FIGURE 2-6   ENTERPRISE APPLICATION ARCHITECTURE

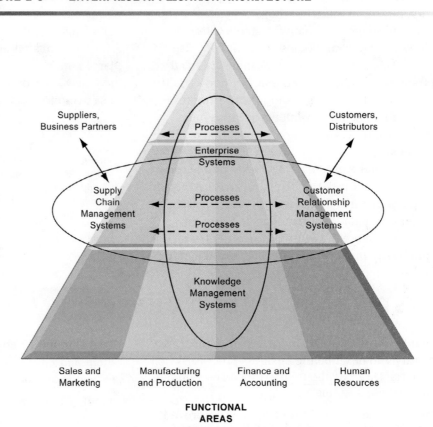

Enterprise applications automate processes that span multiple business functions and organizational levels and may extend outside the organization.

Knowledge Management Systems Some firms perform better than others because they have better knowledge about how to create, produce, and deliver products and services. This firm knowledge is difficult to imitate, unique, and can be leveraged into long-term strategic benefits. **Knowledge management systems (KMS)** enable organizations to better manage processes for capturing and applying knowledge and expertise. These systems collect all relevant knowledge and experience in the firm, and make it available wherever and whenever it is needed to improve business processes and management decisions. They also link the firm to external sources of knowledge.

We examine enterprise systems and systems for supply chain management and customer relationship management in greater detail in Chapter 9. We discuss collaboration systems that support knowledge management in this chapter and cover other types of knowledge management applications in Chapter 11.

Intranets and Extranets

Enterprise applications create deep-seated changes in the way the firm conducts its business, offering many opportunities to integrate important business data into a single system. They are often costly and difficult to implement. Intranets and extranets deserve mention here as alternative tools for increasing integration and expediting the flow of information within the firm, and with customers ad suppliers.

Intranets are simply internal company Web sites that are accessible only by employees. The term "intranet" refers to the fact that it is an internal network, in contrast to the Internet, which is a public network linking organizations and other external networks. Intranets use the same technologies and techniques as the larger Internet, and they often are simply a private access area in a larger company Web site. Likewise with extranets. Extranets are company Web sites that are accessible to authorized vendors and suppliers, and often used to coordinate the movement of supplies to the firm's production apparatus.

For example, Six Flags, which operates 19 theme parks throughout North America, maintains an intranet for its 2,500 full-time employees that provides company-related news and information on each park's day-to-day operations, including weather forecasts, performance schedules, and details about groups and celebrities visiting the parks. The company also uses an extranet to broadcast information about schedule changes and park events to its 30,000 seasonal employees. We describe the technology for intranets and extranets in more detail in Chapter 7.

E-BUSINESS, E-COMMERCE, AND E-GOVERNMENT

The systems and technologies we have just described are transforming firms' relationships with customers, employees, suppliers, and logistic partners into digital relationships using networks and the Internet. So much business is now enabled by or based upon digital networks that we use the terms "electronic business" and "electronic commerce" frequently throughout this text.

Electronic business, or **e-business**, refers to the use of digital technology and the Internet to execute the major business processes in the enterprise. E-business includes activities for the internal management of the firm and for coordination with suppliers and other business partners. It also includes **electronic commerce**, or **e-commerce**.

E-commerce is the part of e-business that deals with the buying and selling of goods and services over the Internet. It also encompasses activities supporting those market transactions, such as advertising, marketing, customer support, security, delivery, and payment.

The technologies associated with e-business have also brought about similar changes in the public sector. Governments on all levels are using Internet technology to deliver information and services to citizens, employees, and businesses with which they work. **E-government** refers to the application of the Internet and networking technologies to digitally enable government and public sector agencies' relationships with citizens, businesses, and other arms of government.

In addition to improving delivery of government services, e-government makes government operations more efficient and also empowers citizens by giving them easier access to information and the ability to network electronically with other citizens. For example, citizens in some states can renew their driver's licenses or apply for unemployment benefits online, and the Internet has become a powerful tool for instantly mobilizing interest groups for political action and fund-raising.

2.3 SYSTEMS FOR COLLABORATION AND TEAMWORK

With all these systems and information, you might wonder how is it possible to make sense out of them? How do people working in firms pull it all together, work towards common goals, and coordinate plans and actions? Information systems can't make decisions, hire or fire people, sign contracts, agree on deals, or adjust the price of goods to the marketplace. In addition to the types of systems we have just described, businesses need special systems to support collaboration and teamwork.

WHAT IS COLLABORATION?

Collaboration is working with others to achieve shared and explicit goals. Collaboration focuses on task or mission accomplishment and usually takes place in a business, or other organization, and between businesses. You collaborate with a colleague in Tokyo having expertise on a topic about which you know nothing. You collaborate with many colleagues in publishing a company blog. If you're in a law firm, you collaborate with accountants in an accounting firm in servicing the needs of a client with tax problems.

Collaboration can be short-lived, lasting a few minutes, or longer term, depending on the nature of the task and the relationship among participants. Collaboration can be one-to-one or many-to-many.

Employees may collaborate in informal groups that are not a formal part of the business firm's organizational structure or they may be organized into formal teams. Teams are part of the organization's business structure for getting things done. **Teams** have a specific mission that someone in the business assigned to them. They have a job to complete. The members of the team need to collaborate on the accomplishment of specific tasks and collectively achieve the team mission. The team mission might be to "win the game," or "increase online sales by 10%," or "prevent insulating foam from falling off a space shuttle." Teams are often short-lived, depending on the problems they tackle and the length of time needed to find a solution and accomplish the mission.

Collaboration and teamwork are more important today than ever for a variety of reasons.

- *Changing nature of work.* The nature of work has changed from factory manufacturing and pre-computer office work where each stage in the production process occurred independently of one another, and was coordinated by supervisors. Work was organized into silos. Within a silo, work passed from one machine tool station to another, from one desktop to another, until the finished product was completed. Today, the kinds of jobs we have require much closer coordination and interaction among the parties involved in producing the service or product. A recent report from the consulting firm McKinsey and Company argued that 41 percent of the U.S. labor force is now composed of jobs where interaction (talking, e-mailing, presenting, and persuading) is the primary value-adding activity. Even in factories, workers today often work in production groups, or pods.

- *Growth of professional work.* "Interaction" jobs tend to be professional jobs in the service sector that require close coordination, and collaboration. Professional jobs require substantial education, and the sharing of information and opinions to get work done. Each actor on the job brings specialized expertise to the problem, and all the actors need to take one another into account in order to accomplish the job.

- *Changing organization of the firm.* For most of the industrial age, managers organized work in a hierarchical fashion. Orders came down the hierarchy, and responses moved back up the hierarchy. Today, work is organized into groups and teams, who are expected to develop their own methods for accomplishing the task. Senior managers observe and measure results, but are much less likely to issue detailed orders or operating procedures. In part this is because expertise has been pushed down in the organization, as have decision-making powers.

- *Changing scope of the firm.* The work of the firm has changed from a single location to multiple locations—offices or factories throughout a region, a nation, or even around the globe. For instance, Henry Ford developed the first mass-production automobile plant at a single Dearborn, Michigan factory. In 2010, Ford expected to produce about 3 million automobiles and employ over 200,000 employees at 90 plants and facilities worldwide. With this kind of global presence, the need for close coordination of design, production, marketing, distribution, and service obviously takes on new importance and scale. Large global companies need to have teams working on a global basis.

- *Emphasis on innovation.* Although we tend to attribute innovations in business and science to great individuals, these great individuals are most likely working with a team of brilliant colleagues, and all have been preceded by a long line of earlier innovators and innovations. Think of Bill Gates and Steve Jobs (founders of Microsoft and Apple), both of whom are highly regarded innovators, and both of whom built strong collaborative teams to nurture and support innovation in their firms. Their initial innovations derived from close collaboration with colleagues and partners. Innovation, in other words, is a group and social process, and most innovations derive from collaboration among individuals in a lab, a business, or government agencies. Strong collaborative practices and technologies are believed to increase the rate and quality of innovation.

- *Changing culture of work and business.* Most research on collaboration supports the notion that diverse teams produce better outputs, faster, than individuals working on their own. Popular notions of the crowd ("crowdsourcing," and the "wisdom of crowds") also provide cultural support for collaboration and teamwork.

BUSINESS BENEFITS OF COLLABORATION AND TEAMWORK

There are many articles and books that have been written about collaboration, some of them by business executives and consultants, and a great many by academic researchers in a variety of businesses. Nearly all of this research is anecdotal. Nevertheless, among both business and academic communities there is a general belief that the more a business firm is "collaborative," the more successful it will be, and that collaboration within and among firms is more essential than in the past.

A recent global survey of business and information systems managers found that investments in collaboration technology produced organizational improvements that returned over four times the amount of the investment, with the greatest benefits for sales, marketing, and research and development functions (Frost and White, 2009). Another study of the value of collaboration also found that the overall economic benefit of collaboration was significant: for every word seen by an employee in e-mails from others, $70 of additional revenue was generated (Aral, Brynjolfsson, and Van Alstyne, 2007).

Table 2-2 summarizes some of the benefits of collaboration identified by previous writers and scholars. Figure 2-7 graphically illustrates how collaboration is believed to impact business performance.

While there are many presumed benefits to collaboration, you really need a supportive business firm culture and the right business processes before you can achieve meaningful collaboration. You also need a healthy investment in collaborative technologies. We now examine these requirements.

TABLE 2-2 BUSINESS BENEFITS OF COLLABORATION

BENEFIT	RATIONALE
Productivity	People working together can complete a complex task faster than the same number of people working in isolation from one another. There will be fewer errors.
Quality	People working collaboratively can communicate errors, and correct actions faster, when they work together than if they work in isolation. Can lead to a reduction in buffers and time delay among production units.
Innovation	People working collaboratively in groups can come up with more innovative ideas for products, services, and administration than the same number working in isolation from one another.
Customer service	People working together in teams can solve customer complaints and issues faster and more effectively than if they were working in isolation from one another.
Financial performance (profitability, sales, and sales growth)	As a result of all of the above, collaborative firms have superior sales growth and financial performance.

FIGURE 2-7   REQUIREMENTS FOR COLLABORATION

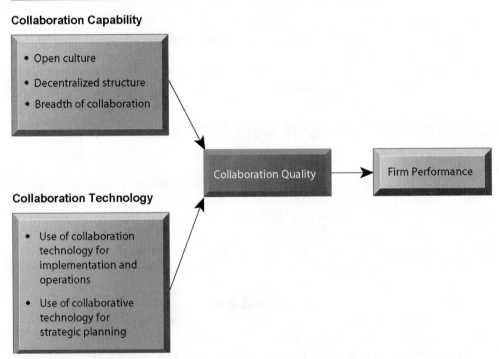

Successful collaboration requires an appropriate organizational structure and culture, along with appropriate collaboration technology.

BUILDING A COLLABORATIVE CULTURE AND BUSINESS PROCESSES

Collaboration won't take place spontaneously in a business firm, especially if there is no supportive culture or business processes. Business firms, especially large firms, had in the past a reputation for being "command and control" organizations where the top leaders thought up all the really important matters, and then ordered lower-level employees to execute senior management plans. The job of middle management supposedly was to pass messages back and forth, up and down the hierarchy.

Command and control firms required lower-level employees to carry out orders without asking too many questions, with no responsibility to improve processes, and with no rewards for teamwork or team performance. If your workgroup needed help from another work group, that was something for the bosses to figure out. You never communicated horizontally, always vertically, so management could control the process. As long as employees showed up for work, and performed the job satisfactorily, that's all that was required. Together, the expectations of management and employees formed a culture, a set of assumptions about common goals and how people should behave. Many business firms still operate this way.

A collaborative business culture and business processes are very different. Senior managers are responsible for achieving results but rely on teams of employees to achieve and implement the results. Policies, products, designs, processes, and systems are much more dependent on teams at all levels of the organization to devise, to create, and to build products and services. Teams are rewarded for their performance, and individuals are rewarded for their performance in a team. The function of middle managers is to build the teams, coordinate their work, and monitor their performance. In a collaborative culture, senior management establishes collaboration and teamwork as vital to the organization, and it actually implements collaboration for the senior ranks of the business as well.

TOOLS AND TECHNOLOGIES FOR COLLABORATION AND TEAMWORK

A collaborative, team-oriented culture won't produce benefits if there are no information systems in place to enable collaboration. Currently there are hundreds of tools designed to deal with the fact that, in order to succeed in our jobs, we are all dependent on one another, our fellow employees, customers, suppliers, and managers. Table 2-3 lists the most important types of collaboration software tools. Some high-end tools like IBM Lotus Notes are expensive, but powerful enough for global firms. Others are available online for free (or with premium versions for a modest fee) and are suitable for small businesses. Let's look more closely at some of these tools.

TABLE 2-3 FIFTEEN CATEGORIES OF COLLABORATIVE SOFTWARE TOOLS

E-mail and instant messaging	White boarding
Collaborative writing	Web presenting
Collaborative reviewing/editing	Work scheduling
Event scheduling	Document sharing (including wikis)
File sharing	Mind mapping
Screen sharing	Large audience Webinars
Audio conferencing	Co-browsing
Video conferencing	

Source: mindmeister.com, 2009.

E-mail and Instant Messaging (IM)

E-mail and instant messaging have been embraced by corporations as a major communication and collaboration tool supporting interaction jobs. Their software operates on computers, cell phones, and other wireless handheld devices and includes features for sharing files as well as transmitting messages. Many instant messaging systems allow users to engage in real-time conversations with multiple participants simultaneously. Gartner technology consultants predict that within a few years, instant messaging will be the "de facto tool" for voice, video, and text chat for 95 percent of employees in big companies.

Social Networking

We've all visited social networking sites such as MySpace and Facebook, which feature tools to help people share their interests and interact. Social networking tools are quickly becoming a corporate tool for sharing ideas and collaborating among interaction-based jobs in the firm. Social networking sites such as Linkedin.com provide networking services to business professionals, while other niche sites have sprung up to serve lawyers, doctors, engineers, and even dentists. IBM built a Community Tools component into its Lotus Notes collaboration software to add social networking features. Users are able to submit questions to others in the company and receive answers via instant messaging.

Wikis

Wikis are a type of Web site that makes it easy for users to contribute and edit text content and graphics without any knowledge of Web page development or programming techniques. The most well-known wiki is Wikipedia, the largest collaboratively edited reference project in the world. It relies on volunteers, makes no money, and accepts no advertising. Wikis are ideal tools for storing and sharing company knowledge and insights. Enterprise software vendor SAP AG has a wiki that acts as a base of information for people outside the company, such as customers and software developers who build programs that interact with SAP software. In the past, those people asked and sometimes answered questions in an informal way on SAP online forums, but that was an inefficient system, with people asking and answering the same questions over and over.

At Intel Corporation, employees built their own internal wiki, and it has been edited over 100,000 times and viewed more than 27 million times by Intel employees. The most common search is for the meaning of Intel acronyms such as EASE for "employee access support environment" and POR for "plan of record." Other popular resources include a page about software engineering processes at the company. Wikis are destined to become the major repository for unstructured corporate knowledge in the next five years in part because they are so much less costly than formal knowledge management systems and they can be much more dynamic and current.

Virtual Worlds

Virtual worlds, such as Second Life, are online 3-D environments populated by "residents" who have built graphical representations of themselves known as avatars. Organizations such as IBM and INSEAD, an international business school with campuses in France and Singapore, are using this virtual world to house online meetings, training sessions, and "lounges." Real-world people represented by avatars meet, interact, and exchange ideas at these virtual locations. Communication takes place in the form of text messages similar to instant messages.

Internet-Based Collaboration Environments

There are now suites of software products providing multi-function platforms for workgroup collaboration among teams of employees who work together from many different locations. Numerous collaboration tools are available, but the most widely used are Internet-based audio conferencing and video conferencing systems, online software services such as Google Apps/Google Sites, and corporate collaboration systems such as Lotus Notes and Microsoft SharePoint.

Virtual Meeting Systems For many businesses, including investment banking, accounting, law, technology services, and management consulting, extensive travel is a fact of life. The expenses incurred by business travel have been steadily rising in recent years, primarily due to increasing energy costs. In an effort to reduce travel expenses, many companies, both large and small, are adopting videoconferencing and Web conferencing technologies.

Companies such as Heinz, General Electric, Pepsico, and Wachovia are using virtual meeting systems for product briefings, training courses, strategy sessions, and even inspirational chats.

An important feature of leading-edge high-end videoconferencing systems is **telepresence** technology, an integrated audio and visual environment that allows a person to give the appearance of being present at a location other than his or her true physical location. The Interactive Session on Management describes telepresence and other technologies for hosting these "virtual" meetings. You can also find video cases on this topic.

Google Apps/Google Sites One of the most widely used "free" online services for collaboration is Google Apps/Google Sites. Google Sites allows users to quickly create online, group-editable Web sites. Google Sites is one part of the larger Google Apps suite of tools. Google Sites users can design and populate Web sites in minutes and, without any advanced technical skills, post a variety of files including calendars, text, spreadsheets, and videos for private, group, or public viewing and editing.

Google Apps works with Google Sites and includes the typical desktop productivity office software tools (word processing, spreadsheets, presentation, contact management, messaging, and mail). A Premier edition charging businesses $50 per year for each user offers 25 gigabytes of mail storage, a 99.9-percent uptime guarantee for e-mail, tools to integrate with the firm's existing infrastructure, and 24/7 phone support. Table 2-4 describes some of the capabilities of Google Apps/Google Sites.

TABLE 2-4 GOOGLE APPS/GOOGLE SITES COLLABORATION FEATURES

GOOGLE APPS/GOOGLE SITES CAPABILITY	DESCRIPTION
Google Calendar	Private and shared calendars; multiple calendars
Google Gmail	Google's free online e-mail service, with mobile access capabilities
Google Talk	Instant messaging, text and voice chat
Google Docs	Online word processing, presentation, spreadsheet, and drawing software; online editing and sharing
Google Sites	Team collaboration sites for sharing documents, schedules, calendars; searching documents and creating group wikis
Google Video	Private hosted video sharing
Google Groups	User-created groups with mailing lists, shared calendars, documents, sites, and video; searchable archives

INTERACTIVE SESSION: MANAGEMENT

VIRTUAL MEETINGS: SMART MANAGEMENT

Instead of taking that 6:30 A.M. plane to make a round of meetings in Dallas, wouldn't it be great if you could attend these events without leaving your desktop? Today you can, thanks to technologies for videoconferencing and for hosting online meetings over the Web. A June 2008 report issued by the Global e-Sustainability Initiative and the Climate Group estimated that up to 20 percent of business travel could be replaced by virtual meeting technology.

A videoconference allows individuals at two or more locations to communicate simultaneously through two-way video and audio transmissions. The critical feature of videoconferencing is the digital compression of audio and video streams by a device called a codec. Those streams are then divided into packets and transmitted over a network or the Internet. Until recently, the technology was plagued by poor audio and video performance, and its cost was prohibitively high for all but the largest and most powerful corporations. Most companies deemed videoconferencing a poor substitute for face-to-face meetings.

However, vast improvements in videoconferencing and associated technologies have renewed interest in this way of working. Videoconferencing is now growing at an annual rate of 30 percent. Proponents of the technology claim that it does more than simply reduce costs. It allows for "better" meetings as well: it's easier to meet with partners, suppliers, subsidiaries, and colleagues from within the office or around the world on a more frequent basis, which in most cases simply cannot be reasonably accomplished through travel. You can also meet with contacts that you wouldn't be able to meet at all without videoconferencing technology.

For example, Rip Curl, a Costa Mesa, California, producer of surfing equipment, uses videoconferencing to help its designers, marketers, and manufacturers collaborate on new products. Executive recruiting firm Korn/Ferry International uses video interviews to screen potential candidates before presenting them to clients.

Today's state-of-the-art videoconferencing systems display sharp high-definition TV images. The top-of-the-line videoconferencing technology is known as telepresence. Telepresence strives to make users feel as if they are actually present in a location different

from their own. You can sit across a table from a large screen showing someone who looks quite real and life-size, but may be in Brussels or Hong Kong. Only the handshake and exchange of business cards are missing. Telepresence products provide the highest-quality videoconferencing available on the market to date. Cisco Systems has installed telepresence systems in more than 500 organizations around the world. Prices for fully equipped telepresence rooms can run to $500,000.

Companies able to afford this technology report large savings. For example, technology consulting firm Accenture reports that it eliminated expenditures for 240 international trips and 120 domestic flights in a single month. The ability to reach customers and partners is also dramatically increased. Other business travelers report tenfold increases in the number of customers and partners they are able to reach for a fraction of the previous price per person. MetLife, which installed Cisco Telepresence in three dedicated conference rooms in Chicago, New York, and New Jersey, claims that the technology not only saved time and expense but also helped the company meet its "green" environmental goals of reducing carbon emissions by 20 percent in 2010.

Videoconferencing products have not traditionally been feasible for small businesses, but another company, LifeSize, has introduced an affordable line of products as low as $5,000. Overall, the product is easy to use and will allow many smaller companies to use a high-quality videoconferencing product.

There are even some free Internet-based options like Skype videoconferencing and ooVoo. These products are of lower quality than traditional videoconferencing products, and they are proprietary, meaning they can only talk to others using that very same system. Most videoconferencing and telepresence products are able to interact with a variety of other devices. Higher-end systems include features like multi-party conferencing, video mail with unlimited storage, no long-distance fees, and a detailed call history.

Companies of all sizes are finding Web-based online meeting tools such as WebEx, Microsoft Office Live Meeting, and Adobe Acrobat Connect especially helpful for training and sales presentations. These products enable participants to share documents and presentations in conjunction with audioconferencing and live video via Webcam. Cornerstone Information Systems, a Bloomington, Indiana, business software company with 60 employees, cut its travel costs by 60 percent and the average time to close a new sale by 30 percent by performing many product demonstrations online.

Before setting up videoconferencing or telepresence, it's important for a company to make sure it really needs the technology to ensure that it will be a profitable venture. Companies should determine how their employees conduct meetings, how they communicate and with what technologies, how much travel they do, and their network's capabilities. There are still plenty of times when face-to-face interaction is more desirable, and often traveling to meet a client is essential for cultivating clients and closing sales.

Videoconferencing figures to have an impact on the business world in other ways, as well. More employees may be able to work closer to home and balance their work and personal lives more efficiently; traditional office environments and corporate headquarters may shrink or disappear; and freelancers, contractors, and workers from other countries will become a larger portion of the global economy.

Sources: Joe Sharkey, "Setbacks in the Air Add to Lure of Virtual Meetings, *The New York Times*, April 26, 2010; Bob Evans, "Pepsi Picks Cisco for Huge TelePresence Deal," February 2, 2010; Esther Schein, "Telepresence Catching On, But Hold On to Your Wallet," *Computerworld*, January 22, 2010; Christopher Musico, "Web Conferencing: Calling Your Conference to Order," *Customer Relationship Management*, February 2009; and Brian Nadel, "3 Videoconferencing Services Pick Up Where Your Travel Budget Leaves Off," *Computerworld*, January 6, 2009; Johna Till Johnson, "Videoconferencing Hits the Big Times.... For Real," *Computerworld*, May 28, 2009.

CASE STUDY QUESTIONS

1. One consulting firm has predicted that video and Web conferencing will make business travel extinct. Do you agree? Why or why not?

2. What is the distinction between videoconferencing and telepresence?

3. What are the ways in which videoconferencing provides value to a business? Would you consider it smart management? Explain your answer.

4. If you were in charge of a small business, would you choose to implement videoconferencing? What factors would you consider in your decision?

MIS IN ACTION

Explore the WebEx Web site (www.webex.com) and answer the following questions:

1. List and describe its capabilities for small-medium and large businesses. How useful is WebEx? How can it help companies save time and money?

2. Compare WebEx video capabilities with the videoconferencing capabilities described in this case.

3. Describe the steps you would take to prepare for a Web conference as opposed to a face-to-face conference.

Google has developed an additional Web-based platform for real-time collaboration and communication called Google Wave. "Waves" are "equal parts conversation and document," in which any participant of a wave can reply anywhere in the message, edit the content, and add or remove participants at any point in the process. Users are able to see responses from other participants on their "wave" while typing occurs, accelerating the pace of discussion.

For example, Clear Channel Radio in Greensboro, North Carolina, used Google Wave for an on air and online promotion that required input from sales people, the sales manager, the station program director, the station promotions director, the online content coordinator, and the Web manager. Without Google Wave, these people would have used numerous back and forth e-mails, sent graphics files to each other for approval, and spent large amounts of time tracking people down by phone. Wave helped them complete the entire project in just a fraction of time it would normally have taken (Boulton, 2010).

Microsoft SharePoint Microsoft SharePoint is the most widely adopted collaboration system for small and medium-sized firms that use Microsoft server and networking products. Some larger firms have adopted it as well. SharePoint is a browser-based collaboration and document management platform, combined with a powerful search engine that is installed on corporate servers.

SharePoint has a Web-based interface and close integration with everyday tools such as Microsoft Office desktop software products. Microsoft's strategy is to take advantage of its "ownership" of the desktop through its Microsoft Office and Windows products. For Microsoft, the path towards enterprise-wide collaboration starts with the Office desktop and Microsoft network servers. SharePoint software makes it possible for employees to share their Office documents and collaborate on projects using Office documents as the foundation.

SharePoint products and technologies provide a platform for Web-based collaboration at the enterprise level. SharePoint can be used to host Web sites that organize and store information in one central location to enable teams to coordinate work activities, collaborate on and publish documents, maintain task lists, implement workflows, and share information via wikis, blogs, and Twitter-style status updates. Because SharePoint stores and organizes information in one place, users can find relevant information quickly and efficiently while working together closely on tasks, projects, and documents.

Here is a list of SharePoint's major capabilities:

- Provides a single workspace for teams to coordinate schedules, organize documents, and participate in discussions, within the organization or over an extranet.

- Facilitates creation and management of documents with the ability to control versions, view past revisions, enforce document-specific security, and maintain document libraries.

- Provides announcements, alerts, and discussion boards to inform users when actions are required or changes are made to existing documentation or information.

- Supports personalized content and both personal and public views of documents and applications.

- Provides templates for blogs and wikis to help teams share information and brainstorm.

- Provides tools to manage document libraries, lists, calendars, tasks, and discussion boards offline, and to synchronize changes when reconnected to the network.

- Provides enterprise search tools for locating people, expertise, and content.

Sony Electronics, a leading provider of consumer and professional electronics products with more 170,000 employees around the world, uses Microsoft Office SharePoint Server 2010 to improve information access, enhance collaboration, and make better use of experts inside the company. Sony uses SharePoint's wiki tools to capture and organize employees' insights and comments into a company-wide body of knowledge, and its people search feature to identify employees with expertise about specific projects and research areas. The company also used SharePoint to create a central file-sharing repository. This helps employees collaboratively write, edit, and exchange documents and eliminates the need to e-mail documents back and forth. All of these improvements have cut development time on key projects from three to six months to three to six weeks. (Microsoft, 2010).

Lotus Notes For very large firms (Fortune 1000 and Russell 2000 firms), the most widely used collaboration tool is IBM's Lotus Notes. Lotus Notes was an early example of groupware, a collaborative software system with capabilities for sharing calendars, collective writing and editing, shared database access, and electronic meetings, with each participant able to see and display information from others and other activities. Notes is now Web-enabled with enhancements for social networking (Lotus Connections) and a scripting and application development environment so that users can build custom applications to suit their unique needs.

IBM Software Group defines Lotus Notes as an "integrated desktop client option for accessing business e-mail, calendars, and applications on an IBM Lotus Domino server." The Notes software installed on the user's client computer allows the machine to be used as a platform for e-mail, instant messaging (working with Lotus Sametime), Web browsing, and calendar/resource reservation work, as well as for interacting with collaborative applications. Today, Notes also provides blogs, wikis, RSS aggregators, CRM, and help desk systems.

Thousands of employees at hundreds of large firms such as Toshiba, Air France, and Global Hyatt Corporation use IBM Lotus Notes as their primary collaboration and teamwork tools. Firmwide installations of Lotus Notes at a large Fortune 1000 firm may cost millions of dollars a year and require extensive support from the corporate information systems department. Although online tools like the Google collaboration services described earlier do not require installation on corporate servers or much support from the corporate IS staff, they are not as powerful as those found in Lotus Notes. It is unclear whether they could scale to the size of a global firm (at least for now). Very large firms adopt IBM Lotus Notes because Notes promises higher levels of security and reliability, and the ability to retain control over sensitive corporate information.

For example, EuroChem, the largest agrochemical company in Russia and one of Europe's top three fertilizer producers, used Lotus Notes to create a single standard platform for collaboration and document management. The software facilitates cooperation and collaboration among geographically dispersed regional production centers and provides a secure automated platform for document exchange. With Lotus Notes, EuroChem is able to register and control all documents, to establish routing paths for document approval, and to maintain a full history of all movements and changes. Security features allow the company to create a personalized work environment for each user and to prevent unauthorized users from accessing sensitive information (IBM, 2009).

Large firms in general do not feel secure using popular online software services for "strategic" applications because of the implicit security concerns. However, most experts believe that these concerns will diminish as experience with online tools grows, and the sophistication of online software service suppliers increases to protect security and reduce vulnerability. Table 2-5 describes additional online collaboration tools.

Checklist for Managers: Evaluating and Selecting Collaboration Software Tools

With so many collaboration tools and services available, how do you choose the right collaboration technology for your firm? To answer this question, you need a framework for understanding just what problems these tools are designed to solve. One framework that has been helpful for us to talk about collaboration tools is the time/space collaboration matrix developed in the early 1990s by a number of collaborative work scholars (Figure 2-8).

TABLE 2-5 OTHER POPULAR ONLINE COLLABORATION TOOLS

TOOL	DESCRIPTION
Socialtext	An enterprise server-based collaboration environment which provides social networking, Twitter-like micro-blogging , wiki workspaces, with integrated weblogs, distributed spreadsheets, and a personal home page for every user. Delivered in a variety of hosted cloud services, as well as on-site appliances to provide enterprise customers with flexible deployment options that meet their security requirements.
Zoho	Collecting and collaborating on text, line drawings, images, Web pages, video, RSS feeds. Project management (includes task management, work flow, reports, time tracking, forums, and file sharing). Free or monthly charge for premium service.
BlueTie	Online collaboration with e-mail, scheduling, to-do lists, contact management, file sharing. $4.99 per user per month.
Basecamp	Sharing to-do lists, files, message boards, milestone tracking. Free for a single project, $24/month for 15 projects with 5 gigabytes of storage.
Onehub	Sharing documents, calendars, Web bookmarks; e-mail integration and IM. Manage hub resources; bulletin board.
WorkZone	Collaboration with file sharing; project management; customization; security.

Socialtext's enterprise social networking products-including microblogging, blogs, wikis, profiles and social spreadsheets-enable employees to share vital information and work together in real-time. Built on a flexible, Web-oriented architecture, Socialtext integrates with virtually any traditional system of record, such as CRM and ERP, enabling companies to discuss, collaborate, and take action on key business processes.

FIGURE 2-8   THE TIME/SPACE COLLABORATION TOOL MATRIX

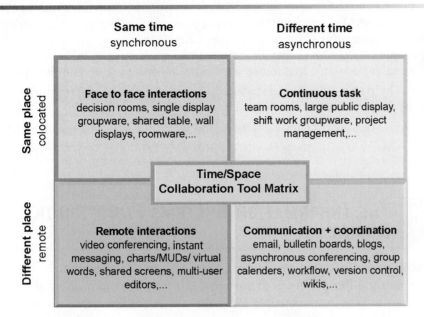

Collaboration technologies can be classified in terms of whether they support interactions at the same or different time or place, and whether these interactions are remote or co-located.

The time/space matrix focuses on two dimensions of the collaboration problem: time and space. For instance, you need to collaborate with people in different time zones and you cannot all meet at the same time. Midnight in New York is noon in Bombay, so this makes it difficult to have a video-conference (the people in New York are too tired). Time is clearly an obstacle to collaboration on a global scale.

Place (location) also inhibits collaboration in large global or even national and regional firms. Assembling people for a physical meeting is made difficult by the physical dispersion of distributed firms (firms with more than one location), the cost of travel, and the time limitations of managers.

The collaboration technologies we have just described are ways of overcoming the limitations of time and space. Using this time/space framework will help you to choose the most appropriate collaboration and teamwork tools for your firm. Note that some tools are applicable in more than one time/place scenario. For example, Internet collaboration suites such as Lotus Notes have capabilities for both synchronous (instant messaging, electronic meeting tools) and asynchronous (e-mail, wikis, document editing) interactions.

Here's a "to-do" list to get started. If you follow these six steps, you should be led to investing in the correct collaboration software for your firm at a price you can afford, and within your risk tolerance.

1. What are the collaboration challenges facing the firm in terms of time and space? Locate your firm in the time/space matrix. Your firm can occupy more than one cell in the matrix. Different collaboration tools will be needed for each situation.
2. Within each cell of the matrix where your firm faces challenges, exactly what kinds of solutions are available? Make a list of vendor products.
3. Analyze each of the products in terms of their cost and benefits to your firm. Be sure to include the costs of training in your cost estimates, and the costs of involving the information systems department if needed.

4. Identify the risks to security and vulnerability involved with each of the products. Is your firm willing to put proprietary information into the hands of external service providers over the Internet? Is your firm willing to risk its important operations to systems controlled by other firms? What are the financial risks facing your vendors? Will they be here in three to five years? What would be the cost of making a switch to another vendor in the event the vendor firm fails?

5. Seek the help of potential users to identify implementation and training issues. Some of these tools are easier to use than others.

6. Make your selection of candidate tools, and invite the vendors to make presentations.

2.4 THE INFORMATION SYSTEMS FUNCTION IN BUSINESS

We've seen that businesses need information systems to operate today and that they use many different kinds of systems. But who is responsible for running these systems? Who is responsible for making sure the hardware, software, and other technologies used by these systems are running properly and are up to date? End users manage their systems from a business standpoint, but managing the technology requires a special information systems function.

In all but the smallest of firms, the **information systems department** is the formal organizational unit responsible for information technology services. The information systems department is responsible for maintaining the hardware, software, data storage, and networks that comprise the firm's IT infrastructure. We describe IT infrastructure in detail in Chapter 5.

THE INFORMATION SYSTEMS DEPARTMENT

The information systems department consists of specialists, such as programmers, systems analysts, project leaders, and information systems managers. **Programmers** are highly trained technical specialists who write the software instructions for computers. **Systems analysts** constitute the principal liaisons between the information systems groups and the rest of the organization. It is the systems analyst's job to translate business problems and requirements into information requirements and systems. **Information systems managers** are leaders of teams of programmers and analysts, project managers, physical facility managers, telecommunications managers, or database specialists. They are also managers of computer operations and data entry staff. Also, external specialists, such as hardware vendors and manufacturers, software firms, and consultants, frequently participate in the day-to-day operations and long-term planning of information systems.

In many companies, the **information systems department** is headed by a **chief information officer (CIO)**. The CIO is a senior manager who oversees the use of information technology in the firm. Today's CIOs are expected to have a strong business background as well as information systems expertise and to play a leadership role in integrating technology into the firm's business strategy. Large firms today also have positions for a chief security officer, chief knowledge officer, and chief privacy officer, all of whom work closely with the CIO.

The **chief security officer (CSO)** is in charge of information systems security for the firm and is responsible for enforcing the firm's information security policy (see Chapter 8). (Sometimes this position is called the chief information security officer [CISO] where information systems security is separated from physical security.) The CSO is responsible for educating and training users and information systems specialists about security, keeping management aware of security threats and breakdowns, and maintaining the tools and policies chosen to implement security.

Information systems security and the need to safeguard personal data have become so important that corporations collecting vast quantities of personal data have established positions for a **chief privacy officer (CPO)**. The CPO is responsible for ensuring that the company complies with existing data privacy laws.

The **chief knowledge officer (CKO)** is responsible for the firm's knowledge management program. The CKO helps design programs and systems to find new sources of knowledge or to make better use of existing knowledge in organizational and management processes.

End users are representatives of departments outside of the information systems group for whom applications are developed. These users are playing an increasingly large role in the design and development of information systems.

In the early years of computing, the information systems group was composed mostly of programmers who performed highly specialized but limited technical functions. Today, a growing proportion of staff members are systems analysts and network specialists, with the information systems department acting as a powerful change agent in the organization. The information systems department suggests new business strategies and new information-based products and services, and coordinates both the development of the technology and the planned changes in the organization.

ORGANIZING THE INFORMATION SYSTEMS FUNCTION

There are many types of business firms, and there are many ways in which the IT function is organized within the firm. A very small company will not have a formal information systems group. It might have one employee who is responsible for keeping its networks and applications running, or it might use consultants for these services. Larger companies will have a separate information systems department, which may be organized along several different lines, depending on the nature and interests of the firm. Our Learning Track describes alternative ways of organizing the information systems function within the business.

The question of how the information systems department should be organized is part of the larger issue of IT governance. **IT governance** includes the strategy and policies for using information technology within an organization. It specifies the decision rights and framework for accountability to ensure that the use of information technology supports the organization's strategies and objectives. How much should the information systems function be centralized? What decisions must be made to ensure effective management and use of information technology, including the return on IT investments? Who should make these decisions? How will these decisions be made and monitored? Firms with superior IT governance will have clearly thought out the answers (Weill and Ross, 2004).

2.5   HANDS-ON MIS PROJECTS

The projects in this section give you hands-on experience analyzing opportunities to improve business processes with new information system applications, using a spreadsheet to improve decision making about suppliers, and using Internet software to plan efficient transportation routes.

Management Decision Problems

1. Don's Lumber Company on the Hudson River is one of the oldest retail lumberyards in New York State. It features a large selection of materials for flooring, decks, moldings, windows, siding, and roofing. The prices of lumber and other building materials are constantly changing. When a customer inquires about the price on pre-finished wood flooring, sales representatives consult a manual price sheet and then call the supplier for the most recent price. The supplier in turn uses a manual price sheet, which has been updated each day. Often the supplier must call back Don's sales reps because the company does not have the newest pricing information immediately on hand. Assess the business impact of this situation, describe how this process could be improved with information technology, and identify the decisions that would have to be made to implement a solution. Who would make those decisions?

2. Henry's Hardware is a small family business in Sacramento, California. The owners must use every square foot of store space as profitably as possible. They have never kept detailed inventory or sales records. As soon as a shipment of goods arrives, the items are immediately placed on store shelves. Invoices from suppliers are only kept for tax purposes. When an item is sold, the item number and price are rung up at the cash register. The owners use their own judgment in identifying items that need to be reordered. What is the business impact of this situation? How could information systems help the owners run their business? What data should these systems capture? What decisions could the systems improve?

Improving Decision Making: Using a Spreadsheet to Select Suppliers

Software skills: Spreadsheet date functions, data filtering, DAVERAGE function
Business skills: Analyzing supplier performance and pricing

In this exercise, you will learn how to use spreadsheet software to improve management decisions about selecting suppliers. You will start with raw transactional data about suppliers organized as a large spreadsheet list. You will use the spreadsheet software to filter the data based on several different criteria to select the best suppliers for your company.

You run a company that manufactures aircraft components. You have many competitors who are trying to offer lower prices and better service to customers, and you are trying to determine whether you can benefit from better supply chain management. In myMISlab, you will find a spreadsheet file that contains a list of all of the items that your firm has ordered from its suppliers during the past three months. A sample is shown below, but the Web site may have a more recent version of this spreadsheet for this exercise. The fields in the spreadsheet file include vendor name, vendor identification number, purchaser's order number, item identification number and item description (for each item ordered from the vendor), cost per item, number of units of the item ordered (quantity), total cost of each

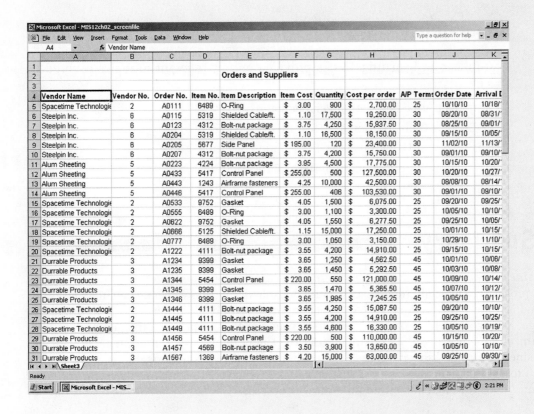

order, vendor's accounts payable terms, order date, and actual arrival date for each order.

Prepare a recommendation of how you can use the data in this spreadsheet database to improve your decisions about selecting suppliers. Some criteria to consider for identifying preferred suppliers include the supplier's track record for on-time deliveries, suppliers offering the best accounts payable terms, and suppliers offering lower pricing when the same item can be provided by multiple suppliers. Use your spreadsheet software to prepare reports to support your recommendations.

Achieving Operational Excellence: Using Internet Software to Plan Efficient Transportation Routes

In this exercise, you will use the same online software tool that businesses use to map out their transportation routes and select the most efficient route. The MapQuest (www.mapquest.com) Web site includes interactive capabilities for planning a trip. The software on this Web site can calculate the distance between two points and provide itemized driving directions to any location.

You have just started working as a dispatcher for Cross-Country Transport, a new trucking and delivery service based in Cleveland, Ohio. Your first assignment is to plan a delivery of office equipment and furniture from Elkhart, Indiana (at the corner of E. Indiana Ave. and Prairie Street) to Hagerstown, Maryland (corner of Eastern Blvd. N. and Potomac Ave.). To guide your trucker, you need to know the most efficient route between the two cities. Use MapQuest to find the route that is the shortest distance between the two cities. Use MapQuest again to find the route that takes the least time. Compare the results. Which route should Cross-Country use?

LEARNING TRACK MODULES

The following Learning Tracks provide content relevant to topics covered in this chapter:

1. Systems from a Functional Perspective
2. IT Enables Collaboration and Teamwork
3. Challenges of Using Business Information Systems
4. Organizing the Information Systems Function

Review Summary

1. **What are business processes? How are they related to information systems?**

   A business process is a logically related set of activities that defines how specific business tasks are performed, and it represents a unique way in which an organization coordinates work, information, and knowledge. Managers need to pay attention to business processes because they determine how well the organization can execute its business, and they may be a source of strategic advantage. There are business processes specific to each of the major business functions, but many business processes are cross-functional. Information systems automate parts of business processes, and they can help organizations redesign and streamline these processes.

2. **How do systems serve the different management groups in a business?**

   Systems serving operational management are transaction processing systems (TPS), such as payroll or order processing, that track the flow of the daily routine transactions necessary to conduct business. Management information systems (MIS) produce reports serving middle management by condensing information from TPS, and these are not highly analytical. Decision-support systems (DSS) support management decisions that are unique and rapidly changing using advanced analytical models. All of these types of systems provide business intelligence that helps managers and enterprise employees make more informed decisions. These systems for business intelligence serve multiple levels of management, and include executive support systems (ESS) for senior management that provide data in the form of graphs, charts, and dashboards delivered via portals using many sources of internal and external information.

3. **How do systems that link the enterprise improve organizational performance?**

   Enterprise applications are designed to coordinate multiple functions and business processes. Enterprise systems integrate the key internal business processes of a firm into a single software system to improve coordination and decision making. Supply chain management systems help the firm manage its relationship with suppliers to optimize the planning, sourcing, manufacturing, and delivery of products and services. Customer relationship management (CRM) systems coordinate the business processes surrounding the firm's customers. Knowledge management systems enable firms to optimize the creation, sharing, and distribution of knowledge. Intranets and extranets are private corporate networks based on Internet technology that assemble information from disparate systems. Extranets make portions of private corporate intranets available to outsiders.

4. **Why are systems for collaboration and teamwork so important and what technologies do they use?**

   Collaboration is working with others to achieve shared and explicit goals. Collaboration and teamwork have become increasingly important in business because of globalization, the decentralization of decision making, and growth in jobs where interaction is the primary value-adding activity. Collaboration is believed to enhance innovation, productivity, quality, and customer service. Effective collaboration today requires a supportive organizational culture as well as information systems and tools for collaborative work. Collaboration tools include e-mail and instant messaging, wikis, videoconferencing systems, virtual worlds, social networking systems, cell phones, and Internet collaboration platforms such as Google Apps/Sites, Microsoft SharePoint, and Lotus Notes.

5. *What is the role of the information systems function in a business?*

The information systems department is the formal organizational unit responsible for information technology services. It is responsible for maintaining the hardware, software, data storage, and networks that comprise the firm's IT infrastructure. The department consists of specialists, such as programmers, systems analysts, project leaders, and information systems managers, and is often headed by a CIO.

Key Terms

Business intelligence, 49
Chief information officer (CIO), 68
Chief knowledge officer (CKO), 69
Chief privacy officer (CPO), 69
Chief security officer (CSO), 68
Collaboration, 56
Customer relationship management (CRM)
  systems, 53
Decision-support systems (DSS), 48
Digital dashboard, 50
Electronic business (e-business), 55
Electronic commerce (e-commerce), 55
E-government, 55
End users, 69
Enterprise applications, 51

Enterprise systems, 51
Executive support systems (ESS), 50
Information systems department, 68
Information systems managers, 68
Interorganizational system, 53
IT governance, 69
Knowledge management systems (KMS), 54
Management information systems (MIS), 47
Portal, 50
Programmers, 68
Supply chain management (SCM) systems, 53
Systems analysts, 68
Teams, 56
Telepresence, 61
Transaction processing systems (TPS), 45

Review Questions

1. What are business processes? How are they related to information systems?

   • Define business processes and describe the role they play in organizations.

   • Describe the relationship between information systems and business processes.

2. How do systems serve the various levels of management in a business?

   • Describe the characteristics of transaction processing systems (TPS) and the roles they play in a business.

   • Describe the characteristics of management information systems (MIS) and explain how MIS differ from TPS and from DSS.

   • Describe the characteristics of decision-support systems (DSS) and how they benefit businesses.

   • Describe the characteristics of executive support systems (ESS) and explain how these systems differ from DSS.

3. How do systems that link the enterprise improve organizational performance?

   • Explain how enterprise applications improve organizational performance.

   • Define enterprise systems, supply chain management systems, customer relationship management systems, and knowledge management systems and describe their business benefits.

   • Explain how intranets and extranets help firms integrate information and business processes.

4. Why are systems for collaboration and teamwork so important and what technologies do they use?

- Define collaboration and teamwork and explain why they have become so important in business today.
- List and describe the business benefits of collaboration.
- Describe a supportive organizational culture and business processes for collaboration.
- List and describe the various types of collaboration and communication systems.

5. What is the role of the information systems function in a business?
   - Describe how the information systems function supports a business.
   - Compare the roles played by programmers, systems analysts, information systems managers, the chief information officer (CIO), chief security officer (CSO), and chief knowledge officer (CKO).

Discussion Questions

1. How could information systems be used to support the order fulfillment process illustrated in Figure 2-1? What are the most important pieces of information these systems should capture? Explain your answer.

2. Identify the steps that are performed in the process of selecting and checking out a book from your college library and the information that flows among these activities. Diagram the process. Are there any ways this process could be improved to improve the performance of your library or your school? Diagram the improved process.

3. How might the BMW Oracle team have used collaboration systems to improve the design and performance of the America's Cup sailboat USA? Which system features would be the most important for these tasks?

Video Cases

Video Cases and Instructional Videos illustrating some of the concepts in this chapter are available. Contact your instructor to access these videos.

Collaboration and Teamwork: Describing Management Decisions and Systems

With a team of three or four other students, find a description of a manager in a corporation in *BusinessWeek*, *Fortune*, *The Wall Street Journal*, or another business publication or do your research on the Web. Gather information about what the manager's company does and the role he or she plays in the company. Identify the organizational level and business function where this manager works. Make a list of the kinds of decisions this manager has to make and the kind of information the manager would need for those decisions. Suggest how information systems could supply this information. If possible, use Google Sites to post links to Web pages, team communication announcements, and work assignments. Try to use Google Docs to develop a presentation of your findings for the class.

Collaboration and Innovation at Procter & Gamble
CASE STUDY

Look in your medicine cabinet. No matter where you live in the world, odds are that you'll find many Procter & Gamble products that you use every day. P&G is the largest manufacturer of consumer products in the world, and one of the top 10 largest companies in the world by market capitalization. The company is known for its successful brands, as well as its ability to develop new brands and maintain its brands' popularity with unique business innovations. Popular P&G brands include Pampers, Tide, Bounty, Folgers, Pringles, Charmin, Swiffer, Crest, and many more. The company has approximately 140,000 employees in more than 80 countries, and its leading competitor is Britain-based Unilever. Founded in 1837 and headquartered in Cincinnati, Ohio, P&G has been a mainstay in the American business landscape for well over 150 years. In 2009, it had $79 billion in revenue and earned a $13.2 billion profit.

P&G's business operations are divided into three main units: Beauty Care, Household Care, and Health and Well-Being, each of which are further subdivided into more specific units. In each of these divisions, P&G has three main focuses as a business. It needs to maintain the popularity of its existing brands, via advertising and marketing; it must extend its brands to related products by developing new products under those brands; and it must innovate and create new brands entirely from scratch. Because so much of P&G's business is built around brand creation and management, it's critical that the company facilitate collaboration between researchers, marketers, and managers. And because P&G is such a big company, and makes such a wide array of products, achieving these goals is a daunting task.

P&G spends 3.4 percent of revenue on innovation, which is more than twice the industry average of 1.6 percent. Its research and development teams consist of 8,000 scientists spread across 30 sites globally. Though the company has an 80 percent "hit" rate on ideas that lead to products, making truly innovative and groundbreaking new products is very difficult in an extremely competitive field like consumer products. What's more, the creativity of bigger companies like P&G has been on the decline, with the top consumer goods companies accounting for only 5 per-

cent of patents filed on home care products in the early 2000s.

Finding better ways to innovate and develop new ideas is critical in a marketplace like consumer goods, and for any company as large as P&G, finding methods of collaboration that are effective across the enterprise can be difficult. That's why P&G has been active in implementing information systems that foster effective collaboration and innovation. The social networking and collaborative tools popularized by Web 2.0 have been especially attractive to P&G management, starting at the top with former CEO A.G. Lafley. Lafley was succeeded by Robert McDonald in 2010, but has been a major force in revitalizing the company.

When Lafley became P&G's CEO in 2000, he immediately asserted that by the end of the decade, the company would generate half of its new product ideas using sources from outside the company, both as a way to develop groundbreaking innovations more quickly and to reduce research and development costs. At the time, Lafley's proclamation was considered to be visionary, but in the past 10 years, P&G has made good on his promise.

The first order of business for P&G was to develop alternatives to business practices that were not sufficiently collaborative. The biggest culprit, says Joe Schueller, Innovation Manager for P&G's Global Business Services division, was perhaps an unlikely one: e-mail. Though it's ostensibly a tool for communication, e-mail is not a sufficiently collaborative way to share information; senders control the flow of information, but may fail to send mail to colleagues who most need to see it, and colleagues that don't need to see certain e-mails will receive mailings long after they've lost interest. Blogs and other collaborative tools, on the other hand, are open to anyone interested in their content, and attract comments from interested users.

However, getting P&G employees to actually use these newer products in place of e-mail has been a struggle for Schueller. Employees have resisted the changes, insisting that newer collaborative tools represent more work on top of e-mail, as opposed to a better alternative. People are accustomed to e-mail, and there's significant organizational inertia against switching to a new way of doing things. Some P&G

processes for sharing knowledge were notoriously inefficient. For instance, some researchers used to write up their experiments using Microsoft Office applications, then print them out and glue them page by page into notebooks. P&G was determined to implement more efficient and collaborative methods of communication to supplant some of these outdated processes.

To that end, P&G launched a total overhaul of its collaboration systems, led by a suite of Microsoft products. The services provided include unified communications (which integrates services for voice transmission, data transmission, instant messaging, e-mail, and electronic conferencing), Microsoft Live Communications Server functionality, Web conferencing with Live Meeting, and content management with SharePoint. According to P&G, over 80,000 employees use instant messaging, and 20,000 use Microsoft Outlook, which provides tools for e-mail, calendaring, task management, contact management, note taking, and Web browsing. Outlook works with Microsoft Office SharePoint Server to support multiple users with shared mailboxes and calendars, SharePoint lists, and meeting schedules.

The presence of these tools suggests more collaborative approaches are taking hold. Researchers use the tools to share the data they've collected on various brands; marketers can more effectively access the data they need to create more highly targeted ad campaigns; and managers are more easily able to find the people and data they need to make critical business decisions.

Companies like P&G are finding that one vendor simply isn't enough to satisfy their diverse needs. That introduces a new challenges: managing information and applications across multiple platforms. For example, P&G found that Google search was inadequate because it doesn't always link information from within the company, and its reliance on keywords for its searches isn't ideal for all of the topics for which employees might search. P&G decided to implement a new search product from start-up Connectbeam, which allows employees to share bookmarks and tag content with descriptive words that appear in future searches, and facilitates social networks of coworkers to help them find and share information more effectively.

The results of the initiative have been immediate. For example, when P&G executives traveled to meet with regional managers, there was no way to integrate all the reports and discussions into a single document. One executive glued the results of experiments into Word documents and passed them out at a conference. Another executive manually entered his data and speech into PowerPoint slides, and then e-mailed the file to his colleagues. One result was that the same file ended up in countless individual mailboxes. Now, P&G's IT department can create a Microsoft SharePoint page where that executive can post all of his presentations. Using SharePoint, the presentations are stored in a single location, but are still accessible to employees and colleagues in other parts of the company. Another collaborative tool, InnovationNet, contains over 5 million research-related documents in digital format accessible via a browser-based portal. That's a far cry from experiments glued in notebooks.

One concern P&G had when implementing these collaborative tools was that if enough employees didn't use them, the tools would be much less useful for those that did use them. Collaboration tools are like business and social networks–the more people connect to the network, the greater the value to all participants. Collaborative tools grow in usefulness as more and more workers contribute their information and insights. They also allow employees quicker access to the experts within the company that have needed information and knowledge. But these benefits are contingent on the lion's share of company employees using the tools.

Another major innovation for P&G was its large-scale adoption of Cisco TelePresence conference rooms at many locations across the globe. For a company as large as P&G, telepresence is an excellent way to foster collaboration between employees across not just countries, but continents. In the past, telepresence technologies were prohibitively expensive and overly prone to malfunction. Today, the technology makes it possible to hold high-definition meetings over long distances. P&G boasts the world's largest rollout of Cisco TelePresence technology.

P&G's biggest challenge in adopting the technology was to ensure that the studios were built to particular specifications in each of the geographically diverse locations where they were installed. Cisco accomplished this, and now P&G's estimates that 35 percent of its employees use telepresence regularly. In some locations, usage is as high as 70 percent. Benefits of telepresence include significant travel savings, more efficient flow of ideas, and quicker decision making. Decisions that once took days now take minutes.

Laurie Heltsley, P&G's director of global business services, noted that the company has saved $4 for every $1 invested in the 70 high-end telepresence systems it has installed over the past few years.

These high-definition systems are used four times as often as the company's earlier versions of videoconferencing systems.

Sources: Joe Sharkey, "Setbacks in the Air Add to Lure of Virtual Meetings," *The New York Times*, April 26, 2010; Matt Hamblen, "Firms Use Collaboration Tools to Tap the Ultimate IP-Worker Ideas," *Computerworld*, September 2, 2009; "Computerworld Honors Program: P&G", 2008; www.pg.com, accessed May 18, 2010; "Procter & Gamble Revolutionizes Collaboration with Cisco TelePresence," www.cisco.com, accessed May 18, 2010; "IT's Role in Collaboration at Procter &Gamble," *Information Week*, February 1, 2007.

CASE STUDY QUESTIONS

1. What is Procter & Gamble's business strategy? What is the relationship of collaboration and innovation to that business strategy?

2. How is P&G using collaboration systems to execute its business model and business strategy? List and describe the collaboration systems and technologies it is using and the benefits of each.

3. Why were some collaborative technologies slow to catch on at P&G?

4. Compare P&G's old and new processes for writing up and distributing the results of a research experiment.

5. Why is telepresence such a useful collaborative tool for a company like P&G?

6. Can you think of other ways P&G could use collaboration to foster innovation?

Chapter 3

Information Systems, Organizations, and Strategy

VERIZON OR AT&T—WHICH COMPANY HAS THE BEST DIGITAL STRATEGY?

Verizon and AT&T are the two largest telecommunications companies in the United States. In addition to voice communication, their customers use their networks to surf the Internet; send e-mail, text, and video messages; share photos; watch videos and high-definition TV; and conduct videoconferences around the globe. All of these products and services are digital.

Competition in this industry is exceptionally intense and fast-changing. Both companies are trying to outflank one another by refining their wireless, landline, and high-speed Internet networks and expanding the range of products, applications, and services available to customers. Wireless services are the most profitable. AT&T is staking its growth on the wireless market by aggressively marketing leading-edge high-end devices such as the iPhone. Verizon has bet on the reliability, power, and range of its wireless and landline networks and its renowned customer service.

For a number of years, Verizon has tried to blunt competition by making heavy technology investments in both its landline and wireless networks. Its wireless network is considered the most far-reaching and reliable in the United States. Verizon is now pouring billions of dollars into a rollout of fourth-generation (4G) cellular technology capable of supporting highly data-intensive applications such as downloading large streams of video and music through smart phones and other network appliances. Returns from Verizon's 4G investment are still uncertain.

Verizon's moves appear more risky financially than AT&T's, because its up-front costs are so high. AT&T's strategy is more conservative. Why not partner with other companies to capitalize on their technology innovations? That was the rationale for AT&T contracting with Apple Computer to be the exclusive network for its iPhone. Even though AT&T subsidizes some of the iPhone's cost to consumers, the iPhone's streamlined design, touch screen, exclusive access to the iTunes music service, and over 250,000 downloadable applications have made it an instant hit. AT&T has also sought to provide cellular services for other network appliances such as Amazon's Kindle e-book reader and netbooks.

The iPhone has been AT&T's primary growth engine, and the Apple relationship made the carrier the U.S. leader in the smartphone carrier marketspace. AT&T has over 43 percent of U.S. smartphone customers, compared with 23 percent for Verizon. Smart-phone customers are highly desirable because they typically pay higher monthly rates for wireless data service plans.

The iPhone became so wildly popular that users overstrained AT&T's networks, leaving many in dense urban areas such as New York and San Francisco with sluggish service or dropped calls. To handle the surging demand, AT&T could upgrade its wireless network, but that would cripple profits. Experts contend that AT&T would have to spend $5 billion to $7 billion to bring its network up to

Verizon's quality. To curb excessive use, AT&T moved to a tiered pricing model for new iPhone users, with data charges based on how much data customers actually use.

Adding to AT&T's woes, its monopoly on the iPhone may be ending. Apple reached an agreement with Verizon in 2010 to make an iPhone that is compatible with Verizon's network. Allowing Verizon to offer iPhone service will more than double Apple's market for this device, but will undoubtedly drive some AT&T iPhone customers to Verizon in the hope of finding better network service. Verizon is further hedging its bets by offering leading-edge smartphones based on Google's Android operating system that compete well against the iPhone. With or without the iPhone, if Verizon's Android phone sales continue to accelerate, the competitive balance will shift again.

Sources: Roger Cheng, "For Telecom Firms, Smartphones Rule," *The Wall Street Journal,* July 19, 2010; Brad Stone and Jenna Wortham, "Even Without iPhone, Verizon Is Gaining," *The New York Times,* July 15, 2010; Roben Farzad, "AT&T's iPhone Mess," *Bloomberg Businessweek,* April 25, 2010; Niraj Sheth, "AT&T Prepares Network for Battle," *The Wall Street Journal,* March 31, 2010; and Amol Sharma, "AT&T, Verizon Make Different Calls," *The Wall Street Journal,* January 28, 2009.

The story of Verizon and AT&T illustrates some of the ways that information systems help businesses compete—and also the challenges of sustaining a competitive advantage. The telecommunications industry in which both companies operate is extremely crowded and competitive, with telecommunications companies vying with cable companies, new upstarts, and each other to provide a wide array of digital services as well as voice transmission. To meet the challenges of surviving and prospering in this environment, each of these companies focused on a different competitive strategy using information technology.

The chapter-opening diagram calls attention to important points raised by this case and this chapter. Both companies identified opportunities to use information technology to offer new products and services. AT&T offered enhanced wireless services for the iPhone, while Verizon initially focused on high-capacity, high-quality network services. AT&T's strategy emphasized keeping costs low while capitalizing on innovations from other technology vendors. Verizon's strategy involved high up-front costs to build a high-capacity network infrastructure, and it also focused on providing a high level of network reliability and customer service.

This case study clearly shows how difficult it is to sustain a competitive advantage. Exclusive rights to use the highly popular iPhone on its network brought AT&T millions of new customers and enhanced its competitive position. But its competitive advantage is likely to erode if it is forced to invest heavily to upgrade its networks, if Apple allows Verizon to offer a version of the iPhone, or if Verizon smartphones are competitive with the iPhone. Changes in service pricing plans may also affect the competitive balance among the various wireless carriers.

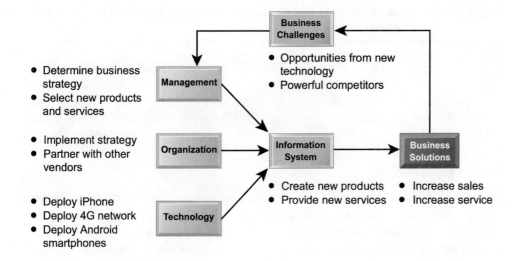

- Determine business strategy
- Select new products and services

- Implement strategy
- Partner with other vendors

- Deploy iPhone
- Deploy 4G network
- Deploy Android smartphones

Business Challenges
- Opportunities from new technology
- Powerful competitors

Management

Organization

Technology

Information System

Business Solutions

- Create new products
- Provide new services

- Increase sales
- Increase service

3.1 ORGANIZATIONS AND INFORMATION SYSTEMS

Information systems and organizations influence one another. Information systems are built by managers to serve the interests of the business firm. At the same time, the organization must be aware of and open to the influences of information systems to benefit from new technologies.

The interaction between information technology and organizations is complex and is influenced by many mediating factors, including the organization's structure, business processes, politics, culture, surrounding environment, and management decisions (see Figure 3-1). You will need to understand how information systems can change social and work life in your firm. You will not be able to design new systems successfully or understand existing systems without understanding your own business organization.

FIGURE 3-1   THE TWO-WAY RELATIONSHIP BETWEEN ORGANIZATIONS AND INFORMATION TECHNOLOGY

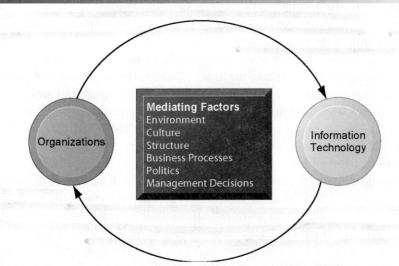

Organizations

Mediating Factors
Environment
Culture
Structure
Business Processes
Politics
Management Decisions

Information Technology

This complex two-way relationship is mediated by many factors, not the least of which are the decisions made—or not made—by managers. Other factors mediating the relationship include the organizational culture, structure, politics, business processes, and environment.

FIGURE 3-2   **THE TECHNICAL MICROECONOMIC DEFINITION OF THE ORGANIZATION**

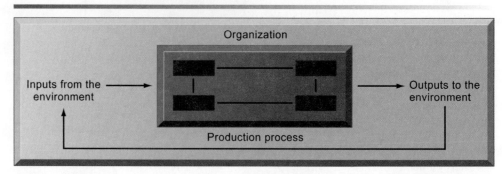

In the microeconomic definition of organizations, capital and labor (the primary production factors provided by the environment) are transformed by the firm through the production process into products and services (outputs to the environment). The products and services are consumed by the environment, which supplies additional capital and labor as inputs in the feedback loop.

As a manager, you will be the one to decide which systems will be built, what they will do, and how they will be implemented. You may not be able to anticipate all of the consequences of these decisions. Some of the changes that occur in business firms because of new information technology (IT) investments cannot be foreseen and have results that may or may not meet your expectations. Who would have imagined fifteen years ago, for instance, that e-mail and instant messaging would become a dominant form of business communication and that many managers would be inundated with more than 200 e-mail messages each day?

WHAT IS AN ORGANIZATION?

An **organization** is a stable, formal social structure that takes resources from the environment and processes them to produce outputs. This technical definition focuses on three elements of an organization. Capital and labor are primary production factors provided by the environment. The organization (the firm) transforms these inputs into products and services in a production function. The products and services are consumed by environments in return for supply inputs (see Figure 3-2).

An organization is more stable than an informal group (such as a group of friends that meets every Friday for lunch) in terms of longevity and routineness. Organizations are formal legal entities with internal rules and procedures that must abide by laws. Organizations are also social structures because they are a collection of social elements, much as a machine has a structure—a particular arrangement of valves, cams, shafts, and other parts.

This definition of organizations is powerful and simple, but it is not very descriptive or even predictive of real-world organizations. A more realistic behavioral definition of an organization is that it is a collection of rights, privileges, obligations, and responsibilities that is delicately balanced over a period of time through conflict and conflict resolution (see Figure 3-3).

In this behavioral view of the firm, people who work in organizations develop customary ways of working; they gain attachments to existing relationships; and they make arrangements with subordinates and superiors about how work will be done, the amount of work that will be done, and under

FIGURE 3-3   **THE BEHAVIORAL VIEW OF ORGANIZATIONS**

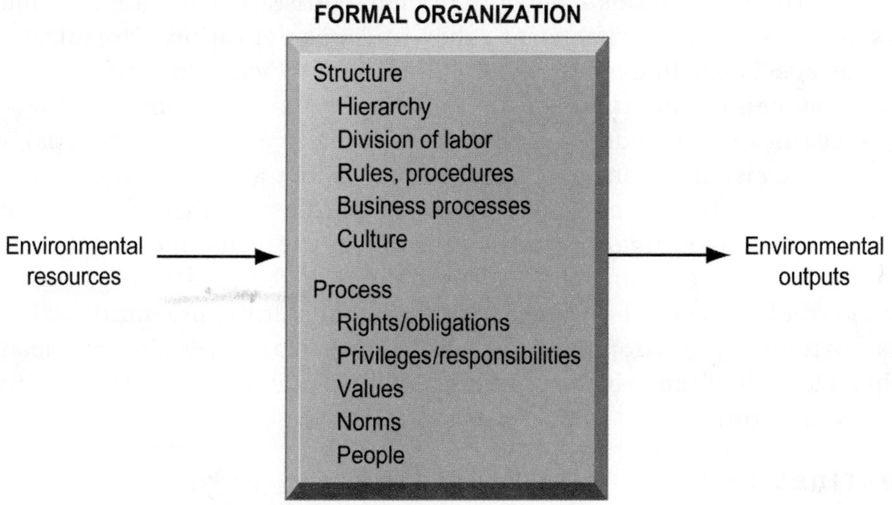

FORMAL ORGANIZATION

Structure
  Hierarchy
  Division of labor
  Rules, procedures
  Business processes
  Culture

Process
  Rights/obligations
  Privileges/responsibilities
  Values
  Norms
  People

Environmental resources → [FORMAL ORGANIZATION] → Environmental outputs

The behavioral view of organizations emphasizes group relationships, values, and structures.

what conditions work will be done. Most of these arrangements and feelings are not discussed in any formal rulebook.

How do these definitions of organizations relate to information systems technology? A technical view of organizations encourages us to focus on how inputs are combined to create outputs when technology changes are introduced into the company. The firm is seen as infinitely malleable, with capital and labor substituting for each other quite easily. But the more realistic behavioral definition of an organization suggests that building new information systems, or rebuilding old ones, involves much more than a technical rearrangement of machines or workers—that some information systems change the organizational balance of rights, privileges, obligations, responsibilities, and feelings that have been established over a long period of time.

Changing these elements can take a long time, be very disruptive, and requires more resources to support training and learning. For instance, the length of time required to implement effectively a new information system is much longer than usually anticipated simply because there is a lag between implementing a technical system and teaching employees and managers how to use the system.

Technological change requires changes in who owns and controls information, who has the right to access and update that information, and who makes decisions about whom, when, and how. This more complex view forces us to look at the way work is designed and the procedures used to achieve outputs.

The technical and behavioral definitions of organizations are not contradictory. Indeed, they complement each other: The technical definition tells us how thousands of firms in competitive markets combine capital, labor, and information technology, whereas the behavioral model takes us inside the individual firm to see how that technology affects the organization's inner workings. Section 3.2 describes how each of these definitions of organizations can help explain the relationships between information systems and organizations.

FEATURES OF ORGANIZATIONS

All modern organizations have certain characteristics. They are bureaucracies with clear-cut divisions of labor and specialization. Organizations arrange specialists in a hierarchy of authority in which everyone is accountable to someone and authority is limited to specific actions governed by abstract rules or procedures. These rules create a system of impartial and universal decision making. Organizations try to hire and promote employees on the basis of technical qualifications and professionalism (not personal connections). The organization is devoted to the principle of efficiency: maximizing output using limited inputs. Other features of organizations include their business processes, organizational culture, organizational politics, surrounding environments, structure, goals, constituencies, and leadership styles. All of these features affect the kinds of information systems used by organizations.

Routines and Business Processes

All organizations, including business firms, become very efficient over time because individuals in the firm develop **routines** for producing goods and services. Routines—sometimes called *standard operating procedures*—are precise rules, procedures, and practices that have been developed to cope with virtually all expected situations. As employees learn these routines, they become highly productive and efficient, and the firm is able to reduce its costs over time as efficiency increases. For instance, when you visit a doctor's office, receptionists have a well-developed set of routines for gathering basic information from you; nurses have a different set of routines for preparing you for an interview with a doctor; and the doctor has a well-developed set of routines for diagnosing you. *Business processes*, which we introduced in Chapters 1 and 2, are collections of such routines. A business firm in turn is a collection of business processes (Figure 3-4).

Organizational Politics

People in organizations occupy different positions with different specialties, concerns, and perspectives. As a result, they naturally have divergent viewpoints about how resources, rewards, and punishments should be distributed. These differences matter to both managers and employees, and they result in political struggle for resources, competition, and conflict within every organization. Political resistance is one of the great difficulties of bringing about organizational change—especially the development of new information systems. Virtually all large information systems investments by a firm that bring about significant changes in strategy, business objectives, business processes, and procedures become politically charged events. Managers that know how to work with the politics of an organization will be more successful than less-skilled managers in implementing new information systems. Throughout this book, you will find many examples of where internal politics defeated the best-laid plans for an information system.

Organizational Culture

All organizations have bedrock, unassailable, unquestioned (by the members) assumptions that define their goals and products. Organizational culture encompasses this set of assumptions about what products the organization should produce, how it should produce them, where, and for whom. Generally, these cultural assumptions are taken totally for granted

FIGURE 3-4   ROUTINES, BUSINESS PROCESSES, AND FIRMS

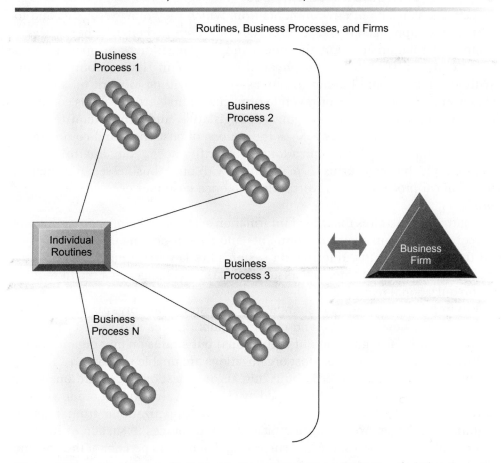

Routines, Business Processes, and Firms

All organizations are composed of individual routines and behaviors, a collection of which make up a business process. A collection of business processes make up the business firm. New information system applications require that individual routines and business processes change to achieve high levels of organizational performance.

and are rarely publicly announced or spoken about. Business processes—the actual way business firms produce value—are usually ensconced in the organization's culture.

You can see organizational culture at work by looking around your university or college. Some bedrock assumptions of university life are that professors know more than students, the reason students attend college is to learn, and classes follow a regular schedule. Organizational culture is a powerful unifying force that restrains political conflict and promotes common understanding, agreement on procedures, and common practices. If we all share the same basic cultural assumptions, agreement on other matters is more likely.

At the same time, organizational culture is a powerful restraint on change, especially technological change. Most organizations will do almost anything to avoid making changes in basic assumptions. Any technological change that threatens commonly held cultural assumptions usually meets a great deal of resistance. However, there are times when the only sensible way for a firm to move forward is to employ a new technology that directly opposes an existing organizational culture. When this occurs, the technology is often stalled while the culture slowly adjusts.

Organizational Environments

Organizations reside in environments from which they draw resources and to which they supply goods and services. Organizations and environments have a reciprocal relationship. On the one hand, organizations are open to, and dependent on, the social and physical environment that surrounds them. Without financial and human resources—people willing to work reliably and consistently for a set wage or revenue from customers—organizations could not exist. Organizations must respond to legislative and other requirements imposed by government, as well as the actions of customers and competitors. On the other hand, organizations can influence their environments. For example, business firms form alliances with other businesses to influence the political process; they advertise to influence customer acceptance of their products.

Figure 3-5 illustrates the role of information systems in helping organizations perceive changes in their environments and also in helping organizations act on their environments. Information systems are key instruments for *environmental scanning*, helping managers identify external changes that might require an organizational response.

Environments generally change much faster than organizations. New technologies, new products, and changing public tastes and values (many of which result in new government regulations) put strains on any organization's culture, politics, and people. Most organizations are unable to adapt to a rapidly changing environment. Inertia built into an organization's standard operating procedures, the political conflict raised by changes to the existing order, and the threat to closely held cultural values inhibit organizations from making significant changes. Young firms typically lack resources to sustain even short periods of troubled times. It is not surprising that only 10 percent of the Fortune 500 companies in 1919 still exist today.

FIGURE 3-5   ENVIRONMENTS AND ORGANIZATIONS HAVE A RECIPROCAL RELATIONSHIP

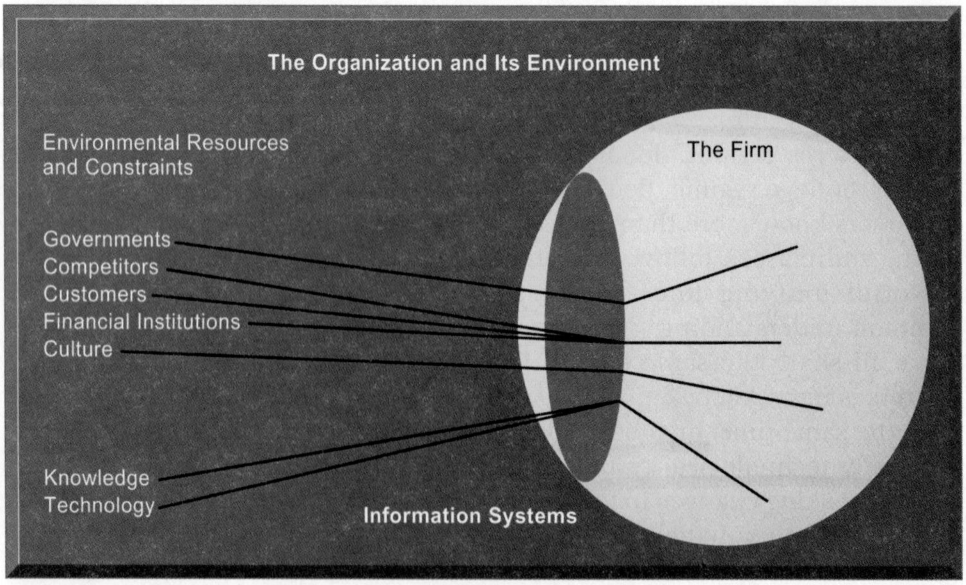

Environments shape what organizations can do, but organizations can influence their environments and decide to change environments altogether. Information technology plays a critical role in helping organizations perceive environmental change and in helping organizations act on their environment.

Disruptive Technologies: Riding the Wave. Sometimes a technology and resulting business innovation comes along to radically change the business landscape and environment. These innovations are loosely called "disruptive." (Christensen, 2003). What makes a technology disruptive? In some cases, **disruptive technologies** are substitute products that perform as well or better (often much better) than anything currently produced. The car substituted for the horse-drawn carriage; the word processor for typewriters; the Apple iPod for portable CD players; digital photography for process film photography.

In these cases, entire industries are put out of business. In other cases, disruptive technologies simply extend the market, usually with less functionality and much less cost, than existing products. Eventually they turn into low-cost competitors for whatever was sold before. Disk drives are an example: small hard disk drives used in PCs extended the market for disk drives by offering cheap digital storage for small files. Eventually, small PC hard disk drives became the largest segment of the disk drive marketplace.

Some firms are able to create these technologies and ride the wave to profits; others learn quickly and adapt their business; still others are obliterated because their products, services, and business models become obsolete. They may be very efficient at doing what no longer needs to be done! There are also cases where no firms benefit, and all the gains go to consumers (firms fail to capture any profits). Table 3-1 describes just a few disruptive technologies from the past.

Disruptive technologies are tricky. Firms that invent disruptive technologies as "first movers" do not always benefit if they lack the resources to exploit the

TABLE 3-1 DISRUPTIVE TECHNOLOGIES: WINNERS AND LOSERS

TECHNOLOGY	DESCRIPTION	WINNERS AND LOSERS
Microprocessor chips (1971)	Thousands and eventually millions of transistors on a silicon chip	Microprocessor firms win (Intel, Texas Instruments) while transistor firms (GE) decline.
Personal computers (1975)	Small, inexpensive, but fully functional desktop computers	PC manufacturers (HP, Apple, IBM), and chip manufacturers prosper (Intel), while mainframe (IBM) and minicomputer (DEC) firms lose.
PC word processing software (1979)	Inexpensive, limited but functional text editing and formatting for personal computers	PC and software manufacturers (Microsoft, HP, Apple) prosper, while the typewriter industry disappears.
World Wide Web (1989)	A global database of digital files and "pages" instantly available	Owners of online content and news benefit, while traditional publishers (newspapers, magazines, broadcast television) lose.
Internet music services (1998)	Repositories of downloadable music on the Web with acceptable fidelity	Owners of online music collections (MP3.com, iTunes), telecommunications providers who own Internet backbone (AT&T, Verizon), local Internet service providers win, while record label firms and music retailers lose (Tower Records).
PageRank algorithm	A method for ranking Web pages in terms of their popularity to supplement Web search by key terms	Google is the winner (they own the patent), while traditional key word search engines (Alta Vista) lose.
Software as Web service	Using the Internet to provide remote access to online software	Online software services companies (Salesforce.com) win, while traditional "boxed" software companies (Microsoft, SAP, Oracle) lose.

technology or fail to see the opportunity. The MITS Altair 8800 is widely regarded as the first PC, but its inventors did not take advantage of their first-mover status. Second movers, so-called "fast followers" such as IBM and Microsoft, reaped the rewards. Citibank's ATMs revolutionized retail banking, but they were copied by other banks. Now all banks use ATMs, with the benefits going mostly to the consumers. Google was not a first mover in search, but an innovative follower that was able to maintain rights to a powerful new search algorithm called PageRank. So far it has been able to hold onto its lead while most other search engines have faded down to small market shares.

Organizational Structure

Organizations all have a structure or shape. Mintzberg's classification, described in Table 3-2, identifies five basic kinds of organizational structure (Mintzberg, 1979).

The kind of information systems you find in a business firm—and the nature of problems with these systems—often reflects the type of organizational structure. For instance, in a professional bureaucracy such as a hospital it is not unusual to find parallel patient record systems operated by the administration, another by doctors, and another by other professional staff such as nurses and social workers. In small entrepreneurial firms you will often find poorly designed systems developed in a rush that often outgrow their usefulness quickly. In huge multidivisional firms operating in hundreds of locations you will often find there is not a single integrating information system, but instead each locale or each division has its set of information systems.

Other Organizational Features

Organizations have goals and use different means to achieve them. Some organizations have coercive goals (e.g., prisons); others have utilitarian goals (e.g., businesses). Still others have normative goals (universities, religious

TABLE 3-2 ORGANIZATIONAL STRUCTURES

ORGANIZATIONAL TYPE	DESCRIPTION	EXAMPLES
Entrepreneurial structure	Young, small firm in a fast-changing environment. It has a simple structure and is managed by an entrepreneur serving as its single chief executive officer.	Small start-up business
Machine bureaucracy	Large bureaucracy existing in a slowly changing environment, producing standard products. It is dominated by a centralized management team and centralized decision making.	Midsize manufacturing firm
Divisionalized bureaucracy	Combination of multiple machine bureaucracies, each producing a different product or service, all topped by one central headquarters.	Fortune 500 firms, such as General Motors
Professional bureaucracy	Knowledge-based organization where goods and services depend on the expertise and knowledge of professionals. Dominated by department heads with weak centralized authority.	Law firms, school systems, hospitals
Adhocracy	Task force organization that must respond to rapidly changing environments. Consists of large groups of specialists organized into short-lived multidisciplinary teams and has weak central management.	Consulting firms, such as the Rand Corporation

groups). Organizations also serve different groups or have different constituencies, some primarily benefiting their members, others benefiting clients, stockholders, or the public. The nature of leadership differs greatly from one organization to another—some organizations may be more democratic or authoritarian than others. Another way organizations differ is by the tasks they perform and the technology they use. Some organizations perform primarily routine tasks that can be reduced to formal rules that require little judgment (such as manufacturing auto parts), whereas others (such as consulting firms) work primarily with nonroutine tasks.

3.2 HOW INFORMATION SYSTEMS IMPACT ORGANIZATIONS AND BUSINESS FIRMS

Information systems have become integral, online, interactive tools deeply involved in the minute-to-minute operations and decision making of large organizations. Over the last decade, information systems have fundamentally altered the economics of organizations and greatly increased the possibilities for organizing work. Theories and concepts from economics and sociology help us understand the changes brought about by IT.

ECONOMIC IMPACTS

From the point of view of economics, IT changes both the relative costs of capital and the costs of information. Information systems technology can be viewed as a factor of production that can be substituted for traditional capital and labor. As the cost of information technology decreases, it is substituted for labor, which historically has been a rising cost. Hence, information technology should result in a decline in the number of middle managers and clerical workers as information technology substitutes for their labor (Laudon, 1990).

As the cost of information technology decreases, it also substitutes for other forms of capital such as buildings and machinery, which remain relatively expensive. Hence, over time we should expect managers to increase their investments in IT because of its declining cost relative to other capital investments.

IT also obviously affects the cost and quality of information and changes the economics of information. Information technology helps firms contract in size because it can reduce transaction costs—the costs incurred when a firm buys on the marketplace what it cannot make itself. According to **transaction cost theory**, firms and individuals seek to economize on transaction costs, much as they do on production costs. Using markets is expensive because of costs such as locating and communicating with distant suppliers, monitoring contract compliance, buying insurance, obtaining information on products, and so forth (Coase, 1937; Williamson, 1985). Traditionally, firms have tried to reduce transaction costs through vertical integration, by getting bigger, hiring more employees, and buying their own suppliers and distributors, as both General Motors and Ford used to do.

Information technology, especially the use of networks, can help firms lower the cost of market participation (transaction costs), making it worthwhile for firms to contract with external suppliers instead of using internal sources. As a result, firms can shrink in size (numbers of employees) because it is far less expensive to outsource work to a competitive marketplace rather than hire employees.

For instance, by using computer links to external suppliers, the Chrysler Corporation can achieve economies by obtaining more than 70 percent of its parts from the outside. Information systems make it possible for companies such as Cisco Systems and Dell Inc. to outsource their production to contract manufacturers such as Flextronics instead of making their products themselves.

Figure 3-6 shows that as transaction costs decrease, firm size (the number of employees) should shrink because it becomes easier and cheaper for the firm to contract for the purchase of goods and services in the marketplace rather than to make the product or offer the service itself. Firm size can stay constant or contract even as the company increases its revenues. For example, when Eastman Chemical Company split off from Kodak in 1994, it had $3.3 billion in revenue and 24,000 full-time employees. In 2009, it generated over $5 billion in revenue with only 10,000 employees.

Information technology also can reduce internal management costs. According to **agency theory**, the firm is viewed as a "nexus of contracts" among self-interested individuals rather than as a unified, profit-maximizing entity (Jensen and Meckling, 1976). A principal (owner) employs "agents" (employees) to perform work on his or her behalf. However, agents need constant supervision and management; otherwise, they will tend to pursue their own interests rather than those of the owners. As firms grow in size and scope, agency costs or coordination costs rise because owners must expend more and more effort supervising and managing employees.

Information technology, by reducing the costs of acquiring and analyzing information, permits organizations to reduce agency costs because it becomes easier for managers to oversee a greater number of employees. Figure 3-7 shows that by reducing overall management costs, information technology enables firms to increase revenues while shrinking the number of middle managers and clerical workers. We have seen examples in earlier chapters where information technology expanded the power and scope of small organizations by enabling them to perform coordinating activities such as processing orders or keeping track of inventory with very few clerks and managers.

FIGURE 3-6   THE TRANSACTION COST THEORY OF THE IMPACT OF INFORMATION TECHNOLOGY ON THE ORGANIZATION

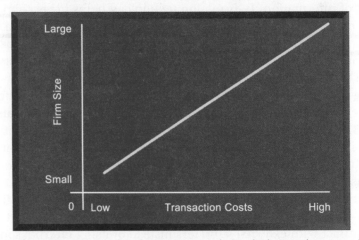

When the costs of participating in markets (transaction costs) were high, it made sense to build large firms and do everything inside the firm. But IT reduces the firm's market transaction costs. This means firms can outsource work using the market, reduce their employee head count, and still grow revenues, relying more on outsourcing firms and external contractors.

FIGURE 3-7   **THE AGENCY COST THEORY OF THE IMPACT OF INFORMATION TECHNOLOGY ON THE ORGANIZATION**

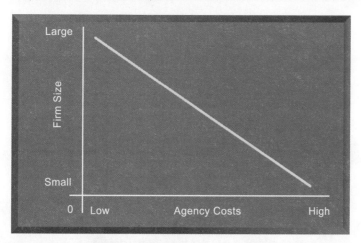

Agency costs are the costs of managing a firm's employees. IT reduces agency costs making management more efficient. Fewer managers are needed to manage employees.   IT makes it possible to build very large global firms and to run them efficiently without greatly expanding management. Without IT, very large global firms would be difficult to operate because they would be very expensive to manage.

Because IT reduces both agency and transaction costs for firms, we should expect firm size to shrink over time as more capital is invested in IT. Firms should have fewer managers, and we expect to see revenue per employee increase over time.

ORGANIZATIONAL AND BEHAVIORAL IMPACTS

Theories based in the sociology of complex organizations also provide some understanding about how and why firms change with the implementation of new IT applications.

IT Flattens Organizations

Large, bureaucratic organizations, which primarily developed before the computer age, are often inefficient, slow to change, and less competitive than newly created organizations. Some of these large organizations have downsized, reducing the number of employees and the number of levels in their organizational hierarchies.

Behavioral researchers have theorized that information technology facilitates flattening of hierarchies by broadening the distribution of information to empower lower-level employees and increase management efficiency (see Figure 3-8). IT pushes decision-making rights lower in the organization because lower-level employees receive the information they need to make decisions without supervision. (This empowerment is also possible because of higher educational levels among the workforce, which give employees the capabilities to make intelligent decisions.) Because managers now receive so much more accurate information on time, they become much faster at making decisions, so fewer managers are required. Management costs decline as a percentage of revenues, and the hierarchy becomes much more efficient.

These changes mean that the management span of control has also been broadened, enabling high-level managers to manage and control more workers

FIGURE 3-8   FLATTENING ORGANIZATIONS

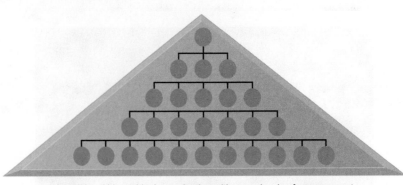

A traditional hierarchical organization with many levels of management

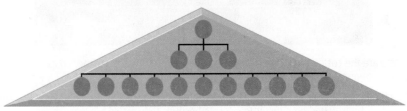

An organization that has been "flattened" by removing layers of management

Information systems can reduce the number of levels in an organization by providing managers with information to supervise larger numbers of workers and by giving lower-level employees more decision-making authority.

spread over greater distances. Many companies have eliminated thousands of middle managers as a result of these changes.

Postindustrial Organizations

Postindustrial theories based more on history and sociology than economics also support the notion that IT should flatten hierarchies. In postindustrial societies, authority increasingly relies on knowledge and competence, and not merely on formal positions. Hence, the shape of organizations flattens because professional workers tend to be self-managing, and decision making should become more decentralized as knowledge and information become more widespread throughout the firm (Drucker, 1988).

Information technology may encourage task force-networked organizations in which groups of professionals come together—face to face or electronically—for short periods of time to accomplish a specific task (e.g., designing a new automobile); once the task is accomplished, the individuals join other task forces. The global consulting service Accenture is an example. It has no operational headquarters and no formal branches. Many of its 190,000 employees move from location to location to work on projects at client locations in 49 different countries.

Who makes sure that self-managed teams do not head off in the wrong direction? Who decides which person works on which team and for how long? How can managers evaluate the performance of someone who is constantly rotating from team to team? How do people know where their careers are headed? New approaches for evaluating, organizing, and informing workers are required, and not all companies can make virtual work effective.

Understanding Organizational Resistance to Change

Information systems inevitably become bound up in organizational politics because they influence access to a key resource—namely, information. Information systems can affect who does what to whom, when, where, and how in an organization. Many new information systems require changes in personal, individual routines that can be painful for those involved and require retraining and additional effort that may or may not be compensated. Because information systems potentially change an organization's structure, culture, business processes, and strategy, there is often considerable resistance to them when they are introduced.

There are several ways to visualize organizational resistance. Leavitt (1965) used a diamond shape to illustrate the interrelated and mutually adjusting character of technology and organization (see Figure 3-9). Here, changes in technology are absorbed, deflected, and defeated by organizational task arrangements, structures, and people. In this model, the only way to bring about change is to change the technology, tasks, structure, and people simultaneously. Other authors have spoken about the need to "unfreeze" organizations before introducing an innovation, quickly implementing it, and "refreezing" or institutionalizing the change (Alter and Ginzberg, 1978; Kolb, 1970).

Because organizational resistance to change is so powerful, many information technology investments flounder and do not increase productivity. Indeed, research on project implementation failures demonstrates that the most common reason for failure of large projects to reach their objectives is not the failure of the technology, but organizational and political resistance to change. Chapter 14 treats this issue in detail. Therefore, as a manger involved in future IT investments, your ability to work with people and organizations is just as important as your technical awareness and knowledge.

THE INTERNET AND ORGANIZATIONS

The Internet, especially the World Wide Web, has an important impact on the relationships between many firms and external entities, and even on the

FIGURE 3-9   ORGANIZATIONAL RESISTANCE AND THE MUTUALLY ADJUSTING RELATIONSHIP BETWEEN TECHNOLOGY AND THE ORGANIZATION

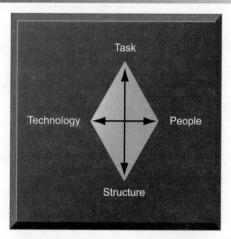

Implementing information systems has consequences for task arrangements, structures, and people. According to this model, to implement change, all four components must be changed simultaneously.
Source: Leavitt (1965).

organization of business processes inside a firm. The Internet increases the accessibility, storage, and distribution of information and knowledge for organizations. In essence, the Internet is capable of dramatically lowering the transaction and agency costs facing most organizations. For instance, brokerage firms and banks in New York can now deliver their internal operating procedures manuals to their employees at distant locations by posting them on the corporate Web site, saving millions of dollars in distribution costs. A global sales force can receive nearly instant product price information updates using the Web or instructions from management sent by e-mail. Vendors of some large retailers can access retailers' internal Web sites directly to find up-to-the-minute sales information and to initiate replenishment orders instantly.

Businesses are rapidly rebuilding some of their key business processes based on Internet technology and making this technology a key component of their IT infrastructures. If prior networking is any guide, one result will be simpler business processes, fewer employees, and much flatter organizations than in the past.

IMPLICATIONS FOR THE DESIGN AND UNDERSTANDING OF INFORMATION SYSTEMS

To deliver genuine benefits, information systems must be built with a clear understanding of the organization in which they will be used. In our experience, the central organizational factors to consider when planning a new system are the following:

- The environment in which the organization must function
- The structure of the organization: hierarchy, specialization, routines, and business processes
- The organization's culture and politics
- The type of organization and its style of leadership
- The principal interest groups affected by the system and the attitudes of workers who will be using the system
- The kinds of tasks, decisions, and business processes that the information system is designed to assist

3.3 USING INFORMATION SYSTEMS TO ACHIEVE COMPETITIVE ADVANTAGE

In almost every industry you examine, you will find that some firms do better than most others. There's almost always a stand-out firm. In the automotive industry, Toyota is considered a superior performer. In pure online retail, Amazon is the leader, in off-line retail Walmart, the largest retailer on earth, is the leader. In online music, Apple's iTunes is considered the leader with more than 75 percent of the downloaded music market, and in the related industry of digital music players, the iPod is the leader. In Web search, Google is considered the leader.

Firms that "do better" than others are said to have a competitive advantage over others: They either have access to special resources that others do not, or they are able to use commonly available resources more efficiently—usually because of superior knowledge and information assets. In any event, they do better in terms of revenue growth, profitability, or productivity growth (efficiency), all of which ultimately in the long run translate into higher stock market valuations than their competitors.

But why do some firms do better than others and how do they achieve competitive advantage? How can you analyze a business and identify its strategic advantages? How can you develop a strategic advantage for your own business? And how do information systems contribute to strategic advantages? One answer to that question is Michael Porter's competitive forces model.

PORTER'S COMPETITIVE FORCES MODEL

Arguably, the most widely used model for understanding competitive advantage is Michael Porter's **competitive forces model** (see Figure 3-10). This model provides a general view of the firm, its competitors, and the firm's environment. Earlier in this chapter, we described the importance of a firm's environment and the dependence of firms on environments. Porter's model is all about the firm's general business environment. In this model, five competitive forces shape the fate of the firm.

Traditional Competitors

All firms share market space with other competitors who are continuously devising new, more efficient ways to produce by introducing new products and services, and attempting to attract customers by developing their brands and imposing switching costs on their customers.

New Market Entrants

In a free economy with mobile labor and financial resources, new companies are always entering the marketplace. In some industries, there are very low barriers to entry, whereas in other industries, entry is very difficult. For instance, it is fairly easy to start a pizza business or just about any small retail business, but it is much more expensive and difficult to enter the computer chip business, which has very high capital costs and requires significant expertise and knowledge that is hard to obtain. New companies have several possible

FIGURE 3-10 PORTER'S COMPETITIVE FORCES MODEL

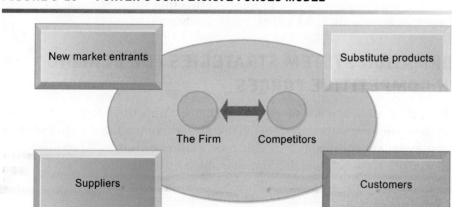

In Porter's competitive forces model, the strategic position of the firm and its strategies are determined not only by competition with its traditional direct competitors but also by four other forces in the industry's environment: new market entrants, substitute products, customers, and suppliers.

advantages: They are not locked into old plants and equipment, they often hire younger workers who are less expensive and perhaps more innovative, they are not encumbered by old worn-out brand names, and they are "more hungry" (more highly motivated) than traditional occupants of an industry. These advantages are also their weakness: They depend on outside financing for new plants and equipment, which can be expensive; they have a less-experienced workforce; and they have little brand recognition.

Substitute Products and Services

In just about every industry, there are substitutes that your customers might use if your prices become too high. New technologies create new substitutes all the time. Even oil has substitutes: Ethanol can substitute for gasoline in cars; vegetable oil for diesel fuel in trucks; and wind, solar, coal, and hydro power for industrial electricity generation. Likewise, the Internet telephone service can substitute for traditional telephone service, and fiber-optic telephone lines to the home can substitute for cable TV lines. And, of course, an Internet music service that allows you to download music tracks to an iPod is a substitute for CD-based music stores. The more substitute products and services in your industry, the less you can control pricing and the lower your profit margins.

Customers

A profitable company depends in large measure on its ability to attract and retain customers (while denying them to competitors), and charge high prices. The power of customers grows if they can easily switch to a competitor's products and services, or if they can force a business and its competitors to compete on price alone in a transparent marketplace where there is little **product differentiation**, and all prices are known instantly (such as on the Internet). For instance, in the used college textbook market on the Internet, students (customers) can find multiple suppliers of just about any current college textbook. In this case, online customers have extraordinary power over used-book firms.

Suppliers

The market power of suppliers can have a significant impact on firm profits, especially when the firm cannot raise prices as fast as can suppliers. The more different suppliers a firm has, the greater control it can exercise over suppliers in terms of price, quality, and delivery schedules. For instance, manufacturers of laptop PCs almost always have multiple competing suppliers of key components, such as keyboards, hard drives, and display screens.

INFORMATION SYSTEM STRATEGIES FOR DEALING WITH COMPETITIVE FORCES

What is a firm to do when it is faced with all these competitive forces? And how can the firm use information systems to counteract some of these forces? How do you prevent substitutes and inhibit new market entrants? There are four generic strategies, each of which often is enabled by using information technology and systems: low-cost leadership, product differentiation, focus on market niche, and strengthening customer and supplier intimacy.

Low-Cost Leadership

Use information systems to achieve the lowest operational costs and the lowest prices. The classic example is Walmart. By keeping prices low and shelves well

Supermarkets and large retail stores such as Walmart use sales data captured at the checkout counter to determine which items have sold and need to be reordered. Walmart's continuous replenishment system transmits orders to restock directly to its suppliers. The system enables Walmart to keep costs low while fine-tuning its merchandise to meet customer demands.

stocked using a legendary inventory replenishment system, Walmart became the leading retail business in the United States. Walmart's continuous replenishment system sends orders for new merchandise directly to suppliers as soon as consumers pay for their purchases at the cash register. Point-of-sale terminals record the bar code of each item passing the checkout counter and send a purchase transaction directly to a central computer at Walmart headquarters. The computer collects the orders from all Walmart stores and transmits them to suppliers. Suppliers can also access Walmart's sales and inventory data using Web technology.

Because the system replenishes inventory with lightning speed, Walmart does not need to spend much money on maintaining large inventories of goods in its own warehouses. The system also enables Walmart to adjust purchases of store items to meet customer demands. Competitors, such as Sears, have been spending 24.9 percent of sales on overhead. But by using systems to keep operating costs low, Walmart pays only 16.6 percent of sales revenue for overhead. (Operating costs average 20.7 percent of sales in the retail industry.)

Walmart's continuous replenishment system is also an example of an **efficient customer response system**. An efficient customer response system directly links consumer behavior to distribution and production and supply chains. Walmart's continuous replenishment system provides such an efficient customer response.

Product Differentiation

Use information systems to enable new products and services, or greatly change the customer convenience in using your existing products and services. For instance, Google continuously introduces new and unique search services on its Web site, such as Google Maps. By purchasing PayPal, an electronic payment system, in 2003, eBay made it much easier for customers to pay sellers and expanded use of its auction marketplace. Apple created the iPod, a unique portable digital music player, plus a unique online Web music service where songs can be purchased for $.69 to $1.29 each. Apple has continued to innovate with its multimedia iPhone, iPad tablet computer, and iPod video player. The chapter-opening case describes how AT&T's business strategy is trying to piggyback off such digital innovations.

Manufacturers and retailers are using information systems to create products and services that are customized and personalized to fit the precise specifications of individual customers. For example, Nike sells customized sneakers through its NIKEiD program on its Web site. Customers are able to select the type of shoe, colors, material, outsoles, and even a logo of up to 8 characters. Nike transmits the orders via computers to specially-equipped plants in China and Korea. The sneakers cost only $10 extra and take about three weeks to reach the customer. This ability to offer individually tailored products or services using the same production resources as mass production is called **mass customization**.

Table 3-3 lists a number of companies that have developed IT-based products and services that other firms have found difficult to copy, or at least a long time to copy.

Focus on Market Niche

Use information systems to enable a specific market focus, and serve this narrow target market better than competitors. Information systems support this strategy by producing and analyzing data for finely tuned sales and marketing techniques. Information systems enable companies to analyze customer buying patterns, tastes, and preferences closely so that they efficiently pitch advertising and marketing campaigns to smaller and smaller target markets.

The data come from a range of sources—credit card transactions, demographic data, purchase data from checkout counter scanners at supermarkets and retail stores, and data collected when people access and interact with Web sites. Sophisticated software tools find patterns in these large pools of data and infer rules from them to guide decision making. Analysis of such data drives one-to-one marketing that creates personal messages based on individualized preferences. For example, Hilton Hotels' OnQ system analyzes detailed data collected on active guests in all of its properties to determine the preferences of each guest and each guest's profitability. Hilton uses this information to give its most profitable customers additional privileges, such as late check-outs. Contemporary customer relationship management (CRM) systems feature analytical capabilities for this type of intensive data analysis (see Chapters 2 and 9).

TABLE 3-3 IT-ENABLED NEW PRODUCTS AND SERVICES PROVIDING COMPETITIVE ADVANTAGE

Amazon: One-click shopping	Amazon holds a patent on one-click shopping that it licenses to other online retailers.
Online music: Apple iPod and iTunes	The iPod is an integrated handheld player backed up with an online library of over 13 million songs
Golf club customization: Ping	Customers can select from more than 1 million different golf club options; a build-to-order system ships their customized clubs within 48 hours.
Online bill payment: CheckFree.com	Fifty-two million households pay bills online in 2010.
Online person-to-person payment: PayPal.com	PayPal enables the transfer of money between individual bank accounts and between bank accounts and credit card accounts.

The Interactive Session on Organizations describes how skillfully credit card companies are able to use this strategy to predict their most profitable cardholders. The companies gather vast quantities of data about consumer purchases and other behaviors and mine these data to construct detailed profiles that identify cardholders who might be good or bad credit risks. These practices have enhanced credit card companies' profitability, but are they in consumers' best interests?

Strengthen Customer and Supplier Intimacy

Use information systems to tighten linkages with suppliers and develop intimacy with customers. Chrysler Corporation uses information systems to facilitate direct access by suppliers to production schedules, and even permits suppliers to decide how and when to ship supplies to Chrysler factories. This allows suppliers more lead time in producing goods. On the customer side, Amazon.com keeps track of user preferences for book and CD purchases, and can recommend titles purchased by others to its customers. Strong linkages to customers and suppliers increase **switching costs** (the cost of switching from one product to a competing product), and loyalty to your firm.

Table 3-4 summarizes the competitive strategies we have just described. Some companies focus on one of these strategies, but you will often see companies pursuing several of them simultaneously. For example, Dell tries to emphasize low cost as well as the ability to customize its personal computers.

THE INTERNET'S IMPACT ON COMPETITIVE ADVANTAGE

Because of the Internet, the traditional competitive forces are still at work, but competitive rivalry has become much more intense (Porter, 2001). Internet technology is based on universal standards that any company can use, making it easy for rivals to compete on price alone and for new competitors to enter the market. Because information is available to everyone, the Internet raises the bargaining power of customers, who can quickly find the lowest-cost provider on the Web. Profits have been dampened. Table 3-5 summarizes some of the potentially negative impacts of the Internet on business firms identified by Porter.

TABLE 3-4 FOUR BASIC COMPETITIVE STRATEGIES

STRATEGY	DESCRIPTION	EXAMPLE
Low-cost leadership	Use information systems to produce products and services at a lower price than competitors while enhancing quality and level of service	Walmart
Product differentiation	Use information systems to differentiate products, and enable new services and products	Google, eBay, Apple, Lands' End
Focus on market niche	Use information systems to enable a focused strategy on a single market niche; specialize	Hilton Hotels, Harrah's
Customer and supplier intimacy	Use information systems to develop strong ties and loyalty with customers and suppliers	Chrysler Corporation Amazon.com

INTERACTIVE SESSION: ORGANIZATIONS

HOW MUCH DO CREDIT CARD COMPANIES KNOW ABOUT YOU?

When Kevin Johnson returned from his honeymoon, a letter from American Express was waiting for him. The letter informed Johnson that AmEx was slashing his credit limit by 60 percent. Why? Not because Johnson missed a payment or had bad credit. The letter stated: "Other customers who have used their card at establishments where you recently shopped, have a poor repayment history with American Express." Johnson had started shopping at Walmart. Welcome to the new era of credit card profiling.

Every time you make a purchase with a credit card, a record of that sale is logged into a massive data repository maintained by the card issuer. Each purchase is assigned a four-digit category code that describes the type of purchase that was made. There are separate codes for grocery stores, fast food restaurants, doctors, bars, bail and bond payments, and dating and escort services. Taken together, these codes allow credit card companies to learn a great deal about each of its customers at a glance.

Credit card companies use these data for multiple purposes. First, they use them to target future promotions for additional products more accurately. Users that purchase airline tickets might receive promotions for frequent flyer miles, for example. The data help card issuers guard against credit card fraud by identifying purchases that appear unusual compared to a cardholder's normal purchase history. The card companies also flag users who frequently charge more than their credit limit or demonstrate erratic spending habits. Lastly, these records are used by law enforcement agencies to track down criminals.

Credit card holders with debt, the ones who never fully pay off their balances entirely and thus have to pay monthly interest charges and other fees, have been a major source of profit for credit card issuers. However, the recent financial crisis and credit crunch have turned them into a mounting liability because so many people are defaulting on their payments and even filing for bankruptcy. So the credit card companies are now focusing on mining credit card data to predict cardholders posing the highest risk.

Using mathematical formulas and insights from behavioral science, these companies are developing more fine-grained profiles to help them get inside the heads of their customers. The data provide new insights about the relationship of certain types of purchases to a customer's ability or inability to pay off credit card balances and other debt. The card-issuing companies now use this information to deny credit card applications or shrink the amount of credit available to high-risk customers.

These companies are generalizing based on certain types of purchases that may unfairly characterize responsible cardholders as risky. Purchases of secondhand clothing, bail bond services, massages, or gambling might cause card issuers to identify you as a risk, even if you maintain your balance responsibly from month to month. Other behaviors that raise suspicion: using your credit card to get your tires re-treaded, to pay for drinks at a bar, to pay for marriage counseling, or to obtain a cash advance. Charged speeding tickets raise suspicion because they may indicate an irrational or impulsive personality. In light of the sub-prime mortgage crisis, credit card companies have even begun to consider individuals from Florida, Nevada, California, and other states hardest hit by foreclosures to be risks simply by virtue of their state of residence.

The same fine-grained profiling also identifies the most reliable credit-worthy cardholders. For example, the credit card companies found that people who buy high-quality bird seed and snow rakes to sweep snow off of their roofs are very likely to pay their debts and never miss payments. Credit card companies are even using their detailed knowledge of cardholder behavior to establish personal connections with the clients that owe them money and convince them to pay off their balances.

One 49-year old woman from Missouri in the throes of a divorce owed $40,000 to various credit card companies at one point, including $28,000 to Bank of America. A Bank of America customer service representative studied the woman's profile and spoke to her numerous times, even pointing out one instance where she was erroneously charged twice. The representative forged a bond with the cardholder, and as a result she paid back the entire $28,000 she owed, (even though she failed to repay much of the remainder that she owed to other credit card companies.)

This example illustrates something the credit card companies now know: when cardholders feel more comfortable with companies, as a result of a good relationship with a customer service rep or for any other reason, they're more likely to pay their debts.

It's common practice for credit card companies to use this information to get a better idea of consumer trends, but should they be able to use it to preemptively deny credit or adjust terms of agreements? Law enforcement is not permitted to profile individuals, but it appears that credit card companies are doing just that.

In June 2008, the FTC filed a lawsuit against CompuCredit, a sub-prime credit card marketer. CompuCredit had been using a sophisticated behavioral scoring model to identify customers who they considered to have risky purchasing behaviors and lower these customers' credit limits. CompuCredit settled the suit by crediting $114 million to the accounts of these supposedly risky customers and paid a $2.5 million penalty.

Congress is investigating the extent to which credit card companies use profiling to determine interest rates and policies for their cardholders. The new credit card reform law signed by President

Barack Obama in May 2009 requires federal regulators to investigate this. Regulators must also determine whether minority cardholders were adversely profiled by these criteria. The new legislation also bars card companies from raising interest rates at any time and for any reason on their customers.

Going forward, you're likely to receive far fewer credit card solicitations in the mail and fewer offers of interest-free cards with rates that skyrocket after an initial grace period. You'll also see fewer policies intended to trick or deceive customers, like cash-back rewards for unpaid balances, which actually encourage cardholders not to pay what they owe. But the credit card companies say that to compensate for these changes, they'll need to raise rates across the board, even for good customers.

Sources: Betty Schiffman, "Who Knows You Better? Your Credit Card Company or Your Spouse?" Daily Finance, April 13, 2010; Charles Duhigg, "What Does Your Credit-Card Company Know about You?" *The New York Times*, June 17, 2009; and CreditCards.com, "Can Your Lifestyle Hurt Your Credit?" MSN Money, June 30, 2009.Boudette.

CASE STUDY QUESTIONS

1. What competitive strategy are the credit card companies pursuing? How do information systems support that strategy?

2. What are the business benefits of analyzing customer purchase data and constructing behavioral profiles?

3. Are these practices by credit card companies ethical? Are they an invasion of privacy? Why or why not?

MIS IN ACTION

1. If you have a credit card, make a detailed list of all of your purchases for the past six months. Then write a paragraph describing what credit card companies learned about your interests and behavior from these purchases.

2. How would this information benefit the credit card companies? What other companies would be interested?

TABLE 3-5 IMPACT OF THE INTERNET ON COMPETITIVE FORCES AND INDUSTRY STRUCTURE

COMPETITIVE FORCE	IMPACT OF THE INTERNET
Substitute products or services	Enables new substitutes to emerge with new approaches to meeting needs and performing functions
Customers' bargaining power	Availability of global price and product information shifts bargaining power to customers
Suppliers' bargaining power	Procurement over the Internet tends to raise bargaining power over suppliers; suppliers can also benefit from reduced barriers to entry and from the elimination of distributors and other intermediaries standing between them and their users
Threat of new entrants	The Internet reduces barriers to entry, such as the need for a sales force, access to channels, and physical assets; it provides a technology for driving business processes that makes other things easier to do
Positioning and rivalry among existing competitors	Widens the geographic market, increasing the number of competitors, and reducing differences among competitors; makes it more difficult to sustain operational advantages; puts pressure to compete on price

The Internet has nearly destroyed some industries and has severely threatened more. For instance, the printed encyclopedia industry and the travel agency industry have been nearly decimated by the availability of substitutes over the Internet. Likewise, the Internet has had a significant impact on the retail, music, book, retail brokerage, software, telecommunications, and newspaper industries.

However, the Internet has also created entirely new markets, formed the basis for thousands of new products, services, and business models, and provided new opportunities for building brands with very large and loyal customer bases. Amazon, eBay, iTunes, YouTube, Facebook, Travelocity, and Google are examples. In this sense, the Internet is "transforming" entire industries, forcing firms to change how they do business.

The Interactive Session on Technology provides more detail on the transformation of the content and media industries. For most forms of media, the Internet has posed a threat to business models and profitability. Growth in book sales other than textbooks and professional publications has been sluggish, as new forms of entertainment continue to compete for consumers' time. Newspapers and magazines have been hit even harder, as their readerships diminish, their advertisers shrink, and more people get their news for free online. The television and film industries have been forced to deal with pirates who are robbing them of some of their profits.

When Apple announced the launch of its new iPad tablet computer, leaders in all of these media saw not only a threat but also a significant opportunity. In fact, the iPad and similar mobile devices may be the savior—if traditional media can strike the right deal with technology providers like Apple and Google. And the iPad may be a threat for companies that fail to adjust their business models to a new method of providing content to users.

THE BUSINESS VALUE CHAIN MODEL

Although the Porter model is very helpful for identifying competitive forces and suggesting generic strategies, it is not very specific about what exactly to do, and it does not provide a methodology to follow for achieving competitive advantages. If your goal is to achieve operational excellence, where do you start? Here's where the business value chain model is helpful.

The **value chain model** highlights specific activities in the business where competitive strategies can best be applied (Porter, 1985) and where information systems are most likely to have a strategic impact. This model identifies specific, critical leverage points where a firm can use information technology most effectively to enhance its competitive position. The value chain model views the firm as a series or chain of basic activities that add a margin of value to a firm's products or services. These activities can be categorized as either primary activities or support activities (see Figure 3-11 on p. 105).

Primary activities are most directly related to the production and distribution of the firm's products and services, which create value for the customer. Primary activities include inbound logistics, operations, outbound logistics, sales and marketing, and service. Inbound logistics includes receiving and storing materials for distribution to production. Operations transforms inputs into finished products. Outbound logistics entails storing and distributing finished products. Sales and marketing includes promoting and selling the firm's products. The service activity includes maintenance and repair of the firm's goods and services.

INTERACTIVE SESSION: TECHNOLOGY

IS THE IPAD A DISRUPTIVE TECHNOLOGY?

Tablet computers have come and gone several times before, but the iPad looks like it will be different. It has a gorgeous 10-inch color display, a persistent Wi-Fi Internet connection, potential use of high-speed cellular networks, functionality from over 250,000 applications available on Apple's App Store, and the ability to deliver video, music, text, social networking applications, and video games. Its entry-level price is just $499. The challenge for Apple is to convince potential users that they need a new, expensive gadget with the functionality that the iPad provides. This is the same challenge faced by the iPhone when it was first announced. As it turned out, the iPhone was a smashing success that decimated the sales of traditional cell phones throughout the world. Will the iPad do likewise as a disruptive technology for the media and content industries? It looks like it is on its way.

The iPad has some appeal to mobile business users, but most experts believe it will not supplant laptops or netbooks. It is in the publishing and media industries where its disruptive impact will first be felt.

The iPad and similar devices (including the Kindle Reader) will force many existing media businesses to change their business models significantly. These companies may need to stop investing in their traditional delivery platforms (like newsprint) and increase their investments in the new digital platform. The iPad will spur people to watch TV on the go, rather than their television set at home, and to read their books, newspapers, and magazines online rather than in print.

Publishers are increasingly interested in e-books as a way to revitalize stagnant sales and attract new readers. The success of Amazon's Kindle has spurred growth in e-book sales to over $91 million wholesale in the first quarter of 2010. Eventually, e-books could account for 25 to 50 percent of all books sold. Amazon, the technology platform provider and the largest distributor of books in the world, has exercised its new power by forcing publishers to sell e-books at $9.95, a price too low for publishers to profit. Publishers are now refusing to supply new books to Amazon unless it raises prices, and Amazon is starting to comply.

The iPad enters this marketplace ready to compete with Amazon over e-book pricing and distribution. Amazon has committed itself to offering the lowest possible prices, but Apple has appealed to publishers by announcing its intention to offer a tiered pricing system, giving publishers the opportunity to participate more actively in the pricing of their books. Apple has agreed with publishers to charge $12 to $14 for e-books, and to act as an agent selling books (with a 30% fee on all e-book sales) rather than a book distributor. Publishers like this arrangement, but worry about long-term pricing expectations, hoping to avoid a scenario where readers come to expect $9.99 e-books as the standard.

Textbook publishers are also eager to establish themselves on the iPad. Many of the largest textbook publishers have struck deals with software firms like ScrollMotion, Inc. to adapt their books for e-book readers. In fact, Apple CEO Steve Jobs designed the iPad with use in schools in mind, and interest on the part of schools in technology like the iPad has been strong. ScrollMotion already has experience using the Apple application platform for the iPhone, so the company is uniquely qualified to convert existing files provided by publishers into a format readable by the iPad and to add additional features, like a dictionary, glossary, quizzes, page numbers, a search function, and high-quality images.

Newspapers are also excited about the iPad, which represents a way for them to continue charging for all of the content that they have been forced to make available online. If the iPad becomes as popular as other hit products from Apple, consumers are more likely to pay for content using that device. The successes of the App Store on the iPhone and of the iTunes music store attest to this. But the experience of the music industry with iTunes also gives all print media reason to worry. The iTunes music store changed the consumer perception of albums and music bundles. Music labels used to make more money selling 12 songs on an album than they did selling popular singles. Now consumers have drastically reduced their consumption of albums, preferring to purchase and download one song at a time. A similar fate may await print newpapers, which are bundles of news articles, many of which are unread.

Apple has also approached TV networks and movie studios about offering access to some of their top shows and movies for a monthly fee, but as of yet the bigger media companies have not responded to Apple's overture. Of course, if the iPad becomes sufficiently popular, that will change, but currently media networks would prefer not to endanger their strong and lucrative partnerships with cable and satellite TV providers. (See the chapter-ending case study.)

And what about Apple's own business model? Apple previously believed content was less important than the popularity of its devices. Now, Apple understands that it needs high-quality content from all the types of media it offers on its devices to be truly successful. The company's new goal is to make deals with each media industry to distribute the content that users want to watch at a price agreed to by the content owners and the platform owners (Apple). The old attitudes of Apple ("Rip, burn,

distribute"), which were designed to sell devices are a thing of the past. In this case of disruptive technology, even the disruptors have been forced to change their behaviors.

Sources: Ken Auletta, "Publish or Perish," *The New Yorker*, April 26, 2010; Yukari Iwatani Kane and Sam Schechner, "Apple Races to Strike Content Deals Ahead of IPad Release," *The Wall Street Journal*, March 18, 2010; Motoko Rich, "Books on iPad Offer Publishers a Pricing Edge," *The New York Times*, January 28, 2010; Jeffrey A. Trachtenberg and Yukari Iwatani Kane, "Textbook Firms Ink Deals for iPad," *The Wall Street Journal*, February 2, 2010; Nick Bilton, "Three Reasons Why the IPad Will Kill Amazon's Kindle," *The New York Times*, January 27, 2010; Jeffrey A Trachtenberg, "Apple Tablet Portends Rewrite for Publishers," *The Wall Street Journal*, January 26, 2010; Brad Stone and Stephanie Clifford, "With Apple Tablet, Print Media Hope for a Payday," *The New York Times*, January 26, 2010; Yukari Iwatani Kane, "Apple Takes Big Gamble on New iPad," *The Wall Street Journal*, January 25, 2010; and Anne Eisenberg, "Devices to Take Textbooks Beyond Text," *The New York Times*, December 6, 2009.

CASE STUDY QUESTIONS

1. Evaluate the impact of the iPad using Porter's competitive forces model.

2. What makes the iPad a disruptive technology? Who are likely to be the winners and losers if the iPad becomes a hit? Why?

3. Describe the effects that the iPad is likely to have on the business models of Apple, content creators, and distributors.

MIS IN ACTION

Visit Apple's site for the iPad and the Amazon.com site for the Kindle. Review the features and specifications of each device. Then answer the following questions:

1. How powerful is the iPad? How useful is it for reading books, newspapers or magazines, for surfing the Web, and for watching video? Can you identify any shortcomings of the device?

2. Compare the capabilities of the Kindle to the iPad. Which is a better device for reading books? Explain your answer.

3. Would you like to use an iPad or Kindle for the books you use in your college courses or read for pleasure instead of traditional print publications? Why or why not?

Support activities make the delivery of the primary activities possible and consist of organization infrastructure (administration and management), human resources (employee recruiting, hiring, and training), technology (improving products and the production process), and procurement (purchasing input).

Now you can ask at each stage of the value chain, "How can we use information systems to improve operational efficiency, and improve customer and supplier intimacy?" This will force you to critically examine how you perform value-adding activities at each stage and how the business processes might be improved. You can also begin to ask how information systems can be used to improve the relationship with customers and with suppliers who lie outside the firm's value chain but belong to the firm's extended value chain where they are absolutely critical to your success. Here, supply chain management systems

FIGURE 3-11   THE VALUE CHAIN MODEL

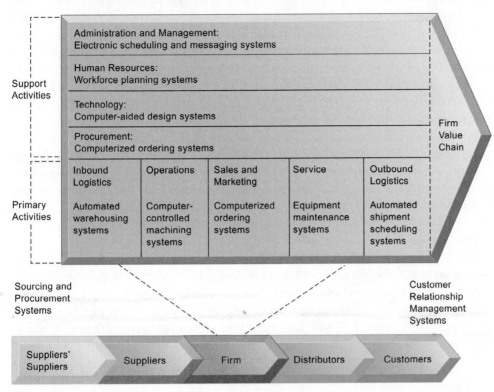

This figure provides examples of systems for both primary and support activities of a firm and of its value partners that can add a margin of value to a firm's products or services.

that coordinate the flow of resources into your firm, and customer relationship management systems that coordinate your sales and support employees with customers, are two of the most common system applications that result from a business value chain analysis. We discuss these enterprise applications in detail later in Chapter 9.

Using the business value chain model will also cause you to consider benchmarking your business processes against your competitors or others in related industries, and identifying industry best practices. **Benchmarking** involves comparing the efficiency and effectiveness of your business processes against strict standards and then measuring performance against those standards. Industry **best practices** are usually identified by consulting companies, research organizations, government agencies, and industry associations as the most successful solutions or problem-solving methods for consistently and effectively achieving a business objective.

Once you have analyzed the various stages in the value chain at your business, you can come up with candidate applications of information systems. Then, once you have a list of candidate applications, you can decide which to develop first. By making improvements in your own business value chain that your competitors might miss, you can achieve competitive advantage by attaining operational excellence, lowering costs, improving profit margins, and forging a closer relationship with customers and suppliers. If your competitors are making similar improvements, then at least you will not be at a competitive disadvantage—the worst of all cases!

Extending the Value Chain: The Value Web

Figure 3-11 shows that a firm's value chain is linked to the value chains of its suppliers, distributors, and customers. After all, the performance of most firms depends not only on what goes on inside a firm but also on how well the firm coordinates with direct and indirect suppliers, delivery firms (logistics partners, such as FedEx or UPS), and, of course, customers.

How can information systems be used to achieve strategic advantage at the industry level? By working with other firms, industry participants can use information technology to develop industry-wide standards for exchanging information or business transactions electronically, which force all market participants to subscribe to similar standards. Such efforts increase efficiency, making product substitution less likely and perhaps raising entry costs—thus discouraging new entrants. Also, industry members can build industry-wide, IT-supported consortia, symposia, and communications networks to coordinate activities concerning government agencies, foreign competition, and competing industries.

Looking at the industry value chain encourages you to think about how to use information systems to link up more efficiently with your suppliers, strategic partners, and customers. Strategic advantage derives from your ability to relate your value chain to the value chains of other partners in the process. For instance, if you are Amazon.com, you want to build systems that:

- Make it easy for suppliers to display goods and open stores on the Amazon site
- Make it easy for customers to pay for goods
- Develop systems that coordinate the shipment of goods to customers
- Develop shipment tracking systems for customers

Internet technology has made it possible to create highly synchronized industry value chains called value webs. A **value web** is a collection of independent firms that use information technology to coordinate their value chains to produce a product or service for a market collectively. It is more customer driven and operates in a less linear fashion than the traditional value chain.

Figure 3-12 shows that this value web synchronizes the business processes of customers, suppliers, and trading partners among different companies in an industry or in related industries. These value webs are flexible and adaptive to changes in supply and demand. Relationships can be bundled or unbundled in response to changing market conditions. Firms will accelerate time to market and to customers by optimizing their value web relationships to make quick decisions on who can deliver the required products or services at the right price and location.

SYNERGIES, CORE COMPETENCIES, AND NETWORK-BASED STRATEGIES

A large corporation is typically a collection of businesses. Often, the firm is organized financially as a collection of strategic business units and the returns to the firm are directly tied to the performance of all the strategic business units. Information systems can improve the overall performance of these business units by promoting synergies and core competencies.

FIGURE 3-12 **THE VALUE WEB**

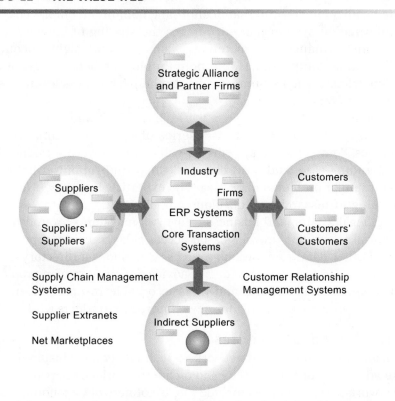

The value web is a networked system that can synchronize the value chains of business partners within an industry to respond rapidly to changes in supply and demand.

Synergies

The idea of synergies is that when the output of some units can be used as inputs to other units, or two organizations pool markets and expertise, these relationships lower costs and generate profits. Recent bank and financial firm mergers, such as the merger of JP Morgan Chase and Bank of New York as well as Bank of America and Countrywide Financial Corporation occurred precisely for this purpose.

One use of information technology in these synergy situations is to tie together the operations of disparate business units so that they can act as a whole. For example, acquiring Countrywide Financial enabled Bank of America to extend its mortgage lending business and to tap into a large pool of new customers who might be interested in its credit card, consumer banking, and other financial products. Information systems would help the merged companies consolidate operations, lower retailing costs, and increase cross-marketing of financial products.

Enhancing Core Competencies

Yet another way to use information systems for competitive advantage is to think about ways that systems can enhance core competencies. The argument is that the performance of all business units will increase insofar as these business units develop, or create, a central core of competencies. A **core competency** is an activity for which a firm is a world-class leader. Core competencies may involve being the world's best miniature parts designer, the best package delivery service, or the best thin-film manufacturer. In general, a core competency relies on knowledge that is gained over many years of practical field experience with a technology. This practical knowledge is typically supplemented with a long-term research effort and committed employees.

Any information system that encourages the sharing of knowledge across business units enhances competency. Such systems might encourage or enhance existing competencies and help employees become aware of new external knowledge; such systems might also help a business leverage existing competencies to related markets.

For example, Procter & Gamble, a world leader in brand management and consumer product innovation, uses a series of systems to enhance its core competencies. Some of these systems for collaboration were introduced in the Chapter 2 ending case study. An intranet called InnovationNet helps people working on similar problems share ideas and expertise. InnovationNet connects those working in research and development (R&D), engineering, purchasing, marketing, legal affairs, and business information systems around the world, using a portal to provide browser-based access to documents, reports, charts, videos, and other data from various sources. It includes a directory of subject matter experts who can be tapped to give advice or collaborate on problem solving and product development, and links to outside research scientists and entrepreneurs who are searching for new, innovative products worldwide.

Network-Based Strategies

The availability of Internet and networking technology have inspired strategies that take advantage of firms' abilities to create networks or network with each other. Network-based strategies include the use of network economics, a virtual company model, and business ecosystems.

Network Economics. Business models based on a network may help firms strategically by taking advantage of **network economics**. In traditional economics—the economics of factories and agriculture—production experiences diminishing returns. The more any given resource is applied to production, the lower the marginal gain in output, until a point is reached where the additional inputs produce no additional outputs. This is the law of diminishing returns, and it is the foundation for most of modern economics.

In some situations, the law of diminishing returns does not work. For instance, in a network, the marginal costs of adding another participant are about zero, whereas the marginal gain is much larger. The larger the number of subscribers in a telephone system or the Internet, the greater the value to all participants because each user can interact with more people. It is not much more expensive to operate a television station with 1,000 subscribers than with 10 million subscribers. The value of a community of people grows with size, whereas the cost of adding new members is inconsequential.

From this network economics perspective, information technology can be strategically useful. Internet sites can be used by firms to build communities of users—like-minded customers who want to share their experiences. This builds customer loyalty and enjoyment, and builds unique ties to customers. EBay, the giant online auction site, and iVillage, an online community for women, are examples. Both businesses are based on networks of millions of users, and both companies have used the Web and Internet communication tools to build communities. The more people offering products on eBay, the more valuable the eBay site is to everyone because more products are listed, and more competition among suppliers lowers prices. Network economics also provides strategic benefits to commercial software vendors. The value of their software and complementary software products increases as more people use them, and there is a larger installed base to justify continued use of the product and vendor support.

Virtual Company Model. Another network-based strategy uses the model of a virtual company to create a competitive business. A **virtual company**, also known as a virtual organization, uses networks to link people, assets, and ideas, enabling it to ally with other companies to create and distribute products and services without being limited by traditional organizational boundaries or physical locations. One company can use the capabilities of another company without being physically tied to that company. The virtual company model is useful when a company finds it cheaper to acquire products, services, or capabilities from an external vendor or when it needs to move quickly to exploit new market opportunities and lacks the time and resources to respond on its own.

Fashion companies, such as GUESS, Ann Taylor, Levi Strauss, and Reebok, enlist Hong Kong-based Li & Fung to manage production and shipment of their garments. Li & Fung handles product development, raw material sourcing, production planning, quality assurance, and shipping. Li & Fung does not own any fabric, factories, or machines, outsourcing all of its work to a network of more than 7,500 suppliers in 37 countries all over the world. Customers place orders to Li & Fung over its private extranet. Li & Fung then sends instructions to appropriate raw material suppliers and factories where the clothing is produced. The Li & Fung extranet tracks the entire production process for each order.

Working as a virtual company keeps Li & Fung flexible and adaptable so that it can design and produce the products ordered by its clients in short order to keep pace with rapidly changing fashion trends.

Business Ecosystems: Keystone and Niche Firms. The Internet and the emergence of digital firms call for some modification of the industry competitive forces model. The traditional Porter model assumes a relatively static industry environment; relatively clear-cut industry boundaries; and a relatively stable set of suppliers, substitutes, and customers, with the focus on industry players in a market environment. Instead of participating in a single industry, some of today's firms are much more aware that they participate in industry sets—collections of industries that provide related services and products (see Figure 3-13). **Business ecosystem** is another term for these loosely coupled but interdependent networks of suppliers, distributors, outsourcing firms, transportation service firms, and technology manufacturers (Iansiti and Levien, 2004).

The concept of a business ecosystem builds on the idea of the value web described earlier, the main difference being that cooperation takes place across many industries rather than many firms. For instance, both Microsoft and Walmart provide platforms composed of information systems, technologies, and services that thousands of other firms in different industries use to enhance their own capabilities. Microsoft has estimated that more than 40,000 firms use its Windows platform to deliver their own products, support Microsoft products, and extend the value of Microsoft's own firm. Walmart's order entry and inventory management system is a platform used by thousands of suppliers to obtain real-time access to customer demand, track shipments, and control inventories.

Business ecosystems can be characterized as having one or a few keystone firms that dominate the ecosystem and create the platforms used by other niche firms. Keystone firms in the Microsoft ecosystem include Microsoft and technology producers such as Intel and IBM. Niche firms include thousands of software

FIGURE 3-13   AN ECOSYSTEM STRATEGIC MODEL

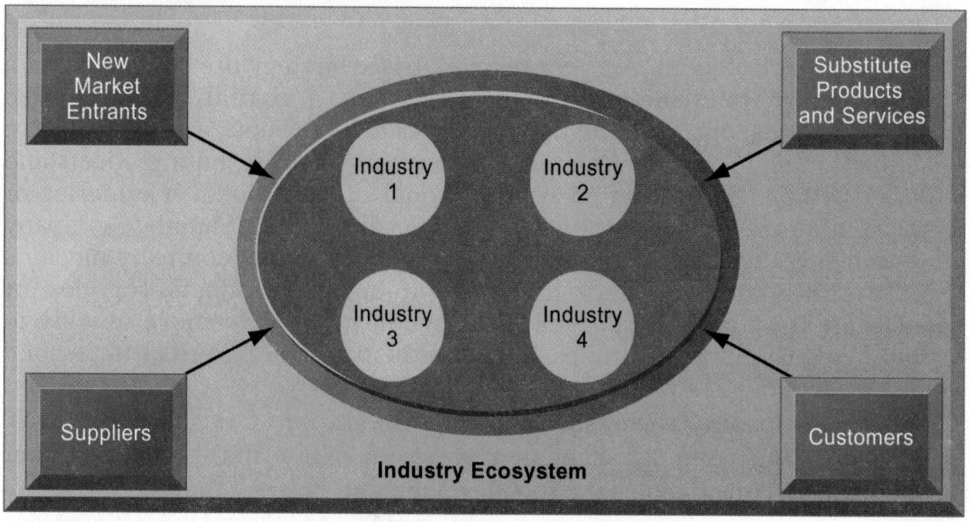

The digital firm era requires a more dynamic view of the boundaries among industries, firms, customers, and suppliers, with competition occurring among industry sets in a business ecosystem. In the ecosystem model, multiple industries work together to deliver value to the customer. IT plays an important role in enabling a dense network of interactions among the participating firms.

application firms, software developers, service firms, networking firms, and consulting firms that both support and rely on the Microsoft products.

Information technology plays a powerful role in establishing business ecosystems. Obviously, many firms use information systems to develop into keystone firms by building IT-based platforms that other firms can use. In the digital firm era, we can expect greater emphasis on the use of IT to build industry ecosystems because the costs of participating in such ecosystems will fall and the benefits to all firms will increase rapidly as the platform grows.

Individual firms should consider how their information systems will enable them to become profitable niche players in larger ecosystems created by keystone firms. For instance, in making decisions about which products to build or which services to offer, a firm should consider the existing business ecosystems related to these products and how it might use IT to enable participation in these larger ecosystems.

A powerful, current example of a rapidly expanding ecosystem is the mobile Internet platform. In this ecosystem there are four industries: device makers (Apple iPhone, RIM BlackBerry, Motorola, LG, and others), wireless telecommunication firms (AT&T, Verizon, T-Mobile, Sprint, and others), independent software applications providers (generally small firms selling games, applications, and ring tones), and Internet service providers (who participate as providers of Internet service to the mobile platform).

Each of these industries has its own history, interests, and driving forces. But these elements come together in a sometimes cooperative, and sometimes competitive, new industry we refer to as the mobile digital platform ecosystem. More than other firms, Apple has managed to combine these industries into a system. It is Apple's mission to sell physical devices (iPhones) that are nearly as powerful as today's personal computers. These devices work only with a high-speed broadband network supplied by the wireless phone carriers. In order to attract a large customer base, the iPhone had to be more than just a cell phone. Apple differentiated this product by making it a "smart phone," one capable of running thousands of different, useful applications. Apple could not develop all these applications itself. Instead it relies on generally small, independent software developers to provide these applications, which can be purchased at the iTunes store. In the background is the Internet service provider industry, which makes money whenever iPhone users connect to the Internet.

3.4 USING SYSTEMS FOR COMPETITIVE ADVANTAGE: MANAGEMENT ISSUES

Strategic information systems often change the organization as well as its products, services, and operating procedures, driving the organization into new behavioral patterns. Successfully using information systems to achieve a competitive advantage is challenging and requires precise coordination of technology, organizations, and management.

SUSTAINING COMPETITIVE ADVANTAGE

The competitive advantages that strategic systems confer do not necessarily last long enough to ensure long-term profitability. Because competitors can retaliate and copy strategic systems, competitive advantage is not always sustainable. Markets, customer expectations, and technology change; globalization has made these changes even more rapid and unpredictable. The Internet can make competitive advantage disappear very quickly because virtually all companies can use this technology. Classic strategic systems, such as American Airlines's SABRE computerized reservation system, Citibank's ATM system, and FedEx's package tracking system, benefited by being the first in their industries. Then rival systems emerged. Amazon.com was an e-commerce leader but now faces competition from eBay, Yahoo, and Google. Information systems alone cannot provide an enduring business advantage. Systems originally intended to be strategic frequently become tools for survival, required by every firm to stay in business, or they may inhibit organizations from making the strategic changes essential for future success.

ALIGNING IT WITH BUSINESS OBJECTIVES

The research on IT and business performance has found that (a) the more successfully a firm can align information technology with its business goals, the more profitable it will be, and (b) only one-quarter of firms achieve alignment of IT with the business. About half of a business firm's profits can be explained by alignment of IT with business (Luftman, 2003).

Most businesses get it wrong: Information technology takes on a life of its own and does not serve management and shareholder interests very well. Instead of business people taking an active role in shaping IT to the enterprise, they ignore it, claim not to understand IT, and tolerate failure in the IT area as just a nuisance to work around. Such firms pay a hefty price in poor performance. Successful firms and managers understand what IT can do and how it works, take an active role in shaping its use, and measure its impact on revenues and profits.

Management Checklist: Performing a Strategic Systems Analysis

To align IT with the business and use information systems effectively for competitive advantage, managers need to perform a strategic systems analysis. To identify the types of systems that provide a strategic advantage to their firms, managers should ask the following questions:

1. What is the structure of the industry in which the firm is located?

   • What are some of the competitive forces at work in the industry? Are there new entrants to the industry? What is the relative power of suppliers, customers, and substitute products and services over prices?

   • Is the basis of competition quality, price, or brand?

   • What are the direction and nature of change within the industry? From where are the momentum and change coming?

   • How is the industry currently using information technology? Is the organization behind or ahead of the industry in its application of information systems?

2. What are the business, firm, and industry value chains for this particular firm?

   • How is the company creating value for the customer—through lower prices and transaction costs or higher quality? Are there any places in the value chain where the business could create more value for the customer and additional profit for the company?

   • Does the firm understand and manage its business processes using the best practices available? Is it taking maximum advantage of supply chain management, customer relationship management, and enterprise systems?

   • Does the firm leverage its core competencies?

   • Is the industry supply chain and customer base changing in ways that benefit or harm the firm?

   • Can the firm benefit from strategic partnerships and value webs?

   • Where in the value chain will information systems provide the greatest value to the firm?

3. Have we aligned IT with our business strategy and goals?

   • Have we correctly articulated our business strategy and goals?

   • Is IT improving the right business processes and activities to promote this strategy?

   • Are we using the right metrics to measure progress toward those goals?

MANAGING STRATEGIC TRANSITIONS

Adopting the kinds of strategic systems described in this chapter generally requires changes in business goals, relationships with customers and suppliers, and business processes. These sociotechnical changes, affecting both social and technical elements of the organization, can be considered **strategic transitions**—a movement between levels of sociotechnical systems.

Such changes often entail blurring of organizational boundaries, both external and internal. Suppliers and customers must become intimately linked and may share each other's responsibilities. Managers will need to devise new business processes for coordinating their firms' activities with those of customers, suppliers, and other organizations. The organizational change requirements surrounding new information systems are so important that they merit attention throughout this text. Chapter 14 examines organizational change issues in more detail.

3.5 HANDS-ON MIS PROJECTS

The projects in this section give you hands-on experience identifying information systems to support a business strategy, analyzing organizational factors affecting the information systems of merging companies, using a database to improve decision making about business strategy, and using Web tools to configure and price an automobile.

Management Decision Problems

1. Macy's, Inc., through its subsidiaries, operates approximately 800 department stores in the United States. Its retail stores sell a range of merchandise, including adult and children's apparel, accessories, cosmetics, home furnishings, and housewares. Senior management has decided that Macy's needs to tailor merchandise more to local tastes, that the colors, sizes, brands, and styles of clothing and other merchandise should be based on the sales patterns in each individual Macy's store. For example, stores in Texas might stock clothing in larger sizes and brighter colors than those in New York, or the Macy's on Chicago's State Street might include a greater variety of makeup shades to attract trendier shoppers. How could information systems help Macy's management implement this new strategy? What pieces of data should these systems collect to help management make merchandising decisions that support this strategy?

2. Today's US Airways is the result of a merger between US Airways and America West Airlines. Before the merger, US Airways dated back to 1939 and had very traditional business processes, a lumbering bureaucracy, and a rigid information systems function that had been outsourced to Electronic Data Systems. America West was formed in 1981 and had a younger workforce, a more freewheeling entrepreneurial culture, and managed its own information systems. The merger was designed to create synergies from US Airways' experience and strong network on the east coast of the United States with America West's low-cost structure, information systems, and routes in the western United States. What features of organizations should management have considered as it merged the two companies and their information systems? What decisions need to be made to make sure the strategy works?

Improving Decision Making: Using a Database to Clarify Business Strategy

Software skills: Database querying and reporting; database design
Business skills: Reservation systems; customer analysis

In this exercise, you'll use database software to analyze the reservation transactions for a hotel and use that information to fine-tune the hotel's business strategy and marketing activities.

The Presidents' Inn is a small three-story hotel on the Atlantic Ocean in Cape May, New Jersey, a popular northeastern U.S. resort. Ten rooms overlook side streets, 10 rooms have bay windows that offer limited views of the ocean, and the remaining 10 rooms in the front of the hotel face the ocean. Room rates are based on room choice, length of stay, and number of guests per room. Room rates are the same for one to four guests. Fifth and sixth guests must pay an additional $20 charge each per day. Guests staying for seven days or more receive a 10-percent discount on their daily room rates.

Business has grown steadily during the past 10 years. Now totally renovated, the inn uses a romantic weekend package to attract couples, a vacation package to attract young families, and a weekday discount package to attract business travelers. The owners currently use a manual reservation and bookkeeping system, which has caused many problems. Sometimes two families have been booked in the same room at the same time. Management does not have immediate data about the hotel's daily operations and income.

In MyMISLab, you will find a database for hotel reservation transactions developed in Microsoft Access. A sample is shown below, but the Web site may have a more recent version of this database for this exercise.

Develop some reports that provide information to help management make the business more competitive and profitable. Your reports should answer the following questions:

- What is the average length of stay per room type?
- What is the average number of visitors per room type?
- What is the base income per room (i.e., length of visit multiplied by the daily rate) during a specified period of time?
- What is the strongest customer base?

After answering these questions, write a brief report describing what the database information reveals about the current business situation. Which specific business strategies might be pursued to increase room occupancy and revenue? How could the database be improved to provide better information for strategic decisions?

Improving Decision Making: Using Web Tools to Configure and Price an Automobile

Software skills: Internet-based software
Business skills: Researching product information and pricing

In this exercise, you'll use software at Web sites for selling cars to find product information about a car of your choice and use that information to make an important purchase decision. You'll also evaluate two of these sites as selling tools.

You are interested in purchasing a new Ford Focus. (If you are personally interested in another car, domestic or foreign, investigate that one instead.) Go to the Web site of CarsDirect (www.carsdirect.com) and begin your investigation. Locate the Ford Focus. Research the various specific automobiles available in that model and determine which you prefer. Explore the full details about the specific car, including pricing, standard features, and options. Locate and read at least two reviews if possible. Investigate the safety of that model

ID	Guest First Name	Guest Last Name	Room	Room Type	Arrival Date	Departure Date	No of Guests
1	Barry	Lloyd	Hayes	Bay-window	12/1/2010	12/4/2010	2
2	Michael	Lunsford	Cleveland	Ocean	12/1/2010	12/9/2010	3
3	Kim	Kyuong	Coolidge	Bay-window	12/4/2010	12/7/2010	1
4	Edward	Holt	Washington	Ocean	12/1/2010	12/3/2010	4
5	Thomas	Collins	Lincoln	Ocean	12/9/2010	12/13/2010	2
6	Paul	Bodkin	Coolidge	Bay-window	12/1/2010	12/3/2010	2
7	Randall	Battenburg	Washington	Ocean	12/4/2010	12/12/2010	2
8	Calvin	Nowotney	Lincoln	Ocean	12/2/2010	12/4/2010	1
9	Homer	Gonzalez	Lincoln	Ocean	12/5/2010	12/7/2010	5
10	David	Sanchez	Jefferson	Bay-window	12/5/2010	12/7/2010	2
11	Buster	Whisler	Jackson	Ocean	12/5/2010	12/8/2010	2
12	Julia	Martines	Reagan	Bay-window	12/10/2010	12/15/2010	1
13	Samuel	Kim	Truman	Side	12/20/2010	12/30/2010	3
14	Arthur	Gottfried	Garfield	Side	12/13/2010	12/15/2010	2
15	Darlene	Shore	Arthur	Ocean	12/24/2010	12/31/2010	5

Record: ◄ ◄ 1 of 30 ► ►► ►* No Filter Search

based on the U.S. government crash tests performed by the National Highway Traffic Safety Administration if those test results are available. Explore the features for locating a vehicle in inventory and purchasing directly. Finally, explore the other capabilities of the CarsDirect site for financing.

Having recorded or printed the information you need from CarsDirect for your purchase decision, surf the Web site of the manufacturer, in this case Ford (www.ford.com). Compare the information available on Ford's Web site with that of CarsDirect for the Ford Focus. Be sure to check the price and any incentives being offered (which may not agree with what you found at CarsDirect). Next, find a local dealer on the Ford site so that you can view the car before making your purchase decision. Explore the other features of Ford's Web site.

Try to locate the lowest price for the car you want in a local dealer's inventory. Which site would you use to purchase your car? Why? Suggest improvements for the sites of CarsDirect and Ford.

LEARNING TRACK MODULE

The following Learning Track provides content relevant to topics covered in this chapter.

1. The Changing Business Environment for Information Technology

Review Summary

1. *Which features of organizations do managers need to know about to build and use information systems successfully? What is the impact of information systems on organizations?*

    All modern organizations are hierarchical, specialized, and impartial, using explicit routines to maximize efficiency. All organizations have their own cultures and politics arising from differences in interest groups, and they are affected by their surrounding environment. Organizations differ in goals, groups served, social roles, leadership styles, incentives, types of tasks performed, and type of structure. These features help explain differences in organizations' use of information systems.

    Information systems and the organizations in which they are used interact with and influence each other. The introduction of a new information system will affect organizational structure, goals, work design, values, competition between interest groups, decision making, and day-to-day behavior. At the same time, information systems must be designed to serve the needs of important organizational groups and will be shaped by the organization's structure, business processes, goals, culture, politics, and management. Information technology can reduce transaction and agency costs, and such changes have been accentuated in organizations using the Internet. New systems disrupt established patterns of work and power relationships, so there is often considerable resistance to them when they are introduced.

2. *How does Porter's competitive forces model help companies develop competitive strategies using information systems?*

    In Porter's competitive forces model, the strategic position of the firm, and its strategies, are determined by competition with its traditional direct competitors, but they are also greatly affected by new market entrants, substitute products and services, suppliers, and customers. Information systems help companies compete by maintaining low costs, differentiating products or services, focusing on market niche, strengthening ties with customers and suppliers, and increasing barriers to market entry with high levels of operational excellence.

3. *How do the value chain and value web models help businesses identify opportunities for strategic information system applications?*

The value chain model highlights specific activities in the business where competitive strategies and information systems will have the greatest impact. The model views the firm as a series of primary and support activities that add value to a firm's products or services. Primary activities are directly related to production and distribution, whereas support activities make the delivery of primary activities possible. A firm's value chain can be linked to the value chains of its suppliers, distributors, and customers. A value web consists of information systems that enhance competitiveness at the industry level by promoting the use of standards and industry-wide consortia, and by enabling businesses to work more efficiently with their value partners.

4. *How do information systems help businesses use synergies, core competencies, and network-based strategies to achieve competitive advantage?*

Because firms consist of multiple business units, information systems achieve additional efficiencies or enhance services by tying together the operations of disparate business units. Information systems help businesses leverage their core competencies by promoting the sharing of knowledge across business units. Information systems facilitate business models based on large networks of users or subscribers that take advantage of network economics. A virtual company strategy uses networks to link to other firms so that a company can use the capabilities of other companies to build, market, and distribute products and services. In business ecosystems, multiple industries work together to deliver value to the customer. Information systems support a dense network of interactions among the participating firms.

5. *What are the challenges posed by strategic information systems and how should they be addressed?*

Implementing strategic systems often requires extensive organizational change and a transition from one sociotechnical level to another. Such changes are called strategic transitions and are often difficult and painful to achieve. Moreover, not all strategic systems are profitable, and they can be expensive to build. Many strategic information systems are easily copied by other firms so that strategic advantage is not always sustainable.

Key Terms

Agency theory, 90
Benchmarking, 105
Best practices, 105
Business ecosystem, 109
Competitive forces model, 95
Core competency, 107
Disruptive technologies, 87
Efficient customer response system, 97
Mass customization, 98
Network economics, 108
Organization, 82

Primary activities, 102
Product differentiation, 96
Routines, 84
Strategic transitions, 112
Support activities, 104
Switching costs, 99
Transaction cost theory, 89
Value chain model, 102
Value web, 106
Virtual company, 109

Review Questions

1. Which features of organizations do managers need to know about to build and use information systems successfully? What is the impact of information systems on organizations?

   • Define an organization and compare the technical definition of organizations with the behavioral definition.

   • Identify and describe the features of organizations that help explain differences in organizations' use of information systems.

   • Describe the major economic theories that help explain how information systems affect organizations.

   • Describe the major behavioral theories that help explain how information systems affect organizations.

   • Explain why there is considerable organizational resistance to the introduction of information systems.

- Describe the impact of the Internet and disruptive technologies on organizations.

2. How does Porter's competitive forces model help companies develop competitive strategies using information systems?
   - Define Porter's competitive forces model and explain how it works.
   - Describe what the competitive forces model explains about competitive advantage.
   - List and describe four competitive strategies enabled by information systems that firms can pursue.
   - Describe how information systems can support each of these competitive strategies and give examples.
   - Explain why aligning IT with business objectives is essential for strategic use of systems.

3. How do the value chain and value web models help businesses identify opportunities for strategic information system applications?
   - Define and describe the value chain model.
   - Explain how the value chain model can be used to identify opportunities for information systems.
   - Define the value web and show how it is related to the value chain.

- Explain how the value web helps businesses identify opportunities for strategic information systems.
- Describe how the Internet has changed competitive forces and competitive advantage.

4. How do information systems help businesses use synergies, core competences, and network-based strategies to achieve competitive advantage?
   - Explain how information systems promote synergies and core competencies.
   - Describe how promoting synergies and core competencies enhances competitive advantage.
   - Explain how businesses benefit by using network economics.
   - Define and describe a virtual company and the benefits of pursuing a virtual company strategy.

5. What are the challenges posed by strategic information systems and how should they be addressed?
   - List and describe the management challenges posed by strategic information systems.
   - Explain how to perform a strategic systems analysis.

Discussion Questions

1. It has been said that there is no such thing as a sustainable strategic advantage. Do you agree? Why or why not?

2. It has been said that the advantage that leading-edge retailers such as Dell and Walmart have over their competition isn't technology; it's their management. Do you agree? Why or why not?

3. What are some of the issues to consider in determining whether the Internet would provide your business with a competitive advantage?

Video Cases

Video Cases and Instructional Videos illustrating some of the concepts in this chapter are available. Contact your instructor to access these videos.

Collaboration and Teamwork: Identifying Opportunities for Strategic Information Systems

With your team of three or four students, select a company described in *The Wall Street Journal*, *Fortune*, *Forbes*, or another business publication. Visit the company's Web site to find additional information about that company and to see how the firm is using the Web. On the basis of this information, analyze the business. Include a description of the organization's features, such as important business processes, culture, structure, and environment, as well as its business strategy. Suggest strategic information systems appropriate for that particular business, including those based on Internet technology, if appropriate. If possible, use Google Sites to post links to Web pages, team communication announcements, and work assignments; to brainstorm; and to work collaboratively on project documents. Try to use Google Docs to develop a presentation of your findings for the class.

Will TV Succumb to the Internet?
CASE STUDY

The Internet has transformed the music industry. Sales of CDs in retail music stores have been steadily declining while sales of songs downloaded through the Internet to iPods and other portable music players are skyrocketing. Moreover, the music industry is still contending with millions of people illegally downloading songs for free. Will the television industry experience a similar fate?

Widespread use of high-speed Internet access, powerful PCs with high-resolution display screens, iPhones, iPads, other mobile handhelds, and leading-edge file-sharing services have made downloading of video content from movies and television shows faster and easier than ever. Free and often illegal downloads of some TV shows are abundant. But the Internet is also providing new ways for television studios to distribute and sell their content, and they are trying to take advantage of that opportunity.

YouTube, which started up in February 2005, quickly became the most popular video-sharing Web site in the world. Even though YouTube's original mission was to provide an outlet for amateur filmmakers, clips of copyrighted Hollywood movies and television shows soon proliferated on the YouTube Web site. It is difficult to gauge how much proprietary content from TV shows winds up on YouTube without the studios' permission. Viacom claimed in a 2008 lawsuit that over 150,000 unauthorized clips of its copyrighted television programs had appeared on YouTube.

YouTube tries to discourage its users from posting illegal clips by limiting the length of videos to 10 minutes each and by removing videos when requested by their copyright owner. YouTube has also implemented Video ID filtering and digital finger-printing technology that allows copyright owners to compare the digital fingerprints of their videos with material on YouTube and then flag infringing material. Using this technology, it is able to filter many unauthorized videos before they appear on the YouTube Web site. If infringing videos do make it online, they can be tracked using Video ID.

The television industry is also striking back by embracing the Internet as another delivery system for its content. Television broadcast networks such as NBC Universal, Fox, and CNN have put television shows on their own Web sites. In March 2007, NBC Universal, News Corp (the owner of Fox Broadcasting), and ABC Inc. formed Hulu.com, a Web site offering streaming video of television shows and movies from NBC, Fox, ABC, Comedy Central, PBS, USA Network, Bravo, FX, Speed, Sundance, Oxygen, Onion News Network, and other networks. Hulu also syndicates its hosting to other sites, including AOL, MSN, Facebook, MySpace, Yahoo!, and Fancast.com, and allows users to embed Hulu clips in their Web site. The site is supported by advertising commercials, and much of its content is free to viewers. CBS's TV.com and Joost are other popular Web television sites.

Content from all of these sites is viewable over iPhones. Hulu has blocked services such as Boxee that try to bring Hulu to TV screens, because that would draw subscribers away from cable and satellite companies, diminishing their revenue.

According to Hulu CEO Jason Kilar, Hulu has successfully brought online TV into the mainstream. It dominates the market for online full-episode TV viewing, with more than 44 million monthly visitors, according to the online measurement firm comScore. Monthly video streams more than tripled in 2009, reaching over 900 million by January 2010.

What if there are so many TV shows available for free on the Web that "Hulu households" cancel their cable subscriptions to watch free TV online? Cable service operators have begun worrying, especially when the cable networks posted some of their programming on the Web. By 2010, nearly 800,000 U.S. households had "cut the cord," dumping their cable, satellite, or high-speed television services from telecom companies such as Verizon's FiOS or AT&T's U-verse. In their place, they turned to Web-based videos from services such as Hulu, downloadable shows from iTunes, by-mail video subscription services such as Netflix, or even old-style over-the-air broadcast programming. Although the "cord cutters" represent less than 1 percent of the 100 million U.S. households subscribing to a cable/satellite/telco television service, the number of cord-cutting U.S. households is predicted to double to about 1.6 million. What if this trend continues?

In July 2009, cable TV operator Comcast Corporation began a trial program to bring some of Time Warner's network shows, including TBS's *My Boys* and TNT's *The Closer,* to the Web. Other cable networks, including A&E and the History Channel, participated in the Comcast test.

By making more television shows available online, but only for cable subscribers, the cable networks hope to preserve and possibly expand the cable TV subscription model in an increasingly digital world. "The vision is you can watch your favorite network's programming on any screen," noted Time Warner Chief Executive Jeff Bewkes. The system used in the Comcast-Time Warner trial is interoperable with cable service providers' systems to authenticate subscribers.

The same technology might also allow cable firms to provide demographic data for more targeted ads and perhaps more sophisticated advertising down the road. Cable programmers stand to earn more advertising revenue from their online content because viewers can't skip ads on TV programs streamed from the Web as they do with traditional TV. Web versions of some television shows in the Comcast–Time Warner trial program, including TNT's *The Closer*, will carry the same number of ads as seen on traditional TV, which amounts to more than four times the ad load on many Internet sites, including Hulu. Many hour-long shows available online are able to accommodate five or six commercial breaks, each with a single 30-second ad. NBC Universal Digital Entertainment has even streamed episodes of series, including *The Office*, with two ads per break. According to research firm eMarketer, these Web-video ads will generate $1.5 billion in ad revenue in 2010 and $2.1 billion in 2011.

For all its early success, Hulu is experiencing growing pains. Although it had generated more than $100 million in advertising revenue within two years, it is still unprofitable. Hulu's content suppliers receive 50 to 70 percent of the advertising revenue Hulu generates from their videos. Some of these media companies have complained that this revenue is very meager, even though use of Hulu has skyrocketed. One major supplier, Viacom, withdrew its programming from Hulu after failing to reach a satisfactory agreement on revenue-sharing, depriving Hulu viewers of such popular shows as *The Daily Show with Jon Stewart* and *The Colbert Report*.

Other companies supplying Hulu's content have pressured the company to earn even more advertising dollars and to set up a subscription service requiring consumers to pay a monthly fee to watch at least some of the shows on the site. On June 29, 2010, Hulu launched such a service, called HuluPlus. For $9.99 per month, paid subscribers get the entire current season of *Glee*, *The Office*, *House* and other shows from broadcasters ABC, Fox, and NBC, as well as all the past seasons of several series. Hulu will

continue to show a few recent episodes for free online. Paying subscribers will get the same number of ads as users of the free Web site in order to keep the subscription cost low. Paying subscribers are also able watch shows in high definition and on multiple devices, including mobile phones and videogame consoles as well as television screens.

Will all of this work out for the cable industry? It's still too early to tell. Although the cable programming companies want an online presence to extend their brands, they don't want to cannibalize TV subscriptions or viewership ratings that generate advertising revenue. Customers accustomed to YouTube and Hulu may rebel if too many ads are shown online. According to Oppenheimer analyst Tim Horan, cable companies will start feeling the impact of customers canceling subscriptions to view online video and TV by 2012. Edward Woo, an Internet and digital media analyst for Wedbush Morgan Securities in Los Angeles, predicts that in a few years, "it should get extremely interesting." Hulu and other Web TV and video sites will have much deeper content, and the technology to deliver that content to home viewers will be more advanced.

Sources: Ryan Nakashima, "Hulu Launches $10 Video Subscription Service," Associated Press, June 29, 2010; Ben Patterson, "Nearly 800,000 U.S. TV Households 'Cut the Cord,' Report Says," Yahoo! News, April 13, 2010; Brian Stelter and Brad Stone, "Successes (and Some Growing Pains) at Hulu, " *The New York Times*, March 31, 2010; Brian Stelter, "Viacom and Hulu Part Ways," *The New York Times*, March 2, 2010; Reinhardt Krause, "Cable TV Leaders Plot Strategy Vs. Free Programs on the Web," *Investors Business Daily*, August 18, 2009; Sam Schechner and Vishesh Kumar, "TV Shows Bring Ads Online," *The Wall Street Journal*, July 16, 2009; and Kevin Hunt, "The Coming TV-Delivery War: Cable vs. Internet," *The Montana Standard*, July 18, 2009.

CASE STUDY QUESTIONS

1. What competitive forces have challenged the television industry? What problems have these forces created?
2. Describe the impact of disruptive technology on the companies discussed in this case.
3. How have the cable programming and delivery companies responded to the Internet?
4. What management, organization, and technology issues must be addressed to solve the cable industry's problems?
5. Have the cable companies found a successful new business model to compete with the Internet? Why or why not?
6. If more television programs were available online, would you cancel your cable subscription? Why or why not?

Chapter 4

Ethical and Social Issues in Information Systems

BEHAVIORAL TARGETING AND YOUR PRIVACY: YOU'RE THE TARGET

Ever get the feeling somebody is trailing you on the Web, watching your every click? Wonder why you start seeing display ads and pop-ups just after you've been scouring the Web for a car, a dress, or cosmetic product? Well, you're right: your behavior is being tracked, and you are being targeted on the Web so that you are exposed to certain ads and not others. The Web sites you visit track the search engine queries you enter, pages visited, Web content viewed, ads clicked, videos watched, content shared, and the products you purchase. Google is the largest Web tracker, monitoring thousands of Web sites. As one wag noted, Google knows more about you than your mother does. In March 2009, Google began displaying ads on thousands of Google-related Web sites based on their previous online activities. To parry a growing public resentment of behavioral targeting, Google said it would give users the ability to see and edit the information that it has compiled about their interests for the purposes of behavioral targeting.

Behavioral targeting seeks to increase the efficiency of online ads by using information that Web visitors reveal about themselves online, and if possible, combine this with offline identity and consumption information gathered by companies such as Acxiom. One of the original promises of the Web was that it can deliver a marketing message tailored to each consumer based on this data, and then measure the results in terms of click-throughs and purchases. The technology used to implement online tracking is a combination of cookies, Flash cookies, and Web beacons (also called Web bugs). Web beacons are small programs placed on your computer when you visit any of thousands of Web sites. They report back to servers operated by the beacon owners the domains and Web pages you visited, what ads you clicked on, and other online behaviors. A recent study of 20 million Web pages published by 2 million domains found Google, Yahoo, Amazon, YouTube, Photobucket, and Flickr among the top 10 Web-bugging sites. Google alone accounts for 20% of all Web bugs. The average home landing page at the top 100 Web domains has over 50 tracking cookies and bugs. And you thought you were surfing alone?

Firms are experimenting with more precise targeting methods. Snapple used behavioral targeting methods (with the help of an online ad firm Tacoda) to identify the types of people attracted to Snapple Green Tea. Answer: people who like the arts and literature, travel internationally, and visit health sites. Microsoft offers MSN advertisers access to personal data derived from 270 million worldwide Windows Live users. The goal of Web beacons and bugs is even more granular: these tools can be used to identify your personal interests and behaviors so precisely targeted ads can be shown to you.

The growth in the power, reach, and scope of behavioral targeting has drawn the attention of privacy groups and the Federal Trade Commission (FTC). Currently, Web tracking is unregulated. In November 2007, the FTC opened hearings to consider proposals from privacy advocates to develop a "do not track list," to develop visual online cues to alert people to tracking, and to allow people to opt out. In the Senate, hearings on behavioral targeting were held throughout 2009 and the first half of 2010 with attention shifting to the privacy of personal location information. While Google, Microsoft, and Yahoo pleaded for legislation to protect them

from consumer lawsuits, the FTC refused to consider new legislation to protect the privacy of Internet users. Instead, the FTC proposed industry self-regulation. In 2009, a consortium of advertising firms (the Network Advertising Initiative) responded positively to FTC-proposed principles to regulate online behavioral advertising. In 2010, Congressional committees pressed leading Internet firms to allow users more opportunities to turn off tracking tools, and to make users aware on entry to a page that they are being tracked. In June 2010, the FTC announced it is examining Facebook Inc.'s efforts to protect user privacy.

All of these regulatory efforts emphasize transparency, user control over their information, security, and the temporal stability of privacy promises (unannounced and sudden changes in information privacy may not be allowed).

Perhaps the central ethical and moral question is understanding what rights individuals have in their own personally identifiable Internet profiles. Are these "ownership" rights, or merely an "interest" in an underlying asset? How much privacy are we willing to give up in order to receive more relevant ads? Surveys suggest that over 70 percent of Americans do not want to receive targeted ads.

Sources: "Web Bug Report," SecuritySpace, July, 2010; Miguel Helft, "Technology Coalition Seeks Stronger Privacy Laws," *New York Times*, March 30, 2010; "Study Finds Behaviorally-Targeted Ads More Than Twice As Valuable, Twice as Effective As Non-targeted Online Ads," Network Advertising Initiative, March 24, 2010; Steve Lohr, "Redrawing the Route to Online Privacy," New York Times, February 28, 2010; "The Collection and Use of Location Information for Commercial Purposes Hearings," U.S. House of Representatives, Committee on Energy and Commerce, Subcommittee on Commerce, Trade and Consumer Protection, February 24, 2010; Tom Krazit, "Groups Call for New Checks on Behavioral Ad Data," CNET News, September 1, 2009; Robert Mitchell, "What Google Knows About You," *Computerworld*, May 11, 2009; Stephanie Clifford, "Many See Privacy on Web as Big Issue, Survey Says," *The New York Times*, March 16, 2009; Miguel Helft, "Google to Offer Ads Based on Interests," *The New York Times*, March 11, 2009; and David Hallerman, "Behavioral Targeting: Marketing Trends," *eMarketer*, June 2008.

The growing use of behavioral targeting techniques described in the chapter-opening case shows that technology can be a double-edged sword. It can be the source of many benefits (by showing you ads relevant to your interests) but it can also create new opportunities for invading your privacy, and enabling the reckless use of that information in a variety of decisions about you.

The chapter-opening diagram calls attention to important points raised by this case and this chapter. Online advertising titans like Google, Microsoft, and Yahoo are all looking for ways to monetize their huge collections of online behavioral data. While search engine marketing is arguably the most effective form of advertising in history, banner display ad marketing is highly inefficient because it displays ads to everyone regardless of their interests. Hence the search engine marketers cannot charge much for display ad space. However, by tracking the online movements of 200 million U.S. Internet users, they can develop a very clear picture of who you are, and use that information to show you ads that might be of interest to you. This would make the marketing process more efficient, and more profitable for all the parties involved.

But this solution also creates an ethical dilemma, pitting the monetary interests of the online advertisers and search engines against the interests of individuals to maintain a sense of control over their personal information and their privacy. Two closely held values are in conflict here. As a manager, you will need to be sensitive to both the negative and positive impacts of information systems for your firm, employees, and customers. You will need to learn how to resolve ethical dilemmas involving information systems.

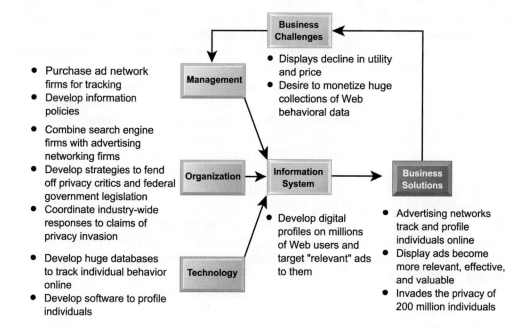

- Purchase ad network firms for tracking
- Develop information policies
- Combine search engine firms with advertising networking firms
- Develop strategies to fend off privacy critics and federal government legislation
- Coordinate industry-wide responses to claims of privacy invasion
- Develop huge databases to track individual behavior online
- Develop software to profile individuals

Business Challenges

- Displays decline in utility and price
- Desire to monetize huge collections of Web behavioral data

Management

Organization

Information System
- Develop digital profiles on millions of Web users and target "relevant" ads to them

Technology

Business Solutions
- Advertising networks track and profile individuals online
- Display ads become more relevant, effective, and valuable
- Invades the privacy of 200 million individuals

4.1 UNDERSTANDING ETHICAL AND SOCIAL ISSUES RELATED TO SYSTEMS

I n the past 10 years, we have witnessed, arguably, one of the most ethically challenging periods for U.S. and global business. Table 4-1 provides a small sample of recent cases demonstrating failed ethical judgment by senior and middle managers. These lapses in management ethical and business judgment occurred across a broad spectrum of industries.

In today's new legal environment, managers who violate the law and are convicted will most likely spend time in prison. U.S. federal sentencing guidelines adopted in 1987 mandate that federal judges impose stiff sentences on business

TABLE 4-1 RECENT EXAMPLES OF FAILED ETHICAL JUDGMENT BY SENIOR MANAGERS

Lehman Brothers (2008–2010)	One of the oldest American investment banks collapses in 2008. Lehman used information systems and accounting sleight of hand to conceal its bad investments. Lehman also engaged in deceptive tactics to shift investments off its books.
WG Trading Co. (2010)	Paul Greenwood, hedge fund manager and general partner at WG Trading, pled guilty to defrauding investors of $554 million over 13 years; Greenwood has forfeited $331 million to the government and faces up to 85 years in prison.
Minerals Management Service (U.S. Department of the Interior) (2010)	Managers accused of accepting gifts and other favors from oil companies, letting oil company rig employees write up inspection reports, and failing to enforce existing regulations on offshore Gulf drilling rigs. Employees systematically falsified information record systems.
Pfizer, Eli Lilly, and AstraZeneca (2009)	Major pharmaceutical firms paid billions of dollars to settle U.S. federal charges that executives fixed clinical trials for antipsychotic and pain killer drugs, marketed them inappropriately to children, and claimed unsubstantiated benefits while covering up negative outcomes. Firms falsified information in reports and systems.
Galleon Group (2009)	Founder of the Galleon Group criminally charged with trading on insider information, paying $250 million to Wall Street banks, and in return received market information that other investors did not get.
Siemens (2009)	The world's largest engineering firm paid over $4 billion to German and U.S. authorities for a decades-long, world-wide bribery scheme approved by corporate executives to influence potential customers and governments. Payments concealed from normal reporting accounting systems.

executives based on the monetary value of the crime, the presence of a conspiracy to prevent discovery of the crime, the use of structured financial transactions to hide the crime, and failure to cooperate with prosecutors (U.S. Sentencing Commission, 2004).

Although in the past business firms would often pay for the legal defense of their employees enmeshed in civil charges and criminal investigations, now firms are encouraged to cooperate with prosecutors to reduce charges against the entire firm for obstructing investigations. These developments mean that, more than ever, as a manager or an employee, you will have to decide for yourself what constitutes proper legal and ethical conduct.

Although these major instances of failed ethical and legal judgment were not masterminded by information systems departments, information systems were instrumental in many of these frauds. In many cases, the perpetrators of these crimes artfully used financial reporting information systems to bury their decisions from public scrutiny in the vain hope they would never be caught. We deal with the issue of control in information systems in Chapter 8. In this chapter, we talk about the ethical dimensions of these and other actions based on the use of information systems.

Ethics refers to the principles of right and wrong that individuals, acting as free moral agents, use to make choices to guide their behaviors. Information systems raise new ethical questions for both individuals and societies because they create opportunities for intense social change, and thus threaten existing distributions of power, money, rights, and obligations. Like other technologies, such as steam engines, electricity, the telephone, and the radio, information technology can be used to achieve social progress, but it can also be used to commit crimes and threaten cherished social values. The development of information technology will produce benefits for many and costs for others.

Ethical issues in information systems have been given new urgency by the rise of the Internet and electronic commerce. Internet and digital firm technologies make it easier than ever to assemble, integrate, and distribute information, unleashing new concerns about the appropriate use of customer information, the protection of personal privacy, and the protection of intellectual property.

Other pressing ethical issues raised by information systems include establishing accountability for the consequences of information systems, setting standards to safeguard system quality that protects the safety of the individual and society, and preserving values and institutions considered essential to the quality of life in an information society. When using information systems, it is essential to ask, "What is the ethical and socially responsible course of action?"

A MODEL FOR THINKING ABOUT ETHICAL, SOCIAL, AND POLITICAL ISSUES

Ethical, social, and political issues are closely linked. The ethical dilemma you may face as a manager of information systems typically is reflected in social and political debate. One way to think about these relationships is given in Figure 4-1. Imagine society as a more or less calm pond on a summer day, a delicate ecosystem in partial equilibrium with individuals and with social and political institutions. Individuals know how to act in this pond because social institutions (family, education, organizations) have developed well-honed rules of behavior, and these are supported by laws developed in the political sector that prescribe behavior and promise sanctions for violations. Now toss a rock into the center of the pond. What happens? Ripples, of course.

FIGURE 4-1   THE RELATIONSHIP BETWEEN ETHICAL, SOCIAL, AND POLITICAL ISSUES IN AN INFORMATION SOCIETY

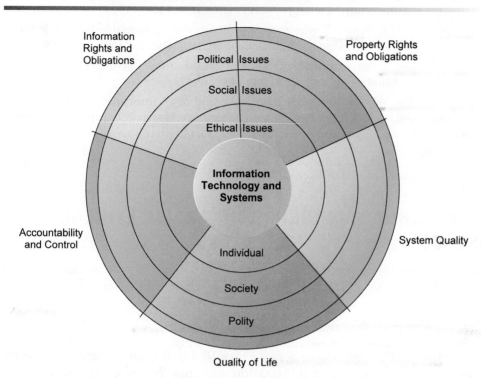

The introduction of new information technology has a ripple effect, raising new ethical, social, and political issues that must be dealt with on the individual, social, and political levels. These issues have five moral dimensions: information rights and obligations, property rights and obligations, system quality, quality of life, and accountability and control.

Imagine instead that the disturbing force is a powerful shock of new information technology and systems hitting a society more or less at rest. Suddenly, individual actors are confronted with new situations often not covered by the old rules. Social institutions cannot respond overnight to these ripples—it may take years to develop etiquette, expectations, social responsibility, politically correct attitudes, or approved rules. Political institutions also require time before developing new laws and often require the demonstration of real harm before they act. In the meantime, you may have to act. You may be forced to act in a legal gray area.

We can use this model to illustrate the dynamics that connect ethical, social, and political issues. This model is also useful for identifying the main moral dimensions of the information society, which cut across various levels of action—individual, social, and political.

FIVE MORAL DIMENSIONS OF THE INFORMATION AGE

The major ethical, social, and political issues raised by information systems include the following moral dimensions:

Information rights and obligations. What **information rights** do individuals and organizations possess with respect to themselves? What can they protect?

Property rights and obligations. How will traditional intellectual property rights be protected in a digital society in which tracing and accounting for ownership are difficult and ignoring such property rights is so easy?

Accountability and control. Who can and will be held accountable and liable for the harm done to individual and collective information and property rights?

System quality. What standards of data and system quality should we demand to protect individual rights and the safety of society?

Quality of life. What values should be preserved in an information- and knowledge-based society? Which institutions should we protect from violation? Which cultural values and practices are supported by the new information technology?

We explore these moral dimensions in detail in Section 4.3.

KEY TECHNOLOGY TRENDS THAT RAISE ETHICAL ISSUES

Ethical issues long preceded information technology. Nevertheless, information technology has heightened ethical concerns, taxed existing social arrangements, and made some laws obsolete or severely crippled. There are four key technological trends responsible for these ethical stresses and they are summarized in Table 4-2.

The doubling of computing power every 18 months has made it possible for most organizations to use information systems for their core production processes. As a result, our dependence on systems and our vulnerability to system errors and poor data quality have increased. Social rules and laws have not yet adjusted to this dependence. Standards for ensuring the accuracy and reliability of information systems (see Chapter 8) are not universally accepted or enforced.

Advances in data storage techniques and rapidly declining storage costs have been responsible for the multiplying databases on individuals—employees, customers, and potential customers—maintained by private and public organizations. These advances in data storage have made the routine violation of individual privacy both cheap and effective. Massive data storage systems are inexpensive enough for regional and even local retailing firms to use in identifying customers.

Advances in data analysis techniques for large pools of data are another technological trend that heightens ethical concerns because companies and government agencies are able to find out highly detailed personal information

TABLE 4-2 TECHNOLOGY TRENDS THAT RAISE ETHICAL ISSUES

TREND	IMPACT
Computing power doubles every 18 months	More organizations depend on computer systems for critical operations.
Data storage costs rapidly declining	Organizations can easily maintain detailed databases on individuals.
Data analysis advances	Companies can analyze vast quantities of data gathered on individuals to develop detailed profiles of individual behavior.
Networking advances	Copying data from one location to another and accessing personal data from remote locations are much easier.

Credit card purchases can make personal information available to market researchers, telemarketers, and direct-mail companies. Advances in information technology facilitate the invasion of privacy.

about individuals. With contemporary data management tools (see Chapter 5), companies can assemble and combine the myriad pieces of information about you stored on computers much more easily than in the past.

Think of all the ways you generate computer information about yourself—credit card purchases, telephone calls, magazine subscriptions, video rentals, mail-order purchases, banking records, local, state, and federal government records (including court and police records), and visits to Web sites. Put together and mined properly, this information could reveal not only your credit information but also your driving habits, your tastes, your associations, and your political interests.

Companies with products to sell purchase relevant information from these sources to help them more finely target their marketing campaigns. Chapters 3 and 6 describe how companies can analyze large pools of data from multiple sources to rapidly identify buying patterns of customers and suggest individual responses. The use of computers to combine data from multiple sources and create electronic dossiers of detailed information on individuals is called **profiling**.

For example, several thousand of the most popular Web sites allow DoubleClick (owned by Google), an Internet advertising broker, to track the activities of their visitors in exchange for revenue from advertisements based on visitor information DoubleClick gathers. DoubleClick uses this information to create a profile of each online visitor, adding more detail to the profile as the visitor accesses an associated DoubleClick site. Over time, DoubleClick can create a detailed dossier of a person's spending and computing habits on the Web that is sold to companies to help them target their Web ads more precisely.

ChoicePoint gathers data from police, criminal, and motor vehicle records; credit and employment histories; current and previous addresses; professional licenses; and insurance claims to assemble and maintain electronic dossiers on almost every adult in the United States. The company sells this personal information to businesses and government agencies. Demand for personal data is so enormous that data broker businesses such as ChoicePoint are flourishing.

A new data analysis technology called **nonobvious relationship awareness (NORA)** has given both the government and the private sector even more powerful profiling capabilities. NORA can take information about people from many disparate sources, such as employment applications, telephone records, customer listings, and "wanted" lists, and correlate relationships to find obscure hidden connections that might help identify criminals or terrorists (see Figure 4-2).

NORA technology scans data and extracts information as the data are being generated so that it could, for example, instantly discover a man at an airline ticket counter who shares a phone number with a known terrorist before that person boards an airplane. The technology is considered a valuable tool for homeland security but does have privacy implications because it can provide such a detailed picture of the activities and associations of a single individual.

Finally, advances in networking, including the Internet, promise to greatly reduce the costs of moving and accessing large quantities of data and open the possibility of mining large pools of data remotely using small desktop machines, permitting an invasion of privacy on a scale and with a precision heretofore unimaginable.

FIGURE 4-2   NONOBVIOUS RELATIONSHIP AWARENESS (NORA)

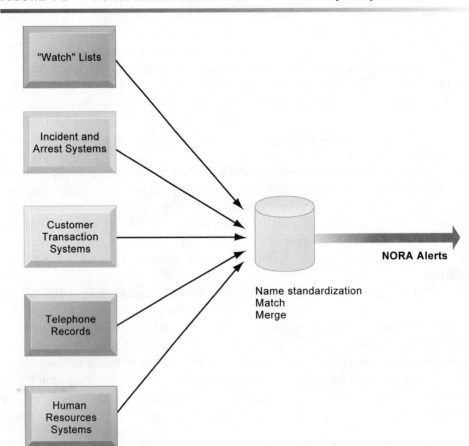

NORA technology can take information about people from disparate sources and find obscure, nonobvious relationships. It might discover, for example, that an applicant for a job at a casino shares a telephone number with a known criminal and issue an alert to the hiring manager.

4.2   ETHICS IN AN INFORMATION SOCIETY

Ethics is a concern of humans who have freedom of choice. Ethics is about individual choice: When faced with alternative courses of action, what is the correct moral choice? What are the main features of ethical choice?

BASIC CONCEPTS: RESPONSIBILITY, ACCOUNTABILITY, AND LIABILITY

Ethical choices are decisions made by individuals who are responsible for the consequences of their actions. **Responsibility** is a key element of ethical action. Responsibility means that you accept the potential costs, duties, and obligations for the decisions you make. **Accountability** is a feature of systems and social institutions: It means that mechanisms are in place to determine who took responsible action, and who is responsible. Systems and institutions in which it is impossible to find out who took what action are inherently incapable of ethical analysis or ethical action. **Liability** extends the concept of responsibility further to the area of laws. Liability is a feature of political systems in which a body of laws is in place that permits individuals to recover the damages done to them by other actors, systems, or organizations. **Due process** is a related feature of law-governed societies and is a process in which laws are known and understood, and there is an ability to appeal to higher authorities to ensure that the laws are applied correctly.

These basic concepts form the underpinning of an ethical analysis of information systems and those who manage them. First, information technologies are filtered through social institutions, organizations, and individuals. Systems do not have impacts by themselves. Whatever information system impacts exist are products of institutional, organizational, and individual actions and behaviors. Second, responsibility for the consequences of technology falls clearly on the institutions, organizations, and individual managers who choose to use the technology. Using information technology in a socially responsible manner means that you can and will be held accountable for the consequences of your actions. Third, in an ethical, political society, individuals and others can recover damages done to them through a set of laws characterized by due process.

ETHICAL ANALYSIS

When confronted with a situation that seems to present ethical issues, how should you analyze it? The following five-step process should help:

1. *Identify and describe clearly the facts.* Find out who did what to whom, and where, when, and how. In many instances, you will be surprised at the errors in the initially reported facts, and often you will find that simply getting the facts straight helps define the solution. It also helps to get the opposing parties involved in an ethical dilemma to agree on the facts.

2. *Define the conflict or dilemma and identify the higher-order values involved.* Ethical, social, and political issues always reference higher values. The parties to a dispute all claim to be pursuing higher values (e.g., freedom, privacy, protection of property, and the free enterprise system). Typically, an ethical issue involves a dilemma: two diametrically opposed courses of action that support worthwhile values. For example, the chapter-ending case study illustrates two competing values: the need to improve health care record keeping and the need to protect individual privacy.

3. *Identify the stakeholders.* Every ethical, social, and political issue has stakeholders: players in the game who have an interest in the outcome, who have invested in the situation, and usually who have vocal opinions. Find out the identity of these groups and what they want. This will be useful later when designing a solution.

4. *Identify the options that you can reasonably take.* You may find that none of the options satisfy all the interests involved, but that some options do a better job than others. Sometimes arriving at a good or ethical solution may not always be a balancing of consequences to stakeholders.

5. *Identify the potential consequences of your options.* Some options may be ethically correct but disastrous from other points of view. Other options may work in one instance but not in other similar instances. Always ask yourself, "What if I choose this option consistently over time?"

CANDIDATE ETHICAL PRINCIPLES

Once your analysis is complete, what ethical principles or rules should you use to make a decision? What higher-order values should inform your judgment? Although you are the only one who can decide which among many ethical principles you will follow, and how you will prioritize them, it is helpful to consider some ethical principles with deep roots in many cultures that have survived throughout recorded history:

1. Do unto others as you would have them do unto you (the **Golden Rule**). Putting yourself into the place of others, and thinking of yourself as the object of the decision, can help you think about fairness in decision making.

2. If an action is not right for everyone to take, it is not right for anyone **(Immanuel Kant's Categorical Imperative)**. Ask yourself, "If everyone did this, could the organization, or society, survive?"

3. If an action cannot be taken repeatedly, it is not right to take at all **(Descartes' rule of change)**. This is the slippery-slope rule: An action may bring about a small change now that is acceptable, but if it is repeated, it would bring unacceptable changes in the long run. In the vernacular, it might be stated as "once started down a slippery path, you may not be able to stop."

4. Take the action that achieves the higher or greater value **(Utilitarian Principle)**. This rule assumes you can prioritize values in a rank order and understand the consequences of various courses of action.

5. Take the action that produces the least harm or the least potential cost **(Risk Aversion Principle)**. Some actions have extremely high failure costs of very low probability (e.g., building a nuclear generating facility in an urban area) or extremely high failure costs of moderate probability (speeding and automobile accidents). Avoid these high-failure-cost actions, paying greater attention to high-failure-cost potential of moderate to high probability.

6. Assume that virtually all tangible and intangible objects are owned by someone else unless there is a specific declaration otherwise. (This is the **ethical "no free lunch" rule.**) If something someone else has created is useful to you, it has value, and you should assume the creator wants compensation for this work.

Actions that do not easily pass these rules deserve close attention and a great deal of caution. The appearance of unethical behavior may do as much harm to you and your company as actual unethical behavior.

PROFESSIONAL CODES OF CONDUCT

When groups of people claim to be professionals, they take on special rights and obligations because of their special claims to knowledge, wisdom, and respect. Professional codes of conduct are promulgated by associations of professionals, such as the American Medical Association (AMA), the American Bar Association (ABA), the Association of Information Technology Professionals (AITP), and the Association for Computing Machinery (ACM). These professional groups take responsibility for the partial regulation of their professions by determining entrance qualifications and competence. Codes of ethics are promises by professions to regulate themselves in the general interest of society. For example, avoiding harm to others, honoring property rights (including intellectual property), and respecting privacy are among the General Moral Imperatives of the ACM's Code of Ethics and Professional Conduct.

SOME REAL-WORLD ETHICAL DILEMMAS

Information systems have created new ethical dilemmas in which one set of interests is pitted against another. For example, many of the large telephone companies in the United States are using information technology to reduce the sizes of their workforces. Voice recognition software reduces the need for human operators by enabling computers to recognize a customer's responses to a series of computerized questions. Many companies monitor what their employees are doing on the Internet to prevent them from wasting company resources on non-business activities.

In each instance, you can find competing values at work, with groups lined up on either side of a debate. A company may argue, for example, that it has a right to use information systems to increase productivity and reduce the size of its workforce to lower costs and stay in business. Employees displaced by information systems may argue that employers have some responsibility for their welfare. Business owners might feel obligated to monitor employee e-mail and Internet use to minimize drains on productivity. Employees might believe they should be able to use the Internet for short personal tasks in place of the telephone. A close analysis of the facts can sometimes produce compromised solutions that give each side "half a loaf." Try to apply some of the principles of ethical analysis described to each of these cases. What is the right thing to do?

4.3 THE MORAL DIMENSIONS OF INFORMATION SYSTEMS

In this section, we take a closer look at the five moral dimensions of information systems first described in Figure 4-1. In each dimension, we identify the ethical, social, and political levels of analysis and use real-world examples to illustrate the values involved, the stakeholders, and the options chosen.

INFORMATION RIGHTS: PRIVACY AND FREEDOM IN THE INTERNET AGE

Privacy is the claim of individuals to be left alone, free from surveillance or interference from other individuals or organizations, including the state. Claims to privacy are also involved at the workplace: Millions of employees are subject to electronic and other forms of high-tech surveillance (Ball, 2001). Information technology and systems threaten individual claims to privacy by making the invasion of privacy cheap, profitable, and effective.

The claim to privacy is protected in the U.S., Canadian, and German constitutions in a variety of different ways and in other countries through various statutes. In the United States, the claim to privacy is protected primarily by the First Amendment guarantees of freedom of speech and association, the Fourth Amendment protections against unreasonable search and seizure of one's personal documents or home, and the guarantee of due process.

Table 4-3 describes the major U.S. federal statutes that set forth the conditions for handling information about individuals in such areas as credit reporting, education, financial records, newspaper records, and electronic communications. The Privacy Act of 1974 has been the most important of these laws, regulating the federal government's collection, use, and disclosure of information. At present, most U.S. federal privacy laws apply only to the federal government and regulate very few areas of the private sector.

Most American and European privacy law is based on a regime called **Fair Information Practices (FIP)** first set forth in a report written in 1973 by a federal government advisory committee (U.S. Department of Health, Education, and Welfare, 1973). FIP is a set of principles governing the collection and use of information about individuals. FIP principles are based on the notion of a mutuality of interest between the record holder and the individual. The individual has an interest in engaging in a transaction, and the record keeper—usually a business or government agency-requires information about the individual to support the transaction. Once information is gathered, the individual maintains an interest in the record, and the record may not be used to support other activities without the individual's consent. In 1998, the FTC restated and extended the original FIP to provide guidelines for protecting online privacy. Table 4-4 describes the FTC's Fair Information Practice principles.

The FTC's FIP principles are being used as guidelines to drive changes in privacy legislation. In July 1998, the U.S. Congress passed the Children's Online Privacy Protection Act (COPPA), requiring Web sites to obtain parental permission before collecting information on children under the age of 13. (This law is

TABLE 4-3 FEDERAL PRIVACY LAWS IN THE UNITED STATES

GENERAL FEDERAL PRIVACY LAWS	PRIVACY LAWS AFFECTING PRIVATE INSTITUTIONS
Freedom of Information Act of 1966 as Amended (5 USC 552)	Fair Credit Reporting Act of 1970
Privacy Act of 1974 as Amended (5 USC 552a)	Family Educational Rights and Privacy Act of 1974
Electronic Communications Privacy Act of 1986	Right to Financial Privacy Act of 1978
Computer Matching and Privacy Protection Act of 1988	Privacy Protection Act of 1980
Computer Security Act of 1987	Cable Communications Policy Act of 1984
Federal Managers Financial Integrity Act of 1982	Electronic Communications Privacy Act of 1986
Driver's Privacy Protection Act of 1994	Video Privacy Protection Act of 1988
E-Government Act of 2002	The Health Insurance Portability and Accountability Act of 1996 (HIPAA)
	Children's Online Privacy Protection Act (COPPA) of 1998
	Financial Modernization Act (Gramm-Leach-Bliley Act) of 1999

TABLE 4-4 FEDERAL TRADE COMMISSION FAIR INFORMATION PRACTICE PRINCIPLES

1. Notice/awareness (core principle). Web sites must disclose their information practices before collecting data. Includes identification of collector; uses of data; other recipients of data; nature of collection (active/inactive); voluntary or required status; consequences of refusal; and steps taken to protect confidentiality, integrity, and quality of the data.

2. Choice/consent (core principle). There must be a choice regime in place allowing consumers to choose how their information will be used for secondary purposes other than supporting the transaction, including internal use and transfer to third parties.

3. Access/participation. Consumers should be able to review and contest the accuracy and completeness of data collected about them in a timely, inexpensive process.

4. Security. Data collectors must take responsible steps to assure that consumer information is accurate and secure from unauthorized use.

5. Enforcement. There must be in place a mechanism to enforce FIP principles. This can involve self-regulation, legislation giving consumers legal remedies for violations, or federal statutes and regulations.

in danger of being overturned.) The FTC has recommended additional legislation to protect online consumer privacy in advertising networks that collect records of consumer Web activity to develop detailed profiles, which are then used by other companies to target online ads. Other proposed Internet privacy legislation focuses on protecting the online use of personal identification numbers, such as social security numbers; protecting personal information collected on the Internet that deals with individuals not covered by COPPA; and limiting the use of data mining for homeland security.

In February 2009, the FTC began the process of extending its fair information practices doctrine to behavioral targeting. The FTC held hearings to discuss its program for voluntary industry principles for regulating behavioral targeting. The online advertising trade group Network Advertising Initiative (discussed later in this section), published its own self-regulatory principles that largely agreed with the FTC. Nevertheless, the government, privacy groups, and the online ad industry are still at loggerheads over two issues. Privacy advocates want both an opt-in policy at all sites and a national Do Not Track list. The industry opposes these moves and continues to insist on an opt-out capability being the only way to avoid tracking (Federal Trade Commission, 2009). Nevertheless, there is an emerging consensus among all parties that greater transparency and user control (especially making opt-out of tracking the default option) is required to deal with behavioral tracking.

Privacy protections have also been added to recent laws deregulating financial services and safeguarding the maintenance and transmission of health information about individuals. The Gramm-Leach-Bliley Act of 1999, which repeals earlier restrictions on affiliations among banks, securities firms, and insurance companies, includes some privacy protection for consumers of financial services. All financial institutions are required to disclose their policies and practices for protecting the privacy of nonpublic personal information and to allow customers to opt out of information-sharing arrangements with nonaffiliated third parties.

The Health Insurance Portability and Accountability Act (HIPAA) of 1996, which took effect on April 14, 2003, includes privacy protection for medical records. The law gives patients access to their personal medical records maintained by health care providers, hospitals, and health insurers, and the right to authorize how protected information about themselves can be used or disclosed. Doctors, hospitals, and other health care providers must limit the disclosure of personal information about patients to the minimum amount necessary to achieve a given purpose.

The European Directive on Data Protection

In Europe, privacy protection is much more stringent than in the United States. Unlike the United States, European countries do not allow businesses to use personally identifiable information without consumers' prior consent. On October 25, 1998, the European Commission's Directive on Data Protection went into effect, broadening privacy protection in the European Union (EU) nations. The directive requires companies to inform people when they collect information about them and disclose how it will be stored and used. Customers must provide their informed consent before any company can legally use data about them, and they have the right to access that information, correct it, and request that no further data be collected. **Informed consent** can be defined as consent given with knowledge of all the facts needed to make a rational decision. EU member nations must translate these principles into their own laws and cannot transfer personal data to countries, such as the United States, that do not have similar privacy protection regulations.

Working with the European Commission, the U.S. Department of Commerce developed a safe harbor framework for U.S. firms. A **safe harbor** is a private, self-regulating policy and enforcement mechanism that meets the objectives of government regulators and legislation but does not involve government regulation or enforcement. U.S. businesses would be allowed to use personal data from EU countries if they develop privacy protection policies that meet EU standards. Enforcement would occur in the United States using self-policing, regulation, and government enforcement of fair trade statutes.

Internet Challenges to Privacy

Internet technology has posed new challenges for the protection of individual privacy. Information sent over this vast network of networks may pass through many different computer systems before it reaches its final destination. Each of these systems is capable of monitoring, capturing, and storing communications that pass through it.

It is possible to record many online activities, including what searches have been conducted, which Web sites and Web pages have been visited, the online content a person has accessed, and what items that person has inspected or purchased over the Web. Much of this monitoring and tracking of Web site visitors occurs in the background without the visitor's knowledge. It is conducted not just by individual Web sites but by advertising networks such as Microsoft Advertising, Yahoo, and DoubleClick that are capable of tracking all browsing behavior at thousands of Web sites. Tools to monitor visits to the World Wide Web have become popular because they help businesses determine who is visiting their Web sites and how to better target their offerings. (Some firms also monitor the Internet usage of their employees to see how they are using company network resources.) The commercial demand for this personal information is virtually insatiable.

Web sites can learn the identities of their visitors if the visitors voluntarily register at the site to purchase a product or service or to obtain a free service, such as information. Web sites can also capture information about visitors without their knowledge using cookie technology.

Cookies are small text files deposited on a computer hard drive when a user visits Web sites. Cookies identify the visitor's Web browser software and track visits to the Web site. When the visitor returns to a site that has stored a cookie, the Web site software will search the visitor's computer, find the cookie, and know what that person has done in the past. It may also update the cookie, depending on the activity during the visit. In this way, the site can customize its contents for each visitor's interests. For example, if you purchase a book on Amazon.com and return later from the same browser, the site will welcome you by name and recommend other books of interest based on your past purchases. DoubleClick, described earlier in this chapter, uses cookies to build its dossiers with details of online purchases and to examine the behavior of Web site visitors. Figure 4-3 illustrates how cookies work.

Web sites using cookie technology cannot directly obtain visitors' names and addresses. However, if a person has registered at a site, that information can be combined with cookie data to identify the visitor. Web site owners can also combine the data they have gathered from cookies and other Web site monitoring tools with personal data from other sources, such as offline data collected from surveys or paper catalog purchases, to develop very detailed profiles of their visitors.

There are now even more subtle and surreptitious tools for surveillance of Internet users. Marketers use Web beacons as another tool to monitor online behavior. **Web beacons**, also called *Web bugs*, are tiny objects invisibly embedded in e-mail messages and Web pages that are designed to monitor the behavior of the user visiting a Web site or sending e-mail. The Web beacon captures and transmits information such as the IP address of the user's computer, the time a Web page was viewed and for how long, the type of Web browser that retrieved the beacon, and previously set cookie values. Web beacons are placed on popular Web sites by "third party" firms who pay the Web sites a fee for access to their audience. Typical popular Web sites contain 25–35 Web beacons.

Other **spyware** can secretly install itself on an Internet user's computer by piggybacking on larger applications. Once installed, the spyware calls out to Web sites to send banner ads and other unsolicited material to the user, and it can also report the user's movements on the Internet to other computers. More information is available about intrusive software in Chapter 8.

About 75 percent of global Internet users use Google search and other services, making Google the world's largest collector of online user data. Whatever Google does with its data has an enormous impact on online privacy. Most experts

FIGURE 4-3   HOW COOKIES IDENTIFY WEB VISITORS

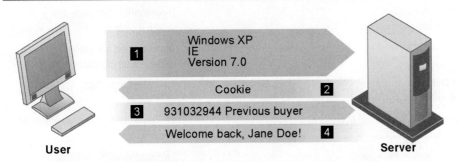

1. The Web server reads the user's Web browser and determines the operating system, browser name, version number, Internet address, and other information.
2. The server transmits a tiny text file with user identification information called a cookie, which the user's browser receives and stores on the user's computer hard drive.
3. When the user returns to the Web site, the server requests the contents of any cookie it deposited previously in the user's computer.
4. The Web server reads the cookie, identifies the visitor, and calls up data on the user.

Cookies are written by a Web site on a visitor's hard drive. When the visitor returns to that Web site, the Web server requests the ID number from the cookie and uses it to access the data stored by that server on that visitor. The Web site can then use these data to display personalized information.

believe that Google possesses the largest collection of personal information in the world—more data on more people than any government agency. Table 4-5 lists the major Google services that collect user data and how Google uses these data.

For a number of years, Google has been using behavioral targeting to help it display more relevant ads based on users' search activities. One of its programs enables advertisers to target ads based on the search histories of Google users, along with any other information the user submits to Google that Google can obtain, such as age, demographics, region, and other Web activities (such as blogging). An additional program allows Google to help advertisers select keywords and design ads for various market segments based on search histories, such as helping a clothing Web site create and test ads targeted at teenage females.

Google has also been scanning the contents of messages received by users of its free Web-based e-mail service called Gmail. Ads that users see when they read their e-mail are related to the subjects of these messages. Profiles are developed on individual users based on the content in their e-mail. Google now displays targeted ads on YouTube and on Google mobile applications, and its DoubleClick ad network serves up targeted banner ads.

In the past, Google refrained from capitalizing too much on the data it collected, considered the best source of data about user interests on the Internet. But with the emergence of rivals such as Facebook who are aggressively tracking and selling online user data, Google has decided to do more to profit from its user data.

The United States has allowed businesses to gather transaction information generated in the marketplace and then use that information for other marketing purposes without obtaining the informed consent of the individual whose information is being used. U.S. e-commerce sites are largely content to publish statements on their Web sites informing visitors about how their information will be used. Some have added opt-out selection boxes to these information policy statements. An **opt-out** model of informed consent permits the collection of personal information until the consumer specifically requests that the

TABLE 4-5 HOW GOOGLE USES THE DATA IT COLLECTS

GOOGLE FEATURE	DATA COLLECTED	USE
Google Search	Google search topics Users' Internet addresses	Targeting text ads placed in search results
Gmail	Contents of e-mail messages	Targeting text ads placed next to the e-mail messages
DoubleClick	Data about Web sites visited on Google's ad network	Targeting banner ads
YouTube	Data about videos uploaded and downloaded; some profile data	Targeting ads for Google display-ad network
Mobile Maps with My Location	User's actual or approximate location	Targeting mobile ads based on user's ZIP code
Google Toolbar	Web-browsing data and search history	No ad use at present
Google Buzz	Users' Google profile data and connections	No ad use at present
Google Chrome	Sample of address-bar entries when Google is the default search engine	No ad use at present
Google Checkout	User's name, address, transaction details	No ad use at present
Google Analytics	Traffic data from Web sites using Google's Analytics service	No ad use at present

data not be collected. Privacy advocates would like to see wider use of an **opt-in** model of informed consent in which a business is prohibited from collecting any personal information unless the consumer specifically takes action to approve information collection and use.

The online industry has preferred self-regulation to privacy legislation for protecting consumers. In 1998, the online industry formed the Online Privacy Alliance to encourage self-regulation to develop a set of privacy guidelines for its members. The group promotes the use of online seals, such as that of TRUSTe, certifying Web sites adhering to certain privacy principles. Members of the advertising network industry, including Google's DoubleClick, have created an additional industry association called the Network Advertising Initiative (NAI) to develop its own privacy policies to help consumers opt out of advertising network programs and provide consumers redress from abuses.

Individual firms like AOL, Yahoo!, and Google have recently adopted policies on their own in an effort to address public concern about tracking people online. AOL established an opt-out policy that allows users of its site to not be tracked. Yahoo follows NAI guidelines and also allows opt-out for tracking and Web beacons (Web bugs). Google has reduced retention time for tracking data.

In general, most Internet businesses do little to protect the privacy of their customers, and consumers do not do as much as they should to protect themselves. Many companies with Web sites do not have privacy policies. Of the companies that do post privacy polices on their Web sites, about half do not monitor their sites to ensure they adhere to these policies. The vast majority of online customers claim they are concerned about online privacy, but less than half read the privacy statements on Web sites (Laudon and Traver, 2010).

In one of the more insightful studies of consumer attitudes towards Internet privacy, a group of Berkeley students conducted surveys of online users, and of complaints filed with the Federal Trade Commission involving privacy issues. Here are some of their results. User concerns: people feel they have no control over the information collected about them, and they don't know who to complain to. Web site practices: Web sites collect all this information, but do not let users have access; the policies are unclear; they share data with "affiliates" but never identify who the affiliates are and how many there are. (MySpace, owned by NewsCorp, has over 1,500 affiliates with whom it shares online information.) Web bug trackers: they are ubiquitous and we are not informed they are on the pages we visit. The results of this study and others suggest that consumers are not saying "Take my privacy, I don't care, send me the service for free." They are saying "We want access to the information, we want some controls on what can be collected, what is done with the information, the ability to opt out of the entire tracking enterprise, and some clarity on what the policies really are, and we don't want those policies changed without our participation and permission." (The full report is available at knowprivacy.org.)

Technical Solutions

In addition to legislation, new technologies are available to protect user privacy during interactions with Web sites. Many of these tools are used for encrypting e-mail, for making e-mail or surfing activities appear anonymous, for preventing client computers from accepting cookies, or for detecting and eliminating spyware.

There are now tools to help users determine the kind of personal data that can be extracted by Web sites. The Platform for Privacy Preferences, known as P3P, enables automatic communication of privacy policies between an e-commerce site and its visitors. **P3P** provides a standard for communicating a Web site's

Web sites are posting their privacy policies for visitors to review. The TRUSTe seal designates Web sites that have agreed to adhere to TRUSTe's established privacy principles of disclosure, choice, access, and security.

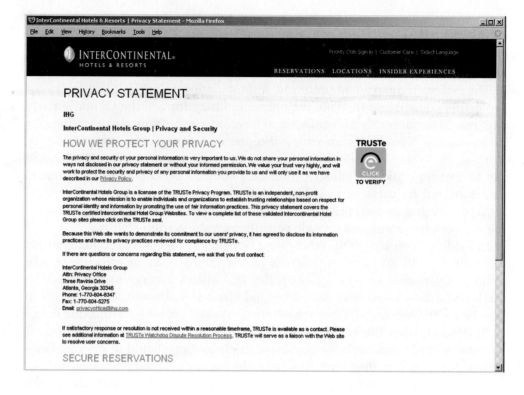

privacy policy to Internet users and for comparing that policy to the user's preferences or to other standards, such as the FTC's FIP guidelines or the European Directive on Data Protection. Users can use P3P to select the level of privacy they wish to maintain when interacting with the Web site.

The P3P standard allows Web sites to publish privacy policies in a form that computers can understand. Once it is codified according to P3P rules, the privacy policy becomes part of the software for individual Web pages (see Figure 4-4). Users of Microsoft Internet Explorer Web browsing software can access and read the P3P site's privacy policy and a list of all cookies coming from the site. Internet Explorer enables users to adjust their computers to screen out all cookies or let in selected cookies based on specific levels of privacy. For example, the "Medium" level accepts cookies from first-party host sites that have opt-in or opt-out policies but rejects third-party cookies that use personally identifiable information without an opt-in policy.

However, P3P only works with Web sites of members of the World Wide Web Consortium who have translated their Web site privacy policies into P3P format. The technology will display cookies from Web sites that are not part of the consortium, but users will not be able to obtain sender information or privacy statements. Many users may also need to be educated about interpreting company privacy statements and P3P levels of privacy. Critics point out that only a small percentage of the most popular Web sites use P3P, most users do not understand their browser's privacy settings, and there is no enforcement of P3P standards—companies can claim anything about their privacy policies.

PROPERTY RIGHTS: INTELLECTUAL PROPERTY

Contemporary information systems have severely challenged existing laws and social practices that protect private intellectual property. **Intellectual property** is considered to be intangible property created by individuals or

FIGURE 4-4   THE P3P STANDARD

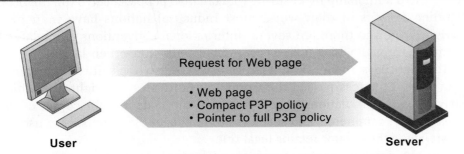

User   **Server**

1. The user with P3P Web browsing software requests a Web page.
2. The Web server returns the Web page along with a compact version of the Web site's policy and a pointer to the full P3P policy. If the Web site is not P3P compliant, no P3P data are returned.
3. The user's Web browsing software compares the response from the Web site with the user's privacy preferences. If the Web site does not have a P3P policy or the policy does not match the privacy levels established by the user, it warns the user or rejects the cookies from the Web site. Otherwise, the Web page loads normally.

P3P enables Web sites to translate their privacy policies into a standard format that can be read by the user's Web browser software. The browser software evaluates the Web site's privacy policy to determine whether it is compatible with the user's privacy preferences.

corporations. Information technology has made it difficult to protect intellectual property because computerized information can be so easily copied or distributed on networks. Intellectual property is subject to a variety of protections under three different legal traditions: trade secrets, copyright, and patent law.

Trade Secrets

Any intellectual work product—a formula, device, pattern, or compilation of data—used for a business purpose can be classified as a **trade secret**, provided it is not based on information in the public domain. Protections for trade secrets vary from state to state. In general, trade secret laws grant a monopoly on the ideas behind a work product, but it can be a very tenuous monopoly.

Software that contains novel or unique elements, procedures, or compilations can be included as a trade secret. Trade secret law protects the actual ideas in a work product, not only their manifestation. To make this claim, the creator or owner must take care to bind employees and customers with nondisclosure agreements and to prevent the secret from falling into the public domain.

The limitation of trade secret protection is that, although virtually all software programs of any complexity contain unique elements of some sort, it is difficult to prevent the ideas in the work from falling into the public domain when the software is widely distributed.

Copyright

Copyright is a statutory grant that protects creators of intellectual property from having their work copied by others for any purpose during the life of the author plus an additional 70 years after the author's death. For corporate-owned works, copyright protection lasts for 95 years after their initial creation. Congress has extended copyright protection to books, periodicals, lectures, dramas, musical compositions, maps, drawings, artwork of any kind, and motion pictures. The intent behind copyright laws has been to encourage creativity and authorship by ensuring that creative people receive the financial and other benefits of their work. Most industrial nations have their own copyright laws, and there are several international conventions and bilateral agreements through which nations coordinate and enforce their laws.

In the mid-1960s, the Copyright Office began registering software programs, and in 1980, Congress passed the Computer Software Copyright Act, which clearly provides protection for software program code and for copies of the original sold in commerce, and sets forth the rights of the purchaser to use the software while the creator retains legal title.

Copyright protects against copying of entire programs or their parts. Damages and relief are readily obtained for infringement. The drawback to copyright protection is that the underlying ideas behind a work are not protected, only their manifestation in a work. A competitor can use your software, understand how it works, and build new software that follows the same concepts without infringing on a copyright.

"Look and feel" copyright infringement lawsuits are precisely about the distinction between an idea and its expression. For instance, in the early 1990s, Apple Computer sued Microsoft Corporation and Hewlett-Packard for infringement of the expression of Apple's Macintosh interface, claiming that the defendants copied the expression of overlapping windows. The defendants countered that the idea of overlapping windows can be expressed only in a single way and, therefore, was not protectable under the merger doctrine of copyright law. When ideas and their expression merge, the expression cannot be copyrighted.

In general, courts appear to be following the reasoning of a 1989 case—*Brown Bag Software vs. Symantec Corp.*—in which the court dissected the elements of software alleged to be infringing. The court found that similar concept, function, general functional features (e.g., drop-down menus), and colors are not protectable by copyright law (*Brown Bag Software vs. Symantec Corp.*, 1992).

Patents

A **patent** grants the owner an exclusive monopoly on the ideas behind an invention for 20 years. The congressional intent behind patent law was to ensure that inventors of new machines, devices, or methods receive the full financial and other rewards of their labor and yet make widespread use of the invention possible by providing detailed diagrams for those wishing to use the idea under license from the patent's owner. The granting of a patent is determined by the United States Patent and Trademark Office and relies on court rulings.

The key concepts in patent law are originality, novelty, and invention. The Patent Office did not accept applications for software patents routinely until a 1981 Supreme Court decision that held that computer programs could be a part of a patentable process. Since that time, hundreds of patents have been granted and thousands await consideration.

The strength of patent protection is that it grants a monopoly on the underlying concepts and ideas of software. The difficulty is passing stringent criteria of nonobviousness (e.g., the work must reflect some special understanding and contribution), originality, and novelty, as well as years of waiting to receive protection.

Challenges to Intellectual Property Rights

Contemporary information technologies, especially software, pose severe challenges to existing intellectual property regimes and, therefore, create significant ethical, social, and political issues. Digital media differ from books, periodicals, and other media in terms of ease of replication; ease of transmission; ease of alteration; difficulty in classifying a software work as a program, book, or even music; compactness—making theft easy; and difficulties in establishing uniqueness.

The proliferation of electronic networks, including the Internet, has made it even more difficult to protect intellectual property. Before widespread use of networks, copies of software, books, magazine articles, or films had to be stored on physical media, such as paper, computer disks, or videotape, creating some hurdles to distribution. Using networks, information can be more widely reproduced and distributed. The Seventh Annual Global Software Piracy Study conducted by the International Data Corporation and the Business Software Alliance reported that the rate of global software piracy climbed to 43 percent in 2009, representing $51 billion in global losses from software piracy. Worldwide, for every $100 worth of legitimate software sold that year, an additional $75 worth was obtained illegally (Business Software Alliance, 2010).

The Internet was designed to transmit information freely around the world, including copyrighted information. With the World Wide Web in particular, you can easily copy and distribute virtually anything to thousands and even millions of people around the world, even if they are using different types of computer systems. Information can be illicitly copied from one place and distributed through other systems and networks even though these parties do not willingly participate in the infringement.

Individuals have been illegally copying and distributing digitized MP3 music files on the Internet for a number of years. File-sharing services such as Napster, and later Grokster, Kazaa, and Morpheus, sprung up to help users locate and swap digital music files, including those protected by copyright. Illegal file sharing became so widespread that it threatened the viability of the music recording industry. The recording industry won some legal battles for shutting these services down, but has not been able to halt illegal file sharing entirely. As more and more homes adopt high-speed Internet access, illegal file sharing of videos will pose similar threats to the motion picture industry.

Mechanisms are being developed to sell and distribute books, articles, and other intellectual property legally on the Internet, and the **Digital Millennium Copyright Act (DMCA)** of 1998 is providing some copyright protection. The DMCA implemented a World Intellectual Property Organization Treaty that makes it illegal to circumvent technology-based protections of copyrighted materials. Internet service providers (ISPs) are required to take down sites of copyright infringers that they are hosting once they are notified of the problem.

Microsoft and other major software and information content firms are represented by the Software and Information Industry Association (SIIA), which lobbies for new laws and enforcement of existing laws to protect intellectual property around the world. The SIIA runs an antipiracy hotline for individuals to report piracy activities, offers educational programs to help organizations combat software piracy, and has published guidelines for employee use of software.

ACCOUNTABILITY, LIABILITY, AND CONTROL

Along with privacy and property laws, new information technologies are challenging existing liability laws and social practices for holding individuals and institutions accountable. If a person is injured by a machine controlled, in part, by software, who should be held accountable and, therefore, held liable? Should a public bulletin board or an electronic service, such as America Online, permit the transmission of pornographic or offensive material (as broadcasters), or should they be held harmless against any liability for what users transmit (as is true of common carriers, such as the telephone system)? What about the Internet? If you outsource your information processing, can you hold the external vendor liable for injuries done to your customers? Some real-world examples may shed light on these questions.

Computer-Related Liability Problems

During the last week of September 2009, thousands of customers of TD Bank, one of the largest banks in North America, scrambled to find their payroll checks, social security checks, and savings and checking account balances. The bank's 6.5 million customers were temporarily out of funds because of a computer glitch. The problems were caused by a failed effort to integrate systems of TD Bank and Commerce Bank. A spokesperson for TD Bank, said that "while the overall integration of the systems went well, there have been some speed-bumps in the final stages, as you might expect with a project of this size and complexity." (Vijayan, 2009). Who is liable for any economic harm caused to individuals or businesses that could not access their full account balances in this period?

This case reveals the difficulties faced by information systems executives who ultimately are responsible for any harm done by systems developed by their staffs. In general, insofar as computer software is part of a machine, and the machine injures someone physically or economically, the producer of the software and the operator can be held liable for damages. Insofar as the software acts like a book, storing and displaying information, courts have been reluctant to hold authors, publishers, and booksellers liable for contents (the exception being instances of fraud or defamation), and hence courts have been wary of holding software authors liable for booklike software.

In general, it is very difficult (if not impossible) to hold software producers liable for their software products that are considered to be like books, regardless of the physical or economic harm that results. Historically, print publishers, books, and periodicals have not been held liable because of fears that liability claims would interfere with First Amendment rights guaranteeing freedom of expression.

What about software as a service? ATM machines are a service provided to bank customers. Should this service fail, customers will be inconvenienced and perhaps harmed economically if they cannot access their funds in a timely manner. Should liability protections be extended to software publishers and operators of defective financial, accounting, simulation, or marketing systems?

Software is very different from books. Software users may develop expectations of infallibility about software; software is less easily inspected than a book, and it is more difficult to compare with other software products for quality; software claims actually to perform a task rather than describe a task, as a book does; and people come to depend on services essentially based on software. Given the centrality of software to everyday life, the chances are excellent that liability law will extend its reach to include software even when the software merely provides an information service.

Telephone systems have not been held liable for the messages transmitted because they are regulated common carriers. In return for their right to provide telephone service, they must provide access to all, at reasonable rates, and achieve acceptable reliability. But broadcasters and cable television stations are subject to a wide variety of federal and local constraints on content and facilities. Organizations can be held liable for offensive content on their Web sites, and online services, such as America Online, might be held liable for postings by their users. Although U.S. courts have increasingly exonerated Web sites and ISPs for posting material by third parties, the threat of legal action still has a chilling effect on small companies or individuals who cannot afford to take their cases to trial.

SYSTEM QUALITY: DATA QUALITY AND SYSTEM ERRORS

The debate over liability and accountability for unintentional consequences of system use raises a related but independent moral dimension: What is an acceptable, technologically feasible level of system quality? At what point should system managers say, "Stop testing, we've done all we can to perfect this software. Ship it!" Individuals and organizations may be held responsible for avoidable and foreseeable consequences, which they have a duty to perceive and correct. And the gray area is that some system errors are foreseeable and correctable only at very great expense, an expense so great that pursuing this level of perfection is not feasible economically—no one could afford the product.

For example, although software companies try to debug their products before releasing them to the marketplace, they knowingly ship buggy products because the time and cost of fixing all minor errors would prevent these products from ever being released. What if the product was not offered on the marketplace, would social welfare as a whole not advance and perhaps even decline? Carrying this further, just what is the responsibility of a producer of computer services—should it withdraw the product that can never be perfect, warn the user, or forget about the risk (let the buyer beware)?

Three principal sources of poor system performance are (1) software bugs and errors, (2) hardware or facility failures caused by natural or other causes, and (3) poor input data quality. A Chapter 8 Learning Track discusses why zero defects in software code of any complexity cannot be achieved and why the seriousness of remaining bugs cannot be estimated. Hence, there is a technological barrier to perfect software, and users must be aware of the potential for catastrophic failure. The software industry has not yet arrived at testing standards for producing software of acceptable but not perfect performance.

Although software bugs and facility catastrophes are likely to be widely reported in the press, by far the most common source of business system failure is data quality. Few companies routinely measure the quality of their data, but individual organizations report data error rates ranging from 0.5 to 30 percent.

QUALITY OF LIFE: EQUITY, ACCESS, AND BOUNDARIES

The negative social costs of introducing information technologies and systems are beginning to mount along with the power of the technology. Many of these negative social consequences are not violations of individual rights or property crimes. Nevertheless, these negative consequences can be extremely harmful to individuals, societies, and political institutions. Computers and information technologies potentially can destroy valuable elements of our culture and society even while they bring us benefits. If there is a balance of good and bad consequences of using information systems, who do we hold responsible for the bad consequences? Next, we briefly examine some of the negative social consequences of systems, considering individual, social, and political responses.

Balancing Power: Center Versus Periphery

An early fear of the computer age was that huge, centralized mainframe computers would centralize power at corporate headquarters and in the nation's capital, resulting in a Big Brother society, as was suggested in George Orwell's novel *1984*. The shift toward highly decentralized computing, coupled with an ideology of empowerment of thousands of workers, and the decentralization of decision making to lower organizational levels, have reduced the fears of power centralization in institutions. Yet much of the empowerment described in popular business magazines is trivial. Lower-level employees may be empowered to make minor decisions, but the key policy decisions may be as centralized as in the past.

Rapidity of Change: Reduced Response Time to Competition

Information systems have helped to create much more efficient national and international markets. The now-more-efficient global marketplace has reduced the normal social buffers that permitted businesses many years to adjust to competition. Time-based competition has an ugly side: The business you work for may not have enough time to respond to global competitors and may be wiped out in a year, along with your job. We stand the risk of developing a "just-in-time society" with "just-in-time jobs" and "just-in-time" workplaces, families, and vacations.

Maintaining Boundaries: Family, Work, and Leisure

Parts of this book were produced on trains and planes, as well as on vacations and during what otherwise might have been "family" time. The danger to ubiquitous computing, telecommuting, nomad computing, and the "do anything anywhere" computing environment is that it is actually coming true. The traditional boundaries that separate work from family and just plain leisure have been weakened.

Although authors have traditionally worked just about anywhere (typewriters have been portable for nearly a century), the advent of information

Although some people enjoy the convenience of working at home, the "do anything anywhere" computing environment can blur the traditional boundaries between work and family time.

systems, coupled with the growth of knowledge-work occupations, means that more and more people are working when traditionally they would have been playing or communicating with family and friends. The work umbrella now extends far beyond the eight-hour day.

Even leisure time spent on the computer threatens these close social relationships. Extensive Internet use, even for entertainment or recreational purposes, takes people away from their family and friends. Among middle school and teenage children, it can lead to harmful anti-social behavior, such as the recent upsurge in cyberbullying.

Weakening these institutions poses clear-cut risks. Family and friends historically have provided powerful support mechanisms for individuals, and they act as balance points in a society by preserving private life, providing a place for people to collect their thoughts, allowing people to think in ways contrary to their employer, and dream.

Dependence and Vulnerability

Today, our businesses, governments, schools, and private associations, such as churches, are incredibly dependent on information systems and are, therefore, highly vulnerable if these systems fail. With systems now as ubiquitous as the telephone system, it is startling to remember that there are no regulatory or standard-setting forces in place that are similar to telephone, electrical, radio, television, or other public utility technologies. The absence of standards and the criticality of some system applications will probably call forth demands for national standards and perhaps regulatory oversight.

Computer Crime and Abuse

New technologies, including computers, create new opportunities for committing crime by creating new valuable items to steal, new ways to steal them, and new ways to harm others. **Computer crime** is the commission of illegal acts through the use of a computer or against a computer system. Computers or computer systems can be the object of the crime (destroying a company's computer center or a company's computer files), as well as the instrument of a crime (stealing computer lists by illegally gaining access to a computer system using a home computer). Simply accessing a computer system without authorization or with intent to do harm, even by accident, is now a federal crime.

Computer abuse is the commission of acts involving a computer that may not be illegal but that are considered unethical. The popularity of the Internet and e-mail has turned one form of computer abuse—spamming—into a serious problem for both individuals and businesses. **Spam** is junk e-mail sent by an organization or individual to a mass audience of Internet users who have expressed no interest in the product or service being marketed. Spammers tend to market pornography, fraudulent deals and services, outright scams, and other products not widely approved in most civilized societies. Some countries have passed laws to outlaw spamming or to restrict its use. In the United States, it is still legal if it does not involve fraud and the sender and subject of the e-mail are properly identified.

Spamming has mushroomed because it only costs a few cents to send thousands of messages advertising wares to Internet users. According to Sophos, a leading vendor of security software, spam accounted for 97 percent of all business e-mail during the second quarter of 2010 (Schwartz, 2010). Spam costs for businesses are very high (estimated at over $50 billion per year) because of the computing and network resources consumed by billions of unwanted e-mail messages and the time required to deal with them.

Internet service providers and individuals can combat spam by using spam filtering software to block suspicious e-mail before it enters a recipient's e-mail inbox. However, spam filters may block legitimate messages. Spammers know how to skirt around filters by continually changing their e-mail accounts, by incorporating spam messages in images, by embedding spam in e-mail attachments and electronic greeting cards, and by using other people's computers that have been hijacked by botnets (see Chapter 7). Many spam messages are sent from one country while another country hosts the spam Web site.

Spamming is more tightly regulated in Europe than in the United States. On May 30, 2002, the European Parliament passed a ban on unsolicited commercial messaging. Electronic marketing can be targeted only to people who have given prior consent.

The U.S. CAN-SPAM Act of 2003, which went into effect on January 1, 2004, does not outlaw spamming but does ban deceptive e-mail practices by requiring commercial e-mail messages to display accurate subject lines, identify the true senders, and offer recipients an easy way to remove their names from e-mail lists. It also prohibits the use of fake return addresses. A few people have been prosecuted under the law, but it has had a negligible impact on spamming. Although Facebook and MySpace have won judgments against spammers, most critics argue the law has too many loopholes and is not effectively enforced (Associated Press, 2009).

Another negative impact of computer technology is the rising danger from people using cell phones to send text messages while driving. Many states have outlawed this behavior, but it has been difficult to eradicate. The Interactive Session on Organizations explores this topic.

Employment: Trickle-Down Technology and Reengineering Job Loss

Reengineering work is typically hailed in the information systems community as a major benefit of new information technology. It is much less frequently noted that redesigning business processes could potentially cause millions of mid-level managers and clerical workers to lose their jobs. One economist has raised the possibility that we will create a society run by a small "high tech elite of corporate professionals . . . in a nation of the permanently unemployed" (Rifkin, 1993).

Other economists are much more sanguine about the potential job losses. They believe relieving bright, educated workers from reengineered jobs will result in these workers moving to better jobs in fast-growth industries. Missing from this equation are unskilled, blue-collar workers and older, less well-educated middle managers. It is not clear that these groups can be retrained easily for high-quality (high-paying) jobs. Careful planning and sensitivity to employee needs can help companies redesign work to minimize job losses.

Equity and Access: Increasing Racial and Social Class Cleavages

Does everyone have an equal opportunity to participate in the digital age? Will the social, economic, and cultural gaps that exist in the United States and other societies be reduced by information systems technology? Or will the cleavages be increased, permitting the better off to become even more better off relative to others?

These questions have not yet been fully answered because the impact of systems technology on various groups in society has not been thoroughly studied. What is known is that information, knowledge, computers, and access

INTERACTIVE SESSION: ORGANIZATIONS

THE PERILS OF TEXTING

Cell phones have become a staple of modern society. Nearly everyone has them, and people carry and use them at all hours of the day. For the most part, this is a good thing: the benefits of staying connected at any time and at any location are considerable. But if you're like most Americans, you may regularly talk on the phone or even text while at the wheel of a car. This dangerous behavior has resulted in increasing numbers of accidents and fatalities caused by cell phone usage. The trend shows no sign of slowing down.

In 2003, a federal study of 10,000 drivers by the National Highway Traffic Safety Administration (NHTSA) set out to determine the effects of using cell phones behind the wheel. The results were conclusive: talking on the phone is equivalent to a 10-point reduction in IQ and a .08 blood alcohol level, which law enforcement considers intoxicated. Hands-free sets were ineffective in eliminating risk, the study found, because the conversation itself is what distracts drivers, not holding the phone. Cell phone use caused 955 fatalities and 240,000 accidents in 2002. Related studies indicated that drivers that talked on the phone while driving increased their crash risk fourfold, and drivers that texted while driving increased their crash risk by a whopping 23 times.

Since that study, mobile device usage has grown by an order of magnitude, worsening this already dangerous situation. The number of wireless subscribers in America has increased by around 1,000 percent since 1995 to nearly 300 million overall in 2010, and Americans' usage of wireless minutes increased by approximately 6,000 percent. This increase in cell phone usage has been accompanied by an upsurge in phone-related fatalities and accidents: In 2010, it's estimated that texting caused 5,870 fatalities and 515,000 accidents, up considerably from prior years. These figures are roughly half of equivalent statistics for drunk driving. Studies show that drivers know that using the phone while driving is one of the most dangerous things you can do on the road, but refuse to admit that it's dangerous when they themselves do it.

Of users that text while driving, the more youthful demographic groups, such as the 18–29 age group, are by far the most frequent texters. About three quarters of Americans in this age group regularly text, compared to just 22 percent of the 35–44 age group. Correspondingly, the majority of accidents involving mobile device use behind the wheel involve young adults. Among this age group, texting behind the wheel is just one of a litany of problems raised by frequent texting: anxiety, distraction, failing grades, repetitive stress injuries, and sleep deprivation are just some of the other problems brought about by excessive use of mobile devices. Teenagers are particularly prone to using cell phones to text because they want to know what's happening to their friends and are anxious about being socially isolated.

Analysts predict that over 800 billion text messages will be sent in 2010. Texting is clearly here to stay, and in fact has supplanted phone calls as the most commonly used method of mobile communication. People are unwilling to give up their mobile devices because of the pressures of staying connected. Neurologists have found that the neural response to multitasking by texting while driving suggests that people develop addictions to the digital devices they use most, getting quick bursts of adrenaline, without which driving becomes boring.

There are interests opposed to legislation prohibiting cell phone use in cars. A number of legislators believe that it's not state or federal government's role to prohibit poor decision making. Auto makers, and some safety researchers, are arguing that with the proper technology and under appropriate conditions, communicating from a moving vehicle is a manageable risk. Louis Tijerina, a veteran of the NHTSA and Ford Motor Co. researcher, notes that even as mobile phone subscriptions have surged to over 250 million during the past decade, the death rate from accidents on the highways has fallen.

Nevertheless, lawmakers are increasingly recognizing the need for more powerful legislation barring drivers from texting behind the wheel. Many states have made inroads with laws prohibiting texting while operating vehicles. In Utah, drivers crashing while texting can receive 15 years in prison, by far the toughest sentence for texting while driving in the nation when the legislation was enacted. Utah's law assumes that drivers understand the risks of texting while driving, whereas in other states, prosecutors must prove that the driver knew about the risks of texting while driving before doing so.

Utah's tough law was the result of a horrifying accident in which a speeding college student, texting at the wheel, rear-ended a car in front. The car lost control, entered the opposite side of the road, and was hit head-on by a pickup truck hauling a trailer, killing the driver instantly. In September 2008, a train engineer in California was texting within a minute prior to the most fatal train accident in almost two decades. Californian authorities responded by banning the use of cell phones by train workers while on duty.

In total, 31 states have banned texting while driving in some form, and most of those states have a full ban for phone users of all ages. The remaining states are likely to follow suit in coming years as well. President Obama also banned texting while driving for all federal government employees in October 2009. Still, there's more work to be done to combat this dangerous and life-threatening practice.

Sources: Paulo Salazar, "Banning Texting While Driving," WCBI.com, August 7, 2010; Jerry Hirsch, "Teen Drivers Dangerously Divide Their Attention," *Los Angeles Times*, August 3, 2010; www.drivinglaws.org, accessed July 2010; www.drivinglaws.org, accessed July 7, 2010; Matt Richtel, "Driver Texting Now an Issue in the Back Seat," *The New York Times*, September 9, 2009; Matt Richtel, "Utah Gets Tough With Texting Drivers," *The New York Times*, August 29, 2009; Matt Richtel, "In Study, Texting Lifts Crash Risk by Large Margin," *The New York Times*, July 28, 2009; Matt Richtel, "Drivers and Legislators Dismiss Cellphone Risks," *The New York Times*, July 19, 2009; Tom Regan, "Some Sobering Stats on Texting While Driving," *The Christian Science Monitor*, May 28, 2009; Katie Hafner, "Texting May be Taking a Toll on Teenagers," *The New York Times*, May 26, 2009; and Tara Parker-Pope, "Texting Until Their Thumbs Hurt," The *New York Times*, May 26, 2009.

CASE STUDY QUESTIONS

1. Which of the five moral dimensions of information systems identified in this text is involved in this case?

2. What are the ethical, social, and political issues raised by this case?

3. Which of the ethical principles described in the text are useful for decision making about texting while driving?

MIS IN ACTION

1. Many people at state and local levels are calling for a federal law against texting while driving. Use a search engine to explore what steps the federal government has taken to discourage texting while driving.

2. Most people are not aware of the widespread impact of texting while driving across the United States. Do a search on "texting while driving." Examine all the search results for the first two pages. Enter the information into a two-column table. In the left column put the locality of the report and year. In the right column give a brief description of the search result, e.g., accident, report, court judgment, etc. What can you conclude from these search results and table?

to these resources through educational institutions and public libraries are inequitably distributed along ethnic and social class lines, as are many other information resources. Several studies have found that certain ethnic and income groups in the United States are less likely to have computers or online Internet access even though computer ownership and Internet access have soared in the past five years. Although the gap is narrowing, higher-income families in each ethnic group are still more likely to have home computers and Internet access than lower-income families in the same group.

A similar **digital divide** exists in U.S. schools, with schools in high-poverty areas less likely to have computers, high-quality educational technology programs, or Internet access availability for their students. Left uncorrected, the digital divide could lead to a society of information haves, computer literate and skilled, versus a large group of information have-nots, computer illiterate and unskilled. Public interest groups want to narrow this digital divide by making digital information services—including the Internet—available to virtually everyone, just as basic telephone service is now.

Health Risks: RSI, CVS, and Technostress

The most common occupational disease today is **repetitive stress injury (RSI)**. RSI occurs when muscle groups are forced through repetitive actions often with high-impact loads (such as tennis) or tens of thousands of repetitions under low-impact loads (such as working at a computer keyboard).

The single largest source of RSI is computer keyboards. The most common kind of computer-related RSI is **carpal tunnel syndrome (CTS)**, in which pressure on the median nerve through the wrist's bony structure, called a carpal tunnel, produces pain. The pressure is caused by constant repetition of keystrokes: in a single shift, a word processor may perform 23,000 keystrokes. Symptoms of carpal tunnel syndrome include numbness, shooting pain, inability to grasp objects, and tingling. Millions of workers have been diagnosed with carpal tunnel syndrome.

RSI is avoidable. Designing workstations for a neutral wrist position (using a wrist rest to support the wrist), proper monitor stands, and footrests all contribute to proper posture and reduced RSI. Ergonomically correct keyboards are also an option. These measures should be supported by frequent rest breaks and rotation of employees to different jobs.

RSI is not the only occupational illness computers cause. Back and neck pain, leg stress, and foot pain also result from poor ergonomic designs of workstations. **Computer vision syndrome (CVS)** refers to any eyestrain condition related to display screen use in desktop computers, laptops, e-readers, smartphones, and hand-held video games. CVS affects about 90 percent of people who spend three hours or more per day at a computer (Beck, 2010). Its symptoms, which are usually temporary, include headaches, blurred vision, and dry and irritated eyes.

The newest computer-related malady is **technostress**, which is stress induced by computer use. Its symptoms include aggravation, hostility toward humans, impatience, and fatigue. According to experts, humans working continuously with computers come to expect other humans and human institutions to behave like computers, providing instant responses, attentiveness, and

Repetitive stress injury (RSI) is the leading occupational disease today. The single largest cause of RSI is computer keyboard work.

an absence of emotion. Technostress is thought to be related to high levels of job turnover in the computer industry, high levels of early retirement from computer-intense occupations, and elevated levels of drug and alcohol abuse.

The incidence of technostress is not known but is thought to be in the millions and growing rapidly in the United States. Computer-related jobs now top the list of stressful occupations based on health statistics in several industrialized countries.

To date, the role of radiation from computer display screens in occupational disease has not been proved. Video display terminals (VDTs) emit nonionizing electric and magnetic fields at low frequencies. These rays enter the body and have unknown effects on enzymes, molecules, chromosomes, and cell membranes. Long-term studies are investigating low-level electromagnetic fields and birth defects, stress, low birth weight, and other diseases. All manufacturers have reduced display screen emissions since the early 1980s, and European countries, such as Sweden, have adopted stiff radiation emission standards.

In addition to these maladies, computer technology may be harming our cognitive functions. Although the Internet has made it much easier for people to access, create, and use information, some experts believe that it is also preventing people from focusing and thinking clearly. The Interactive Session on Technology highlights the debate that has emerged about this problem.

The computer has become a part of our lives—personally as well as socially, culturally, and politically. It is unlikely that the issues and our choices will become easier as information technology continues to transform our world. The growth of the Internet and the information economy suggests that all the ethical and social issues we have described will be heightened further as we move into the first digital century.

INTERACTIVE SESSION: TECHNOLOGY

TOO MUCH TECHNOLOGY?

Do you think that the more information managers receive, the better their decisions? Well, think again. Most of us can no longer imagine the world without the Internet and without our favorite gadgets, whether they're iPads, smartphones, laptops, or cell phones. However, although these devices have brought about a new era of collaboration and communication, they also have introduced new concerns about our relationship with technology. Some researchers suggest that the Internet and other digital technologies are fundamentally changing the way we think—and not for the better. Is the Internet actually making us "dumber," and have we reached a point where we have too much technology? Or does the Internet offer so many new opportunities to discover information that it's actually making us "smarter." And, by the way, how do we define "dumber" and "smarter" in an Internet age?

Wait a second, you're saying. How could this be? The Internet is an unprecedented source for acquiring and sharing all types of information. Creating and disseminating media has never been easier. Resources like Wikipedia and Google have helped to organize knowledge and make that knowledge accessible to the world, and they would not have been possible without the Internet. And other digital media technologies have become indispensable parts of our lives. At first glance, it's not clear how such advancements could do anything but make us smarter.

In response to this argument, several authorities claim that making it possible for millions of people to create media—written blogs, photos, videos—has understandably lowered the quality of media. Bloggers very rarely do original reporting or research but instead copy it from professional resources. YouTube videos contributed by newbies to video come nowhere near the quality of professional videos. Newspapers struggle to stay in business while bloggers provide free content of inconsistent quality.

But similar warnings were issued in response to the development of the printing press. As Gutenberg's invention spread throughout Europe, contemporary literature exploded in popularity, and much of it was considered mediocre by intellectuals of the era. But rather than being destroyed, it was simply in the early stages of fundamental change. As people came to grips with the new technology and

the new norms governing it, literature, newspapers, scientific journals, fiction, and non-fiction all began to contribute to the intellectual climate instead of detracting from it. Today, we can't imagine a world without print media.

Advocates of digital media argue that history is bound to repeat itself as we gain familiarity with the Internet and other newer technologies. The scientific revolution was galvanized by peer review and collaboration enabled by the printing press. According to many digital media supporters, the Internet will usher in a similar revolution in publishing capability and collaboration, and it will be a resounding success for society as a whole.

This may all be true, but from a cognitive standpoint, the effects of the Internet and other digital devices might not be so positive. New studies suggest that digital technologies are damaging our ability to think clearly and focus. Digital technology users develop an inevitable desire to multitask, doing several things at once while using their devices.

Although TV, the Internet, and video games are effective at developing our visual processing ability, research suggests that they detract from our ability to think deeply and retain information. It's true that the Internet grants users easy access to the world's information, but the medium through which that information is delivered is hurting our ability to think deeply and critically about what we read and hear. You'd be "smarter" (in the sense of being able to give an account of the content) by reading a book rather than viewing a video on the same topic while texting with your friends.

Using the Internet lends itself to multitasking. Pages are littered with hyperlinks to other sites; tabbed browsing allows us to switch rapidly between two windows; and we can surf the Web while watching TV, instant messaging friends, or talking on the phone. But the constant distractions and disruptions that are central to online experiences prevent our brains from creating the neural connections that constitute full understanding of a topic. Traditional print media, by contrast, makes it easier to fully concentrate on the content with fewer interruptions.

A recent study conducted by a team of researchers at Stanford found that multitaskers are not only more easily distracted, but were also surprisingly poor at multitasking compared to people who rarely do so themselves. The team also found that multitaskers receive a jolt of excitement when confronted with a new piece of information or a new call, message, or e-mail.

The cellular structure of the brain is highly adaptable and adjusts to the tools we use, so multitaskers quickly become dependent on the excitement they experience when confronted with something new. This means that multitaskers continue to be easily distracted, even if they're totally unplugged from the devices they most often use.

Eyal Ophir, a cognitive scientist on the research team at Stanford, devised a test to measure this phenomenon. Subjects self-identifying as multitaskers were asked to keep track of red rectangles in series of images. When blue rectangles were introduced, multitaskers struggled to recognize whether or not the red rectangles had changed position from image to image. Normal testers significantly outperformed the multitaskers. Less than three percent of multitaskers (called "supertaskers") are able to manage multiple information streams at once; for the vast majority of us, multitasking does not result in greater productivity.

Neuroscientist Michael Merzenich argues that our brains are being 'massively remodeled' by our constant and ever-growing usage of the Web. And it's not just the Web that's contributing to this trend. Our ability to focus is also being undermined by the constant distractions provided by smart phones and other digital technology. Television and video games are no exception. Another study showed that when presented with two identical TV shows, one of which

had a news crawl at the bottom, viewers retained much more information about the show without the news crawl. The impact of these technologies on children may be even greater than the impact on adults, because their brains are still developing, and they already struggle to set proper priorities and resist impulses.

The implications of recent research on the impact of Web 2.0 "social" technologies for management decision making are significant. As it turns out, the "always-connected" harried executive scurrying through airports and train stations, holding multiple voice and text conversations with clients and co-workers on sometimes several mobile devices, might not be a very good decision maker. In fact, the quality of decision making most likely falls as the quantity of digital information increases through multiple channels, and managers lose their critical thinking capabilities. Likewise, in terms of management productivity, studies of Internet use in the workplace suggest that Web 2.0 social technologies offer managers new opportunities to waste time rather than focus on their responsibilities. Checked your Facebook page today? Clearly we need to find out more about the impacts of mobile and social technologies on management work.

Sources: Randall Stross, "Computers at Home: Educational Hope vs. Teenage Reality," *The New York Times*, July 9, 2010; Matt Richtel, "Hooked on Gadgets, and Paying a Mental Price," *The New York Times*, June 6, 2010; Clay Shirky, "Does the Internet Make you Smarter?" *The Wall Street Journal*, June 4, 2010; Nicholas Carr, "Does the Internet Make you Dumber?" *The Wall Street Journal*, June 5, 2010; Ofer Malamud and Christian Pop-Echeles, "Home Computer Use and the Development of Human Capital," January 2010; and "Is Technology Producing a Decline in Critical Thinking and Analysis?" Science Daily, January 29, 2009.

CASE STUDY QUESTIONS

1. What are some of the arguments for and against the use of digital media?

2. How might the brain affected by constant digital media usage?

3. Do you think these arguments outweigh the positives of digital media usage? Why or why not?

4. What additional concerns are there for children using digital media? Should children under 8 use computers and cellphones? Why or why not?

MIS IN ACTION

1. Make a daily log for 1 week of all the activities you perform each day using digital technology (such as cell phones, computers, television, etc.) and the amount of time you spend on each. Note the occasions when you are multitasking. On average, how much time each day do you spend using digital technology? How much of this time do you spend multitasking? Do you think your life is too technology-intense? Justify your response.

4.4 HANDS-ON MIS PROJECTS

The projects in this section give you hands-on experience in analyzing the privacy implications of using online data brokers, developing a corporate policy for employee Web usage, using blog creation tools to create a simple blog, and using Internet newsgroups for market research.

Management Decision Problems

1. USAData's Web site is linked to massive databases that consolidate personal data on millions of people. Anyone with a credit card can purchase marketing lists of consumers broken down by location, age, income level, and interests. If you click on Consumer Leads to order a consumer mailing list, you can find the names, addresses, and sometimes phone numbers of potential sales leads residing in a specific location and purchase the list of those names. One could use this capability to obtain a list, for example, of everyone in Peekskill, New York, making $150,000 or more per year. Do data brokers such as USAData raise privacy issues? Why or why not? If your name and other personal information were in this database, what limitations on access would you want in order to preserve your privacy? Consider the following data users: government agencies, your employer, private business firms, other individuals.

2. As the head of a small insurance company with six employees, you are concerned about how effectively your company is using its networking and human resources. Budgets are tight, and you are struggling to meet payrolls because employees are reporting many overtime hours. You do not believe that the employees have a sufficiently heavy work load to warrant working longer hours and are looking into the amount of time they spend on the Internet.

WEB USAGE REPORT FOR THE WEEK ENDING JANUARY 9, 2010.

USER NAME	MINUTES ONLINE	WEB SITE VISITED
Kelleher, Claire	45	www.doubleclick.net
Kelleher, Claire	107	www.yahoo.com
Kelleher, Claire	96	www.insweb.com
McMahon, Patricia	83	www.itunes.com
McMahon, Patricia	44	www.insweb.com
Milligan, Robert	112	www.youtube.com
Milligan, Robert	43	www.travelocity.com
Olivera, Ernesto	40	www.CNN.com
Talbot, Helen	125	www.etrade.com
Talbot, Helen	27	www.nordstrom.com
Talbot, Helen	35	www.yahoo.com
Talbot, Helen	73	www.ebay.com
Wright, Steven	23	www.facebook.com
Wright, Steven	15	www.autobytel.com

Each employee uses a computer with Internet access on the job. You requested the preceding weekly report of employee Web usage from your information systems department.

- Calculate the total amount of time each employee spent on the Web for the week and the total amount of time that company computers were used for this purpose. Rank the employees in the order of the amount of time each spent online.

- Do your findings and the contents of the report indicate any ethical problems employees are creating? Is the company creating an ethical problem by monitoring its employees' use of the Internet?
- Use the guidelines for ethical analysis presented in this chapter to develop a solution to the problems you have identified.

Achieving Operational Excellence: Creating a Simple Blog

Software skills: Blog creation
Business skills: Blog and Web page design

In this project, you'll learn how to build a simple blog of your own design using the online blog creation software available at Blogger.com. Pick a sport, hobby, or topic of interest as the theme for your blog. Name the blog, give it a title, and choose a template for the blog. Post at least four entries to the blog, adding a label for each posting. Edit your posts, if necessary. Upload an image, such as a photo from your hard drive or the Web to your blog. (Google recommends Open Photo, Flickr: Creative Commons, or Creative Commons Search as sources for photos. Be sure to credit the source for your image.) Add capabilities for other registered users, such as team members, to comment on your blog. Briefly describe how your blog could be useful to a company selling products or services related to the theme of your blog. List the tools available to Blogger (including Gadgets) that would make your blog more useful for business and describe the business uses of each. Save your blog and show it to your instructor.

Improving Decision Making: Using Internet Newsgroups for Online Market Research

Software Skills: Web browser software and Internet newsgroups
Business Skills: Using Internet newsgroups to identify potential customers

This project will help develop your Internet skills in using newsgroups for marketing. It will also ask you to think about the ethical implications of using information in online discussion groups for business purposes.

You are producing hiking boots that you sell through a few stores at this time. You think your boots are more comfortable than those of your competition. You believe you can undersell many of your competitors if you can significantly increase your production and sales. You would like to use Internet discussion groups interested in hiking, climbing, and camping both to sell your boots and to make them well known. Visit groups.google.com, which stores discussion postings from many thousands of newsgroups. Through this site you can locate all relevant newsgroups and search them by keyword, author's name, forum, date, and subject. Choose a message and examine it carefully, noting all the information you can obtain, including information about the author.

- How could you use these newsgroups to market your boots?
- What ethical principles might you be violating if you use these messages to sell your boots? Do you think there are ethical problems in using newsgroups this way? Explain your answer.
- Next use Google or Yahoo.com to search the hiking boots industry and locate sites that will help you develop other new ideas for contacting potential customers.
- Given what you have learned in this and previous chapters, prepare a plan to use newsgroups and other alternative methods to begin attracting visitors to your site.

LEARNING TRACK MODULES

The following Learning Tracks provide content relevant to the topics covered in this chapter:

1. Developing a Corporate Code of Ethics for Information Systems
2. Creating a Web Page

Review Summary

1. **What ethical, social, and political issues are raised by information systems?**

    Information technology is introducing changes for which laws and rules of acceptable conduct have not yet been developed. Increasing computing power, storage, and networking capabilities—including the Internet—expand the reach of individual and organizational actions and magnify their impacts. The ease and anonymity with which information is now communicated, copied, and manipulated in online environments pose new challenges to the protection of privacy and intellectual property. The main ethical, social, and political issues raised by information systems center around information rights and obligations, property rights and obligations, accountability and control, system quality, and quality of life.

2. **What specific principles for conduct can be used to guide ethical decisions?**

    Six ethical principles for judging conduct include the Golden Rule, Immanuel Kant's Categorical Imperative, Descartes' rule of change, the Utilitarian Principle, the Risk Aversion Principle, and the ethical "no free lunch" rule. These principles should be used in conjunction with an ethical analysis.

3. **Why do contemporary information systems technology and the Internet pose challenges to the protection of individual privacy and intellectual property?**

    Contemporary data storage and data analysis technology enables companies to easily gather personal data about individuals from many different sources and analyze these data to create detailed electronic profiles about individuals and their behaviors. Data flowing over the Internet can be monitored at many points. Cookies and other Web monitoring tools closely track the activities of Web site visitors. Not all Web sites have strong privacy protection policies, and they do not always allow for informed consent regarding the use of personal information. Traditional copyright laws are insufficient to protect against software piracy because digital material can be copied so easily and transmitted to many different locations simultaneously over the Internet.

4. **How have information systems affected everyday life?**

    Although computer systems have been sources of efficiency and wealth, they have some negative impacts. Computer errors can cause serious harm to individuals and organizations. Poor data quality is also responsible for disruptions and losses for businesses. Jobs can be lost when computers replace workers or tasks become unnecessary in reengineered business processes. The ability to own and use a computer may be exacerbating socioeconomic disparities among different racial groups and social classes. Widespread use of computers increases opportunities for computer crime and computer abuse. Computers can also create health problems, such as RSI, computer vision syndrome, and technostress.

Key Terms

Accountability, 129
Carpal tunnel syndrome (CTS), 149
Computer abuse, 145
Computer crime, 145
Computer vision syndrome (CVS), 149
Cookies, 134
Copyright, 139
Descartes' rule of change, 130
Digital divide, 148
Digital Millennium Copyright Act (DMCA), 141
Due process, 129

Ethical "no free lunch" rule, 130
Ethics, 124
Fair Information Practices (FIP), 132
Golden Rule, 130
Immanuel Kant's Categorical Imperative, 130
Information rights, 125
Informed consent, 134
Intellectual property, 138
Liability, 129
Nonobvious relationship awareness (NORA), 128
Opt-in, 137
Opt-out, 136

Review Questions

1. What ethical, social, and political issues are raised by information systems?

   - Explain how ethical, social, and political issues are connected and give some examples.
   - List and describe the key technological trends that heighten ethical concerns.
   - Differentiate between responsibility, accountability, and liability.

2. What specific principles for conduct can be used to guide ethical decisions?

   - List and describe the five steps in an ethical analysis.
   - Identify and describe six ethical principles.

3. Why do contemporary information systems technology and the Internet pose challenges to the protection of individual privacy and intellectual property?

   - Define privacy and fair information practices.
   - Explain how the Internet challenges the protection of individual privacy and intellectual property.
   - Explain how informed consent, legislation, industry self-regulation, and technology tools help protect the individual privacy of Internet users.
   - List and define the three different regimes that protect intellectual property rights.

4. How have information systems affected everyday life?

   - Explain why it is so difficult to hold software services liable for failure or injury.
   - List and describe the principal causes of system quality problems.
   - Name and describe four quality-of-life impacts of computers and information systems.
   - Define and describe technostress and RSI and explain their relationship to information technology.

Discussion Questions

1. Should producers of software-based services, such as ATMs, be held liable for economic injuries suffered when their systems fail?

2. Should companies be responsible for unemployment caused by their information systems? Why or why not?

3. Discuss the pros and cons of allowing companies to amass personal data for behavioral targeting.

Video Cases

Video Cases and Instructional Videos illustrating some of the concepts in this chapter are available. Contact your instructor to access these videos.

Collaboration and Teamwork: Developing a Corporate Ethics Code

With three or four of your classmates, develop a corporate ethics code that addresses both employee privacy and the privacy of customers and users of the corporate Web site. Be sure to consider e-mail privacy and employer monitoring of worksites, as well as corporate use of information about employees concerning their off-the-job behavior (e.g.,

lifestyle, marital arrangements, and so forth). If possible, use Google Sites to post links to Web pages, team communication announcements, and work assignments; to brainstorm; and to work collaboratively on project documents. Try to use Google Docs to develop your solution and presentation for the class.

When Radiation Therapy Kills
CASE STUDY

When new expensive medical therapies come along, promising to cure people of illness, one would think that the manufacturers, doctors, and technicians, along with the hospitals and state oversight agencies, would take extreme caution in their application and use. Often this is not the case. Contemporary radiation therapy offers a good example of society failing to anticipate and control the negative impacts of a technology powerful enough to kill people.

For individuals and their families suffering through a battle with cancer, technical advancements in radiation treatment represent hope and a chance for a healthy, cancer-free life. But when these highly complex machines used to treat cancers go awry or when medical technicians and doctors fail to follow proper safety procedures, it results in suffering worse than the ailments radiation aims to cure. A litany of horror stories underscores the consequences when hospitals fail to provide safe radiation treatment to cancer patients. In many of these horror stories, poor software design, poor human-machine interfaces, and lack of proper training are root causes of the problems.

The deaths of Scott Jerome-Parks and Alexandra Jn-Charles, both patients of New York City hospitals, are prime examples of radiation treatments going awry. Jerome-Parks worked in southern Manhattan near the site of the World Trade Center attacks, and suspected that the tongue cancer he developed later was related to toxic dust that he came in contact with after the attacks. His prognosis was uncertain at first, but he had some reason to be optimistic, given the quality of the treatment provided by state-of-the-art linear accelerators at St. Vincent's Hospital, which he selected for his treatment. But after receiving erroneous dosages of radiation several times, his condition drastically worsened.

For the most part, state-of-the-art linear accelerators do in fact provide effective and safe care for cancer patients, and Americans safely receive an increasing amount of medical radiation each year. Radiation helps to diagnose and treat all sorts of cancers, saving many patients' lives in the process, and is administered safely to over half of all cancer patients. Whereas older machines were only capable of imaging a tumor in two dimensions and projecting straight beams of radiation, newer linear accelerators are capable of modeling cancerous tumors in three dimensions and shaping beams of radiation to conform to those shapes.

One of the most common issues with radiation therapy is finding ways to destroy cancerous cells while preserving healthy cells. Using this beam-shaping technique, radiation doesn't pass through as much healthy tissue to reach the cancerous areas. Hospitals advertised their new accelerators as being able to treat previously untreatable cancers because of the precision of the beam-shaping method. Using older machinery, cancers that were too close to important bodily structures were considered too dangerous to treat with radiation due to the imprecision of the equipment.

How, then, are radiation-related accidents increasing in frequency, given the advances in linear acceleration technology? In the cases of Jerome-Parks and Jn-Charles, a combination of machine malfunctions and user error led to these frightening mistakes. Jerome-Parks's brain stem and neck were exposed to excessive dosages of radiation on three separate occasions because of a computer error. The linear accelerator used to treat Jerome-Parks is known as a multi-leaf collimator, a newer, more powerful model that uses over a hundred metal "leaves" to adjust the shape and strength of the beam. The St. Vincent's hospital collimator was made by Varian Medical Systems, a leading supplier of radiation equipment.

Dr. Anthony M. Berson, St. Vincent's chief radiation oncologist, reworked Mr. Jerome Parks's radiation treatment plan to give more protection to his teeth. Nina Kalach, the medical physicist in charge of implementing Jerome-Parks's radiation treatment plan, used Varian software to revise the plan. State records show that as Ms. Kalach was trying to save her work, the computer began seizing up, displaying an error message. The error message asked if Ms. Kalach wanted to save her changes before the program aborted and she responded that she did. Dr. Berson approved the plan.

Six minutes after another computer crash, the first of several radioactive beams was turned on, followed by several additional rounds of radiation the next few days. After the third treatment, Ms. Kalach ran a test to verify that the treatment plan was carried out as prescribed, and found that the multileaf collimator, which was supposed to focus the beam precisely on Mr. Jerome Parks's tumor, was wide open. The patient's entire neck had been exposed and Mr. Jerome-Parks had seven times the prescribed dose of radiation.

As a result of the radiation overdose, Mr. Jerome-Parks's experienced deafness and near-blindness, ulcers in his mouth and throat, persistent nausea, and severe pain. His teeth were falling out, he couldn't swallow, and he was eventually unable to breathe. He died soon after, at the age of 43.

Jn-Charles's case was similarly tragic. A 32-year old mother of two from Brooklyn, she was diagnosed with an aggressive form of breast cancer, but her outlook seemed good after breast surgery and chemotherapy, with only 28 days of radiation treatments left to perform. However, the linear accelerator used at the Brooklyn hospital where Jn-Charles was treated was not a multi-leaf collimator, but instead a slightly older model, which uses a device known as a "wedge" to prevent radiation from reaching unintended areas of the body.

On the day of her 28th and final session, technicians realized that something had gone wrong. Jn-Charles's skin had slowly begun to peel and seemed to resist healing. When the hospital looked into the treatment to see why this could have happened, they discovered that the linear accelerator lacked the crucial command to insert the wedge, which must be programmed by the user. Technicians had failed to notice error messages on their screens indicating the missing wedge during each of the 27 sessions. This meant that Jn-Charles had been exposed to almost quadruple the normal amount of radiation during each of those 27 visits.

Ms. Jn-Charles's radiation overdose created a wound that would not heal despite numerous sessions in a hyperbaric chamber and multiple surgeries. Although the wound closed up over a year later, she died shortly afterwards.

It might seem that the carelessness or laziness of the medical technicians who administered treatment is primarily to blame in these cases, but other factors have contributed just as much. The complexity of new linear accelerator technology has not been accompanied with appropriate updates in software, training, safety procedures, and staffing. St. Vincent's hospital stated that system crashes similar to those involved in the improper therapy for Mr. Jerome-Parks "are not uncommon with the Varian software, and these issues have been communicated to Varian on numerous occasions."

Manufacturers of these machines boast that they can safely administer radiation treatment to more and more patients each day, but hospitals are rarely able to adjust their staffing to handle those workloads or increase the amount of training technicians receive before using newer machines. Medical technicians incorrectly assume that the new systems and software are going to work correctly, but in reality they have not been tested over long periods of time.

Many of these errors could have been detected if the machine operators were paying attention. In fact, many of the reported errors involve mistakes as simple and as egregious as treating patients for the wrong cancers; in one example, a brain cancer patient received radiation intended for breast cancer. Today's linear accelerators also lack some of the necessary safeguards given the amounts of radiation that they can deliver. For example, many linear accelerators are unable to alert users when a dosage of radiation far exceeds the necessary amount to effectively damage a cancerous tumor. Though responsibility ultimately rests with the technician, software programmers may not have designed their product with the technician's needs in mind.

Though the complexity of newer machines has exposed the inadequacy of the safety procedures hospitals employ for radiation treatments, the increasing number of patients receiving radiation due to the speed and increased capability of these machines has created other problems. Technicians at many of the hospitals reporting radiation-related errors reported being chronically overworked, often dealing with over a hundred patients per day. These already swamped medical technicians are not forced to check over the settings of the linear accelerators that they are handling, and errors that are introduced to the computer systems early on are difficult to detect. As a result, the same erroneous treatment may be administered repeatedly, until the technicians and doctors have a reason to check it. Often, the reason is a seriously injured patient.

Further complicating the issue is the fact that the total number of radiation-related accidents each year is essentially unknown. No single agency exists to collect data across the country on these accidents, and many states don't even require that accidents be reported. Even in states that do, hospitals are often reluctant to report errors that they've made, fearful that it will scare potential patients away, affecting their bottom lines. Some instances of hospital error are difficult to detect, since radiation-related cancer may appear a long while after the faulty treatment, and under-radiation doesn't result in any observable injury. Even in New York, which has one of the strictest accident reporting requirements in place and keeps reporting hospitals anonymous to encourage them to share their data, a significant portion of errors go unreported—perhaps even a majority of errors.

The problem is certainly not unique to New York. In New Jersey, 36 patients were over-radiated at a single hospital by an inexperienced team of technicians, and the mistakes continued for months in the absence of a system that detected treatment errors. Patients in Louisiana, Texas, and California repeatedly received incorrect dosages that led to other crippling ailments. Nor is the issue unique to the United States. In Panama, 28 patients at the National Cancer Institute received overdoses of radiation for various types of cancers. Doctors had ordered medical physicists to add a fifth "block," or metal sheet similar to the "leaves" in a multi-leaf collimator, to their linear accelerators, which were only designed to support four blocks. When the staff attempted to get the machine software to work with the extra block, the results were miscalculated dosages and over-radiated patients.

The lack of a central U.S. reporting and regulatory agency for radiation therapy means that in the event of a radiation-related mistake, all of the groups involved are able to avoid ultimate responsibility. Medical machinery and software manufacturers claim that it's the doctors and medical technicians' responsibility to properly use the machines, and the hospitals' responsibility to properly budget time and resources for training. Technicians claim that they are understaffed and overworked, and that there are no procedures in place to check their work and no time to do so even if there were. Hospitals claim that the newer machinery lacks the proper fail-safe mechanisms and that there is no room on already limited budgets for the training that equipment manufacturers claim is required.

Currently, the responsibility for regulating these incidents falls upon the states, which vary widely in their enforcement of reporting. Many states require no reporting at all, but even in a state like Ohio,

which requires reporting of medical mistakes within 15 days of the incident, these rules are routinely broken. Moreover, radiation technicians do not require a license in Ohio, as they do in many other states.

Dr. Fred A. Mettler, Jr., a radiation expert who has investigated radiation accidents worldwide, notes that "while there are accidents, you wouldn't want to scare people to death where they don't get needed radiation therapy." And it bears repeating that the vast majority of the time, radiation works, and saves some people from terminal cancer. But technicians, hospitals, equipment and software manufacturers, and regulators all need to collaborate to create a common set of safety procedures, software features, reporting standards, and certification requirements for technicians in order to reduce the number of radiation accidents.

Sources: Walt Bogdanich, "Medical Group Urges New Rules on Radiation," *The New York Times*, February 4, 2010; "As Technology Surges, Radiation Safeguards Lag," *The New York Times*, January 27, 2010; "Radiation Offers New Cures, and Ways to Do Harm," *The New York Times*, January 24, 2010; and "Case Studies: When Medical Radiation Goes Awry," *The New York Times*, January 21, 2010.

CASE STUDY QUESTIONS

1. What concepts in the chapter are illustrated in this case? What ethical issues are raised by radiation technology?

2. What management, organization, and technology factors were responsible for the problems detailed in this case? Explain the role of each.

3. Do you feel that any of the groups involved with this issue (hospital administrators, technicians, medical equipment and software manufacturers) should accept the majority of the blame for these incidents? Why or why not?

4. How would a central reporting agency that gathered data on radiation-related accidents help reduce the number of radiation therapy errors in the future?

5. If you were in charge of designing electronic software for a linear accelerator, what are some features you would include? Are there any features you would avoid?

PART TWO

Information Technology Infrastructure

Part Two provides the technical foundation for understanding information systems by examining hardware, software, database, and networking technologies along with tools and techniques for security and control. This part answers questions such as: What technologies do businesses today need to accomplish their work? What do I need to know about these technologies to make sure they enhance the performance of the firm? How are these technologies likely to change in the future? What technologies and procedures are required to ensure that systems are reliable and secure?

Chapter 5

IT Infrastructure and Emerging Technologies

BART SPEEDS UP WITH A NEW IT INFRASTRUCTURE

The Bay Area Rapid Transit (BART) is a heavy-rail public transit system that connects San Francisco to Oakland, California, and other neighboring cities to the east and south. BART has provided fast, reliable transportation for more than 35 years and now carries more than 346,000 passengers each day over 104 miles of track and 43 stations. It provides an alternative to driving on bridges and highways, decreasing travel time and the number of cars on the Bay Area's congested roads. It is the fifth busiest rapid transit system in the United States.

BART recently embarked on an ambitious modernization effort to overhaul stations, deploy new rail cars, and extend routes. This modernization effort also encompassed BART's information technology infrastructure. BART's information systems were no longer state-of-the art, and they were starting to affect its ability to provide good service. Aging homegrown financial and human resources systems could no longer provide information rapidly enough for making timely decisions, and they were too unreliable to support its 24/7 operations.

BART upgraded both its hardware and software. It replaced old legacy mainframe applications with Oracle's PeopleSoft Enterprise applications running on HP Integrity blade servers and the Oracle Enterprise Linux operating system. This configuration provides more flexibility and room to grow because BART is able to run the PeopleSoft software in conjunction with new applications it could not previously run.

BART wanted to create a high-availability IT infrastructure using grid computing where it could match computing and storage capacity more closely to actual demand. BART chose to run its applications on a cluster of servers using a grid architecture. Multiple operating environments share capacity and computing resources that can be provisioned, distributed, and redistributed as needed over the grid.

In most data centers, a distinct server is deployed for each application, and each server typically uses only a fraction of its capacity. BART uses virtualization to run multiple applications on the same server, increasing server capacity utilization to 50 percent or higher. This means fewer servers can be used to accomplish the same amount of work.

With blade servers, if BART needs more capacity, it can add another server to the main system. Energy usage is minimized because BART does not have to purchase computing capacity it doesn't need and the blade servers' stripped down modular design minimizes the use of physical space and energy.

By using less hardware and using existing computing resources more efficiently, BART's grid environment saves power and cooling costs. Consolidating applications onto a shared grid of server capacity is expected to reduce energy usage by about 20 percent.

Sources: David Baum, "Speeding into the Modern Age," *Profit*, February 2010; www.bart.gov, accessed June 5, 2010; and Steve Clouther, "The San Francisco Bay Area Rapid Transit Uses IBM Technology to Improve Safety and Reliability," ARC Advisory Group, October 7, 2009.

BART has been widely praised as a successful modern rapid transit system, but its operations and ability to grow where needed were hampered by an outdated IT infrastructure. BART's management felt the best solution was to invest in new hardware and software technologies that were more cost-effective, efficient, and energy-saving.

The chapter-opening diagram calls attention to important points raised by this case and this chapter. Management realized that in order to keep providing the level of service expected by Bay Area residents, it had to modernize its operations, including the hardware and software used for running the organization. The IT infrastructure investments it made had to support BART's business goals and contribute to improving its performance. Other goals included reducing costs and also "green" goals of reducing power and materials consumption.

By replacing its legacy software and computers with blade servers on a grid and more modern business software, BART was able to reduce wasted computer resources not used for processing, use existing resources more efficiently, and cut costs and power consumption. New software tools make it much easier to develop new applications and services. BART's IT infrastucture is easier to manage and capable of scaling to accommodate growing processing loads and new business opportunities. This case shows that the right hardware and software investments not only improve business performance but can also contribute to important social goals, such as conservation of power and materials.

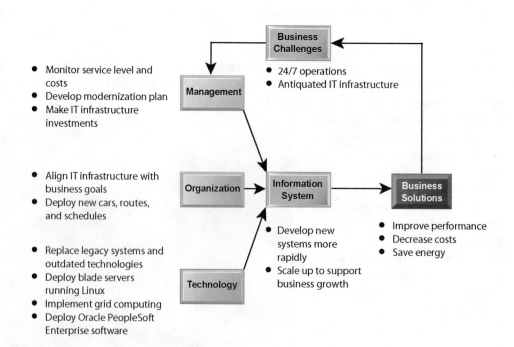

5.1   IT INFRASTRUCTURE

I n Chapter 1, we defined *information technology (IT) infrastructure* as the shared technology resources that provide the platform for the firm's specific information system applications. IT infrastructure includes investment in hardware, software, and services—such as consulting, education, and training—that are shared across the entire firm or across entire business units in the firm. A firm's IT infrastructure provides the foundation for serving customers, working with vendors, and managing internal firm business processes (see Figure 5-1).

Supplying U.S. firms with IT infrastructure (hardware and software) in 2010 is estimated to be a $1 trillion industry when telecommunications, networking equipment, and telecommunications services (Internet, telephone, and data transmission) are included. This does not include IT and related business process consulting services, which would add another $800 billion. Investments in infrastructure account for between 25 and 50 percent of information technology expenditures in large firms, led by financial services firms where IT investment is well over half of all capital investment (Weill et al., 2002).

DEFINING IT INFRASTRUCTURE

IT infrastructure consists of a set of physical devices and software applications that are required to operate the entire enterprise. But IT infrastructure is also a set of firmwide services budgeted by management and comprising both human and technical capabilities. These services include the following:

FIGURE 5-1   **CONNECTION BETWEEN THE FIRM, IT INFRASTRUCTURE, AND BUSINESS CAPABILITIES**

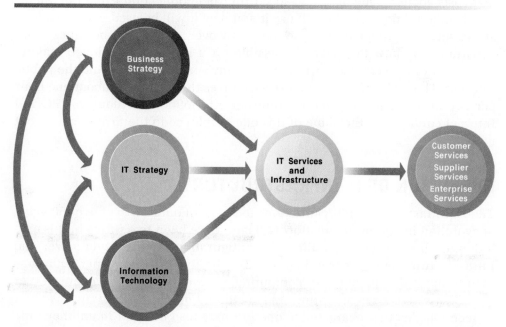

The services a firm is capable of providing to its customers, suppliers, and employees are a direct function of its IT infrastructure. Ideally, this infrastructure should support the firm's business and information systems strategy. New information technologies have a powerful impact on business and IT strategies, as well as the services that can be provided to customers.

- Computing platforms used to provide computing services that connect employees, customers, and suppliers into a coherent digital environment, including large mainframes, midrange computers, desktop and laptop computers, and mobile handheld devices.

- Telecommunications services that provide data, voice, and video connectivity to employees, customers, and suppliers.

- Data management services that store and manage corporate data and provide capabilities for analyzing the data.

- Application software services that provide enterprise-wide capabilities such as enterprise resource planning, customer relationship management, supply chain management, and knowledge management systems that are shared by all business units.

- Physical facilities management services that develop and manage the physical installations required for computing, telecommunications, and data management services.

- IT management services that plan and develop the infrastructure, coordinate with the business units for IT services, manage accounting for the IT expenditure, and provide project management services.

- IT standards services that provide the firm and its business units with policies that determine which information technology will be used, when, and how.

- IT education services that provide training in system use to employees and offer managers training in how to plan for and manage IT investments.

- IT research and development services that provide the firm with research on potential future IT projects and investments that could help the firm differentiate itself in the marketplace.

This "service platform" perspective makes it easier to understand the business value provided by infrastructure investments. For instance, the real business value of a fully loaded personal computer operating at 3 gigahertz that costs about $1,000 or a high-speed Internet connection is hard to understand without knowing who will use it and how it will be used. When we look at the services provided by these tools, however, their value becomes more apparent: The new PC makes it possible for a high-cost employee making $100,000 a year to connect to all the company's major systems and the public Internet. The high-speed Internet service saves this employee about one hour per day in reduced wait time for Internet information. Without this PC and Internet connection, the value of this one employee to the firm might be cut in half.

EVOLUTION OF IT INFRASTRUCTURE

The IT infrastructure in organizations today is an outgrowth of over 50 years of evolution in computing platforms. There have been five stages in this evolution, each representing a different configuration of computing power and infrastructure elements (see Figure 5-2). The five eras are general-purpose mainframe and minicomputer computing, personal computers, client/server networks, enterprise computing, and cloud and mobile computing.

Technologies that characterize one era may also be used in another time period for other purposes. For example, some companies still run traditional mainframe systems or use mainframe computers as massive servers supporting large Web sites and corporate enterprise applications.

FIGURE 5-2   ERAS IN IT INFRASTRUCTURE EVOLUTION

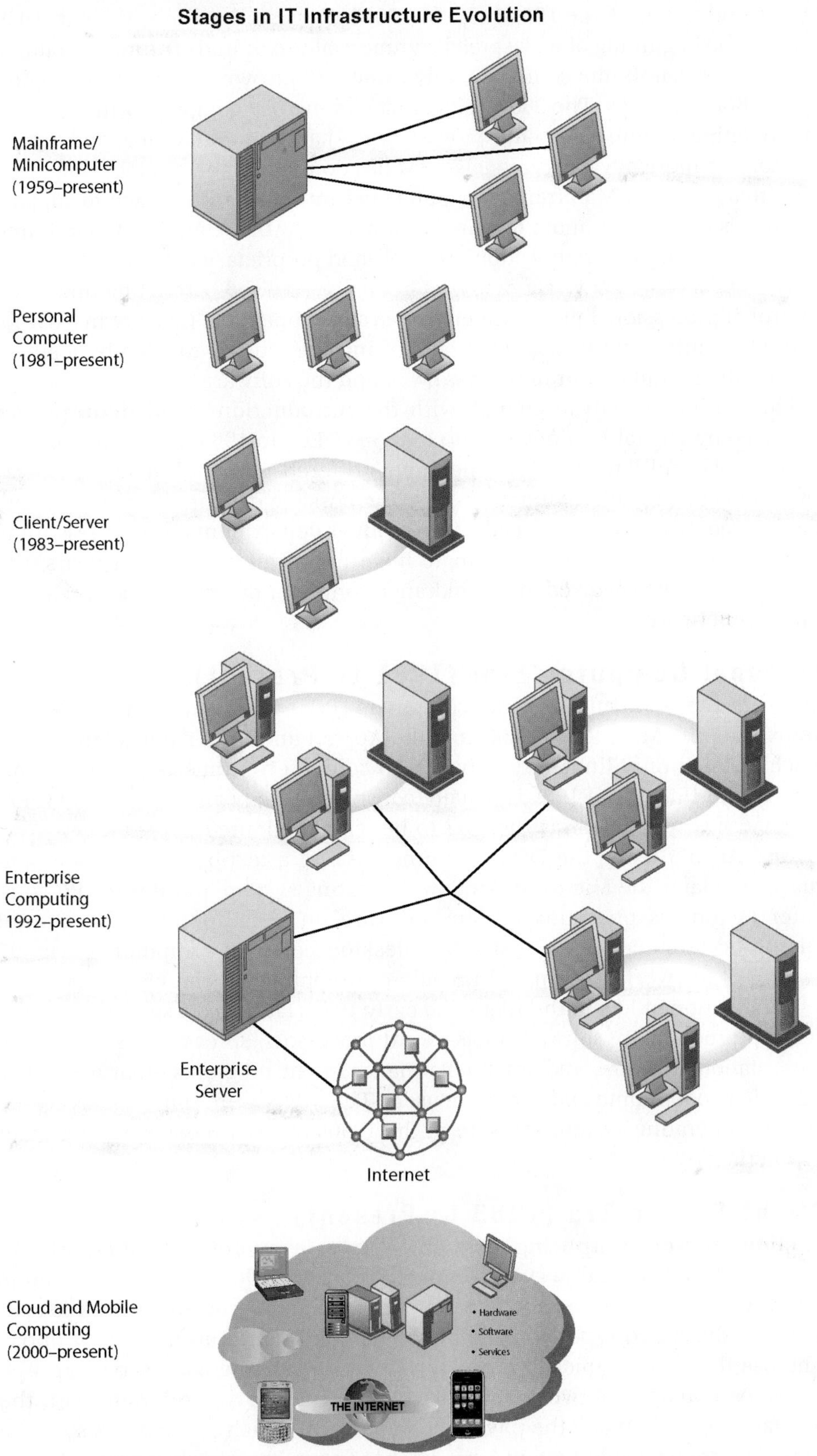

Illustrated here are the typical computing configurations characterizing each of the five eras of IT infrastructure evolution.

General-Purpose Mainframe and Minicomputer Era: (1959 to Present)

The introduction of the IBM 1401 and 7090 transistorized machines in 1959 marked the beginning of widespread commercial use of **mainframe** computers. In 1965, the mainframe computer truly came into its own with the introduction of the IBM 360 series. The 360 was the first commercial computer with a powerful operating system that could provide time sharing, multitasking, and virtual memory in more advanced models. IBM has dominated mainframe computing from this point on. Mainframe computers became powerful enough to support thousands of online remote terminals connected to the centralized mainframe using proprietary communication protocols and proprietary data lines.

The mainframe era was a period of highly centralized computing under the control of professional programmers and systems operators (usually in a corporate data center), with most elements of infrastructure provided by a single vendor, the manufacturer of the hardware and the software.

This pattern began to change with the introduction of **minicomputers** produced by Digital Equipment Corporation (DEC) in 1965. DEC minicomputers (PDP-11 and later the VAX machines) offered powerful machines at far lower prices than IBM mainframes, making possible decentralized computing, customized to the specific needs of individual departments or business units rather than time sharing on a single huge mainframe. In recent years, the minicomputer has evolved into a midrange computer or midrange server and is part of a network.

Personal Computer Era: (1981 to Present)

Although the first truly personal computers (PCs) appeared in the 1970s (the Xerox Alto, the MITS Altair 8800, and the Apple I and II, to name a few), these machines had only limited distribution to computer enthusiasts. The appearance of the IBM PC in 1981 is usually considered the beginning of the PC era because this machine was the first to be widely adopted by American businesses. At first using the DOS operating system, a text-based command language, and later the Microsoft Windows operating system, the **Wintel PC** computer (Windows operating system software on a computer with an Intel microprocessor) became the standard desktop personal computer. Today, 95 percent of the world's estimated 1.5 billion computers use the Wintel standard.

Proliferation of PCs in the 1980s and early 1990s launched a spate of personal desktop productivity software tools—word processors, spreadsheets, electronic presentation software, and small data management programs—that were very valuable to both home and corporate users. These PCs were standalone systems until PC operating system software in the 1990s made it possible to link them into networks.

Client/Server Era (1983 to Present)

In **client/server computing**, desktop or laptop computers called **clients** are networked to powerful **server** computers that provide the client computers with a variety of services and capabilities. Computer processing work is split between these two types of machines. The client is the user point of entry, whereas the server typically processes and stores shared data, serves up Web pages, or manages network activities. The term "server" refers to both the software application and the physical computer on which the network software runs. The server could be a mainframe, but today, server computers typically are more powerful versions of personal computers, based on inexpensive chips and often using multiple processors in a single computer box.

The simplest client/server network consists of a client computer networked to a server computer, with processing split between the two types of machines. This is called a *two-tiered client/server architecture*. Whereas simple client/server networks can be found in small businesses, most corporations have more complex, **multitiered** (often called **N-tier**) **client/server architectures** in which the work of the entire network is balanced over several different levels of servers, depending on the kind of service being requested (see Figure 5-3).

For instance, at the first level, a **Web server** will serve a Web page to a client in response to a request for service. Web server software is responsible for locating and managing stored Web pages. If the client requests access to a corporate system (a product list or price information, for instance), the request is passed along to an **application server**. Application server software handles all application operations between a user and an organization's back-end business systems. The application server may reside on the same computer as the Web server or on its own dedicated computer. Chapters 6 and 7 provide more detail on other pieces of software that are used in multitiered client/server architectures for e-commerce and e-business.

Client/server computing enables businesses to distribute computing work across a series of smaller, inexpensive machines that cost much less than minicomputers or centralized mainframe systems. The result is an explosion in computing power and applications throughout the firm.

Novell NetWare was the leading technology for client/server networking at the beginning of the client/server era. Today, Microsoft is the market leader with its **Windows** operating systems (Windows Server, Windows 7, Windows Vista, and Windows XP).

Enterprise Computing Era (1992 to Present)

In the early 1990s, firms turned to networking standards and software tools that could integrate disparate networks and applications throughout the firm into an enterprise-wide infrastructure. As the Internet developed into a trusted communications environment after 1995, business firms began seriously using the *Transmission Control Protocol/Internet Protocol (TCP/IP)* networking

FIGURE 5-3   A MULTITIERED CLIENT/SERVER NETWORK (N-TIER)

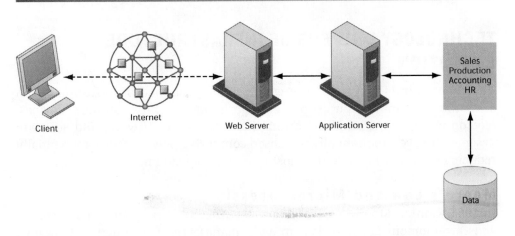

In a multitiered client/server network, client requests for service are handled by different levels of servers.

standard to tie their disparate networks together. We discuss TCP/IP in detail in Chapter 7.

The resulting IT infrastructure links different pieces of computer hardware and smaller networks into an enterprise-wide network so that information can flow freely across the organization and between the firm and other organizations. It can link different types of computer hardware, including mainframes, servers, PCs, mobile phones, and other handheld devices, and it includes public infrastructures such as the telephone system, the Internet, and public network services. The enterprise infrastructure also requires software to link disparate applications and enable data to flow freely among different parts of the business, such as enterprise applications (see Chapters 2 and 9) and Web services (discussed in Section 5.4).

Cloud and Mobile Computing Era (2000 to Present)

The growing bandwidth power of the Internet has pushed the client/server model one step further, towards what is called the "Cloud Computing Model." **Cloud computing** refers to a model of computing that provides access to a shared pool of computing resources (computers, storage, applications, and services), over a network, often the Internet. These "clouds" of computing resources can be accessed on an as-needed basis from any connected device and location. Currently, cloud computing is the fastest growing form of computing, with global revenue expected to reach close to $89 billion in 2011 and nearly $149 billion by 2014 according to Gartner Inc. technology consultants (Cheng and Borzo, 2010; Veverka, 2010).

Thousands or even hundreds of thousands computers are located in cloud data centers, where they can be accessed by desktop computers, laptop computers, netbooks, entertainment centers, mobile devices, and other client machines linked to the Internet, with both personal and corporate computing increasingly moving to mobile platforms. IBM, HP, Dell, and Amazon operate huge, scalable cloud computing centers that provide computing power, data storage, and high-speed Internet connections to firms that want to maintain their IT infrastructures remotely. Software firms such as Google, Microsoft, SAP, Oracle, and Salesforce.com sell software applications as services delivered over the Internet.

We discuss cloud computing in more detail in Section 5.3. The Learning Tracks include a table on Stages in IT Infrastructure Evolution, which compares each era on the infrastructure dimensions introduced.

TECHNOLOGY DRIVERS OF INFRASTRUCTURE EVOLUTION

The changes in IT infrastructure we have just described have resulted from developments in computer processing, memory chips, storage devices, telecommunications and networking hardware and software, and software design that have exponentially increased computing power while exponentially reducing costs. Let's look at the most important developments.

Moore's Law and Microprocessing Power

In 1965, Gordon Moore, the directory of Fairchild Semiconductor's Research and Development Laboratories, an early manufacturer of integrated circuits, wrote in *Electronics* magazine that since the first microprocessor chip was introduced in 1959, the number of components on a chip with the smallest manufacturing costs per component (generally transistors) had doubled each year. This assertion became the foundation of **Moore's Law**. Moore later reduced the rate of growth to a doubling every two years.

This law would later be interpreted in multiple ways. There are at least three variations of Moore's Law, none of which Moore ever stated: (1) the power of microprocessors doubles every 18 months; (2) computing power doubles every 18 months; and (3) the price of computing falls by half every 18 months.

Figure 5-4 illustrates the relationship between number of transistors on a microprocessor and millions of instructions per second (MIPS), a common measure of processor power. Figure 5-5 shows the exponential decline in the cost of transistors and rise in computing power. In 2010 for instance, and Intel 8-Core Xeon processor contains 2.3 billion transistors.

Exponential growth in the number of transistors and the power of processors coupled with an exponential decline in computing costs is likely to continue. Chip manufacturers continue to miniaturize components. Today's transistors should no longer be compared to the size of a human hair but rather to the size of a virus.

By using nanotechnology, chip manufacturers can even shrink the size of transistors down to the width of several atoms. **Nanotechnology** uses individual atoms and molecules to create computer chips and other devices that are thousands of times smaller than current technologies permit. Chip manufacturers are trying to develop a manufacturing process that could produce nanotube processors economically (Figure 5-6). IBM has just started making microprocessors in a production setting using this technology.

The Law of Mass Digital Storage

A second technology driver of IT infrastructure change is the Law of Mass Digital Storage. The world produces as much as 5 exabytes of unique information per year (an exabyte is a billion gigabytes, or 10^{18} bytes). The amount of digital information is roughly doubling every year (Lyman and Varian, 2003). Fortunately, the cost of storing digital information is falling at an exponential

FIGURE 5-4   MOORE'S LAW AND MICROPROCESSOR PERFORMANCE

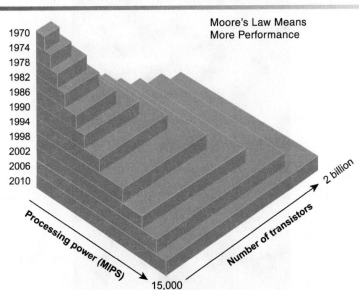

Packing over 2 billion transistors into a tiny microprocessor has exponentially increased processing power. Processing power has increased to over 500,000 MIPS (millions of instructions per second).

Sources: Intel, 2010; authors' estimate.

FIGURE 5-5   FALLING COST OF CHIPS

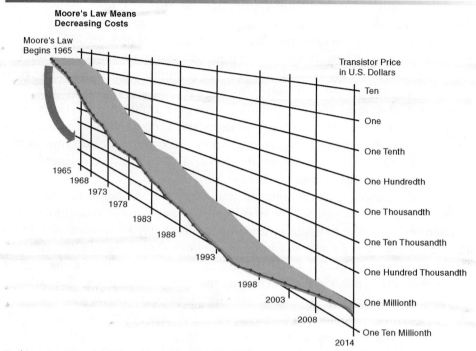

Packing more transistors into less space has driven down transistor cost dramatically as well as the cost of the products in which they are used.

Source: Intel, 2010; authors' estimates.

FIGURE 5-6   EXAMPLES OF NANOTUBES

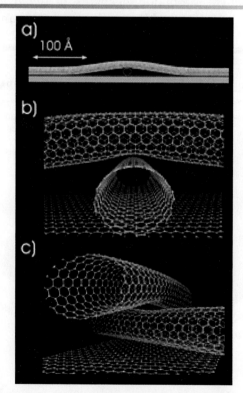

Nanotubes are tiny tubes about 10,000 times thinner than a human hair. They consist of rolled up sheets of carbon hexagons and have the potential uses as minuscule wires or in ultrasmall electronic devices and are very powerful conductors of electrical current.

rate of 100 percent a year. Figure 5-7 shows that the number of kilobytes that can be stored on magnetic media for $1 from 1950 to the present roughly doubled every 15 months.

Metcalfe's Law and Network Economics

Moore's Law and the Law of Mass Storage help us understand why computing resources are now so readily available. But why do people want more computing and storage power? The economics of networks and the growth of the Internet provide some answers.

Robert Metcalfe—inventor of Ethernet local area network technology—claimed in 1970 that the value or power of a network grows exponentially as a function of the number of network members. Metcalfe and others point to the *increasing returns to scale* that network members receive as more and more people join the network. As the number of members in a network grows linearly, the value of the entire system grows exponentially and continues to grow forever as members increase. Demand for information technology has been driven by the social and business value of digital networks, which rapidly multiply the number of actual and potential links among network members.

Declining Communications Costs and the Internet

A fourth technology driver transforming IT infrastructure is the rapid decline in the costs of communication and the exponential growth in the size of the

FIGURE 5-7   THE COST OF STORING DATA DECLINES EXPONENTIALLY 1950–2010

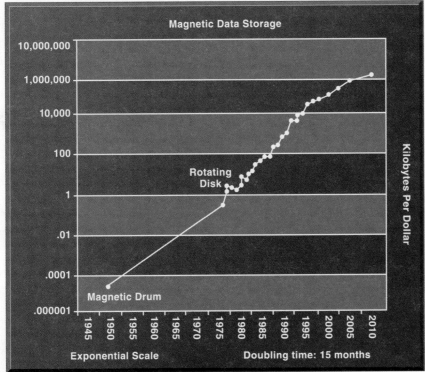

Since the first magnetic storage device was used in 1955, the cost of storing a kilobyte of data has fallen exponentially, doubling the amount of digital storage for each dollar expended every 15 months on average.

Sources: Kurzweil 2003; authors' estimates.

Internet. An estimated 1.8 billion people worldwide now have Internet access (Internet World Stats, 2010). Figure 5-8 illustrates the exponentially declining cost of communication both over the Internet and over telephone networks (which increasingly are based on the Internet). As communication costs fall toward a very small number and approach 0, utilization of communication and computing facilities explodes.

To take advantage of the business value associated with the Internet, firms must greatly expand their Internet connections, including wireless connectivity, and greatly expand the power of their client/server networks, desktop clients, and mobile computing devices. There is every reason to believe these trends will continue.

Standards and Network Effects

Today's enterprise infrastructure and Internet computing would be impossible—both now and in the future—without agreements among manufacturers and widespread consumer acceptance of **technology standards**. Technology standards are specifications that establish the compatibility of products and the ability to communicate in a network (Stango, 2004).

Technology standards unleash powerful economies of scale and result in price declines as manufacturers focus on the products built to a single standard. Without these economies of scale, computing of any sort would be far more expensive than is currently the case. Table 5-1 describes important standards that have shaped IT infrastructure.

Beginning in the 1990s, corporations started moving toward standard computing and communications platforms. The Wintel PC with the Windows operating system and Microsoft Office desktop productivity applications became the standard desktop and mobile client computing platform. Widespread adoption of Unix as the enterprise server operating system of choice made possible the replacement of proprietary and expensive mainframe infrastructures. In telecommunications, the Ethernet standard enabled PCs to connect together in small local area networks (LANs; see Chapter 7), and the TCP/IP standard enabled these LANs to be connected into firm-wide networks, and ultimately, to the Internet.

FIGURE 5-8   EXPONENTIAL DECLINES IN INTERNET COMMUNICATIONS COSTS

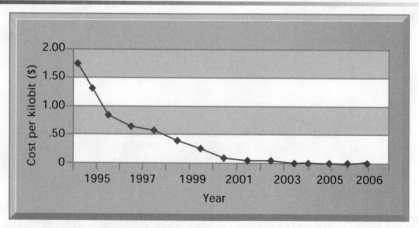

One reason for the growth in the Internet population is the rapid decline in Internet connection and overall communication costs. The cost per kilobit of Internet access has fallen exponentially since 1995. Digital subscriber line (DSL) and cable modems now deliver a kilobit of communication for a retail price of around 2 cents.

Source: Authors.

TABLE 5-1 SOME IMPORTANT STANDARDS IN COMPUTING

STANDARD	SIGNIFICANCE
American Standard Code for Information Interchange (ASCII) (1958)	Made it possible for computer machines from different manufacturers to exchange data; later used as the universal language linking input and output devices such as keyboards and mice to computers. Adopted by the American National Standards Institute in 1963.
Common Business Oriented Language (COBOL) (1959)	An easy-to-use software language that greatly expanded the ability of programmers to write business-related programs and reduced the cost of software. Sponsored by the Defense Department in 1959.
Unix (1969–1975)	A powerful multitasking, multiuser, portable operating system initially developed at Bell Labs (1969) and later released for use by others (1975). It operates on a wide variety of computers from different manufacturers. Adopted by Sun, IBM, HP, and others in the 1980s, it became the most widely used enterprise-level operating system.
Transmission Control Protocol/Internet Protocol (TCP/IP) (1974)	Suite of communications protocols and a common addressing scheme that enables millions of computers to connect together in one giant global network (the Internet). Later, it was used as the default networking protocol suite for local area networks and intranets. Developed in the early 1970s for the U.S. Department of Defense.
Ethernet (1973)	A network standard for connecting desktop computers into local area networks that enabled the widespread adoption of client/server computing and local area networks, and further stimulated the adoption of personal computers.
IBM/Microsoft/Intel Personal Computer (1981)	The standard Wintel design for personal desktop computing based on standard Intel processors and other standard devices, Microsoft DOS, and later Windows software. The emergence of this standard, low-cost product laid the foundation for a 25-year period of explosive growth in computing throughout all organizations around the globe. Today, more than 1 billion PCs power business and government activities every day.
World Wide Web (1989–1993)	Standards for storing, retrieving, formatting, and displaying information as a worldwide web of electronic pages incorporating text, graphics, audio, and video enables creation of a global repository of billions of Web pages.

5.2 INFRASTRUCTURE COMPONENTS

IT infrastructure today is composed of seven major components. Figure 5-9 illustrates these infrastructure components and the major vendors within each component category. These components constitute investments that must be coordinated with one another to provide the firm with a coherent infrastructure.

In the past, technology vendors supplying these components were often in competition with one another, offering purchasing firms a mixture of incompatible, proprietary, partial solutions. But increasingly the vendor firms have been forced by large customers to cooperate in strategic partnerships with one another. For instance, a hardware and services provider such as IBM cooperates with all the major enterprise software providers, has strategic relationships with system integrators, and promises to work with whichever database products its client firms wish to use (even though it sells its own database management software called DB2).

COMPUTER HARDWARE PLATFORMS

U.S. firms will spend about $109 billion in 2010 on computer hardware. This component includes client machines (desktop PCs, mobile computing devices such as netbooks and laptops but not including iPhones or BlackBerrys) and

FIGURE 5-9   **THE IT INFRASTRUCTURE ECOSYSTEM**

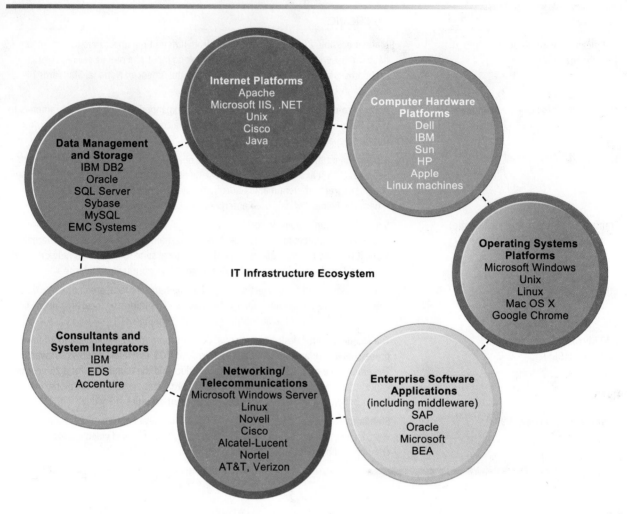

There are seven major components that must be coordinated to provide the firm with a coherent IT infrastructure. Listed here are major technologies and suppliers for each component.

server machines. The client machines use primarily Intel or AMD micro-processors. In 2010, there will be about 90 million PCs sold to U.S. customers (400 million worldwide) (Gartner, 2010).

The server market uses mostly Intel or AMD processors in the form of blade servers in racks, but also includes Sun SPARC microprocessors and IBM POWER chips specially designed for server use. **Blade servers**, which we discussed in the chapter-opening case, are ultrathin computers consisting of a circuit board with processors, memory, and network connections that are stored in racks. They take up less space than traditional box-based servers. Secondary storage may be provided by a hard drive in each blade server or by external mass-storage drives.

The marketplace for computer hardware has increasingly become concentrated in top firms such as IBM, HP, Dell, and Sun Microsystems (acquired by Oracle), and three chip producers: Intel, AMD, and IBM. The industry has collectively settled on Intel as the standard processor, with major exceptions in the server market for Unix and Linux machines, which might use Sun or IBM Unix processors.

Mainframes have not disappeared. The mainframe market has actually grown steadily over the last decade, although the number of providers has dwindled to one: IBM. IBM has also repurposed its mainframe systems so they can be used as giant servers for massive enterprise networks and corporate Web sites. A single IBM mainframe can run up to 17,000 instances of Linux or Windows server software and is capable of replacing thousands of smaller blade servers (see the discussion of virtualization in Section 5.3).

OPERATING SYSTEM PLATFORMS

In 2010, Microsoft Windows comprises about 75 percent of the server operating system market, with 25 percent of corporate servers using some form of the **Unix** operating system or **Linux**, an inexpensive and robust open source relative of Unix. Microsoft Windows Server is capable of providing enterprise-wide operating system and network services, and appeals to organizations seeking Windows-based IT infrastructures (IDC, 2010).

Unix and Linux are scalable, reliable, and much less expensive than mainframe operating systems. They can also run on many different types of processors. The major providers of Unix operating systems are IBM, HP, and Sun, each with slightly different and partially incompatible versions.

At the client level, 90 percent of PCs use some form of Microsoft Windows **operating system** (such as Windows 7, Windows Vista, or Windows XP) to manage the resources and activities of the computer. However, there is now a much greater variety of operating systems than in the past, with new operating systems for computing on handheld mobile digital devices or cloud-connected computers.

Google's **Chrome OS** provides a lightweight operating system for cloud computing using netbooks. Programs are not stored on the user's PC but are used over the Internet and accessed through the Chrome Web browser. User data resides on servers across the Internet. Microsoft has introduced the *Windows Azure* operating system for its cloud services and platform. **Android** is a mobile operating system developed by Android, Inc. (purchased by Google) and later the Open Handset Alliance as a flexible, upgradeable mobile device platform.

Conventional client operating system software is designed around the mouse and keyboard, but increasingly becoming more natural and intuitive by using touch technology. *IPhone OS*, the operating system for the phenomenally popular Apple iPad, iPhone, and iPod Touch, features a **multitouch** interface, where users use their fingers to manipulate objects on the screen. The Interactive Session on Technology explores the implications of using multitouch to interact with the computer.

ENTERPRISE SOFTWARE APPLICATIONS

In addition to software for applications used by specific groups or business units, U.S. firms will spend about $165 billion in 2010 on software for enterprise applications that are treated as components of IT infrastructure. We introduced the various types of enterprise applications in Chapter 2, and Chapter 9 provides a more detailed discussion of each.

The largest providers of enterprise application software are SAP and Oracle (which acquired PeopleSoft). Also included in this category is middleware software supplied by vendors such as BEA for achieving firmwide integration by linking the firm's existing application systems. Microsoft is attempting to move into the lower ends of this market by focusing on small and medium-sized businesses that have not yet implemented enterprise applications.

INTERACTIVE SESSION: TECHNOLOGY

NEW TO THE TOUCH

When Steve Jobs first demonstrated "the pinch"—the two-finger gesture for zooming in and out of photos and Web pages on the iPhone, he not only shook up the mobile phone industry—the entire digital world took notice. The Apple iPhone's multitouch features dramatized new ways of using touch to interact with software and devices.

Touch interfaces are not new. People use them every day to get money from ATMs or to check into flights at airport kiosks. Academic and commercial researchers have been working on multitouch technology for years. What Apple did was to make multitouch more exciting and relevant, popularizing it just as it did in the 1980s with the mouse and the graphical user interface. (These had also been invented elsewhere.)

Multitouch interfaces are potentially more versatile than single-touch interfaces. They allow you to use one or more fingers to perform special gestures that manipulate lists or objects on a screen without moving a mouse, pressing buttons, turning scroll wheels, or striking keys. They take different actions depending on how many fingers they detect and which gestures a user performs. Multitouch gestures are easier to remember than commands because they are based on ingrained human movements that do not have to be learned, scientists say.

The iPhone's Multi-Touch display and software lets you control everything using only your fingers. A panel underneath the display's glass cover senses your touch using electrical fields. It then transmits that information to a LCD screen below it. Special software recognizes multiple simultaneous touch points, (as opposed to the single-touch screen, which recognizes only one touch point.) You can quickly move back and forth through a series of Web pages or photos by "swiping," or placing three fingers on the screen and moving them rapidly sideways. By pinching the image, you can shrink or expand a photo.

Apple has made a concerted effort to provide multitouch features in all of its product categories, but many other consumer technology companies have adopted multitouch for some of their products. Synaptics, a leading supplier of touchpads for laptop makers who compete with Apple, has announced that it is incorporating several multitouch features into its touchpads.

Microsoft's Windows 7 operating system sports multitouch features: When you pair Windows 7 with a touch-screen PC, you can browse online newspapers, flick through photo albums, and shuffle files and folders using nothing but your fingers. To zoom in on something on the screen of a multitouch-compatible PC, you would place two fingers on the screen and spread them apart. To right-click a file, touch it with one finger and tap the screen with a second.

A number of Microsoft Windows PCs have touch screens, with a few Windows laptops emulating some of the multitouch features of Apple computers and handhelds. Microsoft's Surface computer runs on Windows 7 and lets its business customers use multitouch in a table-top display. Customers of hotels, casinos, and retail stores will be able to use multitouch finger gestures to move around digital objects such as photos, to play games, and to browse through product options. The Dell Latitude XT tablet PC uses multitouch, which is helpful to people who can't grasp a mouse and want the functionality of a traditional PC. They can use a finger or a stylus instead. The Android operating system for smartphones has native support for multi-touch, and handsets such as the HTC Desire, Nexus One, and the Motorola Droid have this capability.

Hewlett-Packard (HP) now has laptops and desktops that use touch technology. Its TouchSmart computer lets you use two fingers at once to manipulate images on the screen or to make on-screen gestures designating specific commands without using cursors or scroll bars. To move an object, you touch it with a finger and drag it to its new location. Sliding your finger up and down or sideways smoothly scrolls the display.

The TouchSmart makes it possible for home users to engage in a new type of casual computing—putting on music while preparing dinner, quickly searching for directions before leaving the house, or leaving written, video, or audio memos for family members. Both consumers and businesses have found other uses as well. According to Alan Reed, HP's vice president and general manager for Business Desktops, "There is untapped potential for touch technology in the business marketplace to engage users in a way that has never been done before."

Chicago's O'Hare Airport integrated a group of TouchSmart PCs into "Explore Chicago" tourist kiosks, allowing visitors to check out a virtual Visitor's Center. TouchSmart computing helped an autistic student to speak to and communicate with others for the first time in the 14 years of his life. Without using the TouchSmart PC's wireless keyboard and mouse, users can hold video chats with remote workers through a built-in Webcam and microphone, access e-mail and the Internet, and manage contacts, calendar items, and photos.

Touch-enabled PCs could also appeal to elementary schools seeking an easy-to-use computer for students in early grades, or a wall-mountable information kiosk-type device for parents and visitors. Customers might use touch to place orders with a retailer, conduct virtual video service calls, or to teach or utilize social networking for business.

It's too early to know if the new multitouch interface will ever be as popular as the mouse-driven graphical user interface. Although putting ones fingers on the screen is the ultimate measure of "cool" in the cell phone market, a "killer application" for touch on the PC has not yet emerged. But it's already evident that touch has real advantages on devices where a mouse isn't possible or convenient to use, or the decades-old interface of menus and folders is too cumbersome.

Sources: Claire Cain Miller, "To Win Over Today's Users, Gadgets Have to be Touchable," *The New York Times*, September 1, 2010; Katherine Boehret, "Apple Adds Touches to Its Mac Desktops," *The Wall Street Journal*, August 4, 2010; Ashlee Vance, " Tech Industry Catches Its Breath," *The New York Times*, February 17, 2010; Kathy Sandler, "The Future of Touch," *The Wall Street Journal*, June 2, 2009; Suzanne Robitaille, "Multitouch to the Rescue?" Suite101.com, January 22, 2009; and Eric Lai, "HP Aims TouchSmart Desktop PC at Businesses," *Computerworld*, August 1, 2009.

CASE STUDY QUESTIONS

1. What problems does multitouch technology solve?
2. What are the advantages and disadvantages of a multitouch interface? How useful is it? Explain.
3. Describe three business applications that would benefit from a multitouch interface.
4. What management, organization, and technology issues must be addressed if you or your business was considering systems and computers with multitouch interfaces?

MIS IN ACTION

1. Describe what you would do differently on your PC if it had multitouch capabilities. How much difference would multitouch make in the way you use your computer?

DATA MANAGEMENT AND STORAGE

Enterprise database management software is responsible for organizing and managing the firm's data so that they can be efficiently accessed and used. Chapter 6 describes this software in detail. The leading database software providers are IBM (DB2), Oracle, Microsoft (SQL Server), and Sybase (Adaptive Server Enterprise), which supply more than 90 percent of the U.S. database software marketplace. MySQL is a Linux open source relational database product now owned by Oracle Corporation.

The physical data storage market is dominated by EMC Corporation for large-scale systems, and a small number of PC hard disk manufacturers led by Seagate, Maxtor, and Western Digital.

Digital information is estimated to be growing at 1.2 zettabytes a year. All the tweets, blogs, videos, e-mails, and Facebook postings as well as traditional corporate data add up in 2010 to several thousand Libraries of Congress (EMC Corporation, 2010).

With the amount of new digital information in the world growing so rapidly, the market for digital data storage devices has been growing at more than 15 percent annually over the last five years. In addition to traditional disk arrays and tape libraries, large firms are turning to network-based storage technologies. **Storage area networks (SANs)** connect multiple storage devices on a separate high-speed network dedicated to storage. The SAN creates a large central pool of storage that can be rapidly accessed and shared by multiple servers.

NETWORKING/TELECOMMUNICATIONS PLATFORMS

U.S. firms spend $100 billon a year on networking and telecommunications hardware and a huge $700 billion on networking services (consisting mainly of telecommunications and telephone company charges for voice lines and Internet access; these are not included in this discussion). Chapter 7 is devoted to an in-depth description of the enterprise networking environment, including the Internet. Windows Server is predominantly used as a local area network operating system, followed by Linux and Unix. Large enterprise wide area networks primarily use some variant of Unix. Most local area networks, as well as wide area enterprise networks, use the TCP/IP protocol suite as a standard (see Chapter 7).

The leading networking hardware providers are Cisco, Alcatel-Lucent, Nortel, and Juniper Networks. Telecommunications platforms are typically provided by telecommunications/telephone services companies that offer voice and data connectivity, wide area networking, wireless services, and Internet access. Leading telecommunications service vendors include AT&T and Verizon (see the Chapter 3 opening case). This market is exploding with new providers of cellular wireless, high-speed Internet, and Internet telephone services.

INTERNET PLATFORMS

Internet platforms overlap with, and must relate to, the firm's general networking infrastructure and hardware and software platforms. U.S. firms spent an estimated $40 billion annually on Internet-related infrastructure. These expenditures were for hardware, software, and management services to support a firm's Web site, including Web hosting services, routers, and cabling or wireless equipment. A **Web hosting service** maintains a large Web server, or series of servers, and provides fee-paying subscribers with space to maintain their Web sites.

The Internet revolution created a veritable explosion in server computers, with many firms collecting thousands of small servers to run their Internet operations. Since then there has been a steady push toward server consolidation, reducing the number of server computers by increasing the size and power of each. The Internet hardware server market has become increasingly concentrated in the hands of IBM, Dell, and HP/Compaq, as prices have fallen dramatically.

The major Web software application development tools and suites are supplied by Microsoft (Microsoft Expression Web, SharePoint Designer, and the Microsoft .NET family of development tools); Oracle-Sun (Sun's Java is the most widely used tool for developing interactive Web applications on both the server and client sides); and a host of independent software developers, including Adobe (Flash and text tools like Acrobat), and Real Media (media software). Chapter 7 describes the components of the firm's Internet platform in greater detail.

CONSULTING AND SYSTEM INTEGRATION SERVICES

Today, even a large firm does not have the staff, the skills, the budget, or the necessary experience to deploy and maintain its entire IT infrastructure. Implementing a new infrastructure requires (as noted in Chapters 3 and 14) significant changes in business processes and procedures, training and education, and software integration. Leading consulting firms providing this expertise include Accenture, IBM Global Services, HP Enterprise Services, Infosys, and Wipro Technologies.

Software integration means ensuring the new infrastructure works with the firm's older, so-called legacy systems and ensuring the new elements of the infrastructure work with one another. **Legacy systems** are generally older transaction processing systems created for mainframe computers that continue to be used to avoid the high cost of replacing or redesigning them. Replacing these systems is cost prohibitive and generally not necessary if these older systems can be integrated into a contemporary infrastructure.

5.3   CONTEMPORARY HARDWARE PLATFORM TRENDS

The exploding power of computer hardware and networking technology has dramatically changed how businesses organize their computing power, putting more of this power on networks and mobile handheld devices. We look at seven hardware trends: the emerging mobile digital platform, grid computing, virtualization, cloud computing, green computing, high-performance/power-saving processors, and autonomic computing.

THE EMERGING MOBILE DIGITAL PLATFORM

Chapter 1 pointed out that new mobile digital computing platforms have emerged as alternatives to PCs and larger computers. Cell phones and smartphones such as the BlackBerry and iPhone have taken on many functions of handheld computers, including transmission of data, surfing the Web, transmitting e-mail and instant messages, displaying digital content, and exchanging data with internal corporate systems. The new mobile platform also includes small low-cost lightweight subnotebooks called **netbooks** optimized for wireless communication and Internet access, with core computing functions such as word processing; tablet computers such as the iPad; and digital e-book readers such as Amazon's Kindle with some Web access capabilities.

In a few years, smartphones, netbooks, and tablet computers will be the primary means of accessing the Internet, with business computing moving increasingly from PCs and desktop machines to these mobile devices. For example, senior executives at General Motors are using smartphone applications that drill down into vehicle sales information, financial performance, manufacturing metrics, and project management status. At medical device maker Astra Tech, sales reps use their smartphones to access Salesforce.com customer relationship management (CRM) applications and sales data, checking data on sold and returned products and overall revenue trends before meeting with customers.

GRID COMPUTING

Grid computing, involves connecting geographically remote computers into a single network to create a virtual supercomputer by combining the computational power of all computers on the grid. Grid computing takes advantage of the fact that most computers use their central processing units on average only 25 percent of the time for the work they have been assigned, leaving these idle resources available for other processing tasks. Grid computing was impossible until high-speed Internet connections enabled firms to connect remote machines economically and move enormous quantities of data.

Grid computing requires software programs to control and allocate resources on the grid. Client software communicates with a server software application. The server software breaks data and application code into chunks that are then parceled out to the grid's machines. The client machines perform their traditional tasks while running grid applications in the background.

The business case for using grid computing involves cost savings, speed of computation, and agility, as noted in the chapter-opening case. The chapter-opening case shows that by running its applications on clustered servers on a grid, BART eliminated unused computer resources, used existing resources more efficiently, and reduced costs and power consumption.

VIRTUALIZATION

Virtualization is the process of presenting a set of computing resources (such as computing power or data storage) so that they can all be accessed in ways that are not restricted by physical configuration or geographic location. Virtualization enables a single physical resource (such as a server or a storage device) to appear to the user as multiple logical resources. For example, a server or mainframe can be configured to run many instances of an operating system so that it acts like many different machines. Virtualization also enables multiple physical resources (such as storage devices or servers) to appear as a single logical resource, as would be the case with storage area networks or grid computing. Virtualization makes it possible for a company to handle its computer processing and storage using computing resources housed in remote locations. VMware is the leading virtualization software vendor for Windows and Linux servers. Microsoft offers its own Virtual Server product and has built virtualization capabilities into the newest version of Windows Server.

Business Benefits of Virtualization

By providing the ability to host multiple systems on a single physical machine, virtualization helps organizations increase equipment utilization rates, conserving data center space and energy usage. Most servers run at just 15-20 percent of capacity, and virtualization can boost server utilization rates to 70 percent or higher. Higher utilization rates translate into fewer computers required to process the same amount of work, as illustrated by BART's experience with virtualization in the chapter-opening case.

In addition to reducing hardware and power expenditures, virtualization allows businesses to run their legacy applications on older versions of an operating system on the same server as newer applications. Virtualization also facilitates centralization and consolidation of hardware administration. It is now possible for companies and individuals to perform all of their computing work using a virtualized IT infrastructure, as is the case with cloud computing. We now turn to this topic.

CLOUD COMPUTING

Earlier in this chapter, we introduced cloud computing, in which firms and individuals obtain computer processing, storage, software, and other services as a pool of virtualized resources over a network, primarily the Internet. These resources are made available to users, based on their needs, irrespective of their physical location or the location of the users themselves. The U.S. National Institute of Standards and Technology (NIST) defines cloud computing as having the following essential characteristics (Mell and Grance, 2009):

- **On-demand self-service:** Individuals can obtain computing capabilities such as server time or network storage on their own.

- **Ubiquitous network access:** Individuals can use standard network and Internet devices, including mobile platforms, to access cloud resources.

- **Location independent resource pooling:** Computing resources are pooled to serve multiple users, with different virtual resources dynamically assigned according to user demand. The user generally does not know where the computing resources are located.

- **Rapid elasticity:** Computing resources can be rapidly provisioned, increased, or decreased to meet changing user demand.

- **Measured service:** Charges for cloud resources are based on amount of resources actually used.

Cloud computing consists of three different types of services:

- **Cloud infrastructure as a service:** Customers use processing, storage, networking, and other computing resources from cloud service providers to run their information systems. For example, Amazon uses the spare capacity of its IT infrastructure to provide a broadly based cloud environment selling IT infrastructure services. These include its Simple Storage Service (S3) for storing customers' data and its Elastic Compute Cloud (EC2) service for running their applications. Users pay only for the amount of computing and storage capacity they actually use.

- **Cloud platform as a service:** Customers use infrastructure and programming tools hosted by the service provider to develop their own applications. For example, IBM offers a Smart Business Application Development & Test service for software development and testing on the IBM Cloud. Another example is Salesforce.com's Force.com, described in the chapter-ending case study, which allows developers to build applications that are hosted on its servers as a service.

- **Cloud software as a service:** Customers use software hosted by the vendor on the vendor's hardware and delivered over a network. Leading examples are Google Apps, which provides common business applications online and Salesforce.com, which also leases CRM and related software services over the Internet. Both charge users an annual subscription fee, although Google Apps also has a pared-down free version. Users access these applications from a Web browser, and the data and software are maintained on the providers' remote servers.

A cloud can be private or public. A **public cloud** is maintained by an external service provider, such as Amazon Web Services, accessed through the Internet, and available to the general public. A **private cloud** is a proprietary network or a data center that ties together servers, storage, networks, data, and applications as a set of virtualized services that are shared by users inside a company. Like public clouds, private clouds are able to allocate storage, computing power, or other resources seamlessly to provide computing resources on an as-needed basis. Financial institutions and health care providers are likely to gravitate toward private clouds because these organizations handle so much sensitive financial and personal data. We discuss cloud security issues in Chapter 8.

Since organizations using cloud computing generally do not own the infrastructure, they do not have to make large investments in their own hardware and software. Instead, they purchase their computing services from remote providers and pay only for the amount of computing power they actually use (**utility computing**) or are billed on a monthly or annual subscription basis. The term **on-demand computing** has also been used to describe such services.

For example, Envoy Media Group, a direct-marketing firm that offers highly-targeted media campaigns across multiple channels, including TV, radio, and Internet, hosts its entire Web presence on Azimuth Web Services. The "pay as you go" pricing structure allows the company to quickly and painlessly add servers where they are needed without large investments in hardware. Cloud computing reduced costs about 20 percent because Envoy no longer had to maintain its own hardware or IT personnel.

Cloud computing has some drawbacks. Unless users make provisions for storing their data locally, the responsibility for data storage and control is in the hands of the provider. Some companies worry about the security risks related to entrusting their critical data and systems to an outside vendor that also works with other companies. There are also questions of system reliability. Companies expect their systems to be available 24/7 and do not want to suffer any loss of business capability if their IT infrastructures malfunction. When Amazon's cloud went down in December 2009, subscribers on the U.S. east coast were unable to use their systems for several hours. Another limitation of cloud computing is the possibility of making users dependent on the cloud computing provider.

There are some who believe that cloud computing represents a sea change in the way computing will be performed by corporations as business computing shifts out of private data centers into cloud services (Carr, 2008). This remains a matter of debate. Cloud computing is more immediately appealing to small and medium-sized businesses that lack resources to purchase and own their own hardware and software. However, large corporations have huge investments in complex proprietary systems supporting unique business processes, some of which give them strategic advantages. For them, the most likely scenario is a hybrid computing model where firms use their own infrastructure for their most essential core activities and adopt public cloud computing for less-critical systems or for additional processing capacity during peak business periods. Cloud computing will gradually shift firms from having a fixed infrastructure capacity toward a more flexible infrastructure, some of it owned by the firm, and some of it rented from giant computer centers owned by computer hardware vendors.

GREEN COMPUTING

By curbing hardware proliferation and power consumption, virtualization has become one of the principal technologies for promoting green computing. **Green computing** or **green IT**, refers to practices and technologies for designing, manufacturing, using, and disposing of computers, servers, and associated devices such as monitors, printers, storage devices, and networking and communications systems to minimize impact on the environment.

Reducing computer power consumption has been a very high "green" priority. As companies deploy hundreds or thousands of servers, many are spending almost as much on electricity to power and cool their systems as they did on purchasing the hardware. The U.S. Environmental Protection Agency estimates that data centers will use more than 2 percent of all U.S. electrical power by 2011. Information technology is believed to contribute about 2 percent of the world's greenhouse gases. Cutting power consumption in data centers has become both a serious business and environmental challenge. The Interactive Session on Organizations examines this problem.

AUTONOMIC COMPUTING

With large systems encompassing many thousands of networked devices, computer systems have become so complex today that some experts believe they may not be manageable in the future. One approach to dealing with this problem is to employ autonomic computing. **Autonomic computing** is an industry-wide effort to develop systems that can configure themselves, optimize and tune themselves, heal themselves when broken, and protect themselves from outside intruders and self-destruction.

You can glimpse a few of these capabilities in desktop systems. For instance, virus and firewall protection software are able to detect viruses on PCs, automatically defeat the viruses, and alert operators. These programs can be updated automatically as the need arises by connecting to an online virus protection service such as McAfee. IBM and other vendors are starting to build autonomic features into products for large systems.

HIGH-PERFORMANCE AND POWER-SAVING PROCESSORS

Another way to reduce power requirements and hardware sprawl is to use more efficient and power-saving processors. Contemporary microprocessors now feature multiple processor cores (which perform the reading and execution of computer instructions) on a single chip. A **multicore processor** is an integrated circuit to which two or more processor cores have been attached for enhanced performance, reduced power consumption, and more efficient simultaneous processing of multiple tasks. This technology enables two or more processing engines with reduced power requirements and heat dissipation to perform tasks faster than a resource-hungry chip with a single processing core. Today you'll find dual-core and quad-core processors in PCs and servers with 8-, 10-, 12-, and 16-core processors.

Intel and other chip manufacturers have also developed microprocessors that minimize power consumption. Low power consumption is essential for prolonging battery life in smartphones, netbooks, and other mobile digital devices. You will now find highly power-efficient microprocessors, such as ARM, Apple's A4 processor, and Intel's Atom in netbooks, digital media players, and smartphones. The A4 processor used in the latest version of the iPhone and the iPad consumes approximately 500–800 milliwatts of power, about 1/50 to 1/30 the power consumption of a laptop dual-core processor.

INTERACTIVE SESSION: ORGANIZATIONS

IS GREEN COMPUTING GOOD FOR BUSINESS?

Computer rooms are becoming too hot to handle. Data-hungry tasks such as video on demand, downloading music, exchanging photos, and maintaining Web sites require more and more power-hungry machines. Power and cooling costs for data centers have skyrocketed by more than 800 percent since 1996, with U.S. enterprise data centers predicted to spend twice as much on energy costs as on hardware over the next five years.

The heat generated from rooms full of servers is causing equipment to fail. Some organizations spend more money to keep their data centers cool than they spend to lease the property itself. It's a vicious cycle, as companies must pay to power their servers, and then pay again to keep them cool and operational. Cooling a server requires roughly the same number of kilowatts of energy as running one. All this additional power consumption has a negative impact on the environment and as well as corporate operating costs.

Some of the world's most prominent firms are tackling their power consumption issues with one eye toward saving the environment and the other toward saving dollars. Google and Microsoft are building data centers that take advantage of hydroelectric power. Hewlett-Packard is working on a series of technologies to reduce the carbon footprint of data centers by 75 percent and, with new software and services, to measure energy use and carbon emissions. It reduced its power costs by 20 to 25 percent through consolidation of servers and data centers.

Microsoft's San Antonio data center deploys sensors that measure nearly all power consumption, recycles water used in cooling, and uses internally-developed power management software. Microsoft is also trying to encourage energy-saving software practices by charging business units by the amount of power they consume in the data enter rather than the space they take up on the floor.

None of these companies claim that their efforts will save the world, but they do demonstrate recognition of a growing problem and the commencement of the green computing era. And since these companies' technology and processes are more efficient than most other companies, using their online software services in place of in-house software may also count as a green investment.

PCs typically stay on more than twice the amount of time they are actually being used each day. According to a report by the Alliance to Save Energy, a company with 10,000 personal computer desktops will spend more than $165,000 per year in electricity bills if these machines are left on all night. The group estimates that this practice is wasting around $1.7 billion each year in the United States alone.

Although many companies establish default PC power management settings, about 70 percent of employees turn these settings off. PC power management software from BigFix, 1E NightWatchman, and Verdiem locks PC power settings and automatically powers PCs up right before employees arrive for work in the morning.

Miami-Dade County public schools cut the time its PCs were on from 21 hours to 10.3 hours daily by using BigFix to centrally control PC power settings. City University of New York adopted Verdiem's Surveyor software to turn off its 20,000 PCs when they are inactive at night. Surveyor has trimmed 10 percent from CUNY's power bills, creating an annual savings of around $320,000.

Virtualization is a highly effective tool for cost-effective green computing because it reduces the number of servers and storage resources in the firm's IT infrastructure. Fulton County, Georgia, which provides services for 988,000 citizens, scrutinizes energy usage when purchasing new information technology. It used VMWare virtualization software and a new Fujitsu blade server platform to consolidate underutilized legacy servers so that one machine performs the work that was formerly performed by eight, saving $44,000 per year in power costs. These efforts also created a more up-to-date IT infrastructure.

Experts note that it's important for companies to measure their energy use and inventory and track their information technology assets both before and after they start their green initiatives. Commonly used metrics used by Microsoft and other companies include Power Usage Effectiveness, Data Center Infrastructure Efficiency, and Average Data Efficiency.

It isn't always necessary to purchase new technologies to achieve "green" goals. Organizations can achieve sizable efficiencies by better managing the computing resources they already have.

Health insurer Highmark initially wanted to increase its CPU utilization by 10 percent while reducing power use by 5 percent and eventually by 10 percent. When the company inventoried all of its information technology assets, it found that its information systems staff was hanging onto "dead" servers that served no function but continued to consume power. Unfortunately, many information systems departments still aren't deploying their existing technology resources efficiently or using green measurement tools.

Programs to educate employees in energy conservation may also be necessary. In addition to using

energy-monitoring tools, Honda Motor Corporation trains its data center administrators how to be more energy efficient. For example, it taught them to decommission unused equipment quickly and to use management tools to ensure servers are being optimized.

Sources: Kathleen Lao, "The Green Issue," *Computerworld Canada*, April 2010; Matthew Sarrell, "Greening Your Data Center: The Real Deal," *eWeek*, January 15, 2010; Robert L. Mitchell, "Data Center Density Hits the Wall," *Computerworld*, January 21, 2010; Jim Carlton, "The PC Goes on an Energy Diet," *The Wall Street Journal*, September 8, 2009; and Ronan Kavanagh, "IT Virtualization Helps to Go Green," *Information Management Magazine*, March 2009.

CASE STUDY QUESTIONS

1. What business and social problems does data center power consumption cause?
2. What solutions are available for these problems? Which are environment-friendly?
3. What are the business benefits and costs of these solutions?
4. Should all firms move toward green computing? Why or why not?

MIS IN ACTION

Perform an Internet search on the phrase "green computing" and then answer the following questions:

1. Who are some of the leaders of the green computing movement? Which corporations are leading the way? Which environmental organizations are playing an important role?
2. What are the latest trends in green computing? What kind of impact are they having?
3. What can individuals do to contribute to the green computing movement? Is the movement worthwhile?

5.4 CONTEMPORARY SOFTWARE PLATFORM TRENDS

There are four major themes in contemporary software platform evolution:

- Linux and open source software
- Java and Ajax
- Web services and service-oriented architecture
- Software outsourcing and cloud services

LINUX AND OPEN SOURCE SOFTWARE

Open source software is software produced by a community of several hundred thousand programmers around the world. According to the leading open source professional association, OpenSource.org, open source software is free and can be modified by users. Works derived from the original code must also be free, and the software can be redistributed by the user without additional licensing. Open source software is by definition not restricted to any specific operating system or hardware technology, although most open source software is currently based on a Linux or Unix operating system.

The open source movement has been evolving for more than 30 years and has demonstrated that it can produce commercially acceptable, high-quality software. Popular open source software tools include the Linux operating system, the Apache HTTP Web server, the Mozilla Firefox Web browser, and the Oracle Open Office desktop productivity suite. Open source tools are being used on netbooks as inexpensive alternatives to Microsoft Office. Major hardware and software vendors, including IBM, HP, Dell, Oracle, and SAP, now offer Linux-compatible versions of their products. You can find out more out more about the Open Source Definition from the Open Source Initiative and the history of open source software at the Learning Tracks for this chapter.

Linux

Perhaps the most well known open source software is Linux, an operating system related to Unix. Linux was created by the Finnish programmer Linus Torvalds and first posted on the Internet in August 1991. Linux applications are embedded in cell phones, smartphones, netbooks, and consumer electronics. Linux is available in free versions downloadable from the Internet or in low-cost commercial versions that include tools and support from vendors such as Red Hat.

Although Linux is not used in many desktop systems, it is a major force in local area networks, Web servers, and high-performance computing work, with over 20 percent of the server operating system market. IBM, HP, Intel, Dell, and Oracle-Sun have made Linux a central part of their offerings to corporations.

The rise of open source software, particularly Linux and the applications it supports, has profound implications for corporate software platforms: cost reduction, reliability and resilience, and integration, because Linux works on all the major hardware platforms from mainframes to servers to clients.

SOFTWARE FOR THE WEB: JAVA AND AJAX

Java is an operating system-independent, processor-independent, object-oriented programming language that has become the leading interactive environment for the Web. Java was created by James Gosling and the Green Team at Sun Microsystems in 1992. In November 13, 2006, Sun released much of Java as open source software under the terms of the GNU General Public License (GPL), completing the process on May 8, 2007.

The Java platform has migrated into cellular phones, smartphones, automobiles, music players, game machines, and finally, into set-top cable television systems serving interactive content and pay-per-view services. Java software is designed to run on any computer or computing device, regardless of the specific microprocessor or operating system the device uses. For each of the computing environments in which Java is used, Sun created a Java Virtual Machine that interprets Java programming code for that machine. In this manner, the code is written once and can be used on any machine for which there exists a Java Virtual Machine.

Java developers can create small applet programs that can be embedded in Web pages and downloaded to run on a Web browser. A **Web browser** is an easy-to-use software tool with a graphical user interface for displaying Web pages and for accessing the Web and other Internet resources. Microsoft's Internet Explorer, Mozilla Firefox, and Google Chrome browser are examples. At the enterprise level, Java is being used for more complex e-commerce and e-business applications that require communication with an organization's back-end transaction processing systems.

Ajax

Have you ever filled out a Web order form, made a mistake, and then had to start all over gain after a long wait for a new order form page to appear on your computer screen? Or visited a map site, clicked the North arrow once, and waited some time for an entire new page to load? **Ajax** (Asynchronous JavaScript and XML) is another Web development technique for creating interactive Web applications that prevents all of this inconvenience.

Ajax allows a client and server to exchange small pieces of data behind the scene so that an entire Web page does not have to be reloaded each time the user requests a change. So if you click North on a map site, such as Google Maps, the server downloads just that part of the application that changes with no wait for an entirely new map. You can also grab maps in map applications and move the map in any direction without forcing a reload of the entire page. Ajax uses JavaScript programs downloaded to your client to maintain a near-continuous conversation with the server you are using, making the user experience more seamless.

WEB SERVICES AND SERVICE-ORIENTED ARCHITECTURE

Web services refer to a set of loosely coupled software components that exchange information with each other using universal Web communication standards and languages. They can exchange information between two different systems regardless of the operating systems or programming languages on which the systems are based. They can be used to build open standard Web-based applications linking systems of two different organizations, and they can also be used to create applications that link disparate systems within a single company. Web services are not tied to any one operating system or programming language, and different applications can use them to communicate with each other in a standard way without time-consuming custom coding.

The foundation technology for Web services is **XML**, which stands for **Extensible Markup Language**. This language was developed in 1996 by the World Wide Web Consortium (W3C, the international body that oversees the development of the Web) as a more powerful and flexible markup language than hypertext markup language (HTML) for Web pages. **Hypertext Markup Language (HTML)** is a page description language for specifying how text, graphics, video, and sound are placed on a Web page document. Whereas HTML is limited to describing how data should be presented in the form of Web pages, XML can perform presentation, communication, and storage of data. In XML, a number is not simply a number; the XML tag specifies whether the number represents a price, a date, or a ZIP code. Table 5-2 illustrates some sample XML statements.

TABLE 5-2 EXAMPLES OF XML

PLAIN ENGLISH	XML
Subcompact	<AUTOMOBILETYPE="Subcompact">
4 passenger	<PASSENGERUNIT="PASS">4</PASSENGER>
$16,800	<PRICE CURRENCY="USD">$16,800</PRICE>

By tagging selected elements of the content of documents for their meanings, XML makes it possible for computers to manipulate and interpret their data automatically and perform operations on the data without human intervention. Web browsers and computer programs, such as order processing or enterprise resource planning (ERP) software, can follow programmed rules for applying and displaying the data. XML provides a standard format for data exchange, enabling Web services to pass data from one process to another.

Web services communicate through XML messages over standard Web protocols. *SOAP*, which stands for *Simple Object Access Protocol*, is a set of rules for structuring messages that enables applications to pass data and instructions to one another. *WSDL* stands for *Web Services Description Language*; it is a common framework for describing the tasks performed by a Web service and the commands and data it will accept so that it can be used by other applications. *UDDI*, which stands for *Universal Description, Discovery, and Integration*, enables a Web service to be listed in a directory of Web services so that it can be easily located. Companies discover and locate Web services through this directory much as they would locate services in the yellow pages of a telephone book. Using these protocols, a software application can connect freely to other applications without custom programming for each different application with which it wants to communicate. Everyone shares the same standards.

The collection of Web services that are used to build a firm's software systems constitutes what is known as a service-oriented architecture. A **service-oriented architecture (SOA)** is set of self-contained services that communicate with each other to create a working software application. Business tasks are accomplished by executing a series of these services. Software developers reuse these services in other combinations to assemble other applications as needed.

Virtually all major software vendors provide tools and entire platforms for building and integrating software applications using Web services. IBM includes Web service tools in its WebSphere e-business software platform, and Microsoft has incorporated Web services tools in its Microsoft .NET platform.

Dollar Rent A Car's systems use Web services for its online booking system with Southwest Airlines' Web site. Although both companies' systems are based on different technology platforms, a person booking a flight on Southwest.com can reserve a car from Dollar without leaving the airline's Web site. Instead of struggling to get Dollar's reservation system to share data with Southwest's information systems, Dollar used Microsoft .NET Web services technology as an intermediary. Reservations from Southwest are translated into Web services protocols, which are then translated into formats that can be understood by Dollar's computers.

Other car rental companies have linked their information systems to airline companies' Web sites before. But without Web services, these connections had to be built one at a time. Web services provide a standard way for Dollar's computers to "talk" to other companies' information systems without having to build special links to each one. Dollar is now expanding its use of Web services to link directly to the systems of a small tour operator and a large travel reservation system as well as a wireless Web site for cell phones and smartphones. It does not have to write new software code for each new partner's information systems or each new wireless device (see Figure 5-10).

FIGURE 5-10   HOW DOLLAR RENT A CAR USES WEB SERVICES

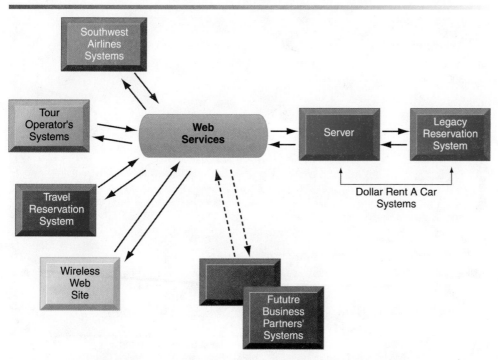

Dollar Rent A Car uses Web services to provide a standard intermediate layer of software to "talk" to other companies' information systems. Dollar Rent A Car can use this set of Web services to link to other companies' information systems without having to build a separate link to each firm's systems.

SOFTWARE OUTSOURCING AND CLOUD SERVICES

Today many business firms continue to operate legacy systems that continue to meet a business need and that would be extremely costly to replace. But they will purchase or rent most of their new software applications from external sources. Figure 5-11 illustrates the rapid growth in external sources of software for U.S. firms.

There are three external sources for software: software packages from a commercial software vendor, outsourcing custom application development to an external vendor, and cloud-based software services and tools.

Software Packages and Enterprise Software

We have already described software packages for enterprise applications as one of the major types of software components in contemporary IT infrastructures. A **software package** is a prewritten commercially available set of software programs that eliminates the need for a firm to write its own software programs for certain functions, such as payroll processing or order handling.

Enterprise application software vendors such as SAP and Oracle-PeopleSoft have developed powerful software packages that can support the primary business processes of a firm worldwide from warehousing, customer relationship management, supply chain management, and finance to human resources. These large-scale enterprise software systems provide a single, integrated, worldwide software system for firms at a cost much less than they would pay if they developed it themselves. Chapter 9 discusses enterprise systems in detail.

FIGURE 5-11   CHANGING SOURCES OF FIRM SOFTWARE

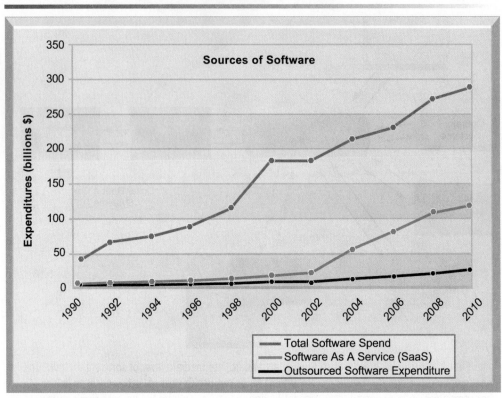

In 2010, U.S. firms will spend over $291 billion on software. About 40 percent of that ($116 billion) will originate outside the firm, either from enterprise software vendors selling firmwide applications or individual application service providers leasing or selling software modules. Another 10 percent ($29 billion) will be provided by SaaS vendors as an online cloud-based service.

Sources: BEA National Income and Product Accounts, 2010; Gartner Group, 2010; author estimates.

Software Outsourcing

Software **outsourcing** enables a firm to contract custom software development or maintenance of existing legacy programs to outside firms, which often operate offshore in low-wage areas of the world. According to the industry analysts, 2010 offshore outsourcing revenues in the United States will be approximately $50 billion, and domestic outsourcing revenues will be $106 billion (Lohr, 2009). The largest expenditure here is paid to domestic U.S. firms providing middleware, integration services, and other software support that are often required to operate larger enterprise systems.

For example, in March 2008, Royal Dutch Shell PLC, the world's third largest oil producer, signed a five-year, $4 billion outsourcing deal with T-Systems International GmbH, AT&T, and Electronic Data Systems (EDS). The agreement assigned AT&T responsibility for networking and telecommunications, T-Systems for hosting and storage, and EDS for end-user computing services and for integration of the infrastructure services. Outsourcing this work has helped Shell cut costs and focus on systems that improve its competitive position in the oil and gas market.

Offshore outsourcing firms have primarily provided lower-level maintenance, data entry, and call center operations. However, with the growing sophistication and experience of offshore firms, particularly in India, more and more new-program development is taking place offshore. Chapter 13 discusses offshore software outsourcing in greater detail.

Cloud-Based Software Services and Tools

In the past, software such as Microsoft Word or Adobe Illustrator came in a box and was designed to operate on a single machine. Today, you're more likely to download the software from the vendor's Web site, or to use the software as a cloud service delivered over the Internet.

Cloud-based software and the data it uses are hosted on powerful servers in massive data centers, and can be accessed with an Internet connection and standard Web browser. In addition to free or low-cost tools for individuals and small businesses provided by Google or Yahoo!, enterprise software and other complex business functions are available as services from the major commercial software vendors. Instead of buying and installing software programs, subscribing companies rent the same functions from these services, with users paying either on a subscription or per-transaction basis. Services for delivering and providing access to software remotely as a Web-based service are now referred to as **software as a service (SaaS)**. A leading example is Salesforce.com, described in the chapter-ending case study, which provides on-demand software services for customer relationship management.

In order to manage their relationship with an outsourcer or technology service provider, firms need a contract that includes a **service level agreement (SLA)**. The SLA is a formal contract between customers and their service providers that defines the specific responsibilities of the service provider and the level of service expected by the customer. SLAs typically specify the nature and level of services provided, criteria for performance measurement, support options, provisions for security and disaster recovery, hardware and software ownership and upgrades, customer support, billing, and conditions for terminating the agreement. We provide a Learning Track on this topic.

Mashups and Apps

The software you use for both personal and business tasks may consist of large self-contained programs, or it may be composed of interchangeable components that integrate freely with other applications on the Internet. Individual users and entire companies mix and match these software components to create their own customized applications and to share information with others. The resulting software applications are called **mashups**. The idea is to take different sources and produce a new work that is "greater than" the sum of its parts. You have performed a mashup if you've ever personalized your Facebook profile or your blog with a capability to display videos or slide shows.

Web mashups combine the capabilities of two or more online applications to create a kind of hybrid that provides more customer value than the original sources alone. For instance, EveryBlock Chicago combines Google Maps with crime data for the city of Chicago. Users can search by location, police beat, or type of crime, and the results are displayed as color-coded map points on a Google Map. Amazon uses mashup technologies to aggregate product descriptions with partner sites and user profiles.

Apps are small pieces of software that run on the Internet, on your computer, or on your cell phone and are generally delivered over the Internet. Google refers to its online services as apps, including the Google Apps suite of desktop productivity tools. But when we talk about apps today, most of the attention goes to the apps that have been developed for the mobile digital platform. It is these apps that turn smartphones and other mobile handheld devices into general-purpose computing tools.

Most of these apps are for the iPhone, Android, and BlackBerry operating system platforms. Many are free or purchased for a small charge, much less than conventional software. There are already over 250,000 apps for the Apple iPhone and iPad platform and over 80,000 that run on smartphones using Google's Android operating system. The success of these mobile platforms depends in large part on the quantity and the quality of the apps they provide. Apps tie the customer to a specific hardware platform: As the user adds more and more apps to his or her mobile phone, the cost of switching to a competing mobile platform rises.

At the moment, the most commonly downloaded apps are games (65%), followed by news and weather (56%), maps/navigation (55%), social networking (54%), music (46%), and video/movies (25%). But there are also serious apps for business users that make it possible to create and edit documents, connect to corporate systems, schedule and participate in meetings, track shipments, and dictate voice messages (see the Chapter 1 Interactive Session on Management). There are also a huge number of e-commerce apps for researching and buying goods and services online.

5.5 MANAGEMENT ISSUES

Creating and managing a coherent IT infrastructure raises multiple challenges: dealing with platform and technology change (including cloud and mobile computing), management and governance, and making wise infrastructure investments.

DEALING WITH PLATFORM AND INFRASTRUCTURE CHANGE

As firms grow, they often quickly outgrow their infrastructure. As firms shrink, they can get stuck with excessive infrastructure purchased in better times. How can a firm remain flexible when most of the investments in IT infrastructure are fixed-cost purchases and licenses? How well does the infrastructure scale? **Scalability** refers to the ability of a computer, product, or system to expand to serve a large number of users without breaking down. New applications, mergers and acquisitions, and changes in business volume all impact computer workload and must be considered when planning hardware capacity.

Firms using mobile computing and cloud computing platforms will require new policies and procedures for managing these platforms. They will need to inventory all of their mobile devices in business use and develop policies and tools for tracking, updating, and securing them and for controlling the data and applications that run on them. Firms using cloud computing and SaaS will need to fashion new contractual arrangements with remote vendors to make sure that the hardware and software for critical applications are always available when needed and that they meet corporate standards for information security. It is up to business management to determine acceptable levels of computer response time and availability for the firm's mission-critical systems to maintain the level of business performance they expect.

MANAGEMENT AND GOVERNANCE

A long-standing issue among information system managers and CEOs has been the question of who will control and manage the firm's IT infrastructure. Chapter 2 introduced the concept of IT governance and described some issues it addresses. Other important questions about IT governance are: Should departments and divisions have the responsibility of making their own information technology decisions or should IT infrastructure be centrally controlled and managed? What is the relationship between central information systems management and business unit information systems management? How will infrastructure costs be allocated among business units? Each organization will need to arrive at answers based on its own needs.

MAKING WISE INFRASTRUCTURE INVESTMENTS

IT infrastructure is a major investment for the firm. If too much is spent on infrastructure, it lies idle and constitutes a drag on firm financial performance. If too little is spent, important business services cannot be delivered and the firm's competitors (who spent just the right amount) will outperform the under-investing firm. How much should the firm spend on infrastructure? This question is not easy to answer.

A related question is whether a firm should purchase and maintain its own IT infrastructure components or rent them from external suppliers, including those offering cloud services. The decision either to purchase your own IT assets or rent them from external providers is typically called the *rent-versus-buy* decision.

Cloud computing may be a low-cost way to increase scalability and flexibility, but firms should evaluate this option carefully in light of security requirements and impact on business processes and work flows. In some instances, the cost of renting software adds up to more than purchasing and maintaining an application in-house. Yet there may be benefits to using SaaS if it allows the company to focus on core business issues instead of technology challenges.

Total Cost of Ownership of Technology Assets

The actual cost of owning technology resources includes the original cost of acquiring and installing hardware and software, as well as ongoing administration costs for hardware and software upgrades, maintenance, technical support, training, and even utility and real estate costs for running and housing the technology. The **total cost of ownership (TCO)** model can be used to analyze these direct and indirect costs to help firms determine the actual cost of specific technology implementations. Table 5-3 describes the most important TCO components to consider in a TCO analysis.

When all these cost components are considered, the TCO for a PC might run up to three times the original purchase price of the equipment. Although the purchase price of a wireless handheld for a corporate employee may run several hundred dollars, the TCO for each device is much higher, ranging from $1,000 to $3,000, according to various consultant estimates. Gains in productivity and efficiency from equipping employees with mobile computing devices must be balanced against increased costs from integrating these devices into the firm's IT infrastructure and from providing technical support. Other cost components include fees for wireless airtime, end-user training, help desk support, and software for special applications. Costs are higher if the mobile devices run many different applications or need to be integrated into back-end systems such as enterprise applications.

Hardware and software acquisition costs account for only about 20 percent of TCO, so managers must pay close attention to administration costs to understand the full cost of the firm's hardware and software. It is possible to reduce some of these administration costs through better management. Many large firms are

TABLE 5-3 TOTAL COST OF OWNERSHIP (TCO) COST COMPONENTS

INFRASTRUCTURE COMPONENT	COST COMPONENTS
Hardware acquisition	Purchase price of computer hardware equipment, including computers, terminals, storage, and printers
Software acquisition	Purchase or license of software for each user
Installation	Cost to install computers and software
Training	Cost to provide training for information systems specialists and end users
Support	Cost to provide ongoing technical support, help desks, and so forth
Maintenance	Cost to upgrade the hardware and software
Infrastructure	Cost to acquire, maintain, and support related infrastructure, such as networks and specialized equipment (including storage backup units)
Downtime	Cost of lost productivity if hardware or software failures cause the system to be unavailable for processing and user tasks
Space and energy	Real estate and utility costs for housing and providing power for the technology

saddled with redundant, incompatible hardware and software because their departments and divisions have been allowed to make their own technology purchases.

In addition to switching to cloud services, these firms could reduce their TCO through greater centralization and standardization of their hardware and software resources. Companies could reduce the size of the information systems staff required to support their infrastructure if the firm minimizes the number of different computer models and pieces of software that employees are allowed to use. In a centralized infrastructure, systems can be administered from a central location and troubleshooting can be performed from that location.

Competitive Forces Model for IT Infrastructure Investment

Figure 5-12 illustrates a competitive forces model you can use to address the question of how much your firm should spend on IT infrastructure.

Market demand for your firm's services. Make an inventory of the services you currently provide to customers, suppliers, and employees. Survey each group, or hold focus groups to find out if the services you currently offer are meeting the needs of each group. For example, are customers complaining of slow responses to their queries about price and availability? Are employees complaining about the difficulty of finding the right information for their jobs? Are suppliers complaining about the difficulties of discovering your production requirements?

Your firm's business strategy. Analyze your firm's five-year business strategy and try to assess what new services and capabilities will be required to achieve strategic goals.

Your firm's IT strategy, infrastructure, and cost. Examine your firm's information technology plans for the next five years and assess its alignment with the firm's business plans. Determine the total IT infrastructure costs. You will want to perform a TCO analysis. If your firm has no IT strategy, you will need to devise one that takes into account the firm's five-year strategic plan.

Information technology assessment. Is your firm behind the technology curve or at the bleeding edge of information technology? Both situations are to be avoided. It is usually not desirable to spend resources on advanced technolo-

FIGURE 5-12   **COMPETITIVE FORCES MODEL FOR IT INFRASTRUCTURE**

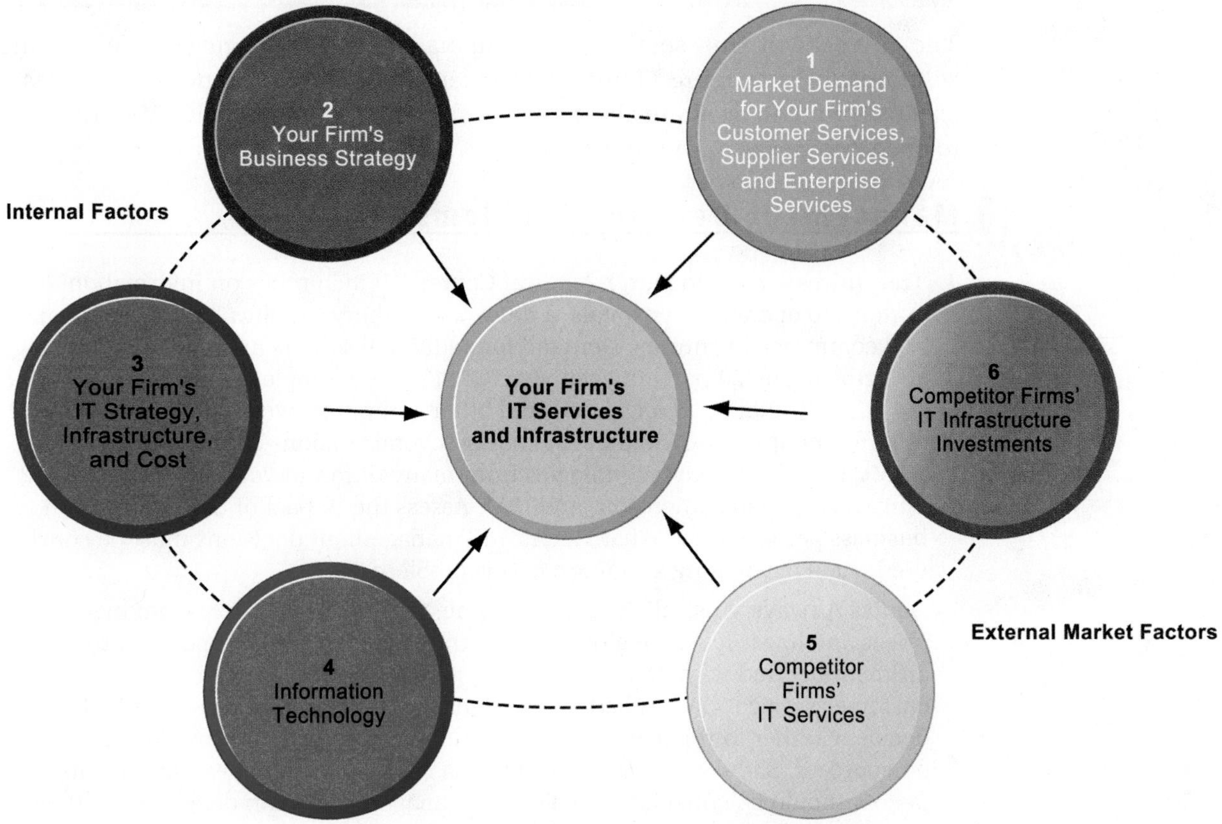

There are six factors you can use to answer the question, "How much should our firm spend on IT infrastructure?"

gies that are still experimental, often expensive, and sometimes unreliable. You want to spend on technologies for which standards have been established and IT vendors are competing on cost, not design, and where there are multiple suppliers. However, you do not want to put off investment in new technologies or allow competitors to develop new business models and capabilities based on the new technologies.

Competitor firm services. Try to assess what technology services competitors offer to customers, suppliers, and employees. Establish quantitative and qualitative measures to compare them to those of your firm. If your firm's service levels fall short, your company is at a competitive disadvantage. Look for ways your firm can excel at service levels.

Competitor firm IT infrastructure investments. Benchmark your expenditures for IT infrastructure against your competitors. Many companies are quite public about their innovative expenditures on IT. If competing firms try to keep IT expenditures secret, you may be able to find IT investment information in public companies' SEC Form 10-K annual reports to the federal government when those expenditures impact a firm's financial results.

Your firm does not necessarily need to spend as much as, or more than, your competitors. Perhaps it has discovered much less-expensive ways of providing services, and this can lead to a cost advantage. Alternatively, your firm may be spending far less than competitors and experiencing commensurate poor performance and losing market share.

5.6   HANDS-ON MIS PROJECTS

The projects in this section give you hands-on experience in developing solutions for managing IT infrastructures and IT outsourcing, using spreadsheet software to evaluate alternative desktop systems, and using Web research to budget for a sales conference.

Management Decision Problems

1. The University of Pittsburgh Medical Center (UPMC) relies on information systems to operate 19 hospitals, a network of other care sites, and international and commercial ventures. Demand for additional servers and storage technology was growing by 20 percent each year. UPMC was setting up a separate server for every application, and its servers and other computers were running a number of different operating systems, including several versions of Unix and Windows. UPMC had to manage technologies from many different vendors, including HP, Sun Microsystems, Microsoft, and IBM. Assess the impact of this situation on business performance. What factors and management decisions must be considered when developing a solution to this problem?

2. Qantas Airways, Australia's leading airline, faces cost pressures from high fuel prices and lower levels of global airline traffic. To remain competitive, the airline must find ways to keep costs low while providing a high level of customer service. Qantas had a 30-year-old data center. Management had to decide whether to replace its IT infrastructure with newer technology or outsource it. Should Qantas outsource to a cloud computing vendor? What factors should be considered by Qantas management when deciding whether to outsource? If Qantas decides to outsource, list and describe points that should be addressed in a service level agreement.

Improving Decision Making: Using a Spreadsheet to Evaluate Hardware and Software Options

Software skills: Spreadsheet formulas
Business skills: Technology pricing

In this exercise, you will use spreadsheet software to calculate the cost of desktop systems, printers, and software.

You have been asked to obtain pricing information on hardware and software for an office of 30 people. Using the Internet, get pricing for 30 PC desktop systems (monitors, computers, and keyboards) manufactured by Lenovo, Dell, and HP/Compaq as listed at their respective corporate Web sites. (For the purposes of this exercise, ignore the fact that desktop systems usually come with preloaded software packages.) Also obtain pricing on 15 desktop printers manufactured by HP, Canon, and Dell. Each desktop system must satisfy the minimum specifications shown in the following table:

MINIMUM DESKTOP SPECIFICATIONS

Processor speed	3 GHz
Hard drive	350 GB
RAM	3 GB
DVD-ROM drive	16 x
Monitor (diagonal measurement)	18 inches

Each desktop printer must satisfy the minimum specifications shown in the following table:

MINIMUM MONOCHROME PRINTER SPECIFICATIONS

Print speed (black and white)	20 pages per minute
Print resolution	600 × 600
Network ready?	Yes
Maximum price/unit	$700

After pricing the desktop systems and printers, obtain pricing on 30 copies of the most recent versions of Microsoft Office, Lotus SmartSuite, and Oracle Open Office desktop productivity packages, and on 30 copies of Microsoft Windows 7 Professional. The application software suite packages come in various versions, so be sure that each package contains programs for word processing, spreadsheets, database, and presentations.

Prepare a spreadsheet showing your research results for the desktop systems, for the printers, and for the software. Use your spreadsheet software to determine the desktop system, printer, and software combination that will offer both the best performance and pricing per worker. Because every two workers will share one printer (15 printers/30 systems), assume only half a printer cost per worker in the spreadsheet. Assume that your company will take the standard warranty and service contract offered by each product's manufacturer.

Improving Decision Making: Using Web Research to Budget for a Sales Conference

Software skills: Internet-based software
Business skills: Researching transportation and lodging costs

The Foremost Composite Materials Company is planning a two-day sales conference for October 15–16, starting with a reception on the evening of October 14. The conference consists of all-day meetings that the entire sales force, numbering 125 sales representatives and their 16 managers, must attend. Each sales representative requires his or her own room, and the company needs two common meeting rooms, one large enough to hold the entire sales force plus visitors (200 total) and the other able to hold half the force. Management has set a budget of $120,000 for the representatives' room rentals. The hotel must also have such services as overhead and computer projectors as well as business center and banquet facilities. It also should have facilities for the company reps to be able to work in their rooms and to enjoy themselves in a swimming pool or gym facility. The company would like to hold the conference in either Miami or Marco Island, Florida.

Foremost usually likes to hold such meetings in Hilton- or Marriott-owned hotels. Use the Hilton and Marriott Web sites to select a hotel in whichever of these cities that would enable the company to hold its sales conference within its budget.

Visit the two sites' homepages, and search them to find a hotel that meets Foremost's sales conference requirements. Once you have selected the hotel, locate flights arriving the afternoon prior to the conference because the attendees will need to check in the day before and attend your reception the evening prior to the conference. Your attendees will be coming from Los Angeles (54), San Francisco (32), Seattle (22), Chicago (19), and Pittsburgh (14). Determine costs of each airline ticket from these cities. When you are finished, create a budget for the conference. The budget will include the cost of each airline ticket, the room cost, and $60 per attendee per day for food.

- What was your final budget?
- Which did you select as the best hotel for the sales conference and why?

LEARNING TRACK MODULES

The following Learning Tracks provide content relevant to topics covered in this chapter:

1. How Computer Hardware and Software Work
2. Service Level Agreements
3. The Open Source Software Initiative
4. Comparing Stages in IT Infrastructure Evolution
5. Cloud Computing

Review Summary

1. **What is IT infrastructure and what are its components?**

   IT infrastructure is the shared technology resources that provide the platform for the firm's specific information system applications. IT infrastructure includes hardware, software, and services that are shared across the entire firm. Major IT infrastructure components include computer hardware platforms, operating system platforms, enterprise software platforms, networking and telecommunications platforms, database management software, Internet platforms, and consulting services and systems integrators.

2. **What are the stages and technology drivers of IT infrastructure evolution?**

   The five stages of IT infrastructure evolution are: the mainframe era, the personal computer era, the client/server era, the enterprise computing era, and the cloud and mobile computing era. Moore's Law deals with the exponential increase in processing power and decline in the cost of computer technology, stating that every 18 months the power of microprocessors doubles and the price of computing falls in half. The Law of Mass Digital Storage deals with the exponential decrease in the cost of storing data, stating that the number of kilobytes of data that can be stored on magnetic media for $1 roughly doubles every 15 months. Metcalfe's Law helps shows that a network's value to participants grows exponentially as the network takes on more members. Also driving exploding computer use is the rapid decline in costs of communication and growing agreement in the technology industry to use computing and communications standards.

3. **What are the current trends in computer hardware platforms?**

   Increasingly, computing is taking place on a mobile digital platform. Grid computing involves connecting geographically remote computers into a single network to create a computational grid that combines the computing power of all the computers on the network. Virtualization organizes computing resources so that their use is not restricted by physical configuration or geographic location. In cloud computing, firms and individuals obtain computing power and software as services over a network, including the Internet, rather than purchasing and installing the hardware and software on their own computers. A multicore processor is a microprocessor to which two or more processing cores have been attached for enhanced performance. Green computing includes practices and technologies for producing, using, and disposing of information technology hardware to minimize negative impact on the environment. In autonomic computing, computer systems have capabilities for automatically configuring and repairing themselves. Power-saving processors dramatically reduce power consumption in mobile digital devices.

4. **What are the current trends in software platforms?**

   Open source software is produced and maintained by a global community of programmers and is often downloadable for free. Linux is a powerful, resilient open source operating system that can run on multiple hardware platforms and is used widely to run Web servers. Java is an operating-system– and hardware-independent programming language that is the leading interactive programming environment for the Web. Web services are loosely coupled software components based on open Web standards that work with any application software and operating system. They can be used as components of Web-based applications linking the systems of two different organizations or to link disparate systems of a single company. Companies are purchasing their new software applications from outside sources, including software packages, by outsourcing custom application development to an external vendor (that may be offshore), or by renting online software services (SaaS). Mashups combine two different software services to create new software applications and services. Apps are small pieces of software that run on the Internet, on a computer, or on a mobile phone and are generally delivered over the Internet.

5. *What are the challenges of managing IT infrastructure and management solutions?*

Major challenges include dealing with platform and infrastructure change, infrastructure management and governance, and making wise infrastructure investments. Solution guidelines include using a competitive forces model to determine how much to spend on IT infrastructure and where to make strategic infrastructure investments, and establishing the total cost of ownership (TCO) of information technology assets. The total cost of owning technology resources includes not only the original cost of computer hardware and software but also costs for hardware and software upgrades, maintenance, technical support, and training.

Key Terms

Ajax, 189
Android, 177
Application server, 169
Apps, 193
Autonomic computing, 185
Blade servers, 176
Chrome OS, 177
Clients, 168
Client/server computing, 168
Cloud computing, 170
Extensible Markup Language (XML), 189
Green computing, 184
Grid computing, 182
Hypertext Markup Language (HTML), 189
Java, 188
Legacy systems, 181
Linux, 177
Mainframe, 168
Mashup, 193
Minicomputers, 168
Moore's Law, 171
Multicore processor, 185
Multitiered (N-tier) client/server architecture, 169
Multitouch, 177
Nanotechnology, 171

Netbook, 181
On-demand computing, 184
Open source software, 187
Operating system, 177
Outsourcing, 192
Private cloud, 183
Public cloud, 183
SaaS (Software as a Service), 193
Scalability, 194
Service level agreement (SLA), 193
Server, 168
Service-oriented architecture (SOA), 190
Software package, 191
Storage area network (SAN), 180
Technology standards, 174
Total cost of ownership (TCO), 195
Unix, 177
Utility computing, 184
Virtualization, 182
Web browser, 188
Web hosting service, 180
Web server, 169
Web services, 189
Windows, 169
Wintel PC, 168

Review Questions

1. What is IT infrastructure and what are its components?

   - Define IT infrastructure from both a technology and a services perspective.
   - List and describe the components of IT infrastructure that firms need to manage.

2. What are the stages and technology drivers of IT infrastructure evolution?

   - List each of the eras in IT infrastructure evolution and describe its distinguishing characteristics.
   - Define and describe the following: Web server, application server, multitiered client/server architecture.
   - Describe Moore's Law and the Law of Mass Digital Storage.
   - Describe how network economics, declining communications costs, and technology standards affect IT infrastructure.

3. What are the current trends in computer hardware platforms?

   - Describe the evolving mobile platform, grid computing, and cloud computing.

   - Explain how businesses can benefit from autonomic computing, virtualization, green computing, and multicore processors.

4. What are the current trends in software platforms?

   - Define and describe open source software and Linux and explain their business benefits.
   - Define Java and Ajax and explain why they are important.
   - Define and describe Web services and the role played by XML.
   - Name and describe the three external sources for software.
   - Define and describe software mashups and apps.

5. What are the challenges of managing IT infrastructure and management solutions?

   - Name and describe the management challenges posed by IT infrastructure.
   - Explain how using a competitive forces model and calculating the TCO of technology assets help firms make good infrastructure investments.

Discussion Questions

1. Why is selecting computer hardware and software for the organization an important management decision? What management, organization, and technology issues should be considered when selecting computer hardware and software?

2. Should organizations use software service providers for all their software needs? Why or why not? What management, organization, and technology factors should be considered when making this decision?

3. What are the advantages and disadvantages of cloud computing?

Video Cases

Video Cases and Instructional Videos illustrating some of the concepts in this chapter are available. Contact your instructor to access these videos.

Collaboration and Teamwork: Evaluating Server and Mobile Operating Systems

Form a group with three or four of your classmates. Choose server or mobile operating systems to evaluate. You might research and compare the capabilities and costs of Linux versus the most recent version of the Windows operating system or Unix. Alternatively, you could compare the capabilities of the Android mobile operating system with the most recent version

of the iPhone operating system (iOS). If possible, use Google Sites to post links to Web pages, team communication announcements, and work assignments; to brainstorm; and to work collaboratively on project documents. Try to use Google Docs to develop a presentation of your findings for the class.

Salesforce.Com: Cloud Services Go Mainstream
CASE STUDY

Salesforce.com, one of the most disruptive technology companies of the past few years, has single-handedly shaken up the software industry with its innovative business model and resounding success. Salesforce provides customer relationship management (CRM) and other application software solutions in the form of software as a service leased over the Internet, as opposed to software bought and installed on machines locally.

The company was founded in 1999 by former Oracle executive Marc Benioff, and has since grown to over 3,900 employees, 82,400 corporate customers, and 2.1 million subscribers. It earned $1.3 billion in revenue in 2009, making it one of the top 50 software companies in the world. Salesforce attributes its success to the many benefits of its on-demand model of software distribution.

The on-demand model eliminates the need for large up-front hardware and software investments in systems and lengthy implementations on corporate computers. Subscriptions start as low as $9 per user per month for the pared-down Group version for small sales and marketing teams, with monthly subscriptions for more advanced versions for large enterprises starting around $65 per user.

For example, the Minneapolis-based Haagen-Dazs Shoppe owned by Nestle USA calculated it would have had to spend $65,000 for a custom-designed database to help management stay in contact with the company's retail franchises. The company only had to pay $20,000 to establish service with Salesforce, plus a monthly charge of $125 per month for 20 users to use wireless handhelds or the Web to remotely monitor all the Haagen-Dazs franchises across the United States.

Salesforce.com implementations take three months at the longest, and usually less than a month. There is no hardware for subscribers to purchase, scale, and maintain. There are no operating systems, database servers, or application servers to install, no consultants and staff, and no expensive licensing and maintenance fees. The system is accessible via a standard Web browser, with some functions accessible by mobile handheld devices. Salesforce.com continually updates its software behind the scenes. There are tools for customizing some features of the software to support a company's unique business processes. Subscribers can leave if business turns sour or a better system comes along. If they lay people off, they can cut down on the number of Salesforce subscriptions they buy.

Salesforce faces significant challenges as it continues to grow and refine its business. The first challenge comes from increased competition, both from traditional industry leaders and new challengers hoping to replicate Salesforce's success. Microsoft, SAP, and Oracle have rolled out subscription-based versions of their CRM products in response to Salesforce. Smaller competitors like NetSuite, Salesboom.com, and RightNow also have made some inroads against Salesforce's market share.

Salesforce still has plenty of catching up to do to reach the size and market share of its larger competitors. As recently as 2007, SAP's market share was nearly four times as large as Salesforce's, and IBM's customer base includes 9,000 software companies that run their applications on their software and that are likelier to choose a solution offered by IBM over Salesforce.

Salesforce needs to continually prove to customers that it is reliable and secure enough to remotely handle their corporate data and applications. The company has experienced a number of service outages. For example, on January 6, 2009, a core network device failed and prevented data in Europe, Japan, and North America from being processed for 38 minutes. Over 177 million transactions were affected. While most of Salesforce's customers accept that IT services provided through the cloud are going to be available slightly less than full time, some customers and critics used the outage as an opportunity to question the soundness of the entire concept of cloud computing. In February 2009, a similar outage occurred, affecting Europe and as well as North America a few hours later.

Thus far, Salesforce has experienced only one security breach. In November 2007, a Salesforce employee was tricked into divulging his corporate password to scammers, exposing Salesforce's customer list. Salesforce clients were subjected to a barrage of highly targeted scams and hacking attempts that appeared authentic. Although this incident raised a red flag, many customers reported that Salesforce's handling of the situation was satisfactory. All of Salesforce's major customers regularly send auditors to Salesforce to check security.

Another challenge for Salesforce is to expand its business model into other areas. Salesforce is currently used mostly by sales staff needing to keep track of leads and customer lists. One way the company is trying to provide additional functionality is through a partnership with Google and more specifically Google Apps. Salesforce is combining its services with Gmail, Google Docs, Google Talk, and Google Calendar to allow its customers to accomplish more tasks via the Web. Salesforce and Google both hope that their Salesforce.com for Google Apps initiative will galvanize further growth in on-demand software.

Salesforce has also partnered with Apple to distribute its applications for use on the iPhone. The company hopes that it can tap into the large market of iPhone users, pitching the ability to use Salesforce applications any time, anywhere. And Salesforce introduced a development tool for integrating with Facebook's social network to enable customers to build applications that call functions at the Facebook site. (In early 2010, Salesforce introduced its own social networking application called Chatter, which enables employees to create profiles and make status updates that appear in colleagues' news feeds, similar to Facebook and Twitter.)

In order to grow its revenues to the levels that industry observers and Wall Street eventually expects Salesforce is changing its focus from selling a suite of software applications to providing a broader cloud computing "platform" on which many software companies deliver applications. As CEO Marc Benioff put it, over the past decade, "we focused on software as a service...In the next decade, Salesforce.com will really be focused on the platform as a service."

The company has intensified its efforts to provide cloud computing offerings to its customers. The new Salesforce.com Web site places much more emphasis on cloud computing, grouping products into three types of clouds: the Sales Cloud, the Service Cloud, and the Custom Cloud. The Sales and Service clouds consist of applications meant to improve sales and customer service, respectively, but the Custom Cloud is another name for the Force.com application development platform, where customers can develop their own applications for use within the broader Salesforce network.

Force.com provides a set of development tools and IT services that enable users to customize their Salesforce customer relationship management applications or to build entirely new applications and run them "in the cloud" on Salesforce's data center infrastructure. Salesforce opened up Force.com to other independent software developers and listed their programs on its AppExchange.

Using AppExchange, small businesses can go online and easily download over 950 software applications, some add-ons to Salesforce.com and others that are unrelated, even in non-customer-facing functions such as human resources. Force.com Sites, based on the Force.com development environment, enables users to develop Web pages and register domain names. Pricing is based on site traffic.

Salesforce's cloud infrastructure includes two data centers in the United States and a third in Singapore, with others in Europe and Japan planned for the future. Salesforce has additionally partnered with Amazon to enable Force.com customers to tap into Amazon's cloud computing services (Elastic Compute Cloud and Simple Storage Service.) Amazon's services would handle the "cloudburst computing" tasks of Force.com applications that require extra processing power or storage capacity.

An International Data Center (IDC) report estimated that the Force.com platform enables users to build and run business applications and Web sites five times faster and at half the cost of non-cloud alternatives. For instance, RehabCare, a national provider of medical rehabilitation services, used Force.com to build a mobile iPhone patient admission application for clinicians. RehabCare's information systems team built a prototype application within four days that runs on the Force.com platform. It would have taken six months to build a similar mobile application using Microsoft development tools. About 400 clinicians now use the app.

Author Solutions, a self-publishing company based in Bloomington, Minnesota, uses the Force.com platform to host the applications driving its operations. It reports saving up to 75 percent from not having to maintain and manage its own data center, e-commerce, and workflow applications, and the ability to scale as it business mushroomed. Workflow modifications that once took 30 to 120 hours are accomplished in one-fourth the time. The time and cost for adding a new product, which used to take 120 to 240 hours (and cost $6,000 to $12,000) has been reduced by 75 percent. The new platform is able to handle 30 percent more work volume than the old systems with the same number of employees.

The question is whether the audience for Salesforce's AppExchange and Force.com platforms will prove large enough to deliver the level of growth Salesforce wants. It still isn't clear whether the company will generate the revenue it needs to provide cloud computing services on the same scale as Google or Amazon and also make its cloud computing investments pay off.

Some analysts believe the platform may not be attractive to larger companies for their application needs. Yet another challenge is providing constant availability. Salesforce.com subscribers depend on the service being available 24/7. But thanks to the previously described outages, many companies have rethought their dependency on software as a service. Salesforce.com provides tools to assure customers about its system reliability and also offers PC applications that tie into their services so users can work offline.

Still, a number of companies are reluctant to jump on the SaaS and cloud computing bandwagon. Moreover, it is still not clear whether software delivered over the Web will cost less in the long run. According to Gartner consultants analyst Rob DiSisto, it may be cheaper to subscribe to Salesforce.com's software services for the first few years, but what happens after that? Will the expense of upgrading and managing on-demand software become higher than the fees companies are paying to own and host their own software?

Sources: "How Salesforce.com Brings Success to the Cloud," *IT BusinessEdge.com*, accessed June 10, 2010; Lauren McKay, "Salesforce.com Extends Chatter Across the Cloud," *CRM Magazine*, April 14, 2010; Jeff Cogswell, "Salesforce.com Assembles an Array of Development Tools for Force.com," *eWeek*, February 15, 2010; Mary Hayes Weier, "Why Force.com Is Important to Cloud Computing," *Information Week*, November 23, 2009; Jessi Hempel, "Salesforce Hits Stride," CNN Money.com, March 2, 2009; Clint Boulton, "Salesforce.com Network Device Failure Shuts Thousands Out of SaaS Apps," *eWeek*, January 7, 2009; J. Nicholas Hoover, "Service Outages Force Cloud Adopters to Rethink Tactics," *Information Week*, August 18/25, 2008; and Charles Babcock, "Salesforce Ascends Beyond Software As Service," *Information Week*, November 10, 2008.

CASE STUDY QUESTIONS

1. How does Salesforce.com use cloud computing?

2. What are some of the challenges facing Salesforce as it continues its growth? How well will it be able to meet those challenges?

3. What kinds of businesses could benefit from switching to Salesforce and why?

4. What factors would you take into account in deciding whether to use Saleforce.com for your business?

5. Could a company run its entire business using Salesforce.com, Force.com, and App Exchange? Explain your answer.

Chapter 6

Foundations of Business Intelligence: Databases and Information Management

RR DONNELLEY TRIES TO MASTER ITS DATA

Right now you are most likely using an RR Donnelley product. Chicago-based RR Donnelley is a giant commercial printing and service company providing printing services, forms and labels, direct mail, and other services. This textbook probably came off its presses. The company's recent expansion has been fueled by a series of acquisitions, including commercial printer Moore Wallace in 2005 and printing and supply chain management company Banta in January 2007. RR Donnelley's revenue has jumped from $2.4 billion in 2003 to over $9.8 billion today.

However, all that growth created information management challenges. Each acquired company had its own systems and its own set of customer, vendor, and product data. Coming from so many different sources, the data were often inconsistent, duplicated, or incomplete. For example, different units of the business might each have a different meaning for the entity "customer." One might define "customer" as a specific billing location, while another might define "customer" as the legal parent entity of a company. Donnelley had to use time-consuming manual processes to reconcile the data stored in multiple systems in order to get a clear enterprise-wide picture of each of its customers, since they might be doing business with several different units of the company. These conditions heightened inefficiencies and costs.

RR Donnelley had become so big that it was impractical to store the information from all of its units in a single system. But Donnelley still needed a clear single set of data that was accurate and consistent for the entire enterprise. To solve this problem, RR Donnelley turned to master data management (MDM). MDM seeks to ensure that an organization does not use multiple versions of the same piece of data in different parts of its operations by merging disparate records into a single authenticated master file. Once the master file is in place, employees and applications access a single consolidated view of the company's data. It is especially useful for companies such as Donnelley that have data integration problems as a result of mergers and acquisitions.

Implementing MDM is a multi-step process that includes business process analysis, data cleansing, data consolidation and reconciliation, and data migration into a master file of all the company's data. Companies must identify what group in the company "owns" each piece of data and is responsible for resolving inconsistent definitions of data and other discrepancies. Donnelley launched its MDM program in late 2005 and began creating a single set of identifiers for its customer and vendor data. The company opted for a registry model using Purisma's Data Hub in which customer data continue to reside in the system where they originate but are registered in a master "hub" and cross-referenced so applications can find the data. The data in their source system are not touched.

Nearly a year later, Donnelley brought up its

Customer Master Data Store, which integrates the data from numerous systems from Donnelley acquisitions. Data that are outdated, incomplete, or incorrectly formatted are corrected or eliminated. A registry points to where the source data are stored. By having a single consistent enterprise-wide set of data with common definitions and standards, management is able to easily find out what kind of business and how much business it has with a particular customer to identify top customers and sales opportunities. And when Donelley acquires a company, it can quickly see a list of overlapping customers.

Sources: John McCormick, "Mastering Data at R.R. Donnelley," *Information Management Magazine*, March 2009; www.rrdonnelley.com, accessed June 10, 2010; and www.purisma.com, accessed June 10, 2010.

R R Donnelley's experience illustrates the importance of data management for businesses. Donnelley has experienced phenomenal growth, primarily through acquisitions. But its business performance depends on what it can or cannot do with its data. How businesses store, organize, and manage their data has a tremendous impact on organizational effectiveness.

The chapter-opening diagram calls attention to important points raised by this case and this chapter. Management decided that the company needed to centralize the management of the company's data. Data about customers, vendors, products, and other important entities had been stored in a number of different systems and files where they could not be easily retrieved and analyzed. They were often redundant and inconsistent, limiting their usefulness. Management was unable to obtain an enterprise-wide view of all of its customers at all of its acquisitions to market its products and services and provide better service and support.

In the past, RR Donnelley had used heavily manual paper processes to reconcile its inconsistent and redundant data and manage its information from an enterprise-wide perspective. This solution was no longer viable as the organization grew larger. A more appropriate solution was to identify, consolidate, cleanse, and standardize customer and other data in a single master data management registry. In addition to using appropriate technology, Donnelley had to correct and reorganize the data into a standard format and establish rules, responsibilities, and procedures for updating and using the data.

A master data management system helps RR Donnelley boost profitability by making it easier to identify customers and sales opportunities. It also improves operational efficiency and decision making by having more accurate and complete customer data available and reducing the time required to reconcile redundant and inconsistent data.

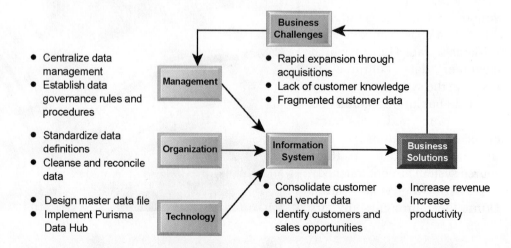

6.1   ORGANIZING DATA IN A TRADITIONAL FILE ENVIRONMENT

An effective information system provides users with accurate, timely, and relevant information. Accurate information is free of errors. Information is timely when it is available to decision makers when it is needed. Information is relevant when it is useful and appropriate for the types of work and decisions that require it.

You might be surprised to learn that many businesses don't have timely, accurate, or relevant information because the data in their information systems have been poorly organized and maintained. That's why data management is so essential. To understand the problem, let's look at how information systems arrange data in computer files and traditional methods of file management.

FILE ORGANIZATION TERMS AND CONCEPTS

A computer system organizes data in a hierarchy that starts with bits and bytes and progresses to fields, records, files, and databases (see Figure 6-1). A bit represents the smallest unit of data a computer can handle. A group of bits, called a byte, represents a single character, which can be a letter, a

FIGURE 6-1   THE DATA HIERARCHY

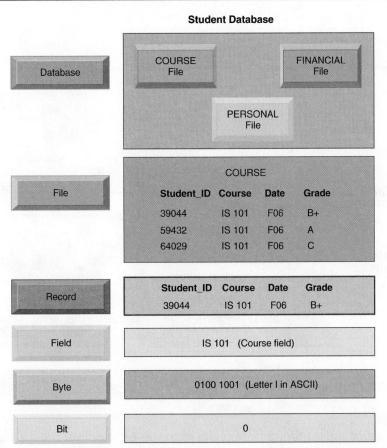

A computer system organizes data in a hierarchy that starts with the bit, which represents either a 0 or a 1. Bits can be grouped to form a byte to represent one character, number, or symbol. Bytes can be grouped to form a field, and related fields can be grouped to form a record. Related records can be collected to form a file, and related files can be organized into a database.

number, or another symbol. A grouping of characters into a word, a group of words, or a complete number (such as a person's name or age) is called a **field**. A group of related fields, such as the student's name, the course taken, the date, and the grade, comprises a **record**; a group of records of the same type is called a **file**.

For example, the records in Figure 6-1 could constitute a student course file. A group of related files makes up a **database**. The student course file illustrated in Figure 6-1 could be grouped with files on students' personal histories and financial backgrounds to create a student database.

A record describes an entity. An **entity** is a person, place, thing, or event on which we store and maintain information. Each characteristic or quality describing a particular entity is called an **attribute**. For example, Student_ID, Course, Date, and Grade are attributes of the entity COURSE. The specific values that these attributes can have are found in the fields of the record describing the entity COURSE.

PROBLEMS WITH THE TRADITIONAL FILE ENVIRONMENT

In most organizations, systems tended to grow independently without a company-wide plan. Accounting, finance, manufacturing, human resources, and sales and marketing all developed their own systems and data files. Figure 6-2 illustrates the traditional approach to information processing.

FIGURE 6-2   **TRADITIONAL FILE PROCESSING**

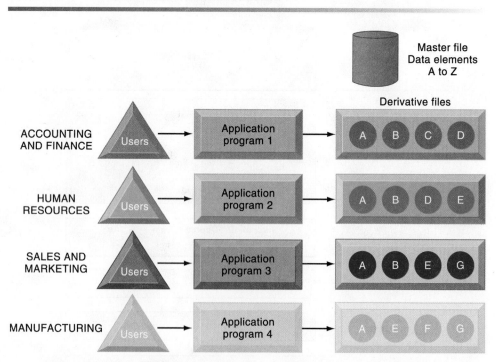

The use of a traditional approach to file processing encourages each functional area in a corporation to develop specialized applications. Each application requires a unique data file that is likely to be a subset of the master file. These subsets of the master file lead to data redundancy and inconsistency, processing inflexibility, and wasted storage resources.

Each application, of course, required its own files and its own computer program to operate. For example, the human resources functional area might have a personnel master file, a payroll file, a medical insurance file, a pension file, a mailing list file, and so forth until tens, perhaps hundreds, of files and programs existed. In the company as a whole, this process led to multiple master files created, maintained, and operated by separate divisions or departments. As this process goes on for 5 or 10 years, the organization is saddled with hundreds of programs and applications that are very difficult to maintain and manage. The resulting problems are data redundancy and inconsistency, program-data dependence, inflexibility, poor data security, and an inability to share data among applications.

Data Redundancy and Inconsistency

Data redundancy is the presence of duplicate data in multiple data files so that the same data are stored in more than place or location. Data redundancy occurs when different groups in an organization independently collect the same piece of data and store it independently of each other. Data redundancy wastes storage resources and also leads to **data inconsistency**, where the same attribute may have different values. For example, in instances of the entity COURSE illustrated in Figure 6-1, the Date may be updated in some systems but not in others. The same attribute, Student_ID, may also have different names in different systems throughout the organization. Some systems might use Student_ID and others might use ID, for example.

Additional confusion might result from using different coding systems to represent values for an attribute. For instance, the sales, inventory, and manufacturing systems of a clothing retailer might use different codes to represent clothing size. One system might represent clothing size as "extra large," whereas another might use the code "XL" for the same purpose. The resulting confusion would make it difficult for companies to create customer relationship management, supply chain management, or enterprise systems that integrate data from different sources.

Program-Data Dependence

Program-data dependence refers to the coupling of data stored in files and the specific programs required to update and maintain those files such that changes in programs require changes to the data. Every traditional computer program has to describe the location and nature of the data with which it works. In a traditional file environment, any change in a software program could require a change in the data accessed by that program. One program might be modified from a five-digit to a nine-digit ZIP code. If the original data file were changed from five-digit to nine-digit ZIP codes, then other programs that required the five-digit ZIP code would no longer work properly. Such changes could cost millions of dollars to implement properly.

Lack of Flexibility

A traditional file system can deliver routine scheduled reports after extensive programming efforts, but it cannot deliver ad hoc reports or respond to unanticipated information requirements in a timely fashion. The information required by ad hoc requests is somewhere in the system but may be too expensive to retrieve. Several programmers might have to work for weeks to put together the required data items in a new file.

Poor Security

Because there is little control or management of data, access to and dissemination of information may be out of control. Management may have no way of knowing who is accessing or even making changes to the organization's data.

Lack of Data Sharing and Availability

Because pieces of information in different files and different parts of the organization cannot be related to one another, it is virtually impossible for information to be shared or accessed in a timely manner. Information cannot flow freely across different functional areas or different parts of the organization. If users find different values of the same piece of information in two different systems, they may not want to use these systems because they cannot trust the accuracy of their data.

6.2 | THE DATABASE APPROACH TO DATA MANAGEMENT

Database technology cuts through many of the problems of traditional file organization. A more rigorous definition of a **database** is a collection of data organized to serve many applications efficiently by centralizing the data and controlling redundant data. Rather than storing data in separate files for each application, data are stored so as to appear to users as being stored in only one location. A single database services multiple applications. For example, instead of a corporation storing employee data in separate information systems and separate files for personnel, payroll, and benefits, the corporation could create a single common human resources database.

DATABASE MANAGEMENT SYSTEMS

A **database management system (DBMS)** is software that permits an organization to centralize data, manage them efficiently, and provide access to the stored data by application programs. The DBMS acts as an interface between application programs and the physical data files. When the application program calls for a data item, such as gross pay, the DBMS finds this item in the database and presents it to the application program. Using traditional data files, the programmer would have to specify the size and format of each data element used in the program and then tell the computer where they were located.

The DBMS relieves the programmer or end user from the task of understanding where and how the data are actually stored by separating the logical and physical views of the data. The *logical view* presents data as they would be perceived by end users or business specialists, whereas the *physical view* shows how data are actually organized and structured on physical storage media.

The database management software makes the physical database available for different logical views required by users. For example, for the human resources database illustrated in Figure 6-3, a benefits specialist might require a view consisting of the employee's name, social security number, and health insurance coverage. A payroll department member might need data such as the employee's name, social security number, gross pay, and net pay. The data for

FIGURE 6-3   **HUMAN RESOURCES DATABASE WITH MULTIPLE VIEWS**

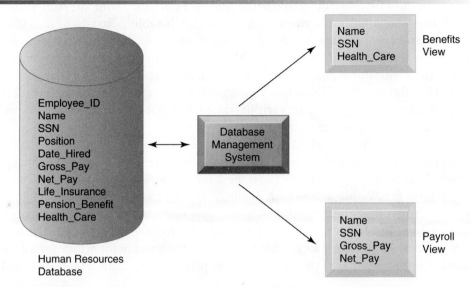

A single human resources database provides many different views of data, depending on the information requirements of the user. Illustrated here are two possible views, one of interest to a benefits specialist and one of interest to a member of the company's payroll department.

all these views are stored in a single database, where they can be more easily managed by the organization.

How a DBMS Solves the Problems of the Traditional File Environment

A DBMS reduces data redundancy and inconsistency by minimizing isolated files in which the same data are repeated. The DBMS may not enable the organization to eliminate data redundancy entirely, but it can help control redundancy. Even if the organization maintains some redundant data, using a DBMS eliminates data inconsistency because the DBMS can help the organization ensure that every occurrence of redundant data has the same values. The DBMS uncouples programs and data, enabling data to stand on their own. Access and availability of information will be increased and program development and maintenance costs reduced because users and programmers can perform ad hoc queries of data in the database. The DBMS enables the organization to centrally manage data, their use, and security.

Relational DBMS

Contemporary DBMS use different database models to keep track of entities, attributes, and relationships. The most popular type of DBMS today for PCs as well as for larger computers and mainframes is the **relational DBMS**. Relational databases represent data as two-dimensional tables (called relations). Tables may be referred to as files. Each table contains data on an entity and its attributes. Microsoft Access is a relational DBMS for desktop systems, whereas DB2, Oracle Database, and Microsoft SQL Server are relational DBMS for large mainframes and midrange computers. MySQL is a popular open-source DBMS, and Oracle Database Lite is a DBMS for small handheld computing devices.

Let's look at how a relational database organizes data about suppliers and parts (see Figure 6-4). The database has a separate table for the entity SUPPLIER and a table for the entity PART. Each table consists of a grid of columns and rows of data. Each individual element of data for each entity is stored as a separate field, and each field represents an attribute for that entity. Fields in a relational database are also called columns. For the entity SUPPLIER, the supplier identification number, name, street, city, state, and ZIP code are stored as separate fields within the SUPPLIER table and each field represents an attribute for the entity SUPPLIER.

The actual information about a single supplier that resides in a table is called a row. Rows are commonly referred to as records, or in very technical terms, as **tuples**. Data for the entity PART have their own separate table.

The field for Supplier_Number in the SUPPLIER table uniquely identifies each record so that the record can be retrieved, updated, or sorted and it is called a **key field**. Each table in a relational database has one field that is designated as its **primary key**. This key field is the unique identifier for all the information in any row of the table and this primary key cannot be duplicated.

FIGURE 6-4 **RELATIONAL DATABASE TABLES**

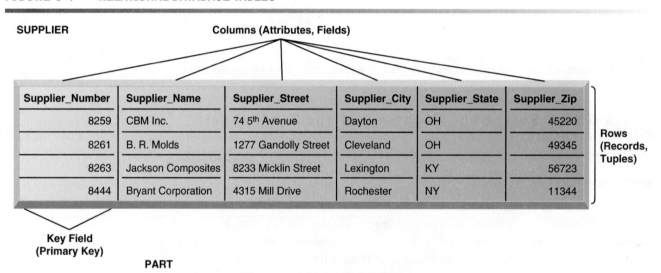

A relational database organizes data in the form of two-dimensional tables. Illustrated here are tables for the entities SUPPLIER and PART showing how they represent each entity and its attributes. Supplier_Number is a primary key for the SUPPLIER table and a foreign key for the PART table.

Supplier_Number is the primary key for the SUPPLIER table and Part_Number is the primary key for the PART table. Note that Supplier_Number appears in both the SUPPLIER and PART tables. In the SUPPLIER table, Supplier_Number is the primary key. When the field Supplier_Number appears in the PART table it is called a **foreign key** and is essentially a lookup field to look up data about the supplier of a specific part.

Operations of a Relational DBMS

Relational database tables can be combined easily to deliver data required by users, provided that any two tables share a common data element. Suppose we wanted to find in this database the names of suppliers who could provide us with part number 137 or part number 150. We would need information from two tables: the SUPPLIER table and the PART table. Note that these two files have a shared data element: Supplier_Number.

In a relational database, three basic operations, as shown in Figure 6-5, are used to develop useful sets of data: select, join, and project. The *select* operation creates a subset consisting of all records in the file that meet stated criteria. Select creates, in other words, a subset of rows that meet certain criteria. In our example, we want to select records (rows) from the PART table where the Part_Number equals 137 or 150. The *join* operation combines relational tables to provide the user with more information than is available in individual tables. In our example, we want to join the now-shortened PART table (only parts 137 or 150 will be presented) and the SUPPLIER table into a single new table.

The *project* operation creates a subset consisting of columns in a table, permitting the user to create new tables that contain only the information required. In our example, we want to extract from the new table only the following columns: Part_Number, Part_Name, Supplier_Number, and Supplier_Name.

Object-Oriented DBMS

Many applications today and in the future require databases that can store and retrieve not only structured numbers and characters but also drawings, images, photographs, voice, and full-motion video. DBMS designed for organizing structured data into rows and columns are not well suited to handling graphics-based or multimedia applications. Object-oriented databases are better suited for this purpose.

An **object-oriented DBMS** stores the data and procedures that act on those data as objects that can be automatically retrieved and shared. Object-oriented database management systems (OODBMS) are becoming popular because they can be used to manage the various multimedia components or Java applets used in Web applications, which typically integrate pieces of information from a variety of sources.

Although object-oriented databases can store more complex types of information than relational DBMS, they are relatively slow compared with relational DBMS for processing large numbers of transactions. Hybrid **object-relational DBMS** systems are now available to provide capabilities of both object-oriented and relational DBMS.

Databases in the Cloud

Suppose your company wants to use cloud computing services. Is there a way to manage data in the cloud? The answer is a qualified "Yes." Cloud computing providers offer database management services, but these services typically have less functionality than their on-premises counterparts. At the moment,

FIGURE 6-5 THE THREE BASIC OPERATIONS OF A RELATIONAL DBMS

PART

Part_Number	Part_Name	Unit_Price	Supplier_Number
137	Door latch	22.00	8259
145	Side mirror	12.00	8444
150	Door molding	6.00	8263
152	Door lock	31.00	8259
155	Compressor	54.00	8261
178	Door handle	10.00	8259

Select Part_Number = 137 or 150

SUPPLIER

Supplier_Number	Supplier_Name	Supplier_Street	Supplier_City	Supplier_State	Supplier_Zip
8259	CBM Inc.	74 5th Avenue	Dayton	OH	45220
8261	B. R. Molds	1277 Gandolly Street	Cleveland	OH	49345
8263	Jackson Components	8233 Micklin Street	Lexington	KY	56723
8444	Bryant Corporation	4315 Mill Drive	Rochester	NY	11344

Join by Supplier_Number

Part_Number	Part_Name	Supplier_Number	Supplier_Name
137	Door latch	8259	CBM Inc.
150	Door molding	8263	Jackson Components

Project selected columns

The select, join, and project operations enable data from two different tables to be combined and only selected attributes to be displayed.

the primary customer base for cloud-based data management consists of Web-focused start-ups or small to medium-sized businesses looking for database capabilities at a lower price than a standard relational DBMS.

Amazon Web Services has both a simple non-relational database called SimpleDB and a Relational Database Service, which is based on an online implementation of the MySQL open source DBMS. Amazon Relational Database Service (Amazon RDS) offers the full range of capabilities of MySQL. Pricing is based on usage. (Charges run from 11 cents per hour for a small database using 1.7 GB of server memory to $3.10 per hour for a large database using 68 GB of server memory.) There are also charges for the volume of data stored, the number of input-output requests, the amount of data written to the database, and the amount of data read from the database.

Amazon Web Services additionally offers Oracle customers the option to license Oracle Database 11g, Oracle Enterprise Manager, and Oracle Fusion Middleware to run on the Amazon EC2 (Elastic Cloud Compute) platform.

Microsoft SQL Azure Database is a cloud-based relational database service based on Microsoft's SQL Server DBMS. It provides a highly available, scalable database service hosted by Microsoft in the cloud. SQL Azure Database helps reduce costs by integrating with existing software tools and providing symmetry with on-premises and cloud databases.

TicketDirect, which sells tickets to concerts, sporting events, theater performances, and movies in Australia and New Zealand, adopted the SQL Azure Database cloud platform in order to improve management of peak system loads during major ticket sales. It migrated its data to the SQL Azure database. By moving to a cloud solution, TicketDirect is able to scale its computing resources in response to real-time demand while keeping costs low.

CAPABILITIES OF DATABASE MANAGEMENT SYSTEMS

A DBMS includes capabilities and tools for organizing, managing, and accessing the data in the database. The most important are its data definition language, data dictionary, and data manipulation language.

DBMS have a **data definition** capability to specify the structure of the content of the database. It would be used to create database tables and to define the characteristics of the fields in each table. This information about the database would be documented in a data dictionary. A **data dictionary** is an automated or manual file that stores definitions of data elements and their characteristics.

Microsoft Access has a rudimentary data dictionary capability that displays information about the name, description, size, type, format, and other properties of each field in a table (see Figure 6-6). Data dictionaries for large corporate databases may capture additional information, such as usage, ownership (who in the organization is responsible for maintaining the data), authorization; security, and the individuals, business functions, programs, and reports that use each data element.

Querying and Reporting

DBMS includes tools for accessing and manipulating information in databases. Most DBMS have a specialized language called a **data manipulation language** that is used to add, change, delete, and retrieve the data in the database. This language contains commands that permit end users and programming specialists to extract data from the database to satisfy information requests and develop applications. The most prominent data manipulation language today is **Structured Query Language**, or **SQL**. Figure 6-7 illustrates the SQL query that

FIGURE 6-6   MICROSOFT ACCESS DATA DICTIONARY FEATURES

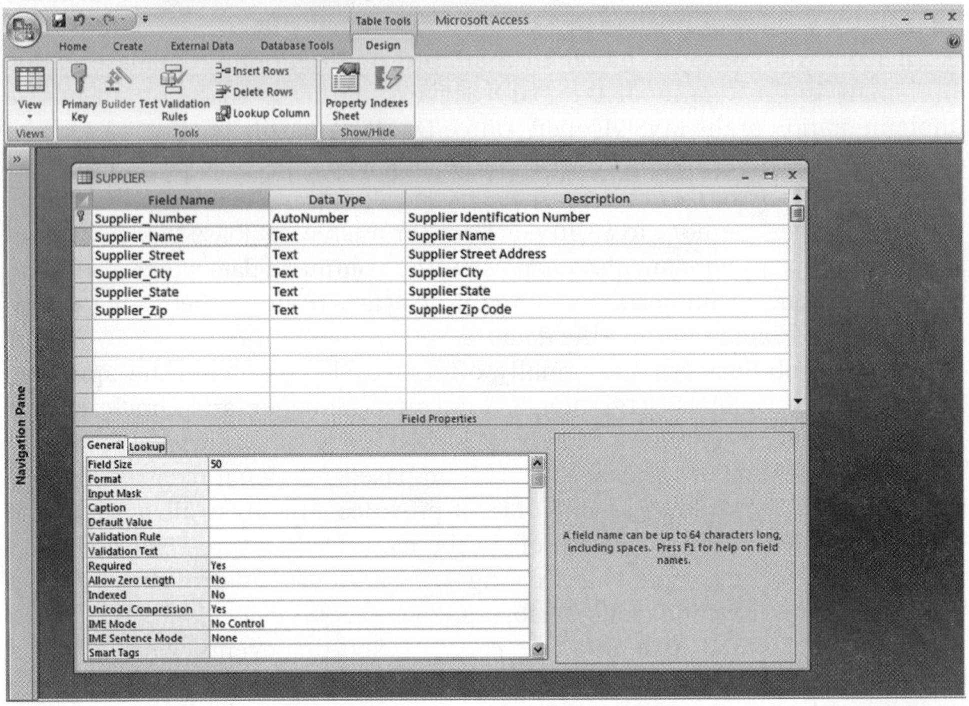

Microsoft Access has a rudimentary data dictionary capability that displays information about the size, format, and other characteristics of each field in a database. Displayed here is the information maintained in the SUPPLIER table. The small key icon to the left of Supplier_Number indicates that it is a key field.

would produce the new resultant table in Figure 6-5. You can find out more about how to perform SQL queries in our Learning Tracks for this chapter.

Users of DBMS for large and midrange computers, such as DB2, Oracle, or SQL Server, would employ SQL to retrieve information they needed from the database. Microsoft Access also uses SQL, but it provides its own set of user-friendly tools for querying databases and for organizing data from databases into more polished reports.

In Microsoft Access, you will find features that enable users to create queries by identifying the tables and fields they want and the results, and then selecting the rows from the database that meet particular criteria. These actions in turn are translated into SQL commands. Figure 6-8 illustrates how

FIGURE 6-7   EXAMPLE OF AN SQL QUERY

```
SELECT PART.Part_Number, PART.Part_Name, SUPPLIER.Supplier_Number,
SUPPLIER.Supplier_Name
FROM PART, SUPPLIER
WHERE PART.Supplier_Number = SUPPLIER.Supplier_Number AND
Part_Number = 137 OR Part_Number = 150;
```

Illustrated here are the SQL statements for a query to select suppliers for parts 137 or 150. They produce a list with the same results as Figure 6-5.

FIGURE 6-8   AN ACCESS QUERY

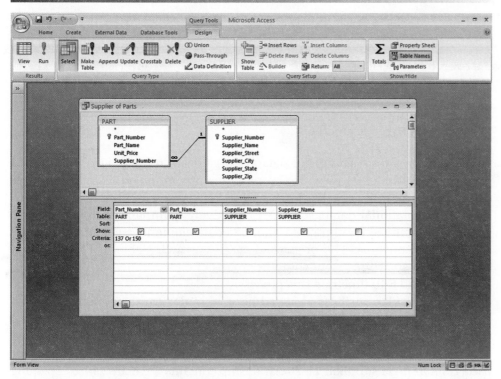

Illustrated here is how the query in Figure 6-7 would be constructed using Microsoft Access query-building tools. It shows the tables, fields, and selection criteria used for the query.

the same query as the SQL query to select parts and suppliers would be constructed using the Microsoft query-building tools.

Microsoft Access and other DBMS include capabilities for report generation so that the data of interest can be displayed in a more structured and polished format than would be possible just by querying. Crystal Reports is a popular report generator for large corporate DBMS, although it can also be used with Access. Access also has capabilities for developing desktop system applications. These include tools for creating data entry screens, reports, and developing the logic for processing transactions.

DESIGNING DATABASES

To create a database, you must understand the relationships among the data, the type of data that will be maintained in the database, how the data will be used, and how the organization will need to change to manage data from a company-wide perspective. The database requires both a conceptual design and a physical design. The conceptual, or logical, design of a database is an abstract model of the database from a business perspective, whereas the physical design shows how the database is actually arranged on direct-access storage devices.

Normalization and Entity-Relationship Diagrams

The conceptual database design describes how the data elements in the database are to be grouped. The design process identifies relationships among data elements and the most efficient way of grouping data elements together to meet business information requirements. The process also identifies redundant data elements and the groupings of data elements required for specific

FIGURE 6-9   AN UNNORMALIZED RELATION FOR ORDER

ORDER (Before Normalization)

Order_ Number	Order_ Date	Part_ Number	Part_ Name	Unit_ Price	Part_ Quantity	Supplier_ Number	Supplier_ Name	Supplier_ Street	Supplier_ City	Supplier_ State	Supplier_ Zip

An unnormalized relation contains repeating groups. For example, there can be many parts and suppliers for each order. There is only a one-to-one correspondence between Order_Number and Order_Date.

application programs. Groups of data are organized, refined, and streamlined until an overall logical view of the relationships among all the data in the database emerges.

To use a relational database model effectively, complex groupings of data must be streamlined to minimize redundant data elements and awkward many-to-many relationships. The process of creating small, stable, yet flexible and adaptive data structures from complex groups of data is called **normalization**. Figures 6-9 and 6-10 illustrate this process.

In the particular business modeled here, an order can have more than one part but each part is provided by only one supplier. If we build a relation called ORDER with all the fields included here, we would have to repeat the name and address of the supplier for every part on the order, even though the order is for parts from a single supplier. This relationship contains what are called repeating data groups because there can be many parts on a single order to a given supplier. A more efficient way to arrange the data is to break down ORDER into smaller relations, each of which describes a single entity. If we go step by step and normalize the relation ORDER, we emerge with the relations illustrated in Figure 6-10. You can find out more about normalization, entity-relationship diagramming, and database design in the Learning Tracks for this chapter.

Relational database systems try to enforce **referential integrity** rules to ensure that relationships between coupled tables remain consistent. When one table has a foreign key that points to another table, you may not add a record to the table with the foreign key unless there is a corresponding record in the linked table. In the database we examined earlier in this chapter, the foreign key

FIGURE 6-10   NORMALIZED TABLES CREATED FROM ORDER

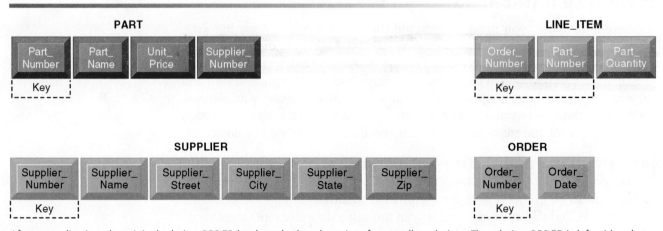

After normalization, the original relation ORDER has been broken down into four smaller relations. The relation ORDER is left with only two attributes and the relation LINE_ITEM has a combined, or concatenated, key consisting of Order_Number and Part_Number.

Supplier_Number links the PART table to the SUPPLIER table. We may not add a new record to the PART table for a part with Supplier_Number 8266 unless there is a corresponding record in the SUPPLIER table for Supplier_Number 8266. We must also delete the corresponding record in the PART table if we delete the record in the SUPPLIER table for Supplier_Number 8266. In other words, we shouldn't have parts from nonexistent suppliers!

Database designers document their data model with an **entity-relationship diagram**, illustrated in Figure 6-11. This diagram illustrates the relationship between the entities SUPPLIER, PART, LINE_ITEM, and ORDER. The boxes represent entities. The lines connecting the boxes represent relationships. A line connecting two entities that ends in two short marks designates a one-to-one relationship. A line connecting two entities that ends with a crow's foot topped by a short mark indicates a one-to-many relationship. Figure 6-11 shows that one ORDER can contain many LINE_ITEMs. (A PART can be ordered many times and appear many times as a line item in a single order.) Each PART can have only one SUPPLIER, but many PARTs can be provided by the same SUPPLIER.

It can't be emphasized enough: If the business doesn't get its data model right, the system won't be able to serve the business well. The company's systems will not be as effective as they could be because they'll have to work with data that may be inaccurate, incomplete, or difficult to retrieve. Understanding the organization's data and how they should be represented in a database is perhaps the most important lesson you can learn from this course.

For example, Famous Footwear, a shoe store chain with more than 800 locations in 49 states, could not achieve its goal of having "the right style of shoe in the right store for sale at the right price" because its database was not properly designed for rapidly adjusting store inventory. The company had an Oracle relational database running on an IBM AS/400 midrange computer, but the database was designed primarily for producing standard reports for management rather than for reacting to marketplace changes. Management could not obtain precise data on specific items in inventory in each of its stores. The company had to work around this problem by building a new database where the sales and inventory data could be better organized for analysis and inventory management.

6.3 USING DATABASES TO IMPROVE BUSINESS PERFORMANCE AND DECISION MAKING

Businesses use their databases to keep track of basic transactions, such as paying suppliers, processing orders, keeping track of customers, and paying employees. But they also need databases to provide information that will help the company

FIGURE 6-11   AN ENTITY-RELATIONSHIP DIAGRAM

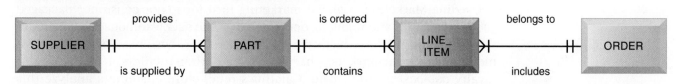

This diagram shows the relationships between the entities SUPPLIER, PART, LINE_ITEM, and ORDER that might be used to model the database in Figure 6-10.

run the business more efficiently, and help managers and employees make better decisions. If a company wants to know which product is the most popular or who is its most profitable customer, the answer lies in the data.

For example, by analyzing data from customer credit card purchases, Louise's Trattoria, a Los Angeles restaurant chain, learned that quality was more important than price for most of its customers, who were college-educated and liked fine wine. Acting on this information, the chain introduced vegetarian dishes, more seafood selections, and more expensive wines, raising sales by more than 10 percent.

In a large company, with large databases or large systems for separate functions, such as manufacturing, sales, and accounting, special capabilities and tools are required for analyzing vast quantities of data and for accessing data from multiple systems. These capabilities include data warehousing, data mining, and tools for accessing internal databases through the Web.

DATA WAREHOUSES

Suppose you want concise, reliable information about current operations, trends, and changes across the entire company If you worked in a large company, obtaining this might be difficult because data are often maintained in separate systems, such as sales, manufacturing, or accounting. Some of the data you need might be found in the sales system, and other pieces in the manufacturing system. Many of these systems are older legacy systems that use outdated data management technologies or file systems where information is difficult for users to access.

You might have to spend an inordinate amount of time locating and gathering the data you need, or you would be forced to make your decision based on incomplete knowledge. If you want information about trends, you might also have trouble finding data about past events because most firms only make their current data immediately available. Data warehousing addresses these problems.

What Is a Data Warehouse?

A **data warehouse** is a database that stores current and historical data of potential interest to decision makers throughout the company. The data originate in many core operational transaction systems, such as systems for sales, customer accounts, and manufacturing, and may include data from Web site transactions. The data warehouse consolidates and standardizes information from different operational databases so that the information can be used across the enterprise for management analysis and decision making.

Figure 6-12 illustrates how a data warehouse works. The data warehouse makes the data available for anyone to access as needed, but it cannot be altered. A data warehouse system also provides a range of ad hoc and standardized query tools, analytical tools, and graphical reporting facilities. Many firms use intranet portals to make the data warehouse information widely available throughout the firm.

Catalina Marketing, a global marketing firm for major consumer packaged goods companies and retailers, operates a gigantic data warehouse that includes three years of purchase history for 195 million U.S. customer loyalty program members at supermarkets, pharmacies, and other retailers. It is the largest loyalty database in the world. Catalina's retail store customers analyze this database of customer purchase histories to determine individual customers' buying preferences. When a shopper checks out at the cash register of one of

FIGURE 6-12   COMPONENTS OF A DATA WAREHOUSE

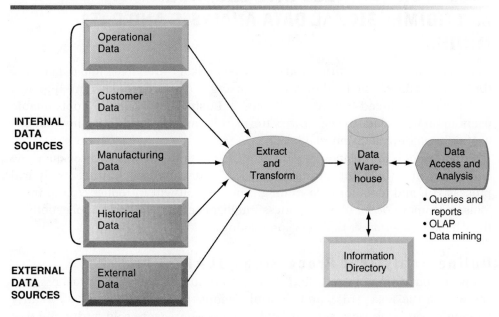

The data warehouse extracts current and historical data from multiple operational systems inside the organization. These data are combined with data from external sources and reorganized into a central database designed for management reporting and analysis. The information directory provides users with information about the data available in the warehouse.

Catalina's retail customers, the purchase is instantly analyzed along with that customer's buying history in the data warehouse to determine what coupons that customer will receive at checkout along with a receipt.

The U.S. Internal Revenue Service (IRS) maintains a Compliance Data Warehouse that consolidates taxpayer data that had been fragmented among many different legacy systems, including personal information about taxpayers and archived tax returns. These systems had been designed to process tax return forms efficiently but their data were very difficult to query and analyze. The Compliance Data Warehouse integrates taxpayer data from many disparate sources into a relational structure, which makes querying and analysis much easier. With a complete and comprehensive picture of taxpayers, the warehouse helps IRS analysts and staff identify people who are most likely to cheat on their income tax payments and respond rapidly to taxpayer queries.

Data Marts

Companies often build enterprise-wide data warehouses, where a central data warehouse serves the entire organization, or they create smaller, decentralized warehouses called data marts. A **data mart** is a subset of a data warehouse in which a summarized or highly focused portion of the organization's data is placed in a separate database for a specific population of users. For example, a company might develop marketing and sales data marts to deal with customer information. Before implementing an enterprise-wide data warehouse, bookseller Barnes & Noble maintained a series of data marts—one for point-of-sale data in retail stores, another for college bookstore sales, and a third for online sales. A data mart typically focuses on a single subject area or line of business, so it usually can be constructed more rapidly and at lower cost than an enterprise-wide data warehouse.

TOOLS FOR BUSINESS INTELLIGENCE: MULTIDIMENSIONAL DATA ANALYSIS AND DATA MINING

Once data have been captured and organized in data warehouses and data marts, they are available for further analysis using tools for business intelligence, which we introduced briefly in Chapter 2. Business intelligence tools enable users to analyze data to see new patterns, relationships, and insights that are useful for guiding decision making.

Principal tools for business intelligence include software for database querying and reporting, tools for multidimensional data analysis (online analytical processing), and tools for data mining. This section will introduce you to these tools, with more detail about business intelligence analytics and applications in the Chapter 12 discussion of decision making.

Online Analytical Processing (OLAP)

Suppose your company sells four different products—nuts, bolts, washers, and screws—in the East, West, and Central regions. If you wanted to ask a fairly straightforward question, such as how many washers were sold during the past quarter, you could easily find the answer by querying your sales database. But what if you wanted to know how many washers sold in each of your sales regions and compare actual results with projected sales?

To obtain the answer, you would need **online analytical processing (OLAP)**. OLAP supports multidimensional data analysis, enabling users to view the same data in different ways using multiple dimensions. Each aspect of information—product, pricing, cost, region, or time period—represents a different dimension. So, a product manager could use a multidimensional data analysis tool to learn how many washers were sold in the East in June, how that compares with the previous month and the previous June, and how it compares with the sales forecast. OLAP enables users to obtain online answers to ad hoc questions such as these in a fairly rapid amount of time, even when the data are stored in very large databases, such as sales figures for multiple years.

Figure 6-13 shows a multidimensional model that could be created to represent products, regions, actual sales, and projected sales. A matrix of actual sales can be stacked on top of a matrix of projected sales to form a cube with six faces. If you rotate the cube 90 degrees one way, the face showing will be product versus actual and projected sales. If you rotate the cube 90 degrees again, you will see region versus actual and projected sales. If you rotate 180 degrees from the original view, you will see projected sales and product versus region. Cubes can be nested within cubes to build complex views of data. A company would use either a specialized multidimensional database or a tool that creates multidimensional views of data in relational databases.

Data Mining

Traditional database queries answer such questions as, "How many units of product number 403 were shipped in February 2010?" OLAP, or multidimensional analysis, supports much more complex requests for information, such as "Compare sales of product 403 relative to plan by quarter and sales region for the past two years." With OLAP and query-oriented data analysis, users need to have a good idea about the information for which they are looking.

Data mining is more discovery-driven. Data mining provides insights into corporate data that cannot be obtained with OLAP by finding hidden patterns and

FIGURE 6-13   MULTIDIMENSIONAL DATA MODEL

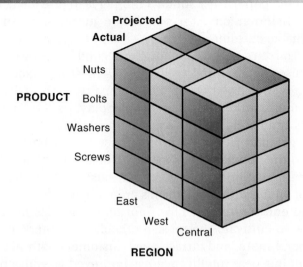

The view that is showing is product versus region. If you rotate the cube 90 degrees, the face will show product versus actual and projected sales. If you rotate the cube 90 degrees again, you will see region versus actual and projected sales. Other views are possible.

relationships in large databases and inferring rules from them to predict future behavior. The patterns and rules are used to guide decision making and forecast the effect of those decisions. The types of information obtainable from data mining include associations, sequences, classifications, clusters, and forecasts.

- *Associations* are occurrences linked to a single event. For instance, a study of supermarket purchasing patterns might reveal that, when corn chips are purchased, a cola drink is purchased 65 percent of the time, but when there is a promotion, cola is purchased 85 percent of the time. This information helps managers make better decisions because they have learned the profitability of a promotion.

- In *sequences*, events are linked over time. We might find, for example, that if a house is purchased, a new refrigerator will be purchased within two weeks 65 percent of the time, and an oven will be bought within one month of the home purchase 45 percent of the time.

- *Classification* recognizes patterns that describe the group to which an item belongs by examining existing items that have been classified and by inferring a set of rules. For example, businesses such as credit card or telephone companies worry about the loss of steady customers. Classification helps discover the characteristics of customers who are likely to leave and can provide a model to help managers predict who those customers are so that the managers can devise special campaigns to retain such customers.

- *Clustering* works in a manner similar to classification when no groups have yet been defined. A data mining tool can discover different groupings within data, such as finding affinity groups for bank cards or partitioning a database into groups of customers based on demographics and types of personal investments.

- Although these applications involve predictions, *forecasting* uses predictions in a different way. It uses a series of existing values to forecast what other values will be. For example, forecasting might find patterns in data to help managers estimate the future value of continuous variables, such as sales figures.

These systems perform high-level analyses of patterns or trends, but they can also drill down to provide more detail when needed. There are data mining applications for all the functional areas of business, and for government and scientific work. One popular use for data mining is to provide detailed analyses of patterns in customer data for one-to-one marketing campaigns or for identifying profitable customers.

For example, Harrah's Entertainment, the second-largest gambling company in its industry, uses data mining to identify its most profitable customers and generate more revenue from them. The company continually analyzes data about its customers gathered when people play its slot machines or use Harrah's casinos and hotels. Harrah's marketing department uses this information to build a detailed gambling profile, based on a particular customer's ongoing value to the company. For instance, data mining lets Harrah's know the favorite gaming experience of a regular customer at one of its Midwest riverboat casinos, along with that person's preferences for room accomodations, restaurants, and entertainment. This information guides management decisions about how to cultivate the most profitable customers, encourage those customers to spend more, and attract more customers with high revenue-generating potential. Business intelligence has improved Harrah's profits so much that it has become the centerpiece of the firm's business strategy.

Predictive analytics use data mining techniques, historical data, and assumptions about future conditions to predict outcomes of events, such as the probability a customer will respond to an offer or purchase a specific product. For example, the U.S. division of The Body Shop International plc used predictive analytics with its database of catalog, Web, and retail store customers to identify customers who were more likely to make catalog purchases. That information helped the company build a more precise and targeted mailing list for its catalogs, improving the response rate for catalog mailings and catalog revenues.

Text Mining and Web Mining

Business intelligence tools deal primarily with data that have been structured in databases and files. However, unstructured data, most in the form of text files, is believed to account for over 80 percent of an organization's useful information. E-mail, memos, call center transcripts, survey responses, legal cases, patent descriptions, and service reports are all valuable for finding patterns and trends that will help employees make better business decisions. **Text mining** tools are now available to help businesses analyze these data. These tools are able to extract key elements from large unstructured data sets, discover patterns and relationships, and summarize the information. Businesses might turn to text mining to analyze transcripts of calls to customer service centers to identify major service and repair issues.

Text mining is a relatively new technology, but what's really new are the myriad ways in which unstructured data are being generated by consumers and the business uses for these data. The Interactive Session on Technology explores some of these business applications of text mining.

The Web is another rich source of valuable information, some of which can now be mined for patterns, trends, and insights into customer behavior. The discovery and analysis of useful patterns and information from the World Wide Web is called **Web mining**. Businesses might turn to Web mining to help them understand customer behavior, evaluate the effectiveness of a particular Web site, or quantify the success of a marketing campaign. For instance, marketers use Google Trends and Google Insights for Search services, which track the popularity of various words and phrases used in Google search queries, to learn what people are interested in and what they are interested in buying.

INTERACTIVE SESSION: TECHNOLOGY

WHAT CAN BUSINESSES LEARN FROM TEXT MINING?

Text mining is the discovery of patterns and relationships from large sets of unstructured data—the kind of data we generate in e-mails, phone conversations, blog postings, online customer surveys, and tweets. The mobile digital platform has amplified the explosion in digital information, with hundreds of millions of people calling, texting, searching, "apping" (using applications), buying goods, and writing billions of e-mails on the go.

Consumers today are more than just consumers: they have more ways to collaborate, share information, and influence the opinions of their friends and peers, and the data they create in doing so have significant value to businesses. Unlike structured data, which are generated from events such as completing a purchase transaction, unstructured data have no distinct form. Nevertheless, managers believe such data may offer unique insights into customer behavior and attitudes that were much more difficult to determine years ago.

For example, in 2007, JetBlue experienced unprecedented levels of customer discontent in the wake of a February ice storm that resulted in widespread flight cancellations and planes stranded on Kennedy Airport runways. The airline received 15,000 e-mails per day from customers during the storm and immediately afterwards, up from its usual daily volume of 400. The volume was so much larger than usual that JetBlue had no simple way to read everything its customers were saying.

Fortunately, the company had recently contracted with Attensity, a leading vendor of text analytics software, and was able to use the software to analyze all of the e-mail it had received within two days. According to JetBlue research analyst Bryan Jeppsen, Attensity Analyze for Voice of the Customer (VoC) enabled JetBlue to rapidly extract customer sentiments, preferences, and requests it couldn't find any other way. This tool uses a proprietary technology to automatically identify facts, opinions, requests, trends, and trouble spots from the unstructured text of survey responses, service notes, e-mail messages, Web forums, blog entries, news articles, and other customer communications. The technology is able to accurately and automatically identify the many different "voices" customers use to express their feedback (such as a negative voice, positive voice, or conditional voice), which helps organiza-

tions pinpoint key events and relationships, such as intent to buy, intent to leave, or customer "wish" events. It can reveal specific product and service issues, reactions to marketing and public relations efforts, and even buying signals.

Attensity's software integrated with JetBlue's other customer analysis tools, such as Satmetrix's Net Promoter metrics, which classifies customers into groups that are generating positive, negative, or no feedback about the company. Using Attensity's text analytics in tandem with these tools, JetBlue developed a customer bill of rights that addressed the major issues customers had with the company.

Hotel chains like Gaylord Hotels and Choice Hotels are using text mining software to glean insights from thousands of customer satisfaction surveys provided by their guests. Gaylord Hotels is using Clarabridge's text analytics solution delivered via the Internet as a hosted software service to gather and analyze customer feedback from surveys, e-mail, chat messaging, staffed call centers, and online forums associated with guests' and meeting planners' experiences at the company's convention resorts. The Clarabridge software sorts through the hotel chain's customer surveys and gathers positive and negative comments, organizing them into a variety of categories to reveal less obvious insights. For example, guests complained about many things more frequently than noisy rooms, but complaints of noisy rooms were most frequently correlated with surveys indicating an unwillingness to return to the hotel for another stay.

Analyzing customer surveys used to take weeks, but now takes only days, thanks to the Clarabridge software. Location managers and corporate executives have also used findings from text mining to influence decisions on building improvements.

Wendy's International adopted Clarabridge software to analyze nearly 500,000 messages it collects each year from its Web-based feedback forum, call center notes, e-mail messages, receipt-based surveys, and social media. The chain's customer satisfaction team had previously used spreadsheets and keyword searches to review customer comments, a very slow manual approach. Wendy's management was looking for a better tool to speed analysis, detect emerging issues, and pinpoint troubled areas of the business at the store, regional, or corporate level.

The Clarabridge technology enables Wendy's to track customer experiences down to the store level within minutes. This timely information helps store, regional, and corporate managers spot and address problems related to meal quality, cleanliness, and speed of service.

Text analytics software caught on first with government agencies and larger companies with information systems departments that had the means to properly use the complicated software, but Clarabridge is now offering a version of its product geared towards small businesses. The technology has already caught on with law enforcement, search tool interfaces, and "listening platforms" like Nielsen Online. Listening platforms are text mining tools that focus on brand management, allowing companies to determine how consumers feel about their brand and take steps to respond to negative sentiment.

Structured data analysis won't be rendered obsolete by text analytics, but companies that are able to use both methods to develop a clearer picture of their customers' attitudes will have an easier time establishing and building their brand and gleaning insights that will enhance profitability.

Sources: Doug Henschen, "Wendy's Taps Text Analytics to Mine Customer Feedback," *Information Week*, March 23, 2010; David Stodder," How Text Analytics Drive Customer Insight" *Information Week*, February 1, 2010; Nancy David Kho, "Customer Experience and Sentiment Analysis," *KMWorld*, February 1, 2010; Siobhan Gorman, "Details of Einstein Cyber-Shield Disclosed by White House," *The Wall Street Journal*, March 2, 2010; www.attensity.com, accessed June 16, 2010; and www.clarabridge.com, accessed June 17, 2010.

CASE STUDY QUESTIONS

1. What challenges does the increase in unstructured data present for businesses?

2. How does text-mining improve decision-making?

3. What kinds of companies are most likely to benefit from text mining software? Explain your answer.

4. In what ways could text mining potentially lead to the erosion of personal information privacy? Explain.

MIS IN ACTION

Visit a Web site such as QVC.com or TripAdvisor.com detailing products or services that have customer reviews. Pick a product, hotel, or other service with at least 15 customer reviews and read those reviews, both positive and negative. How could Web content mining help the offering company improve or better market this product or service? What pieces of information should highlighted?

Web mining looks for patterns in data through content mining, structure mining, and usage mining. Web content mining is the process of extracting knowledge from the content of Web pages, which may include text, image, audio, and video data. Web structure mining extracts useful information from the links embedded in Web documents. For example, links pointing to a document indicate the popularity of the document, while links coming out of a document indicate the richness or perhaps the variety of topics covered in the document. Web usage mining examines user interaction data recorded by a Web server whenever requests for a Web site's resources are received. The usage data records the user's behavior when the user browses or makes transactions on the Web site and collects the data in a server log. Analyzing such data can help companies determine the value of particular customers, cross marketing strategies across products, and the effectiveness of promotional campaigns.

DATABASES AND THE WEB

Have you ever tried to use the Web to place an order or view a product catalog? If so, you were probably using a Web site linked to an internal corporate database. Many companies now use the Web to make some of the information in their internal databases available to customers and business partners.

Suppose, for example, a customer with a Web browser wants to search an online retailer's database for pricing information. Figure 6-14 illustrates how that customer might access the retailer's internal database over the Web. The user accesses the retailer's Web site over the Internet using Web browser software on his or her client PC. The user's Web browser software requests data from the organization's database, using HTML commands to communicate with the Web server.

Because many back-end databases cannot interpret commands written in HTML, the Web server passes these requests for data to software that translates HTML commands into SQL so that they can be processed by the DBMS working with the database. In a client/server environment, the DBMS resides on a dedicated computer called a **database server**. The DBMS receives the SQL requests and provides the required data. The middleware transfers information from the organization's internal database back to the Web server for delivery in the form of a Web page to the user.

Figure 6-14 shows that the middleware working between the Web server and the DBMS is an application server running on its own dedicated computer (see Chapter 5). The application server software handles all application operations, including transaction processing and data access, between browser-based computers and a company's back-end business applications or databases. The application server takes requests from the Web server, runs the business logic to process transactions based on those requests, and provides connectivity to the organization's back-end systems or databases. Alternatively, the software for handling these operations could be a custom program or a CGI script. A CGI script is a compact program using the *Common Gateway Interface (CGI)* specification for processing data on a Web server.

There are a number of advantages to using the Web to access an organization's internal databases. First, Web browser software is much easier to use than proprietary query tools. Second, the Web interface requires few or no changes to the internal database. It costs much less to add a Web interface in front of a legacy system than to redesign and rebuild the system to improve user access.

Accessing corporate databases through the Web is creating new efficiencies, opportunities, and business models. ThomasNet.com provides an up-to-date online directory of more than 600,000 suppliers of industrial products, such as chemicals, metals, plastics, rubber, and automotive equipment. Formerly called Thomas Register, the company used to send out huge paper catalogs with this information. Now it provides this information to users online via its Web site and has become a smaller, leaner company.

Other companies have created entirely new businesses based on access to large databases through the Web. One is the social networking site MySpace, which helps users stay connected with each other or meet new people.

FIGURE 6-14   LINKING INTERNAL DATABASES TO THE WEB

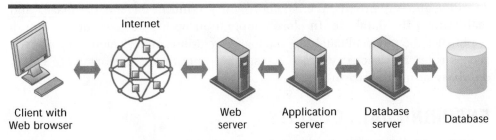

Users access an organization's internal database through the Web using their desktop PCs and Web browser software.

MySpace features music, comedy, videos, and "profiles" with information supplied by 122 million users about their age, hometown, dating preferences, marital status, and interests. It maintains a massive database to house and manage all of this content. Facebook uses a similar database.

6.4   MANAGING DATA RESOURCES

Setting up a database is only a start. In order to make sure that the data for your business remain accurate, reliable, and readily available to those who need it, your business will need special policies and procedures for data management.

ESTABLISHING AN INFORMATION POLICY

Every business, large and small, needs an information policy. Your firm's data are an important resource, and you don't want people doing whatever they want with them. You need to have rules on how the data are to be organized and maintained, and who is allowed to view the data or change them.

An **information policy** specifies the organization's rules for sharing, disseminating, acquiring, standardizing, classifying, and inventorying information. Information policy lays out specific procedures and accountabilities, identifying which users and organizational units can share information, where information can be distributed, and who is responsible for updating and maintaining the information. For example, a typical information policy would specify that only selected members of the payroll and human resources department would have the right to change and view sensitive employee data, such as an employee's salary or social security number, and that these departments are responsible for making sure that such employee data are accurate.

If you are in a small business, the information policy would be established and implemented by the owners or managers. In a large organization, managing and planning for information as a corporate resource often requires a formal data administration function. **Data administration** is responsible for the specific policies and procedures through which data can be managed as an organizational resource. These responsibilities include developing information policy, planning for data, overseeing logical database design and data dictionary development, and monitoring how information systems specialists and end-user groups use data.

You may hear the term **data governance** used to describe many of these activities. Promoted by IBM, data governance deals with the policies and processes for managing the availability, usability, integrity, and security of the data employed in an enterprise, with special emphasis on promoting privacy, security, data quality, and compliance with government regulations.

A large organization will also have a database design and management group within the corporate information systems division that is responsible for defining and organizing the structure and content of the database, and maintaining the database. In close cooperation with users, the design group establishes the physical database, the logical relations among elements, and the access rules and security procedures. The functions it performs are called **database administration**.

ENSURING DATA QUALITY

A well-designed database and information policy will go a long way toward ensuring that the business has the information it needs. However, additional steps must be taken to ensure that the data in organizational databases are accurate and remain reliable.

What would happen if a customer's telephone number or account balance were incorrect? What would be the impact if the database had the wrong price for the product you sold or your sales system and inventory system showed different prices for the same product? Data that are inaccurate, untimely, or inconsistent with other sources of information lead to incorrect decisions, product recalls, and financial losses. Inaccurate data in criminal justice and national security databases might even subject you to unnecessarily surveillance or detention, as described in the chapter-ending case study.

According to Forrester Research, 20 percent of U.S. mail and commercial package deliveries were returned because of incorrect names or addresses. Gartner Inc. reported that more than 25 percent of the critical data in large Fortune 1000 companies' databases is inaccurate or incomplete, including bad product codes and product descriptions, faulty inventory descriptions, erroneous financial data, incorrect supplier information, and incorrect employee data. (Gartner, 2007).

Think of all the times you've received several pieces of the same direct mail advertising on the same day. This is very likely the result of having your name maintained multiple times in a database. Your name may have been misspelled or you used your middle initial on one occasion and not on another or the information was initially entered onto a paper form and not scanned properly into the system. Because of these inconsistencies, the database would treat you as different people! We often receive redundant mail addressed to Laudon, Lavdon, Lauden, or Landon.

If a database is properly designed and enterprise-wide data standards established, duplicate or inconsistent data elements should be minimal. Most data quality problems, however, such as misspelled names, transposed numbers, or incorrect or missing codes, stem from errors during data input. The incidence of such errors is rising as companies move their businesses to the Web and allow customers and suppliers to enter data into their Web sites that directly update internal systems.

Before a new database is in place, organizations need to identify and correct their faulty data and establish better routines for editing data once their database is in operation. Analysis of data quality often begins with a **data quality audit**, which is a structured survey of the accuracy and level of completeness of the data in an information system. Data quality audits can be performed by surveying entire data files, surveying samples from data files, or surveying end users for their perceptions of data quality.

Data cleansing, also known as *data scrubbing*, consists of activities for detecting and correcting data in a database that are incorrect, incomplete, improperly formatted, or redundant. Data cleansing not only corrects errors but also enforces consistency among different sets of data that originated in separate information systems. Specialized data-cleansing software is available to automatically survey data files, correct errors in the data, and integrate the data in a consistent company-wide format.

Data quality problems are not just business problems. They also pose serious problems for individuals, affecting their financial condition and even their jobs. The Interactive Session on Organizations describes some of these impacts, as it details the data quality problems found in the companies that collect and report consumer credit data. As you read this case, look for the management, organization, and technology factors behind this problem, and whether existing solutions are adequate.

INTERACTIVE SESSION: ORGANIZATIONS

CREDIT BUREAU ERRORS—BIG PEOPLE PROBLEMS

You've found the car of your dreams. You have a good job and enough money for a down payment. All you need is an auto loan for $14,000. You have a few credit card bills, which you diligently pay off each month. But when you apply for the loan you're turned down. When you ask why, you're told you have an overdue loan from a bank you've never heard of. You've just become one of the millions of people who have been victimized by inaccurate or outdated data in credit bureaus' information systems.

Most data on U.S. consumers' credit histories are collected and maintained by three national credit reporting agencies: Experian, Equifax, and TransUnion. These organizations collect data from various sources to create a detailed dossier of an individual's borrowing and bill paying habits. This information helps lenders assess a person's credit worthiness, the ability to pay back a loan, and can affect the interest rate and other terms of a loan, including whether a loan will be granted in the first place. It can even affect the chances of finding or keeping a job: At least one-third of employers check credit reports when making hiring, firing, or promotion decisions.

U.S. credit bureaus collect personal information and financial data from a variety of sources, including creditors, lenders, utilities, debt collection agencies, and the courts. These data are aggregated and stored in massive databases maintained by the credit bureaus. The credit bureaus then sell this information to other companies to use for credit assessment.

The credit bureaus claim they know which credit cards are in each consumer's wallet, how much is due on the mortgage, and whether the electric bill is paid on time. But if the wrong information gets into their systems, whether through identity theft or errors transmitted by creditors, watch out! Untangling the mess can be almost impossible.

The bureaus understand the importance of providing accurate information to both lenders and consumers. But they also recognize that their own systems are responsible for many credit-report errors. Some mistakes occur because of the procedures for matching loans to individual credit reports.

The sheer volume of information being transmitted from creditors to credit bureaus increases the likelihood of mistakes. Experian, for example, updates 30 million credit reports each day and roughly 2 billion credit reports each month. It matches the identifying personal information in a credit application or credit account with the identifying personal information in a consumer credit file. Identifying personal information includes items such as name (first name, last name and middle initial), full current address and ZIP code, full previous address and ZIP code, and social security number. The new credit information goes into the consumer credit file that it best matches.

The credit bureaus rarely receive information that matches in all the fields in credit files, so they have to determine how much variation to allow and still call it a match. Imperfect data lead to imperfect matches. A consumer might provide incomplete or inaccurate information on a credit application. A creditor might submit incomplete or inaccurate information to the credit bureaus. If the wrong person matches better than anyone else, the data could unfortunately go into the wrong account.

Perhaps the consumer didn't write clearly on the account application. Name variations on different credit accounts can also result in less-than-perfect matches. Take the name Edward Jeffrey Johnson. One account may say Edward Johnson. Another may say Ed Johnson. Another might say Edward J. Johnson. Suppose the last two digits of Edward's social security number get transposed—more chance for mismatches.

If the name or social security number on another person's account partially matches the data in your file, the computer might attach that person's data to your record. Your record might likewise be corrupted if workers in companies supplying tax and bankruptcy data from court and government records accidentally transpose a digit or misread a document.

The credit bureaus claim it is impossible for them to monitor the accuracy of the 3.5 billion pieces of credit account information they receive each month. They must continually contend with bogus claims from consumers who falsify lender information or use shady credit-repair companies that challenge all the negative information on a credit report regardless of its validity. To separate the good from the bad, the credit bureaus use an automated e-OSCAR (Electronic Online Solution for Complete and Accurate Reporting) system to forward consumer disputes to lenders for verification.

If your credit report showed an error, the bureaus usually do not contact the lender directly to correct the information. To save money, the bureaus send consumer protests and evidence to a data processing center run by a third-party contractor. These contractors rapidly summarize every complaint with a short comment and 2-digit code from a menu of 26 options. For example, the code A3 designates "belongs to another individual with a similar name." These summaries are often too brief to include the background banks need to understand a complaint.

Although this system fixes large numbers of errors (data are updated or corrected for 72 percent of disputes), consumers have few options if the system fails. Consumers who file a second dispute without providing new information might have their dispute dismissed as "frivolous." If the consumer tries to contact the lender that made the error on their own, banks have no obligation to investigate the dispute—unless it's sent by a credit bureau.

Sources: Dennis McCafferty, "Bad Credit Could Cost You a Job," *Baseline*, June 7, 2010; Kristen McNamara, "Bad Credit Derails Job Seekers," *The Wall Street Journal*, March 16, 2010; Anne Kadet, Lucy Lazarony, "Your Name Can Mess Up Your Credit Report, Bankrate.com, accessed July 1, 2009; "Credit Report Fix a Headache," *Atlanta Journal-Constitution*, June 14, 2009; and "Why Credit Bureaus Can't Get It Right," *Smart Money*, March 2009.

CASE STUDY QUESTIONS

1. Assess the business impact of credit bureaus' data quality problems for the credit bureaus, for lenders, for individuals.

2. Are any ethical issues raised by credit bureaus' data quality problems? Explain your answer.

3. Analyze the management, organization, and technology factors responsible for credit bureaus' data quality problems.

4. What can be done to solve these problems?

MIS IN ACTION

Go to the Experian Web site (www.experian.com) and explore the site, with special attention to its services for businesses and small businesses. Then answer the following questions:

1. List and describe five services for businesses and explain how each uses consumer data. Describe the kinds of businesses that would use these services.

2. Explain how each of these services is affected by inaccurate consumer data.

6.5   HANDS-ON MIS PROJECTS

The projects in this section give you hands-on experience in analyzing data quality problems, establishing company-wide data standards, creating a database for inventory management, and using the Web to search online databases for overseas business resources.

Management Decision Problems

1. Emerson Process Management, a global supplier of measurement, analytical, and monitoring instruments and services based in Austin, Texas, had a new data warehouse designed for analyzing customer activity to improve service and marketing that was full of inaccurate and redundant data. The data in the warehouse came from numerous transaction processing systems in Europe, Asia, and other locations around the world. The team that designed the warehouse had assumed that sales groups in all these areas would enter customer names and addresses the same way, regardless of their location. In fact, cultural differences combined with complications from absorbing companies that Emerson had acquired led to multiple ways of entering quotes, billing, shipping, and other data. Assess the potential business impact of these data quality problems. What decisions have to be made and steps taken to reach a solution?

2. Your industrial supply company wants to create a data warehouse where management can obtain a single corporate-wide view of critical sales information to identify best-selling products in specific geographic areas, key customers, and sales trends. Your sales and product information are stored in several different systems: a divisional sales system running on a Unix server and a corporate sales system running on an IBM mainframe. You would like to create a single standard format that consolidates these data from both systems. The following format has been proposed.

PRODUCT_ID	PRODUCT_DESCRIPTION	COST_PER_UNIT	UNITS_SOLD	SALES_REGION	DIVISION	CUSTOMER_ID

The following are sample files from the two systems that would supply the data for the data warehouse:

CORPORATE SALES SYSTEM

PRODUCT_ID	PRODUCT_DESCRIPTION	UNIT_COST	UNITS_SOLD	SALES_TERRITORY	DIVISION
60231	Bearing, 4"	5.28	900,245	Northeast	Parts
85773	SS assembly unit	12.45	992,111	Midwest	Parts

MECHANICAL PARTS DIVISION SALES SYSTEM

PROD_NO	PRODUCT_DESCRIPTION	COST_PER_UNIT	UNITS_SOLD	SALES_REGION	CUSTOMER_ID
60231	4" Steel bearing	5.28	900,245	N.E.	Anderson
85773	SS assembly unit	12.45	992,111	M.W.	Kelly Industries

- What business problems are created by not having these data in a single standard format?

- How easy would it be to create a database with a single standard format that could store the data from both systems? Identify the problems that would have to be addressed.

- Should the problems be solved by database specialists or general business managers? Explain.

- Who should have the authority to finalize a single company-wide format for this information in the data warehouse?

Achieving Operational Excellence: Building a Relational Database for Inventory Management

Software skills: Database design, querying, and reporting
Business skills: Inventory management

Businesses today depend on databases to provide reliable information about items in inventory, items that need restocking, and inventory costs. In this exercise, you'll use database software to design a database for managing inventory for a small business.

Sylvester's Bike Shop, located in San Francisco, California, sells road, mountain, hybrid, leisure, and children's bicycles. Currently, Sylvester's purchases bikes from three suppliers but plans to add new suppliers in the near future. This rapidly growing business needs a database system to manage this information.

Initially, the database should house information about suppliers and products. The database will contain two tables: a supplier table and a product table. The reorder level refers to the number of items in inventory that triggers a decision to order more items to prevent a stockout. (In other words, if the number of units of a particular item in inventory falls below the reorder level, the item should be reordered.) The user should be able to perform several queries and produce several managerial reports based on the data contained in the two tables.

Using the information found in the tables in MyMISLab, build a simple relational database for Sylvester's. Once you have built the database, perform the following activities:

- Prepare a report that identifies the five most expensive bicycles. The report should list the bicycles in descending order from most expensive to least expensive, the quantity on hand for each, and the markup percentage for each.

- Prepare a report that lists each supplier, its products, the quantities on hand, and associated reorder levels. The report should be sorted alphabetically by supplier. Within each supplier category, the products should be sorted alphabetically.

- Prepare a report listing only the bicycles that are low in stock and need to be reordered. The report should provide supplier information for the items identified.

- Write a brief description of how the database could be enhanced to further improve management of the business. What tables or fields should be added? What additional reports would be useful?

Improving Decision Making: Searching Online Databases for Overseas Business Resources

Software skills: Online databases
Business skills: Researching services for overseas operations

Internet users have access to many thousands of Web-enabled databases with information on services and products in faraway locations. This project develops skills in searching these online databases.

Your company is located in Greensboro, North Carolina, and manufactures office furniture of various types. You have recently acquired several new customers in Australia, and a study you commissioned indicates that, with a presence there, you could greatly increase your sales. Moreover, your study indicates that you could do even better if you actually manufactured many of your products locally (in Australia). First, you need to set up an office in Melbourne to establish a presence, and then you need to begin importing from the United States. You then can plan to start producing locally.

You will soon be traveling to the area to make plans to actually set up an office, and you want to meet with organizations that can help you with your operation. You will need to engage people or organizations that offer many services necessary for you to open your office, including lawyers, accountants, import-export experts, telecommunications equipment and support, and even trainers who can help you to prepare your future employees to work for you. Start by searching for U.S. Department of Commerce advice on doing business in Australia. Then try the following online databases to locate companies that you would like to meet with during your coming trip: Australian Business Register (abr.business.gov.au/), Australia Trade Now (australiatradenow.com/), and the Nationwide Business Directory of Australia (www.nationwide.com.au). If necessary, you could also try search engines such as Yahoo and Google. Then perform the following activities:

- List the companies you would contact to interview on your trip to determine whether they can help you with these and any other functions you think vital to establishing your office.

- Rate the databases you used for accuracy of name, completeness, ease of use, and general helpfulness.

- What does this exercise tell you about the design of databases?

LEARNING TRACK MODULES

The following Learning Tracks provide content relevant to topics covered in this chapter:

1. Database Design, Normalization, and Entity-Relationship Diagramming
2. Introduction to SQL
3. Hierarchical and Network Data Models

Review Summary

1. ***What are the problems of managing data resources in a traditional file environment and how are they solved by a database management system?***

    Traditional file management techniques make it difficult for organizations to keep track of all of the pieces of data they use in a systematic way and to organize these data so that they can be easily accessed. Different functional areas and groups were allowed to develop their own files independently. Over time, this traditional file management environment creates problems such as data redundancy and inconsistency, program-data dependence, inflexibility, poor security, and lack of data sharing and availability. A database management system (DBMS) solves these problems with software that permits centralization of data and data management so that businesses have a single consistent source for all their data needs. Using a DBMS minimizes redundant and inconsistent files.

2. ***What are the major capabilities of DBMS and why is a relational DBMS so powerful?***

    The principal capabilities of a DBMS includes a data definition capability, a data dictionary capability, and a data manipulation language. The data definition capability specifies the structure and content of the database. The data dictionary is an automated or manual file that stores information about the data in the database, including names, definitions, formats, and descriptions of data elements. The data manipulation language, such as SQL, is a specialized language for accessing and manipulating the data in the database.

    The relational database is the primary method for organizing and maintaining data today in information systems because it is so flexible and accessible. It organizes data in two-dimensional tables called relations with rows and columns. Each table contains data about an entity and its attributes. Each row represents a record and each column represents an attribute or field. Each table also contains a key field to uniquely identify each record for retrieval or manipulation. Relational database tables can be combined easily to deliver data required by users, provided that any two tables share a common data element.

3. ***What are some important database design principles?***

    Designing a database requires both a logical design and a physical design. The logical design models the database from a business perspective. The organization's data model should reflect its key business processes and decision-making requirements. The process of creating small, stable, flexible, and adaptive data structures from complex groups of data when designing a relational database is termed normalization. A well-designed relational database will not have many-to-many relationships, and all attributes for a specific entity will only apply to that entity. It will try to enforce referential integrity rules to ensure that relationships between coupled tables remain consistent. An entity-relationship diagram graphically depicts the relationship between entities (tables) in a relational database.

4. ***What are the principal tools and technologies for accessing information from databases to improve business performance and decision making?***

    Powerful tools are available to analyze and access the information in databases. A data warehouse consolidates current and historical data from many different operational systems in a central database designed for reporting and analysis. Data warehouses support multidimensional data analysis, also known as online analytical processing (OLAP). OLAP represents relationships among data as a multidimensional structure, which can be visualized as cubes of data and cubes within cubes of data, enabling more sophisticated data analysis. Data mining analyzes large pools of data, including the contents of data warehouses, to find patterns and rules that can be used to predict future behavior and guide decision making. Text mining tools help businesses analyze large unstructured data sets consisting of text. Web mining tools focus on analysis of useful patterns and information from the World Wide Web, examining the structure of Web sites and activities of Web site users as well as the contents of Web pages. Conventional databases can be linked via middleware to the Web or a Web interface to facilitate user access to an organization's internal data.

5. *Why are information policy, data administration, and data quality assurance essential for managing the firm's data resources?*

Developing a database environment requires policies and procedures for managing organizational data as well as a good data model and database technology. A formal information policy governs the maintenance, distribution, and use of information in the organization. In large corporations, a formal data administration function is responsible for information policy, as well as for data planning, data dictionary development, and monitoring data usage in the firm.

Data that are inaccurate, incomplete, or inconsistent create serious operational and financial problems for businesses because they may create inaccuracies in product pricing, customer accounts, and inventory data, and lead to inaccurate decisions about the actions that should be taken by the firm. Firms must take special steps to make sure they have a high level of data quality. These include using enterprise-wide data standards, databases designed to minimize inconsistent and redundant data, data quality audits, and data cleansing software.

Key Terms

Attribute, 210

Data administration, 230

Data cleansing, 231

Data definition, 217

Data dictionary, 217

Data governance, 230

Data inconsistency, 211

Data manipulation language, 217

Data mart, 223

Data mining, 224

Data quality audit, 231

Data redundancy, 211

Data warehouse, 222

Database, 210

Database (rigorous definition), 212

Database administration, 230

Database management system (DBMS), 212

Database server, 229

Entity, 210

Entity-relationship diagram, 221

Field, 210

File, 210

Foreign key, 215

Information policy, 230

Key field, 214

Normalization, 219

Object-oriented DBMS, 215

Object-relational DBMS, 215

Online analytical processing (OLAP), 224

Predictive analytics, 226

Primary key, 210

Program-data dependence, 211

Record, 214

Referential integrity, 220

Relational DBMS, 213

Structured Query Language (SQL), 217

Text mining, 226

Tuple, 214

Web mining, 226

Review Questions

1. What are the problems of managing data resources in a traditional file environment and how are they solved by a database management system?

   - List and describe each of the components in the data hierarchy.
   - Define and explain the significance of entities, attributes, and key fields.
   - List and describe the problems of the traditional file environment.
   - Define a database and a database management system and describe how it solves the problems of a traditional file environment.

2. What are the major capabilities of DBMS and why is a relational DBMS so powerful?

   - Name and briefly describe the capabilities of a DBMS.
   - Define a relational DBMS and explain how it organizes data.
   - List and describe the three operations of a relational DBMS.

3. What are some important database design principles?

   - Define and describe normalization and referential integrity and explain how they contribute to a well-designed relational database.
   - Define and describe an entity-relationship diagram and explain its role in database design.

4. What are the principal tools and technologies for accessing information from databases to improve business performance and decision making?

   - Define a data warehouse, explaining how it works and how it benefits organizations.
   - Define business intelligence and explain how it is related to database technology.
   - Describe the capabilities of online analytical processing (OLAP).
   - Define data mining, describing how it differs from OLAP and the types of information it provides.
   - Explain how text mining and Web mining differ from conventional data mining.
   - Describe how users can access information from a company's internal databases through the Web.

5. Why are information policy, data administration, and data quality assurance essential for managing the firm's data resources?

   - Describe the roles of information policy and data administration in information management.
   - Explain why data quality audits and data cleansing are essential.

Discussion Questions

1. It has been said that you do not need database management software to create a database environment. Discuss.

2. To what extent should end users be involved in the selection of a database management system and database design?

3. What are the consequences of an organization not having an information policy?

Video Cases

Video Cases and Instructional Videos illustrating some of the concepts in this chapter are available. Contact your instructor to access these videos.

Collaboration and Teamwork: Identifying Entities and Attributes in an Online Database

With your team of three or four students, select an online database to explore, such as AOL Music, iGo.com, or the Internet Movie Database (IMDb). Explore one of these Web sites to see what information it provides. Then list the entities and attributes that the company running the Web site must keep track of in its databases. Diagram the relationship between the entities you have identified. If possible, use Google Sites to post links to Web pages, team communication announcements, and work assignments; to brainstorm; and to work collaboratively on project documents. Try to use Google Docs to develop a presentation of your findings for the class.

The Terror Watch List Database's Troubles Continue
CASE STUDY

In the aftermath of the 9-11 attacks, the FBI's Terrorist Screening Center, or TSC, was established to consolidate information about suspected terrorists from multiple government agencies into a single list to enhance inter-agency communication. A database of suspected terrorists known as the terrorist watch list was created. Multiple U.S. government agencies had been maintaining separate lists and these agencies lacked a consistent process to share relevant information.

Records in the TSC database contain sensitive but unclassified information on terrorist identities, such as name and date of birth, that can be shared with other screening agencies. Classified information about the people in the watch list is maintained in other law enforcement and intelligence agency databases. Records for the watchlist database are provided by two sources: The National Counterterrorism Center (NCTC) managed by the Office of the Director of National Intelligence provides identifying information on individuals with ties to international terrorism. The FBI provides identifying information on individuals with ties to purely domestic terrorism.

These agencies collect and maintain terrorist information and nominate individuals for inclusion in the TSC's consolidated watch list. They are required to follow strict procedures established by the head of the agency concerned and approved by the U.S. Attorney General. TSC staff must review each record submitted before it is added to the database. An individual will remain on the watch list until the respective department or agency that nominated that person to the list determines that the person should be removed from the list and deleted from the database

The TSC watch list database is updated daily with new nominations, modifications to existing records, and deletions. Since its creation, the list has ballooned to 400,000 people, recorded as 1.1 million names and aliases, and is continuing to grow at a rate of 200,000 records each year. Information on the list is distributed to a wide range of government agency systems for use in efforts to deter or detect the movements of known or suspected terrorists.

Recipient agencies include the FBI, CIA, National Security Agency (NSA), Transportation Security Administration (TSA), Department of Homeland Security, State Department, Customs and Border Protection, Secret Service, U.S. Marshals Service, and the White House. Airlines use data supplied by the TSA system in their NoFly and Selectee lists for prescreening passengers, while the U.S. Customs and Border Protection system uses the watchlist data to help screen travelers entering the United States. The State Department system screens applicants for visas to enter the United States and U.S. residents applying for passports, while state and local law enforcement agencies use the FBI system to help with arrests, detentions, and other criminal justice activities. Each of these agencies receives the subset of data in the watch list that pertains to its specific mission.

When an individual makes an airline reservation, arrives at a U.S. port of entry, applies for a U.S. visa, or is stopped by state or local police within the United States, the frontline screening agency or airline conducts a name-based search of the individual against the records from the terrorist watch list database. When the computerized name-matching system generates a "hit" (a potential name match) against a watch list record, the airline or agency will review each potential match. Matches that are clearly positive or exact matches that are inconclusive (uncertain or difficult to verify) are referred to the applicable screening agency's intelligence or operations center and to the TSC for closer examination. In turn, TSC checks its databases and other sources, including classified databases maintained by the NCTC and FBI to confirm whether the individual is a positive, negative, or inconclusive match to the watch list record. TSC creates a daily report summarizing all positive matches to the watch list and distributes them to numerous federal agencies.

The process of consolidating information from disparate agencies has been a slow and painstaking one, requiring the integration of at least 12 different databases. Two years after the process of integration took place, 10 of the 12 databases had been processed. The remaining two databases (the U.S. Immigration and Customs Enforcement's Automatic Biometric Identification System and the FBI's Integrated Automated Fingerprint Identification System) are both fingerprint databases. There is still more work to be done to optimize the list's usefulness.

Reports from both the Government Accountability Office and the Office of the Inspector General assert that the list contains inaccuracies and that government departmental policies for nomination and removal from the lists are not uniform. There has also been public outcry resulting from the size of the list and well-publicized incidents of obvious non-terrorists finding that they are included on the list.

Information about the process for inclusion on the list must necessarily be carefully protected if the list is to be effective against terrorists. The specific criteria for inclusion are not public knowledge. We do know, however, that government agencies populate their watch lists by performing wide sweeps of information gathered on travelers, using many misspellings and alternate variations of the names of suspected terrorists. This often leads to the inclusion of people who do not belong on watch lists, known as "false positives." It also results in some people being listed multiple times under different spellings of their names.

While these selection criteria may be effective for tracking as many potential terrorists as possible, they also lead to many more erroneous entries on the list than if the process required more finely tuned information to add new entries. Notable examples of 'false positives' include Michael Hicks, an 8-year-old New Jersey Cub Scout who is continually stopped at the airport for additional screening and the late senator Ted Kennedy, who had been repeatedly delayed in the past because his name resembles an alias once used by a suspected terrorist. Like Kennedy, Hicks may have been added because his name is the same or similar to a different suspected terrorist.

These incidents call attention to the quality and accuracy of the data in the TSC consolidated terrorist watch list. In June 2005, a report by the Department of Justice's Office of the Inspector General found inconsistent record counts, duplicate records, and records that lacked data fields or had unclear sources for their data. Although TSC subsequently enhanced its efforts to identify and correct incomplete or inaccurate watch list records, the Inspector General noted in September 2007 that TSC management of the watch list still showed some weaknesses.

Given the option between a list that tracks every potential terrorist at the cost of unnecessarily tracking some innocents, and a list that fails to track many terrorists in an effort to avoid tracking innocents, many would choose the list that tracked every terrorist despite the drawbacks. But to make matters worse for those already inconvenienced by wrongful inclusion on the list, there is currently no simple and quick redress process for innocents that hope to remove themselves from it.

The number of requests for removal from the watch list continues to mount, with over 24,000 requests recorded (about 2,000 each month) and only 54 percent of them resolved. The average time to process a request in 2008 was 40 days, which was not (and still is not) fast enough to keep pace with the number of requests for removal coming in. As a result, law-abiding travelers that inexplicably find themselves on the watch list are left with no easy way to remove themselves from it.

In February 2007, the Department of Homeland Security instituted its Traveler Redress Inquiry Program (TRIP) to help people that have been erroneously added to terrorist watch lists remove themselves and avoid extra screening and questioning. John Anderson's mother claimed that despite her best efforts, she was unable to remove her son from the watch lists. Senator Kennedy reportedly was only able to remove himself from the list by personally bringing up the matter to Tom Ridge, then the Director of the Department of Homeland Security.

Security officials say that mistakes such as the one that led to Anderson and Kennedy's inclusion on no-fly and consolidated watch lists occur due to the matching of imperfect data in airline reservation systems with imperfect data on the watch lists. Many airlines don't include gender, middle name, or date of birth in their reservations records, which increases the likelihood of false matches.

One way to improve screening and help reduce the number of people erroneously marked for additional investigation would be to use a more sophisticated system involving more personal data about individuals on the list. The TSA is developing just such a system, called "Secure Flight," but it has been continually delayed due to privacy concerns regarding the sensitivity and safety of the data it would collect. Other similar surveillance programs and watch lists, such as the NSA's attempts to gather information about suspected terrorists, have drawn criticism for potential privacy violations.

Additionally, the watch list has drawn criticism because of its potential to promote racial profiling and discrimination. Some allege that they were included by virtue of their race and ethnic descent, such as David Fathi, an attorney for the ACLU of Iranian descent, and Asif Iqbal, a U.S. citizen of Pakistani decent with the same name as a Guantanamo detainee. Outspoken critics of U.S. foreign policy, such as some elected officials and university professors, have also found themselves on the list.

A report released in May 2009 by Department of Justice Inspector General Glenn A. Fine found that the FBI had incorrectly kept nearly 24,000 people on its own watch list that supplies data to the terrorist watch list on the basis of outdated or irrelevant information. Examining nearly 69,000 referrals to the FBI list, the report found that 35 percent of those people remained on the list despite inadequate justification. Even more worrisome, the list did not contain the names of people who should have been listed because of their terrorist ties.

FBI officials claim that the bureau has made improvements, including better training, faster processing of referrals, and requiring field office supervisors to review watch-list nominations for accuracy and completeness. But this watch list and the others remain imperfect tools. In early 2008, it was revealed that 20 known terrorists were not correctly listed on the consolidated watch list. (Whether these individuals were able to enter the U.S. as a result is unclear.)

Umar Farouk Abdulmutallab, the Nigerian who unsuccessfully tried to detonate plastic explosives on the Northwest Airlines flight from Amsterdam to Detroit on Christmas Day 2009, had not made it onto the no-fly list. Although Abdulmutallab's father had reported concern over his son's radicalization to the U.S. State Department, the Department did not revoke Adbulmutallab's visa because his name was misspelled in the visa database, so he was allowed to enter the United States. Faisal Shahzad, the Times Square car bomber, was apprehended on May 3, 2010, only moments before his Emirates airline flight to Dubai and Pakistan was about to take off. The airline had failed to check a last-minute update to the no-fly list that had added Shahzad's name.

Sources: Scott Shane, "Lapses Allowed Suspect to Board Plane," *The New York Times*, May 4, 2010; Mike McIntire, "Ensnared by Error on Growing U.S. Watch List," *The New York Times*, April 6, 2010; Eric Lipton, Eric Schmitt, and Mark Mazzetti, "Review of Jet Bomb Plot Shows More Missed Clues," *The New York Times*, January 18, 2010; Lizette Alvarez, "Meet Mikey, 8: U.S. Has Him on Watch List," *The New York Times*, January 14, 2010; Eric Lichtblau, "Justice Dept. Finds Flaws in F.B.I. Terror List," *The New York Times*, May 7, 2009; Bob Egelko, "Watch-list Name Confusion Causes Hardship," *San Francisco Chronicle*, March 20, 2008; "Reports Cite Lack of Uniform Policy for Terrorist Watch List," *The Washington Post*, March 18, 2008; Siobhan Gorman, "NSA's Domestic Spying Grows as Agency Sweeps Up Data," *The Wall Street Journal*, March 10, 2008; Ellen Nakashima, and Scott McCartney, "When Your Name is Mud at the Airport," *The Wall Street Journal*, January 29, 2008.

CASE STUDY QUESTIONS

1. What concepts in this chapter are illustrated in this case?

2. Why was the consolidated terror watch list created? What are the benefits of the list?

3. Describe some of the weaknesses of the watch list. What management, organization, and technology factors are responsible for these weaknesses?

4. How effective is the system of watch lists described in this case study? Explain your answer.

5. If you were responsible for the management of the TSC watch list database, what steps would you take to correct some of these weaknesses?

6. Do you believe that the terror watch list represents a significant threat to individuals' privacy or Constitutional rights? Why or why not?

Chapter 7

Telecommunications, the Internet, and Wireless Technology

Interactive Sessions:

The Battle Over Net Neutrality

Monitoring Employees on Networks: Unethical or Good Business?

HYUNDAI HEAVY INDUSTRIES CREATES A WIRELESS SHIPYARD

What's it like to be the world's largest shipbuilder? Ask Hyundai Heavy Industries (HHI), headquartered in Ulsan, South Korea, which produces 10 percent of the world's ships. HHI produces tankers, bulk carriers, containerships, gas and chemical carriers, ship engines, offshore oil and gas drilling platforms, and undersea pipelines.

Coordinating and optimizing the production of so many different products, is obviously a daunting task. The company has already invested nearly $50 million in factory planning software to help manage this effort. But HHI's "factory" encompasses 11 square kilometers (4.2 square miles) stretching over land and sea, including nine drydocks, the largest of which spans more than seven football fields to support construction of four vessels simultaneously. Over 12,000 workers build up to 30 ships at one time, using millions of parts ranging in size from small rivets to five-story buildings.

This production environment proved too large and complex to easily track the movement of parts and inventory in real time as these events were taking place. Without up-to-the-minute data, the efficiencies from enterprise resource planning software are very limited. To make matters worse, the recent economic downturn hit HHI especially hard, as world trading and shipping plummeted. Orders for new ships in 2009 plunged to 7.9 million compensated gross tons (CGT, a measurement of vessel size), down from 150 million CGT the previous year. In this economic environment, Hyundai Heavy was looking for new ways to reduce expenses and streamline production.

HHI's solution was a high-speed wireless network across the entire shipyard, which was built by KT Corp., South Korea's largest telecommunications firm. It is able to transmit data at a rate of 4 megabits per second, about four times faster than the typical cable modem delivering high-speed Internet service to U.S. households. The company uses radio sensors to track the movement of parts as they move from fabrication shop to the side of a drydock and then onto a ship under construction. Workers on the ship use notebook computers or handheld mobile phones to access plans and engage in two-way video conversations with ship designers in the office, more than a mile away.

In the past, workers who were inside a vessel below ground or below sea level had to climb topside to use a phone or walkie-talkie when they had to talk to someone about a problem. The new wireless network is connected to the electric lines in the ship, which convey digital data to Wi-Fi wireless transmitters placed around the hull during construction. Workers' Internet phones, webcams, and PCs are linked to the Wi-Fi system, so workers can use Skype VoIP to call their colleagues on the surface. Designers in an office building a mile from the construction site use the webcams to investigate problems.

On the shipyard roads, 30 transporter trucks fitted to receivers connected to the wireless network update their location every 20 seconds to a control room. This helps dispatchers

match the location of transporters with orders for parts, shortening the trips each truck makes. All of the day's movements are finished by 6 P.M. instead of 8 P.M. By making operations more efficient and reducing labor costs, the wireless technology is expected to save Hyundai Heavy $40 million annually.

Sources: Evan Ramstad, "High-Speed Wireless Transforms a Shipyard," *The Wall Street Journal*, March 15, 2010 and "Hyundai Heavy Plans Wireless Shipyard," *The Korea Herald*, March 30, 2010.

Hyundai Heavy Industries's experience illustrates some of the powerful capabilities and opportunities provided by contemporary networking technology. The company used wireless networking technology to connect designers, laborers, ships under construction, and transportation vehicles to accelerate communication and coordination, and cut down on the time, distance, or number of steps required to perform a task.

The chapter-opening diagram calls attention to important points raised by this case and this chapter. Hyundai Heavy Industries produces ships and other products that are very labor-intensive and sensitive to changes in global economic conditions. Its production environment is large, complex, and extremely difficult to coordinate and manage. The company needs to keep operating costs as low as possible. HHI's shipyard extends over a vast area, and it was extremely difficult to monitor and coordinate different projects and work teams.

Management decided that wireless technology provided a solution and arranged for the deployment of a wireless network throughout the entire shipyard. The network also links the yard to designers in HHI's office a mile away. The network made it much easier to track parts and production activities and to optimize the movements of transporter trucks. HHI had to redesign its production and other work processes to take advantage of the new technology.

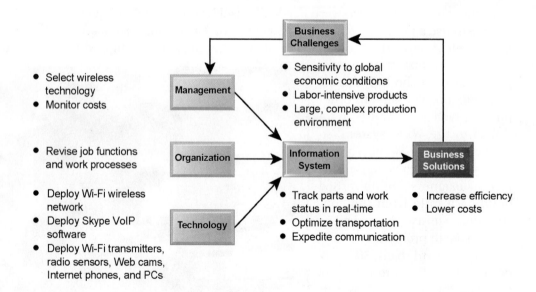

7.1   TELECOMMUNICATIONS AND NETWORKING IN TODAY'S BUSINESS WORLD

I f you run or work in a business, you can't do without networks. You need to communicate rapidly with your customers, suppliers, and employees. Until about 1990, businesses used the postal system or telephone system with voice or fax for communication. Today, however, you and your employees use computers and e-mail, the Internet, cell phones, and mobile computers connected to wireless networks for this purpose. Networking and the Internet are now nearly synonymous with doing business.

NETWORKING AND COMMUNICATION TRENDS

Firms in the past used two fundamentally different types of networks: telephone networks and computer networks. Telephone networks historically handled voice communication, and computer networks handled data traffic. Telephone networks were built by telephone companies throughout the twentieth century using voice transmission technologies (hardware and software), and these companies almost always operated as regulated monopolies throughout the world. Computer networks were originally built by computer companies seeking to transmit data between computers in different locations.

Thanks to continuing telecommunications deregulation and information technology innovation, telephone and computer networks are converging into a single digital network using shared Internet-based standards and equipment. Telecommunications providers today, such as AT&T and Verizon, offer data transmission, Internet access, cellular telephone service, and television programming as well as voice service. (See the Chapter 3 opening case.) Cable companies, such as Cablevision and Comcast, now offer voice service and Internet access. Computer networks have expanded to include Internet telephone and limited video services. Increasingly, all of these voice, video, and data communications are based on Internet technology.

Both voice and data communication networks have also become more powerful (faster), more portable (smaller and mobile), and less expensive. For instance, the typical Internet connection speed in 2000 was 56 kilobits per second, but today more than 60 percent of U.S. Internet users have high-speed **broadband** connections provided by telephone and cable TV companies running at 1 to 15 million bits per second. The cost for this service has fallen exponentially, from 25 cents per kilobit in 2000 to a tiny fraction of a cent today.

Increasingly, voice and data communication, as well as Internet access, are taking place over broadband wireless platforms, such as cell phones, mobile handheld devices, and PCs in wireless networks. In a few years, more than half the Internet users in the United States will use smartphones and mobile netbooks to access the Internet. In 2010, 84 million Americans accessed the Internet through mobile devices, and this number is expected to double by 2014 (eMarketer, 2010).

WHAT IS A COMPUTER NETWORK?

If you had to connect the computers for two or more employees together in the same office, you would need a computer network. Exactly what is a network? In its simplest form, a network consists of two or more connected computers.

Figure 7-1 illustrates the major hardware, software, and transmission components used in a simple network: a client computer and a dedicated server computer, network interfaces, a connection medium, network operating system software, and either a hub or a switch.

Each computer on the network contains a network interface device called a **network interface card (NIC)**. Most personal computers today have this card built into the motherboard. The connection medium for linking network components can be a telephone wire, coaxial cable, or radio signal in the case of cell phone and wireless local area networks (Wi-Fi networks).

The **network operating system (NOS)** routes and manages communications on the network and coordinates network resources. It can reside on every computer in the network, or it can reside primarily on a dedicated server computer for all the applications on the network. A server computer is a computer on a network that performs important network functions for client computers, such as serving up Web pages, storing data, and storing the network operating system (and hence controlling the network). Server software such as Microsoft Windows Server, Linux, and Novell Open Enterprise Server are the most widely used network operating systems.

Most networks also contain a switch or a hub acting as a connection point between the computers. **Hubs** are very simple devices that connect network components, sending a packet of data to all other connected devices. A **switch** has more intelligence than a hub and can filter and forward data to a specified destination on the network.

What if you want to communicate with another network, such as the Internet? You would need a router. A **router** is a communications processor used to route packets of data through different networks, ensuring that the data sent gets to the correct address.

FIGURE 7-1 **COMPONENTS OF A SIMPLE COMPUTER NETWORK**

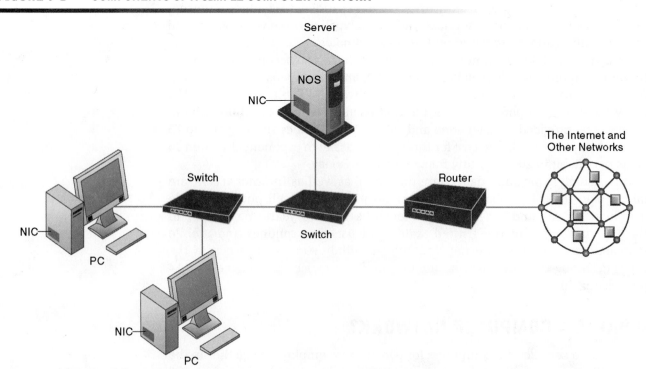

Illustrated here is a very simple computer network, consisting of computers, a network operating system residing on a dedicated server computer, cable (wiring) connecting the devices, network interface cards (NICs), switches, and a router.

Networks in Large Companies

The network we've just described might be suitable for a small business. But what about large companies with many different locations and thousands of employees? As a firm grows, and collects hundreds of small local area networks, these networks can be tied together into a corporate-wide networking infrastructure. The network infrastructure for a large corporation consists of a large number of these small local area networks linked to other local area networks and to firmwide corporate networks. A number of powerful servers support a corporate Web site, a corporate intranet, and perhaps an extranet. Some of these servers link to other large computers supporting back-end systems.

Figure 7-2 provides an illustration of these more complex, larger scale corporate-wide networks. Here you can see that the corporate network infrastructure supports a mobile sales force using cell phones and smartphones, mobile employees linking to the company Web site, internal company networks using mobile wireless local area networks (Wi-Fi networks), and a videoconferencing system to support managers across the world. In addition to these computer networks, the firm's infrastructure usually includes a separate telephone network that handles most voice data. Many firms are dispensing with their traditional telephone networks and using Internet telephones that run on their existing data networks (described later).

As you can see from this figure, a large corporate network infrastructure uses a wide variety of technologies—everything from ordinary telephone service and corporate data networks to Internet service, wireless Internet, and cell phones.

FIGURE 7-2   CORPORATE NETWORK INFRASTRUCTURE

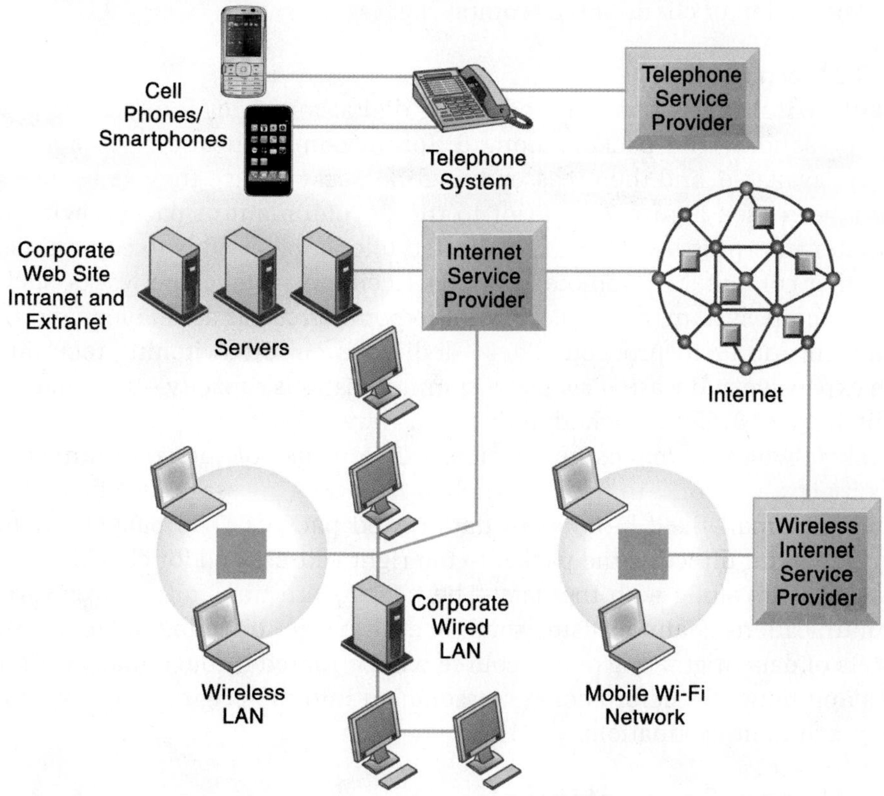

Today's corporate network infrastructure is a collection of many different networks from the public switched telephone network, to the Internet, to corporate local area networks linking workgroups, departments, or office floors.

One of the major problems facing corporations today is how to integrate all the different communication networks and channels into a coherent system that enables information to flow from one part of the corporation to another, and from one system to another. As more and more communication networks become digital, and based on Internet technologies, it will become easier to integrate them.

KEY DIGITAL NETWORKING TECHNOLOGIES

Contemporary digital networks and the Internet are based on three key technologies: client/server computing, the use of packet switching, and the development of widely used communications standards (the most important of which is Transmission Control Protocol/Internet Protocol, or TCP/IP) for linking disparate networks and computers.

Client/Server Computing

We introduced client/server computing in Chapter 5. Client/server computing is a distributed computing model in which some of the processing power is located within small, inexpensive client computers, and resides literally on desktops, laptops, or in handheld devices. These powerful clients are linked to one another through a network that is controlled by a network server computer. The server sets the rules of communication for the network and provides every client with an address so others can find it on the network.

Client/server computing has largely replaced centralized mainframe computing in which nearly all of the processing takes place on a central large mainframe computer. Client/server computing has extended computing to departments, workgroups, factory floors, and other parts of the business that could not be served by a centralized architecture. The Internet is the largest implementation of client/server computing.

Packet Switching

Packet switching is a method of slicing digital messages into parcels called packets, sending the packets along different communication paths as they become available, and then reassembling the packets once they arrive at their destinations (see Figure 7-3). Prior to the development of packet switching, computer networks used leased, dedicated telephone circuits to communicate with other computers in remote locations. In circuit-switched networks, such as the telephone system, a complete point-to-point circuit is assembled, and then communication can proceed. These dedicated circuit-switching techniques were expensive and wasted available communications capacity—the circuit was maintained regardless of whether any data were being sent.

Packet switching makes much more efficient use of the communications capacity of a network. In packet-switched networks, messages are first broken down into small fixed bundles of data called packets. The packets include information for directing the packet to the right address and for checking transmission errors along with the data. The packets are transmitted over various communications channels using routers, each packet traveling independently. Packets of data originating at one source will be routed through many different paths and networks before being reassembled into the original message when they reach their destinations.

TCP/IP and Connectivity

In a typical telecommunications network, diverse hardware and software components need to work together to transmit information. Different com-

FIGURE 7-3   PACKED-SWITCHED NETWORKS AND PACKET COMMUNICATIONS

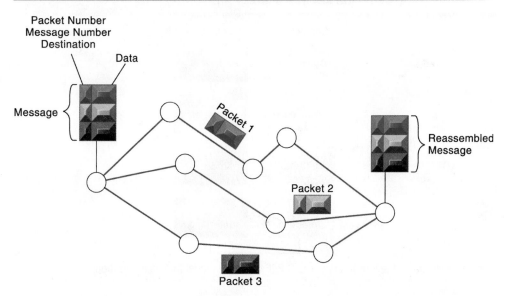

Data are grouped into small packets, which are transmitted independently over various communications channels and reassembled at their final destination.

ponents in a network communicate with each other only by adhering to a common set of rules called protocols. A **protocol** is a set of rules and procedures governing transmission of information between two points in a network.

In the past, many diverse proprietary and incompatible protocols often forced business firms to purchase computing and communications equipment from a single vendor. But today, corporate networks are increasingly using a single, common, worldwide standard called **Transmission Control Protocol/ Internet Protocol (TCP/IP)**. TCP/IP was developed during the early 1970s to support U.S. Department of Defense Advanced Research Projects Agency (DARPA) efforts to help scientists transmit data among different types of computers over long distances.

TCP/IP uses a suite of protocols, the main ones being TCP and IP. TCP refers to the Transmission Control Protocol (TCP), which handles the movement of data between computers. TCP establishes a connection between the computers, sequences the transfer of packets, and acknowledges the packets sent. IP refers to the Internet Protocol (IP), which is responsible for the delivery of packets and includes the disassembling and reassembling of packets during transmission. Figure 7-4 illustrates the four-layered Department of Defense reference model for TCP/IP.

1. Application layer. The Application layer enables client application programs to access the other layers and defines the protocols that applications use to exchange data. One of these application protocols is the Hypertext Transfer Protocol (HTTP), which is used to transfer Web page files.

2. Transport layer. The Transport layer is responsible for providing the Application layer with communication and packet services. This layer includes TCP and other protocols.

3. Internet layer. The Internet layer is responsible for addressing, routing, and packaging data packets called IP datagrams. The Internet Protocol is one of the protocols used in this layer.

FIGURE 7-4 **THE TRANSMISSION CONTROL PROTOCOL/INTERNET PROTOCOL (TCP/IP) REFERENCE MODEL**

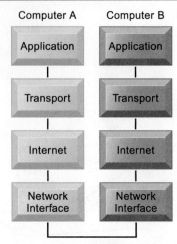

This figure illustrates the four layers of the TCP/IP reference model for communications

4. Network Interface layer. At the bottom of the reference model, the Network Interface layer is responsible for placing packets on and receiving them from the network medium, which could be any networking technology.

Two computers using TCP/IP are able to communicate even if they are based on different hardware and software platforms. Data sent from one computer to the other passes downward through all four layers, starting with the sending computer's Application layer and passing through the Network Interface layer. After the data reach the recipient host computer, they travel up the layers and are reassembled into a format the receiving computer can use. If the receiving computer finds a damaged packet, it asks the sending computer to retransmit it. This process is reversed when the receiving computer responds.

7.2 COMMUNICATIONS NETWORKS

Let's look more closely at alternative networking technologies available to businesses.

SIGNALS: DIGITAL VS. ANALOG

There are two ways to communicate a message in a network: either using an analog signal or a digital signal. An *analog signal* is represented by a continuous waveform that passes through a communications medium and has been used for voice communication. The most common analog devices are the telephone handset, the speaker on your computer, or your iPod earphone, all of which create analog wave forms that your ear can hear.

A *digital signal* is a discrete, binary waveform, rather than a continuous waveform. Digital signals communicate information as strings of two discrete states: one bit and zero bits, which are represented as on-off electrical pulses. Computers use digital signals and require a modem to convert these digital signals into analog signals that can be sent over (or received from) telephone lines, cable lines, or wireless media that use analog signals (see Figure 7-5). **Modem** stands for

FIGURE 7-5   FUNCTIONS OF THE MODEM

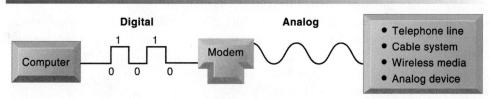

A modem is a device that translates digital signals into analog form (and vice versa) so that computers can transmit data over analog networks such as telephone and cable networks.

modulator-demodulator. Cable modems connect your computer to the Internet using a cable network. DSL modems connect your computer to the Internet using a telephone company's land line network. Wireless modems perform the same function as traditional modems, connecting your computer to a wireless network that could be a cell phone network, or a Wi-Fi network. Without modems, computers could not communicate with one another using analog networks (which include the telephone system and cable networks).

TYPES OF NETWORKS

There are many different kinds of networks and ways of classifying them. One way of looking at networks is in terms of their geographic scope (see Table 7-1).

Local Area Networks

If you work in a business that uses networking, you are probably connecting to other employees and groups via a local area network. A **local area network (LAN)** is designed to connect personal computers and other digital devices within a half-mile or 500-meter radius. LANs typically connect a few computers in a small office, all the computers in one building, or all the computers in several buildings in close proximity. LANs also are used to link to long-distance wide area networks (WANs, described later in this section) and other networks around the world using the Internet.

Review Figure 7-1, which could serve as a model for a small LAN that might be used in an office. One computer is a dedicated network file server, providing users with access to shared computing resources in the network, including software programs and data files.

The server determines who gets access to what and in which sequence. The router connects the LAN to other networks, which could be the Internet or another corporate network, so that the LAN can exchange information with networks external to it. The most common LAN operating systems are Windows, Linux, and Novell. Each of these network operating systems supports TCP/IP as their default networking protocol.

TABLE 7-1 TYPES OF NETWORKS

TYPE	AREA
Local area network (LAN)	Up to 500 meters (half a mile); an office or floor of a building
Campus area network (CAN)	Up to 1,000 meters (a mile); a college campus or corporate facility
Metropolitan area network (MAN)	A city or metropolitan area
Wide area network (WAN)	A transcontinental or global area

Ethernet is the dominant LAN standard at the physical network level, specifying the physical medium to carry signals between computers, access control rules, and a standardized set of bits used to carry data over the system. Originally, Ethernet supported a data transfer rate of 10 megabits per second (Mbps). Newer versions, such as Fast Ethernet and Gigabit Ethernet, support data transfer rates of 100 Mbps and 1 gigabits per second (Gbps), respectively, and are used in network backbones.

The LAN illustrated in Figure 7-1 uses a client/server architecture where the network operating system resides primarily on a single file server, and the server provides much of the control and resources for the network. Alternatively, LANs may use a peer-to-peer architecture. A **peer-to-peer** network treats all processors equally and is used primarily in small networks with 10 or fewer users. The various computers on the network can exchange data by direct access and can share peripheral devices without going through a separate server.

In LANs using the Windows Server family of operating systems, the peer-to-peer architecture is called the *workgroup network model,* in which a small group of computers can share resources, such as files, folders, and printers, over the network without a dedicated server. The Windows *domain network model*, in contrast, uses a dedicated server to manage the computers in the network.

Larger LANs have many clients and multiple servers, with separate servers for specific services, such as storing and managing files and databases (file servers or database servers), managing printers (print servers), storing and managing e-mail (mail servers), or storing and managing Web pages (Web servers).

Sometimes LANs are described in terms of the way their components are connected together, or their **topology**. There are three major LAN topologies: star, bus, and ring (see Figure 7-6).

In a **star topology**, all devices on the network connect to a single hub. Figure 7-6 illustrates a simple star topology in which all network traffic flows through the hub. In an *extended star network*, multiple layers of hubs are organized into a hierarchy.

In a **bus topology**, one station transmits signals, which travel in both directions along a single transmission segment. All of the signals are broadcast in both directions to the entire network. All machines on the network receive the same signals, and software installed on the client computers enables each client to listen for messages addressed specifically to it. The bus topology is the most common Ethernet topology.

A **ring topology** connects network components in a closed loop. Messages pass from computer to computer in only one direction around the loop, and only one station at a time may transmit. The ring topology is primarily found in older LANs using Token Ring networking software.

Metropolitan and Wide Area Networks

Wide area networks (WANs) span broad geographical distances—entire regions, states, continents, or the entire globe. The most universal and powerful WAN is the Internet. Computers connect to a WAN through public networks, such as the telephone system or private cable systems, or through leased lines or satellites. A **metropolitan area network (MAN)** is a network that spans a metropolitan area, usually a city and its major suburbs. Its geographic scope falls between a WAN and a LAN.

FIGURE 7-6   **NETWORK TOPOLOGIES**

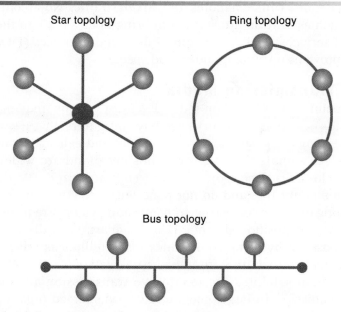

The three basic network topologies are the star, bus, and ring.

PHYSICAL TRANSMISSION MEDIA

Networks use different kinds of physical transmission media, including twisted wire, coaxial cable, fiber optics, and media for wireless transmission. Each has advantages and limitations. A wide range of speeds is possible for any given medium depending on the software and hardware configuration.

Twisted Wire

Twisted wire consists of strands of copper wire twisted in pairs and is an older type of transmission medium. Many of the telephone systems in buildings had twisted wires installed for analog communication, but they can be used for digital communication as well. Although an older physical transmission medium, the twisted wires used in today's LANs, such as CAT5, can obtain speeds up to 1 Gbps. Twisted-pair cabling is limited to a maximum recommended run of 100 meters (328 feet).

Coaxial Cable

Coaxial cable, similar to that used for cable television, consists of thickly insulated copper wire that can transmit a larger volume of data than twisted wire. Cable was used in early LANs and is still used today for longer (more than 100 meters) runs in large buildings. Coaxial has speeds up to 1 Gbps.

Fiber Optics and Optical Networks

Fiber-optic cable consists of bound strands of clear glass fiber, each the thickness of a human hair. Data are transformed into pulses of light, which are sent through the fiber-optic cable by a laser device at rates varying from 500 kilobits to several trillion bits per second in experimental settings. Fiber-optic cable is considerably faster, lighter, and more durable than wire media, and is well suited to systems requiring transfers of large volumes of data. However, fiber-optic cable is more expensive than other physical transmission media and harder to install.

Until recently, fiber-optic cable had been used primarily for the high-speed network backbone, which handles the major traffic. Now cellular phone companies such as Verizon are starting to bring fiber lines into the home for new types of services, such as Verizon's Fiber Optic Services (FiOS) Internet service that provides up 50 Mbps download speeds.

Wireless Transmission Media

Wireless transmission is based on radio signals of various frequencies. There are three kinds of wireless networks used by computers: microwave, cellular, and Wi-Fi. **Microwave** systems, both terrestrial and celestial, transmit high-frequency radio signals through the atmosphere and are widely used for high-volume, long-distance, point-to-point communication. Microwave signals follow a straight line and do not bend with the curvature of the earth. Therefore, long-distance terrestrial transmission systems require that transmission stations be positioned about 37 miles apart. Long-distance transmission is also possible by using communication satellites as relay stations for microwave signals transmitted from terrestrial stations.

Communication satellites use microwave transmission and are typically used for transmission in large, geographically dispersed organizations that would be difficult to network using cabling media or terrestrial microwave, as well as for home Internet service, especially in rural areas. For instance, the global energy company BP p.l.c. uses satellites for real-time data transfer of oil field exploration data gathered from searches of the ocean floor. Using geosynchronous satellites, exploration ships transfer these data to central computing centers in the United States for use by researchers in Houston, Tulsa, and suburban Chicago. Figure 7-7 illustrates how this system works. Satellites are also used for home television and Internet service. The two major satellite Internet providers (Dish Network and DirectTV) have about 30 million subscribers, and about 17 percent of all U.S. households access the Internet using satellite services (eMarketer, 2010).

FIGURE 7-7   **BP'S SATELLITE TRANSMISSION SYSTEM**

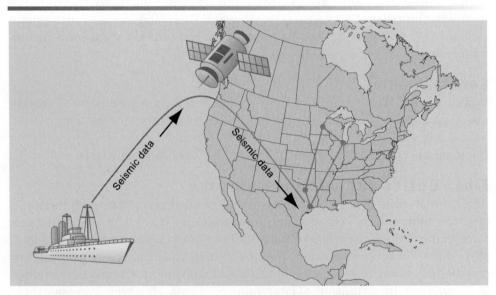

Communication satellites help BP transfer seismic data between oil exploration ships and research centers in the United States.

Cellular systems also use radio waves and a variety of different protocols to communicate with radio antennas (towers) placed within adjacent geographic areas called cells. Communications transmitted from a **cell phone** to a local cell pass from antenna to antenna—cell to cell—until they reach their final destination.

Wireless networks are supplanting traditional wired networks for many applications and creating new applications, services, and business models. In Section 7.4, we provide a detailed description of the applications and technology standards driving the "wireless revolution."

Transmission Speed

The total amount of digital information that can be transmitted through any telecommunications medium is measured in bits per second (bps). One signal change, or cycle, is required to transmit one or several bits; therefore, the transmission capacity of each type of telecommunications medium is a function of its frequency. The number of cycles per second that can be sent through that medium is measured in **hertz**—one hertz is equal to one cycle of the medium.

The range of frequencies that can be accommodated on a particular telecommunications channel is called its **bandwidth**. The bandwidth is the difference between the highest and lowest frequencies that can be accommodated on a single channel. The greater the range of frequencies, the greater the bandwidth and the greater the channel's transmission capacity.

7.3 THE GLOBAL INTERNET

We all use the Internet, and many of us can't do without it. It's become an indispensable personal and business tool. But what exactly is the Internet? How does it work, and what does Internet technology have to offer for business? Let's look at the most important Internet features.

WHAT IS THE INTERNET?

The Internet has become the world's most extensive, public communication system that now rivals the global telephone system in reach and range. It's also the world's largest implementation of client/server computing and internet-working, linking millions of individual networks all over the world. This global network of networks began in the early 1970s as a U.S. Department of Defense network to link scientists and university professors around the world.

Most homes and small businesses connect to the Internet by subscribing to an Internet service provider. An **Internet service provider (ISP)** is a commercial organization with a permanent connection to the Internet that sells temporary connections to retail subscribers. EarthLink, NetZero, AT&T, and Time Warner are ISPs. Individuals also connect to the Internet through their business firms, universities, or research centers that have designated Internet domains.

There are a variety of services for ISP Internet connections. Connecting via a traditional telephone line and modem, at a speed of 56.6 kilobits per second (Kbps) used to be the most common form of connection worldwide, but it has been largely replaced by broadband connections. Digital subscriber line (DSL), cable, satellite Internet connections, and T lines provide these broadband services.

Digital subscriber line (DSL) technologies operate over existing telephone lines to carry voice, data, and video at transmission rates ranging from 385 Kbps all the way up to 9 Mbps. **Cable Internet connections** provided by cable television vendors use digital cable coaxial lines to deliver high-speed Internet access to homes and businesses. They can provide high-speed access to the Internet of up to 15 Mbps. In areas where DSL and cable services are unavailable, it is possible to access the Internet via satellite, although some satellite Internet connections have slower upload speeds than other broadband services.

T1 and T3 are international telephone standards for digital communication. They are leased, dedicated lines suitable for businesses or government agencies requiring high-speed guaranteed service levels. **T1 lines** offer guaranteed delivery at 1.54 Mbps, and T3 lines offer delivery at 45 Mbps. The Internet does not provide similar guaranteed service levels, but simply "best effort."

INTERNET ADDRESSING AND ARCHITECTURE

The Internet is based on the TCP/IP networking protocol suite described earlier in this chapter. Every computer on the Internet is assigned a unique **Internet Protocol (IP) address**, which currently is a 32-bit number represented by four strings of numbers ranging from 0 to 255 separated by periods. For instance, the IP address of www.microsoft.com is 207.46.250.119.

When a user sends a message to another user on the Internet, the message is first decomposed into packets using the TCP protocol. Each packet contains its destination address. The packets are then sent from the client to the network server and from there on to as many other servers as necessary to arrive at a specific computer with a known address. At the destination address, the packets are reassembled into the original message.

The Domain Name System

Because it would be incredibly difficult for Internet users to remember strings of 12 numbers, the **Domain Name System (DNS)** converts domain names to IP addresses. The **domain name** is the English-like name that corresponds to the unique 32-bit numeric IP address for each computer connected to the Internet. DNS servers maintain a database containing IP addresses mapped to their corresponding domain names. To access a computer on the Internet, users need only specify its domain name.

DNS has a hierarchical structure (see Figure 7-8). At the top of the DNS hierarchy is the root domain. The child domain of the root is called a top-level domain, and the child domain of a top-level domain is called is a second-level domain. Top-level domains are two- and three-character names you are familiar with from surfing the Web, for example, .com, .edu, .gov, and the various country codes such as .ca for Canada or .it for Italy. Second-level domains have two parts, designating a top-level name and a second-level name—such as buy.com, nyu.edu, or amazon.ca. A host name at the bottom of the hierarchy designates a specific computer on either the Internet or a private network.

The most common domain extensions currently available and officially approved are shown in the following list. Countries also have domain names such as .uk, .au, and .fr (United Kingdom, Australia, and France, respectively), and there is a new class of "internationalized" top level domains that use non-English characters (ICANN, 2010). In the future, this list will expand to include many more types of organizations and industries.

FIGURE 7-8   THE DOMAIN NAME SYSTEM

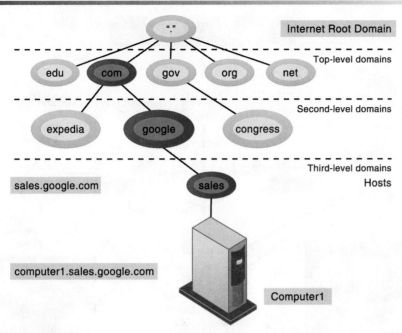

Domain Name System is a hierarchical system with a root domain, top-level domains, second-level domains, and host computers at the third level.

.com	Commercial organizations/businesses
.edu	Educational institutions
.gov	U.S. government agencies
.mil	U.S. military
.net	Network computers
.org	Nonprofit organizations and foundations
.biz	Business firms
.info	Information providers

Internet Architecture and Governance

Internet data traffic is carried over transcontinental high-speed backbone networks that generally operate today in the range of 45 Mbps to 2.5 Gbps (see Figure 7-9). These trunk lines are typically owned by long-distance telephone companies (called *network service providers*) or by national governments. Local connection lines are owned by regional telephone and cable television companies in the United States that connect retail users in homes and businesses to the Internet. The regional networks lease access to ISPs, private companies, and government institutions.

Each organization pays for its own networks and its own local Internet connection services, a part of which is paid to the long-distance trunk line owners. Individual Internet users pay ISPs for using their service, and they generally pay a flat subscription fee, no matter how much or how little they use the Internet. A debate is now raging on whether this arrangement should continue or whether heavy Internet users who download large video and music files should pay more for the bandwidth they consume. The Interactive Session on Organizations explores this topic, as it examines the pros and cons of network neutrality.

FIGURE 7-9   INTERNET NETWORK ARCHITECTURE

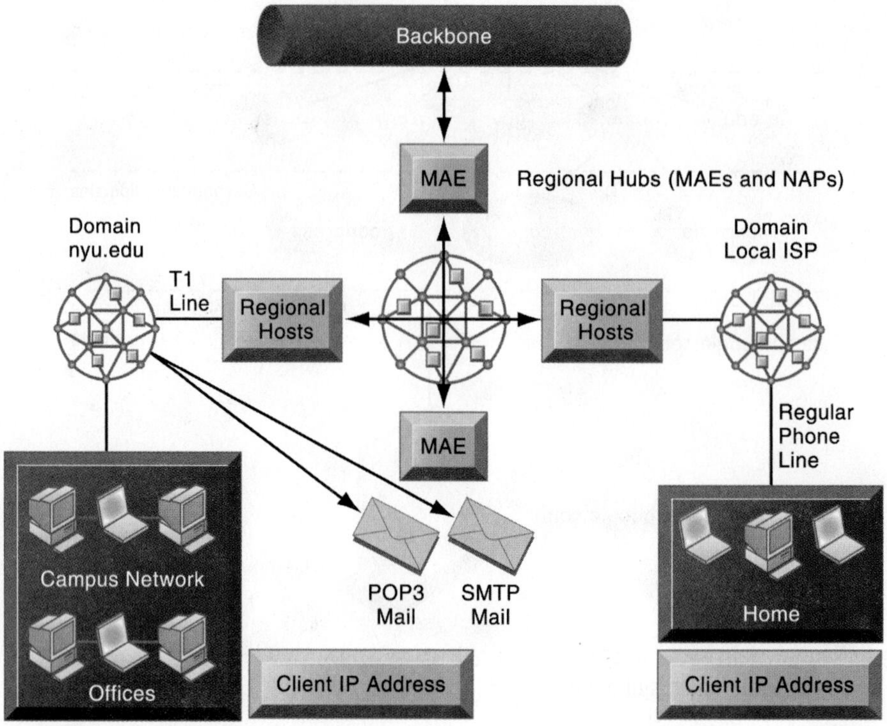

The Internet backbone connects to regional networks, which in turn provide access to Internet service providers, large firms, and government institutions. Network access points (NAPs) and metropolitan-area exchanges (MAEs) are hubs where the backbone intersects regional and local networks and where backbone owners connect with one another.

No one "owns" the Internet, and it has no formal management. However, worldwide Internet policies are established by a number of professional organizations and government bodies, including the Internet Architecture Board (IAB), which helps define the overall structure of the Internet; the Internet Corporation for Assigned Names and Numbers (ICANN), which assigns IP addresses; and the World Wide Web Consortium (W3C), which sets Hypertext Markup Language and other programming standards for the Web.

These organizations influence government agencies, network owners, ISPs, and software developers with the goal of keeping the Internet operating as efficiently as possible. The Internet must also conform to the laws of the sovereign nation-states in which it operates, as well as the technical infrastructures that exist within the nation-states. Although in the early years of the Internet and the Web there was very little legislative or executive interference, this situation is changing as the Internet plays a growing role in the distribution of information and knowledge, including content that some find objectionable.

The Future Internet: IPv6 and Internet2

The Internet was not originally designed to handle the transmission of massive quantities of data and billions of users. Because many corporations and governments have been given large blocks of millions of IP addresses to accommodate current and future workforces, and because of sheer Internet population growth, the world will run out of available IP addresses using the existing addressing convention by 2012 or 2013. Under development is a new version of the IP addressing schema called *Internet Protocol version 6 (IPv6)*, which contains 128-bit addresses (2 to the power of 128), or more than a quadrillion possible unique addresses.

Internet2 and Next-Generation Internet (NGI) are consortia representing 200 universities, private businesses, and government agencies in the United States that are working on a new, robust, high-bandwidth version of the Internet. They have established several new high-performance backbone networks with bandwidths reaching as much as 100 Gbps. Internet2 research groups are developing and implementing new technologies for more effective routing practices; different levels of service, depending on the type and importance of the data being transmitted; and advanced applications for distributed computation, virtual laboratories, digital libraries, distributed learning, and tele-immersion. These networks do not replace the public Internet, but they do provide test beds for leading-edge technology that may eventually migrate to the public Internet.

INTERNET SERVICES AND COMMUNICATION TOOLS

The Internet is based on client/server technology. Individuals using the Internet control what they do through client applications on their computers, such as Web browser software. The data, including e-mail messages and Web pages, are stored on servers. A client uses the Internet to request information from a particular Web server on a distant computer, and the server sends the requested information back to the client over the Internet. Chapters 5 and 6 describe how Web servers work with application servers and database servers to access information from an organization's internal information systems applications and their associated databases. Client platforms today include not only PCs and other computers but also cell phones, small handheld digital devices, and other information appliances.

Internet Services

A client computer connecting to the Internet has access to a variety of services. These services include e-mail, electronic discussion groups, chatting and instant messaging, **Telnet**, **File Transfer Protocol (FTP)**, and the Web. Table 7-2 provides a brief description of these services.

Each Internet service is implemented by one or more software programs. All of the services may run on a single server computer, or different services may

TABLE 7-2 MAJOR INTERNET SERVICES

CAPABILITY	FUNCTIONS SUPPORTED
E-mail	Person-to-person messaging; document sharing
Chatting and instant messaging	Interactive conversations
Newsgroups	Discussion groups on electronic bulletin boards
Telnet	Logging on to one computer system and doing work on another
File Transfer Protocol (FTP)	Transferring files from computer to computer
World Wide Web	Retrieving, formatting, and displaying information (including text, audio, graphics, and video) using hypertext links

INTERACTIVE SESSION: ORGANIZATIONS

THE BATTLE OVER NET NEUTRALITY

What kind of Internet user are you? Do you primarily use the Net to do a little e-mail and look up phone numbers? Or are you online all day, watching YouTube videos, downloading music files, or playing massively multiplayer online games? If you're the latter, you are consuming a great deal of bandwidth, and hundreds of millions of people like you might start to slow the Internet down. YouTube consumed as much bandwidth in 2007 as the entire Internet did in 2000. That's one of the arguments being made today for charging Internet users based on the amount of transmission capacity they use.

If user demand for the Internet overwhelms network capacity, the Internet might not come to a screeching halt, but users would be faced with very sluggish download speeds and slow performance of YouTube, Facebook, and other data-heavy services. (Heavy use of iPhones in urban areas such as New York and San Francisco has already degraded service on the AT&T wireless network. AT&T reports that 3 percent of its subscriber base accounts for 40 percent of its data traffic.)

Other researchers believe that as digital traffic on the Internet grows, even at a rate of 50 percent per year, the technology for handling all this traffic is advancing at an equally rapid pace.

In addition to these technical issues, the debate about metering Internet use centers around the concept of network neutrality. Network neutrality is the idea that Internet service providers must allow customers equal access to content and applications, regardless of the source or nature of the content. Presently, the Internet is indeed neutral: all Internet traffic is treated equally on a first-come, first-served basis by Internet backbone owners.

However, telecommunications and cable companies are unhappy with this arrangement. They want to be able to charge differentiated prices based on the amount of bandwidth consumed by content being delivered over the Internet. These companies believe that differentiated pricing is "the fairest way" to finance necessary investments in their network infrastructures.

Internet service providers point to the upsurge in piracy of copyrighted materials over the Internet. Comcast, the second largest Internet service provider in the United States, reported that illegal file sharing of copyrighted material was consuming 50 percent of its network capacity. In 2008, the company slowed down transmission of BitTorrent files, used extensively for piracy and illegal sharing of copyrighted materials, including video. The Federal Communications Commission (FCC) ruled that Comcast had to stop slowing peer-to-peer traffic in the name of network management. Comcast then filed a lawsuit challenging the FCC's authority to enforce network neutrality. In April 2010, a federal appeals court ruled in favor of Comcast that the FCC did not have the authority to regulate how an Internet provider manages its network.

Advocates of net neutrality are pushing Congress to find ways to regulate the industry to prevent network providers from adopting Comcast-like practices. The strange alliance of net neutrality advocates includes MoveOn.org, the Christian Coalition, the American Library Association, every major consumer group, many bloggers and small businesses, and some large Internet companies like Google and Amazon.

Net neutrality advocates argue that the risk of censorship increases when network operators can selectively block or slow access to certain content such as Hulu videos or access to competing low-cost services such as Skype and Vonage. There are already many examples of Internet providers restricting access to sensitive materials (such as Pakistan's government blocking access to anti-Muslim sites and YouTube as a whole in response to content it deemed defamatory to Islam.)

Proponents of net neutrality also argue that a neutral Internet encourages everyone to innovate without permission from the phone and cable companies or other authorities, and this level playing field has spawned countless new businesses. Allowing unrestricted information flow becomes essential to free markets and democracy as commerce and society increasingly move online.

Network owners believe regulation to enforce net neutrality will impede U.S. competitiveness by stifling innovation, discouraging capital expenditures for new networks, and curbing their networks' ability to cope with the exploding demand for Internet and wireless traffic. U.S. Internet service lags behind many other nations in overall speed, cost, and quality of service, adding credibility to this argument.

And with enough options for Internet access, regulation would not be essential for promoting net neutrality. Dissatisfied consumers could simply switch to providers who enforce net neutrality and allow unlimited Internet use.

Since the Comcast ruling was overturned, FCC efforts to support net neutrality have been in a holding pattern as it searches for some means of regulating broadband Internet service within the constraints of current law and current court rulings. One proposal is to reclassify broadband Internet transmission as a telecommunications service so the FCC could apply decades-old regulations for traditional telephone networks.

In August 2010, Verizon and Google issued a policy statement proposing that regulators enforce net neutrality on wired connections, but not on wireless networks, which are becoming the dominant Internet platform. The proposal was an effort to define some sort of middle ground that would safeguard net neutrality while giving carriers the flexibility they needed to manage their networks and generate revenue from them. None of the major players in the net neutrality debate showed support and both sides remain dug in.

Sources: Joe Nocera, "The Struggle for What We Already Have," *The New York Times*, September 4, 2010; Claire Cain Miller, "Web Plan is Dividing Companies," *The New York Times*, August 11, 2010; Wayne Rash, "Net Neutrality Looks Dead in the Clutches of Congress," *eWeek*, June 13 2010; Amy Schatz and Spencer E. Ante, "FCC Web Rules Create Pushback," *The Wall Street Journal*, May 6, 2010; Amy Schatz, "New U.S. Push to Regulate Internet Access," *The Wall Street Journal*, May 5, 2010; and Joanie Wexler: "Net Neutrality: Can We Find Common Ground?" *Network World*, April 1, 2009.

CASE STUDY QUESTIONS

1. What is network neutrality? Why has the Internet operated under net neutrality up to this point in time?

2. Who's in favor of net neutrality? Who's opposed? Why?

3. What would be the impact on individual users, businesses, and government if Internet providers switched to a tiered service model?

4. Are you in favor of legislation enforcing network neutrality? Why or why not?

MIS IN ACTION

1. Visit the Web site of the Open Internet Coalition and select five member organizations. Then visit the Web site of each of these organizations or surf the Web to find out more information about each. Write a short essay explaining why each organization is in favor of network neutrality.

2. Calculate how much bandwidth you consume when using the Internet every day. How many e-mails do you send daily and what is the size of each? (Your e-mail program may have e-mail file size information.) How many music and video clips do you download daily and what is the size of each? If you view YouTube often, surf the Web to find out the size of a typical YouTube file. Add up the number of e-mail, audio, and video files you transmit or receive on a typical day.

be allocated to different machines. Figure 7-10 illustrates one way that these services can be arranged in a multitiered client/server architecture.

E-mail enables messages to be exchanged from computer to computer, with capabilities for routing messages to multiple recipients, forwarding messages, and attaching text documents or multimedia files to messages. Although some organizations operate their own internal electronic mail systems, most e-mail today is sent through the Internet. The costs of e-mail is far lower than equivalent voice, postal, or overnight delivery costs, making the Internet a very inexpensive and rapid communications medium. Most e-mail messages arrive anywhere in the world in a matter of seconds.

Nearly 90 percent of U.S. workplaces have employees communicating interactively using **chat** or instant messaging tools. Chatting enables two or

FIGURE 7-10   CLIENT/SERVER COMPUTING ON THE INTERNET

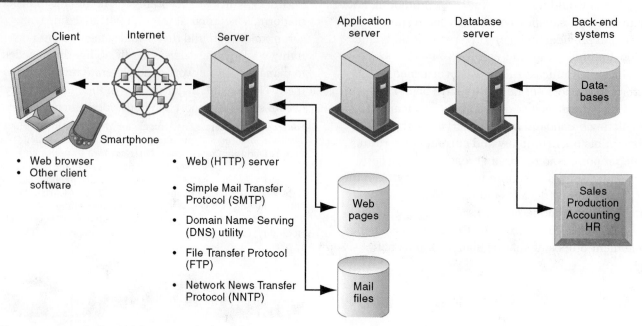

Client computers running Web browser and other software can access an array of services on servers over the Internet. These services may all run on a single server or on multiple specialized servers.

more people who are simultaneously connected to the Internet to hold live, interactive conversations. Chat systems now support voice and video chat as well as written conversations. Many online retail businesses offer chat services on their Web sites to attract visitors, to encourage repeat purchases, and to improve customer service.

Instant messaging is a type of chat service that enables participants to create their own private chat channels. The instant messaging system alerts the user whenever someone on his or her private list is online so that the user can initiate a chat session with other individuals. Instant messaging systems for consumers include Yahoo! Messenger, Google Talk, and Windows Live Messenger. Companies concerned with security use proprietary instant messaging systems such as Lotus Sametime.

Newsgroups are worldwide discussion groups posted on Internet electronic bulletin boards on which people share information and ideas on a defined topic, such as radiology or rock bands. Anyone can post messages on these bulletin boards for others to read. Many thousands of groups exist that discuss almost all conceivable topics.

Employee use of e-mail, instant messaging, and the Internet is supposed to increase worker productivity, but the accompanying Interactive Session on Management shows that this may not always be the case. Many company managers now believe they need to monitor and even regulate their employees' online activity. But is this ethical? Although there are some strong business reasons why companies may need to monitor their employees' e-mail and Web activities, what does this mean for employee privacy?

Voice over IP

The Internet has also become a popular platform for voice transmission and corporate networking. **Voice over IP (VoIP)** technology delivers voice information in digital form using packet switching, avoiding the tolls charged by local and long-distance telephone networks (see Figure 7-11). Calls that would ordinarily be transmitted over public telephone networks travel over the corporate network based on the Internet Protocol, or the public Internet. Voice calls can be made and received with a computer equipped with a microphone and speakers or with a VoIP-enabled telephone.

Cable firms such as Time Warner and Cablevision provide VoIP service bundled with their high-speed Internet and cable offerings. Skype offers free VoIP worldwide using a peer-to-peer network, and Google has its own free VoIP service.

Although there are up-front investments required for an IP phone system, VoIP can reduce communication and network management costs by 20 to 30 percent. For example, VoIP saves Virgin Entertainment Group $700,000 per year in long-distance bills. In addition to lowering long-distance costs and eliminating monthly fees for private lines, an IP network provides a single voice-data infrastructure for both telecommunications and computing services. Companies no longer have to maintain separate networks or provide support services and personnel for each different type of network.

Another advantage of VoIP is its flexibility. Unlike the traditional telephone network, phones can be added or moved to different offices without rewiring or reconfiguring the network. With VoIP, a conference call is arranged by a simple click-and-drag operation on the computer screen to select the names of the conferees. Voice mail and e-mail can be combined into a single directory.

Unified Communications

In the past, each of the firm's networks for wired and wireless data, voice communications, and videoconferencing operated independently of each other and had to be managed separately by the information systems department. Now, however, firms are able to merge disparate communications modes into a single universally accessible service using unified communications technology. **Unified communications** integrates disparate channels for voice communications, data communications, instant messaging, e-mail, and electronic conferencing into a single experience where users can seamlessly switch back and forth between different communication modes. Presence technology shows whether a person is available to receive a call. Companies will need to examine

FIGURE 7-11   HOW VOICE OVER IP WORKS

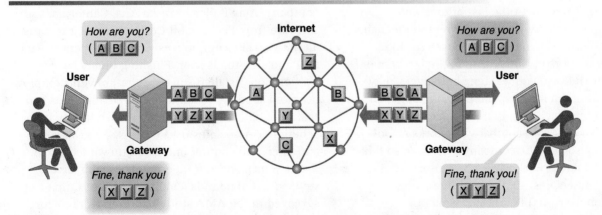

An VoIP phone call digitizes and breaks up a voice message into data packets that may travel along different routes before being reassembled at the final destination. A processor nearest the call's destination, called a gateway, arranges the packets in the proper order and directs them to the telephone number of the receiver or the IP address of the receiving computer.

INTERACTIVE SESSION: MANAGEMENT

MONITORING EMPLOYEES ON NETWORKS: UNETHICAL OR GOOD BUSINESS?

When you were at work, how many minutes (or hours) did you spend on Facebook today? Did you send personal e-mail or visit some sports Web sites? If so, you're not alone. According to a Nucleus Research study, 77 percent of workers with Facebook accounts use them during work hours. An IDC Research study shows that as much as 40 percent of Internet surfing occurring during work hours is personal, while other studies report as many as 90 percent of employees receive or send personal e-mail at work.

This behavior creates serious business problems. Checking e-mail, responding to instant messages, and sneaking in a brief YouTube video create a series of nonstop interruptions that divert employee attention from the job tasks they are supposed to be performing. According to Basex, a New York City business research company, these distractions take up as much as 28 percent of the average U.S. worker's day and result in $650 billion in lost productivity each year!

Many companies have begun monitoring their employee use of e-mail, blogs, and the Internet, sometimes without their knowledge. A recent American Management Association (AMA) survey of 304 U.S. companies of all sizes found that 66 percent of these companies monitor employee e-mail messages and Web connections. Although U.S. companies have the legal right to monitor employee Internet and e-mail activity while they are at work, is such monitoring unethical, or is it simply good business?

Managers worry about the loss of time and employee productivity when employees are focusing on personal rather than company business. Too much time on personal business, on the Internet or not, can mean lost revenue. Some employees may even be billing time they spend pursuing personal interests online to clients, thus overcharging them.

If personal traffic on company networks is too high, it can also clog the company's network so that legitimate business work cannot be performed. Schemmer Associates, an architecture firm in Omaha, Nebraska, and Potomac Hospital in Woodridge, Virginia, found that computing resources were limited by a lack of bandwidth caused by employees using corporate Internet connections to watch and download video files.

When employees use e-mail or the Web (including social networks) at employer facilities or with employer equipment, anything they do, including anything illegal, carries the company's name. Therefore, the employer can be traced and held liable. Management in many firms fear that racist, sexually explicit, or other potentially offensive material accessed or traded by their employees could result in adverse publicity and even lawsuits for the firm. Even if the company is found not to be liable, responding to lawsuits could cost the company tens of thousands of dollars.

Companies also fear leakage of confidential information and trade secrets through e-mail or blogs. A recent survey conducted by the American Management Association and the ePolicy Institute found that 14 percent of the employees polled admitted they had sent confidential or potentially embarrassing company e-mails to outsiders.

U.S. companies have the legal right to monitor what employees are doing with company equipment during business hours. The question is whether electronic surveillance is an appropriate tool for maintaining an efficient and positive workplace. Some companies try to ban all personal activities on corporate networks—zero tolerance. Others block employee access to specific Web sites or social sites or limit personal time on the Web.

For example, Enterprise Rent-A-Car blocks employee access to certain social sites and monitors the Web for employees' online postings about the company. Ajax Boiler in Santa Ana, California, uses software from SpectorSoft Corporation that records all the Web sites employees visit, time spent at each site, and all e-mails sent. Flushing Financial Corporation installed software that prevents employees from sending e-mail to specified addresses and scans e-mail attachments for sensitive information. Schemmer Associates uses OpenDNS to categorize and filter Web content and block unwanted video.

Some firms have fired employees who have stepped out of bounds. One-third of the companies surveyed in the AMA study had fired workers for misusing the Internet on the job. Among managers who fired employees for Internet misuse, 64 percent did so because the employees' e-mail contained inappropriate or offensive language, and more than 25 percent fired workers for excessive personal use of e-mail.

No solution is problem free, but many consultants believe companies should write corporate policies on employee e-mail and Internet use. The policies should include explicit ground rules that state, by position or level, under what circumstances employees can use company facilities for e-mail, blogging, or Web surfing. The policies should also inform employees whether these activities are monitored and explain why.

IBM now has "social computing guidelines" that cover employee activity on sites such as Facebook and Twitter. The guidelines urge employees not to conceal their identities, to remember that they are personally responsible for what they publish, and to refrain from discussing controversial topics that are not related to their IBM role.

The rules should be tailored to specific business needs and organizational cultures. For example, although some companies may exclude all employees from visiting sites that have explicit sexual material, law firm or hospital employees may require access to these sites. Investment firms will need to allow many of their employees access to other investment sites. A company dependent on widespread information sharing, innovation, and independence could very well find that monitoring creates more problems than it solves.

Sources: Joan Goodchild, "Not Safe for Work: What's Acceptable for Office Computer Use," *CIO Australia*, June 17, 2010; Sarah E. Needleman, "Monitoring the Monitors," *The Wall Street Journal*, August 16, 2010; Michelle Conline and Douglas MacMillan, "Web 2.0: Managing Corporate Reputations," *Business Week*, May 20, 2009; James Wong, "Drafting Trouble-Free Social Media Policies," Law.com, June 15, 2009; and Maggie Jackson, "May We Have Your Attention, Please?" *Business Week*, June 23, 2008.

CASE STUDY QUESTIONS

1. Should managers monitor employee e-mail and Internet usage? Why or why not?
2. Describe an effective e-mail and Web use policy for a company.
3. Should managers inform employees that their Web behavior is being monitored? Or should managers monitor secretly? Why or why not?

MIS IN ACTION

Explore the Web site of online employee monitoring software such as Websense, Barracuda Networks, MessageLabs, or SpectorSoft, and answer the following questions:

1. What employee activities does this software track? What can an employer learn about an employee by using this software?
2. How can businesses benefit from using this software?
3. How would you feel if your employer used this software where you work to monitor what you are doing on the job? Explain your response.

how work flows and business processes will be altered by this technology in order to gauge its value.

CenterPoint Properties, a major Chicago area industrial real estate company, used unified communications technology to create collaborative Web sites for each of its real estate deals. Each Web site provides a single point for accessing structured and unstructured data. Integrated presence technology lets team members e-mail, instant message, call, or videoconference with one click.

Virtual Private Networks

What if you had a marketing group charged with developing new products and services for your firm with members spread across the United States? You would want to be able to e-mail each other and communicate with the home office without any chance that outsiders could intercept the communications. In the past, one answer to this problem was to work with large private networking firms who offered secure, private, dedicated networks to customers. But this was an expensive solution. A much less-expensive solution is to create a virtual private network within the public Internet.

A **virtual private network (VPN)** is a secure, encrypted, private network that has been configured within a public network to take advantage of the economies of scale and management facilities of large networks, such as the Internet (see Figure 7-12). A VPN provides your firm with secure, encrypted communications at a much lower cost than the same capabilities offered by traditional non-Internet providers who use their private networks to secure communications. VPNs also provide a network infrastructure for combining voice and data networks.

Several competing protocols are used to protect data transmitted over the public Internet, including *Point-to-Point Tunneling Protocol (PPTP)*. In a process called tunneling, packets of data are encrypted and wrapped inside IP packets. By adding this wrapper around a network message to hide its content, business firms create a private connection that travels through the public Internet.

THE WEB

You've probably used the Web to download music, to find information for a term paper, or to obtain news and weather reports. The Web is the most popular Internet service. It's a system with universally accepted standards for storing, retrieving, formatting, and displaying information using a client/server architecture. Web pages are formatted using hypertext with embedded links that connect documents to one another and that also link pages to other objects, such as sound, video, or animation files. When you click a graphic and a video clip plays, you have clicked a hyperlink. A typical **Web site** is a collection of Web pages linked to a home page.

Hypertext

Web pages are based on a standard Hypertext Markup Language (HTML), which formats documents and incorporates dynamic links to other documents and pictures stored in the same or remote computers (see Chapter 5). Web

FIGURE 7-12   A VIRTUAL PRIVATE NETWORK USING THE INTERNET

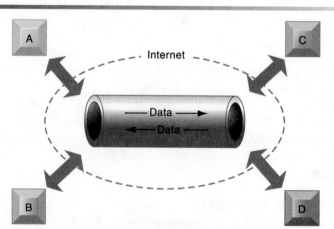

This VPN is a private network of computers linked using a secure "tunnel" connection over the Internet. It protects data transmitted over the public Internet by encoding the data and "wrapping" them within the Internet Protocol (IP). By adding a wrapper around a network message to hide its content, organizations can create a private connection that travels through the public Internet.

pages are accessible through the Internet because Web browser software operating your computer can request Web pages stored on an Internet host server using the **Hypertext Transfer Protocol (HTTP)**. HTTP is the communications standard used to transfer pages on the Web. For example, when you type a Web address in your browser, such as www.sec.gov, your browser sends an HTTP request to the sec.gov server requesting the home page of sec.gov.

HTTP is the first set of letters at the start of every Web address, followed by the domain name, which specifies the organization's server computer that is storing the document. Most companies have a domain name that is the same as or closely related to their official corporate name. The directory path and document name are two more pieces of information within the Web address that help the browser track down the requested page. Together, the address is called a **uniform resource locator (URL)**. When typed into a browser, a URL tells the browser software exactly where to look for the information. For example, in the URL *http://www.megacorp.com/content/features/082610.html*, *http* names the protocol used to display Web pages, www.megacorp.com is the domain name, *content/features* is the directory path that identifies where on the domain Web server the page is stored, and *082610.html* is the document name and the name of the format it is in (it is an HTML page).

Web Servers

A Web server is software for locating and managing stored Web pages. It locates the Web pages requested by a user on the computer where they are stored and delivers the Web pages to the user's computer. Server applications usually run on dedicated computers, although they can all reside on a single computer in small organizations.

The most common Web server in use today is Apache HTTP Server, which controls 54 percent of the market. Apache is an open source product that is free of charge and can be downloaded from the Web. Microsoft Internet Information Services is the second most commonly used Web server, with a 25 percent market share.

Searching for Information on the Web

No one knows for sure how many Web pages there really are. The surface Web is the part of the Web that search engines visit and about which information is recorded. For instance, Google visited about 100 billion pages in 2010, and this reflects a large portion of the publicly accessible Web page population. But there is a "deep Web" that contains an estimated 900 billion additional pages, many of them proprietary (such as the pages of *The Wall Street Journal Online*, which cannot be visited without an access code) or that are stored in protected corporate databases.

Search Engines Obviously, with so many Web pages, finding specific Web pages that can help you or your business, nearly instantly, is an important problem. The question is, how can you find the one or two pages you really want and need out of billions of indexed Web pages? **Search engines** attempt to solve the problem of finding useful information on the Web nearly instantly, and, arguably, they are the "killer app" of the Internet era. Today's search engines can sift through HTML files, files of Microsoft Office applications, PDF files, as well as audio, video, and image files. There are hundreds of different search engines in the world, but the vast majority of search results are supplied by three top providers: Google, Yahoo!, and Microsoft's Bing search engine.

Web search engines started out in the early 1990s as relatively simple software programs that roamed the nascent Web, visiting pages and gathering information about the content of each page. The first search engines were simple keyword indexes of all the pages they visited, leaving the user with lists of pages that may not have been truly relevant to their search.

In 1994, Stanford University computer science students David Filo and Jerry Yang created a hand-selected list of their favorite Web pages and called it "Yet Another Hierarchical Officious Oracle," or Yahoo!. Yahoo! was not initially a search engine but rather an edited selection of Web sites organized by categories the editors found useful, but it has since developed its own search engine capabilities.

In 1998, Larry Page and Sergey Brin, two other Stanford computer science students, released their first version of Google. This search engine was different: Not only did it index each Web page's words but it also ranked search results based on the relevance of each page. Page patented the idea of a page ranking system (PageRank System), which essentially measures the popularity of a Web page by calculating the number of sites that link to that page as well as the number of pages which it links to. Brin contributed a unique Web crawler program that indexed not only keywords on a page but also combinations of words (such as authors and the titles of their articles). These two ideas became the foundation for the Google search engine. Figure 7-13 illustrates how Google works.

Search engine Web sites are so popular that many people use them as their home page, the page where they start surfing the Web(see Chapter 10). As useful as they are, no one expected search engines to be big money makers. Today, however, search engines are the foundation for the fastest growing form of marketing and advertising, search engine marketing.

Search engines have become major shopping tools by offering what is now called **search engine marketing**. When users enter a search term at Google, Bing, Yahoo!, or any of the other sites serviced by these search engines, they receive two types of listings: sponsored links, for which advertisers have paid to be listed (usually at the top of the search results page), and unsponsored "organic" search results. In addition, advertisers can purchase small text boxes on the side of search results pages. The paid, sponsored advertisements are the fastest growing form of Internet advertising and are powerful new marketing tools that precisely match consumer interests with advertising messages at the right moment. Search engine marketing monetizes the value of the search process. In 2010, search engine marketing generated $12.3 billion in revenue, half of all online advertising ($25.6 billion). Ninety seven percent of Google's annual revenue of $23.6 billion comes from search engine marketing (eMarketer, 2010).

Because search engine marketing is so effective, companies are starting to optimize their Web sites for search engine recognition. The better optimized the page is, the higher a ranking it will achieve in search engine result listings. **Search engine optimization (SEO)** is the process of improving the quality and volume of Web traffic to a Web site by employing a series of techniques that help a Web site achieve a higher ranking with the major search engines when certain keywords and phrases are put in the search field. One technique is to make sure that the keywords used in the Web site description match the keywords likely to be used as search terms by prospective customers. For example, your Web site is more likely to be among the first ranked by search engines

FIGURE 7-13   HOW GOOGLE WORKS

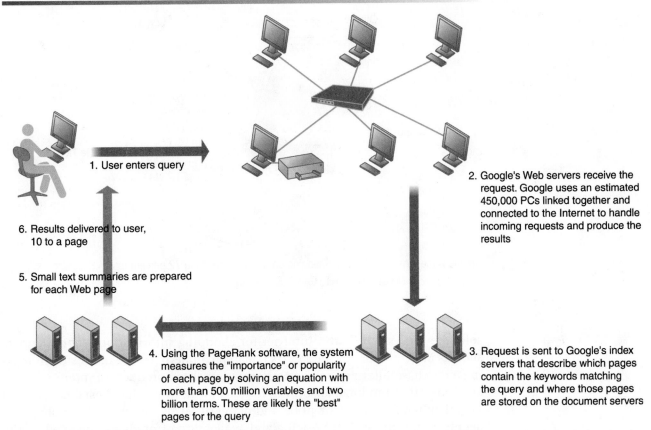

1. User enters query

6. Results delivered to user, 10 to a page

5. Small text summaries are prepared for each Web page

2. Google's Web servers receive the request. Google uses an estimated 450,000 PCs linked together and connected to the Internet to handle incoming requests and produce the results

4. Using the PageRank software, the system measures the "importance" or popularity of each page by solving an equation with more than 500 million variables and two billion terms. These are likely the "best" pages for the query

3. Request is sent to Google's index servers that describe which pages contain the keywords matching the query and where those pages are stored on the document servers

The Google search engine is continuously crawling the Web, indexing the content of each page, calculating its popularity, and storing the pages so that it can respond quickly to user requests to see a page. The entire process takes about one-half second.

if it uses the keyword "lighting" rather than "lamps" if most prospective customers are searching for "lighting." It is also advantageous to link your Web site to as many other Web sites as possible because search engines evaluate such links to determine the popularity of a Web page and how it is linked to other content on the Web. The assumption is the more links there are to a Web site, the more useful the Web site must be.

In 2010, about 110 million people each day in the United States alone used a search engine, producing over 17 billion searches a month. There are hundreds of search engines, but the top three (Google, Yahoo!, and Bing) account for over 90 percent of all searches (see Figure 7-14).

Although search engines were originally built to search text documents, the explosion in online video and images has created a demand for search engines that can quickly find specific videos. The words "dance," "love," "music," and "girl" are all exceedingly popular in titles of YouTube videos, and searching on these keywords produces a flood of responses even though the actual contents of the video may have nothing to do with the search term. Searching videos is challenging because computers are not very good or quick at recognizing digital images. Some search engines have started indexing movie scripts so it will be possible to search on dialogue to find a movie. Blinkx.com is a popular video search service and Google has added video search capabilities.

Intelligent Agent Shopping Bots Chapter 11 describes the capabilities of software agents with built-in intelligence that can gather or filter information and perform other tasks to assist users. **Shopping bots** use intelligent agent

FIGURE 7-14   TOP U.S. WEB SEARCH ENGINES

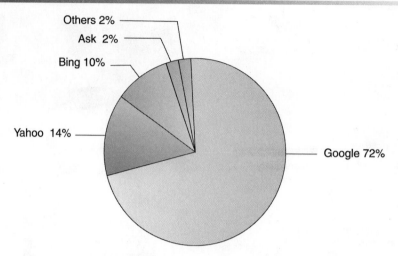

Google is the most popular search engine on the Web, handling 72 percent of all Web searches.
Sources: Based on data from SeoConsultants.com, August 28, 2010.

software for searching the Internet for shopping information. Shopping bots such as MySimon or Google Product Search can help people interested in making a purchase filter and retrieve information about products of interest, evaluate competing products according to criteria the users have established, and negotiate with vendors for price and delivery terms. Many of these shopping agents search the Web for pricing and availability of products specified by the user and return a list of sites that sell the item along with pricing information and a purchase link.

Web 2.0

Today's Web sites don't just contain static content—they enable people to collaborate, share information, and create new services and content online. These second-generation interactive Internet-based services are referred to as **Web 2.0**. If you have shared photos over the Internet at Flickr or another photo site, posted a video to YouTube, created a blog, used Wikipedia, or added a widget to your Facebook page, you've used some of these Web 2.0 services.

Web 2.0 has four defining features: interactivity, real-time user control, social participation (sharing), and user-generated content. The technologies and services behind these features include cloud computing, software mashups and widgets, blogs, RSS, wikis, and social networks.

Mashups and widgets, which we introduced in Chapter 5, are software services that enable users and system developers to mix and match content or software components to create something entirely new. For example, Yahoo's photo storage and sharing site Flickr combines photos with other information about the images provided by users and tools to make it usable within other programming environments.

These software applications run on the Web itself instead of the desktop. With Web 2.0, the Web is not just a collection of destination sites, but a source of data and services that can be combined to create applications users need. Web 2.0 tools and services have fueled the creation of social networks and other online communities where people can interact with one another in the manner of their choosing.

A **blog**, the popular term for a Weblog, is a personal Web site that typically contains a series of chronological entries (newest to oldest) by its author, and links to related Web pages. The blog may include a *blogroll* (a collection of links to other blogs) and *trackbacks* (a list of entries in other blogs that refer to a post on the first blog). Most blogs allow readers to post comments on the blog entries as well. The act of creating a blog is often referred to as "blogging." Blogs are either hosted by a third-party site such as Blogger.com, LiveJournal.com, TypePad.com, and Xanga.com, or prospective bloggers can download software such as Movable Type to create a blog that is housed by the user's ISP.

Blog pages are usually variations on templates provided by the blogging service or software. Therefore, millions of people without HTML skills of any kind can post their own Web pages and share content with others. The totality of blog-related Web sites is often referred to as the **blogosphere**. Although blogs have become popular personal publishing tools, they also have business uses (see Chapters 9 and 10).

If you're an avid blog reader, you might use RSS to keep up with your favorite blogs without constantly checking them for updates. **RSS**, which stands for Rich Site Summary or Really Simple Syndication, syndicates Web site content so that it can be used in another setting. RSS technology pulls specified content from Web sites and feeds it automatically to users' computers, where it can be stored for later viewing.

To receive an RSS information feed, you need to install aggregator or news reader software that can be downloaded from the Web. (Most current Web browsers include RSS reading capabilities.) Alternatively, you can establish an account with an aggregator Web site. You tell the aggregator to collect all updates from a given Web page, or list of pages, or gather information on a given subject by conducting Web searches at regular intervals. Once subscribed, you automatically receive new content as it is posted to the specified Web site.

A number of businesses use RSS internally to distribute updated corporate information. Wells Fargo uses RSS to deliver news feeds that employees can customize to see the business news of greatest relevance to their jobs. RSS feeds are so popular that online publishers are developing ways to present advertising along with content.

Blogs allow visitors to add comments to the original content, but they do not allow visitors to change the original posted material. **Wikis**, in contrast, are collaborative Web sites where visitors can add, delete, or modify content on the site, including the work of previous authors. Wiki comes from the Hawaiian word for "quick."

Wiki software typically provides a template that defines layout and elements common to all pages, displays user-editable software program code, and then renders the content into an HTML-based page for display in a Web browser. Some wiki software allows only basic text formatting, whereas other tools allow the use of tables, images, or even interactive elements, such as polls or games. Most wikis provide capabilities for monitoring the work of other users and correcting mistakes.

Because wikis make information sharing so easy, they have many business uses. For example, Motorola sales representatives use wikis for sharing sales information. Instead of developing a different pitch for every client, reps reuse the information posted on the wiki. The U.S. Department of Homeland Security's National Cyber Security Center deployed a wiki to facilitate collaboration among federal agencies on cybersecurity. NCSC and other agencies use the wiki for real-time information sharing on threats, attacks, and responses and as a repository for technical and standards information.

Social networking sites enable users to build communities of friends and professional colleagues. Members each typically create a "profile," a Web page for posting photos, videos, MP3 files, and text, and then share these profiles with others on the service identified as their "friends" or contacts. Social networking sites are highly interactive, offer real-time user control, rely on user-generated content, and are broadly based on social participation and sharing of content and opinions. Leading social networking sites include Facebook, MySpace (with 500 million and 180 million global members respectively in 2010), and LinkedIn (for professional contacts).

For many, social networking sites are the defining Web 2.0 application, and one that will radically change how people spend their time online; how people communicate and with whom; how business people stay in touch with customers, suppliers, and employees; how providers of goods and services learn about their customers; and how advertisers reach potential customers. The large social networking sites are also morphing into application development platforms where members can create and sell software applications to other members of the community. Facebook alone had over 1 million developers who created over 550,000 applications for gaming, video sharing, and communicating with friends and family. We talk more about business applications of social networking in Chapters 2 and 10, and you can find social networking discussions in many other chapters of the text. You can also find a more detailed discussion of Web 2.0 in our Learning Tracks.

Web 3.0: The Future Web

Every day about 110 million Americans enter 500 million queries search engines. How many of these 500 million queries produce a meaningful result (a useful answer in the first three listings)? Arguably, fewer than half. Google, Yahoo, Microsoft, and Amazon are all trying to increase the odds of people finding meaningful answers to search engine queries. But with over 100 billion Web pages indexed, the means available for finding the information you really want are quite primitive, based on the words used on the pages, and the relative popularity of the page among people who use those same search terms. In other words, it's hit and miss.

To a large extent, the future of the Web involves developing techniques to make searching the 100 billion public Web pages more productive and meaningful for ordinary people. Web 1.0 solved the problem of obtaining access to information. Web 2.0 solved the problem of sharing that information with others, and building new Web experiences. **Web 3.0** is the promise of a future Web where all this digital information, all these contacts, can be woven together into a single meaningful experience.

Sometimes this is referred to as the **Semantic Web**. "Semantic" refers to meaning. Most of the Web's content today is designed for humans to read and for computers to display, not for computer programs to analyze and manipulate. Search engines can discover when a particular term or keyword appears in a Web document, but they do not really understand its meaning or how it relates to other information on the Web. You can check this out on Google by entering two searches. First, enter "Paris Hilton". Next, enter "Hilton in Paris". Because Google does not understand ordinary English, it has no idea that you are interested in the Hilton Hotel in Paris in the second search. Because it cannot understand the meaning of pages it has indexed, Google's search engine returns the most popular pages for those queries where "Hilton" and "Paris" appear on the pages.

First described in a 2001 *Scientific American* article, the Semantic Web is a collaborative effort led by the World Wide Web Consortium to add a layer of meaning atop the existing Web to reduce the amount of human involvement in searching for and processing Web information (Berners-Lee et al., 2001).

Views on the future of the Web vary, but they generally focus on ways to make the Web more "intelligent," with machine-facilitated understanding of information promoting a more intuitive and effective user experience. For instance, let's say you want to set up a party with your tennis buddies at a local restaurant Friday night after work. One problem is that you had earlier scheduled to go to a movie with another friend. In a Semantic Web 3.0 environment, you would be able to coordinate this change in plans with the schedules of your tennis buddies, the schedule of your movie friend, and make a reservation at the restaurant all with a single set of commands issued as text or voice to your handheld smartphone. Right now, this capability is beyond our grasp.

Work proceeds slowly on making the Web a more intelligent experience, in large part because it is difficult to make machines, including software programs, that are truly intelligent like humans. But there are other views of the future Web. Some see a 3-D Web where you can walk through pages in a 3-D environment. Others point to the idea of a pervasive Web that controls everything from the lights in your living room, to your car's rear view mirror, not to mention managing your calendar and appointments.

Other complementary trends leading toward a future Web 3.0 include more widespread use of cloud computing and SaaS business models, ubiquitous connectivity among mobile platforms and Internet access devices, and the transformation of the Web from a network of separate siloed applications and content into a more seamless and interoperable whole. These more modest visions of the future Web 3.0 are more likely to be realized in the near term.

7.4 THE WIRELESS REVOLUTION

If you have a cell phone, do you use it for taking and sending photos, sending text messages, or downloading music clips? Do you take your laptop to class or to the library to link up to the Internet? If so, you're part of the wireless revolution! Cell phones, laptops, and small handheld devices have morphed into portable computing platforms that let you perform some of the computing tasks you used to do at your desk.

Wireless communication helps businesses more easily stay in touch with customers, suppliers, and employees and provides more flexible arrangements for organizing work. Wireless technology has also created new products, services, and sales channels, which we discuss in Chapter 10.

If you require mobile communication and computing power or remote access to corporate systems, you can work with an array of wireless devices, including cell phones, **smartphones**, and wireless-enabled personal computers. We introduced smartphones in our discussions of the mobile digital platform in Chapters 1 and 5. In addition to voice transmission, they feature capabilities for e-mail, messaging, wireless Internet access, digital photography, and personal information management. The features of the iPhone and BlackBerry illustrate the extent to which cellphones have evolved into small mobile computers.

CELLULAR SYSTEMS

Digital cellular service uses several competing standards. In Europe and much of the rest of the world outside the United Sates, the standard is Global System for Mobile Communication (GSM). GSM's strength is its international roaming capability. There are GSM cell phone systems in the United States, including T-Mobile and AT&T Wireless.

The major standard in the United States is Code Division Multiple Access (CDMA), which is the system used by Verizon and Sprint. CDMA was developed by the military during World War II. It transmits over several frequencies, occupies the entire spectrum, and randomly assigns users to a range of frequencies over time.

Earlier generations of cellular systems were designed primarily for voice and limited data transmission in the form of short text messages. Wireless carriers now offer more powerful cellular networks called third-generation or **3G** networks, with transmission speeds ranging from 144 Kbps for mobile users in, say, a car, to more than 2 Mbps for stationary users. This is sufficient transmission capacity for video, graphics, and other rich media, in addition to voice, making 3G networks suitable for wireless broadband Internet access. Many of the cellular handsets available today are 3G-enabled.

High-speed cellular networks are widely used in Japan, South Korea, Taiwan, Hong Kong, Singapore, and parts of northern Europe. In U.S. locations without full 3G coverage, U.S. cellular carriers have upgraded their networks to support higher-speed transmission, enabling cell phones to be used for Web access, music downloads, and other broadband services. PCs equipped with a special card can use these broadband cellular services for anytime, anywhere wireless Internet access.

The next evolution in wireless communication, called **4G networks**, is entirely packet switched and capable of 100 Mbps transmission speed (which can reach 1 Gbps under optimal conditions), with premium quality and high security. Voice, data, and high-quality streaming video will be available to users anywhere, anytime. Pre-4G technologies currently include Long Term Evolution (LTE) and the mobile WiMax. (See the discussion of WiMax in the following section.) You can find out more about cellular generations in the Learning Tracks for this chapter.

WIRELESS COMPUTER NETWORKS AND INTERNET ACCESS

If you have a laptop computer, you might be able to use it to access the Internet as you move from room to room in your dorm, or table to table in your university library. An array of technologies provide high-speed wireless access to the Internet for PCs and other wireless handheld devices as well as for cell phones. These new high-speed services have extended Internet access to numerous locations that could not be covered by traditional wired Internet services.

Bluetooth

Bluetooth is the popular name for the 802.15 wireless networking standard, which is useful for creating small **personal area networks (PANs)**. It links up to eight devices within a 10-meter area using low-power, radio-based communication and can transmit up to 722 Kbps in the 2.4-GHz band.

Wireless phones, pagers, computers, printers, and computing devices using Bluetooth communicate with each other and even operate each other without direct user intervention (see Figure 7-15). For example, a person could direct a notebook computer to send a document file wirelessly to a printer. Bluetooth connects wireless keyboards and mice to PCs or cell phones to earpieces without wires. Bluetooth has low-power requirements, making it appropriate for battery-powered handheld computers, cell phones, or PDAs.

Although Bluetooth lends itself to personal networking, it has uses in large corporations. For example, FedEx drivers use Bluetooth to transmit the delivery data captured by their handheld PowerPad computers to cellular transmitters, which forward the data to corporate computers. Drivers no longer need to spend time docking their handheld units physically in the transmitters, and Bluetooth has saved FedEx $20 million per year.

Wi-Fi and Wireless Internet Access

The 802.11 set of standards for wireless LANs and wireless Internet access is also known as **Wi-Fi**. The first of these standards to be widely adopted was 802.11b, which can transmit up to 11 Mbps in the unlicensed 2.4-GHz band and has an effective distance of 30 to 50 meters. The 802.11g standard can transmit up to 54 Mbps in the 2.4-GHz range. 802.11n is capable of transmitting over 100 Mbps. Today's PCs and netbooks have built-in support for Wi-Fi, as do the iPhone, iPad, and other smartphones.

In most Wi-Fi communication, wireless devices communicate with a wired LAN using access points. An access point is a box consisting of a radio receiver/transmitter and antennas that links to a wired network, router, or hub. Mobile access points such as Virgin Mobile's MiFi use the existing cellular network to create Wi-Fi connections.

FIGURE 7-15   A BLUETOOTH NETWORK (PAN)

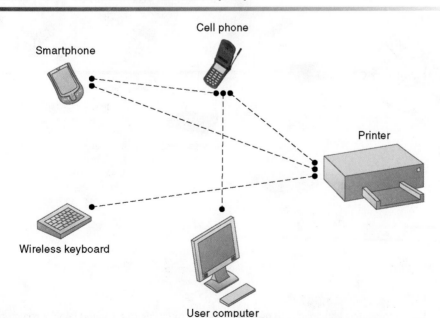

Bluetooth enables a variety of devices, including cell phones, smartphones, wireless keyboards and mice, PCs, and printers, to interact wirelessly with each other within a small 30-foot (10-meter) area. In addition to the links shown, Bluetooth can be used to network similar devices to send data from one PC to another, for example.

Figure 7-16 illustrates an 802.11 wireless LAN that connects a small number of mobile devices to a larger wired LAN and to the Internet. Most wireless devices are client machines. The servers that the mobile client stations need to use are on the wired LAN. The access point controls the wireless stations and acts as a bridge between the main wired LAN and the wireless LAN. (A bridge connects two LANs based on different technologies.) The access point also controls the wireless stations.

The most popular use for Wi-Fi today is for high-speed wireless Internet service. In this instance, the access point plugs into an Internet connection, which could come from a cable TV line or DSL telephone service. Computers within range of the access point use it to link wirelessly to the Internet.

Hotspots typically consist of one or more access points providing wireless Internet access in a public place. Some hotspots are free or do not require any additional software to use; others may require activation and the establishment of a user account by providing a credit card number over the Web.

Businesses of all sizes are using Wi-Fi networks to provide low-cost wireless LANs and Internet access. Wi-Fi hotspots can be found in hotels, airport lounges, libraries, cafes, and college campuses to provide mobile access to the Internet. Dartmouth College is one of many campuses where students now use Wi-Fi for research, course work, and entertainment.

Wi-Fi technology poses several challenges, however. One is Wi-Fi's security features, which make these wireless networks vulnerable to intruders. We provide more detail about Wi-Fi security issues in Chapter 8.

FIGURE 7-16   AN 802.11 WIRELESS LAN

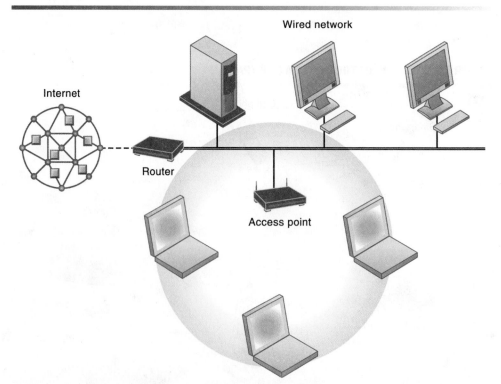

Mobile laptop computers equipped with network interface cards link to the wired LAN by communicating with the access point. The access point uses radio waves to transmit network signals from the wired network to the client adapters, which convert them into data that the mobile device can understand. The client adapter then transmits the data from the mobile device back to the access point, which forwards the data to the wired network.

Another drawback of Wi-Fi networks is susceptibility to interference from nearby systems operating in the same spectrum, such as wireless phones, microwave ovens, or other wireless LANs. However, wireless networks based on the 802.11n standard are able to solve this problem by using multiple wireless antennas in tandem to transmit and receive data and technology called *MIMO* (multiple input multiple output) to coordinate multiple simultaneous radio signals.

WiMax

A surprisingly large number of areas in the United States and throughout the world do not have access to Wi-Fi or fixed broadband connectivity. The range of Wi-Fi systems is no more than 300 feet from the base station, making it difficult for rural groups that don't have cable or DSL service to find wireless access to the Internet.

The IEEE developed a new family of standards known as WiMax to deal with these problems. **WiMax**, which stands for Worldwide Interoperability for Microwave Access, is the popular term for IEEE Standard 802.16. It has a wireless access range of up to 31 miles and transmission speed of up to 75 Mbps.

WiMax antennas are powerful enough to beam high-speed Internet connections to rooftop antennas of homes and businesses that are miles away. Cellular handsets and laptops with WiMax capabilities are appearing in the marketplace. Mobile WiMax is one of the pre-4G network technologies we discussed earlier in this chapter. Clearwire, which is owned by Sprint-Nextel, is using WiMax technology as the foundation for the 4G networks it is deploying throughout the United States.

RFID AND WIRELESS SENSOR NETWORKS

Mobile technologies are creating new efficiencies and ways of working throughout the enterprise. In addition to the wireless systems we have just described, radio frequency identification systems and wireless sensor networks are having a major impact.

Radio Frequency Identification (RFID)

Radio frequency identification (RFID) systems provide a powerful technology for tracking the movement of goods throughout the supply chain. RFID systems use tiny tags with embedded microchips containing data about an item and its location to transmit radio signals over a short distance to RFID readers. The RFID readers then pass the data over a network to a computer for processing. Unlike bar codes, RFID tags do not need line-of-sight contact to be read.

The RFID tag is electronically programmed with information that can uniquely identify an item plus other information about the item, such as its location, where and when it was made, or its status during production. Embedded in the tag is a microchip for storing the data. The rest of the tag is an antenna that transmits data to the reader.

The reader unit consists of an antenna and radio transmitter with a decoding capability attached to a stationary or handheld device. The reader emits radio waves in ranges anywhere from 1 inch to 100 feet, depending on its power output, the radio frequency employed, and surrounding environmental conditions. When an RFID tag comes within the range of the reader, the tag is activated and starts sending data. The reader captures these data, decodes them, and sends them back over a wired or wireless network to a host computer for further processing (see Figure 7-17). Both RFID tags and antennas come in a variety of shapes and sizes.

Active RFID tags are powered by an internal battery and typically enable data to be rewritten and modified. Active tags can transmit for hundreds of feet but may cost several dollars per tag. Automated toll-collection systems such as New York's E-ZPass use active RFID tags.

Passive RFID tags do not have their own power source and obtain their operating power from the radio frequency energy transmitted by the RFID reader. They are smaller, lighter, and less expensive than active tags, but only have a range of several feet.

In inventory control and supply chain management, RFID systems capture and manage more detailed information about items in warehouses or in production than bar coding systems. If a large number of items are shipped together, RFID systems track each pallet, lot, or even unit item in the shipment. This technology may help companies such as Walmart improve receiving and storage operations by improving their ability to "see" exactly what stock is stored in warehouses or on retail store shelves.

Walmart has installed RFID readers at store receiving docks to record the arrival of pallets and cases of goods shipped with RFID tags. The RFID reader reads the tags a second time just as the cases are brought onto the sales floor from backroom storage areas. Software combines sales data from Walmart's point-of-sale systems and the RFID data regarding the number of cases brought out to the sales floor. The program determines which items will soon be depleted and automatically generates a list of items to pick in the warehouse to replenish store shelves before they run out. This information helps Walmart reduce out-of-stock items, increase sales, and further shrink its costs.

The cost of RFID tags used to be too high for widespread use, but now it is less than 10 cents per passive tag in the United States. As the price decreases, RFID is starting to become cost-effective for some applications.

In addition to installing RFID readers and tagging systems, companies may need to upgrade their hardware and software to process the massive amounts of data produced by RFID systems—transactions that could add up to tens or hundreds of terabytes.

FIGURE 7-17   HOW RFID WORKS

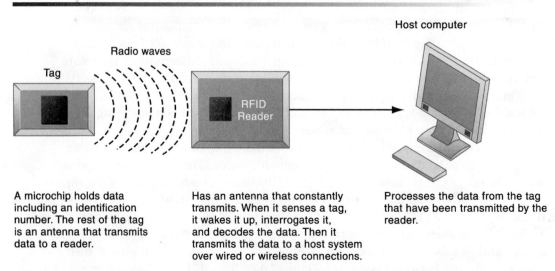

A microchip holds data including an identification number. The rest of the tag is an antenna that transmits data to a reader.

Has an antenna that constantly transmits. When it senses a tag, it wakes it up, interrogates it, and decodes the data. Then it transmits the data to a host system over wired or wireless connections.

Processes the data from the tag that have been transmitted by the reader.

RFID uses low-powered radio transmitters to read data stored in a tag at distances ranging from 1 inch to 100 feet. The reader captures the data from the tag and sends them over a network to a host computer for processing.

Software is used to filter, aggregate, and prevent RFID data from overloading business networks and system applications. Applications often need to be redesigned to accept large volumes of frequently generated RFID data and to share those data with other applications. Major enterprise software vendors, including SAP and Oracle-PeopleSoft, now offer RFID-ready versions of their supply chain management applications.

Wireless Sensor Networks

If your company wanted state-of-the art technology to monitor building security or detect hazardous substances in the air, it might deploy a wireless sensor network. **Wireless sensor networks (WSNs)** are networks of interconnected wireless devices that are embedded into the physical environment to provide measurements of many points over large spaces. These devices have built-in processing, storage, and radio frequency sensors and antennas. They are linked into an interconnected network that routes the data they capture to a computer for analysis.

These networks range from hundreds to thousands of nodes. Because wireless sensor devices are placed in the field for years at a time without any maintenance or human intervention, they must have very low power requirements and batteries capable of lasting for years.

Figure 7-18 illustrates one type of wireless sensor network, with data from individual nodes flowing across the network to a server with greater processing power. The server acts as a gateway to a network based on Internet technology.

Wireless sensor networks are valuable in areas such as monitoring environmental changes, monitoring traffic or military activity, protecting property, efficiently operating and managing machinery and vehicles, establishing security perimeters, monitoring supply chain management, or detecting chemical, biological, or radiological material.

FIGURE 7-18   A WIRELESS SENSOR NETWORK

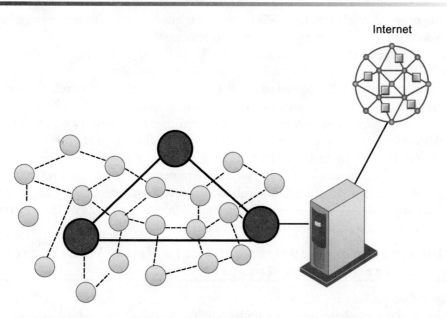

The small circles represent lower-level nodes and the larger circles represent high-end nodes. Lower-level nodes forward data to each other or to higher-level nodes, which transmit data more rapidly and speed up network performance.

7.5   HANDS-ON MIS PROJECTS

The projects in this section give you hands-on experience evaluating and selecting communications technology, using spreadsheet software to improve selection of telecommunications services, and using Web search engines for business research.

Management Decision Problems

1. Your company supplies ceramic floor tiles to Home Depot, Lowe's, and other home improvement stores. You have been asked to start using radio frequency identification tags on each case of tiles you ship to help your customers improve the management of your products and those of other suppliers in their warehouses. Use the Web to identify the cost of hardware, software, and networking components for an RFID system for your company. What factors should be considered? What are the key decisions that have to be made in determining whether your firm should adopt this technology?

2. BestMed Medical Supplies Corporation sells medical and surgical products and equipment from over 700 different manufacturers to hospitals, health clinics, and medical offices. The company employs 500 people at seven different locations in western and midwestern states, including account managers, customer service and support representatives, and warehouse staff. Employees communicate via traditional telephone voice services, e-mail, instant messaging, and cell phones. Management is inquiring about whether the company should adopt a system for unified communications. What factors should be considered? What are the key decisions that have to be made in determining whether to adopt this technology? Use the Web, if necessary, to find out more about unified communications and its costs.

Improving Decision Making: Using Spreadsheet Software to Evaluate Wireless Services

Software skills: Spreadsheet formulas, formatting
Business skills: Analyzing telecommunications services and costs

In this project, you'll use the Web to research alternative wireless services and use spreadsheet software to calculate wireless service costs for a sales force.

You would like to equip your sales force of 35 based in Cincinnati, Ohio, with mobile phones that have capabilities for voice transmission, text messaging, and taking and sending photos. Use the Web to select a wireless service provider that provides nationwide service as well as good service in your home area. Examine the features of the mobile handsets offered by each of these vendors. Assume that each of the 35 salespeople will need to spend three hours per day during business hours (8 a.m. to 6 p.m.) on mobile voice communication, send 30 text messages per day, and five photos per week. Use your spreadsheet software to determine the wireless service and handset that will offer the best pricing per user over a two-year period. For the purposes of this exercise, you do not need to consider corporate discounts.

Achieving Operational Excellence: Using Web Search Engines for Business Research

Software skills: Web search tools
Business skills: Researching new technologies

This project will help develop your Internet skills in using Web search engines for business research.

You want to learn more about ethanol as an alternative fuel for motor vehicles. Use the following search engines to obtain that information: Yahoo!, Google, and Bing. If you wish, try some other search engines as well. Compare the volume and quality of information you find with each search tool. Which tool is the easiest to use? Which produced the best results for your research? Why?

LEARNING TRACK MODULES

The following Learning Tracks provide content relevant to topics covered in this chapter:

1. Computing and Communications Services Provided by Commercial Communications Vendors
2. Broadband Network Services and Technologies
3. Cellular System Generations
4. WAP and I-Mode: Wireless Cellular Standards for Web Access
5. Wireless Applications for Customer Relationship Management, Supply Chain Management, and Health care
6. Web 2.0

Review Summary

1. *What are the principal components of telecommunications networks and key networking technologies?*

    A simple network consists of two or more connected computers. Basic network components include computers, network interfaces, a connection medium, network operating system software, and either a hub or a switch. The networking infrastructure for a large company includes the traditional telephone system, mobile cellular communication, wireless local area networks, video-conferencing systems, a corporate Web site, intranets, extranets, and an array of local and wide area networks, including the Internet.

    Contemporary networks have been shaped by the rise of client/server computing, the use of packet switching, and the adoption of Transmission Control Protocol/Internet Protocol (TCP/IP) as a universal communications standard for linking disparate networks and computers, including the Internet. Protocols provide a common set of rules that enable communication among diverse components in a telecommunications network.

2. *What are the main telecommunications transmission media and types of networks?*

    The principal physical transmission media are twisted copper telephone wire, coaxial copper cable, fiber-optic cable, and wireless transmission. Twisted wire enables companies to use existing wiring for telephone systems for digital communication, although it is relatively slow. Fiber-optic and coaxial cable are used for high-volume transmission but are expensive to install. Microwave and communications satellites are used for wireless communication over long distances.

    local area networks (LANs) connect PCs and other digital devices together within a 500-meter radius and are used today for many corporate computing tasks. Network components may be connected together using a star, bus, or ring topology. Wide area networks (WANs) span broad geographical distances, ranging from several miles to continents, and are private networks that are independently managed. Metropolitan area networks (MANs) span a single urban area.

    Digital subscriber line (DSL) technologies, cable Internet connections, and T1 lines are often used for high-capacity Internet connections.

    Cable Internet connections provide high-speed access to the Web or corporate intranets at speeds of up to 10 Mbps. A T1 line supports a data transmission rate of 1.544 Mbps.

3. *How do the Internet and Internet technology work, and how do they support communication and e-business?*

    The Internet is a worldwide network of networks that uses the client/server model of computing and the TCP/IP network reference model. Every computer on the Internet is assigned a unique numeric IP address. The Domain Name System (DNS) converts IP addresses to more user-friendly domain names. Worldwide Internet policies are established by organizations and government bodies, such as the Internet Architecture Board (IAB) and the World Wide Web Consortium (W3C).

    Major Internet services include e-mail, newgroups, chatting, instant messaging, Telnet, FTP, and the Web. Web pages are based on Hypertext Markup Language (HTML) and can display text, graphics, video, and audio. Web site directories, search engines, and RSS technology help users locate the information they need on the Web. RSS, blogs, social networking, and wikis are features of Web 2.0.

    Firms are also starting to realize economies by using VoIP technology for voice transmission and by using virtual private networks (VPNs) as low-cost alternatives to private WANs.

4. *What are the principal technologies and standards for wireless networking, communication, and Internet access?*

    Cellular networks are evolving toward high-speed, high-bandwidth, digital packet-switched transmission. Broadband 3G networks are capable of transmitting data at speeds ranging from 144 Kbps to more than 2 Mbps. 4G networks capable of transmission speeds that could reach 1 Gbps are starting to be rolled out.

    Major cellular standards include Code Division Multiple Access (CDMA), which is used primarily in the United States, and Global System for Mobile Communications (GSM), which is the standard in Europe and much of the rest of the world.

Standards for wireless computer networks include Bluetooth (802.15) for small personal area networks (PANs), Wi-Fi (802.11) for local area networks (LANs), and WiMax (802.16) for metropolitan area networks (MANs).

5. *Why are radio frequency identification (RFID) and wireless sensor networks valuable for business?*
Radio frequency identification (RFID) systems provide a powerful technology for tracking the movement of goods by using tiny tags with embedded data about an item and its location. RFID readers read the radio signals transmitted by these tags and pass the data over a network to a computer for processing. Wireless sensor networks (WSNs) are networks of interconnected wireless sensing and transmitting devices that are embedded into the physical environment to provide measurements of many points over large spaces.

Key Terms

3G networks, 276

4G networks, 276

Bandwidth, 257

Blog, 273

Blogosphere, 273

Bluetooth, 276

Broadband, 247

Bus topology, 254

Cable Internet connections, 258

Cell phone, 257

Chat, 263

Coaxial cable, 255

Digital subscriber line (DSL), 258

Domain name, 258

Domain Name System (DNS), 258

E-mail, 263

Fiber-optic cable, 255

File Transfer Protocol (FTP), 261

Hertz, 257

Hotspots, 278

Hubs, 248

HypertextTransfer Protocol (HTTP), 269

Instant messaging, 264

Internet Protocol (IP) address, 258

Internet service provider (ISP), 257

Internet2, 261

local area network (LAN), 253

Metropolitan-area network (MAN), 200

Microwave, 256

Modem, 252

Network interface card (NIC), 248

Network operating system (NOS), 248

Packet switching, 250

Peer-to-peer, 254

Personal-area networks (PANs), 276

Protocol, 251

Radio frequency identification (RFID), 279

Ring topology, 254

Router, 248

RSS, 273

Search engines, 270

Search engine marketing, 270

Search engine optimization (SEO), 270

Semantic Web, 274

Shopping bots, 272

Smartphones, 276

Social networking, 274

Star topology, 254

Switch, 248

T1 lines, 258

Telnet, 261

Topology, 254

Transmission Control Protocol/Internet Protocol (TCP/IP), 251

Twisted wire, 255

Unified communications, 265

Uniform resource locator (URL), 269

Virtual private network (VPN), 268

Voice over IP (VoIP), 264

Web 2.0, 272

Web 3.0, 274

Web site, 268

Wide-area networks (WANs), 254

Wi-Fi, 277

Wiki, 273

WiMax, 279

Wireless sensor networks (WSNs), 281

Review Questions

1. What are the principal components of telecommunications networks and key networking technologies?

- Describe the features of a simple network and the network infrastructure for a large company.
- Name and describe the principal technologies and trends that have shaped contemporary telecommunications systems.

2. What are the main telecommunications transmission media and types of networks?

- Name the different types of physical transmission media and compare them in terms of speed and cost.
- Define a LAN, and describe its components and the functions of each component.
- Name and describe the principal network topologies.

3. How do the Internet and Internet technology work, and how do they support communication and e-business?

- Define the Internet, describe how it works, and explain how it provides business value.
- Explain how the Domain Name System (DNS) and IP addressing system work.

- List and describe the principal Internet services.
- Define and describe VoIP and virtual private networks, and explain how they provide value to businesses.
- List and describe alternative ways of locating information on the Web.
- Compare Web 2.0 and Web 3.0.

4. What are the principal technologies and standards for wireless networking, communications, and Internet access?

- Define Bluetooth, Wi-Fi, WiMax, 3G and 4G networks.
- Describe the capabilities of each and for which types of applications each is best suited.

5. Why are RFID and wireless sensor networks (WSNs) valuable for business?

- Define RFID, explain how it works, and describe how it provides value to businesses.
- Define WSNs, explain how they work, and describe the kinds of applications that use them.

Discussion Questions

1. It has been said that within the next few years, smartphones will become the single most important digital device we own. Discuss the implications of this statement.

2. Should all major retailing and manufacturing companies switch to RFID? Why or why not?

3. Compare Wi-Fi and high-speed cellular systems for accessing the Internet. What are the advantages and disadvantages of each?

Video Cases

Video Cases and Instructional Videos illustrating some of the concepts in this chapter are available. Contact your instructor to access these videos.

Collaboration and Teamwork: Evaluating Smartphones

Form a group with three or four of your classmates. Compare the capabilities of Apple's iPhone with a smartphone handset from another vendor with similar features. Your analysis should consider the purchase cost of each device, the wireless networks where each device can operate, service plan and handset costs, and the services available for each device.

You should also consider other capabilities of each device, including the ability to integrate with existing corporate or PC applications. Which device would you select? What criteria would you use to guide your selection? If possible, use Google Sites to post links to Web pages, team communication announcements, and work assignments; to brainstorm; and to work collaboratively on project documents. Try to use Google Docs to develop a - presentation of your findings for the class.

Google, Apple, and Microsoft Struggle for Your Internet Experience
CASE STUDY

In what looks like a college food fight, the three Internet titans—Google, Microsoft, and Apple—are in an epic struggle to dominate your Internet experience. What's at stake is where you search, buy, find your music and videos, and what device you will use to do all these things. The prize is a projected 2015 $400 billion e-commerce marketplace where the major access device will be a mobile smartphone or tablet computer. Each firm generates extraordinary amounts of cash based on different business models. Each firm brings billions of dollars of spare cash to the fight.

In this triangular fight, at one point or another, each firm has befriended one of the other firms to combat the other firm. Two of the firms—Google and Apple—are determined to prevent Microsoft from expanding its dominance beyond the PC desktop. So Google and Apple are friends. But when it comes to mobile phones and apps, Goggle and Apple are enemies: each want to dominate the mobile market. Apple and Microsoft are determined to prevent Google from extending beyond its dominance in search and advertising. So Apple and Microsoft are friends. But when it comes to the mobile marketplace for devices and apps, Apple and Microsoft are enemies. Google and Microsoft are just plain enemies in a variety of battles. Google is trying to weaken Microsoft's PC software dominance, and Microsoft is trying to break into the search advertising market with Bing.

Today the Internet, along with hardware devices and software applications, is going through a major expansion. Mobile devices with advanced functionality and ubiquitous Internet access are rapidly gaining on traditional desktop computing as the most popular form of computing, changing the basis for competition throughout the industry. Research firm Gartner predicts that by 2013, mobile phones will surpass PCs as the way most people access the Internet. Today, mobile devices account for 5 percent of all searches performed on the Internet; in 2016, they are expected to account for 23.5% of searches.

These mobile Internet devices are made possible by a growing cloud of computing capacity available to anyone with a smartphone and Internet connectivity. Who needs a desktop PC anymore when you can listen to music and watch videos 24/7? It's no surprise, then, that today's tech titans are so aggressively battling for control of this brave new mobile world.

Apple, Google, and Microsoft already compete in an assortment of fields. Google has a huge edge in advertising, thanks to its dominance in Internet search. Microsoft's offering, Bing, has grown to about 10 percent of the search market, and the rest essentially belongs to Google. Apple is the leader in mobile software applications, thanks to the popularity of the App Store for its iPhones. Google and Microsoft have less popular app offerings on the Web.

Microsoft is still the leader in PC operating systems and desktop productivity software, but has failed miserably with smartphone hardware and software, mobile computing, cloud-based software apps, its Internet portal, and even its game machines and software. All contribute less than 5 percent to Microsoft's revenue (the rest comes from Windows, Office, and network software). While Windows is still the operating system on 95 percent of the world's 2 billion PCs, Google's Android OS and Apple's iOS are the dominant players in the mobile computing market. The companies also compete in music, Internet browsers, online video, and social networking.

For both Apple and Google, the most critical battleground is mobile computing. Apple has several advantages that will serve it well in the battle for mobile supremacy. It's no coincidence that since the Internet exploded in size and popularity, so too did the company's revenue, which totaled well over $40 billion in 2009. The iMac, iPod, and iPhone have all contributed to the company's enormous success in the Internet era, and the company hopes that the iPad will follow the trend of profitability set by these products. Apple has a loyal user base that has steadily grown and is very likely to buy future product offerings. Apple is hopeful that the iPad will be as successful as the iPhone, which already accounts for over 30 percentof Apple's revenue. So far, the iPad appears to be living up to this expectation.

Part of the reason for the popularity of the Apple iPhone, and for the optimism surrounding Internet-equipped smartphones in general, has been the success of the App Store. A vibrant selection of applications (apps) distinguishes Apple's offerings from its competitors', and gives the company a measurable head start in this marketplace. Apple already offers over 250,000 apps for its devices, and Apple takes a 30% cut of all app sales. Apps greatly enrich the experience of using a mobile device, and without them, the predictions for the future of mobile Internet would not be nearly as bright. Whoever creates the most appealing set of devices and applications will derive a significant competitive advantage over rival companies. Right now, that company is Apple.

But the development of smartphones and mobile Internet is still in its infancy. Google has acted swiftly to enter the battle for mobile supremacy while it can still 'win', irreparably damaging its relationship with Apple, its former ally, in the process. As more people switch to mobile computing as their primary method for accessing the Internet, Google is aggressively following the eyeballs. Google is as strong as the size of its advertising network. With the impending shift towards mobile computing looming, it's no certainty that it will be able to maintain its dominant position in search. That's why the dominant online search company began developing a mobile operating system and its Nexus One entry into the smartphone marketplace. Google hopes to control its own destiny in an increasingly mobile world.

Google's efforts to take on Apple began when it acquired Android, Inc., the developer of the mobile operating system of the same name. Google's original goal was to counter Microsoft's attempts to enter the mobile device market, but Microsoft was largely unsuccessful. Instead, Apple and Research In Motion, makers of the popular BlackBerry series of smartphones, filled the void. Google continued to develop Android, adding features that Apple's offerings lacked, such as the ability to run multiple apps at once. After an initial series of blocky, unappealing prototypes, there are now Android-equipped phones that are functionally and aesthetically competitive with the iPhone. For example, the Motorola Droid was heavily advertised, using the slogan "Everything iDon't...Droid Does."

Google has been particularly aggressive with its entry into the mobile computing market because it is concerned about Apple's preference for 'closed', proprietary standards on its phones. It would like smartphones to have open nonproprietary

platforms where users can freely roam the Web and pull in apps that work on many different devices.

Apple believes devices such as smartphones and tablets should have proprietary standards and be tightly controlled, with customers using applications on these devices that have been downloaded from the its App Store. Thus Apple retains the final say over whether or not its mobile users can access various services on the Web, and that includes services provided by Google. Google doesn't want Apple to be able to block it from providing its services on iPhones, or any other smartphone. A high- profile example of Apple's desire to fend off Google occurred after Google attempted to place its voice mail management program, Google Voice, onto the iPhone. Apple cited privacy concerns in preventing Google's effort.

Soon after, Google CEO Eric Schmidt stepped down from his post on Apple's board of directors. Since Schmidt's departure from Apple's board, the two companies have been in an all-out war. They've battled over high-profile acquisitions, including mobile advertising firm AdMob, which was highly sought after by both companies. AdMob sells banner ads that appear inside mobile applications, and the company is on the cutting edge of developing new methods of mobile advertising. Apple was close to a deal with the start-up when Google swooped in and bought AdMob for $750 million in stock. Google doesn't expect to earn anything close to that in returns from the deal, but it was willing to pay a premium to disrupt Apple's mobile advertising effort.

Undeterred, Apple bought top competitor Quattro Wireless for $275 million in January 2010. It then shuttered the service in September of that year in favor of its own iAd advertising platform. IAd allows developers of the programs in Apple's App Store for the iPhone, iPad, and iPod Touch to embed ads in their software. Apple will sell the ads and give the app developers 60 percent of the ad revenue.

Apple has been more than willing to use similarly combative tactics to slow its competition down. Apple sued HTC, the Taiwanese mobile phone manufacturer of Android-equipped phones, citing patent infringement. Apple CEO Steve Jobs has consistently bashed Google in the press, characterizing the company as a bully and questioning its ethics. Many analysts speculate that Apple may take a shot at Google by teaming up with a partner that would have been unthinkable just a few years ago: Microsoft. News reports suggest that Apple is considering striking a deal with Microsoft to make Bing its default search engine on both the iPhone and Apple's Web browser. This would be a blow to Google, and a boon to Microsoft, which would receive a much needed boost to its fledgling search service.

The struggle between Apple and Google wouldn't matter much if there wasn't so much potential money at stake. Billions of dollars hang in the balance, and the majority of that money will come from advertising. App sales are another important component, especially for Apple. Apple has the edge in selection and quality of apps, but while sales have been brisk, developers have complained that making money is too difficult. A quarter of the 250,000 apps available in early 2010 were free, which makes no money for developers or for Apple but it does bring consumers to the Apple marketplace where they can be sold other apps or entertainment services.

Google in the meantime is moving aggressively to support manufacturers of handsets that run its Android operating system and can access its services online. Apple relies on sales of its devices to remain profitable. It has had no problems with this so far, but Google only needs to spread its advertising networks onto these devices to make a profit. In fact, some analysts speculate that Google envisions a future where mobile phones cost a fraction of what they do today, or are even free, requiring only the advertising revenue generated by the devices to turn a profit. Apple would struggle to remain competitive in this environment. Jobs has kept the Apple garden closed for a simple reason: you need an Apple device to play there.

The three-way struggle between Microsoft, Apple, and Google really has no precedent in the history of computing platforms. In early contests it was typically a single firm that rode the crest of a new technology to become the dominant player. Examples include IBM's dominance of the mainframe market, Digital Euipment's dominance of minicomputers, Microsoft's dominance of PC operating systems and productivity applications, and Cisco Systems' dominance of the Internet router market. In the current struggle are three firms trying to dominate the customer experience on the Internet. Each firm brings certain strengths and weaknesses to the fray. Will a single firm "win," or will all three survive the contest for the consumer Internet experience? It's still too early to tell.

Sources: Jennifer LeClaire, "Quattro Wireless to be Closed as Apple Focuses on IAd," *Top Tech News*, August 20, 2010; Yukari Iwatani Kane and Emily Steel, "Apple Fights Rival Google on New Turf," *The Wall Street Journal*, April 8, 2010; Brad Stone and Miguel Helft, "Apple's Spat with Google Is Getting Personal," *The New York Times*, March 12, 2010; Peter Burrows, "Apple vs. Google," *BusinessWeek*, January 14, 2010; Holman W. Jenkins, Jr., "The Microsofting of Apple?", *The Wall Street Journal*, February 10, 2010; Jessica E. Vascellaro and Ethan Smith, "Google and Microsoft Crank Up Rivalry," *The Wall Street Journal*, October 21, 2009; Jessica E. Vascellaro and Don Clark, "Google Targets Microsoft's Turf," *The Wall Street Journal*, July 9, 2009; Miguel Helft, "Google Set to Acquire AdMob for $750 Million," *The New York Times*, November 10, 2009; Jessica E. Vascellaro, "Google Rolls Out New Tools as it Battles Rival," *The Wall Street Journal*, December 8, 2009; and Jessica E. Vascellaro and Yukari Iwatani Kane, "Apple, Google Rivalry Heats Up," *The Wall Street Journal*, December 10, 2009.

CASE STUDY QUESTIONS

1. Compare the business models and areas of strength of Apple, Google, and Microsoft.

2. Why is mobile computing so important to these three firms? Evaluate the mobile platform offerings of each firm.

3. What is the significance of applications and app stores to the success or failure of mobile computing?

4. Which company and business model do you think will prevail in this epic struggle? Explain your answer.

5. What difference would it make to you as a manager or individual consumer if Apple, Google, or Microsoft dominated the Internet experience? Explain your answer.

Chapter 8

Securing Information Systems

Interactive Sessions:

When Antivirus Software Cripples Your Computers

How Secure Is the Cloud?

YOU'RE ON FACEBOOK? WATCH OUT!

Facebook is the world's largest online social network, and increasingly, the destination of choice for messaging friends, sharing photos and videos, and collecting "eyeballs" for business advertising and market research. But, watch out! It's also a great place for losing your identity or being attacked by malicious software.

How could that be? Facebook has a security team that works hard to counter threats on that site. It uses up-to-date security technology to protect its Web site. But with 500 million users, it can't police everyone and everything. And Facebook makes an extraordinarily tempting target for both mischief-makers and criminals.

Facebook has a huge worldwide user base, an easy-to-use Web site, and a community of users linked to their friends. Its members are more likely to trust messages they receive from friends, even if this communication is not legitimate. Perhaps for these reasons, research from the Kaspersky Labs security firm shows malicious software on social networking sites such as Facebook and MySpace is 10 times more successful at infecting users than e-mail-based attacks. Moreover, IT security firm Sophos reported on February 1, 2010, that Facebook poses the greatest security risk of all the social networking sites.

Here are some examples of what can go wrong:

According to a February 2010 report from Internet security company NetWitness, Facebook served as the primary delivery method for an 18-month-long hacker attack in which Facebook users were tricked into revealing their passwords and downloading a rogue program that steals financial data. A legitimate-looking Facebook e-mail notice asked users to provide information to help the social network update its login system. When the user clicked the "update" button in the e-mail, that person was directed to a bogus Facebook login screen where the user's name was filled in and that person was prompted to provide his or her password. Once the user supplied that information, an "Update Tool," installed the Zeus "Trojan horse" rogue software program designed to steal financial and personal data by surreptitiously tracking users' keystrokes as they enter information into their computers. The hackers, most likely an Eastern European criminal group, stole as many as 68,000 login credentials from 2,400 companies and government agencies for online banking, social networking sites, and e-mail.

The Koobface worm targets Microsoft Windows users of Facebook, Twitter, and other social networking Web sites in order to gather sensitive information from the victims such as credit card numbers. Koobface was first detected in December 2008. It spreads by delivering bogus Facebook messages to people who are "friends" of a Facebook user whose computer has already been infected. Upon receipt, the message directs the recipients to a third-party Web site, where they are prompted to download what is purported to be an update of the Adobe Flash player. If they download and execute the file, Koobface is able to infect their system and use the computer for more malicious work.

For much of May 2010, Facebook members and their

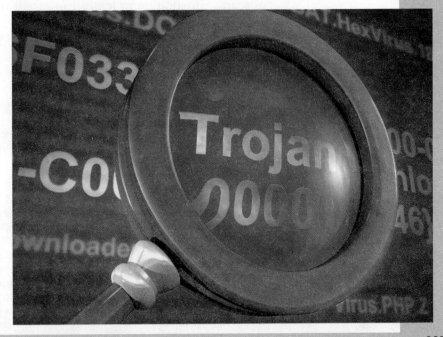

friends were victims of a spam campaign that tries to e-mail unsolicited advertisements and steal Facebook users' login credentials. The attack starts with a message containing a link to a bogus Web page sent by infected users to all of their friends. The message addresses each friend by name and invites that person to click on a link to "the most hilarious video ever." The link transports the user to a rogue Web site mimicking the Facebook login form. When users try to log in, the page redirects back to a Facebook application page that installs illicit adware software, which bombards their computers with all sorts of unwanted ads.

Recovering from these attacks is time-consuming and costly, especially for business firms. A September 2010 study by Panda Security found that one-third of small and medium businesses it surveyed had been hit by malicious software from social networks, and more than a third of these suffered more than $5,000 in losses. Of course, for large businesses, losses from Facebook are much greater.

Sources: Lance Whitney, "Social-Media Malware Hurting Small Businesses," CNET News, September 15, 2010; Raj Dash, "Report: Facebook Served as Primary Distribution Channel for Botnet Army," allfacebook.com, February 18, 2010; Sam Diaz, "Report: Bad Guys Go Social: Facebook Tops Security Risk List," *ZDNet*, February 1, 2010; Lucian Constantin, "Weekend Adware Scam Returns to Facebook," Softpedia, May 29, 2010; Brad Stone, "Viruses that Leave Victims Red in the Facebook," *The New York Times*, December 14, 2009; and Brian Prince, "Social Networks 10 Times as Effective for Hackers, Malware," *eWeek*, May 13, 2009.

The problems created by malicious software on Facebook illustrate some of the reasons why businesses need to pay special attention to information system security. Facebook provides a plethora of benefits to both individuals and businesses. But from a security standpoint, using Facebook is one of the easiest ways to expose a computer system to malicious software—your computer, your friends' computers, and even the computers of Facebook-participating businesses.

The chapter-opening diagram calls attention to important points raised by this case and this chapter. Although Facebook's management has a security policy and security team in place, Facebook has been plagued with many security problems that affect both individuals and businesses. The "social" nature of this site and large number of users make it unusually attractive for criminals and hackers intent on stealing valuable personal and financial information and propagating malicious software. Even though Facebook and its users deploy security technology, they are still vulnerable to new kinds of malicious software attacks and criminal scams. In addition to losses from theft of financial data, the difficulties of eradicating the malicious software or repairing damage caused by identity theft add to operational costs and make both individuals and businesses less effective.

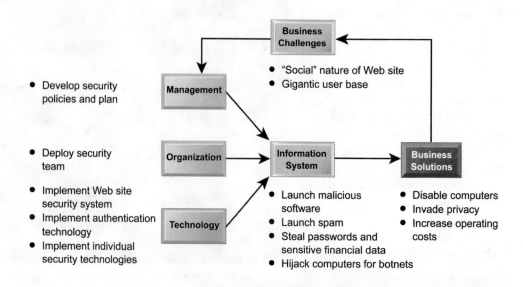

8.1   SYSTEM VULNERABILITY AND ABUSE

C an you imagine what would happen if you tried to link to the Internet without a firewall or antivirus software? Your computer would be disabled in a few seconds, and it might take you many days to recover. If you used the computer to run your business, you might not be able to sell to your customers or place orders with your suppliers while it was down. And you might find that your computer system had been penetrated by outsiders, who perhaps stole or destroyed valuable data, including confidential payment data from your customers. If too much data were destroyed or divulged, your business might never be able to operate!

In short, if you operate a business today, you need to make security and control a top priority. **Security** refers to the policies, procedures, and technical measures used to prevent unauthorized access, alteration, theft, or physical damage to information systems. **Controls** are methods, policies, and organizational procedures that ensure the safety of the organization's assets; the accuracy and reliability of its records; and operational adherence to management standards.

WHY SYSTEMS ARE VULNERABLE

When large amounts of data are stored in electronic form, they are vulnerable to many more kinds of threats than when they existed in manual form. Through communications networks, information systems in different locations are interconnected. The potential for unauthorized access, abuse, or fraud is not limited to a single location but can occur at any access point in the network. Figure 8-1 illustrates the most common threats against contemporary information systems. They can stem from technical, organizational, and environmental factors compounded by poor management decisions. In the multi-tier client/server computing environment illustrated here, vulnerabilities exist at each layer and in the communications between the layers. Users at the client layer can cause harm by introducing errors or by accessing systems without

FIGURE 8-1   CONTEMPORARY SECURITY CHALLENGES AND VULNERABILITIES

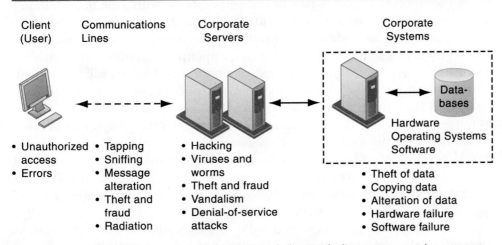

The architecture of a Web-based application typically includes a Web client, a server, and corporate information systems linked to databases. Each of these components presents security challenges and vulnerabilities. Floods, fires, power failures, and other electrical problems can cause disruptions at any point in the network.

authorization. It is possible to access data flowing over networks, steal valuable data during transmission, or alter messages without authorization. Radiation may disrupt a network at various points as well. Intruders can launch denial-of-service attacks or malicious software to disrupt the operation of Web sites. Those capable of penetrating corporate systems can destroy or alter corporate data stored in databases or files.

Systems malfunction if computer hardware breaks down, is not configured properly, or is damaged by improper use or criminal acts. Errors in programming, improper installation, or unauthorized changes cause computer software to fail. Power failures, floods, fires, or other natural disasters can also disrupt computer systems.

Domestic or offshore partnering with another company adds to system vulnerability if valuable information resides on networks and computers outside the organization's control. Without strong safeguards, valuable data could be lost, destroyed, or could fall into the wrong hands, revealing important trade secrets or information that violates personal privacy.

The popularity of handheld mobile devices for business computing adds to these woes. Portability makes cell phones, smartphones, and tablet computers easy to lose or steal. Smartphones share the same security weaknesses as other Internet devices, and are vulnerable to malicious software and penetration from outsiders. In 2009, security experts identified 30 security flaws in software and operating systems of smartphones made by Apple, Nokia, and BlackBerry maker Research in Motion.

Even the apps that have been custom-developed for mobile devices are capable of turning into rogue software. For example, in December 2009, Google pulled dozens of mobile banking apps from its Android Market because they could have been updated to capture customers' banking credentials. Smartphones used by corporate executives may contain sensitive data such as sales figures, customer names, phone numbers, and e-mail addresses. Intruders may be able to access internal corporate networks through these devices.

Internet Vulnerabilities

Large public networks, such as the Internet, are more vulnerable than internal networks because they are virtually open to anyone. The Internet is so huge that when abuses do occur, they can have an enormously widespread impact. When the Internet becomes part of the corporate network, the organization's information systems are even more vulnerable to actions from outsiders.

Computers that are constantly connected to the Internet by cable modems or digital subscriber line (DSL) lines are more open to penetration by outsiders because they use fixed Internet addresses where they can be easily identified. (With dial-up service, a temporary Internet address is assigned for each session.) A fixed Internet address creates a fixed target for hackers.

Telephone service based on Internet technology (see Chapter 7) is more vulnerable than the switched voice network if it does not run over a secure private network. Most Voice over IP (VoIP) traffic over the public Internet is not encrypted, so anyone with a network can listen in on conversations. Hackers can intercept conversations or shut down voice service by flooding servers supporting VoIP with bogus traffic.

Vulnerability has also increased from widespread use of e-mail, instant messaging (IM), and peer-to-peer file-sharing programs. E-mail may contain attachments that serve as springboards for malicious software or unauthorized access to internal corporate systems. Employees may use e-mail messages to transmit valuable trade secrets, financial data, or confidential customer information to unauthorized recipients. Popular IM applications for consumers do not use a secure layer for text messages, so they can be intercepted and read by outsiders during transmission over the public Internet. Instant messaging activity over the Internet can in some cases be used as a back door to an otherwise secure network. Sharing files over peer-to-peer (P2P) networks, such as those for illegal music sharing, may also transmit malicious software or expose information on either individual or corporate computers to outsiders.

Wireless Security Challenges

Is it safe to log onto a wireless network at an airport, library, or other public location? It depends on how vigilant you are. Even the wireless network in your home is vulnerable because radio frequency bands are easy to scan. Both Bluetooth and Wi-Fi networks are susceptible to hacking by eavesdroppers. Although the range of Wi-Fi networks is only several hundred feet, it can be extended up to one-fourth of a mile using external antennae. Local area networks (LANs) using the 802.11 standard can be easily penetrated by outsiders armed with laptops, wireless cards, external antennae, and hacking software. Hackers use these tools to detect unprotected networks, monitor network traffic, and, in some cases, gain access to the Internet or to corporate networks.

Wi-Fi transmission technology was designed to make it easy for stations to find and hear one another. The *service set identifiers (SSIDs)* identifying the access points in a Wi-Fi network are broadcast multiple times and can be picked up fairly easily by intruders' sniffer programs (see Figure 8-2). Wireless networks in many locations do not have basic protections against **war driving**,

FIGURE 8-2   WI-FI SECURITY CHALLENGES

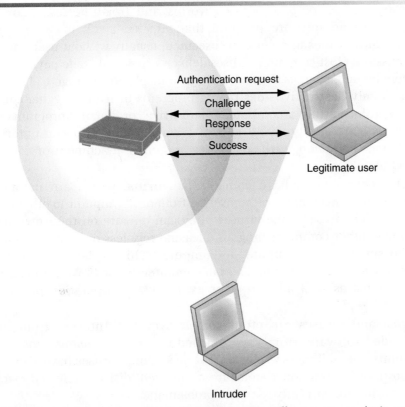

Many Wi-Fi networks can be penetrated easily by intruders using sniffer programs to obtain an address to access the resources of a network without authorization.

in which eavesdroppers drive by buildings or park outside and try to intercept wireless network traffic.

A hacker can employ an 802.11 analysis tool to identify the SSID. (Windows XP, Vista, and 7 have capabilities for detecting the SSID used in a network and automatically configuring the radio NIC within the user's device.) An intruder that has associated with an access point by using the correct SSID is capable of accessing other resources on the network, using the Windows operating system to determine which other users are connected to the network, access their computer hard drives, and open or copy their files.

Intruders also use the information they have gleaned to set up rogue access points on a different radio channel in physical locations close to users to force a user's radio NIC to associate with the rogue access point. Once this association occurs, hackers using the rogue access point can capture the names and passwords of unsuspecting users.

The initial security standard developed for Wi-Fi, called Wired Equivalent Privacy (WEP), is not very effective. WEP is built into all standard 802.11 products, but its use is optional. Many users neglect to use WEP security features, leaving them unprotected. The basic WEP specification calls for an access point and all of its users to share the same 40-bit encrypted password, which can be easily decrypted by hackers from a small amount of traffic. Stronger encryption and authentication systems are now available, such as Wi-Fi Protected Access 2 (WPA2), but users must be willing to install them.

MALICIOUS SOFTWARE: VIRUSES, WORMS, TROJAN HORSES, AND SPYWARE

Malicious software programs are referred to as **malware** and include a variety of threats, such as computer viruses, worms, and Trojan horses. A **computer virus** is a rogue software program that attaches itself to other software programs or data files in order to be executed, usually without user knowledge or permission. Most computer viruses deliver a "payload." The payload may be relatively benign, such as the instructions to display a message or image, or it may be highly destructive—destroying programs or data, clogging computer memory, reformatting a computer's hard drive, or causing programs to run improperly. Viruses typically spread from computer to computer when humans take an action, such as sending an e-mail attachment or copying an infected file.

Most recent attacks have come from **worms**, which are independent computer programs that copy themselves from one computer to other computers over a network. (Unlike viruses, they can operate on their own without attaching to other computer program files and rely less on human behavior in order to spread from computer to computer. This explains why computer worms spread much more rapidly than computer viruses.) Worms destroy data and programs as well as disrupt or even halt the operation of computer networks.

Worms and viruses are often spread over the Internet from files of downloaded software, from files attached to e-mail transmissions, or from compromised e-mail messages or instant messaging. Viruses have also invaded computerized information systems from "infected" disks or infected machines. E-mail worms are currently the most problematic.

Malware targeting mobile devices is not as extensive as that targeting computers, but is spreading nonetheless using e-mail, text messages, Bluetooth, and file downloads from the Web via Wi-Fi or cellular networks. There are now more than 200 viruses and worms targeting mobile phones, such as Cabir, Commwarrior, Frontal.A, and Ikee.B. Frontal.A installs a corrupted file that causes phone failure and prevents the user from rebooting, while Ikee.B turns jailbroken iPhones into botnet-controlled devices. Mobile device viruses pose serious threats to enterprise computing because so many wireless devices are now linked to corporate information systems.

Web 2.0 applications, such as blogs, wikis, and social networking sites such as Facebook and MySpace, have emerged as new conduits for malware or spyware. These applications allow users to post software code as part of the permissible content, and such code can be launched automatically as soon as a Web page is viewed. The chapter-opening case study describes other channels for malware targeting Facebook. In September 2010, hackers exploited a Twitter security flaw to send users to Japanese pornographic sites and automatically generated messages from other accounts (Coopes, 2010).

Table 8-1 describes the characteristics of some of the most harmful worms and viruses that have appeared to date.

Over the past decade, worms and viruses have caused billions of dollars of damage to corporate networks, e-mail systems, and data. According to Consumer Reports' State of the Net 2010 survey, U.S. consumers lost $3.5 billion

TABLE 8-1 EXAMPLES OF MALICIOUS CODE

NAME	TYPE	DESCRIPTION
Conficker (aka Downadup, Downup)	Worm	First detected in November 2008. Uses flaws in Windows software to take over machines and link them into a virtual computer that can be commanded remotely. Has more than 5 million computers worldwide under its control. Difficult to eradicate.
Storm	Worm/ Trojan horse	First identified in January 2007. Spreads via e-mail spam with a fake attachment. Infected up to 10 million computers, causing them to join its zombie network of computers engaged in criminal activity.
Sasser.ftp	Worm	First appeared in May 2004. Spread over the Internet by attacking random IP addresses. Causes computers to continually crash and reboot, and infected computers to search for more victims. Affected millions of computers worldwide, disrupting British Airways flight check-ins, operations of British coast guard stations, Hong Kong hospitals, Taiwan post office branches, and Australia's Westpac Bank. Sasser and its variants caused an estimated $14.8 billion to $18.6 billion in damages worldwide.
MyDoom.A	Worm	First appeared on January 26, 2004. Spreads as an e-mail attachment. Sends e-mail to addresses harvested from infected machines, forging the sender's address. At its peak this worm lowered global Internet performance by 10 percent and Web page loading times by as much as 50 percent. Was programmed to stop spreading after February 12, 2004.
Sobig.F	Worm	First detected on August 19, 2003. Spreads via e-mail attachments and sends massive amounts of mail with forged sender information. Deactivated itself on September 10, 2003, after infecting more than 1 million PCs and doing $5 to $10 billion in damage.
ILOVEYOU	Virus	First detected on May 3, 2000. Script virus written in Visual Basic script and transmitted as an attachment to e-mail with the subject line ILOVEYOU. Overwrites music, image, and other files with a copy of itself and did an estimated $10 billion to $15 billion in damage.
Melissa	Macro virus/ worm	First appeared in March 1999. Word macro script mailing infected Word file to first 50 entries in user's Microsoft Outlook address book. Infected 15 to 29 percent of all business PCs, causing $300 million to $600 million in damage.

because of malware and online scams, and the majority of these losses came from malware (Consumer Reports, 2010).

A **Trojan horse** is a software program that appears to be benign but then does something other than expected, such as the Zeus Trojan described in the chapter-opening case. The Trojan horse is not itself a virus because it does not replicate, but it is often a way for viruses or other malicious code to be introduced into a computer system. The term *Trojan horse* is based on the huge wooden horse used by the Greeks to trick the Trojans into opening the gates to their fortified city during the *Trojan War*. Once inside the city walls, Greek soldiers hidden in the horse revealed themselves and captured the city.

At the moment, **SQL injection attacks** are the largest malware threat. SQL injection attacks take advantage of vulnerabilities in poorly coded Web application software to introduce malicious program code into a company's systems and networks. These vulnerabilities occur when a Web application fails to properly validate or filter data entered by a user on a Web page, which might occur when ordering something online. An attacker uses this input validation error to send a rogue SQL query to the underlying database to access the database, plant malicious code, or access other systems on the network. Large Web applications have hundreds of places for inputting user data, each of which creates an opportunity for an SQL injection attack.

A large number of Web-facing applications are believed to have SQL injection vulnerabilities, and tools are available for hackers to check Web applications for these vulnerabilities. Such tools are able to locate a data entry field on a Web page form, enter data into it, and check the response to see if shows vulnerability to a SQL injection.

Some types of spyware also act as malicious software. These small programs install themselves surreptitiously on computers to monitor user Web surfing activity and serve up advertising. Thousands of forms of spyware have been documented.

Many users find such **spyware** annoying and some critics worry about its infringement on computer users' privacy. Some forms of spyware are especially nefarious. **Keyloggers** record every keystroke made on a computer to steal serial numbers for software, to launch Internet attacks, to gain access to e-mail accounts, to obtain passwords to protected computer systems, or to pick up personal information such as credit card numbers. Other spyware programs reset Web browser home pages, redirect search requests, or slow performance by taking up too much memory. The Zeus Trojan described in the chapter-opening case uses keylogging to steal financial information.

HACKERS AND COMPUTER CRIME

A **hacker** is an individual who intends to gain unauthorized access to a computer system. Within the hacking community, the term *cracker* is typically used to denote a hacker with criminal intent, although in the public press, the terms hacker and cracker are used interchangeably. Hackers and crackers gain unauthorized access by finding weaknesses in the security protections employed by Web sites and computer systems, often taking advantage of various features of the Internet that make it an open system that is easy to use.

Hacker activities have broadened beyond mere system intrusion to include theft of goods and information, as well as system damage and **cybervandalism**, the intentional disruption, defacement, or even destruction of a Web site or corporate information system. For example, cybervandals have turned many of the MySpace "group" sites, which are dedicated to interests such as home beer brewing or animal welfare, into cyber-graffiti walls, filled with offensive comments and photographs.

Spoofing and Sniffing

Hackers attempting to hide their true identities often spoof, or misrepresent, themselves by using fake e-mail addresses or masquerading as someone else. **Spoofing** also may involve redirecting a Web link to an address different from the intended one, with the site masquerading as the intended destination. For example, if hackers redirect customers to a fake Web site that looks almost exactly like the true site, they can then collect and process orders, effectively stealing business as well as sensitive customer information from the true site. We provide more detail on other forms of spoofing in our discussion of computer crime.

A **sniffer** is a type of eavesdropping program that monitors information traveling over a network. When used legitimately, sniffers help identify potential network trouble spots or criminal activity on networks, but when used for criminal purposes, they can be damaging and very difficult to detect. Sniffers enable hackers to steal proprietary information from anywhere on a network, including e-mail messages, company files, and confidential reports.

Denial-of-Service Attacks

In a **denial-of-service (DoS) attack**, hackers flood a network server or Web server with many thousands of false communications or requests for services to crash the network. The network receives so many queries that it cannot keep up with them and is thus unavailable to service legitimate requests. A **distributed denial-of-service (DDoS)** attack uses numerous computers to inundate and overwhelm the network from numerous launch points.

For example, during the 2009 Iranian election protests, foreign activists trying to help the opposition engaged in DDoS attacks against Iran's government. The official Web site of the Iranian government (ahmadinejad.ir) was rendered inaccessible on several occasions.

Although DoS attacks do not destroy information or access restricted areas of a company's information systems, they often cause a Web site to shut down, making it impossible for legitimate users to access the site. For busy e-commerce sites, these attacks are costly; while the site is shut down, customers cannot make purchases. Especially vulnerable are small and midsize businesses whose networks tend to be less protected than those of large corporations.

Perpetrators of DoS attacks often use thousands of "zombie" PCs infected with malicious software without their owners' knowledge and organized into a **botnet**. Hackers create these botnets by infecting other people's computers with bot malware that opens a back door through which an attacker can give instructions. The infected computer then becomes a slave, or zombie, serving a master computer belonging to someone else. Once a hacker infects enough computers, her or she can use the amassed resources of the botnet to launch DDos attacks, phishing campaigns, or unsolicited "spam" e-mail.

The number of computers that are part of botnets is variously estimated to be from 6 to 24 million, with thousands of botnets operating worldwide. The largest botnet attack in 2010 was the Mariposa botnet, which started in Spain and spread across the world. Mariposa had infected and controlled about 12.7 million computers in its efforts to steal credit card numbers and online banking passwords. More than half the Fortune 1000 companies, 40 major banks, and numerous government agencies were infected—and did not know it.

The chapter-ending case study describes multiple waves of DDoS attacks targeting a number of Web sites of government agencies and other organizations in South Korea and the United States in July 2009. The attacker used a botnet controlling over 65,000 computers, and was able to cripple some of these sites for several days. Most of the botnet originated from China, and North Korea. Botnet attacks thought to have originated in Russia were responsible for crippling the Web sites of the Estonian government in April 2007 and the Georgian government in July 2008.

Computer Crime

Most hacker activities are criminal offenses, and the vulnerabilities of systems we have just described make them targets for other types of **computer crime** as well. For example, in early July 2009, U.S. federal agents arrested Sergey Aleynikov, a computer programmer at investment banking firm Goldman Sachs, for stealing proprietary computer programs used in making lucrative rapid-fire trades in the financial markets. The software brought Goldman many millions of dollars of profits per year and, in the wrong hands, could have been used to manipulate financial markets in unfair ways. Computer crime is defined by the U.S. Department of Justice as "any violations of criminal law that involve a knowledge of computer technology for their perpetration, investigation, or prosecution." Table 8-2 provides examples of the computer as a target of crime and as an instrument of crime.

No one knows the magnitude of the computer crime problem—how many systems are invaded, how many people engage in the practice, or the total economic damage. According to the 2009 CSI Computer Crime and Security Survey of 500 companies, participants' average annual loss from computer crime and security attacks was close to $234,000 (Computer Security Institute, 2009). Many companies are reluctant to report computer crimes because the crimes may involve employees, or the company fears that publicizing its vulnerability will hurt its reputation. The most economically damaging kinds of computer crime are

TABLE 8-2 EXAMPLES OF COMPUTER CRIME

COMPUTERS AS TARGETS OF CRIME

Breaching the confidentiality of protected computerized data

Accessing a computer system without authority

Knowingly accessing a protected computer to commit fraud

Intentionally accessing a protected computer and causing damage, negligently or deliberately

Knowingly transmitting a program, program code, or command that intentionally causes damage to a protected computer

Threatening to cause damage to a protected computer

COMPUTERS AS INSTRUMENTS OF CRIME

Theft of trade secrets

Unauthorized copying of software or copyrighted intellectual property, such as articles, books, music, and video

Schemes to defraud

Using e-mail for threats or harassment

Intentionally attempting to intercept electronic communication

Illegally accessing stored electronic communications, including e-mail and voice mail

Transmitting or possessing child pornography using a computer

DoS attacks, introducing viruses, theft of services, and disruption of computer systems.

Identity Theft

With the growth of the Internet and electronic commerce, identity theft has become especially troubling. **Identity theft** is a crime in which an imposter obtains key pieces of personal information, such as social security identification numbers, driver's license numbers, or credit card numbers, to impersonate someone else. The information may be used to obtain credit, merchandise, or services in the name of the victim or to provide the thief with false credentials. According to Javelin Strategy and Research, losses from identity theft rose to $54 billion in 2009, and over 11 million U.S. adults were victims of identity fraud (Javelin Strategy & Research, 2010).

Identify theft has flourished on the Internet, with credit card files a major target of Web site hackers. Moreover, e-commerce sites are wonderful sources of customer personal information—name, address, and phone number. Armed with this information, criminals are able to assume new identities and establish new credit for their own purposes.

One increasingly popular tactic is a form of spoofing called **phishing**. Phishing involves setting up fake Web sites or sending e-mail or text messages that look like those of legitimate businesses to ask users for confidential personal data. The message instructs recipients to update or confirm records by providing social security numbers, bank and credit card information, and other confidential data either by responding to the e-mail message, by entering the information at a bogus Web site, or by calling a telephone number. EBay, PayPal, Amazon.com, Walmart, and a variety of banks, are among the top spoofed companies.

New phishing techniques called evil twins and pharming are harder to detect. **Evil twins** are wireless networks that pretend to offer trustworthy Wi-Fi connections to the Internet, such as those in airport lounges, hotels, or coffee shops. The bogus network looks identical to a legitimate public network. Fraudsters try to capture passwords or credit card numbers of unwitting users who log on to the network.

Pharming redirects users to a bogus Web page, even when the individual types the correct Web page address into his or her browser. This is possible if pharming perpetrators gain access to the Internet address information stored by Internet service providers to speed up Web browsing and the ISP companies have flawed software on their servers that allows the fraudsters to hack in and change those addresses.

In the largest instance of identity theft to date, Alberto Gonzalez of Miami and two Russian co-conspirators penetrated the corporate systems of TJX Corporation, Hannaford Brothers, 7-Eleven, and other major retailers, stealing over 160 million credit and debit card numbers between 2005 and 2008. The group initially planted "sniffer" programs in these companies' computer networks that captured card data as they were being transmitted between computer systems. They later switched to SQL injection attacks, which we introduced earlier in this chapter, to penetrate corporate databases. In March 2010, Gonzalez was sentenced to 20 years in prison. TJX alone spent over $200 million to deal with its data theft, including legal settlements.

The U.S. Congress addressed the threat of computer crime in 1986 with the Computer Fraud and Abuse Act. This act makes it illegal to access a computer system without authorization. Most states have similar laws, and nations in Europe have comparable legislation. Congress also passed the National Information Infrastructure Protection Act in 1996 to make virus distribution and hacker attacks that disable Web sites federal crimes. U.S. legislation, such as the Wiretap Act, Wire Fraud Act, Economic Espionage Act, Electronic Communications Privacy Act, E-mail Threats and Harassment Act, and Child Pornography Act, covers computer crimes involving intercepting electronic communication, using electronic communication to defraud, stealing trade secrets, illegally accessing stored electronic communications, using e-mail for threats or harassment, and transmitting or possessing child pornography.

Click Fraud

When you click on an ad displayed by a search engine, the advertiser typically pays a fee for each click, which is supposed to direct potential buyers to its products. **Click fraud** occurs when an individual or computer program fraudulently clicks on an online ad without any intention of learning more about the advertiser or making a purchase. Click fraud has become a serious problem at Google and other Web sites that feature pay-per-click online advertising.

Some companies hire third parties (typically from low-wage countries) to fraudulently click on a competitor's ads to weaken them by driving up their marketing costs. Click fraud can also be perpetrated with software programs doing the clicking, and botnets are often used for this purpose. Search engines such as Google attempt to monitor click fraud but have been reluctant to publicize their efforts to deal with the problem.

Global Threats: Cyberterrorism and Cyberwarfare

The cybercriminal activities we have described—launching malware, denial-of-service attacks, and phishing probes—are borderless. Computer security firm Sophos reported that 42 percent of the malware it identified in early 2010 originated in the United States, while 11 percent came from China, and 6 percent from Russia (Sophos, 2010). The global nature of the Internet makes it possible for cybercriminals to operate—and to do harm—anywhere in the world.

Concern is mounting that the vulnerabilities of the Internet or other networks make digital networks easy targets for digital attacks by terrorists, foreign intelligence services, or other groups seeking to create widespread disruption and harm. Such cyberattacks might target the software that runs electrical power grids, air traffic control systems, or networks of major banks and financial institutions. At least 20 countries, including China, are believed to be developing offensive and defensive cyberwarfare capabilities. The chapter-ending case study discusses this problem in greater detail.

INTERNAL THREATS: EMPLOYEES

We tend to think the security threats to a business originate outside the organization. In fact, company insiders pose serious security problems. Employees have access to privileged information, and in the presence of sloppy internal security procedures, they are often able to roam throughout an organization's systems without leaving a trace.

Studies have found that user lack of knowledge is the single greatest cause of network security breaches. Many employees forget their passwords to access computer systems or allow co-workers to use them, which compromises the system. Malicious intruders seeking system access sometimes trick employees into revealing their passwords by pretending to be legitimate members of the company in need of information. This practice is called **social engineering**.

Both end users and information systems specialists are also a major source of errors introduced into information systems. End users introduce errors by entering faulty data or by not following the proper instructions for processing data and using computer equipment. Information systems specialists may create software errors as they design and develop new software or maintain existing programs.

SOFTWARE VULNERABILITY

Software errors pose a constant threat to information systems, causing untold losses in productivity. Growing complexity and size of software programs, coupled with demands for timely delivery to markets, have contributed to an increase in software flaws or vulnerabilities For example, a database-related software error prevented millions of JP Morgan Chase retail and small-business customers from accessing their online bank accounts for two days in September 2010 (Dash, 2010).

A major problem with software is the presence of hidden **bugs** or program code defects. Studies have shown that it is virtually impossible to eliminate all bugs from large programs. The main source of bugs is the complexity of decision-making code. A relatively small program of several hundred lines will contain tens of decisions leading to hundreds or even thousands of different paths. Important programs within most corporations are usually much larger, containing tens of thousands or even millions of lines of code, each with many times the choices and paths of the smaller programs.

Zero defects cannot be achieved in larger programs. Complete testing simply is not possible. Fully testing programs that contain thousands of choices and millions of paths would require thousands of years. Even with rigorous testing, you would not know for sure that a piece of software was dependable until the product proved itself after much operational use.

Flaws in commercial software not only impede performance but also create security vulnerabilities that open networks to intruders. Each year security firms identify thousands of software vulnerabilities in Internet and PC software. For instance, in 2009, Symantec identified 384 browser vulnerabilities: 169 in Firefox, 94 in Safari, 45 in Internet Explorer, 41 in Chrome, and 25 in Opera. Some of these vulnerabilities were critical (Symantec, 2010).

To correct software flaws once they are identified, the software vendor creates small pieces of software called **patches** to repair the flaws without disturbing the proper operation of the software. An example is Microsoft's Windows Vista Service Pack 2, released in April 2009, which includes some security enhancements to counter malware and hackers. It is up to users of the software to track these vulnerabilities, test, and apply all patches. This process is called *patch management*.

Because a company's IT infrastructure is typically laden with multiple business applications, operating system installations, and other system services, maintaining patches on all devices and services used by a company is often time-consuming and costly. Malware is being created so rapidly that companies have very little time to respond between the time a vulnerability and a patch are announced and the time malicious software appears to exploit the vulnerability.

The need to respond so rapidly to the torrent of security vulnerabilities even creates defects in the software meant to combat them, including popular antivirus products. What happened in the spring of 2010 to McAfee, a leading vendor of commercial antivirus software is an example, as discussed in the Interactive Session on Management.

INTERACTIVE SESSION: MANAGEMENT

WHEN ANTIVIRUS SOFTWARE CRIPPLES YOUR COMPUTERS

McAfee is a prominent antivirus software and computer security company based in Santa Clara, California. Its popular VirusScan product (now named AntiVirus Plus) is used by companies and individual consumers across the world, driving its revenues of $1.93 billion in 2009.

A truly global company, McAfee has over 6,000 employees across North America, Europe, and Asia. VirusScan and other McAfee security products address endpoint security, network security, and risk and compliance. The company has worked to compile a long track record of good customer service and strong quality assurance.

At 6 a.m. PDT April 21, 2010, McAfee made a blunder that threatened to destroy that track record and prompted the possible departure of hundreds of valued customers. McAfee released what should have been a routine update for its flagship VirusScan product that was intended to deal with a powerful new virus known as 'W32/wecorl.a'. Instead, McAfee's update caused potentially hundreds of thousands of McAfee-equipped machines running Windows XP to crash and fail to reboot. How could McAfee, a company whose focus is saving and preserving computers, commit a gaffe that accomplished the opposite for a significant portion of its client base?

That was the question McAfee's angry clients were asking on the morning of April 21, when their computers were crippled or totally non-functional. The updates mistakenly targeted a critical Windows file, svchost.exe, which hosts other services used by various programs on PCs. Usually, more than one instance of the process is running at any given time, and eliminating them all would cripple any system. Though many viruses, including W32/wecorl.a, disguise themselves using the name svchost.exe to avoid detection, McAfee had never had problems with viruses using that technique before.

To make matters worse, without svchost.exe, Windows computers can't boot properly. VirusScan users applied the update, tried rebooting their systems, and were powerless to act as their systems went haywire, repeatedly rebooting, losing their network capabilities and, worst of all, their ability to detect USB drives, which is the only way of fixing affected computers. Companies using McAfee and

that relied heavily on Windows XP computers struggled to cope with the majority of their machines suddenly failing.

Angry network administrators turned to McAfee for answers, and the company was initially just as confused as its clients regarding how such a monumental slipup could occur. Soon, McAfee determined that the majority of affected machines were using Windows XP Service Pack 3 combined with McAfee VirusScan version 8.7. They also noted that the "Scan Processes on enable" option of VirusScan, off by default in most VirusScan installations, was turned on in the majority of affected computers.

McAfee conducted a more thorough investigation into its mistake and published a FAQ sheet that explained more completely why they had made such a big mistake and which customers were affected. The two most prominent points of failure were as follows: first, users should have received a warning that svchost.exe was going to be quarantined or deleted, instead of automatically disposing of the file. Next, McAfee's automated quality assurance testing failed to detect such a critical error because of what the company called "inadequate coverage of product and operating systems in the test systems used."

The only way tech support staffs working in organizations could fix the problem was to go from computer to computer manually. McAfee released a utility called "SuperDAT Remediation Tool," which had to be downloaded to an unaffected machine, placed on a flash drive, and run in Windows Safe Mode on affected machines. Because affected computers lacked network access, this had to be done one computer at a time until all affected machines were repaired. The total number of machines impacted is not known but it doubtless involved tens of thousands of corporate computers. Needless to say, network administrators and corporate tech support divisions were incensed.

Regarding the flaws in McAfee's quality assurance processes, the company explained in the FAQ that they had not included Windows XP Service Pack 3 with VirusScan version 8.7 in the test configuration of operating systems and McAfee product versions. This explanation flabbergasted many of McAfee's clients and other industry analysts, since XP SP3 is the most widely used desktop PC configuration.

Vista and Windows 7 generally ship with new computers and are rarely installed on functioning XP computers.

Another reason that the problem spread so quickly without detection was the increasing demand for faster antivirus updates. Most companies aggressively deploy their updates to ensure that machines spend as little time exposed to new viruses as possible. McAfee's update reached a large number of machines so quickly without detection because most companies trust their antivirus provider to get it right.

Unfortunately for McAfee, it only takes a single slipup or oversight to cause significant damage to an antivirus company's reputation. McAfee was criticized for its slow response to the crisis and for its initial attempts to downplay the issue's impact on its customers. The company released a

statement claiming that only a small fraction of its customers were affected, but this was soon shown to be false. Two days after the update was released, McAfee executive Barry McPherson finally apologized to customers on the company's blog. Soon after, CEO David DeWalt recorded a video for customers, still available via McAfee's Web site, in which he apologized for and explained the incident.

Sources: Peter Svensson, "McAfee Antivirus Program Goes Berserk, Freezes PCs," Associated Press, April 21, 2010; Gregg Keizer, "McAfee Apologizes for Crippling PCs with Bad Update," *Computerworld*, April 23, 2010 and "McAfee Update Mess Explained," *Computerworld*, April 22, 2010; Ed Bott, "McAfee Admits 'Inadequate' Quality Control Caused PC Meltdown," *ZDNet*, April 22, 2010; and Barry McPherson, "An Update on False Positive Remediation," http://siblog. mcafee.com/support/an-update-on-false-positive-remediation, April 22, 2010.

CASE STUDY QUESTIONS

1. What management, organization, and technology factors were responsible for McAfee's software problem?
2. What was the business impact of this software problem, both for McAfee and for its customers?
3. If you were a McAfee enterprise customer, would you consider McAfee's response to the problem be acceptable? Why or why not?
4. What should McAfee do in the future to avoid similar problems?

MIS IN ACTION

Search online for the apology by Barry McPherson ("Barry McPherson apology") and read the reaction of customers. Do you think McPherson's apology helped or inflamed the situation? What is a "false positive remediation"?

8.2   BUSINESS VALUE OF SECURITY AND CONTROL

Many firms are reluctant to spend heavily on security because it is not directly related to sales revenue. However, protecting information systems is so critical to the operation of the business that it deserves a second look.

Companies have very valuable information assets to protect. Systems often house confidential information about individuals' taxes, financial assets, medical records, and job performance reviews. They also can contain information on corporate operations, including trade secrets, new product development plans, and marketing strategies. Government systems may store information on weapons systems, intelligence operations, and military targets. These information assets have tremendous value, and the repercussions can be devastating if they are lost, destroyed, or placed in the wrong hands. One study estimated that when the security of a large firm is compromised, the company loses approximately 2.1 percent of its market value within two days of the security breach, which translates into an average loss of $1.65 billion in stock market value per incident (Cavusoglu, Mishra, and Raghunathan, 2004).

Inadequate security and control may result in serious legal liability. Businesses must protect not only their own information assets but also those of customers, employees, and business partners. Failure to do so may open the firm to costly litigation for data exposure or theft. An organization can be held liable for needless risk and harm created if the organization fails to take appropriate protective action to prevent loss of confidential information, data corruption, or breach of privacy. For example, BJ's Wholesale Club was sued by the U.S. Federal Trade Commission for allowing hackers to access its systems and steal credit and debit card data for fraudulent purchases. Banks that issued the cards with the stolen data sought $13 million from BJ's to compensate them for reimbursing card holders for the fraudulent purchases. A sound security and control framework that protects business information assets can thus produce a high return on investment. Strong security and control also increase employee productivity and lower operational costs.

LEGAL AND REGULATORY REQUIREMENTS FOR ELECTRONIC RECORDS MANAGEMENT

Recent U.S. government regulations are forcing companies to take security and control more seriously by mandating the protection of data from abuse, exposure, and unauthorized access. Firms face new legal obligations for the retention and storage of electronic records as well as for privacy protection.

If you work in the health care industry, your firm will need to comply with the Health Insurance Portability and Accountability Act (HIPAA) of 1996. **HIPAA** outlines medical security and privacy rules and procedures for simplifying the administration of health care billing and automating the transfer of health care data between health care providers, payers, and plans. It requires members of the health care industry to retain patient information for six years and ensure the confidentiality of those records. It specifies privacy, security, and electronic transaction standards for health care providers handling patient information, providing penalties for breaches of medical privacy, disclosure of patient records by e-mail, or unauthorized network access.

If you work in a firm providing financial services, your firm will need to comply with the Financial Services Modernization Act of 1999, better known as the **Gramm-Leach-Bliley Act** after its congressional sponsors. This act requires financial institutions to ensure the security and confidentiality of customer data. Data must be stored on a secure medium, and special security measures must be enforced to protect such data on storage media and during transmittal.

If you work in a publicly traded company, your company will need to comply with the Public Company Accounting Reform and Investor Protection Act of 2002, better known as the **Sarbanes-Oxley Act** after its sponsors Senator Paul Sarbanes of Maryland and Representative Michael Oxley of Ohio. This Act was designed to protect investors after the financial scandals at Enron, WorldCom, and other public companies. It imposes responsibility on companies and their management to safeguard the accuracy and integrity of financial information that is used internally and released externally. One of the Learning Tracks for this chapter discusses Sarbanes-Oxley in detail.

Sarbanes-Oxley is fundamentally about ensuring that internal controls are in place to govern the creation and documentation of information in financial statements. Because information systems are used to generate, store, and transport such data, the legislation requires firms to consider information systems security and other controls required to ensure the integrity, confidentiality, and accuracy of their data. Each system application that deals with critical financial reporting data requires controls to make sure the data are accurate. Controls to secure the corporate network, prevent unauthorized access to systems and data, and ensure data integrity and availability in the event of disaster or other disruption of service are essential as well.

ELECTRONIC EVIDENCE AND COMPUTER FORENSICS

Security, control, and electronic records management have become essential for responding to legal actions. Much of the evidence today for stock fraud, embezzlement, theft of company trade secrets, computer crime, and many civil cases is in digital form. In addition to information from printed or typewritten pages, legal cases today increasingly rely on evidence represented as digital data stored on portable floppy disks, CDs, and computer hard disk drives, as well as in e-mail, instant messages, and e-commerce transactions over the Internet. E-mail is currently the most common type of electronic evidence.

In a legal action, a firm is obligated to respond to a discovery request for access to information that may be used as evidence, and the company is required by law to produce those data. The cost of responding to a discovery request can be enormous if the company has trouble assembling the required data or the data have been corrupted or destroyed. Courts now impose severe financial and even criminal penalties for improper destruction of electronic documents.

An effective electronic document retention policy ensures that electronic documents, e-mail, and other records are well organized, accessible, and neither retained too long nor discarded too soon. It also reflects an awareness of how to preserve potential evidence for computer forensics. **Computer forensics** is the scientific collection, examination, authentication, preservation, and analysis of data held on or retrieved from computer storage media in such a way that the information can be used as evidence in a court of law. It deals with the following problems:

- Recovering data from computers while preserving evidential integrity
- Securely storing and handling recovered electronic data
- Finding significant information in a large volume of electronic data
- Presenting the information to a court of law

Electronic evidence may reside on computer storage media in the form of computer files and as *ambient data*, which are not visible to the average user. An example might be a file that has been deleted on a PC hard drive. Data that a computer user may have deleted on computer storage media can be recovered through various techniques. Computer forensics experts try to recover such hidden data for presentation as evidence.

An awareness of computer forensics should be incorporated into a firm's contingency planning process. The CIO, security specialists, information systems staff, and corporate legal counsel should all work together to have a plan in place that can be executed if a legal need arises. You can find out more about computer forensics in the Learning Tracks for this chapter.

8.3 ESTABLISHING A FRAMEWORK FOR SECURITY AND CONTROL

Even with the best security tools, your information systems won't be reliable and secure unless you know how and where to deploy them. You'll need to know where your company is at risk and what controls you must have in place to protect your information systems. You'll also need to develop a security policy and plans for keeping your business running if your information systems aren't operational.

INFORMATION SYSTEMS CONTROLS

Information systems controls are both manual and automated and consist of both general controls and application controls. **General controls** govern the design, security, and use of computer programs and the security of data files in general throughout the organization's information technology infrastructure. On the whole, general controls apply to all computerized applications and consist of a combination of hardware, software, and manual procedures that create an overall control environment.

General controls include software controls, physical hardware controls, computer operations controls, data security controls, controls over implementation of system processes, and administrative controls. Table 8-3 describes the functions of each of these controls.

Application controls are specific controls unique to each computerized application, such as payroll or order processing. They include both automated and manual procedures that ensure that only authorized data are completely and accurately processed by that application. Application controls can be classified as (1) input controls, (2) processing controls, and (3) output controls.

Input controls check data for accuracy and completeness when they enter the system. There are specific input controls for input authorization, data conversion, data editing, and error handling. *Processing controls* establish that data are complete and accurate during updating. *Output controls* ensure that

TABLE 8-3 GENERAL CONTROLS

TYPE OF GENERAL CONTROL	DESCRIPTION
Software controls	Monitor the use of system software and prevent unauthorized access of software programs, system software, and computer programs.
Hardware controls	Ensure that computer hardware is physically secure, and check for equipment malfunction. Organizations that are critically dependent on their computers also must make provisions for backup or continued operation to maintain constant service.
Computer operations controls	Oversee the work of the computer department to ensure that programmed procedures are consistently and correctly applied to the storage and processing of data. They include controls over the setup of computer processing jobs and backup and recovery procedures for processing that ends abnormally.
Data security controls	Ensure that valuable business data files on either disk or tape are not subject to unauthorized access, change, or destruction while they are in use or in storage.
Implementation controls	Audit the systems development process at various points to ensure that the process is properly controlled and managed.
Administrative controls	Formalize standards, rules, procedures, and control disciplines to ensure that the organization's general and application controls are properly executed and enforced.

the results of computer processing are accurate, complete, and properly distributed. You can find more detail about application and general controls in our Learning Tracks.

RISK ASSESSMENT

Before your company commits resources to security and information systems controls, it must know which assets require protection and the extent to which these assets are vulnerable. A risk assessment helps answer these questions and determine the most cost-effective set of controls for protecting assets.

A **risk assessment** determines the level of risk to the firm if a specific activity or process is not properly controlled. Not all risks can be anticipated and measured, but most businesses will be able to acquire some understanding of the risks they face. Business managers working with information systems specialists should try to determine the value of information assets, points of vulnerability, the likely frequency of a problem, and the potential for damage. For example, if an event is likely to occur no more than once a year, with a maximum of a $1,000 loss to the organization, it is not be wise to spend $20,000 on the design and maintenance of a control to protect against that event. However, if that same event could occur at least once a day, with a potential loss of more than $300,000 a year, $100,000 spent on a control might be entirely appropriate.

Table 8-4 illustrates sample results of a risk assessment for an online order processing system that processes 30,000 orders per day. The likelihood of each exposure occurring over a one-year period is expressed as a percentage. The next column shows the highest and lowest possible loss that could be expected each time the exposure occurred and an average loss calculated by adding the highest and lowest figures together and dividing by two. The expected annual loss for each exposure can be determined by multiplying the average loss by its probability of occurrence.

This risk assessment shows that the probability of a power failure occurring in a one-year period is 30 percent. Loss of order transactions while power is down could range from $5,000 to $200,000 (averaging $102,500) for each occurrence, depending on how long processing is halted. The probability of embezzlement occurring over a yearly period is about 5 percent, with potential losses ranging from $1,000 to $50,000 (and averaging $25,500) for each occurrence. User errors have a 98 percent chance of occurring over a yearly period, with losses ranging from $200 to $40,000 (and averaging $20,100) for each occurrence.

Once the risks have been assessed, system builders will concentrate on the control points with the greatest vulnerability and potential for loss. In this case, controls should focus on ways to minimize the risk of power failures and user errors because anticipated annual losses are highest for these areas.

TABLE 8-4 ONLINE ORDER PROCESSING RISK ASSESSMENT

EXPOSURE	PROBABILITY OF OCCURRENCE (%)	LOSS RANGE/ AVERAGE ($)	EXPECTED ANNUAL LOSS ($)
Power failure	30%	$5,000–$200,000 ($102,500)	$30,750
Embezzlement	5%	$1,000–$50,000 ($25,500)	$1,275
User error	98%	$200–$40,000 ($20,100)	$19,698

SECURITY POLICY

Once you've identified the main risks to your systems, your company will need to develop a security policy for protecting the company's assets. A **security policy** consists of statements ranking information risks, identifying acceptable security goals, and identifying the mechanisms for achieving these goals. What are the firm's most important information assets? Who generates and controls this information in the firm? What existing security policies are in place to protect the information? What level of risk is management willing to accept for each of these assets? Is it willing, for instance, to lose customer credit data once every 10 years? Or will it build a security system for credit card data that can withstand the once-in-a-hundred-year disaster? Management must estimate how much it will cost to achieve this level of acceptable risk.

The security policy drives policies determining acceptable use of the firm's information resources and which members of the company have access to its information assets. An **acceptable use policy (AUP)** defines acceptable uses of the firm's information resources and computing equipment, including desktop and laptop computers, wireless devices, telephones, and the Internet. The policy should clarify company policy regarding privacy, user responsibility, and personal use of company equipment and networks. A good AUP defines unacceptable and acceptable actions for every user and specifies consequences for noncompliance. For example, security policy at Unilever, the giant multinational consumer goods company, requires every employee equipped with a laptop or mobile handheld device to use a company-specified device and employ a password or other method of identification when logging onto the corporate network.

Security policy also includes provisions for identity management. **Identity management** consists of business processes and software tools for identifying the valid users of a system and controlling their access to system resources. It includes policies for identifying and authorizing different categories of system users, specifying what systems or portions of systems each user is allowed to access, and the processes and technologies for authenticating users and protecting their identities.

Figure 8-3 is one example of how an identity management system might capture the access rules for different levels of users in the human resources function. It specifies what portions of a human resource database each user is permitted to access, based on the information required to perform that person's job. The database contains sensitive personal information such as employees' salaries, benefits, and medical histories.

The access rules illustrated here are for two sets of users. One set of users consists of all employees who perform clerical functions, such as inputting employee data into the system. All individuals with this type of profile can update the system but can neither read nor update sensitive fields, such as salary, medical history, or earnings data. Another profile applies to a divisional manager, who cannot update the system but who can read all employee data fields for his or her division, including medical history and salary. We provide more detail on the technologies for user authentication later on in this chapter.

DISASTER RECOVERY PLANNING AND BUSINESS CONTINUITY PLANNING

If you run a business, you need to plan for events, such as power outages, floods, earthquakes, or terrorist attacks that will prevent your information systems and your business from operating. **Disaster recovery planning** devises

FIGURE 8-3   ACCESS RULES FOR A PERSONNEL SYSTEM

SECURITY PROFILE 1

User: Personnel Dept. Clerk

Location: Division 1

Employee Identification
Codes with This Profile: 00753, 27834, 37665, 44116

Data Field Restrictions	Type of Access
All employee data for Division 1 only	Read and Update
• Medical history data	None
• Salary	None
• Pensionable earnings	None

SECURITY PROFILE 2

User: Divisional Personnel Manager

Location: Division 1

Employee Identification
Codes with This Profile: 27321

Data Field Restrictions	Type of Access
All employee data for Division 1 only	Read Only

These two examples represent two security profiles or data security patterns that might be found in a personnel system. Depending on the access rules, a user would have certain restrictions on access to various systems, locations, or data in an organization.

plans for the restoration of computing and communications services after they have been disrupted. Disaster recovery plans focus primarily on the technical issues involved in keeping systems up and running, such as which files to back up and the maintenance of backup computer systems or disaster recovery services.

For example, MasterCard maintains a duplicate computer center in Kansas City, Missouri, to serve as an emergency backup to its primary computer center in St. Louis. Rather than build their own backup facilities, many firms contract with disaster recovery firms, such as Comdisco Disaster Recovery Services in Rosemont, Illinois, and SunGard Availability Services, headquartered in Wayne, Pennsylvania. These disaster recovery firms provide hot sites housing spare computers at locations around the country where subscribing firms can run their critical applications in an emergency. For example, Champion Technologies, which supplies chemicals used in oil and gas operations, is able to switch its enterprise systems from Houston to a SunGard hot site in Scottsdale, Arizona, in two hours.

Business continuity planning focuses on how the company can restore business operations after a disaster strikes. The business continuity plan identifies critical business processes and determines action plans for handling mission-critical functions if systems go down. For example, Deutsche Bank, which provides investment banking and asset management services in 74 different countries, has a well-developed business continuity plan that it continually updates and refines. It maintains full-time teams in Singapore, Hong Kong, Japan, India, and Australia to coordinate plans addressing loss of facilities, personnel, or critical systems so that the company can continue to operate when a catastrophic event occurs. Deutsche Bank's plan distinguishes between processes critical for business survival and those critical to crisis support and is coordinated with the company's disaster recovery planning for its computer centers.

Business managers and information technology specialists need to work together on both types of plans to determine which systems and business processes are most critical to the company. They must conduct a business impact analysis to identify the firm's most critical systems and the impact a systems outage would have on the business. Management must determine the maximum amount of time the business can survive with its systems down and which parts of the business must be restored first.

THE ROLE OF AUDITING

How does management know that information systems security and controls are effective? To answer this question, organizations must conduct comprehensive and systematic audits. An **MIS audit** examines the firm's overall security environment as well as controls governing individual information systems. The auditor should trace the flow of sample transactions through the system and perform tests, using, if appropriate, automated audit software. The MIS audit may also examine data quality.

Security audits review technologies, procedures, documentation, training, and personnel. A thorough audit will even simulate an attack or disaster to test the response of the technology, information systems staff, and business employees.

The audit lists and ranks all control weaknesses and estimates the probability of their occurrence. It then assesses the financial and organizational impact of each threat. Figure 8-4 is a sample auditor's listing of control weaknesses for a loan system. It includes a section for notifying management of such weaknesses and for management's response. Management is expected to devise a plan for countering significant weaknesses in controls.

8.4 TECHNOLOGIES AND TOOLS FOR PROTECTING INFORMATION RESOURCES

Businesses have an array of technologies for protecting their information resources. They include tools for managing user identities, preventing unauthorized access to systems and data, ensuring system availability, and ensuring software quality.

IDENTITY MANAGEMENT AND AUTHENTICATION

Large and midsize companies have complex IT infrastructures and many different systems, each with its own set of users. Identity management software automates the process of keeping track of all these users and their system privileges, assigning each user a unique digital identity for accessing each system. It also includes tools for authenticating users, protecting user identities, and controlling access to system resources.

To gain access to a system, a user must be authorized and authenticated. **Authentication** refers to the ability to know that a person is who he or she

FIGURE 8-4   SAMPLE AUDITOR'S LIST OF CONTROL WEAKNESSES

Function: Loans Location: Peoria, IL	Prepared by: J. Ericson Date: June 16, 2011		Received by: T. Benson Review date: June 28, 2011	
Nature of Weakness and Impact	Chance for Error/Abuse		Notification to Management	
	Yes/No	Justification	Report date	Management response
User accounts with missing passwords	Yes	Leaves system open to unauthorized outsiders or attackers	5/10/11	Eliminate accounts without passwords
Network configured to allow some sharing of system files	Yes	Exposes critical system files to hostile parties connected to the network	5/10/11	Ensure only required directories are shared and that they are protected with strong passwords
Software patches can update production programs without final approval from Standards and Controls group	No	All production programs require management approval; Standards and Controls group assigns such cases to a temporary production status		

This chart is a sample page from a list of control weaknesses that an auditor might find in a loan system in a local commercial bank. This form helps auditors record and evaluate control weaknesses and shows the results of discussing those weaknesses with management, as well as any corrective actions taken by management.

claims to be. Authentication is often established by using **passwords** known only to authorized users. An end user uses a password to log on to a computer system and may also use passwords for accessing specific systems and files. However, users often forget passwords, share them, or choose poor passwords that are easy to guess, which compromises security. Password systems that are too rigorous hinder employee productivity. When employees must change complex passwords frequently, they often take shortcuts, such as choosing passwords that are easy to guess or writing down their passwords at their workstations in plain view. Passwords can also be "sniffed" if transmitted over a network or stolen through social engineering.

New authentication technologies, such as tokens, smart cards, and biometric authentication, overcome some of these problems. A **token** is a physical device, similar to an identification card, that is designed to prove the identity of a single user. Tokens are small gadgets that typically fit on key rings and display passcodes that change frequently. A **smart card** is a device about the size of a credit card that contains a chip formatted with access permission and other data. (Smart cards are also used in electronic payment systems.) A reader device interprets the data on the smart card and allows or denies access.

Biometric authentication uses systems that read and interpret individual human traits, such as fingerprints, irises, and voices, in order to grant or deny access. Biometric authentication is based on the measurement of a physical or behavioral trait that makes each individual unique. It compares a person's unique characteristics, such as the fingerprints, face, or retinal image, against a

This PC has a biometric fingerprint reader for fast yet secure access to files and networks. New models of PCs are starting to use biometric identification to authenticate users.

stored profile of these characteristics to determine whether there are any differences between these characteristics and the stored profile. If the two profiles match, access is granted. Fingerprint and facial recognition technologies are just beginning to be used for security applications, with many PC laptops equipped with fingerprint identification devices and several models with built-in webcams and face recognition software.

FIREWALLS, INTRUSION DETECTION SYSTEMS, AND ANTIVIRUS SOFTWARE

Without protection against malware and intruders, connecting to the Internet would be very dangerous. Firewalls, intrusion detection systems, and antivirus software have become essential business tools.

Firewalls

Firewalls prevent unauthorized users from accessing private networks. A firewall is a combination of hardware and software that controls the flow of incoming and outgoing network traffic. It is generally placed between the organization's private internal networks and distrusted external networks, such as the Internet, although firewalls can also be used to protect one part of a company's network from the rest of the network (see Figure 8-5).

The firewall acts like a gatekeeper who examines each user's credentials before access is granted to a network. The firewall identifies names, IP addresses, applications, and other characteristics of incoming traffic. It checks this information against the access rules that have been programmed into the system by the network administrator. The firewall prevents unauthorized communication into and out of the network.

In large organizations, the firewall often resides on a specially designated computer separate from the rest of the network, so no incoming request directly accesses private network resources. There are a number of firewall screening technologies, including static packet filtering, stateful inspection, Network Address Translation, and application proxy filtering. They are frequently used in combination to provide firewall protection.

FIGURE 8-5   A CORPORATE FIREWALL

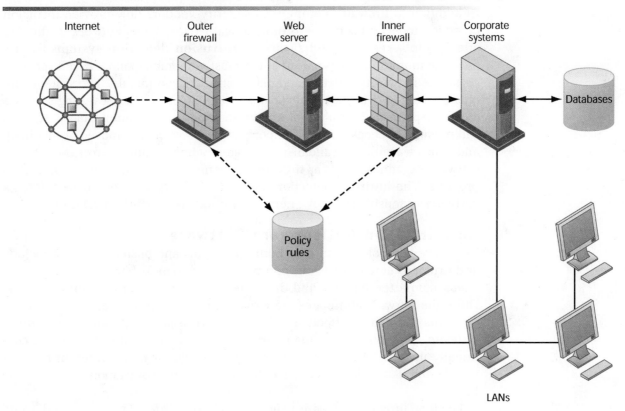

The firewall is placed between the firm's private network and the public Internet or another distrusted network to protect against unauthorized traffic.

Packet filtering examines selected fields in the headers of data packets flowing back and forth between the trusted network and the Internet, examining individual packets in isolation. This filtering technology can miss many types of attacks. *Stateful inspection* provides additional security by determining whether packets are part of an ongoing dialogue between a sender and a receiver. It sets up state tables to track information over multiple packets. Packets are accepted or rejected based on whether they are part of an approved conversation or whether they are attempting to establish a legitimate connection.

Network Address Translation (NAT) can provide another layer of protection when static packet filtering and stateful inspection are employed. NAT conceals the IP addresses of the organization's internal host computer(s) to prevent sniffer programs outside the firewall from ascertaining them and using that information to penetrate internal systems.

Application proxy filtering examines the application content of packets. A proxy server stops data packets originating outside the organization, inspects them, and passes a proxy to the other side of the firewall. If a user outside the company wants to communicate with a user inside the organization, the outside user first "talks" to the proxy application and the proxy application communicates with the firm's internal computer. Likewise, a computer user inside the organization goes through the proxy to talk with computers on the outside.

To create a good firewall, an administrator must maintain detailed internal rules identifying the people, applications, or addresses that are allowed or rejected. Firewalls can deter, but not completely prevent, network penetration by outsiders and should be viewed as one element in an overall security plan.

Intrusion Detection Systems

In addition to firewalls, commercial security vendors now provide intrusion detection tools and services to protect against suspicious network traffic and attempts to access files and databases. **Intrusion detection systems** feature full-time monitoring tools placed at the most vulnerable points or "hot spots" of corporate networks to detect and deter intruders continually. The system generates an alarm if it finds a suspicious or anomalous event. Scanning software looks for patterns indicative of known methods of computer attacks, such as bad passwords, checks to see if important files have been removed or modified, and sends warnings of vandalism or system administration errors. Monitoring software examines events as they are happening to discover security attacks in progress. The intrusion detection tool can also be customized to shut down a particularly sensitive part of a network if it receives unauthorized traffic.

Antivirus and Antispyware Software

Defensive technology plans for both individuals and businesses must include antivirus protection for every computer. **Antivirus software** is designed to check computer systems and drives for the presence of computer viruses. Often the software eliminates the virus from the infected area. However, most antivirus software is effective only against viruses already known when the software was written. To remain effective, the antivirus software must be continually updated. Antivirus products are available for many different types of mobile and handheld devices in addition to servers, workstations, and desktop PCs.

Leading antivirus software vendors, such as McAfee, Symantec, and Trend Micro, have enhanced their products to include protection against spyware. Antispyware software tools such as Ad-Aware, Spybot S&D, and Spyware Doctor are also very helpful.

Unified Threat Management Systems

To help businesses reduce costs and improve manageability, security vendors have combined into a single appliance various security tools, including firewalls, virtual private networks, intrusion detection systems, and Web content filtering and antispam software. These comprehensive security management products are called **unified threat management (UTM)** systems. Although initially aimed at small and medium-sized businesss, UTM products are available for all sizes of networks. Leading UTM vendors include Crossbeam, Fortinet, and Check Point, and networking vendors such as Cisco Systems and Juniper Networks provide some UTM capabilities in their equipment.

SECURING WIRELESS NETWORKS

Despite its flaws, WEP provides some margin of security if Wi-Fi users remember to activate it. A simple first step to thwart hackers is to assign a unique name to your network's SSID and instruct your router not to broadcast it. Corporations can further improve Wi-Fi security by using it in conjunction with virtual private network (VPN) technology when accessing internal corporate data.

In June 2004, the Wi-Fi Alliance industry trade group finalized the 802.11i specification (also referred to as Wi-Fi Protected Access 2 or WPA2) that replaces WEP with stronger security standards. Instead of the static encryption keys used in WEP, the new standard uses much longer keys that continually change, making them harder to crack. It also employs an encrypted authentication system with a central authentication server to ensure that only authorized users access the network.

ENCRYPTION AND PUBLIC KEY INFRASTRUCTURE

Many businesses use encryption to protect digital information that they store, physically transfer, or send over the Internet. **Encryption** is the process of transforming plain text or data into cipher text that cannot be read by anyone other than the sender and the intended receiver. Data are encrypted by using a secret numerical code, called an encryption key, that transforms plain data into cipher text. The message must be decrypted by the receiver.

Two methods for encrypting network traffic on the Web are SSL and S-HTTP. **Secure Sockets Layer (SSL)** and its successor Transport Layer Security (TLS) enable client and server computers to manage encryption and decryption activities as they communicate with each other during a secure Web session. **Secure Hypertext Transfer Protocol (S-HTTP)** is another protocol used for encrypting data flowing over the Internet, but it is limited to individual messages, whereas SSL and TLS are designed to establish a secure connection between two computers.

The capability to generate secure sessions is built into Internet client browser software and servers. The client and the server negotiate what key and what level of security to use. Once a secure session is established between the client and the server, all messages in that session are encrypted.

There are two alternative methods of encryption: symmetric key encryption and public key encryption. In symmetric key encryption, the sender and receiver establish a secure Internet session by creating a single encryption key and sending it to the receiver so both the sender and receiver share the same key. The strength of the encryption key is measured by its bit length. Today, a typical key will be 128 bits long (a string of 128 binary digits).

The problem with all symmetric encryption schemes is that the key itself must be shared somehow among the senders and receivers, which exposes the key to outsiders who might just be able to intercept and decrypt the key. A more secure form of encryption called **public key encryption** uses two keys: one shared (or public) and one totally private as shown in Figure 8-6. The keys are mathematically related so that data encrypted with one key can be decrypted using only the other key. To send and receive messages, communi-

FIGURE 8-6   PUBLIC KEY ENCRYPTION

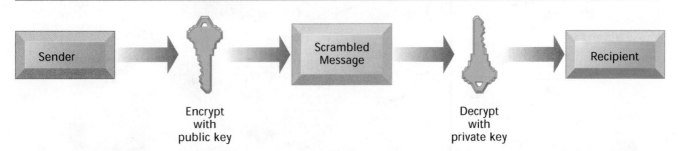

A public key encryption system can be viewed as a series of public and private keys that lock data when they are transmitted and unlock the data when they are received. The sender locates the recipient's public key in a directory and uses it to encrypt a message. The message is sent in encrypted form over the Internet or a private network. When the encrypted message arrives, the recipient uses his or her private key to decrypt the data and read the message.

cators first create separate pairs of private and public keys. The public key is kept in a directory and the private key must be kept secret. The sender encrypts a message with the recipient's public key. On receiving the message, the recipient uses his or her private key to decrypt it.

Digital certificates are data files used to establish the identity of users and electronic assets for protection of online transactions (see Figure 8-7). A digital certificate system uses a trusted third party, known as a certificate authority (CA, or certification authority), to validate a user's identity. There are many CAs in the United States and around the world, including VeriSign, IdenTrust, and Australia's KeyPost.

The CA verifies a digital certificate user's identity offline. This information is put into a CA server, which generates an encrypted digital certificate containing owner identification information and a copy of the owner's public key. The certificate authenticates that the public key belongs to the designated owner. The CA makes its own public key available publicly either in print or perhaps on the Internet. The recipient of an encrypted message uses the CA's public key to decode the digital certificate attached to the message, verifies it was issued by the CA, and then obtains the sender's public key and identification information contained in the certificate. Using this information, the recipient can send an encrypted reply. The digital certificate system would enable, for example, a credit card user and a merchant to validate that their digital certificates were issued by an authorized and trusted third party before they exchange data. **Public key infrastructure (PKI)**, the use of public key cryptography working with a CA, is now widely used in e-commerce.

ENSURING SYSTEM AVAILABILITY

As companies increasingly rely on digital networks for revenue and operations, they need to take additional steps to ensure that their systems and applications

FIGURE 8-7   DIGITAL CERTIFICATES

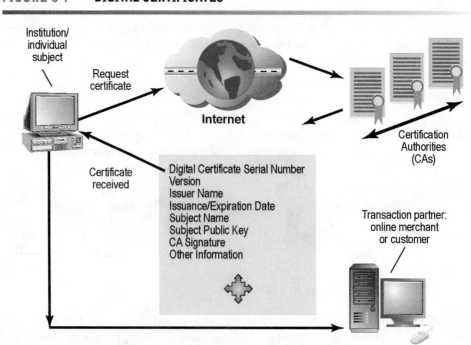

Digital certificates help establish the identity of people or electronic assets. They protect online transactions by providing secure, encrypted, online communication.

are always available. Firms such as those in the airline and financial services industries with critical applications requiring online transaction processing have traditionally used fault-tolerant computer systems for many years to ensure 100-percent availability. In **online transaction processing**, transactions entered online are immediately processed by the computer. Multitudinous changes to databases, reporting, and requests for information occur each instant.

Fault-tolerant computer systems contain redundant hardware, software, and power supply components that create an environment that provides continuous, uninterrupted service. Fault-tolerant computers use special software routines or self-checking logic built into their circuitry to detect hardware failures and automatically switch to a backup device. Parts from these computers can be removed and repaired without disruption to the computer system.

Fault tolerance should be distinguished from **high-availability computing**. Both fault tolerance and high-availability computing try to minimize downtime. **Downtime** refers to periods of time in which a system is not operational. However, high-availability computing helps firms recover quickly from a system crash, whereas fault tolerance promises continuous availability and the elimination of recovery time altogether.

High-availability computing environments are a minimum requirement for firms with heavy e-commerce processing or for firms that depend on digital networks for their internal operations. High-availability computing requires backup servers, distribution of processing across multiple servers, high-capacity storage, and good disaster recovery and business continuity plans. The firm's computing platform must be extremely robust with scalable processing power, storage, and bandwidth.

Researchers are exploring ways to make computing systems recover even more rapidly when mishaps occur, an approach called **recovery-oriented computing**. This work includes designing systems that recover quickly, and implementing capabilities and tools to help operators pinpoint the sources of faults in multi-component systems and easily correct their mistakes.

Controlling Network Traffic: Deep Packet Inspection

Have you ever tried to use your campus network and found it was very slow? It may be because your fellow students are using the network to download music or watch YouTube. Bandwith-consuming applications such as file-sharing programs, Internet phone service, and online video are able to clog and slow down corporate networks, degrading performance. For example, Ball Sate University in Muncie, Indiana, found its network had slowed because a small minority of students were using peer-to-peer file-sharing programs to download movies and music.

A technology called **deep packet inspection (DPI)** helps solve this problem. DPI examines data files and sorts out low-priority online material while assigning higher priority to business-critical files. Based on the priorities established by a network's operators, it decides whether a specific data packet can continue to its destination or should be blocked or delayed while more important traffic proceeds. Using a DPI system from Allot Communications, Ball State was able to cap the amount of file-sharing traffic and assign it a much lower priority. Ball State's preferred network traffic speeded up.

Security Outsourcing

Many companies, especially small businesses, lack the resources or expertise to provide a secure high-availability computing environment on their own. They can outsource many security functions to **managed security service**

providers (MSSPs) that monitor network activity and perform vulnerability testing and intrusion detection. SecureWorks, BT Managed Security Solutions Group, and Symantec are leading providers of MSSP services.

SECURITY ISSUES FOR CLOUD COMPUTING AND THE MOBILE DIGITAL PLATFORM

Although cloud computing and the emerging mobile digital platform have the potential to deliver powerful benefits, they pose new challenges to system security and reliability. We now describe some of these challenges and how they should be addressed.

Security in the Cloud

When processing takes place in the cloud, accountability and responsibility for protection of sensitive data still reside with the company owning that data. Understanding how the cloud computing provider organizes its services and manages the data is critical. The Interactive Session on Technology details some of the cloud security issues that should be addressed.

Cloud users need to confirm that regardless of where their data are stored or transferred, they are protected at a level that meets their corporate requirements. They should stipulate that the cloud provider store and process data in specific jurisdictions according to the privacy rules of those jurisdictions. Cloud clients should find how the cloud provider segregates their corporate data from those of other companies and ask for proof that encryption mechanisms are sound. It's also important to know how the cloud provider will respond if a disaster strikes, whether the provider will be able to completely restore your data, and how long this should take. Cloud users should also ask whether cloud providers will submit to external audits and security certifications. These kinds of controls can be written into the service level agreement (SLA) before to signing with a cloud provider.

Securing Mobile Platforms

If mobile devices are performing many of the functions of computers, they need to be secured like desktops and laptops against malware, theft, accidental loss, unauthorized access, and hacking attempts. Mobile devices accessing corporate systems and data require special protection.

Companies should make sure that their corporate security policy includes mobile devices, with additional details on how mobile devices should be supported, protected, and used. They will need tools to authorize all devices in use; to maintain accurate inventory records on all mobile devices, users, and applications; to control updates to applications; and to lock down lost devices so they can't be compromised. Firms should develop guidelines stipulating approved mobile platforms and software applications as well as the required software and procedures for remote access of corporate systems. Companies will need to ensure that all smartphones are up to date with the latest security patches and antivirus/anti-spam software, and they should encrypt communication whenever possible.

ENSURING SOFTWARE QUALITY

In addition to implementing effective security and controls, organizations can improve system quality and reliability by employing software metrics and rigorous software testing. Software metrics are objective assessments of the system in the form of quantified measurements. Ongoing use of metrics allows

INTERACTIVE SESSION: TECHNOLOGY

HOW SECURE IS THE CLOUD?

New York-based investment banking and financial services firm Cowen and Co. has moved its global sales systems to the cloud using Salesforce.com. So far, Cowen's CIO Daniel Flax is pleased. Using cloud services has helped the company lower upfront technology costs, decrease downtime and support additional services. But he's trying to come to grips with cloud security issues.Cloud computing is indeed cloudy, and this lack of transparency is troubling to many.

One of the biggest risks of cloud computing is that it is highly distributed. Cloud applications and application mash-ups reside in virtual libraries in large remote data centers and server farms that supply business services and data management for multiple corporate clients. To save money and keep costs low, cloud computing providers often distribute work to data centers around the globe where work can be accomplished most efficiently. When you use the cloud, you may not know precisely where your data are being hosted, and you might not even know the country where they are being stored.

The dispersed nature of cloud computing makes it difficult to track unauthorized activity. Virtually all cloud providers use encryption, such as Secure Sockets Layer, to secure the data they handle while the data are being transmitted. But if the data are stored on devices that also store other companies' data, it's important to ensure these stored data are encrypted as well.

Indian Harvest Specialtifoods, a Bemidji, Minnesota-based company that distributes rice, grains, and legumes to restaurants worldwide, relies on cloud software provider NetSuite to ensure that its data sent to the cloud are fully protected. Mike Mullin, Indian Harvest's IT director, feels that using SSL (Secure Sockets Layer) to encrypt the data gives him some level of confidence that the data are secure. He also points out that his company and other users of cloud services need to pay attention to their own security practices, especially access controls. "Your side of the infrastructure is just as vulnerable, if not more vulnerable, than the provider's side," he observes.

One way to deal with these problems is to use a cloud vendor that is a public company, which is required by law to disclose how it manages information. Salesforce.com meets this requirement, with strict processes and guidelines for managing its data centers. "We know our data are in the U.S. and we have a report on the very data centers that we're talking about, " says Flax.

Another alternative is to use a cloud provider that give subscribers the option to choose where their cloud computing work takes place. For example, Terremark Worldwide Inc. is giving its subscriber Agora Games the option to choose where its applications run. Terremark has a Miami facility but is adding other locations. In the past, Agora had no say over where Terremark hosted its applications and data.

Even if your data are totally secure in the cloud, you may not be able to prove it. Some cloud providers don't meet current compliance requirements regarding security, and some of those providers, such as Amazon, have asserted that they don't intend to meet those rules and won't allow compliance auditors on-site.

There are laws restricting where companies can send and store some types of information—personally identifiable information in the European Union (EU), government work in the United Sates or applications that employ certain encryption algorithms. Companies required to meet these regulations involving protected data either in the United States or the EU won't be able to use public cloud providers.

Some of these regulations call for proof that systems are securely managed, which may require confirmation from an independent audit. Large providers are unlikely to allow another company's auditors to inspect their data centers. Microsoft found a way to deal with this problem that may be helpful. The company reduced 26 different types of audits to a list of 200 necessary controls for meeting compliance standards that were applied to its data center environments and services. Microsoft does not give every customer or auditor access to its data centers, but its compliance framework allows auditors to order from a menu of tests and receive the results.

Companies expect their systems to be running 24/7, but cloud providers haven't always been able to provide this level of service. Millions of customers of Salesforce.com suffered a 38-minute outage in early January 2009 and others several years earlier. The January 2009 outage locked more than 900,000 subscribers out of crucial applications and data needed to transact business with customers. More than 300,000 customers using Intuit's online network of small business aplications were unable to access these services for two days in June 2010 following a power outage.

Agreements for services such as Amazon EC2 and Microsoft Azure state that these companies are not going to be held liable for data losses or fines or other legal penalties when companies use their services. Both vendors offer guidance on how to use their cloud platforms securely, and they may still be able to protect data better than some companies' home-grown facilities.

Salesforce.com had been building up and redesigning its infrastructure to ensure better service. The company invested $50 million in Mirrorforce technology, a mirroring system that creates a duplicate database in a separate location and synchronizes the data instantaneously. If one database is disabled, the other takes over. Salesforce.com added two data centers on the East and West coasts in addition to its Silicon Valley facility. The company distributed processing for its larger customers among these centers to balance its database load.

Sources: Seth Fineberg, "A Shadow on the Cloud?" *Information Management*, August, 2010; Ellen Messmer, "Secrecy of Cloud Computing Providers Raises IT Security Risks," *IT World*, July 13, 2010; John Edwards, "Cutting Through the Fog of Cloud Security," *Computerworld*, February 23, 2009; Wayne Rash, "Is Cloud Computing Secure? Prove It," *eWeek*, September 21, 2009; Robert Lemos, "Five Lessons from Microsoft on Cloud Security," *Computerworld*, August 25, 2009; and Mike Fratto, "Cloud Control," *Information Week*, January 26, 2009.

CASE STUDY QUESTIONS

1. What security and control problems are described in this case?

2. What people, organization, and technology factors contribute to these problems?

3. How secure is cloud computing? Explain your answer.

4. If you were in charge of your company's information systems department, what issues would you want to clarify with prospective vendors?

5. Would you entrust your corporate systems to a cloud computing provider? Why or why not?

MIS IN ACTION

Go to www.trust.salesforce.com, then answer the following questions:

1. Click on Security and describe Salesforce.com's security provisions. How helpful are these?

2. Click on Best Practices and describe what subscribing companies can do to tighten security. How helpful are these guidelines?

3. If you ran a business, would you feel confident about using Salesforce.com's on-demand service? Why or why not?

the information systems department and end users to jointly measure the performance of the system and identify problems as they occur. Examples of software metrics include the number of transactions that can be processed in a specified unit of time, online response time, the number of payroll checks printed per hour, and the number of known bugs per hundred lines of program code. For metrics to be successful, they must be carefully designed, formal, objective, and used consistently.

Early, regular, and thorough testing will contribute significantly to system quality. Many view testing as a way to prove the correctness of work they have done. In fact, we know that all sizable software is riddled with errors, and we must test to uncover these errors.

Good testing begins before a software program is even written by using a *walkthrough*—a review of a specification or design document by a small group of people carefully selected based on the skills needed for the particular objectives being tested. Once developers start writing software programs, coding walkthroughs also can be used to review program code. However, code must be tested by computer runs. When errors are discovered, the source is found and eliminated through a process called *debugging*. You can find out more about the various stages of testing required to put an information system into operation in Chapter 13. Our Learning Tracks also contain descriptions of methodologies for developing software programs that also contribute to software quality.

8.5   HANDS-ON MIS PROJECTS

The projects in this section give you hands-on experience analyzing security vulnerabilities, using spreadsheet software for risk analysis, and using Web tools to research security outsourcing services.

Management Decision Problems

1. K2 Network operates online game sites used by about 16 million people in over 100 countries. Players are allowed to enter a game for free, but must buy digital "assets" from K2, such as swords to fight dragons, if they want to be deeply involved. The games can accommodate millions of players at once and are played simultaneously by people all over the world. Prepare a security analysis for this Internet-based business. What kinds of threats should it anticipate? What would be their impact on the business? What steps can it take to prevent damage to its Web sites and continuing operations?

2. A survey of your firm's information technology infrastructure has produced the following security analysis statistics:

SECURITY VULNERABILITIES BY TYPE OF COMPUTING PLATFORM

PLATFORM	NUMBER OF COMPUTERS	HIGH RISK	MEDIUM RISK	LOW RISK	TOTAL VULNERABILITIES
Windows Server (corporate applications)	1	11	37	19	
Windows 7 Enterprise (high-level administrators)	3	56	242	87	
Linux (e-mail and printing services)	1	3	154	98	
Sun Solaris (Unix) (E-commerce and Web servers)	2	12	299	78	
Windows 7 Enterprise user desktops and laptops with office productivity tools that can also be linked to the corporate network running corporate applications and intranet	195	14	16	1,237	

High risk vulnerabilities include non-authorized users accessing applications, guessable passwords, user names matching the password, active user accounts with missing passwords, and the existence of unauthorized programs in application systems.

Medium risk vulnerabilities include the ability of users to shut down the system without being logged on, passwords and screen saver settings that were not established for PCs, and outdated versions of software still being stored on hard drives.

Low risk vulnerabilities include the inability of users to change their passwords, user passwords that have not been changed periodically, and passwords that were smaller than the minimum size specified by the company.

- Calculate the total number of vulnerabilities for each platform. What is the potential impact of the security problems for each computing platform on the organization?
- If you only have one information systems specialist in charge of security, which platforms should you address first in trying to eliminate these vulnerabilities? Second? Third? Last? Why?
- Identify the types of control problems illustrated by these vulnerabilities and explain the measures that should be taken to solve them.
- What does your firm risk by ignoring the security vulnerabilities identified?

Improving Decision Making: Using Spreadsheet Software to Perform a Security Risk Assessment

Software skills: Spreadsheet formulas and charts
Business skills: Risk assessment

This project uses spreadsheet software to calculate anticipated annual losses from various security threats identified for a small company.

Mercer Paints is a small but highly regarded paint manufacturing company located in Alabama. The company has a network in place linking many of its business operations. Although the firm believes that its security is adequate, the recent addition of a Web site has become an open invitation to hackers. Management requested a risk assessment. The risk assessment identified a number of potential exposures. These exposures, their associated probabilities, and average losses are summarized in the following table.

MERCER PAINTS RISK ASSESSMENT

EXPOSURE	PROBABILITY OF OCCURRENCE	AVERAGE LOSS
Malware attack	60%	$75,000
Data loss	12%	$70,000
Embezzlement	3%	$30,000
User errors	95%	$25,000
Threats from hackers	95%	$90,000
Improper use by employees	5%	$5,000
Power failure	15%	$300,000

- In addition to the potential exposures listed, you should identify at least three other potential threats to Mercer Paints, assign probabilities, and estimate a loss range.
- Use spreadsheet software and the risk assessment data to calculate the expected annual loss for each exposure.
- Present your findings in the form of a chart. Which control points have the greatest vulnerability? What recommendations would you make to Mercer Paints? Prepare a written report that summarizes your findings and recommendations.

Improving Decision Making: Evaluating Security Outsourcing Services

Software skills: Web browser and presentation software
Business skills: Evaluating business outsourcing services

Businesses today have a choice of whether to outsource the security function or maintain their own internal staff for this purpose. This project will help develop your Internet skills in using the Web to research and evaluate security outsourcing services.

As an information systems expert in your firm, you have been asked to help management decide whether to outsource security or keep the security function within the firm. Search the Web to find information to help you decide whether to outsource security and to locate security outsourcing services.

- Present a brief summary of the arguments for and against outsourcing computer security for your company.
- Select two firms that offer computer security outsourcing services, and compare them and their services.
- Prepare an electronic presentation for management summarizing your findings. Your presentation should make the case on whether or not your company should outsource computer security. If you believe your company should outsource, the presentation should identify which security outsourcing service should be selected and justify your selection.

LEARNING TRACK MODULES

The following Learning Tracks provide content relevant to topics covered in this chapter:

1. The Booming Job Market in IT Security
2. The Sarbanes-Oxley Act
3. Computer Forensics
4. General and Application Controls for Information Systems
5. Management Challenges of Security and Control
6. Software Vulnerability and Reliability

Review Summary

1. *Why are information systems vulnerable to destruction, error, and abuse?*

   Digital data are vulnerable to destruction, misuse, error, fraud, and hardware or software failures. The Internet is designed to be an open system and makes internal corporate systems more vulnerable to actions from outsiders. Hackers can unleash denial-of-service (DoS) attacks or penetrate corporate networks, causing serious system disruptions. Wi-Fi networks can easily be penetrated by intruders using sniffer programs to obtain an address to access the resources of the network. Computer viruses and worms can disable systems and Web sites. The dispersed nature of cloud computing makes it difficult to track unauthorized activity or to apply controls from afar. Software presents problems because software bugs may be impossible to eliminate and because software vulnerabilities can be exploited by hackers and malicious software. End users often introduce errors.

2. *What is the business value of security and control?*

   Lack of sound security and control can cause firms relying on computer systems for their core business functions to lose sales and productivity. Information assets, such as confidential employee records, trade secrets, or business plans, lose much of their value if they are revealed to outsiders or if they expose the firm to legal liability. New laws, such as HIPAA, the Sarbanes-Oxley Act, and the Gramm-Leach-Bliley Act, require companies to practice stringent electronic records management and adhere to strict standards for security, privacy, and control. Legal actions requiring electronic evidence and computer forensics also require firms to pay more attention to security and electronic records management.

3. *What are the components of an organizational framework for security and control?*

   Firms need to establish a good set of both general and application controls for their information systems. A risk assessment evaluates information assets, identifies control points and control weaknesses, and determines the most cost-effective set of controls. Firms must also develop a coherent corporate security policy and plans for continuing business operations in the event of disaster or disruption. The security policy includes policies for acceptable use and identity management. Comprehensive and systematic MIS auditing helps organizations determine the effectiveness of security and controls for their information systems.

4. *What are the most important tools and technologies for safeguarding information resources?*

   Firewalls prevent unauthorized users from accessing a private network when it is linked to the Internet. Intrusion detection systems monitor private networks from suspicious network traffic and attempts to access corporate systems. Passwords, tokens, smart cards, and biometric authentication are used to authenticate system users. Antivirus software checks computer systems for infections by viruses and worms and often eliminates the malicious software, while antispyware software combats intrusive and harmful spyware programs. Encryption, the coding and scrambling of messages, is a widely used technology for securing electronic transmissions over unprotected networks. Digital certificates combined with public key encryption provide further protection of electronic transactions by authenticating a user's identity. Companies can use fault-tolerant computer systems or create high-availability computing environments to make sure that their information systems are always available. Use of software metrics and rigorous software testing help improve software quality and reliability.

Key Terms

Acceptable use policy (AUP), 310

Antivirus software, 316

Application controls, 308

Authentication, 312

Biometric authentication, 313

Botnet, 299

Bugs, 303

Business continuity planning, 311

Click fraud, 302

Computer crime, 300

Computer forensics, 307

Computer virus, 296

Controls, 293

Cybervandalism, 298

Deep packet inspection (DPI), 319

Denial-of-service (DoS) attack, 299

Review Questions

1. Why are information systems vulnerable to destruction, error, and abuse?
   - List and describe the most common threats against contemporary information systems.
   - Define malware and distinguish among a virus, a worm, and a Trojan horse.
   - Define a hacker and explain how hackers create security problems and damage systems.
   - Define computer crime. Provide two examples of crime in which computers are targets and two examples in which computers are used as instruments of crime.
   - Define identity theft and phishing and explain why identity theft is such a big problem today.
   - Describe the security and system reliability problems created by employees.
   - Explain how software defects affect system reliability and security.

2. What is the business value of security and control?
   - Explain how security and control provide value for businesses.
   - Describe the relationship between security and control and recent U.S. government regulatory requirements and computer forensics.

3. What are the components of an organizational framework for security and control?
   - Define general controls and describe each type of general control.
   - Define application controls and describe each type of application control.
   - Describe the function of risk assessment and explain how it is conducted for information systems.
   - Define and describe the following: security policy, acceptable use policy, and identity management.
   - Explain how MIS auditing promotes security and control.

4. What are the most important tools and technologies for safeguarding information resources?
   - Name and describe three authentication methods.
   - Describe the roles of firewalls, intrusion detection systems, and antivirus software in promoting security.

- Explain how encryption protects information.
- Describe the role of encryption and digital certificates in a public key infrastructure.
- Distinguish between fault-tolerant and high-availability computing, and between disaster

recovery planning and business continuity planning.
- Identify and describe the security problems posed by cloud computing.
- Describe measures for improving software quality and reliability.

Discussion Questions

1. Security isn't simply a technology issue, it's a business issue. Discuss.

2. If you were developing a business continuity plan for your company, where would you start? What aspects of the business would the plan address?

3. Suppose your business had an e-commerce Web site where it sold goods and accepted credit card payments. Discuss the major security threats to this Web site and their potential impact. What can be done to minimize these threats?

Video Cases

Video Cases and Instructional Videos illustrating some of the concepts in this chapter are available. Contact your instructor to access these videos.

Collaboration and Teamwork: Evaluating Security Software Tools

With a group of three or four students, use the Web to research and evaluate security products from two competing vendors, such as antivirus software, firewalls, or antispyware software. For each product, describe its capabilities, for what types of businesses it is best suited, and its cost to purchase and install.

Which is the best product? Why? If possible, use Google Sites to post links to Web pages, team communication announcements, and work assignments; to brainstorm; and to work collaboratively on project documents. Try to use Google Docs to develop a presentation of your findings for the class.

Are We Ready for Cyberwarfare?
CASE STUDY

For most of us, the Internet is a tool we use for e-mail, news, entertainment, socializing, and shopping. But for computer security experts affiliated with government agencies and private contractors, as well as their hacker counterparts from across the globe, the Internet has become a battlefield—a war zone where cyberwarfare is becoming more frequent and hacking techniques are becoming more advanced. Cyberwarfare poses a unique and daunting set of challenges for security experts, not only in detecting and preventing intrusions but also in tracking down perpetrators and bringing them to justice.

Cyberwarfare can take many forms. Often, hackers use botnets, massive networks of computers that they control thanks to spyware and other malware, to launch large-scale DDoS attacks on their target's servers. Other methods allow intruders to access secure computers remotely and copy or delete e-mail and files from the machine, or even to remotely monitor users of a machine using more sophisticated software. For cybercriminals, the benefit of cyberwarfare is that they can compete with traditional superpowers for a fraction of the cost of, for example, building up a nuclear arsenal. Because more and more modern technological infrastructure will rely on the Internet to function, cyberwarriors will have no shortage of targets at which to take aim.

Cyberwarfare also involves defending against these types of attacks. That's a major focus of U.S. intelligence agencies. While the U.S. is currently at the forefront of cyberwarfare technologies, it's unlikely to maintain technological dominance because of the relatively low cost of the technologies needed to mount these types of attacks.

In fact, hackers worldwide have already begun doing so in earnest. In July 2009, 27 American and South Korean government agencies and other organizations were hit by a DDoS attack. An estimated 65,000 computers belonging to foreign botnets flooded the Web sites with access requests. Affected sites included those of the White House, the Treasury, the Federal Trade Commission, the Defense Department, the Secret Service, the New York Stock Exchange, and the Washington Post, in addition to the Korean Defense Ministry, National Assembly, the presidential Blue House, and several others.

The attacks were not sophisticated, but were widespread and prolonged, succeeding in slowing down most of the U.S. sites and forcing several South Korean sites to stop operating. North Korea or pro-North Korean groups were suspected to be behind the attacks, but the Pyongyang government denied any involvement.

The lone positive from the attacks was that only the Web sites of these agencies were affected. However, other intrusions suggest that hackers already have the potential for much more damaging acts of cyberwarfare. The Federal Aviation Administration (FAA), which oversees the airline activity of the United States, has already been subject to successful attacks on its systems, including one in 2006 that partially shut down air-traffic data systems in Alaska.

In 2007 and 2008, computer spies broke into the Pentagon's $300 billion Joint Strike Fighter project. Intruders were able to copy and siphon off several terabytes of data related to design and electronics systems, potentially making it easier to defend against the fighter when it's eventually produced. The intruders entered through vulnerabilities of two or three contractors working on the fighter jet project. Fortunately, computers containing the most sensitive data were not connected to the Internet, and were therefore inaccessible to the intruders. Former U.S. officials say that this attack originated in China, and that China had been making steady progress in developing online-warfare techniques. China rebutted these claims, stating that the U.S. media was subscribing to outdated, Cold War-era thinking in blaming them, and that Chinese hackers were not skilled enough to perpetrate an attack of that magnitude.

In December 2009, hackers reportedly stole a classified PowerPoint slide file detailing U.S. and South Korean strategy for fighting a war against North Korea. In Iraq, insurgents intercepted Predator drone feeds using software they had downloaded from the Internet.

Earlier that year, in April, cyberspies infiltrated the U.S. electrical grid, using weak points where computers on the grid are connected to the Internet, and left behind software programs whose purpose is unclear, but which presumably could be used to disrupt the system. Reports indicated that the spies originated in computer networks in China and Russia. Again, both nations denied the charges.

In response to these and other intrusions, the federal government launched a program called "Perfect Citizen" to detect cyberassaults on private companies running critical infrastructure. The U.S. National Security Agency (NSA) plans to install sensors in computer networks for critical infrastructure that would be activated by unusual activity signalling an impending cyberattack. The initial focus will be older large computer control systems that have since been linked to the Internet, making them more vulnerable to cyber attack. NSA will likely start with electric, nuclear, and air-traffic control systems with the greatest impact on national security.

As of this writing, most federal agencies get passing marks for meeting the requirements of the Federal Information Security Management Act, the most recent set of standards passed into law. But as cyberwarfare technologies develop and become more advanced, the standards imposed by this legislation will likely be insufficient to defend against attacks.

In each incident of cyberwarfare, the governments of the countries suspected to be responsible have roundly denied the charges with no repercussions. How could this be possible? The major reason is that tracing identities of specific attackers through cyberspace is next to impossible, making deniability simple.

The real worry for security experts and government officials is an act of cyberwar against a critical resource, such as the electric grid, financial system, or communications systems. First of all, the U.S. has no clear policy about how the country would respond to that level of a cyberattack. Although the electric grid was accessed by hackers, it hasn't yet actually been attacked. A three-year study of U.S. cybersecurity recommended that such a policy be created and made public. It also suggested that the U.S. attempt to find common ground with other nations to join forces in preventing these attacks.

Secondly, the effects of such an attack would likely be devastating. Mike McConnell, the former director of national intelligence, stated that if even a single large American bank were successfully attacked, "it would have an order-of-magnitude greater impact on the global economy" than the World Trade Center attacks, and that "the ability to threaten the U.S. money supply is the equivalent of today's nuclear weapon." Such an attack would have a catastrophic effect on the U.S. financial system, and by extension, the world economy.

Lastly, many industry analysts are concerned that the organization of our cybersecurity is messy, with no clear leader among our intelligence agencies. Several different agencies, including the Pentagon and the NSA, have their sights on being the leading agency in the ongoing efforts to combat cyberwarfare. In June 2009, Secretary of Defense Robert Gates ordered the creation of the first headquarters designed to coordinate government cybersecurity efforts, called Cybercom. Cybercom was activated in May 2010 with the goal of coordinating the operation and protection of military and Pentagon computer networks in the hopes of resolving this organizational tangle.

In confronting this problem, one critical question has arisen: how much control over enforcing cybersecurity should be given to American spy agencies, since they are prohibited from acting on American soil? Cyberattacks know no borders, so distinguishing between American soil and foreign soil means domestic agencies will be unnecessarily inhibited in their ability to fight cybercrime. For example, if the NSA was investigating the source of a cyberattack on government Web sites, and determined that the attack originated from American servers, under our current laws, it would not be able to investigate further.

Some experts believe that there is no effective way for a domestic agency to conduct computer operations without entering prohibited networks within the United States, or even conducting investigations in countries that are American allies. The NSA has already come under heavy fire for its surveillance actions after 9-11, and this has the potential to raise similar privacy concerns. Preventing terrorist or cyberwar attacks may require examining some e-mail messages from other countries or giving intelligence agencies more access to networks or Internet service providers. There is a need for an open debate about what constitutes a violation of privacy and what is acceptable during 'cyber-wartime', which is essentially all the time. The law may need to be changed to accommodate effective cybersecurity techniques, but it's unclear that this can be done without eroding some privacy rights that we consider essential.

As for these offensive measures, it's unclear as to how strong the United States' offensive capabilities for cyberwarfare are. The government closely guards this information, almost all of which is classified. But former military and intelligence officials indicate that our cyberwarfare capabilities have dramatically increased in sophistication in the past year or two.

And because tracking cybercriminals has proven so difficult, it may be that the best defense is a strong offense.

Sources: "Cyber Task Force Passes Mission to Cyber Command,"Defence Professionals, September 8, 2010; Siobhan Gorman, "U.S. Plans Cyber Shield for Utilities, Companies," *The Wall Street Journal*, July 8, 2010 and "U.S. Hampered in Fighting Cyber Attacks, Report Says," *The Wall Street Journal*, June, 16, 2010; Siobhan Gorman, Yochi Dreazen and August Cole, "Drone Breach Stirs Calls to Fill Cyber Post," *The Wall Street Journal*, December 21, 2009; Sean Gallagher, "New Threats Compel DOD to Rethink Cyber Strategy," *Defense Knowledge Technologies and Net-Enabled Warfare*, January 22, 2010; Lance Whitney, Cyber Command Chief Details Threat to U.S.," *Military Tech*, August 5, 2010; Hoover, J. Nicholas, "Cybersecurity Balancing Act," *Information Week*, April 27, 2009; David E. Sanger, John Markoff, and Thom Shanker, "U.S. Steps Up Effort on Digital Defenses," *The New York Times*, April 28, 2009; John Markoff and Thom Shanker. "Panel Advises Clarifying U.S. Plans on Cyberwar," *The New York Times*, April 30, 2009; Siobhan Gorman and Evan Ramstad, "Cyber Blitz Hits U.S., Korea," *The Wall Street Journal*, July 9, 2009; Lolita C. Baldor, "White House Among Targets of Sweeping Cyber Attack," Associated Press, July 8, 2009; Choe Sang-Hun, "Cyberattacks Hit U.S. and South Korean Web Sites," *The New York Times*, July 9, 2009; Siobhan Gorman, "FAA's Air-Traffic Networks Breached by Hackers," *The Wall Street Journal*, May 7, 2009; Thom Shanker, "New Military Command for Cyberspace," *The New York Times*, June 24, 2009; David E. Sanger and Thom Shanker, "Pentagon Plans New Arm to Wage Wars in Cyberspace," *The New York Times*, May 29, 2009; Siobhan Gorman, August Cole, and Yochi Dreazen, "Computer Spies Breach Fighter-Jet Project," *The Wall Street Journal*, April 21, 2009; Siobhan Gorman, "Electricity Grid in U.S. Penetrated by Spies," *The Wall Street Journal*, April 8, 2009; "Has Power Grid Been Hacked? U.S. Won't Say," Reuters, April 8, 2009; Markoff, John, "Vast Spy System Loots Computers in 103 Countries." *The New York Times,* March 29, 2009; Markoff, John, "Tracking Cyberspies Through the Web Wilderness," *The New York Times*, May 12, 2009.

CASE STUDY QUESTIONS

1. Is cyberwarfare a serious problem? Why or why not?

2. Assess the management, organization, and technology factors that have created this problem.

3. What solutions have been proposed? Do you think they will be effective? Why or why not?

4. Are there other solutions for this problem that should be pursued? What are they?

PART THREE

Key System Applications for the Digital Age

Part Three examines the core information system applications businesses are using today to improve operational excellence and decision making. These applications include enterprise systems; systems for supply chain management, customer relationship management, collaboration, and knowledge management; e-commerce applications; and decision-support systems. This part answers questions such as: How can enterprise applications improve business performance? How do firms use e-commerce to extend the reach of their businesses? How can systems improve collaboration and decision making and help companies make better use of their knowledge assets?

Chapter 9

Achieving Operational Excellence and Customer Intimacy: Enterprise Applications

CANNONDALE LEARNS TO MANAGE A GLOBAL SUPPLY CHAIN

I f you enjoy cycling, you may very well be using a Cannondale bicycle. Cannondale, headquartered in Bethel, Connecticut, is the world-leading manufacturer of high-end bicycles, apparel, footwear, and accessories, with dealers and distributors in more than 66 countries. Cannondale's supply and distribution chains span the globe, and the company must coordinate manufacturing, assembly, and sales/distribution sites in many different countries. Cannondale produces more than 100 different bicycle models each year; 60 percent of these are newly introduced to meet ever-changing customer preferences.

Cannondale offers both make-to-stock and make-to-order models. A typical bicycle requires a 150-day lead time and a four-week manufacturing window, and some models have bills of materials with over 150 parts. (A bill of materials specifies the raw materials, assemblies, components, parts, and quantities of each needed to manufacture a final product.) Cannondale must manage more than 1 million of these bills of materials and more than 200,000 individual parts. Some of these parts come from specialty vendors with even longer lead times and limited production capacity.

Obviously, managing parts availability in a constantly changing product line impacted by volatile customer demand requires a great deal of manufacturing flexibility. Until recently, that flexibility was missing. Cannondale had an antiquated legacy material requirements planning system for planning production, controlling inventory, and managing manufacturing processes that could only produce reports on a weekly basis. By Tuesday afternoon, Monday's reports were already out of date. The company was forced to substitute parts in order to meet demand, and sometimes it lost sales. Cannondale needed a solution that could track the flow of parts more accurately, support its need for flexibility, and work with its existing business systems, all within a restricted budget.

Cannondale selected the Kinaxis RapidResponse on-demand software service as a solution. RapidResponse furnishes accurate and detailed supply chain information via an easy-to-use spreadsheet interface, using data supplied automatically from Cannondale's existing manufacturing systems. Data from operations at multiple sites are assembled in a single place for analysis and decision making. Supply chain participants from different locations are able to model manufacturing and inventory data in "what-if" scenarios to see the impact of alternative actions across the entire supply chain. Old forecasts can be compared to new ones, and the system can evaluate the constraints of a new plan.

Cannondale buyers, planners, master schedulers, sourcers, product managers, customer service, and finance personnel, use RapidResponse for sales reporting, forecasting, monitoring daily inventory availability, and feeding production schedule information to Cannondale's manufacturing and order processing systems. Users are able to see up-to-date information for all sites. Management uses the system daily to examine areas where there are backlogs.

The improved supply chain information from RapidResponse enables Cannondale to respond to customer orders much more rapidly with lower levels of inventory and safety stock. Cycle times and lead times for producing products have also been reduced. The company's dates for promising deliveries are more reliable and accurate.

Sources: Kinaxis Corp., "Cannondale Improves Customer Response Times While Reducing Inventory Using RapidResponse," 2010; www.kinaxis.com, accessed June 21, 2010; and www.cannondale.com, accessed June 21, 2010.

Cannondale's problems with its supply chain illustrate the critical role of supply chain management (SCM) systems in business. Cannondale's business performance was impeded because it could not coordinate its sourcing, manufacturing, and distribution processes. Costs were unnecessarily high because the company was unable to accurately determine the exact amount of each product it needed to fulfill orders and hold just that amount in inventory. Instead, the company resorted to keeping extra "safety stock" on hand "just in case." When products were not available when the customer wanted them, Cannondale lost sales.

The chapter-opening diagram calls attention to important points raised by this case and this chapter. Like many other firms, Cannondale had a complex supply chain and manufacturing processes to coordinate in many different locations. The company had to deal with hundreds and perhaps thousands of suppliers of parts and raw materials. It was not always possible to have just the right amount of each part or component available when it was needed because the company lacked accurate, up-to-date information about parts in inventory and what manufacturing processes needed those parts.

An on-demand supply chain management software service from Kinaxis helped solve this problem. The Kinaxis RapidResponse software takes in data from Cannondale's existing manufacturing systems and assembles data from multiple sites to furnish a single view of Cannondale's supply chain based on up-to-date information. Cannondale staff are able to see exactly what parts are available or on order as well as the status of bikes in production. With better tools for planning, users are able to see the impact of changes in supply and demand so that they can make better decisions about how to respond to these changes. The system has greatly enhanced operational efficiency and decision making.

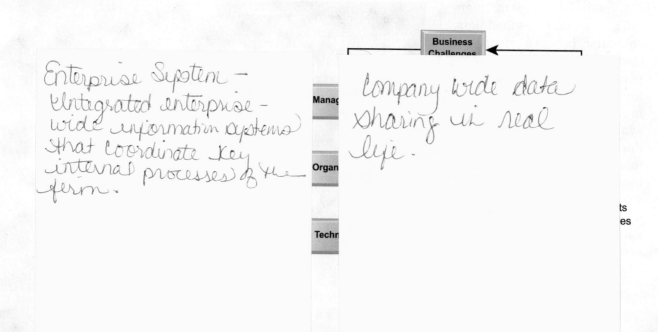

9.1 ENTERPRISE SYSTEMS

Around the globe, companies are increasingly becoming more connected, both internally and with other companies. If you run a business, you'll want to be able to react instantaneously when a customer places a large order or when a shipment from a supplier is delayed. You may also want to know the impact of these events on every part of the business and how the business is performing at any point in time, especially if you're running a large company. Enterprise systems provide the integration to make this possible. Let's look at how they work and what they can do for the firm.

WHAT ARE ENTERPRISE SYSTEMS?

Imagine that you had to run a business based on information from tens or even hundreds of different databases and systems, none of which could speak to one another? Imagine your company had 10 different major product lines, each produced in separate factories, and each with separate and incompatible sets of systems controlling production, warehousing, and distribution.

At the very least, your decision making would often be based on manual hard-copy reports, often out of date, and it would be difficult to really understand what is happening in the business as a whole. Sales personnel might not be able to tell at the time they place an order whether the ordered items are in inventory, and manufacturing could not easily use sales data to plan for new production. You now have a good idea of why firms need a special enterprise system to integrate information.

Chapter 2 introduced enterprise systems, also known as enterprise resource planning (ERP) systems, which are based on a suite of integrated software modules and a common central database. The database collects data from many different divisions and departments in a firm, and from a large number of key business processes in manufacturing and production, finance and accounting, sales and marketing, and human resources, making the data available for applications that support nearly all of an organization's internal business activities. When new information is entered by one process, the information is made immediately available to other business processes (see Figure 9-1).

If a sales representative places an order for tire rims, for example, the system verifies the customer's credit limit, schedules the shipment, identifies the best shipping route, and reserves the necessary items from inventory. If inventory stock were insufficient to fill the order, the system schedules the manufacture of more rims, ordering the needed materials and components from suppliers. Sales and production forecasts are immediately updated. General ledger and corporate cash levels are automatically updated with the revenue and cost information from the order. Users could tap into the system and find out where that particular order was at any minute. Management could obtain information at any point in time about how the business was operating. The system could also generate enterprise-wide data for management analyses of product cost and profitability.

FIGURE 9-1   **HOW ENTERPRISE SYSTEMS WORK**

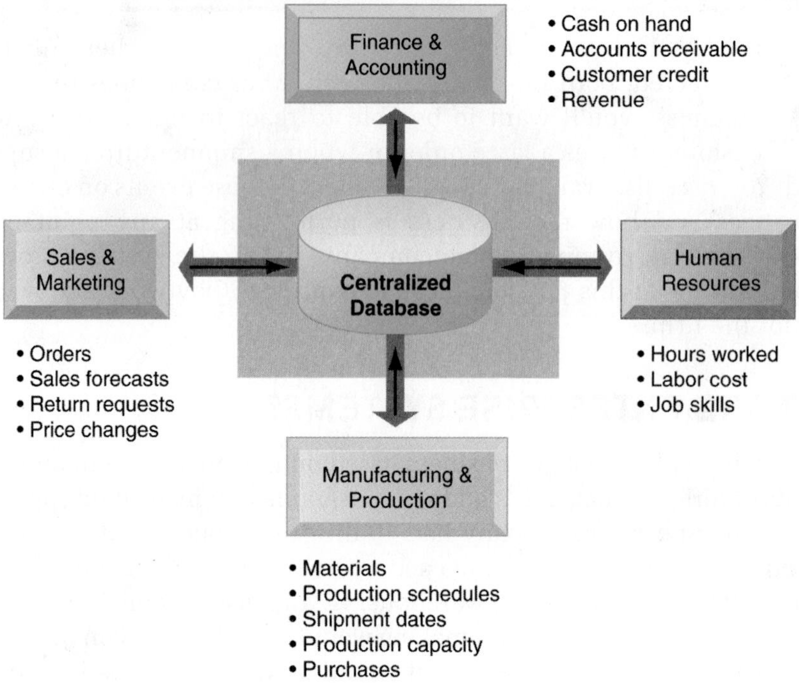

Enterprise systems feature a set of integrated software modules and a central database that enables data to be shared by many different business processes and functional areas throughout the enterprise.

ENTERPRISE SOFTWARE

Enterprise software is built around thousands of predefined business processes that reflect best practices. Table 9-1 describes some of the major business processes supported by enterprise software.

Companies implementing this software must first select the functions of the system they wish to use and then map their business processes to the predefined business processes in the software. (One of our Learning Tracks shows how SAP enterprise software handles the procurement process for a new piece of equipment.) Identifying the organization's business processes to be included in the system and then mapping them to the processes in the enterprise software is often a major effort. A firm would use configuration tables provided by the software to tailor a particular aspect of the system to the way it does business. For example, the firm could use these tables to select whether it wants to track revenue by product line, geographical unit, or distribution channel.

TABLE 9-1 BUSINESS PROCESSES SUPPORTED BY ENTERPRISE SYSTEMS

Financial and accounting processes, including general ledger, accounts payable, accounts receivable, fixed assets, cash management and forecasting, product-cost accounting, cost-center accounting, asset accounting, tax accounting, credit management, and financial reporting.

Human resources processes, including personnel administration, time accounting, payroll, personnel planning and development, benefits accounting, applicant tracking, time management, compensation, workforce planning, performance management, and travel expense reporting.

Manufacturing and production processes, including procurement, inventory management, purchasing, shipping, production planning, production scheduling, material requirements planning, quality control, distribution, transportation execution, and plant and equipment maintenance.

Sales and marketing processes, including order processing, quotations, contracts, product configuration, pricing, billing, credit checking, incentive and commission management, and sales planning.

If the enterprise software does not support the way the organization does business, companies can rewrite some of the software to support the way their business processes work. However, enterprise software is unusually complex, and extensive customization may degrade system performance, compromising the information and process integration that are the main benefits of the system. If companies want to reap the maximum benefits from enterprise software, they must change the way they work to conform to the business processes in the software. To implement a new enterprise system, Tasty Baking Company identified its existing business processes and then translated them into the business processes built into the SAP ERP software it had selected. To ensure it obtained the maximum benefits from the enterprise software, Tasty Baking Company deliberately planned for customizing less than 5 percent of the system and made very few changes to the SAP software itself. It used as many tools and features that were already built into the SAP software as it could. SAP has more than 3,000 configuration tables for its enterprise software.

Leading enterprise software vendors include SAP, Oracle (with its acquisition PeopleSoft) Infor Global Solutions, and Microsoft. There are versions of enterprise software packages designed for small businesses and on-demand versions, including software services delivered over the Web (see the Interactive Session on Technology in Section 9.4). Although initially designed to automate the firm's internal "back-office" business processes, enterprise systems have become more externally-oriented and capable of communicating with customers, suppliers, and other entities.

BUSINESS VALUE OF ENTERPRISE SYSTEMS

Enterprise systems provide value both by increasing operational efficiency and by providing firm-wide information to help managers make better decisions. Large companies with many operating units in different locations have used enterprise systems to enforce standard practices and data so that everyone does business the same way worldwide.

Coca Cola, for instance, implemented a SAP enterprise system to standardize and coordinate important business processes in 200 countries. Lack of standard, company-wide business processes prevented the company from leveraging its worldwide buying power to obtain lower prices for raw materials and from reacting rapidly to market changes.

Enterprise systems help firms respond rapidly to customer requests for information or products. Because the system integrates order, manufacturing, and delivery data, manufacturing is better informed about producing only what customers have ordered, procuring exactly the right amount of components or raw materials to fill actual orders, staging production, and minimizing the time that components or finished products are in inventory.

Alcoa, the world's leading producer of aluminum and aluminum products with operations spanning 41 countries and 500 locations, had initially been organized around lines of business, each of which had its own set of information systems. Many of these systems were redundant and inefficient. Alcoa's costs for executing requisition-to-pay and financial processes were much higher and its cycle times were longer than those of other companies in its industry. (Cycle time refers to the total elapsed time from the beginning to the end of a process.) The company could not operate as a single worldwide entity.

After implementing enterprise software from Oracle, Alcoa eliminated many redundant processes and systems. The enterprise system helped Alcoa reduce requisition-to-pay cycle time by verifying receipt of goods and automatically generating receipts for payment. Alcoa's accounts payable transaction processing dropped 89 percent. Alcoa was able to centralize financial and procurement activities, which helped the company reduce nearly 20 percent of its worldwide costs.

Enterprise systems provide much valuable information for improving management decision making. Corporate headquarters has access to up-to-the-minute data on sales, inventory, and production and uses this information to create more accurate sales and production forecasts. Enterprise software includes analytical tools for using data captured by the system to evaluate overall organizational performance. Enterprise system data have common standardized definitions and formats that are accepted by the entire organization. Performance figures mean the same thing across the company. Enterprise systems allow senior management to easily find out at any moment how a particular organizational unit is performing, determine which products are most or least profitable, and calculate costs for the company as a whole.

For example, Alcoa's enterprise system includes functionality for global human resources management that shows correlations between investment in employee training and quality, measures the company-wide costs of delivering services to employees, and measures the effectiveness of employee recruitment, compensation, and training.

9.2 SUPPLY CHAIN MANAGEMENT SYSTEMS

If you manage a small firm that makes a few products or sells a few services, chances are you will have a small number of suppliers. You could coordinate your supplier orders and deliveries using a telephone and fax machine. But if you manage a firm that produces more complex products and services, then you will have hundreds of suppliers, and your suppliers will each have their own set of suppliers. Suddenly, you are in a situation where you will need to coordinate the activities of hundreds or even thousands of other firms in order to produce your products and services. Supply chain management systems, which we introduced in Chapter 2, are an answer to these problems of supply chain complexity and scale.

THE SUPPLY CHAIN

A firm's **supply chain** is a network of organizations and business processes for procuring raw materials, transforming these materials into intermediate and finished products, and distributing the finished products to customers. It links suppliers, manufacturing plants, distribution centers, retail outlets, and customers to supply goods and services from source through consumption. Materials, information, and payments flow through the supply chain in both directions

Goods start out as raw materials and, as they move through the supply chain, are transformed into intermediate products (also referred to as components or parts), and finally, into finished products. The finished products are shipped to distribution centers and from there to retailers and customers. Returned items flow in the reverse direction from the buyer back to the seller.

Let's look at the supply chain for Nike sneakers as an example. Nike designs, markets, and sells sneakers, socks, athletic clothing, and accessories throughout the world. Its primary suppliers are contract manufacturers with factories in China, Thailand, Indonesia, Brazil, and other countries. These companies fashion Nike's finished products.

Nike's contract suppliers do not manufacture sneakers from scratch. They obtain components for the sneakers—the laces, eyelets, uppers, and soles—from other suppliers and then assemble them into finished sneakers. These suppliers in turn have their own suppliers. For example, the suppliers of soles have suppliers for synthetic rubber, suppliers for chemicals used to melt the rubber for molding, and suppliers for the molds into which to pour the rubber. Suppliers of laces would have suppliers for their thread, for dyes, and for the plastic lace tips.

Figure 9-2 provides a simplified illustration of Nike's supply chain for sneakers; it shows the flow of information and materials among suppliers, Nike, and Nike's distributors, retailers, and customers. Nike's contract manufacturers are its primary suppliers. The suppliers of soles, eyelets, uppers, and laces are the secondary (Tier 2) suppliers. Suppliers to these suppliers are the tertiary (Tier 3) suppliers.

The *upstream* portion of the supply chain includes the company's suppliers, the suppliers' suppliers, and the processes for managing relationships with them. The *downstream* portion consists of the organizations and processes for distributing and delivering products to the final customers. Companies doing manufacturing, such as Nike's contract suppliers of sneakers, also manage their own *internal supply chain* processes for transforming materials, components, and services furnished by their suppliers into finished products or intermediate products (components or parts) for their customers and for managing materials and inventory.

FIGURE 9-2   NIKE'S SUPPLY CHAIN

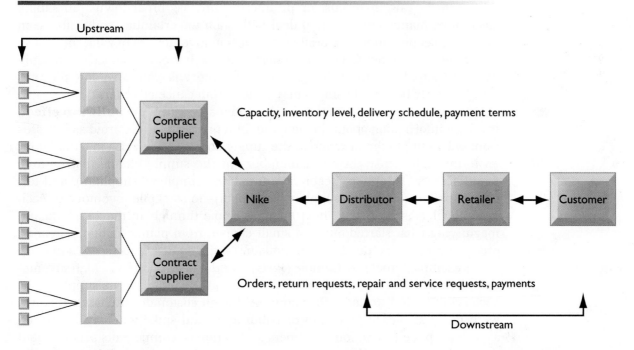

This figure illustrates the major entities in Nike's supply chain and the flow of information upstream and downstream to coordinate the activities involved in buying, making, and moving a product. Shown here is a simplified supply chain, with the upstream portion focusing only on the suppliers for sneakers and sneaker soles.

The supply chain illustrated in Figure 9-2 only shows two contract manufacturers for sneakers and only the upstream supply chain for sneaker soles. Nike has hundreds of contract manufacturers turning out finished sneakers, socks, and athletic clothing, each with its own set of suppliers. The upstream portion of Nike's supply chain would actually comprise thousands of entities. Nike also has numerous distributors and many thousands of retail stores where its shoes are sold, so the downstream portion of its supply chain is also large and complex.

INFORMATION SYSTEMS AND SUPPLY CHAIN MANAGEMENT

Inefficiencies in the supply chain, such as parts shortages, underutilized plant capacity, excessive finished goods inventory, or high transportation costs, are caused by inaccurate or untimely information. For example, manufacturers may keep too many parts in inventory because they do not know exactly when they will receive their next shipments from their suppliers. Suppliers may order too few raw materials because they do not have precise information on demand. These supply chain inefficiencies waste as much as 25 percent of a company's operating costs.

If a manufacturer had perfect information about exactly how many units of product customers wanted, when they wanted them, and when they could be produced, it would be possible to implement a highly efficient **just-in-time strategy**. Components would arrive exactly at the moment they were needed and finished goods would be shipped as they left the assembly line.

In a supply chain, however, uncertainties arise because many events cannot be foreseen—uncertain product demand, late shipments from suppliers, defective parts or raw materials, or production process breakdowns. To satisfy customers, manufacturers often deal with such uncertainties and unforeseen events by keeping more material or products in inventory than what they think they may actually need. The *safety stock* acts as a buffer for the lack of flexibility in the supply chain. Although excess inventory is expensive, low fill rates are also costly because business may be lost from canceled orders.

One recurring problem in supply chain management is the **bullwhip effect**, in which information about the demand for a product gets distorted as it passes from one entity to the next across the supply chain. A slight rise in demand for an item might cause different members in the supply chain—distributors, manufacturers, suppliers, secondary suppliers (suppliers' suppliers), and tertiary suppliers (suppliers' suppliers' suppliers)—to stockpile inventory so each has enough "just in case." These changes ripple throughout the supply chain, magnifying what started out as a small change from planned orders, creating excess inventory, production, warehousing, and shipping costs (see Figure 9-3).

For example, Procter & Gamble (P&G) found it had excessively high inventories of its Pampers disposable diapers at various points along its supply chain because of such distorted information. Although customer purchases in stores were fairly stable, orders from distributors would spike when P&G offered aggressive price promotions. Pampers and Pampers' components accumulated in warehouses along the supply chain to meet demand that did not actually exist. To eliminate this problem, P&G revised its marketing, sales, and supply chain processes and used more accurate demand forecasting

The bullwhip is tamed by reducing uncertainties about demand and supply when all members of the supply chain have accurate and up-to-date information. If all supply chain members share dynamic information about inventory

FIGURE 9-3   THE BULLWHIP EFFECT

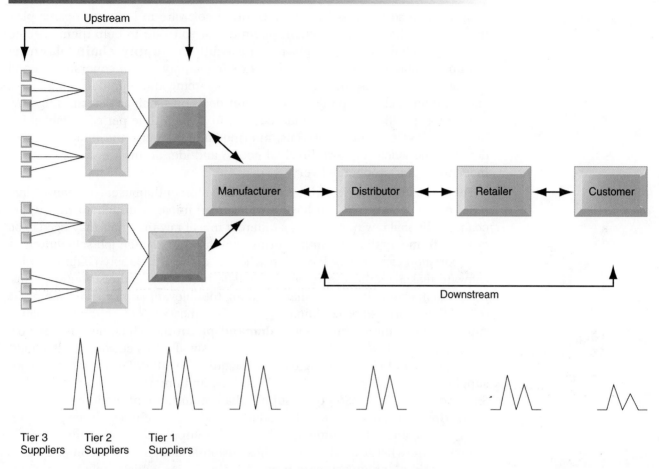

Inaccurate information can cause minor fluctuations in demand for a product to be amplified as one moves further back in the supply chain. Minor fluctuations in retail sales for a product can create excess inventory for distributors, manufacturers, and suppliers.

levels, schedules, forecasts, and shipments, they have more precise knowledge about how to adjust their sourcing, manufacturing, and distribution plans. Supply chain management systems provide the kind of information that helps members of the supply chain make better purchasing and scheduling decisions. Table 9-2 describes how firms benefit from these systems.

TABLE 9-2 HOW INFORMATION SYSTEMS FACILITATE SUPPLY CHAIN MANAGEMENT

INFORMATION FROM SUPPLY CHAIN MANAGEMENT SYSTEMS HELPS FIRMS

Decide when and what to produce, store, and move

Rapidly communicate orders

Track the status of orders

Check inventory availability and monitor inventory levels

Reduce inventory, transportation, and warehousing costs

Track shipments

Plan production based on actual customer demand

Rapidly communicate changes in product design

SUPPLY CHAIN MANAGEMENT SOFTWARE

Supply chain software is classified as either software to help businesses plan their supply chains (supply chain planning) or software to help them execute the supply chain steps (supply chain execution). **Supply chain planning systems** enable the firm to model its existing supply chain, generate demand forecasts for products, and develop optimal sourcing and manufacturing plans. Such systems help companies make better decisions such as determining how much of a specific product to manufacture in a given time period; establishing inventory levels for raw materials, intermediate products, and finished goods; determining where to store finished goods; and identifying the transportation mode to use for product delivery.

For example, if a large customer places a larger order than usual or changes that order on short notice, it can have a widespread impact throughout the supply chain. Additional raw materials or a different mix of raw materials may need to be ordered from suppliers. Manufacturing may have to change job scheduling. A transportation carrier may have to reschedule deliveries. Supply chain planning software makes the necessary adjustments to production and distribution plans. Information about changes is shared among the relevant supply chain members so that their work can be coordinated. One of the most important—and complex— supply chain planning functions is **demand planning**, which determines how much product a business needs to make to satisfy all of its customers' demands. Manugistics and i2 Technologies (both acquired by JDA Software) are major supply chain management software vendors, and enterprise software vendors SAP and Oracle-PeopleSoft offer supply chain management modules.

Whirlpool Corporation, which produces washing; machines, dryers, refrigerators, ovens, and other home appliances, uses supply chain planning systems to make sure what it produces matches customer demand. The company uses supply chain planning software from i2 Technologies, which includes modules for master scheduling, deployment planning, and inventory planning. Whirlpool also installed i2's Web-based tool for Collaborative Planning, Forecasting, and Replenishment (CPFR) for sharing and combining its sales forecasts with those of its major sales partners. Improvements in supply chain planning combined with new state-of-the-art distribution centers helped Whirlpool increase availability of products in stock when customers needed them to 97 percent, while reducing the number of excess finished goods in inventory by 20 percent and forecasting errors by 50 percent (Barrett, 2009).

Supply chain execution systems manage the flow of products through distribution centers and warehouses to ensure that products are delivered to the right locations in the most efficient manner. They track the physical status of goods, the management of materials, warehouse and transportation operations, and financial information involving all parties. Haworth Incorporated's Warehouse Management System (WMS) is an example. Haworth is a world-leading manufacturer and designer of office furniture, with distribution centers in four different states. The WMS tracks and controls the flow of finished goods from Haworth's distribution centers to its customers. Acting on shipping plans for customer orders, the WMS directs the movement of goods based on immediate conditions for space, equipment, inventory, and personnel.

The Interactive Session on Organizations describes how supply chain management software improved decision making and operational performance at Southwest Airlines. This company maintains a competitive edge by combining superb customer service with low costs. Effectively managing its parts inventory is crucial to achieving these goals.

INTERACTIVE SESSION: ORGANIZATIONS

SOUTHWEST AIRLINES TAKES OFF WITH BETTER SUPPLY CHAIN MANAGEMENT

"Weather at our destination is 50 degrees with some broken clouds, but they'll try to have them fixed before we arrive. Thank you, and remember, nobody loves you or your money more than Southwest Airlines."

Crew humor at 30,000 feet? Must be Southwest Airlines. The company is the largest low-fare, high-frequency, point-to-point airline in the world, and largest overall measured by number of passengers per year. Founded in 1971 with four planes serving three cities, the company now operates over 500 aircraft in 68 cities, and has revenues of $10.1 billion. Southwest has the best customer service record among major airlines, the lowest cost structure, and the lowest and simplest fares. The stock symbol is LUV (for Dallas's Love Field where the company is headquartered), but love is the major theme of Southwest's employee and customer relationships. The company has made a profit every year since 1973, one of the few airlines that can make that claim.

Despite a freewheeling, innovative corporate culture, even Southwest needs to get serious about its information systems to maintain profitability. Southwest is just like any other company that needs to manage its supply chain and inventory efficiently. The airline's success has led to continued expansion, and as the company has grown, its legacy information systems have been unable to keep up with the increasingly large amount of data being generated.

One of the biggest problems with Southwest's legacy systems was lack of information visibility. Often, the data that Southwest's managers needed were safely stored on their systems but weren't "visible", or readily available for viewing or use in other systems. Information about what replacement parts were available at a given time was difficult or impossible to acquire, and that affected response times for everything from mechanical problems to part fulfillment.

For Southwest, which prides itself on its excellent customer service, getting passengers from one location to another with minimal delay is critically important. Repairing aircraft quickly is an important part of accomplishing that goal. The company had $325 million in service parts inventory, so any solution that more efficiently handled that inventory

and reduced aircraft groundings would have a strong impact on the airline's bottom line. Richard Zimmerman, Southwest's manager of inventory management, stated that "there's a significant cost when we have to ground aircraft because we ran out of a part. The long-term, cost-effective way to solve that problem was to increase productivity and to ensure that our maintenance crews were supported with the right spare parts, through the right software application."

Southwest's management started looking for a better inventory management solution, and a vendor that was capable of working within the airline's unique corporate culture. After an extensive search, Southwest eventually chose i2 Technologies, a leading supply chain management software and services company that was recently purchased by JDA Software. Southwest implemented the i2 Demand Planner, i2 Service Parts Planner, and i2 Service Budget Optimizer to overhaul its supply chain management and improve data visibility.

I2 Demand Planner improves Southwest's forecasts for all of the part location combinations in its system, and provides better visibility into demand for each part. Planners are able to differentiate among individual parts based on criticality and other dimensions such as demand volume, demand variability, and dollar usage. I2 Service Parts Planner helps Southwest replenish its store of parts and ensures that "the right parts are in the right location at the right time." The software can recommend the best mix of parts for each location that will satisfy the customer service requirements of that location at the lowest cost. If excess inventory builds up in certain service locations, the software will recommend the most cost-efficient way to transfer that excess inventory to locations with parts deficits. I2 Service Budget Optimizer helps Southwest use its historical data of parts usage to generate forecasts of future parts usage.

Together, these solutions gather data from Southwest's legacy systems and provide useful information to Southwest's managers. Most importantly, Southwest can recognize demand shortages before they become problems, thanks to the visibility provided by i2's solutions. Southwest's managers now have a clear and unobstructed view of all of the data up and down the company's supply chain.

By using what-if analysis, planners can quantify the cost to the company of operating at different levels of service. Zimmerman added that i2 "will help us lower inventory costs and keep our cost per air seat mile down to the lowest in the industry. Also, the solutions will help us ensure that the maintenance team can quickly repair the aircraft so that our customers experience minimal delays." The results of the i2 implementation were

increased availability of parts, increased speed and intelligence of decision making, reduced parts inventory by 15 percent, saving the company over $30 million, and increased service levels from 92 percent prior to the implementation to over 95 percent afterwards.

Sources: Chris Lauer, *Southwest Airlines: Corporations That Changed the World*, Greenwood Press, May 2010; www.i2.com, "Ensuring Optimal Parts Inventory at Southwest Airlines," and "Service Parts Management," accessed April 25, 2010; and www.southwest.com, accessed July 1, 2010.

CASE STUDY QUESTIONS

1. Why is parts inventory management so important at Southwest Airlines? What business processes are affected by the airline's ability or inability to have required parts on hand?

2. Why management, organization, and technology factors were responsible for Southwest's problems with inventory management?

3. How did implementing the i2 software change the way Southwest ran its business?

4. Describe two decisions that were improved by implementing the i2 system.

MIS IN ACTION

Visit i2's site (www.i2.com) and learn more about some of the other companies using its software. Pick one of these companies, then answer the following questions:

1. What problem did the company need to address with i2's software?

2. Why did the company select i2 as its software vendor?

3. What were the gains that the company realized as a result of the software implementation?

GLOBAL SUPPLY CHAINS AND THE INTERNET

Before the Internet, supply chain coordination was hampered by the difficulties of making information flow smoothly among disparate internal supply chain systems for purchasing, materials management, manufacturing, and distribution. It was also difficult to share information with external supply chain partners because the systems of suppliers, distributors, or logistics providers were based on incompatible technology platforms and standards. Enterprise and supply chain management systems enhanced with Internet technology supply some of this integration.

A manager will use a Web interface to tap into suppliers' systems to determine whether inventory and production capabilities match demand for the firm's products. Business partners will use Web-based supply chain management tools to collaborate online on forecasts. Sales representatives will access suppliers' production schedules and logistics information to monitor customers' order status.

Global Supply Chain Issues

More and more companies are entering international markets, outsourcing manufacturing operations, and obtaining supplies from other countries as well as selling abroad. Their supply chains extend across multiple countries and regions. There are additional complexities and challenges to managing a global supply chain.

Global supply chains typically span greater geographic distances and time differences than domestic supply chains and have participants from a number of different countries. Although the purchase price of many goods might be lower abroad, there are often additional costs for transportation, inventory (the need for a larger buffer of safety stock), and local taxes or fees. Performance standards may vary from region to region or from nation to nation. Supply chain management may need to reflect foreign government regulations and cultural differences. All of these factors impact how a company takes orders, plans distribution, sizes warehousing, and manages inbound and outbound logistics throughout the global markets it services.

The Internet helps companies manage many aspects of their global supply chains, including sourcing, transportation, communications, and international finance. Today's apparel industry, for example, relies heavily on outsourcing to contract manufacturers in China and other low-wage countries. Apparel companies are starting to use the Web to manage their global supply chain and production issues.

For example, Koret of California, a subsidiary of apparel maker Kellwood Co., uses e-SPS Web-based software to gain end-to-end visibility into its entire global supply chain. E-SPS features Web-based software for sourcing, work-in-progress tracking, production routing, product-development tracking, problem identification and collaboration, delivery-date projections, and production-related inquiries and reports.

As goods are being sourced, produced, and shipped, communication is required among retailers, manufacturers, contractors, agents, and logistics providers. Many, especially smaller companies, still share product information over the phone, via e-mail, or through faxes. These methods slow down the supply chain and also increase errors and uncertainty. With e-SPS, all supply chain members communicate through a Web-based system. If one of Koret's vendors makes a change in the status of a product, everyone in the supply chain sees the change.

In addition to contract manufacturing, globalization has encouraged outsourcing warehouse management, transportation management, and related operations to third-party logistics providers, such as UPS Supply Chain Solutions and Schneider Logistics Services. These logistics services offer Web-based software to give their customers a better view of their global supply chains. Customers are able to check a secure Web site to monitor inventory and shipments, helping them run their global supply chains more efficiently.

Demand-Driven Supply Chains: From Push to Pull Manufacturing and Efficient Customer Response

In addition to reducing costs, supply chain management systems facilitate efficient customer response, enabling the workings of the business to be driven more by customer demand. (We introduced efficient customer response systems in Chapter 3.)

Earlier supply chain management systems were driven by a push-based model (also known as build-to-stock). In a **push-based model**, production master schedules are based on forecasts or best guesses of demand for products, and products are "pushed" to customers. With new flows of information made possible by Web-based tools, supply chain management more easily follows a pull-based model. In a **pull-based model**, also known as a demand-driven model or build-to-order, actual customer orders or purchases trigger events in the supply chain. Transactions to produce and deliver only what customers have ordered move up the supply chain from retailers to distributors to manufacturers and eventually to suppliers. Only products to fulfill these orders move back down the supply chain to the retailer. Manufacturers use only actual order demand information to drive their production schedules and the procurement of components or raw materials, as illustrated in Figure 9-4. Walmart's continuous replenishment system described in Chapter 3 is an example of the pull-based model.

The Internet and Internet technology make it possible to move from sequential supply chains, where information and materials flow sequentially from company to company, to concurrent supply chains, where information flows in many directions simultaneously among members of a supply chain network. Complex supply networks of manufacturers, logistics suppliers, outsourced manufacturers, retailers, and distributors are able to adjust immediately to changes in schedules or orders. Ultimately, the Internet could create a "digital logistics nervous system" throughout the supply chain (see Figure 9-5).

BUSINESS VALUE OF SUPPLY CHAIN MANAGEMENT SYSTEMS

You have just seen how supply chain management systems enable firms to streamline both their internal and external supply chain processes and provide management with more accurate information about what to produce, store, and move. By implementing a networked and integrated supply chain management system, companies match supply to demand, reduce inventory levels, improve delivery service, speed product time to market, and use assets more effectively.

Total supply chain costs represent the majority of operating expenses for many businesses and in some industries approach 75 percent of the total operating budget. Reducing supply chain costs may have a major impact on firm profitability.

In addition to reducing costs, supply chain management systems help increase sales. If a product is not available when a customer wants it, customers often try to purchase it from someone else. More precise control of the supply chain enhances the firm's ability to have the right product available for customer purchases at the right time.

FIGURE 9-4   PUSH- VERSUS PULL-BASED SUPPLY CHAIN MODELS

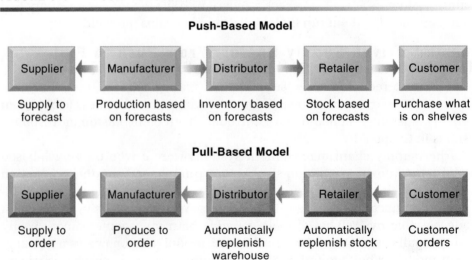

The difference between push- and pull-based models is summarized by the slogan "Make what we sell, not sell what we make."

FIGURE 9-5   THE FUTURE INTERNET-DRIVEN SUPPLY CHAIN

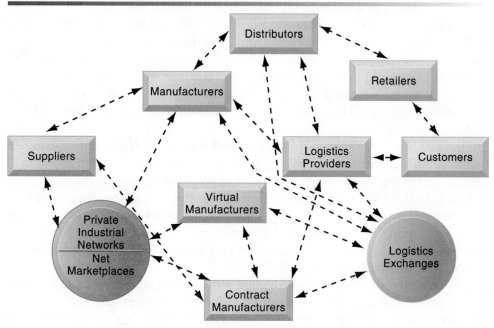

The future Internet-driven supply chain operates like a digital logistics nervous system. It provides multidirectional communication among firms, networks of firms, and e-marketplaces so that entire networks of supply chain partners can immediately adjust inventories, orders, and capacities.

9.3   CUSTOMER RELATIONSHIP MANAGEMENT SYSTEMS

You've probably heard phrases such as "the customer is always right" or "the customer comes first." Today these words ring more true than ever. Because competitive advantage based on an innovative new product or service is often very short lived, companies are realizing that their only enduring competitive strength may be their relationships with their customers. Some say that the basis of competition has switched from who sells the most products and services to who "owns" the customer, and that customer relationships represent a firm's most valuable asset.

WHAT IS CUSTOMER RELATIONSHIP MANAGEMENT?

What kinds of information would you need to build and nurture strong, long-lasting relationships with customers? You'd want to know exactly who your customers are, how to contact them, whether they are costly to service and sell to, what kinds of products and services they are interested in, and how much money they spend on your company. If you could, you'd want to make sure you knew each of your customers well, as if you were running a small-town store. And you'd want to make your good customers feel special.

In a small business operating in a neighborhood, it is possible for business owners and managers to really know their customers on a personal, face-to-face basis. But in a large business operating on a metropolitan, regional, national, or even global basis, it is impossible to "know your customer" in this intimate way. In these kinds of businesses there are too many customers and too many different ways that customers interact with the firm (over the Web, the phone, fax, and in person). It becomes especially difficult to integrate information from all theses sources and to deal with the large numbers of customers.

A large business's processes for sales, service, and marketing tend to be highly compartmentalized, and these departments do not share much essential customer information. Some information on a specific customer might be stored and organized in terms of that person's account with the company. Other pieces of information about the same customer might be organized by products that were purchased. There is no way to consolidate all of this information to provide a unified view of a customer across the company.

This is where customer relationship management systems help. Customer relationship management (CRM) systems, which we introduced in Chapter 2, capture and integrate customer data from all over the organization, consolidate the data, analyze the data, and then distribute the results to various systems and customer touch points across the enterprise. A **touch point** (also known as a contact point) is a method of interaction with the customer, such as telephone, e-mail, customer service desk, conventional mail, Web site, wireless device, or retail store.

Well-designed CRM systems provide a single enterprise view of customers that is useful for improving both sales and customer service. Such systems likewise provide customers with a single view of the company regardless of what touch point the customer uses (see Figure 9-6).

Good CRM systems provide data and analytical tools for answering questions such as these: "What is the value of a particular customer to the firm over his or her lifetime?" "Who are our most loyal customers?" (It can cost six times more to sell to a new customer than to an existing customer.) "Who are our most profitable customers?" and "What do these profitable customers want to buy?" Firms use the answers to these questions to acquire new customers, provide better service and support to existing customers, customize their offerings more precisely to customer preferences, and provide ongoing value to retain profitable customers.

FIGURE 9-6   CUSTOMER RELATIONSHIP MANAGEMENT (CRM)

CRM systems examine customers from a multifaceted perspective. These systems use a set of integrated applications to address all aspects of the customer relationship, including customer service, sales, and marketing.

CUSTOMER RELATIONSHIP MANAGEMENT SOFTWARE

Commercial CRM software packages range from niche tools that perform limited functions, such as personalizing Web sites for specific customers, to large-scale enterprise applications that capture myriad interactions with customers, analyze them with sophisticated reporting tools, and link to other major enterprise applications, such as supply chain management and enterprise systems. The more comprehensive CRM packages contain modules for **partner relationship management (PRM)** and **employee relationship management (ERM)**.

PRM uses many of the same data, tools, and systems as customer relationship management to enhance collaboration between a company and its selling partners. If a company does not sell directly to customers but rather works through distributors or retailers, PRM helps these channels sell to customers directly. It provides a company and its selling partners with the ability to trade information and distribute leads and data about customers, integrating lead generation, pricing, promotions, order configurations, and availability. It also provides a firm with tools to assess its partners' performances so it can make sure its best partners receive the support they need to close more business.

ERM software deals with employee issues that are closely related to CRM, such as setting objectives, employee performance management, performance-based compensation, and employee training. Major CRM application software vendors include Oracle-owned Siebel Systems and PeopleSoft, SAP, Salesforce.com, and Microsoft Dynamics CRM.

Customer relationship management systems typically provide software and online tools for sales, customer service, and marketing. We briefly describe some of these capabilities.

Sales Force Automation (SFA)

Sales force automation modules in CRM systems help sales staff increase their productivity by focusing sales efforts on the most profitable customers, those who are good candidates for sales and services. CRM systems provide sales prospect and contact information, product information, product configuration capabilities, and sales quote generation capabilities. Such software can assemble information about a particular customer's past purchases to help the salesperson make personalized recommendations. CRM software enables sales, marketing, and delivery departments to easily share customer and prospect information. It increases each salesperson's efficiency in reducing the cost per sale as well as the cost of acquiring new customers and retaining old ones. CRM software also has capabilities for sales forecasting, territory management, and team selling.

Customer Service

Customer service modules in CRM systems provide information and tools to increase the efficiency of call centers, help desks, and customer support staff. They have capabilities for assigning and managing customer service requests.

One such capability is an appointment or advice telephone line: When a customer calls a standard phone number, the system routes the call to the correct service person, who inputs information about that customer into the system only once. Once the customer's data are in the system, any service representative can handle the customer relationship. Improved access to consistent and accurate customer information help call centers handle more calls per day and decrease the duration of each call. Thus, call centers and customer service groups achieve greater productivity, reduced transaction time, and higher quality of service at lower cost. The customer is happier because he or she spends less time on the phone restating his or her problem to customer service representatives.

CRM systems may also include Web-based self-service capabilities: The company Web site can be set up to provide inquiring customers personalized support information as well as the option to contact customer service staff by phone for additional assistance.

Marketing

CRM systems support direct-marketing campaigns by providing capabilities for capturing prospect and customer data, for providing product and service information, for qualifying leads for targeted marketing, and for scheduling and tracking direct-marketing mailings or e-mail (see Figure 9-7). Marketing modules also include tools for analyzing marketing and customer data, identifying profitable and unprofitable customers, designing products and services to satisfy specific customer needs and interests, and identifying opportunities for cross-selling.

Cross-selling is the marketing of complementary products to customers. (For example, in financial services, a customer with a checking account might be sold a money market account or a home improvement loan.) CRM tools also help firms manage and execute marketing campaigns at all stages, from planning to determining the rate of success for each campaign.

Figure 9-8 illustrates the most important capabilities for sales, service, and marketing processes that would be found in major CRM software products. Like enterprise software, this software is business-process driven, incorporating hundreds of business processes thought to represent best practices in each of these areas. To achieve maximum benefit, companies need to revise and model their business processes to conform to the best-practice business processes in the CRM software.

Figure 9-9 illustrates how a best practice for increasing customer loyalty through customer service might be modeled by CRM software. Directly

FIGURE 9-7   HOW CRM SYSTEMS SUPPORT MARKETING

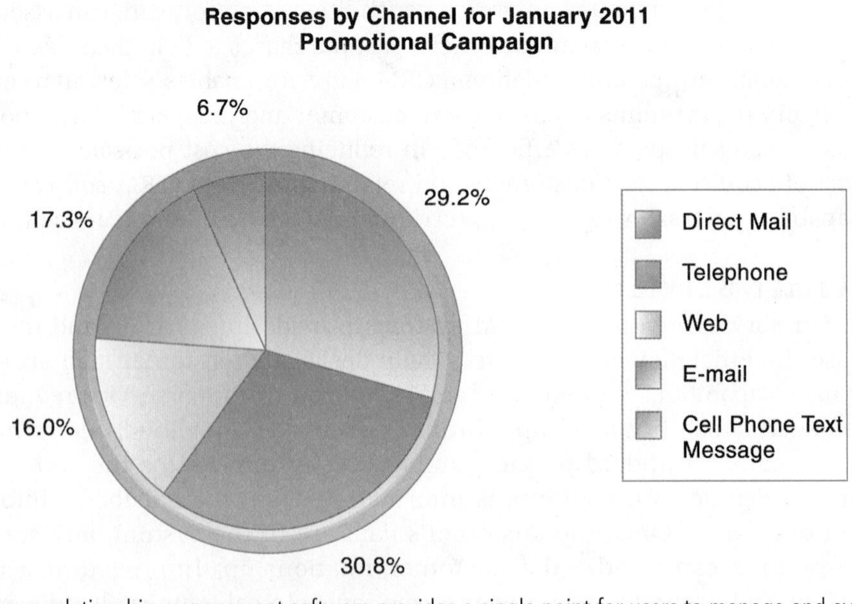

Customer relationship management software provides a single point for users to manage and evaluate marketing campaigns across multiple channels, including e-mail, direct mail, telephone, the Web, and wireless messages.

FIGURE 9-8   CRM SOFTWARE CAPABILITIES

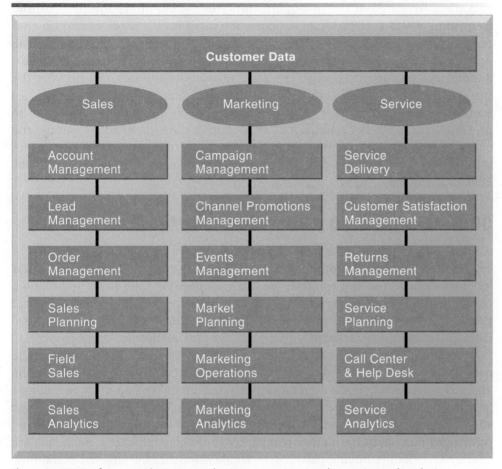

The major CRM software products support business processes in sales, service, and marketing, integrating customer information from many different sources. Included are support for both the operational and analytical aspects of CRM.

FIGURE 9-9   CUSTOMER LOYALTY MANAGEMENT PROCESS MAP

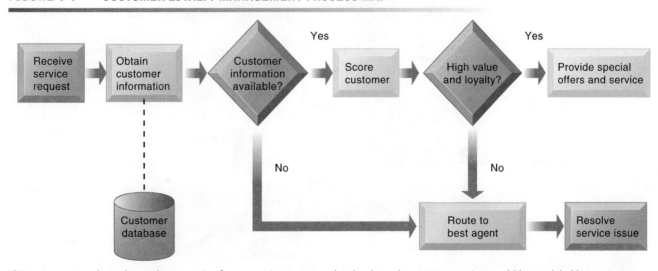

This process map shows how a best practice for promoting customer loyalty through customer service would be modeled by customer relationship management software. The CRM software helps firms identify high-value customers for preferential treatment.

servicing customers provides firms with opportunities to increase customer retention by singling out profitable long-term customers for preferential treatment. CRM software can assign each customer a score based on that person's value and loyalty to the company and provide that information to help call centers route each customer's service request to agents who can best handle that customer's needs. The system would automatically provide the service agent with a detailed profile of that customer that includes his or her score for value and loyalty. The service agent would use this information to present special offers or additional service to the customer to encourage the customer to keep transacting business with the company. You will find more information on other best-practice business processes in CRM systems in our Learning Tracks.

OPERATIONAL AND ANALYTICAL CRM

All of the applications we have just described support either the operational or analytical aspects of customer relationship management. **Operational CRM** includes customer-facing applications, such as tools for sales force automation, call center and customer service support, and marketing automation. **Analytical CRM** includes applications that analyze customer data generated by operational CRM applications to provide information for improving business performance.

Analytical CRM applications are based on data warehouses that consolidate the data from operational CRM systems and customer touch points for use with online analytical processing (OLAP), data mining, and other data analysis techniques (see Chapter 6). Customer data collected by the organization might be combined with data from other sources, such as customer lists for direct-marketing campaigns purchased from other companies or demographic data. Such data are analyzed to identify buying patterns, to create segments for targeted marketing, and to pinpoint profitable and unprofitable customers (see Figure 9-10).

FIGURE 9-10   ANALYTICAL CRM DATA WAREHOUSE

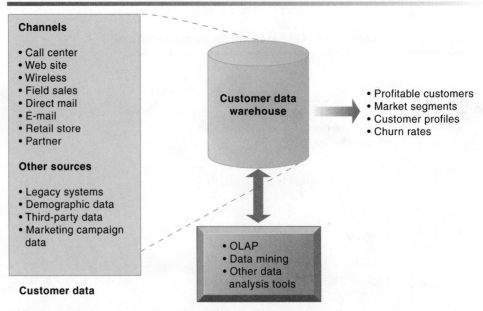

Analytical CRM uses a customer data warehouse and tools to analyze customer data collected from the firm's customer touch points and from other sources.

Another important output of analytical CRM is the customer's lifetime value to the firm. **Customer lifetime value (CLTV)** is based on the relationship between the revenue produced by a specific customer, the expenses incurred in acquiring and servicing that customer, and the expected life of the relationship between the customer and the company.

BUSINESS VALUE OF CUSTOMER RELATIONSHIP MANAGEMENT SYSTEMS

Companies with effective customer relationship management systems realize many benefits, including increased customer satisfaction, reduced direct-marketing costs, more effective marketing, and lower costs for customer acquisition and retention. Information from CRM systems increases sales revenue by identifying the most profitable customers and segments for focused marketing and cross-selling.

Customer churn is reduced as sales, service, and marketing better respond to customer needs. The **churn rate** measures the number of customers who stop using or purchasing products or services from a company. It is an important indicator of the growth or decline of a firm's customer base.

9.4 ENTERPRISE APPLICATIONS: NEW OPPORTUNITIES AND CHALLENGES

Many firms have implemented enterprise systems and systems for supply chain management and customer relationship because they are such powerful instruments for achieving operational excellence and enhancing decision making. But precisely because they are so powerful in changing the way the organization works, they are challenging to implement. Let's briefly examine some of these challenges, as well as new ways of obtaining value from these systems.

ENTERPRISE APPLICATION CHALLENGES

Promises of dramatic reductions in inventory costs, order-to-delivery time, as well as more efficient customer response and higher product and customer profitability make enterprise systems and systems for supply chain management and customer relationship management very alluring. But to obtain this value, you must clearly understand how your business has to change to use these systems effectively.

Enterprise applications involve complex pieces of software that are very expensive to purchase and implement. It might take a large Fortune 500 company several years to complete a large-scale implementation of an enterprise system or a system for SCM or CRM. The total cost for an average large system implementation based on SAP or Oracle software, including software, database tools, consulting fees, personnel costs, training, and perhaps hardware costs, runs over $12 million. The implementation cost of a enterprise system for a small or mid-sized company based on software from a "Tier II" vendor such as Epicor or Lawson averages $3.5 million. (Wailgum, 2009).

Enterprise applications require not only deep-seated technological changes but also fundamental changes in the way the business operates. Companies must make sweeping changes to their business processes to work with the software. Employees must accept new job functions and responsibilities. They must learn how to perform a new set of work activities and understand how the information they enter into the system can affect other parts of the company. This requires new organizational learning.

Supply chain management systems require multiple organizations to share information and business processes. Each participant in the system may have to change some of its processes and the way it uses information to create a system that best serves the supply chain as a whole.

Some firms experienced enormous operating problems and losses when they first implemented enterprise applications because they didn't understand how much organizational change was required. For example, Kmart had trouble getting products to store shelves when it first implemented i2 Technologies supply chain management software in July 2000. The i2 software did not work well with Kmart's promotion-driven business model, which created sharp spikes in drops in demand for products. Overstock.com's order tracking system went down for a full week in October 2005 when the company replaced a home-grown system with an Oracle enterprise system. The company rushed to implement the software, and did not properly synchronize the Oracle software's process for recording customer refunds with its accounts receivable system. These problems contributed to a third-quarter loss of $14.5 million that year.

Enterprise applications also introduce "switching costs." Once you adopt an enterprise application from a single vendor, such as SAP, Oracle, or others, it is very costly to switch vendors, and your firm becomes dependent on the vendor to upgrade its product and maintain your installation.

Enterprise applications are based on organization-wide definitions of data. You'll need to understand exactly how your business uses its data and how the data would be organized in a customer relationship management, supply chain management, or enterprise system. CRM systems typically require some data cleansing work.

Enterprise software vendors are addressing these problems by offering pared-down versions of their software and "fast-start" programs for small and medium-sized businesses and best-practice guidelines for larger companies. Our Interactive Session on Technology describes how on-demand and cloud-based tools deal with this problem as well.

Companies adopting enterprise applications can also save time and money by keeping customizations to the minimum. For example, Kennametal, a $2 billion metal-cutting tools company in Pennsylvania, had spent $10 million over 13 years maintaining an ERP system with over 6,400 customizations. The company is now replacing it with a "plain vanilla," non-customized-version of SAP enterprise software and changing its business processes to conform to the software (Johnson, 2010).

NEXT-GENERATION ENTERPRISE APPLICATIONS

Today, enterprise application vendors are delivering more value by becoming more flexible, Web-enabled, and capable of integration with other systems. Standalone enterprise systems, customer relationship systems, and supply chain management systems are becoming a thing of the past.

The major enterprise software vendors have created what they call *enterprise solutions*, *enterprise suites*, or *e-business suites* to make their customer relationship management, supply chain management, and enterprise systems work closely with each other, and link to systems of customers and suppliers. SAP Business Suite, Oracle's e-Business Suite, and Microsoft's Dynamics suite (aimed at mid-sized companies) are examples, and they now utilize Web services and service-oriented architecture (SOA, see Chapter 5).

SAP's next-generation enterprise applications are based on its enterprise service-oriented architecture. It incorporates service-oriented architecture (SOA) standards and uses its NetWeaver tool as an integration platform linking SAP's own applications and Web services developed by independent software vendors. The goal is to make enterprise applications easier to implement and manage.

For example, the current version of SAP enterprise software combines key applications in finance, logistics and procurement, and human resources administration into a core ERP component. Businesses then extend these applications by linking to function-specific Web services such as employee recruiting or collections management provided by SAP and other vendors. SAP provides over 500 Web services through its Web site.

Oracle also has included SOA and business process management capabilities into its Fusion middleware products. Businesses can use Oracle tools to customize Oracle's applications without breaking the entire application.

Next-generation enterprise applications also include open source and on-demand solutions. Compared to commercial enterprise application software, open source products such as Compiere, Apache Open for Business (OFBiz), and Openbravo are not as mature, nor do they include as much support. However, companies such as small manufacturers are choosing this option because there are no software licensing fees and fees are based on usage. (Support and customization for open source products cost extra.)

SAP now offers an on-demand enterprise software solution called Business ByDesign for small and medium-sized businesses in select countries. For large businesses, SAP's on-site software is the only version available. SAP is, however, hosting function-specific applications (such as e-sourcing and expense management) available by subscription that integrate with customers' on-site SAP Business Suite systems.

The most explosive growth in software as a service (SaaS) offerings has been for customer relationship management. Salesforce.com has been the leader in hosted CRM solutions, but Oracle and SAP have also developed SaaS capabilities. SaaS and cloud-based versions of enterprise systems are starting to be offered by vendors such as NetSuite and Plex Online. Compiere sells both cloud and on-premise versions of its ERP systems. Use of cloud-based enterprise applications is starting to take off, as discussed in the Interactive Session on Technology.

The major enterprise application vendors also offer portions of their products that work on mobile handhelds. You can find out more about this topic in our Learning Track on Wireless Applications for Customer Relationship Management, Supply Chain Management, and Healthcare.

Salesforce.com and Oracle have added Web 2.0 capabilities that enable organizations to identify new ideas more rapidly, improve team productivity, and deepen interactions with customers. For example, Salesforce Ideas enables subscribers to harness the "wisdom of crowds" by allowing their customers to submit and discuss new ideas. Dell Computer deployed this technology as Dell IdeaStorm (dellideastorm.com) to encourage its customers to suggest and vote on new concepts and feature changes in Dell products.

INTERACTIVE SESSION: TECHNOLOGY

ENTERPRISE APPLICATIONS MOVE TO THE CLOUD

You've already read about Salesforce.com in this book. It's the most successful enterprise-scale software as a service (SaaS). Until recently, there were few other SaaS enterprise software applications available on the Internet. Today, that's changed, as a growing number of cloud-based enterprise resource planning (ERP) and customer relationship management (CRM) application providers enter this marketspace. While traditional enterprise software vendors like Oracle are using their well-established position to grab a share of the cloud-based application market, newcomers like RightNow, Compiere, and SugarCRM have found success using some different tactics.

Most companies interested in cloud computing are small to midsize and lack the know-how or financial resources to successfully build and maintain ERP and CRM applications in-house. Others are simply looking to cut costs by moving their applications to the cloud. According to the International Data Corporation (IDC), about 3.2 percent of U.S. small businesses, or about 230,000 businesses, use cloud services. Small-business spending on cloud services increased by 36.2 percent in 2010 to $2.4 billion.

Even larger companies have made the switch to the cloud. For example, camera manufacturer Nikon decided to go with a cloud-based solution as it attempted to merge customer data from 25 disparate sources and applications into a single system. Company officials were hoping to eliminate maintenance and administrative costs, but not at the expense of a storage system that met their requirements, was never out of service, and worked perfectly.

Nikon found its solution with RightNow, a cloud-based CRM provider located in Bozeman, Montana. The company was founded in 1997 and has attracted firms intrigued by its customizable applications, impeccable customer service, and robust infrastructure. Prices start at $110 per user per month and the average deployment time is 45 days.

Nikon had been using several different systems to perform business functions, and was struggling to merge customer data located in a variety of legacy systems. While looking for vendors to help implement a Web-based FAQ system to answer customer questions and provide support on the basis of these data, the company came across RightNow. Nikon found that not only did RightNow have the capabil-

ity to implement that system, but it also had an array of other useful services. When Nikon discovered that it could combine outbound e-mail, contact management, and customer records into a single system in RightNow's cloud, it made the move, expecting to receive a solid return on the investment.

What Nikon got was far more than expected: an astonishing 3,200 percent return on investment (ROI), equivalent to a savings of $14 million after three years! The FAQ system reduced the number of incoming calls to Nikon's customer service staff. More customers found the information they needed on the Web, call response times dropped by 50 percent, and incoming e-mail dropped by 70 percent. While Nikon still hosts its SAP ERP system internally due to its complexity, Nikon switched its entire CRM system to RightNow.

Not all companies experience gains of that magnitude, and cloud computing does have drawbacks. Many companies are concerned about maintaining control of their data and security. Although cloud computing companies are prepared to handle these issues, availability assurances and service-level agreements are uncommon. Companies that manage their CRM apps with a cloud infrastructure have no guarantees that their data will be available at all times, or even that the provider will still exist in the future.

Many smaller companies have taken advantage of a new type of cloud computing known as open source cloud computing. Under this model, cloud vendors make the source code of their applications available to their customers and allow them to make any changes they want on their own. This differs from the traditional model, where cloud vendors offer applications which are customizable, but not at the source code level.

For example, Jerry Skaare, president of O-So-Pure (OSP), a manufacturer of ultraviolet water purification systems, selected the Compiere Cloud Edition versions of ERP software hosted on the Amazon EC2 Cloud virtual environment. OSP had long outgrown its existing ERP system and was held back by inefficient, outdated processes in accounting, inventory, manufacturing, and e-commerce. Compiere ERP provides a complete end-to-end ERP solution that automates processes from accounting to purchasing, order fulfillment, manufacturing, and warehousing.

Compiere uses a model-driven platform that stores business logic in an applications dictionary rather than being hard-coded into software programs. Firms using Compiere are able to customize their applications by creating, modifying, or deleting business logic in the applications dictionary without extensive programming. In contrast to traditional ERP systems that encourage subscribers to modify their business processes to conform to the software, Compiere encourages its subscribers to customize its system to match their unique business needs.

The fact that the Compiere software is open source also makes it easier for users to modify. OSP was attracted to this feature, along with the robust functionality, scalability, and low cost, of the Compiere ERP Cloud Edition. Skaare said that he was comfortable that "the little idiosyncrasies of my company" could be handled by the software. Though Skaare is unlikely to make any changes himself, it's important for him to know that his staff has the option to tweak OSP's ERP applications. Open source cloud computing provides companies that flexibility.

Not to be outdone, established CRM companies like Oracle have moved into SaaS. Pricing starts at $70 per month per user. Oracle may have an edge because its CRM system has so many capabilities and includes embedded tools for forecasting and analytics, including interactive dashboards. Subscribers are able to use these tools to answer questions such as "How efficient is your sales effort?" or "How much are your customers spending?"

Bryant & Stratton College, a pioneer in career education, used Oracle CRM On Demand to create more successful marketing campaigns. Bryant & Stratton analyzed past campaigns for tech-savvy recent high school graduates, as well as older, non-traditional students returning to school later in life. Oracle CRM On Demand tracked advertising to prospective students and determined accurate costs for each lead, admissions application, and registered attending student. This information helped the school determine the true value of each type of marketing program.

Sources: Marta Bright, "Know Who. Know How." *Oracle Magazine*, January/February 2010; Brad Stone, "Companies Slowly Join Cloud-Computing," *The New York Times*, April 28, 2010; and Esther Shein, "Open-source CRM and ERP: New Kids on the Cloud," *Computerworld*, October 30, 2009.

CASE STUDY QUESTIONS

1. What types of companies are most likely to adopt cloud-based ERP and CRM software services? Why? What companies might not be well-suited for this type of software?

2. What are the advantages and disadvantages of using cloud-based enterprise applications?

3. What management, organization, and technology issues should be addressed in deciding whether to use a conventional ERP or CRM system versus a cloud-based version?

MIS IN ACTION

Visit the Web site of RightNow, Compiere, or another competing company offering a cloud-based version of ERP or CRM. Then answer the following questions:

1. What kinds of open source offerings does the company have, if any? Describe some of the features.

2. Toward what types of companies is the company marketing its services?

3. What other services does the company offer?

Enterprise application vendors have also beefed up their business intelligence features to help managers obtain more meaningful information from the massive amounts of data generated by these systems. Rather than requiring users to leave an application and launch separate reporting and analytics tools, the vendors are starting to embed analytics within the context of the application itself. They are also offering complementary analytics products, such as SAP Business Objects and Oracle Business Intelligence Enterprise Edition. We discuss business intelligence analytics in greater detail in Chapter 12.

Service Platforms

Another way of extending enterprise applications is to use them to create service platforms for new or improved business processes that integrate information from multiple functional areas. These enterprise-wide service platforms provide a greater degree of cross-functional integration than the traditional enterprise applications. A **service platform** integrates multiple applications from multiple business functions, business units, or business partners to deliver a seamless experience for the customer, employee, manager, or business partner.

For instance, the order-to-cash process involves receiving an order and seeing it all the way through obtaining payment for the order. This process begins with lead generation, marketing campaigns, and order entry, which are typically supported by CRM systems. Once the order is received, manufacturing is scheduled and parts availability is verified—processes that are usually supported by enterprise software. The order then is handled by processes for distribution planning, warehousing, order fulfillment, and shipping, which are usually supported by supply chain management systems. Finally, the order is billed to the customer, which is handled by either enterprise financial applications or accounts receivable. If the purchase at some point required customer service, customer relationship management systems would again be invoked.

A service such as order-to-cash requires data from enterprise applications and financial systems to be further integrated into an enterprise-wide composite process. To accomplish this, firms need software tools that use existing applications as building blocks for new cross-enterprise processes (see Figure 9-11). Enterprise application vendors provide middleware and tools that use XML and Web services for integrating enterprise applications with older legacy applications and systems from other vendors.

FIGURE 9-11   ORDER-TO-CASH SERVICE

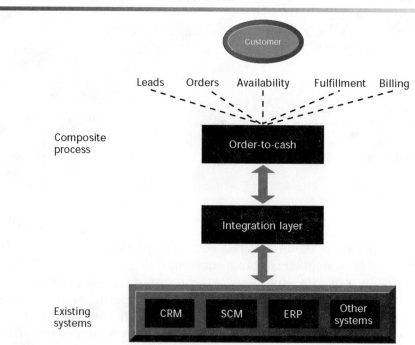

Order-to-cash is a composite process that integrates data from individual enterprise systems and legacy financial applications. The process must be modeled and translated into a software system using application integration tools.

Increasingly, these new services will be delivered through portals. Portal software can integrate information from enterprise applications and disparate in-house legacy systems, presenting it to users through a Web interface so that the information appears to be coming from a single source. For example, Valero Energy, North America's largest refiner, used SAP NetWeaver Portal to create a service for wholesale clients to view their account information all at once. SAP NetWeaver Portal provides an interface to clients' invoice, price, electronic funds, and credit card transaction data stored in SAP's customer relationship management system data warehouse as well as in non-SAP systems.

9.5   HANDS-ON MIS PROJECTS

The projects in this section give you hands-on experience analyzing business process integration, suggesting supply chain management and customer relationship management applications, using database software to manage customer service requests, and evaluating supply chain management business services.

Management Decision Problems

1. Toronto-based Mercedes-Benz Canada, with a network of 55 dealers, did not know enough about its customers. Dealers provided customer data to the company on an ad hoc basis. Mercedes did not force dealers to report this information, and its process for tracking dealers that failed to report was cumbersome. There was no real incentive for dealers to share information with the company. How could CRM and partner relationship management (PRM) systems help solve this problem?

2. Office Depot sells a wide range of office products and services in the United States and internationally, including general office supplies, computer supplies, business machines (and related supplies), and office furniture. The company tries to offer a wider range of office supplies at lower cost than other retailers by using just-in-time replenishment and tight inventory control systems. It uses information from a demand forecasting system and point-of-sale data to replenish its inventory in its 1,600 retail stores. Explain how these systems help Office Depot minimize costs and any other benefits they provide. Identify and describe other supply chain management applications that would be especially helpful to Office Depot.

Improving Decision Making: Using Database Software to Manage Customer Service Requests

Software skills: Database design; querying and reporting
Business skills: Customer service analysis

In this exercise, you'll use database software to develop an application that tracks customer service requests and analyzes customer data to identify customers meriting priority treatment.

Prime Service is a large service company that provides maintenance and repair services for close to 1,200 commercial businesses in New York, New Jersey, and Connecticut. Its customers include businesses of all sizes. Customers with service needs call into its customer service department with requests for repairing heating ducts, broken windows, leaky roofs, broken water pipes, and other problems. The company assigns each request a number and

		ACCT_ID ▾	NAME ▾	ADDR ▾	CITY ▾	STAT ▾	ZIP ▾	DOLLAR_SIZI ▾	CONTACT_FI ▾	CONTACT_L ▾	PHONE ▾	Ad(
		1	Able Association	123 Axion Stre	Albertown	NY	11444-4444	$50,000	Alison	Ableson	(209) 111-1111	
		2	Briggs Bakery	123 Boggs Stre	Brimstone	CT	11200-1234	$94,000	Barry	Berryman	(210) 111-1212	
		3	Constant Carriers	31 Carmine Le	Carver	NJ	20111-1212	$200,000	Carl	Compress	(202) 123-1222	
		4	Darning Drapers	1234 Dante Av(	Driblle	NY	12345-6849	$60,000	Delilah	Dilman	(209) 123-4321	
		5	Eagle Engineers	Eagle Park	Edmonton	CT	11222-2313	$45,000	Eddie	Exeter	(210) 212-2233	
	*	(New)						$0				

Record: I◀ ◀ 1 of 5 ▶ ▶I ▶⧉ 🚫 No Filter Search

writes down the service request number, identification number of the customer account, the date of the request, the type of equipment requiring repair, and a brief description of the problem. The service requests are handled on a first-come-first-served basis. After the service work has been completed, Prime calculates the cost of the work, enters the price on the service request form, and bills the client.

Management is not happy with this arrangement because the most important and profitable clients—those with accounts of more than $70,000—are treated no differently from its clients with small accounts. It would like to find a way to provide its best customers with better service. Management would also like to know which types of service problems occur most frequently so that it can make sure it has adequate resources to address them.

Prime Service has a small database with client account information, which can be found in MyMISLab. A sample is shown above, but the Web site may have a more recent version of this database for this exercise. The database table includes fields for the account ID, company (account) name, street address, city, state, ZIP code, account size (in dollars), contact last name, contact first name, and contact telephone number. The contact is the name of the person in each company who is responsible for contacting Prime about maintenance and repair work.

Use your database software to design a solution that would enable Prime's customer service representatives to identify the most important customers so that they could receive priority service. Your solution will require more than one table. Populate your database with at least 15 service requests. Create several reports that would be of interest to management, such as a list of the highest—and lowest—priority accounts or a report showing the most frequently occurring service problems. Create a report listing service calls that customer service representatives should respond to first on a specific date.

Achieving Operational Excellence: Evaluating Supply Chain Management Services

Software skills: Web browser and presentation software
Business skills: Evaluating supply chain management services

Trucking companies no longer merely carry goods from one place to another. Some also provide supply chain management services to their customers and help them manage their information. In this project, you'll use the Web to research and evaluate two of these business services.

Investigate the Web sites of two companies, UPS Logistics and Schneider Logistics, to see how these companies' services can be used for supply chain management. Then respond to the following questions:

- What supply chain processes can each of these companies support for their clients?

- How can customers use the Web sites of each company to help them with supply chain management?
- Compare the supply chain management services provided by these companies. Which company would you select to help your firm manage its supply chain? Why?

LEARNING TRACK MODULES

The following Learning Tracks provide content relevant to topics covered in this chapter:

1. SAP Business Process Map
2. Business Processes in Supply Chain Management and Supply Chain Metrics
3. Best-Practice Business Processes in CRM Software

Review Summary

1. *How do enterprise systems help businesses achieve operational excellence?*

    Enterprise software is based on a suite of integrated software modules and a common central database. The database collects data from and feeds the data into numerous applications that can support nearly all of an organization's internal business activities. When new information is entered by one process, the information is made available immediately to other business processes.

    Enterprise systems support organizational centralization by enforcing uniform data standards and business processes throughout the company and a single unified technology platform. The firmwide data generated by enterprise systems helps managers evaluate organizational performance.

2. *How do supply chain management systems coordinate planning, production, and logistics with suppliers?*

    Supply chain management systems automate the flow of information among members of the supply chain so they can use it to make better decisions about when and how much to purchase, produce, or ship. More accurate information from supply chain management systems reduces uncertainty and the impact of the bullwhip effect.

    Supply chain management software includes software for supply chain planning and for supply chain execution. Internet technology facilitates the management of global supply chains by providing the connectivity for organizations in different countries to share supply chain information. Improved communication among supply chain members also facilitates efficient customer response and movement toward a demand-driven model.

3. *How do customer relationship management systems help firms achieve customer intimacy?*

    Customer relationship management (CRM) systems integrate and automate customer-facing processes in sales, marketing, and customer service, providing an enterprise-wide view of customers. Companies can use this knowledge when they interact with customers to provide them with better service or to sell new products and services. These systems also identify profitable or nonprofitable customers or opportunities to reduce the churn rate.

The major customer relationship management software packages provide capabilities for both operational CRM and analytical CRM. They often include modules for managing relationships with selling partners (partner relationship management) and for employee relationship management.

4. *What are the challenges posed by enterprise applications?*

Enterprise applications are difficult to implement. They require extensive organizational change, large new software investments, and careful assessment of how these systems will enhance organizational performance. Enterprise applications cannot provide value if they are implemented atop flawed processes or if firms do not know how to use these systems to measure performance improvements. Employees require training to prepare for new procedures and roles. Attention to data management is essential.

5. *How are enterprise applications used in platforms for new cross-functional services?*

Service platforms integrate data and processes from the various enterprise applications (customer relationship management, supply chain management, and enterprise systems), as well as from disparate legacy applications to create new composite business processes. Web services tie various systems together. The new services are delivered through enterprise portals, which can integrate disparate applications so that information appears to be coming from a single source. Open source, mobile, and cloud versioins of some of these products are becoming available.

Key Terms

Analytical CRM, 354	*Operational CRM, 354*
Bullwhip effect, 342	*Partner relationship management (PRM), 351*
Churn rate, 355	*Pull-based model, 347*
Cross-selling, 352	*Push-based model, 347*
Customer lifetime value (CLTV), 355	*Service platform, 360*
Demand planning, 344	*Supply chain, 340*
Employee relationship management (ERM), 351	*Supply chain execution systems, 344*
Enterprise software, 338	*Supply chain planning systems, 344*
Just-in-time strategy, 342	*Touch point, 350*

Review Questions

1. How do enterprise systems help businesses achieve operational excellence?

    - Define an enterprise system and explain how enterprise software works.

    - Describe how enterprise systems provide value for a business.

2. How do supply chain management systems coordinate planning, production, and logistics with suppliers?

    - Define a supply chain and identify each of its components.

    - Explain how supply chain management systems help reduce the bullwhip effect and how they provide value for a business.

    - Define and compare supply chain planning systems and supply chain execution systems.

    - Describe the challenges of global supply chains and how Internet technology can help companies manage them better.

- Distinguish between a push-based and pull-based model of supply chain management and explain how contemporary supply chain management systems facilitate a pull-based model.

3. How do customer relationship management systems help firms achieve customer intimacy?
   - Define customer relationship management and explain why customer relationships are so important today.
   - Describe how partner relationship management (PRM) and employee relationship management (ERM) are related to customer relationship management (CRM).
   - Describe the tools and capabilities of customer relationship management software for sales, marketing, and customer service.

- Distinguish between operational and analytical CRM.

4. What are the challenges posed by enterprise applications?
   - List and describe the challenges posed by enterprise applications.
   - Explain how these challenges can be addressed.

5. How are enterprise applications used in platforms for new cross-functional services?
   - Define a service platform and describe the tools for integrating data from enterprise applications.
   - How are enterprise applications taking advantage of cloud computing, wireless technology, Web 2.0, and open source technology?

Discussion Questions

1. Supply chain management is less about managing the physical movement of goods and more about managing information. Discuss the implications of this statement.

2. If a company wants to implement an enterprise application, it had better do its homework. Discuss the implications of this statement.

3. Which enterprise application should a business install first: ERP, SCM, or CRM? Explain your answer.

Video Cases

Video Cases and Instructional Videos illustrating some of the concepts in this chapter are available. Contact your instructor to access these videos.

Collaboration and Teamwork: Analyzing Enterprise Application Vendors

With a group of three or four students, use the Web to research and evaluate the products of two vendors of enterprise application software. You could compare, for example, the SAP and Oracle enterprise systems, the supply chain management systems from i2 and SAP, or the customer relationship management systems of Oracle's Siebel CRM and Salesforce.com. Use what you have learned from these companies' Web sites to compare the software packages you have selected in terms of business functions supported, technology platforms, cost, and ease of use. Which vendor would you select? Why? Would you select the same vendor for a small business as well as a large one? If possible, use Google Sites to post links to Web pages, team communication announcements, and work assignments; to brainstorm; and to work collaboratively on project documents. Try to use Google Docs to develop a presentation of your findings for the class.

Border States Industries Fuels Rapid Growth with ERP
CASE STUDY

Border States Industries Inc., also known as Border States Electric (BSE), is a wholesale distributor for the construction, industrial, utility, and data communications markets. The company is headquartered in Fargo, North Dakota, and has 57 sales offices in states along the U.S. borders with Canada and Mexico as well as in South Dakota, Wisconsin, Iowa, and Missouri. BSE has 1,400 employees and is wholly employee-owned through its employee stock ownership plan. For the fiscal year ending March 31, 2008, BSE earned revenues of over US $880 million.

BSE's goal is to provide customers with what they need whenever they need it, including providing custom services beyond delivery of products. Thus, the company is not only a wholesale distributor but also a provider of supply chain solutions, with extensive service operations such as logistics, job-site trailers, and kitting (packaging individually separate but related items together as one unit). BSE has distribution agreements with more than 9,000 product vendors.

BSE had relied on its own legacy ERP system called Rigel since 1988 to support its core business processes. However, Rigel had been designed exclusively for electrical wholesalers, and by the mid-1990s, the system could not support BSE's new lines of business and extensive growth.

At that point, BSE's management decided to implement a new ERP system and selected the enterprise software from SAP AG. The ERP solution included SAP's modules for sales and distribution, materials management, financials and controlling, and human resources.

BSE initially budgeted $6 million for the new system, with a start date of November 1, 1998. Senior management worked with IBM and SAP consulting to implement the system. Although close involvement of management was one key ingredient in the systems' success, day-to-day operations suffered while managers were working on the project.

BSE also decided to customize the system extensively. It wrote its own software to enable the ERP system to interface automatically with systems from other vendors, including Taxware Systems, Inc., Innovis Inc., and TOPCALL International GmbH. The Taxware system enabled BSE to comply with the sales tax requirements of all the states and munici-

palities where it conducts business. The Innovis system supported electronic data interchange (EDI) so that BSE could electronically exchange purchase and payment transactions with its suppliers. The TOPCALL system enabled BSE to fax customers and vendors directly from the SAP system.

At the time of this implementation, BSE had no experience with SAP software, and few consultants familiar with the version of the SAP software that BSE was using. Instead of adopting the best-practice business processes embedded in the SAP software, BSE hired consultants to further customize the SAP software to make its new SAP system look like its old Rigel system in certain areas. For example, it tried to make customer invoices resemble the invoices produced by the old Rigel system.

Implementing these changes required so much customization of the SAP software that BSE had to delay the launch date for the new ERP system until February 1, 1999. By that time, continued customization and tuning raised total implementation costs to $9 million (an increase of 50 percent).

Converting and cleansing data from BSE's legacy system took far longer than management had anticipated. The first group of "expert users" were trained too early in the project and had to be retrained when the new system finally went live. BSE never fully tested the system as it would be used in a working production environment before the system actually went live.

For the next five years, BSE continued to use its SAP ERP system successfully as it acquired several small companies and expanded its branch office infrastructure to 24 states. As the business grew further, profits and inventory turns increased. However, the Internet brought about the need for additional changes, as customers sought to transact business with BSE through an e-commerce storefront. BSE automated online credit card processing and special pricing agreements (SPAs) with designated customers. Unfortunately, the existing SAP software did not support these changes, so the company had to process thousands of SPAs manually.

To process a credit card transaction in a branch office, BSE employees had to leave their desks, walk over to a dedicated credit card processing system in the back office, manually enter the credit card numbers, wait for transaction approval, and then return to their workstations to continue processing sales transactions.

In 2004, BSE began upgrading its ERP system to a more recent version of the SAP software. The software included new support for bills of material and kitting, which were not available in the old system. This functionality enabled BSE to provide better support to utility customers because it could prepare kits that could be delivered directly to a site.

This time the company kept customization to a minimum and used the SAP best practices for wholesale distribution embedded in the software. It also replaced TOPCALL with software from Esker for faxing and e-mailing outbound invoices, order acknowledgments, and purchase orders and added capabilities from Vistex Inc. to automate SPA rebate claims processing. BSE processes over 360,000 SPA claims each year, and the Vistex software enabled BSE to reduce rebate fulfillment time to 72 hours and transaction processing time by 63 percent. In the past, it took 15 to 30 days for BSE to receive rebates from vendors.

BSE budgeted $1.6 million and 4.5 months for implementation, which management believed was sufficient for a project of this magnitude. This time there were no problems. The new system went live on its target date and cost only $1.4 million to implement—14 percent below budget.

In late 2006, BSE acquired a large company that was anticipated to increase sales volume by 20 percent each year. This acquisition added 19 new branches to BSE. These new branches were able to run BSE's SAP software within a day after the acquisition had been completed. BSE now tracks 1.5 million unique items with the software.

Since BSE first deployed SAP software in 1998, sales have increased 300 percent, profits have climbed more than 500 percent, 60 percent of accounts payable transactions take place electronically using EDI, and SPA processing has been reduced by 63 percent. The company turns over its inventory more than four times per year. Instead of waiting 15 to 20 days for monthly financial statements, monthly and year-to-date financial results are available within a day after closing the books. Manual work for handling incoming mail, preparing bank deposits, and taking checks physically to the bank has been significantly reduced. Over 60 percent of vendor invoices arrive electronically, which has reduced staff size in accounts payable and the number of transaction errors. Transaction costs are lower.

The number of full-time BSE employees did increase in the information systems area to support the SAP software. BSE had initially expected to have

3 IT staff supporting the system, but needed 8 people when the first ERP implementation went live in 1999 and 11 by 2006 to support additional SAP software and the new acquisition. BSE's information technology (IT) costs rose by approximately $3 million per year after the first SAP implementation. However, sales expanded during the same period, so the increased overhead for the system produced a cost increase of only .5 percent of total sales.

BSE management has pointed out that much of the work that was automated by the ERP systems has been in the accounting department and involved activities that were purely transactional. This has freed up resources for adding more employees who work directly with customers trying to reduce costs and increase sales.

In the past, BSE had maintained much of its data outside its major corporate systems using PC-based Microsoft Access database and Excel spreadsheet software. Management lacked a single company-wide version of corporate data because the data were fragmented into so many different systems. Now the company is standardized on one common platform and the information is always current and available to management. Management can obtain a picture of how the entire business is performing at any moment in time. Since the SAP system makes all of BSE's planning and budgeting data available online, management is able to make better and quicker decisions.

In 2006, Gartner Group Consultants performed an independent evaluation of BSE's ERP implementation. Gartner interviewed top executives and analyzed BSE data on the impact of the ERP system on BSE's business process costs, using costs as a percentage of sales as its final metric for assessing the financial impact of SAP software. Cost categories analyzed included costs of goods sold, overhead and administration, warehousing costs, IT support, and delivery.

Gartner's analysis validated that the SAP software implementation cost from 1998 to 2001 did indeed total $9 million and that this investment was paid back by savings from the new ERP system within 2.5 years. Between 1998 and 2006, the SAP software implemented by BSE produced total savings of $30 million, approximately one-third of BSE's cumulative earnings during the same period. As a percentage of sales, warehouse costs went down 1 percent, delivery costs decreased by .5 percent, and total overhead costs declined by 1.5 percent. Gartner calculated the total return on investment (ROI) for the project between 1998 and 2006 was $3.3 million per year, or 37 percent of the original investment.

BSE is now focusing on providing more support for Internet sales, including online ordering, inventory, order status, and invoice review, all within a SAP software environment. The company implemented SAP NetWeaver Master Data Management to provide tools to manage and maintain catalog data and prepare the data for publication online and in traditional print media. The company is using SAP's Web Dynpro development environment to enable wireless warehouse and inventory management activities to interact with the SAP software. And it is using SAP NetWeaver Business Intelligence software to learn more about customers, their buying habits, and opportunities to cross-sell and upsell products.

Sources: Border States Industries, "Operating System-SAP Software," 2010; Jim Shepherd and Aurelie Cordier, "Wholesale Distributor Uses ERP Solution to Fuel Rapid Growth," AMR Research, 2009; SAP AG, "Border States Industries: SAP Software Empowers Wholesale Distributor," 2008; www.borderstateselectric.com, accessed July 7, 2009; and "Border States (BSE)," 2008 ASUG Impact Award.

CASE STUDY QUESTIONS

1. What problems was Border States Industries encountering as it expanded? What management, organization, and technology factors were responsible for these problems?

2. How easy was it to develop a solution using SAP ERP software? Explain your answer.

3. List and describe the benefits from the SAP software.

4. How much did the new system solution transform the business? Explain your answer.

5. How successful was this solution for BSE? Identify and describe the metrics used to measure the success of the solution.

6. If you had been in charge of SAP's ERP implementations, what would you have done differently?

Chapter 10

E-commerce: Digital Markets, Digital Goods

4FOOD: BURGERS GO SOCIAL

4 Food, a new organic burger restaurant in Manhattan, opened its doors with a promise of delicious food. But what's equally delicious are its plans to drive the business using social networking. This restaurant wants to be much more than a place to dine. It wants to be a vast social networking experience.

Inside the restaurant, located at the corner of Madison Avenue and 40th Street, a 240-square-foot monitor constantly streams Twitter tweets, restaurant information, and Foursquare check-ins. Foursquare is a Web and mobile application that allows registered users to connect with friends and update their location information. Points are awarded for "checking in" at selected restaurants, bars, and other sites. Customers see tweets and status updates and reply to them or add their own messages with their cell phones or other mobile devices using 4Food's free Wi-Fi wireless Internet connection.

This restaurant has multiple options for placing an order. You can give your order to a restaurant employee using an iPad, or you can place the order online yourself. Naturally, 4Food has its own Facebook page, which it uses for social marketing. Tagging its Facebook wall makes you eligible to win an iPad. 4Food offered $20 worth of food to whoever was the first to tweet a picture of himself or herself in front of the restaurant's "tag wall"—a wall in the front of the restaurant inviting people to write "tweets" using a Magic Marker. 4Food also uses social networks for hiring and to promote it's "De-Junk NYC" campaign to promote innovative ideas for improving the city.

But what makes 4Food really stand out is its use of crowdsourcing for both marketing and menu development. This restaurant has an online tool for customers to invent their own sandwiches and other dishes and to give their inventions clever names. Every time someone orders an item invented by another customer, the inventor receives a $.25 in-store credit. With 4Food's list of ingredients, millions of combinations are possible.

Some customers will no doubt use their extensive social networks to promote the burgers they invented. Those with hundreds of thousands of followers on social networks could conceivably earn free burgers for the rest of their lives if they constantly promote 4Food. All of these measures create very low- cost incentives for large numbers of customers to actively promote the restaurant. They also generate word-of-mouth "buzz" with minimal expenditure. All it takes is establishing a presence on social networks and rolling out promotions.

Will 4Food be successful? Competing with 20,000 other New York City restaurants won't be easy. But by using social networking technology to forge ties with customers and giving those customers a stake in the success of products, 4Food hopes to have the recipe for a successful business.

Sources: Mike Elgan, "New York Burger Joint Goes Social, Mobile," Computerworld, May 31, 2010 and www.4food.com, accessed October 22, 2010.

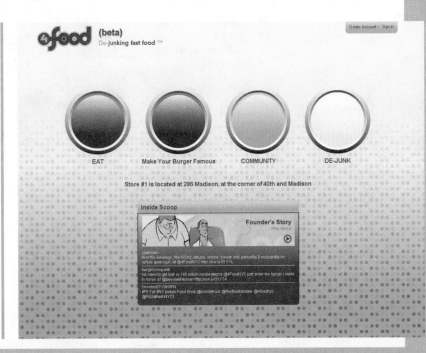

4Food exemplifies the new face of e-commerce. Selling physical goods on the Internet is still important, but much of the excitement and interest now centers around services and social experiences-connecting with friends and family through social networking; sharing photos, video, and music, and ideas; and using social networking to attract customers and design new products and services. 4Food 's business model relies on mobile technology and social networking tools to attract customers, take orders, promote its brand, and use customer feedback to improve its menu offerings.

The chapter-opening diagram calls attention to important points raised by this case and this chapter. The business challenge facing 4Food is that it needs a way to stand out amid 20,000 other restaurants in New York City. E-commerce and social networking technology introduced new opportunities for linking to customers and for distinguishing products and services. 4Food's management decided to base its business model around social technology, and make social networking part of the dining experience. 4Food uses social networking and mobile technology—including Twitter, Foursquare, and Facebook—to attract customers, to process reservations, to promote its brand image, and to solicit customer feedback for improving its menu offerings. By taking advantage of social networking tools, 4Food is able to differentiate itself from other restaurants and promote the business at a very low cost.

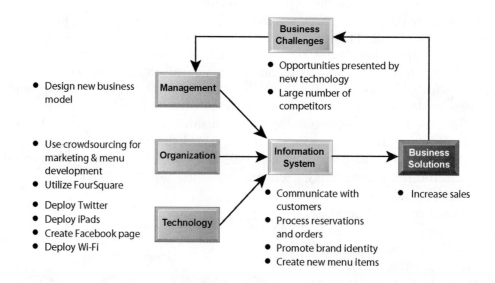

10.1 E-COMMERCE AND THE INTERNET

Have you ever purchased music over the Web or streamed a movie? Have you ever used the Web to search for information about your sneakers before you bought them in a retail store? If so, you've participated in e-commerce. In 2010, 133 million adult Americans bought something online, as did millions of others worldwide. And although most purchases still take place through traditional channels, e-commerce continues to grow rapidly and to transform the way many companies do business. In 2010, e-commerce represents about 6 percent of all retail sales in the United States, and is growing at 12 percent annually (eMarketer, 2010a).

E-COMMERCE TODAY

E-commerce refers to the use of the Internet and the Web to transact business. More formally, e-commerce is about digitally enabled commercial transactions between and among organizations and individuals. For the most part, this means transactions that occur over the Internet and the Web. Commercial transactions involve the exchange of value (e.g., money) across organizational or individual boundaries in return for products and services.

E-commerce began in 1995 when one of the first Internet portals, Netscape.com, accepted the first ads from major corporations and popularized the idea that the Web could be used as a new medium for advertising and sales. No one envisioned at the time what would turn out to be an exponential growth curve for e-commerce retail sales, which doubled and tripled in the early years. E-commerce grew at double-digit rates until the recession of 2008–2009 when growth slowed to a crawl. In 2009, e-commerce revenues were flat (Figure 10-1), not bad considering that traditional retail sales were shrinking by 5 percent annually. In fact, e-commerce during the recession was the only stable segment in retail. Some online retailers forged ahead at a record pace: Amazon's 2009 revenues were up 25 percent over 2008 sales. Despite the recession, in 2010, the

FIGURE 10-1 THE GROWTH OF E-COMMERCE

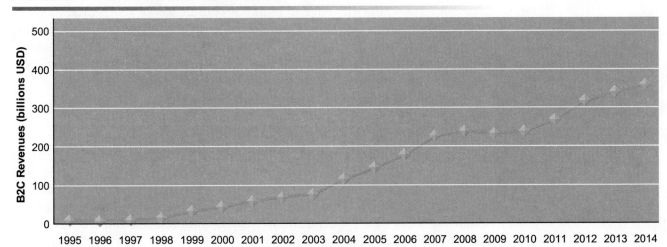

Retail e-commerce revenues grew 15–25 percent per year until the recession of 2008–2009, when they slowed measurably. In 2010, e-commerce revenues are growing again at an estimated 12 percent annually.

number of online buyers increased by 6 percent to 133 million, and the average annual purchase is up 5 percent to $1,139. Amazon's sales grew by 28 percent in the year.

Mirroring the history of many technological innovations, such as the telephone, radio, and television, the very rapid growth in e-commerce in the early years created a market bubble in e-commerce stocks. Like all bubbles, the "dot-com" bubble burst (in March 2001). A large number of e-commerce companies failed during this process. Yet for many others, such as Amazon, eBay, Expedia, and Google, the results have been more positive: soaring revenues, fine-tuned business models that produce profits, and rising stock prices. By 2006, e-commerce revenues returned to solid growth, and have continued to be the fastest growing form of retail trade in the United States, Europe, and Asia.

- Online consumer sales grew to an estimated $225 billion in 2010, an increase of more than 12 percent over 2009 (including travel services and digital downloads), with 133 million people purchasing online and 162 million shopping and gathering information but not necessarily purchasing (eMarketer, 2010a).

- The number of individuals of all ages online in the United States expanded to 221 million in 2010, up from 147 million in 2004. In the world, over 1.9 billion people are now connected to the Internet. Growth in the overall Internet population has spurred growth in e-commerce (eMarketer, 2010b).

- Approximately 80 million households have broadband access to the Internet in 2010, representing about 68 percent of all households.

- About 83 million Americans now access the Internet using a smartphone such as an iPhone, Droid, or BlackBerry. Mobile e-commerce has begun a rapid growth based on apps, ring tones, downloaded entertainment, and location-based services. In a few years, mobile phones will be the most common Internet access device.

- On an average day, an estimated 128 million adult U.S. Internet users go online. About 102 million send e-mail, 81 million use a search engine, and 71 million get news. Around 63 million use a social network, 43 million do online banking, 38 million watch an online video, and 28 million look for information on Wikipedia (Pew Internet & American Life Project, 2010).

- B2B e-commerce-use of the Internet for business-to-business commerce and collaboration among business partners expanded to more than $3.6 trillion.

The e-commerce revolution is still unfolding. Individuals and businesses will increasingly use the Internet to conduct commerce as more products and services come online and households switch to broadband telecommunications. More industries will be transformed by e-commerce, including travel reservations, music and entertainment, news, software, education, and finance. Table 10-1 highlights these new e-commerce developments.

WHY E-COMMERCE IS DIFFERENT

Why has e-commerce grown so rapidly? The answer lies in the unique nature of the Internet and the Web. Simply put, the Internet and e-commerce technologies are much more rich and powerful than previous technology revolutions like radio, television, and the telephone. Table 10-2 describes the unique features of the Internet and Web as a commercial medium. Let's explore each of these unique features in more detail.

Ubiquity

In traditional commerce, a marketplace is a physical place, such as a retail store, that you visit to transact business. E-commerce is ubiquitous, meaning

TABLE 10-1 THE GROWTH OF E-COMMERCE

BUSINESS TRANSFORMATION

- E-commerce remains the fastest growing form of commerce when compared to physical retail stores, services, and entertainment.

- The first wave of e-commerce transformed the business world of books, music, and air travel. In the second wave, nine new industries are facing a similar transformation scenario: marketing and advertising, telecommunications, movies, television, jewelry and luxury goods, real estate, online travel, bill payments, and software.

- The breadth of e-commerce offerings grows, especially in the services economy of social networking, travel, information clearinghouses, entertainment, retail apparel, appliances, and home furnishings.

- The online demographics of shoppers broaden to match that of ordinary shoppers.

- Pure e-commerce business models are refined further to achieve higher levels of profitability, whereas traditional retail brands, such as Sears, JCPenney, L.L.Bean, and Walmart, use e-commerce to retain their dominant retail positions.

- Small businesses and entrepreneurs continue to flood the e-commerce marketplace, often riding on the infrastructures created by industry giants, such as Amazon, Apple, and Google, and increasingly taking advantage of cloud-based computing resources.

- Mobile e-commerce begins to take off in the United States with location-based services and entertainment downloads including e-books.

TECHNOLOGY FOUNDATIONS

- Wireless Internet connections (Wi-Fi, WiMax, and 3G/4G smart phones) grow rapidly.

- Powerful handheld mobile devices support music, Web surfing, and entertainment as well as voice communication. Podcasting and streaming take off as mediums for distribution of video, radio, and user-generated content.

- The Internet broadband foundation becomes stronger in households and businesses as transmission prices fall. More than 80 million households had broadband cable or DSL access to the Internet in 2010 ,about 68 percent of all households in the United States (eMarketer, 2010a).

- Social networking software and sites such as Facebook, MySpace, Twitter, LinkedIn, and thousands of others become a major new platform for e-commerce, marketing, and advertising. Facebook hits 500 million users worldwide, and 180 million in the United States (comScore, 2010).

- New Internet-based models of computing, such as cloud computing, software as a service (SaaS), and Web 2.0 software greatly reduce the cost of e-commerce Web sites.

NEW BUSINESS MODELS EMERGE

- More than half the Internet user population have joined an online social network, contribute to social bookmarking sites, create blogs, and share photos. Together these sites create a massive online audience as large as television that is attractive to marketers.

- The traditional advertising business model is severely disrupted as Google and other technology players such as Microsoft and Yahoo! seek to dominate online advertising, and expand into offline ad brokerage for television and newspapers.

- Newspapers and other traditional media adopt online, interactive models but are losing advertising revenues to the online players despite gaining online readers.

- Online entertainment business models offering television, movies, music, sports, and e-books surge, with cooperation among the major copyright owners in Hollywood and New York with the Internet distributors like Google, YouTube, Facebook, and Microsoft.

that is it available just about everywhere, at all times. It makes it possible to shop from your desktop, at home, at work, or even from your car, using mobile commerce. The result is called a **marketspace**—a marketplace extended beyond traditional boundaries and removed from a temporal and geographic location.

TABLE 10-2 EIGHT UNIQUE FEATURES OF E-COMMERCE TECHNOLOGY

E-commerce Technology Dimension	Business Significance
Ubiquity. Internet/Web technology is available everywhere: at work, at home, and elsewhere via mobile devices.	The marketplace is extended beyond traditional boundaries and is removed from a temporal and geographic location. "Marketspace" anytime, is created; shopping can take place anywhere. Customer convenience is enhanced, and shopping costs are reduced.
Global reach. The technology reaches across national boundaries, around the Earth.	Commerce is enabled across cultural and national boundaries seamlessly and without modification. The marketspace includes, potentially, billions of consumers and millions of businesses worldwide.
Universal standards. There is one set of technology standards, namely Internet standards.	With one set of technical standards across the globe, disparate computer systems can easily communicate with each other.
Richness. Video, audio, and text messages are possible.	Video, audio, and text marketing messages are integrated into a single marketing message and consumer experience.
Interactivity. The technology works through interaction with the user.	Consumers are engaged in a dialog that dynamically adjusts the experience to the individual, and makes the consumer a co-participant in the process of delivering goods to the market.
Information Density. The technology reduces information costs and raises quality.	Information processing, storage, and communication costs drop dramatically, whereas currency, accuracy, and timeliness improve greatly. Information becomes plentiful, cheap, and more accurate.
Personalization/Customization. The technology allows personalized messages to be delivered to individuals as well as groups.	Personalization of marketing messages and customization of products and services are based on individual characteristics.
Social technology. User content generation and social networking.	New Internet social and business models enable user content creation and distribution, and support social networks.

From a consumer point of view, ubiquity reduces **transaction costs**—the costs of participating in a market. To transact business, it is no longer necessary that you spend time or money traveling to a market, and much less mental effort is required to make a purchase.

Global Reach

E-commerce technology permits commercial transactions to cross cultural and national boundaries far more conveniently and cost effectively than is true in traditional commerce. As a result, the potential market size for e-commerce merchants is roughly equal to the size of the world's online population (estimated to be more than 1.9 billion, and growing rapidly) (Internetworldstats.com, 2010).

In contrast, most traditional commerce is local or regional—it involves local merchants or national merchants with local outlets. Television and radio stations and newspapers, for instance, are primarily local and regional institutions with limited, but powerful, national networks that can attract a national audience but not easily cross national boundaries to a global audience.

Universal Standards

One strikingly unusual feature of e-commerce technologies is that the technical standards of the Internet and, therefore, the technical standards for conducting e-commerce are universal standards. They are shared by all nations around the world and enable any computer to link with any other computer regardless of the technology platform each is using. In contrast, most traditional commerce technologies differ from one nation to the next. For instance, television and radio standards differ around the world, as does cell telephone technology.

The universal technical standards of the Internet and e-commerce greatly lower **market entry costs**—the cost merchants must pay simply to bring their goods to market. At the same time, for consumers, universal standards reduce **search costs**—the effort required to find suitable products.

Richness

Information **richness** refers to the complexity and content of a message. Traditional markets, national sales forces, and small retail stores have great richness: They are able to provide personal, face-to-face service using aural and visual cues when making a sale. The richness of traditional markets makes them powerful selling or commercial environments. Prior to the development of the Web, there was a trade-off between richness and reach: The larger the audience reached, the less rich the message. The Web makes it possible to deliver rich messages with text, audio, and video simultaneously to large numbers of people.

Interactivity

Unlike any of the commercial technologies of the twentieth century, with the possible exception of the telephone, e-commerce technologies are interactive, meaning they allow for two-way communication between merchant and consumer. Television, for instance, cannot ask viewers any questions or enter into conversations with them, and it cannot request that customer information be entered into a form. In contrast, all of these activities are possible on an e-commerce Web site. Interactivity allows an online merchant to engage a consumer in ways similar to a face-to-face experience but on a massive, global scale.

Information Density

The Internet and the Web vastly increase **information density**—the total amount and quality of information available to all market participants, consumers, and merchants alike. E-commerce technologies reduce information collection, storage, processing, and communication costs while greatly increasing the currency, accuracy, and timeliness of information.

Information density in e-commerce markets make prices and costs more transparent. **Price transparency** refers to the ease with which consumers can find out the variety of prices in a market; **cost transparency** refers to the ability of consumers to discover the actual costs merchants pay for products.

There are advantages for merchants as well. Online merchants can discover much more about consumers than in the past. This allows merchants to segment the market into groups that are willing to pay different prices and permits the merchants to engage in **price discrimination**—selling the same goods, or nearly the same goods, to different targeted groups at different prices. For instance, an online merchant can discover a consumer's avid interest in expensive, exotic vacations and then pitch high-end vacation plans to that consumer at a premium price, knowing this person is willing to pay extra for such a vacation. At the same time, the online merchant can pitch the same vacation plan at a lower price to a more price-sensitive consumer. Information density also helps merchants differentiate their products in terms of cost, brand, and quality.

Personalization/Customization

E-commerce technologies permit **personalization**: Merchants can target their marketing messages to specific individuals by adjusting the message to a person's name, interests, and past purchases. The technology also permits **customization**—changing the delivered product or service based on a user's preferences or prior behavior. Given the interactive nature of e-commerce technology, much information about the consumer can be gathered in the marketplace at the moment of purchase. With the increase in information density, a great deal of information about the consumer's past purchases and behavior can be stored and used by online merchants.

The result is a level of personalization and customization unthinkable with traditional commerce technologies. For instance, you may be able to shape what you see on television by selecting a channel, but you cannot change the content of the channel you have chosen. In contrast, the *Wall Street Journal* Online allows you to select the type of news stories you want to see first and gives you the opportunity to be alerted when certain events happen.

Social Technology: User Content Generation and Social Networking

In contrast to previous technologies, the Internet and e-commerce technologies have evolved to be much more social by allowing users to create and share with their personal friends (and a larger worldwide community) content in the form of text, videos, music, or photos. Using these forms of communication, users are able to create new social networks and strengthen existing ones.

All previous mass media in modern history, including the printing press, use a broadcast model (one-to-many) where content is created in a central location by experts (professional writers, editors, directors, and producers) and audiences are concentrated in huge numbers to consume a standardized product. The new Internet and e-commerce empower users to create and distribute content on a large scale, and permit users to program their own content consumption. The Internet provides a unique many-to-many model of mass communications.

KEY CONCEPTS IN E-COMMERCE: DIGITAL MARKETS AND DIGITAL GOODS IN A GLOBAL MARKETPLACE

The location, timing, and revenue models of business are based in some part on the cost and distribution of information. The Internet has created a digital marketplace where millions of people all over the world are able to exchange massive amounts of information directly, instantly, and for free. As a result, the Internet has changed the way companies conduct business and increased their global reach.

The Internet reduces information asymmetry. An **information asymmetry** exists when one party in a transaction has more information that is important for the transaction than the other party. That information helps determine their relative bargaining power. In digital markets, consumers and suppliers can "see" the prices being charged for goods, and in that sense digital markets are said to be more "transparent" than traditional markets.

For example, before auto retailing sites appeared on the Web, there was a significant information asymmetry between auto dealers and customers. Only the auto dealers knew the manufacturers' prices, and it was difficult for consumers to shop around for the best price. Auto dealers' profit margins depended on this asymmetry of information. Today's consumers have access to a legion of Web sites providing competitive pricing information, and three-fourths of U.S. auto buyers use the Internet to shop around for the best deal. Thus, the Web has reduced the information asymmetry surrounding an auto purchase. The Internet has also helped businesses seeking to purchase from other businesses reduce information asymmetries and locate better prices and terms.

Digital markets are very flexible and efficient because they operate with reduced search and transaction costs, lower **menu costs** (merchants' costs of changing prices), greater price discrimination, and the ability to change prices dynamically based on market conditions. In **dynamic pricing**, the price of a product varies depending on the demand characteristics of the customer or the supply situation of the seller.

These new digital markets may either reduce or increase switching costs, depending on the nature of the product or service being sold, and they may cause some extra delay in gratification. Unlike a physical market, you can't immediately consume a product such as clothing purchased over the Web (although immediate consumption is possible with digital music downloads and other digital products.)

Digital markets provide many opportunities to sell directly to the consumer, bypassing intermediaries, such as distributors or retail outlets. Eliminating intermediaries in the distribution channel can significantly lower purchase transaction costs. To pay for all the steps in a traditional distribution channel, a product may have to be priced as high as 135 percent of its original cost to manufacture.

Figure 10-2 illustrates how much savings result from eliminating each of these layers in the distribution process. By selling directly to consumers or reducing the number of intermediaries, companies are able to raise profits while charging lower prices. The removal of organizations or business process layers responsible for intermediary steps in a value chain is called **disintermediation**.

FIGURE 10-2   THE BENEFITS OF DISINTERMEDIATION TO THE CONSUMER

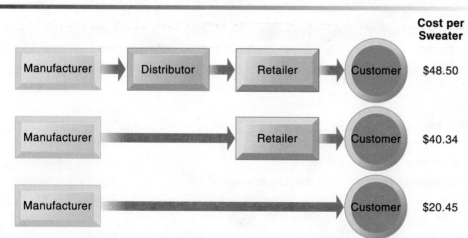

The typical distribution channel has several intermediary layers, each of which adds to the final cost of a product, such as a sweater. Removing layers lowers the final cost to the consumer.

Disintermediation is affecting the market for services. Airlines and hotels operating their own reservation sites online earn more per ticket because they have eliminated travel agents as intermediaries. Table 10-3 summarizes the differences between digital markets and traditional markets.

Digital Goods

The Internet digital marketplace has greatly expanded sales of digital goods. **Digital goods** are goods that can be delivered over a digital network. Music tracks, video, Hollywood movies, software, newspapers, magazines, and books can all be expressed, stored, delivered, and sold as purely digital products. Currently, most of these products are sold as physical goods, for example, CDs, DVDs, newspapers, and hard-copy books. But the Internet offers the possibility of delivering all these products on demand as digital products.

In general, for digital goods, the marginal cost of producing another unit is about zero (it costs nothing to make a copy of a music file). However, the cost of producing the original first unit is relatively high—in fact, it is nearly the total cost of the product because there are few other costs of inventory and distribution. Costs of delivery over the Internet are very low, marketing costs remain the same, and pricing can be highly variable. (On the Internet, the merchant can change prices as often as desired because of low menu costs.)

The impact of the Internet on the market for these kinds of digital goods is nothing short of revolutionary, and we see the results around us every day. Businesses dependent on physical products for sales—such as bookstores, book publishers, music labels, and film studios—face the possibility of declining sales and even destruction of their businesses. Newspapers and magazines are losing readers to the Internet, and losing advertisers even as online newspaper readership soars. Record label companies are losing sales to music download sites and Internet piracy, and music stores are going out of business. Video rental firms, such as Blockbuster (now in bankruptcy), based on a physical DVD market and physical stores, lost sales to Netflix using an Internet catalog and streaming video model. Hollywood studios as well face the prospect that Internet pirates will distribute their product as a digital stream, bypassing Hollywood's monopoly on DVD rentals and sales, which

TABLE 10-3 DIGITAL MARKETS COMPARED TO TRADITIONAL MARKETS

	DIGITAL MARKETS	TRADITIONAL MARKETS
Information asymmetry	Asymmetry reduced	Asymmetry high
Search costs	Low	High
Transaction costs	Low (sometimes virtually nothing)	High (time, travel)
Delayed gratification	High (or lower in the case of a digital good)	Lower: purchase now
Menu costs	Low	High
Dynamic pricing	Low cost, instant	High cost, delayed
Price discrimination	Low cost, instant	High cost, delayed
Market segmentation	Low cost, moderate precision	High cost, less precision
Switching costs	Higher/lower (depending on product characteristics)	High
Network effects	Strong	Weaker
Disintermediation	More possible/likely	Less possible/unlikely

now accounts for more than half of industry film revenues. To date, pirated movies have not seriously threatened Hollywood revenues in part because the major film studios and Internet distributors like YouTube, Amazon, and Apple are learning how to cooperate. Table 10.4 describes digital goods and how they differ from traditional physical goods.

10.2 E-COMMERCE: BUSINESS AND TECHNOLOGY

E-commerce has grown from a few advertisements on early Web portals in 1995, to over 6 percent of all retail sales in 2010 (an estimated $255 billion), surpassing the mail order catalog business. E-commerce is a fascinating combination of business models and new information technologies. Let's start with a basic understanding of the types of e-commerce, and then describe e-commerce business and revenue models. We'll also cover new technologies that help companies reach over 221 million online consumers in the United States, and an estimated 800 million more worldwide.

TYPES OF E-COMMERCE

There are many ways to classify electronic commerce transactions. One is by looking at the nature of the participants in the electronic commerce transaction. The three major electronic commerce categories are business-to-consumer (B2C) e-commerce, business-to-business (B2B) e-commerce, and consumer-to-consumer (C2C) e-commerce.

- **Business-to-consumer (B2C)** electronic commerce involves retailing products and services to individual shoppers. BarnesandNoble.com, which sells books, software, and music to individual consumers, is an example of B2C e-commerce.

- **Business-to-business (B2B)** electronic commerce involves sales of goods and services among businesses. ChemConnect's Web site for buying and selling chemicals and plastics is an example of B2B e-commerce.

- **Consumer-to-consumer (C2C)** electronic commerce involves consumers selling directly to consumers. For example, eBay, the giant Web auction site, enables people to sell their goods to other consumers by auctioning their merchandise off to the highest bidder, or for a fixed price. Craigslist is the most widely used platform used by consumers to buy from and sell directly to others.

TABLE 10-4 HOW THE INTERNET CHANGES THE MARKETS FOR DIGITAL GOODS

	DIGITAL GOODS	TRADITIONAL GOODS
Marginal cost/unit	Zero	Greater than zero , high
Cost of production	High (most of the cost)	Variable
Copying cost	Approximately 0	Greater than zero, high
Distributed delivery cost	Low	High
Inventory cost	Low	High
Marketing cost	Variable	Variable
Pricing	More variable (bundling, random pricing games)	Fixed, based on unit costs

Another way of classifying electronic commerce transactions is in terms of the platforms used by participants in a transaction. Until recently, most e-commerce transactions took place using a personal computer connected to the Internet over wired networks. Two wireless mobile alternatives have emerged: smartphones and dedicated e-readers like the Kindle using cellular networks, and smartphones and small tablet computers using Wi-Fi wireless networks. The use of handheld wireless devices for purchasing goods and services from any location is termed **mobile commerce** or **m-commerce**. Both business-to-business and business-to-consumer e-commerce transactions can take place using m-commerce technology, which we discuss in detail in Section 10.3.

E-COMMERCE BUSINESS MODELS

Changes in the economics of information described earlier have created the conditions for entirely new business models to appear, while destroying older business models. Table 10-5 describes some of the most important Internet business models that have emerged. All, in one way or another, use the Internet to add extra value to existing products and services or to provide the foundation for new products and services.

Portal

Portals such as Google, Bing, Yahoo, MSN, and AOL offer powerful Web search tools as well as an integrated package of content and services, such as news, e-mail, instant messaging, maps, calendars, shopping, music downloads, video streaming, and more, all in one place. Initially, portals were primarily "gateways" to the Internet. Today, however, the portal business model provides a destination site where users start their Web searching and linger to read news, find entertainment, and meet other people, and be exposed to advertising. Portals generate revenue primarily by attracting very large audiences, charging advertisers for ad placement, collecting referral fees for steering customers to other sites, and charging for premium services. In 2010, portals generated an estimated $13.5 billion in revenues. Although there are hundreds of portal/search engine sites, the top five sites (Google, Yahoo, MSN/Bing, AOL, and Ask.com) gather more than 95 percent of the Internet traffic because of their superior brand recognition (eMarketer, 2010e).

E-tailer

Online retail stores, often called **e-tailers**, come in all sizes, from giant Amazon with 2010 revenues of more than $24 billion, to tiny local stores that have Web sites. An e-tailer is similar to the typical bricks-and-mortar storefront, except that customers only need to connect to the Internet to check their inventory and place an order. Altogether, online retail generated about $152 billion in revenues for 2010. The value proposition of e-tailers is to provide convenient, low-cost shopping 24/7, offering large selections and consumer choice. Some e-tailers, such as Walmart.com or Staples.com, referred to as "bricks-and-clicks," are subsidiaries or divisions of existing physical stores and carry the same products. Others, however, operate only in the virtual world, without any ties to physical locations. Amazon, BlueNile.com, and Drugstore.com are examples of this type of e-tailer. Several other variations of e-tailers—such as online versions of direct mail catalogs, online malls, and manufacturer-direct online sales—also exist.

TABLE 10-5 INTERNET BUSINESS MODELS

CATEGORY	DESCRIPTION	EXAMPLES
E-tailer	Sells physical products directly to consumers or to individual businesses.	Amazon RedEnvelope.com
Transaction broker	Saves users money and time by processing online sales transactions and generating a fee each time a transaction occurs.	ETrade.com Expedia
Market creator	Provides a digital environment where buyers and sellers can meet, search for products, display products, and establish prices for those products. Can serve consumers or B2B e-commerce, generating revenue from transaction fees.	eBay Priceline.com ChemConnect.com
Content provider	Creates revenue by providing digital content, such as news, music, photos, or video, over the Web. The customer may pay to access the content, or revenue may be generated by selling advertising space.	WSJ.com GettyImages.com iTunes.com Games.com
Community provider	Provides an online meeting place where people with similar interests can communicate and find useful information.	Facebook MySpace iVillage , Twitter
Portal	Provides initial point of entry to the Web along with specialized content and other services.	Yahoo Bing Google
Service provider	Provides Web 2.0 applications such as photo sharing, video sharing, and user-generated content as services. Provides other services such as online data storage and backup.	Google Apps Photobucket.com Xdrive.com

Content Provider

While e-commerce began as a retail product channel, it has increasingly turned into a global content channel. "Content" is defined broadly to include all forms of intellectual property. **Intellectual property** refers to all forms of human expression that can be put into a tangible medium such as text, CDs, DVDs, or stored on any digital (or other) media, including the Web. Content providers distribute information content, such as digital video, music, photos, text, and artwork, over the Web. The value proposition of online content providers is that consumers can find a wide range of content online, conveniently, and purchase this content inexpensively, to be played, or viewed, on multiple computer devices or smartphones.

Providers do not have to be the creators of the content (although sometimes they are, like Disney.com), and are more likely to be Internet-based distributors of content produced and created by others. For example, Apple sells music tracks at its iTunes Store, but it does not create or commission new music.

The phenomenal popularity of the iTunes Store, and Apple's Internet-connected devices like the iPhone, iPod, and iPad, have enabled new forms of digital content delivery from podcasting to mobile streaming. **Podcasting** is a method of publishing audio or video broadcasts via the Internet, allowing subscribing users to download audio or video files onto their personal computers or portable music players. **Streaming** is a publishing method for music and video files that flows a continuous stream of content to a user's device without being stored locally on the device.

Estimates vary, but total download and subscription media revenues for 2010 are somewhere between $8 billion and $10 billion annually. They are the fastest growing segment within e-commerce, growing at an estimated 20 percent annual rate (eMarketer, 2010b).

Transaction Broker

Sites that process transactions for consumers normally handled in person, by phone, or by mail are transaction brokers. The largest industries using this model are financial services and travel services. The online transaction broker's primary value propositions are savings of money and time, as well as providing an extraordinary inventory of financial products and travel packages, in a single location. Online stock brokers and travel booking services charge fees that are considerably less than traditional versions of these services.

Market Creator

Market creators build a digital environment in which buyers and sellers can meet, display products, search for products, and establish prices. The value proposition of online market creators is that they provide a platform where sellers can easily display their wares and where purchasers can buy directly from sellers. Online auction markets like eBay and Priceline are good examples of the market creator business model. Another example is Amazon's Merchants platform (and similar programs at eBay) where merchants are allowed to set up stores on Amazon's Web site and sell goods at fixed prices to consumers. This is reminiscent of open air markets where the market creator operates a facility (a town square) where merchants and consumers meet. Online market creators will generate about $12 billion in revenues for 2010.

Service Provider

While e-tailers sell products online, service providers offer services online. There's been an explosion in online services. Web 2.0 applications, photo sharing, and online sites for data backup and storage all use a service provider business model. Software is no longer a physical product with a CD in a box, but increasingly software as a service (SaaS) that you subscribe to online rather than purchase from a retailer (see Chapter 5). Google has led the way in developing online software service applications such as Google Apps, Gmail, and online data storage services.

Community Provider

Community providers are sites that create a digital online environment where people with similar interests can transact (buy and sell goods); share interests, photos, videos; communicate with like-minded people; receive interest-related information; and even play out fantasies by adopting online personalities called avatars. The social networking sites Facebook, MySpace, LinkedIn, and Twitter; online communities such as iVillage; and hundreds of other smaller, niche sites such as Doostang and Sportsvite all offer users community-building tools and services. Social networking sites have been the fastest growing Web sites in recent years, often doubling their audience size in a year. However, they are struggling to achieve profitability. The Interactive Session on Organizations explores this topic.

INTERACTIVE SESSION: ORGANIZATIONS

TWITTER SEARCHES FOR A BUSINESS MODEL

Twitter, the social networking site based on 140-character text messages, is the buzz social networking phenomenon of the year. Like all social networking sites, such as Facebook, MySpace, YouTube, Flickr, and others, Twitter provides a platform for users to express themselves by creating content and sharing it with their "followers," who sign up to receive someone's "tweets." And like most social networking sites, Twitter faces the problem of how to make money. As of October 2010, Twitter has failed to generate earnings as its management ponders how best to exploit the buzz and user base it has created.

Twitter began as a Web-based version of popular text messaging services provided by cell phone carriers. Executives in a podcasting company called Odeo were searching for a new revenue-producing product or service. In March 2006, they created a stand-alone, private company called Twitter.

The basic idea was to marry short text messaging on cell phones with the Web and its ability to create social groups. You start by establishing a Twitter account online, and identifying the friends that you would like to receive your messages. By sending a text message called a "tweet" to a short code on your cell phone (40404), you can tell your friends what you are doing, your location, and whatever else you might want to say. You are limited to 140 characters, but there is no installation and no charge. This social network messaging service to keep buddies informed is a smash success.

Coming up with solid numbers for Twitter is not easy because the firm is not releasing any "official" figures. By September 2010, Twitter, according to comScore, had around 30 million unique monthly users in the United States, and perhaps 96 million worldwide, displacing MySpace as the number three global social network (behind Facebook and Microsoft's Live Profile).

The number of individual tweets is also known only by the company. According to the company, by early 2007, Twitter had transmitted 20,000 tweets, which jumped to 60,000 tweets in a few months. During the Iranian rebellion in June 2009, there were reported to be over 200,000 tweets per hour worldwide. In October 2010, Twitter was recording over 1.2 million tweets a month. On the other hand, experts believe that 80 percent of tweets are

generated by only 10 percent of users, and that the median number of tweet readers per tweet is 1 (most tweeters tweet to one follower). Even more disturbing is that Twittter has a 60 percent churn rate: only 40 percent of users remain more than one month. Obviously, many users lose interest in learning about their friends' breakfast menu, and many feel "too connected" to their "friends," who in fact may only be distant acquaintances, if that. On the other hand, celebrities such as Britney Spears have hundreds of thousands of "friends" who follow their activities, making Twitter a marvelous, free public relations tool. Twitter unfortunately does not make a cent on these activities.

The answer to these questions about unique users, numbers of tweets, and churn rate are critical to understanding the business value of Twitter as a firm. To date, Twitter has generated losses and has unknown revenues, but in February 2009, it raised $35 million in a deal that valued the company at $255 million. The following September, Twitter announced it had raised $100 million in additional funding, from private equity firms, previous investors, and mutual fund giant T. Rowe Price, based on a company valuation of a staggering $1 billion!

So how can Twitter make money from its users and their tweets? What's its business model and how might it evolve over time? To start, consider the company's assets and customer value proposition. The main asset is user attention and audience size (eyeballs per day). The value proposition is "get it now" or real-time news on just about anything from the mundane to the monumental. An equally important asset is the database of tweets that contains the comments, observations, and opinions of the audience, and the search engine that mines those tweets for patterns. These are real-time and spontaneous observations.

Yet another asset has emerged in the last year: Twitter is a powerful alternative media platform for the distribution of news, videos, and pictures. Once again, no one predicted that Twitter would be the first to report on terrorist attacks in Mumbai, the landing of a passenger jet in the Hudson River, the Iranian rebellion in June 2009, or the political violence in Bangkok and Kenya in May 2010.

How can these assets be monetized? Advertising, what else! In April 2010, Twitter announced it s first foray into the big-time ad marketplace with Promoted Tweets. Think Twitter search engine: in response to a user's query to Twitter's search function for, say netbooks, a Best Buy ad for netbooks will be displayed. The company claims Promoted Tweets are not really ads because they look like all other tweets, just a part of the tweet stream of messages. These so-called "organic tweets" differ therefore from traditional search engine text ads, or social network ads which are far from organic. So far, Best Buy, Bravo, Red Bull, Sony, Starbucks, and Virgin American have signed up. If this actually works, thousands of companies might sign up to blast messages to millions of subscribers in response to related queries.

A second Twitter monetization effort announced in June 2010 is called Promoted Trends. Trends is a section of the Twitter home page that lets users know what's hot, what a lot of people are talking about. The company claims this is "organic," and a true reflection of what people are tweeting about. Promoted Trends are trends that companies would like to initiate. A company can place a Promoted Trends banner on the bottom of the page and when users click on the banner, they are taken to the follower page for that movie or product. Disney bought Promoted Trends for its film *Toy Story 3*, according to Twitter.

In July 2010, Twitter announced its third initiative of the year: @earlybird accounts, which users can follow to receive special offers. Walt Disney Pictures has used the service to promote *The Sorcerer's Apprentice* by offering twofers (buy one ticket, get another one free). The service could work nicely with so-called real-time or "flash" marketing campaigns in entertainment, fashion, luxury goods, technology, and beauty products. So far, Twitter has over 50,000 @earlybird followers and hopes to reach "influentials," people who shape the purchasing decisions of many others.

Another monetizing service is temporal real-time search. If there's one thing Twitter has uniquely among all the social network sites, it's real-time information. In 2010, Twitter entered into agreements with Google, Microsoft, and Yahoo to permit these search engines to index tweets and make them available to the entire Internet. This service will give free real-time content to the search engines as opposed to archival content. It is unclear who's doing who a service here, and the financial arrangements are not public.

Other large players are experimenting. Dell created a Twitter outlet account, @DellOutlet, and is using it to sell open-box and discontinued computers.

Dell also maintains several customer service accounts. Twitter could charge such accounts a commission on sales because Twitter is acting like an e-commerce sales platform similar to Amazon. Other firms have used their Twitter followers' fan base to market discount air tickets (Jet Blue) and greeting cards (Somecards).

Freemium is another possibility: ask users to pay a subscription fee for premium services such as videos and music downloads. However, it may be too late for this idea because users have come to expect the service to be free. Twitter could charge service providers such as doctors, dentists, lawyers, and hair salons for providing their customers with unexpected appointment availabilities. But Twitter's most likely steady revenue source might be its database of hundreds of millions of real-time tweets. Major firms such as Starbucks, Amazon, Intuit (QuickBooks and Mint.com), and Dell have used Twitter to understand how their customers are reacting to products, services, and Web sites, and then making corrections or changes in those services and products. Twitter is a fabulous listening post on the Internet frontier.

The possibilities are endless, and just about any of the above scenarios offers some solution to the company's problem, which is a lack of revenue (forget about profits). The company is coy about announcing its business model, what one pundit described as hiding behind a "Silicon Valley Mona Lisa smile." These Wall Street pundits are thought to be party poopers in the Valley. In a nod to Apple's iTunes and Amazon's merchant services, Twitter has turned over its messaging capabilities and software platform to others, one of which is CoTweet.com, a company that organizes multiple Twitter exchanges for customers so they can be tracked more easily. Google is selling ad units based around a company's last five tweets (ads are displayed to users who have created or viewed tweets about a company). Twitter is not charging for this service. In the meantime, observers wonder if Twitter is twittering away its assets and may not ever show a profit for its $160 million investment.

Sources: Matthew Shaer, "Twitter Hits 145 Million User Mark, Sees Rise in Mobile Use," Christian Science Monitor, September 3, 2010; Jason Lipshutz, "Lady Gaga to Steal Britney Spears' Twitter Crown," Reuters, August 19, 2010; Emir Afrati, "Twitter's Early Bird Ad Ploy Takes Flight," *Wall Street Journal*, July 14, 2010; Jessica Guynn, "Twitter Tests New Promoted Trends Feature with 'Toy Story 3' from Disney's Pixar," *Los Angeles Times*, June 16, 2010; Erica Naone, "Will Twitter's Ad Strategy Work," *Technology Review*, April 15, 2010; Jessica Vascellaro and Emily Steel, "Twitter Rolls Out Ads," *Wall Street Journal*, April 14, 2010; Brad Stone, "Twitter's Latest Valuation: $1 Billion," *New York Times*, September 24, 2009; Jon Fine, "Twitter Makes a Racket. But Revenues?" *Business Week*, April 9, 2009.

CASE STUDY QUESTIONS   MIS IN ACTION

1. Based on your reading in this chapter, how would you characterize Twitter's business model?

2. If Twitter is to have a revenue model, which of the revenue models described in this chapter would work?

3. What is the most important asset that Twitter has, and how could it monetize this asset?

4. What impact will a high customer churn rate have on Twitter's potential advertising revenue?

1. Go to Twitter.com and enter a search on your favorite (or least favorite) car. Can you find the company's official site? What else do you find? Describe the results and characterize the potential risks and rewards for companies that would like to advertise to Twitter's audience.

2. How would you improve Twitter's Web site to make it more friendly for large advertisers?

3. Teenagers are infrequent users of Twitter because they use their cell phones for texting, and most users are adults 18–34 years of age. Find five users of Twitter and ask them how long they have used the service, are they likely to continue using the service, and how would they feel about banner ads appearing on their Twitter Web screen and phone screens. Are loyal users of Twitter less likely (or more likely) to tolerate advertising on Twitter?

E-COMMERCE REVENUE MODELS

A firm's **revenue model** describes how the firm will earn revenue, generate profits, and produce a superior return on investment. Although there are many different e-commerce revenue models that have been developed, most companies rely on one, or some combination, of the following six revenue models: advertising, sales, subscription, free/freemium, transaction fee, and affiliate.

Advertising Revenue Model

In the **advertising revenue model**, a Web site generates revenue by attracting a large audience of visitors who can then be exposed to advertisements. The advertising model is the most widely used revenue model in e-commerce, and arguably, without advertising revenues, the Web would be a vastly different experience from what it is now. Content on the Web—everything from news to videos and opinions—is "free" to visitors because advertisers pay the production and distribution costs in return for the right to expose visitors to ads. Companies will spend an estimated $240 billion on advertising in 2010, and an estimated $25 billion of that amount on online advertising (in the form of a paid message on a Web site, paid search listing, video, widget, game, or other online medium, such as instant messaging). In the last five years, advertisers have increased online spending and cut outlays on traditional channels such as radio and newspapers. Television advertising has expanded along with online advertising revenues.

Web sites with the largest viewership or that attract a highly specialized, differentiated viewership and are able to retain user attention ("stickiness") are able to charge higher advertising rates. Yahoo, for instance, derives nearly all its revenue from display ads (banner ads) and to a lesser extent search engine text ads. Ninety-eight percent of Google's revenue derives from selling keywords to advertisers in an auction-like market (the AdSense program). The average Facebook user spends over five hours a week on the site, far longer than other portal sites.

Sales Revenue Model

In the **sales revenue model**, companies derive revenue by selling goods, information, or services to customers. Companies such as Amazon (which sells books, music, and other products), LLBean.com, and Gap.com, all have sales revenue models. Content providers make money by charging for downloads of entire files such as music tracks (iTunes Store) or books or for downloading music and/or video streams (Hulu.com TV shows—see Chapter 3). Apple has pioneered and strengthened the acceptance of micropayments. **Micropayment systems** provide content providers with a cost-effective method for processing high volumes of very small monetary transactions (anywhere from $.25 to $5.00 per transaction). MyMISlab has a Learning Track with more detail on micropayment and other e-commerce payment systems.

Subscription Revenue Model

In the **subscription revenue model**, a Web site offering content or services charges a subscription fee for access to some or all of its offerings on an ongoing basis. Content providers often use this revenue model. For instance, the online version of *Consumer Reports* provides access to premium content, such as detailed ratings, reviews, and recommendations, only to subscribers, who have a choice of paying a $5.95 monthly subscription fee or a $26.00 annual fee. Netflix is one of the most successful subscriber sites with more that 15 million subscribers in September 2010. The Wall Street Journal has the largest online subscription newspaper with more than 1 million online subscribers. To be successful, the subscription model requires that the content be perceived as a having high added value, differentiated, and not readily available elsewhere nor easily replicated. Companies successfully offering content or services online on a subscription basis include Match.com and eHarmony (dating services), Ancestry.com and Genealogy.com (genealogy research), Microsoft's Xboxlive.com (video games), and Rhapsody.com (music).

Free/Freemium Revenue Model

In the **free/freemium revenue model**, firms offer basic services or content for free, while charging a premium for advanced or special features. For example, Google offers free applications, but charges for premium services. Pandora, the subscription radio service, offers a free service with limited play time, and a premium service with unlimited play. The Flickr photo-sharing service offers free basic services for sharing photos with friends and family, and also sells a $24.95 "premium" package that provides users unlimited storage, high-definition video storage and playback, and freedom from display advertising. The idea is to attract very large audiences with free services, and then to convert some of this audience to pay a subscription for premium services. One problem with this model is converting people from being "free loaders" into paying customers. "Free" can be a powerful model for losing money.

Transaction Fee Revenue Model

In the **transaction fee revenue model**, a company receives a fee for enabling or executing a transaction. For example, eBay provides an online auction marketplace and receives a small transaction fee from a seller if the seller is successful in selling an item. E*Trade, an online stockbroker, receives transaction fees each time it executes a stock transaction on behalf of a customer. The transaction revenue model enjoys wide acceptance in part because the true cost of using the platform is not immediately apparent to the user.

Affiliate Revenue Model

In the **affiliate revenue model**, Web sites (called "affiliate Web sites") send visitors to other Web sites in return for a referral fee or percentage of the revenue from any resulting sales. For example, MyPoints makes money by connecting companies to potential customers by offering special deals to its members. When members take advantage of an offer and make a purchase, they earn "points" they can redeem for free products and services, and MyPoints receives a referral fee. Community feedback sites such as Epinions and Yelp receive much of their revenue from steering potential customers to Web sites where they make a purchase. Amazon uses affiliates who steer business to the Amazon Web site by placing the Amazon logo on their blogs. Personal blogs may be involved in affiliate marketing. Some bloggers are paid directly by manufacturers, or receive free products, for speaking highly of products and providing links to sales channels.

WEB 2.0: SOCIAL NETWORKING AND THE WISDOM OF CROWDS

One of the fastest growing areas of e-commerce revenues are Web 2.0 online services, which we described in Chapter 7. The most popular Web 2.0 service is social networking, online meeting places where people can meet their friends and their friends' friends. Every day over 60 million Internet users in the United States visit a social networking site like Facebook, MySpace, LinkedIn, and hundreds of others.

Social networking sites link people through their mutual business or personal connections, enabling them to mine their friends (and their friends' friends) for sales leads, job-hunting tips, or new friends. MySpace, Facebook, and Friendster appeal to people who are primarily interested in extending their friendships, while LinkedIn focuses on job networking for professionals.

Social networking sites and online communities offer new possibilities for e-commerce. Networking sites like Facebook and MySpace sell banner, video, and text ads; sell user preference information to marketers; and sell products such as music, videos, and e-books. Corporations set up their own Facebook and MySpace profiles to interact with potential customers. For example, Procter & Gamble set up a MySpace profile page for Crest toothpaste soliciting "friends" for a fictional character called "Miss Irresistable." Business firms can also "listen" to what social networkers are saying about their products, and obtain valuable feedback from consumers. At user-generated content sites like YouTube, high-quality video content is used to display advertising, and Hollywood studios have set up their own channels to market their products. The Interactive Session on Management looks more closely at social networking on Facebook, focusing on its impact on privacy.

At **social shopping** sites like Kaboodle, ThisNext, and Stylehive you can swap shopping ideas with friends. Facebook offers this same service on a voluntary basis. Online communities are also ideal venues to employ viral marketing techniques. Online viral marketing is like traditional word-of-mouth marketing except that the word can spread across an online commu-

FACEBOOK: MANAGING YOUR PRIVACY FOR THEIR PROFIT

Facebook is the largest social networking site in the world. Founded in 2004 by Mark Zuckerberg, the site had over 500 million worldwide users as of October 2010, and has long since surpassed all of its social networking peers. Facebook allows users to create a profile and join various types of self-contained networks, including college-wide, workplace, and regional networks. The site includes a wide array of tools that allow users to connect and interact with other users, including messaging, groups, photo-sharing, and user-created applications.

Although the site is the leader in social networking, it has waged a constant struggle to develop viable methods of generating revenue. Though many investors are still optimistic regarding Facebook's future profitability, it still needs to adjust its business model to monetize the site traffic and personal information it has accumulated.

Like many businesses of its kind, Facebook makes its money through advertising. Facebook represents a unique opportunity for advertisers to reach highly targeted audiences based on their demographic information, hobbies and personal preferences, geographical regions, and other narrowly specified criteria in a comfortable and engaging environment. Businesses both large and small can place advertisements that are fully integrated into primary features of the site or create Facebook pages where users can learn more about and interact with them.

However, many individuals on Facebook aren't interested in sharing their personal information with anyone other than a select group of their friends on the site. This is a difficult issue for Facebook. The company needs to provide a level of privacy that makes their users comfortable, but it's that very privacy that prevents it from gathering as much information as it would like, and the more information Facebook has, the more money it earns. Facebook's goal is to persuade its users to be comfortable sharing information willingly by providing an environment that becomes richer and more entertaining as the amount of information shared increases. In trying to achieve this goal, the site has made a number of missteps, but is improving its handling of users' privacy rights.

The launch of Facebook's Beacon advertising service in 2007 was a lightning rod for criticism of Facebook's handling of its private information.

Beacon was intended to inform users about what their friends were purchasing and what sites they were visiting away from Facebook. Users were angry that Beacon continued to communicate private information even after a user opted out of the service. After significant public backlash and the threat of a class-action lawsuit, Facebook shut down Beacon in September 2009.

Facebook has also drawn criticism for preserving the personal information of people who attempted to remove their profiles from the site. In early 2009, it adjusted its terms of service to assign it ownership rights over the information contained in deleted profiles. In many countries, this practice is illegal, and the user backlash against the move was swift.

In response, Facebook's chief privacy officer, Chris Kelly, presided over a total overhaul of Facebook's privacy policy, which took the form of an open collaboration with some of the most vocal critics of the old policies, including the previously mentioned protest group's founders. In February, Facebook went forward with the new terms after holding a vote open to all Facebook users, 75 percent of whom approved. The site now allows users either to deactivate or to delete their account entirely, and only saves information after deactivation.

In late 2009, tensions between Facebook and its users came to a head when the site rolled out new privacy controls for users, but had adjusted those settings to be public by default. Even users that had previously set their privacy to be "friends-only" for photos and profile information had their content exposed, including the profile of Zuckerberg himself. When asked about the change, Zuckerberg explained that the moves were in response to a shift in social norms towards openness and away from privacy, saying "we decided that these would be the social norms now and we just went for it."

The fallout from the change and is still ongoing, and more privacy problems keep cropping up. In October 2010, Facebook unveiled new features giving users more control over how they share personal information on the site with other users and third-party applications. These include a groups feature allowing users to distinguish specific circles of "friends" and choose what information they want to share with each group and whether the groups are public or private.

Shortly thereafter, a Wall Street Journal investigation found that some of the most popular Facebook applications (apps) had been transmitting user IDs— identifying information which could provide access to people's names and, in some cases, their friends' names—to dozens of advertising and Internet tracking companies. Sharing user IDs is in violation of Facebook's privacy policies.

All these privacy flaps have not diminished advertiser interest. Facebook serves ads on each user's home page and on the sidebars of user profiles. In addition to an image and headline from the advertiser, Facebook ads include the names of any user's friends who have clicked on a button indicating they like the brand or ad. A Nielsen Co. study found that including information about individuals a person knows in an ad boosted recall of the ad by 68 percent and doubled awareness of a brand's message. To determine what ads to serve to particular people, Facebook abstracts profile information into keywords, and advertisers match ads to those keywords. No individual data is shared with any advertiser.

However, it's still unclear how much money is there to be made from advertising on Facebook. The site insists that it doesn't plan to charge its users any kind of fee for site access. Facebook's 2010 revenue was expected to approach $1 billion, which is a far cry from a $33 billion private market valuation. But the site has already become a critical component of the Web's social fabric, and Facebook management insists that it's unworried about profitability in 2010 or the immediate future.

Sources: Emily Steel and Geoffrey A. Fowler, "Facebook in Privacy Breach," *The Wall Street Journal*, October 18, 2010; Jessica E. Vascellaro, "Facebook Makes Gains in Web Ads," *The Wall Street Journal*, May 12, 2010 and "Facebook Grapples with Privacy Issues," *The Wall Street Journal*, May 19, 2010; Geoffrey A. Fowler, "Facebook Fights Privacy Concerns," *The Wall Street Journal*, August 21, 2010 and "Facebook Tweaks Allow Friends to Sort Who They Really 'Like,'" *The Wall Street Journal*, October 5, 2010; Emily Steel and Geoffrey A. Fowler, "Facebook Touts Selling Power of Friendship," *The Wall Street Journal*, July 7, 2010; Brad Stone, "Is Facebook Growing Up Too Fast?" *The New York Times*, March 29, 2009; and CG Lynch, "Facebook's Chief Privacy Officer: Balancing Needs of Users with the Business of Social Networks", *CIO.com*, April 1 2009.

CASE STUDY QUESTIONS

1. What concepts in the chapter are illustrated in this case?

2. Describe the weaknesses of Facebook's privacy policies and features. What management, organization, and technology factors have contributed to those weaknesses?

3. List and describe some of the options that Facebook managers have in balancing privacy and profitability. How can Facebook better safeguard user privacy? What would be the impact on its profitability and business model?

4. Do you anticipate that Facebook will be successful in developing a business model that monetizes their site traffic? Why or why not?

MIS IN ACTION

Visit Facebook's Web site and review the site's privacy policy. Then answer the following questions:

1. To what user information does Facebook retain the rights?

2. What is Facebook's stance regarding information shared via third-party applications developed for the Facebook platform?

3. Did you find the privacy policy to be clear and reasonable? What would you change, if anything?

nity at the speed of light, and go much further geographically than a small network of friends.

The Wisdom of Crowds

Creating sites where thousands, even millions, of people can interact offers business firms new ways to market and advertise, to discover who likes (or hates) their products. In a phenomenon called "**the wisdom of crowds**," some argue that large numbers of people can make better decisions about a wide range of topics or products than a single person or even a small committee of experts (Surowiecki, 2004).

Obviously this is not always the case, but it can happen in interesting ways. In marketing, the wisdom of crowds concept suggests that firms should consult with thousands of their customers first as a way of establishing a relationship with them, and second, to better understand how their products and services are used and appreciated (or rejected). Actively soliciting the comments of your customers builds trust and sends the message to your customers that you care what they are thinking, and that you need their advice.

Beyond merely soliciting advice, firms can be actively helped in solving some business problems using what is called **crowdsourcing**. For instance, in 2006, Netflix announced a contest in which it offered to pay $1 million to the person or team who comes up with a method for improving by 10 percent Netflix's prediction of what movies customers would like as measured against their actual choices. By 2009, Netflix received 44,014 entries from 5,169 teams in 186 countries. The winning team improved a key part of Netflix's business: a recommender system that recommends to its customers what new movies to order based on their personal past movie choices and the choices of millions of other customers who are like them (Howe, 2008; Resnick and Varian, 1997).

Firms can also use the wisdom of crowds in the form of prediction markets. **Prediction markets** are established as peer-to-peer betting markets where participants make bets on specific outcomes of, say, quarterly sales of a new product, designs for new products, or political elections. The world's largest commercial prediction market is Betfair, founded in 2000, where you bet for or against specific outcomes on football games, horse races, and whether or not the Dow Jones will go up or down in a single day. Iowa Electronic Markets (IEM) is an academic market focused on elections. You can place bets on the outcome of local and national elections.

E-COMMERCE MARKETING

While e-commerce and the Internet have changed entire industries and enable new business models, no industry has been more affected than marketing and marketing communications. The Internet provides marketers with new ways of identifying and communicating with millions of potential customers at costs far lower than traditional media, including search engine marketing, data mining, recommender systems, and targeted e-mail. The Internet enables **long tail marketing**. Before the Internet, reaching a large audience was very expensive, and marketers had to focus on attracting the largest number of consumers with popular hit products, whether music, Hollywood movies, books, or cars. In contrast, the Internet allows marketers to inexpensively find potential customers for which demand is very low, people on the far ends of the bell (normal) curve. For instance, the Internet makes it possible to sell independent music profitably to very small audiences. There's always some demand for almost any product. Put a string of such long tail sales together and you have a profitable business.

The Internet also provides new ways—often instantaneous and spontaneous—to gather information from customers, adjust product offerings, and increase customer value. Table 10-6 describes the leading marketing and advertising formats used in e-commerce.

Many e-commerce marketing firms use behavioral targeting techniques to increase the effectiveness of banner, rich media, and video ads. **Behavioral**

TABLE 10-6 ONLINE MARKETING AND ADVERTISING FORMATS (BILLIONS)

MARKETING FORMAT	2010 REVENUE	DESCRIPTION
Search engine	$12.3	Text ads targeted at precisely what the customer is looking for at the moment of shopping and purchasing. Sales oriented.
Display ads	$5.8	Banner ads (pop-ups and leave-behinds) with interactive features; increasingly behaviorally targeted to individual Web activity. Brand development and sales.
Classified	$1.9	Job, real estate, and services ads; interactive, rich media, and personalized to user searches. Sales and branding.
Rich media	$1.57	Animations, games, and puzzles. Interactive, targeted, and entertaining. Branding orientation.
Affiliate and blog marketing	$1.5	Blog and Web site marketing steers customers to parent sites; interactive, personal, and often with video. Sales orientation.
Video	$1.5	Fastest growing format, engaging and entertaining; behaviorally targeted, interactive. Branding and sales.
Sponsorships	$.4	Online games, puzzle, contests, and coupon sites sponsored by firms to promote products. Sales orientation.
E-mail	$.27	Effective, targeted marketing tool with interactive and rich media potential. Sales oriented.

targeting refers to tracking the click-streams (history of clicking behavior) of individuals on thousands of Web sites for the purpose of understanding their interests and intentions, and exposing them to advertisements that are uniquely suited to their behavior. Proponents believe this more precise understanding of the customer leads to more efficient marketing (the firm pays for ads only to those shoppers who are most interested in their products) and larger sales and revenues. Unfortunately, behavioral targeting of millions of Web users also leads to the invasion of personal privacy without user consent (see our discussion in Chapter 4). When consumers lose trust in their Web experience, they tend not to purchase anything.

Behavioral targeting takes place at two levels: at individual Web sites and on various advertising networks that track users across thousands of Web sites. All Web sites collect data on visitor browser activity and store it in a database. They have tools to record the site that users visited prior to coming to the Web site, where these users go when they leave that site, the type of operating system they use, browser information, and even some location data. They also record the specific pages visited on the particular site, the time spent on each page of the site, the types of pages visited, and what the visitors purchased (see Figure 10-3). Firms analyze this information about customer interests and behavior to develop precise profiles of existing and potential customers.

This information enables firms to understand how well their Web site is working, create unique personalized Web pages that display content or ads for products or services of special interest to each user, improve the customer's experience, and create additional value through a better understanding of the shopper (see Figure 10-4). By using personalization technology to modify the Web pages presented to each customer, marketers achieve some of the benefits of using individual salespeople at dramatically lower costs. For instance,

FIGURE 10-3   WEB SITE VISITOR TRACKING

The shopper clicks on the home page. The store can tell that the shopper arrived from the Yahoo! portal at 2:30 PM (which might help determine staffing for customer service centers) and how long she lingered on the home page (which might indicate trouble navigating the site).

The shopper clicks on blouses, clicks to select a woman's white blouse, then clicks to view the same item in pink. The shopper clicks to select this item in a size 10 in pink and clicks to place it in her shopping cart. This information can help the store determine which sizes and colors are most popular.

From the shopping cart page, the shopper clicks to close the browser to leave the Web site without purchasing the blouse. This action could indicate the shopper changed her mind or that she had a problem with the Web site's checkout and payment process. Such behavior might signal that the Web site was not well designed.

E-commerce Web sites have tools to track a shopper's every step through an online store. Close examination of customer behavior at a Web site selling women's clothing shows what the store might learn at each step and what actions it could take to increase sales.

General Motors will show a Chevrolet banner ad to women emphasizing safety and utility, while men will receive different ads emphasizing power and ruggedness.

FIGURE 10-4   WEB SITE PERSONALIZATION

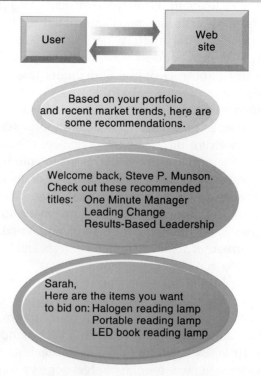

Firms can create unique personalized Web pages that display content or ads for products or services of special interest to individual users, improving the customer experience and creating additional value.

What if you are a large national advertising company with many different clients trying to reach millions of consumers? What if you were a large global manufacturer trying to reach potential consumers for your products? With millions of Web sites, working with each one would be impractical. Advertising networks solve this problem by creating a network of several thousand of the most popular Web sites visited by millions of people, tracking the behavior of these users across the entire network, building profiles of each user, and then selling these profiles to advertisers. Popular Web sites download dozens of Web tracking cookies, bugs, and beacons, which report user online behavior to remote servers without the users' knowledge. Looking for young, single consumers, with college degrees, living in the Northeast, in the 18–34 age range who are interested purchasing a European car? Not a problem. Advertising networks can identify and deliver hundreds of thousands of people who fit this profile and expose them to ads for European cars as they move from one Web site to another. Estimates vary, but behaviorally targeted ads are 10 times more likely to produce a consumer response than a randomly chosen banner or video ad (see Figure 10-5). So-called advertising exchanges use this same technology to auction access to people with very specific profiles to advertisers in a few milliseconds.

B2B E-COMMERCE: NEW EFFICIENCIES AND RELATIONSHIPS

The trade between business firms (business-to-business commerce or B2B) represents a huge marketplace. The total amount of B2B trade in the United States in 2009 was about $12.2 trillion, with B2B e-commerce (online B2B) contributing about $3.6 trillion of that amount (U.S. Census Bureau, 2010; authors' estimates). By 2014, B2B e-commerce should grow to about $5.1 trillion in the United States, assuming an average growth rate of about 7 percent. The process of conducting trade among business firms is complex and requires significant human interven-

FIGURE 10-5   HOW AN ADVERTISING NETWORK SUCH AS DOUBLECLICK WORKS

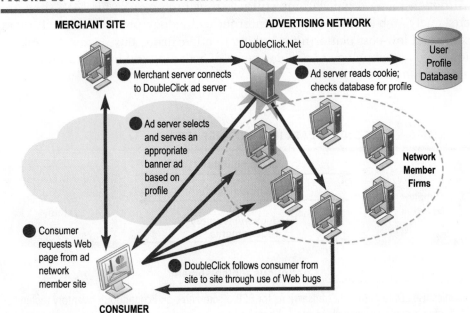

Advertising networks have become controversial among privacy advocates because of their ability to track individual consumers across the Internet. We discuss privacy issues further in Chapter 4.

tion, and therefore, it consumes significant resources. Some firms estimate that each corporate purchase order for support products costs them, on average, at least $100 in administrative overhead. Administrative overhead includes processing paper, approving purchase decisions, spending time using the telephone and fax machines to search for products and arrange for purchases, arranging for shipping, and receiving the goods. Across the economy, this adds up to trillions of dollars annually being spent for procurement processes that could potentially be automated. If even just a portion of inter-firm trade were automated, and parts of the entire procurement process assisted by the Internet, literally trillions of dollars might be released for more productive uses, consumer prices potentially would fall, productivity would increase, and the economic wealth of the nation would expand. This is the promise of B2B e-commerce. The challenge of B2B e-commerce is changing existing patterns and systems of procurement, and designing and implementing new Internet-based B2B solutions.

Business-to-business e-commerce refers to the commercial transactions that occur among business firms. Increasingly, these transactions are flowing through a variety of different Internet-enabled mechanisms. About 80 percent of online B2B e-commerce is still based on proprietary systems for **electronic data interchange (EDI)**. Electronic data interchange enables the computer-to-computer exchange between two organizations of standard transactions such as invoices, bills of lading, shipment schedules, or purchase orders. Transactions are automatically transmitted from one information system to another through a network, eliminating the printing and handling of paper at one end and the inputting of data at the other. Each major industry in the United States and much of the rest of the world has EDI standards that define the structure and information fields of electronic documents for that industry.

EDI originally automated the exchange of documents such as purchase orders, invoices, and shipping notices. Although some companies still use EDI for document automation, firms engaged in just-in-time inventory replenishment and continuous production use EDI as a system for continuous replenishment. Suppliers have online access to selected parts of the purchasing firm's production and delivery schedules and automatically ship materials and goods to meet prespecified targets without intervention by firm purchasing agents (see Figure 10-6).

Although many organizations still use private networks for EDI, they are increasingly Web-enabled because Internet technology provides a much more flexible and low-cost platform for linking to other firms. Businesses are able to extend digital technology to a wider range of activities and broaden their circle of trading partners.

FIGURE 10-6   ELECTRONIC DATA INTERCHANGE (EDI)

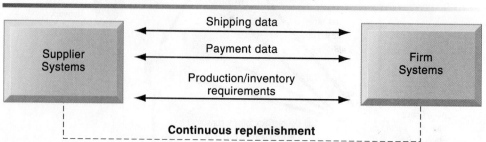

Companies use EDI to automate transactions for B2B e-commerce and continuous inventory replenishment. Suppliers can automatically send data about shipments to purchasing firms. The purchasing firms can use EDI to provide production and inventory requirements and payment data to suppliers.

Take procurement, for example. Procurement involves not only purchasing goods and materials but also sourcing, negotiating with suppliers, paying for goods, and making delivery arrangements. Businesses can now use the Internet to locate the lowest-cost supplier, search online catalogs of supplier products, negotiate with suppliers, place orders, make payments, and arrange transportation. They are not limited to partners linked by traditional EDI networks.

The Internet and Web technology enable businesses to create new electronic storefronts for selling to other businesses with multimedia graphic displays and interactive features similar to those for B2C commerce. Alternatively, businesses can use Internet technology to create extranets or electronic marketplaces for linking to other businesses for purchase and sale transactions.

Private industrial networks typically consist of a large firm using an extranet to link to its suppliers and other key business partners (see Figure 10-7). The network is owned by the buyer, and it permits the firm and designated suppliers, distributors, and other business partners to share product design and development, marketing, production scheduling, inventory management, and unstructured communication, including graphics and e-mail. Another term for a private industrial network is a **private exchange**.

An example is VW Group Supply, which links the Volkswagen Group and its suppliers. VW Group Supply handles 90 percent of all global purchasing for Volkswagen, including all automotive and parts components.

Net marketplaces, which are sometimes called e-hubs, provide a single, digital marketplace based on Internet technology for many different buyers and sellers (see Figure 10-8). They are industry owned or operate as independent intermediaries between buyers and sellers. Net marketplaces generate revenue from purchase and sale transactions and other services provided to clients. Participants in Net marketplaces can establish prices through online negotiations, auctions, or requests for quotations, or they can use fixed prices.

FIGURE 10-7   A PRIVATE INDUSTRIAL NETWORK

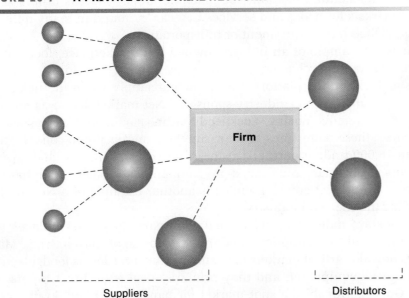

A private industrial network, also known as a private exchange, links a firm to its suppliers, distributors, and other key business partners for efficient supply chain management and other collaborative commerce activities.

FIGURE 10-8   A NET MARKETPLACE

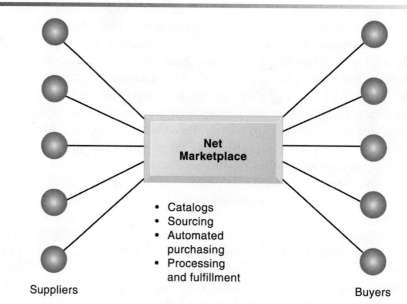

Net marketplaces are online marketplaces where multiple buyers can purchase from multiple sellers.

There are many different types of Net marketplaces and ways of classifying them. Some Net marketplaces sell direct goods and some sell indirect goods. *Direct goods* are goods used in a production process, such as sheet steel for auto body production. *Indirect goods* are all other goods not directly involved in the production process, such as office supplies or products for maintenance and repair. Some Net marketplaces support contractual purchasing based on long-term relationships with designated suppliers, and others support short-term spot purchasing, where goods are purchased based on immediate needs, often from many different suppliers.

Some Net marketplaces serve vertical markets for specific industries, such as automobiles, telecommunications, or machine tools, whereas others serve horizontal markets for goods and services that can be found in many different industries, such as office equipment or transportation.

Exostar is an example of an industry-owned Net marketplace, focusing on long-term contract purchasing relationships and on providing common networks and computing platforms for reducing supply chain inefficiencies. This aerospace and defense industry-sponsored Net marketplace was founded jointly by BAE Systems, Boeing, Lockheed Martin, Raytheon, and Rolls-Royce plc to connect these companies to their suppliers and facilitate collaboration. More than 16,000 trading partners in the commercial, military, and government sectors use Exostar's sourcing, e-procurement, and collaboration tools for both direct and indirect goods. Elemica is another example of a Net marketplace serving the chemical industry.

Exchanges are independently owned third-party Net marketplaces that connect thousands of suppliers and buyers for spot purchasing. Many exchanges provide vertical markets for a single industry, such as food, electronics, or industrial equipment, and they primarily deal with direct inputs. For example, Go2paper enables a spot market for paper, board, and kraft among buyers and sellers in the paper industries from over 75 countries.

Exchanges proliferated during the early years of e-commerce but many have failed. Suppliers were reluctant to participate because the exchanges encouraged competitive bidding that drove prices down and did not offer any long-term relationships with buyers or services to make lowering prices worthwhile. Many essential direct purchases are not conducted on a spot basis because they require contracts and consideration of issues such as delivery timing, customization, and quality of products.

10.3 THE MOBILE DIGITAL PLATFORM AND MOBILE E-COMMERCE

Walk down the street in any major metropolitan area and count how many people are pecking away at their iPhones or BlackBerrys. Ride the trains, fly the planes, and you'll see your fellow travelers reading an online newspaper, watching a video on their phone, or reading a novel on their Kindle. In five years, the majority of Internet users in the United States will rely on mobile devices as their primary device for accessing the Internet. M-commerce has taken off.

In 2010, m-commerce represented less than 10 percent of all e-commerce, with about $5 billion in annual revenues generated by selling music, videos, ring tones, applications, movies, television, and location-based services like local restaurant locators and traffic updates. However, m-commerce is the fastest growing form of e-commerce, with some areas expanding at a rate of 50 percent or more per year, and is estimated to grow to $19 billion in 2014 (see Figure 10-9). In 2010, there were an estimated 5 billion cell phone subscribers worldwide, with over 855 million in China and 300 million in the United States (eMarketer, 2010d).

M-COMMERCE SERVICES AND APPLICATIONS

The main areas of growth in mobile e-commerce are location-based services, about $215 million in revenue in 2010; software application sales at stores such as iTunes (about $1.8 billion); entertainment downloads of ring tones, music, video, and TV shows (about $1 billion); mobile display advertising ($784 million); direct shopping services such as Slifter ($200 million); and e-book sales ($338 million).

FIGURE 10-9   CONSOLIDATED MOBILE COMMERCE REVENUES

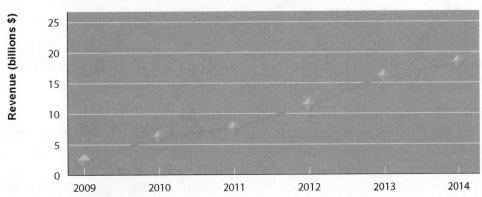

Mobile e-commerce is the fastest growing type of B2C e-commerce although it represents only a small part of all e-commerce in 2010.

M-commerce applications have taken off for services that are time-critical, that appeal to people on the move, or that accomplish a task more efficiently than other methods. They are especially popular in Europe, Japan, South Korea, and other countries with strong wireless broadband infrastructures. The following sections describe some examples.

Location-Based Services

Wikitude.me provides a special kind of browser for smart phones equipped with a built-in global positioning system (GPS) and compass that can identify your precise location and where the phone is pointed. Using information from over 800,000 points of interest available on Wikipedia, plus thousands of other local sites, the browser overlays information about points of interest you are viewing, and displays that information on your smartphone screen, superimposed on a map or photograph that you just snapped. For example, users can point their smart phone cameras towards mountains from a tour bus and see the names and heights of the mountains displayed on the screen. Lost in a European medieval city, or downtown Los Angeles? Open up the Wikitude browser, point your camera at a building, and then find the address and other interesting details. Wikitude.me also allows users to geo-tag the world around them, and then submit the tags to Wikitude in order to share content with other users. In 2010, both Facebook and Twitter launched a Places feature that allows users to let their friends know where they are. These services compete with Foursquare and Gowalla, which allow users to check in at places and broadcast their location to friends.

Loopt is a free social networking application that allows you to share your status and track the location of friends via smartphones such as the iPhone, BlackBerry, and over 100 other mobile devices. Users also have the ability to integrate Loopt with other social networks, including Facebook and Twitter. Loopt has 4 million users. The service doesn't sell information to advertisers, but does post ads based on user location. Loopt's target is to deal with advertisers at the walking level (within 200 to 250 meters).

Foursquare provides a similar service to 4 million registered users, who are able to connect with friends and update their location. Points are awarded for "checking in" at designated venues. Users choose to have their check-ins posted on their accounts on Twitter, Facebook, or both. Users also earn badges by checking in at locations with certain tags, for check-in frequency, or for the time of check-in. More than 3,000 restaurants, bars, and other businesses (including 4Food, described in the chapter-opening case) use Foursquare to attract customers with promotions.

Banking and Financial Services

Banks and credit card companies are rolling out services that let customers manage their accounts from their mobile devices. JPMorgan Chase and Bank of America customers can use their cell phones to check account balances, transfer funds, and pay bills.

Wireless Advertising and Retailing

Although the mobile advertising market is currently small ($784 million), it is rapidly growing (up 17 percent from last year and expected to grow to over $6.2 billion by 2014), as more and more companies seek ways to exploit new databases of location-specific information. Alcatel-Lucent offers a new service to be managed by 1020 Placecast that will identify cell phone users within a specified distance of an advertiser's nearest outlet and notify them about the outlet's address and phone number, perhaps including a link to a coupon or other promotion. 1020 Placecast's clients include Hyatt, FedEx, and Avis Rent A Car.

Yahoo displays ads on its mobile home page for companies such as Pepsi, Procter & Gamble, Hilton, Nissan, and Intel. Google is displaying ads linked to cell phone searches by users of the mobile version of its search engine, while Microsoft offers banner and text advertising on its MSN Mobile portal in the United States. Ads are embedded in games, videos, and other mobile applications.

Shopkick is a mobile application that enables retailers such as Best Buy, Sports Authority, and Macy's to offer coupons to people when they walk into their stores. The shopkick app automatically recognizes when the user has entered a partner retail store, and offers a new virtual currency called "kickbucks," which can be redeemed for Facebook credits, iTunes Gift Cards, travel vouchers, DVD's, or immediate cash-back rewards at any of the partner stores.

In 2010, shoppers ordered about $2.2 billion in physical goods from Web sites via smartphones (over 1 billion of that at Amazon alone). Thirty percent of retailers have m-commerce Web sites—simplified versions of their Web sites that make it possible for shoppers to use cell phones to place orders. Clothing retailers Lilly Pulitzer and Armani Exchange, Home Depot, and 1–800 Flowers are among those companies with specialized apps for m-commerce sales.

Games and Entertainment

Cell phones have developed into portable entertainment platforms. Smartphones like the iPhone and Droid offer downloadable and streaming digital games, movies, TV shows, music, and ringtones.

Users of broadband services from the major wireless vendors can stream on-demand video clips, news clips, and weather reports. MobiTV, offered by Sprint and AT&T Wireless, features live TV programs, including MSNBC and Fox Sports. Film companies are starting to produce short films explicitly designed to play on mobile phones. User-generated content is also appearing in mobile form. Facebook, MySpace, YouTube, and other social networking sites have versions for mobile devices. In 2010, the top 10 most popular apps on Facebook are games, led by Farmville with over 16 million daily users.

10.4 BUILDING AN E-COMMERCE WEB SITE

Building a successful e-commerce site requires a keen understanding of business, technology, and social issues, as well as a systematic approach. A complete treatment of the topic is beyond the scope of this text, and students should consult books devoted to just this topic (Laudon and Traver, 2011). The two most important management challenges in building a successful e-commerce site are (1) developing a clear understanding of your business objectives and (2) knowing how to choose the right technology to achieve those objectives.

PIECES OF THE SITE-BUILDING PUZZLE

Let's assume you are a manager for a medium-sized, industrial parts firm of around 10,000 employees worldwide, operating in eight countries in Europe, Asia, and North America. Senior management has given you a budget of $1 million to build an e-commerce site within one year. The purpose of this site will be to sell and service the firm's 20,000 customers, who are mostly small machine and metal fabricating shops around the world. Where do you start?

First, you must be aware of the main areas where you will need to make decisions. On the organizational and human resources fronts, you will have to bring together a team of individuals who possess the skill sets needed to build and manage a successful e-commerce site. This team will make the key decisions about technology, site design, and social and information policies that will be applied at your site. The entire site development effort must be closely managed if you hope to avoid the disasters that have occurred at some firms.

You will also need to make decisions about your site's hardware, software, and telecommunications infrastructure. The demands of your customers should drive your choices of technology. Your customers will want technology that enables them to find what they want easily, view the product, purchase the product, and then receive the product from your warehouses quickly. You will also have to carefully consider your site's design. Once you have identified the key decision areas, you will need to think about a plan for the project.

BUSINESS OBJECTIVES, SYSTEM FUNCTIONALITY, AND INFORMATION REQUIREMENTS

In planning your Web site you need to answer the question, "What do we want the e-commerce site to do for our business?" The key lesson to be learned here is to let the business decisions drive the technology, not the reverse. This will ensure that your technology platform is aligned with your business. We will assume here that you have identified a business strategy and chosen a business model to achieve your strategic objectives. (Review Chapter 3.) But how do you translate your strategies, business models, and ideas into a working e-commerce site?

Your planning should identify the specific business objectives for your site, and then develop a list of system functionalities and information requirements. Business objectives are simply capabilities you want your site to have. System functionalities are types of information systems capabilities you will need to achieve your business objectives. The information requirements for a system are the information elements that the system must produce in order to achieve the business objectives.

Table 10-7 describes some basic business objectives, system functionalities, and information requirements for a typical e-commerce site. The objectives must be translated into a description of system functionalities and ultimately into a set of precise information requirements. The specific information requirements for a system typically are defined in much greater detail than Table 10-7 indicates (see Chapter 13). The business objectives of an e-commerce site are similar to those of a physical retail store, but they must be provided entirely in digital form, 24 hours a day, 7 days a week.

BUILDING THE WEB SITE: IN-HOUSE VERSUS OUTSOURCING

There are many choices for building and maintaining Web sites. Much depends on how much money you are willing to spend. Choices range from outsourcing the entire Web site development to an external vendor to building everything yourself (in-house). You also have a second decision to make: will you host

TABLE 10-7 SYSTEM ANALYSIS: BUSINESS OBJECTIVES, SYSTEM FUNCTIONALITY, AND INFORMATION REQUIREMENTS FOR A TYPICAL E-COMMERCE SITE

BUSINESS OBJECTIVE	SYSTEM FUNCTIONALITY	INFORMATION REQUIREMENTS
Display goods	Digital catalog	Dynamic text and graphics catalog
Provide product information (content)	Product database	Product description, stocking numbers, inventory levels
Personalize/customize product	Customer on-site tracking	Site log for every customer visit; data mining capability to identify common customer paths and appropriate responses
Execute a transaction payment	Shopping cart/payment system	Secure credit card clearing; multiple options
Accumulate customer information	Customer database	Name, address, phone, and e-mail for all customers; online customer registration
Provide after-sale customer support	Sales database and customer relationship management system (CRM)	Customer ID, product, date, payment, shipment date
Coordinate marketing/advertising	Ad server, e-mail server, e-mail, campaign manager, ad banner manager	Site behavior log of prospects and customers linked to e-mail and banner ad campaigns
Understand marketing effectiveness	Site tracking and reporting system	Number of unique visitors, pages visited, products purchased, identified by marketing campaign
Provide production and supplier links	Inventory management system	Product and inventory levels, supplier ID and contact, order quantity data by product

(operate) the site on your firm's own servers or will you outsource the hosting to a Web host provider? There are some vendors who will design, build, and host your site, while others will either build or host (but not both). Figure 10-10 illustrates the alternatives.

FIGURE 10-10 CHOICES IN BUILDING AND HOSTING WEB SITES

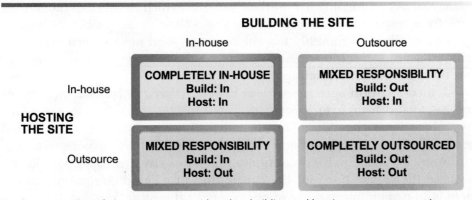

You have a number of alternatives to consider when building and hosting an e-commerce site.

The Building Decision

If you elect to build your own site, there are a range of options. Unless you are fairly skilled, you should use a pre-built template to create the Web site. For example, Yahoo Merchant Solutions, Amazon Stores, and eBay all provide templates that merely require you to input text, graphics, and other data, as well as the infrastructure to run the Web site once it has been created. This is the least costly and simplest solution, but you will be limited to the "look and feel" and functionality provided by the template and infrastructure.

If you have some experience with computers, you might decide to build the site yourself. There is a broad variety of tools, ranging from those that help you build everything truly "from scratch," such as Adobe Dreamweaver, Adobe InDesign, and Microsoft Expression, to top-of-the-line prepackaged site-building tools that can create sophisticated sites customized to your needs.

The decision to build a Web site on your own has a number of risks. Given the complexity of features such as shopping carts, credit card authentication and processing, inventory management, and order processing, development costs are high, as are the risks of doing a poor job. You will be reinventing what other specialized firms have already built, and your staff may face a long, difficult learning curve, delaying your entry to market. Your efforts could fail. On the positive side, you may able to build a site that does exactly what you want, and develop the in-house knowledge to revise the site rapidly if necessitated by a changing business environment.

If you choose more expensive site-building packages, you will be purchasing state-of-the-art software that is well tested. You could get to market sooner. However, to make a sound decision, you will have to evaluate many different software packages and this can take a long time. You may have to modify the packages to fit your business needs and perhaps hire additional outside consultants to do the modifications. Costs rise rapidly as modifications mount. (We discuss this problem in greater detail in Chapter 13.) A $4,000 package can easily become a $40,000 to $60,000 development project.

In the past, bricks-and-mortar retailers typically designed their e-commerce sites themselves (because they already had the skilled staff and IT infrastructure in place to do this). Today, however, larger retailers rely heavily on external vendors to provide sophisticated Web site capabilities, while also maintaining a substantial internal staff. Medium-size start-ups will often purchase a sophisticated package and then modify it to suit their needs. Very small mom-and-pop firms seeking simple storefronts will use templates.

The Hosting Decision

Now let's look at the hosting decision. Most businesses choose to outsource hosting and pay a company to host their Web site, which means that the hosting company is responsible for ensuring the site is "live" or accessible, 24 hours a day. By agreeing to a monthly fee, the business need not concern itself with technical aspects of setting up and maintaining a Web server, telecommunications links, or specialized staffing.

With a **co-location** agreement, your firm purchases or leases a Web server (and has total control over its operation) but locates the server in a vendor's physical facility. The vendor maintains the facility, communications lines, and the machinery. In the age of cloud computing, it is much less expensive to host your Web site in virtualized computing facilities. In this case, you do not purchase the server, but rent the capabilities of a cloud computing center. There is an extraordinary range of prices for cloud hosting, ranging from $4.95 a month, to several hundred thousands of dollars per month depending on the size of the Web site, bandwidth, storage, and support requirements. Very large providers (such as IBM, HP, and Oracle) achieve large economies of scale by establishing huge "server farms" located strategically around the country and the globe. What this means is that the cost of pure hosting has fallen as fast as the fall in server prices, dropping about 50 percent every year.

Web Site Budgets

Simple Web sites can be built and hosted with a first-year cost of $5,000 or less. The Web sites of large firms with high levels of interactivity and linkage to corporate systems cost several million dollars a year to create and operate. For instance, in September 2006, Bluefly, which sells discounted women's and men's designer clothes online, embarked on the process of developing an improved version of its Web site based on software from Art Technology Group (ATG). It launched the new site in August 2008. To date, it has invested over $5.3 million in connection with the redevelopment of the Web site. In 2010, Bluefly had online sales of $81 million, and is growing at 7.5 percent a year. It's e-commerce technology budget is over $8 million a year, roughly 10 percent of its total revenues (Bluefly, Inc., 2010).

Figure 10-11 provides some idea of the relative size of various Web site cost components. In general, the cost of hardware, software, and telecommunications for building and operating a Web site has fallen dramatically (by over 50 percent) since 2000, making it possible for very small entrepreneurs to create fairly sophisticated sites. At the same time, the costs of system maintenance and content creation have risen to make up more than half of typical Web site budgets. Providing content and smooth 24/7 operations are both very labor-intensive.

FIGURE 10-11 COMPONENTS OF A WEB SITE BUDGET

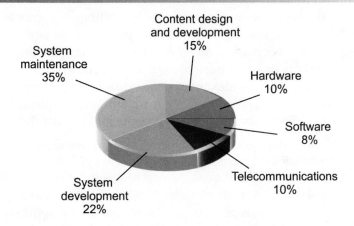

10.5 HANDS-ON MIS PROJECTS

The projects in this section give you hands-on experience developing e-commerce strategies for businesses, using spreadsheet software to research the profitability of an e-commerce company, and using Web tools to research and evaluate e-commerce hosting services.

Management Decision Problems

1. Columbiana is a small, independent island in the Caribbean. It wants to develop its tourist industry and attract more visitors. The island has many

historical buildings, forts, and other sites, along with rain forests and striking mountains. A few first-class hotels and several dozen less-expensive accommodations can be found along its beautiful white sand beaches. The major airlines have regular flights to Columbiana, as do several small airlines. Columbiana's government wants to increase tourism and develop new markets for the country's tropical agricultural products. How can a Web presence help? What Internet business model would be appropriate? What functions should the Web site perform?

2. Explore the Web sites of the following companies: Blue Nile, J.Crew, Circuit City, Black&Decker, Peet's Coffee & Tea, and Priceline. Determine which of these Web sites would benefit most from adding a company-sponsored blog to the Web site. List the business benefits of the blog. Specify the intended audience for the blog. Decide who in the company should author the blog, and select some topics for the blog.

Improving Decision Making: Using Spreadsheet Software to Analyze a Dot-Com Business

Software skills: Spreadsheet downloading, formatting, and formulas
Business skills: Financial statement analysis

Publicly traded companies, including those specializing in e-commerce, are required to file financial data with the U.S. Securities and Exchange Commission. By analyzing this information, you can determine the profitability of an e-commerce company and the viability of its business model.

Pick one e-commerce company on the Internet, for example, Ashford, Buy.com, Yahoo, or Priceline. Study the Web pages that describe the company and explain its purpose and structure. Use the Web to find articles that comment on the company. Then visit the Securities and Exchange Commission's Web site at www.sec.gov and select Filings & Forms to access the company's 10-K (annual report) form showing income statements and balance sheets. Select only the sections of the 10-K form containing the desired portions of financial statements that you need to examine, and download them into your spreadsheet. (MyMISLab provides more detailed instructions on how to download this 10-K data into a spreadsheet.) Create simplified spreadsheets of the company's balance sheets and income statements for the past three years.

- Is the company a dot-com success, borderline business, or failure? What information dictates the basis of your decision? Why? When answering these questions, pay special attention to the company's three-year trends in revenues, costs of sales, gross margins, operating expenses, and net margins.

- Prepare an overhead presentation (with a minimum of five slides), including appropriate spreadsheets or charts, and present your work to your professor and classmates.

Achieving Operational Excellence: Evaluating E-Commerce Hosting Services

Software skills: Web browser software
Business skills: Evaluating e-commerce hosting services

This project will help develop your Internet skills in commercial services for hosting an e-commerce site for a small start-up company.

You would like to set up a Web site to sell towels, linens, pottery, and tableware from Portugal and are examining services for hosting small business

Internet storefronts. Your Web site should be able to take secure credit card payments and to calculate shipping costs and taxes. Initially, you would like to display photos and descriptions of 40 different products. Visit Yahoo! Small Business, GoDaddy, and Volusion and compare the range of e-commerce hosting services they offer to small businesses, their capabilities, and costs. Also examine the tools they provide for creating an e-commerce site. Compare these services and decide which you would use if you were actually establishing a Web store. Write a brief report indicating your choice and explaining the strengths and weaknesses of each.

LEARNING TRACK MODULES

The following Learning Tracks provide content relevant to topics covered in this chapter:

1. Building a Web Page
2. E-commerce Challenges: The Story of Online Groceries
3. Build an E-commerce Business Plan
4. Hot New Careers in E-commerce

Review Summary

1. *What are the unique features of e-commerce, digital markets, and digital goods?*

    E-commerce involves digitally enabled commercial transactions between and among organizations and individuals. Unique features of e-commerce technology include ubiquity, global reach, universal technology standards, richness, interactivity, information density, capabilities for personalization and customization, and social technology.

    Digital markets are said to be more "transparent" than traditional markets, with reduced information asymmetry, search costs, transaction costs, and menu costs, along with the ability to change prices dynamically based on market conditions. Digital goods, such as music, video, software, and books, can be delivered over a digital network. Once a digital product has been produced, the cost of delivering that product digitally is extremely low.

2. *What are the principal e-commerce business and revenue models?*

    E-commerce business models are e-tailers, transaction brokers, market creators, content providers, community providers, service providers, and portals. The principal e-commerce revenue models are advertising, sales, subscription, free/freemium, transaction fee, and affiliate.

3. *How has e-commerce transformed marketing?*

    The Internet provides marketers with new ways of identifying and communicating with millions of potential customers at costs far lower than traditional media. Crowdsourcing utilizing the "wisdom of crowds" helps companies learn from customers in order to improve product offerings and increase customer value. Behavioral targeting techniques increase the effectiveness of banner, rich media, and video ads.

4. *How has e-commerce affected business-to-business transactions?*

    B2B e-commerce generates efficiencies by enabling companies to locate suppliers, solicit bids, place orders, and track shipments in transit electronically. Net marketplaces provide a single, digital marketplace for many buyers and sellers. Private industrial networks link a firm with its suppliers and other strategic business partners to develop highly efficient and responsive supply chains.

5. *What is the role of m-commerce in business, and what are the most important m-commerce applications?*

    M-commerce is especially well-suited for location-based applications, such as finding local hotels and restaurants, monitoring local traffic and weather, and providing personalized location-based marketing. Mobile phones and handhelds are being used for mobile bill payment, banking, securities trading, transportation schedule updates, and downloads of digital content, such as music, games, and video clips. M-commerce requires wireless portals and special digital payment systems that can handle micropayments.

6. *What issues must be addressed when building an e-commerce Web site?*

    Building a successful e-commerce site requires a clear understanding of the business objectives to be achieved by the site and selection of the right technology to achieve those objectives. E-commerce sites can be built and hosted in-house or partially or fully outsourced to external service providers.

Key Terms

Advertising revenue model, 387
Affiliate revenue model, 389
Behavioral marketing, 392
Business-to-business (B2B) electronic commerce, 381
Business-to-consumer (B2C) electronic commerce, 381
Co-location, 404
Community providers, 384
Consumer-to-consumer (C2C) electronic commerce, 381
Cost Transparency, 377
Crowdsourcing, 392
Customization, 378
Digital goods, 380
Disintermediation, 379
Dynamic pricing, 379
Electronic data interchange (EDI), 396
E-tailer, 382
Exchanges, 398
Free/freemium revenue model, 388
Information asymmetry, 378
Information density, 377
Intellectual property, 383
Long tail marketing, 392
Market creator, 384

Market entry costs, 377
Marketspace, 375
Menu costs, 379
Micropayment systems, 388
Mobile commerce (m-commerce), 382
Net marketplaces, 397
Personalization, 378
Podcasting, 383
Prediction market, 392
Price discrimination, 377
Price transparency, 377
Private exchange, 397
Private industrial networks, 397
Revenue model, 387
Richness, 377
Sales revenue model, 388
Search costs, 377
Social shopping, 389
Streaming, 383
Subscription revenue model, 388
Transaction costs, 376
Transaction fee revenue model, 388
Wisdom of crowds, 391

Review Questions

1. What are the unique features of e-commerce, digital markets, and digital goods?

    *   Name and describe four business trends and three technology trends shaping e-commerce today.

    *   List and describe the eight unique features of e-commerce.

    *   Define a digital market and digital goods and describe their distinguishing features.

2. What are the principal e-commerce business and revenue models?

    *   Name and describe the principal e-commerce business models.

    *   Name and describe the e-commerce revenue models.

3. How has e-commerce transformed marketing?

    *   Explain how social networking and the "wisdom of crowds" help companies improve their marketing.

    *   Define behavioral targeting and explain how it works at individual Web sites and on advertising networks.

4. How has e-commerce affected business-to-business transactions?

    *   Explain how Internet technology supports business-to-business electronic commerce.

- Define and describe Net marketplaces and explain how they differ from private industrial networks (private exchanges).

5. What is the role of m-commerce in business, and what are the most important m-commerce applications?

- List and describe important types of m-commerce services and applications.

- Describe some of the barriers to m-commerce.

6. What issues must be addressed when building an e-commerce Web site?

- List and describe each of the factors that go into the building of an e-commerce Web site.

- List and describe four business objectives, four system functionalities, and four information requirements of a typical e-commerce Web site.

- List and describe each of the options for building and hosting e-commerce Web sites.

Discussion Questions

1. How does the Internet change consumer and supplier relationships?

2. The Internet may not make corporations obsolete, but the corporations will have to change their business models. Do you agree? Why or why not?

3. How have social technologies changed e-commerce?

Video Cases

Video Cases and Instructional Videos illustrating some of the concepts in this chapter are available. Contact your instructor to access these videos.

Collaboration and Teamwork: Performing a Competitive Analysis of E-commerce Sites

Form a group with three or four of your classmates. Select two businesses that are competitors in the same industry and that use their Web sites for electronic commerce. Visit these Web sites. You might compare, for example, the Web sites for iTunes and Napster, Amazon and BarnesandNoble.com, or E*Trade and Scottrade. Prepare an evaluation of each business's Web site in terms of its functions, user friendliness, and ability to support the company's business strategy. Which Web site does a better job? Why? Can you make some recommendations to improve these Web sites? If possible, use Google Sites to post links to Web pages, team communication announcements, and work assignments; to brainstorm; and to work collaboratively on project documents. Try to use Google Docs to develop a presentation of your findings for the class.

Amazon vs. Walmart: Which Giant Will Dominate E-commerce?
CASE STUDY

Since arriving on the dot-com scene in 1995, Amazon.com has grown from a small online bookseller to one of the largest retailing companies in the world, and easily the largest e-commerce retailer. The company has come a long way from its roots as a small Internet start-up selling books online. In addition to books, Amazon now sells millions of new, used, and collectible items in categories such as apparel and accessories, electronics, computers, kitchen and housewares, music, DVDs, videos, cameras, office products, toys and baby items, computers, software, travel services, sporting goods, jewelry, and watches. In 2010, sales of electronics and general merchandise comprised the majority of Amazon's sales for the first time.

Amazon.com would like to be "The Walmart of the Web," and it is indeed the Internet's top retailer. But in 2010, another firm emerged as a serious challenger for the title of 'Walmart of the Web': Walmart. Though Walmart is a latecomer to the world of e-commerce, the world's largest retailer appears to have its sights set on Amazon and is ready to battle it out for online e-tailing supremacy.

In contrast with Amazon, Walmart was founded as a traditional, off-line, physical store in 1962, and has grown from a single general store managed by founder Sam Walton to the largest retailer in the world with nearly 8,000 stores worldwide.

Based in Bentonville, Arkansas, Walmart made $405 billion in sales last year, which is about 20 times as much as Amazon. In fact, based on current size alone, the battle between Walmart and Amazon is far from a clash of two similarly powerful titans. Walmart is clearly the bigger and stronger of the two, and for the time being, Amazon is not a big threat to Walmart as a whole.

Amazon, however, is not an easy target. The company has created a recognizable and highly successful brand in online retailing as a mass-market, low-price, high-volume online superstore. It has developed extensive warehousing facilities and an extremely efficient distribution network specifically designed for Web shopping. Its premium shipping service, Amazon Prime, provides "free" two-day shipping at an affordable price (currently only $79 per year), often considered to be a weak point for online retailers. Even without Amazon Prime, designated Super Saver items totaling at least $25.00 ship for free.

Amazon's technology platform is massive and powerful enough to support not only sales of its own items but also those of third-party small and large businesses, which integrate their products into Amazon's Web site and use its order entry and payment systems to process their own sales. (Amazon does not own these products, and shipping is handled by the third party, with Amazon collecting 10-20 percent on the sale). This enables Amazon to offer an even wider array of products than it could carry on its own while keeping inventory costs low and increasing revenue. Amazon has further expanded its product selection via acquisitions such as the 2009 purchase of online shoe shopping site Zappos.com, which earned $1 billion in retail sales in 2008 and gave the company an edge in footwear.

In the third quarter of 2009, when retail sales dipped 4 percent across the board, Amazon's sales increased by 24 percent. Its sales of electronics and general merchandise, which is the most prominent area of competition between Amazon and Walmart, were up 44 percent. And e-commerce is expected to become an increasingly large portion of total retail sales. Some estimates indicate that e-commerce could account for 15 to 20 percent of total retail in the United States within the next decade, as more and more shoppers opt to avoid the hassle of shopping at a physical location in favor of shopping online. If this happens, Amazon is in the best position to benefit. In the meantime, e-commerce has not suffered as much from the recession and is recovering more quickly than traditional retail, giving Walmart more reason for concern.

However, Walmart also brings a strong hand to the table. It is an even larger and more recognizable brand than Amazon. Consumers associate Walmart with the lowest price, which Walmart has the flexibility to offer on any given item because of its size and ability to keep overhead costs to a minimum. Walmart can lose money selling a hot product at extremely low margins and expect to make money on the strength of the large quantities of other items it sells. It also has a legendary continuous inventory replenishment system that starts restocking merchandise as soon as an item reaches the checkout counter. Walmart's efficiency, flexibility, and ability to fine-tune its inventory to carry exactly what customers want have been enduring sources of competitive advantage. Walmart also has a significant physical presence, with stores all across the United States and in many other countries, and its stores provide the instant gratification of shopping, buying an item, and taking it home immediately, as opposed to waiting when ordering from Amazon.

Walmart believes Amazon's Achilles' heel is the costs and delays of shipping online purchases to buyers. Customers who buy some of the more than 1.5 million products on Walmart.com can have them shipped free to a local Walmart, and pick up their purchases at these stores. Internet shoppers may be tempted to pick up other items once they are inside the store. New service desks at the front of some stores make it even easier for shoppers to retrieve their purchases. A Walmart on the outskirts of Chicago is testing a drive-through window, similar to those found at pharmacies and fast-food restaurants, where shoppers can pick up their Internet orders.

In late 2009, Walmart.com began aggressively lowering prices on a wide variety of popular items, making sure in each instance to undercut Amazon's price. The types of items Walmart discounted included books, DVDs, other electronics, and toys. The message was clear: Walmart is not going down without a fight in e-commerce. And Walmart.com executive Raul Vazquez echoed the same thought, saying that Walmart will adjust its prices "as low as we need to" to be the "low-cost leader" on the Web. In other words, the two companies are now locked in a price war, and both sites are determined to win.

The most high profile area where the two companies have done battle is in online book sales. Amazon's Kindle e-book reader may have started the conflict by offering the most popular books in e-book format for just $9.99. Though many publishers have since balked at allowing their books to be sold in the e-book format for that price, the battle has raged on in traditional formats. Several high-profile book releases, such as Stephen King's newest novel, *Under the Dome*, illustrated just how low both companies are willing to go. Walmart lowered its price for the novel to just $10, claiming that it wasn't in response to the $9.99 e-book price. Amazon matched that price shortly thereafter. In response, Walmart dropped the price to $9.00 a few days later. The book's retail cover price is $35 dollars, and its wholesale price is about $17. This means that

both retailers are losing at least $7 on every copy of *Under the Dome* that they sell at that price.

Walmart sees its massive price cuts as a way to gain market share quickly as they enter the online bookselling marketplace at a time when e-book readers and Apple iPhones and iPads make the e-book format popular. Amazon has demonstrated that in the short term it is more than capable of competing with Walmart on price. As of this case's writing, Amazon had raised its price on the *Under the Dome* back up to $17. Walmart's price, of course, was $16.99. The two sites have had similar clashes over many high-profile books, like J.K. Rowling's *Harry Potter and the Half-Blood Prince* and James Patterson's *I, Alex Cross*, the latter selling for $13.00 on Amazon and $12.99 on Walmart.com as of this writing.

The feud between the two sites has spilled over into other types of merchandise. Amazon and Walmart.com have competed over Xbox 360 consoles, popular DVD releases, and other big-ticket electronics. Even popular toys like the perennial top seller Easy-Bake Oven have been caught up in the fray. With the 2009 holiday shopping season in full swing, Walmart dropped its price for the toy from $28 to just $17. Amazon slashed its price to $18 on the very same day.

Amazon claims it doesn't see shipping as a weakness. According to Amazon spokesperson Craig Berman, "Shopping on Amazon means you don't have to fight the crowds. We bring the items to your doorstep. You don't have to fight through traffic or find a parking space." Moreover, Amazon has taken steps recently to speed delivery times. In October, it began offering same-day delivery in seven U.S. cities, at an extra cost to shoppers. By working with carriers and improving its own internal systems, Amazon also started offering second-day deliveries on Saturdays, shaving two days off some orders. And Amazon continues to expand its selection of goods to be as exhaustive as Walmart's. In November 2010, Walmart introduced free shipping for all online orders.

Amazon founder and CEO Jeff Bezos is fond of describing the U.S. retail market as having "room for many winners." Will this hold true for Walmart and Amazon going forward? Walmart remains unchallenged among traditional physical retailers, but will it topple Amazon on the Web? Or will Amazon continue to be the "Walmart" of online retailers? Alternatively, will Walmart end up enlarging the online retail market space, helping Amazon grow in the process?

Sources: Kelly Evans, "How America Now Shops: Online Stores, Dollar Retailers (Watch Out Walmart)," *The Wall Street Journal*, March 23, 2010; Brad Stone, "The Fight Over Who Sets Prices at the Online Mall," *The New York Times*, February 8, 2010; Paul Sharma, "The Music Battle, Replayed with Books," *The Wall Street Journal*, November 24, 2009; Martin Peers, "Rivals Explore Amazon's Territory," *The Wall Street Journal*, January 7, 2010; "Is Wal-Mart Gaining on Amazon.com?" *The Wall Street Journal*, reprinted on MSN Money, December 18, 2009; "Amazon Steps Into Zappos' Shoes," *eMarketer*, July 24, 2009; Brad Stone, "Can Amazon Be the Wal-Mart of the Web?" *The New York Times*, September 20, 2009; and Brad Stone and Stephanie Rosenbloom, "Price War Brews Between Amazon and Wal-Mart," *The New York Times*, November 24, 2009.

CASE STUDY QUESTIONS

1. What concepts in the chapter are illustrated in this case?

2. Analyze Amazon and Walmart.com using the value chain and competitive forces models.

3. What are the management, organization, and technology factors that have contributed to the success of both Wal-Mart and Amazon?

4. Compare Wal-Mart's and Amazon's e-commerce business models. Which is stronger? Explain your answer.

5. Where would you prefer to make your Internet purchases? Amazon or Walmart.com? Why?

Chapter 11

Managing Knowledge

Interactive Sessions:

Augmented Reality: Reality Gets Better

The Flash Crash: Machines Gone Wild?

CANADIAN TIRE KEEPS THE WHEELS ROLLING WITH KNOWLEDGE MANAGEMENT SYSTEMS

Don't be fooled by its modest name: Canadian Tire sells a lot more than tires. This company is actually five interrelated companies consisting of petroleum outlets, financial services, and retail outlets selling automotive, sports, leisure, home products, and apparel. It is also one of Canada's largest companies and most-shopped retailers, with 57,000 employees, and 1,200 stores and gas stations across Canada. The retail outlets are independently owned and operated and are spread across Canada. Canadian Tire also sells merchandise online.

Obviously, a company this big needs efficient and effective ways of communicating with its workforce and dealers, and arming them with up-to-date information to run the business. The company created two different systems for this purpose, a dealer portal and an employee information intranet.

The dealer portal was based on Microsoft Office SharePoint Portal Server, and provided a central online source for merchandise setup information, alerts, best practices, product ordering, and problem resolution. The money saved from reducing daily and weekly mailings to dealers saved the company $1–2 million annually. Customer service improved because the dealers no longer had to wade through thick paper product binders. Now product manuals are all online, and dealers are able to automatically find accurate up-to-date information.

The employee intranet called TIREnet was initially more problematic. It was based on Lotus Notes Domino software and had been poorly designed. Employees complained that the site was disorganized, brimming with outdated and redundant material, and lacked effective search features. People spent more time than necessary searching for administrative and human resource-related documents.

Canadian Tire upgraded TIREnet with a new interface that was more streamlined and intuitive. The foundation for the new TIREnet was Microsoft SharePoint Server, and the company reorganized the internal Web site so that it was easier to use and find information. SharePoint provides an option to freeze specific content, such as human resources documents, so only staff with appropriate clearance can post changes.

Canadian Tire catalogued more than 30,000 documents from the old system and transferred them to the new system. Employees no longer have to browse through TIREnet to locate a document. SharePoint's Enterprise Search technology lets employees search for documents by typing their queries into a search box, instantly providing more up-to-date information for decision making.

It is also much easier to keep documents current. Employees and managers archived up to 50 percent of old TIREnet content that was irrelevant and outdated. Documents are now automatically updated according to who has reviewed each, and the last date each was accessed. This information helps Canadian Tire management identify and remove outdated and time-sensitive material, further reducing the time required to find information.

Sources: Microsoft Canada, "Wheels In Motion," www.microsoft.ca, acessed July 15, 2010; Microsoft Corprpation, "Canadian Tire Major Tire Company Adopts Collaboration System with Faster Search," July 6, 2009; and www.canadiantire.ca, accessed September 28, 2010.

Canadian Tire's experience shows how business performance can benefit by making organizational knowledge more easily available. Facilitating access to knowledge, improving the quality and currency of knowledge, and using that knowledge to improve business processes are vital to success and survival.

The chapter-opening diagram calls attention to important points raised by this case and this chapter. Canadian Tire is a very large and far-flung company with multiple lines of business. It has many different business units and retail dealers with which to communicate knowledge about the operation of the business. Delays in accessing product information impaired dealer efficiency and customer service, while cumbersome processes and tools for accessing information used by employees similarly hampered internal operations.

Canadian Tire developed a successful information-sharing platform for its dealers using Microsoft SharePoint Server, improving dealer operations and customer service. But the knowledge it provided internally to employees was disorganized and out of date. Canadian Tire revamped its TIREnet employee intranet based on Lotus Notes Domino software by switching its technology platform to Microsoft SharePoint Server and by streamlining and simplifying the user interface. It improved its business processes for classifying and storing documents to make them easier to locate using SharePoint search technology. SharePoint has tools for automatically tracking the time and authorship of document updates, which also helps Canadian Tire keep its information up to date. Thanks to better knowledge management, Canadian Tire is operating much more efficiently and effectively.

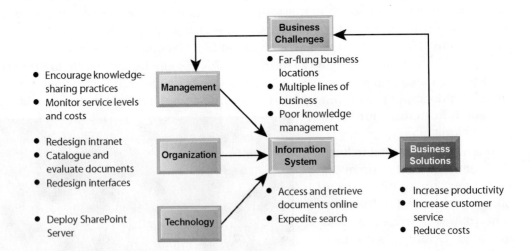

11.1 THE KNOWLEDGE MANAGEMENT LANDSCAPE

Knowledge management and collaboration systems are among the fastest growing areas of corporate and government software investment. The past decade has shown an explosive growth in research on knowledge and knowledge management in the economics, management, and information systems fields.

Knowledge management and collaboration are closely related. Knowledge that cannot be communicated and shared with others is nearly useless. Knowledge becomes useful and actionable when shared throughout the firm. As we described in Chapter 2, collaboration systems include Internet-based collaboration environments like Google Sites and IBM's Lotus Notes, social networking, e-mail and instant messaging, virtual meeting systems, wikis, and virtual worlds. In this chapter, we will be focusing on knowledge management systems, always mindful of the fact that communicating and sharing knowledge are becoming increasingly important.

We live in an information economy in which the major source of wealth and prosperity is the production and distribution of information and knowledge. About 55 percent of the U.S. labor force consists of knowledge and information workers, and 60 percent of the gross domestic product of the United States comes from the knowledge and information sectors, such as finance and publishing.

Knowledge management has become an important theme at many large business firms as managers realize that much of their firm's value depends on the firm's ability to create and manage knowledge. Studies have found that a substantial part of a firm's stock market value is related to its intangible assets, of which knowledge is one important component, along with brands, reputations, and unique business processes. Well-executed knowledge-based projects have been known to produce extraordinary returns on investment, although the impacts of knowledge-based investments are difficult to measure (Gu and Lev, 2001; Blair and Wallman, 2001).

IMPORTANT DIMENSIONS OF KNOWLEDGE

There is an important distinction between data, information, knowledge, and wisdom. Chapter 1 defines **data** as a flow of events or transactions captured by an organization's systems that, by itself, is useful for transacting but little else. To turn data into useful *information*, a firm must expend resources to organize data into categories of understanding, such as monthly, daily, regional, or store-based reports of total sales. To transform information into **knowledge**, a firm must expend additional resources to discover patterns, rules, and contexts where the knowledge works. Finally, **wisdom** is thought to be the collective and individual experience of applying knowledge to the solution of problems. Wisdom involves where, when, and how to apply knowledge.

Knowledge is both an individual attribute and a collective attribute of the firm. Knowledge is a cognitive, even a physiological, event that takes place inside peoples' heads. It is also stored in libraries and records, shared in lectures, and stored by firms in the form of business processes and employee know-how. Knowledge residing in the minds of employees that has not been documented is called **tacit knowledge**, whereas knowledge that has been documented is called **explicit knowledge**. Knowledge can reside in e-mail, voice mail, graphics, and unstructured documents as well as structured documents. Knowledge is generally believed to have a location, either in the minds of humans or in specific business processes. Knowledge is "sticky" and not universally applicable or easily moved. Finally, knowledge is thought to be situational and contextual. For example, you must know when to perform a procedure as well as how to perform it. Table 11-1 reviews these dimensions of knowledge.

We can see that knowledge is a different kind of firm asset from, say, buildings and financial assets; that knowledge is a complex phenomenon; and that there are many aspects to the process of managing knowledge. We can also recognize that knowledge-based core competencies of firms—the two or three things that an organization does best—are key organizational assets. Knowing how to do things effectively and efficiently in ways that other organizations cannot duplicate is a primary source of profit and competitive advantage that cannot be purchased easily by competitors in the marketplace.

For instance, having a unique build-to-order production system constitutes a form of knowledge and perhaps a unique asset that other firms cannot copy easily. With knowledge, firms become more efficient and effective in their use of scarce resources. Without knowledge, firms become less efficient and less effective in their use of resources and ultimately fail.

Organizational Learning and Knowledge Management

Like humans, organizations create and gather knowledge using a variety of organizational learning mechanisms. Through collection of data, careful measurement of planned activities, trial and error (experiment), and feedback from customers and the environment in general, organizations gain experience. Organizations that learn adjust their behavior to reflect that learning by creating new business processes and by changing patterns of management decision

TABLE 11-1 IMPORTANT DIMENSIONS OF KNOWLEDGE

KNOWLEDGE IS A FIRM ASSET

Knowledge is an intangible asset.

The transformation of data into useful information and knowledge requires organizational resources.

Knowledge is not subject to the law of diminishing returns as are physical assets, but instead experiences network effects as its value increases as more people share it.

KNOWLEDGE HAS DIFFERENT FORMS

Knowledge can be either tacit or explicit (codified).

Knowledge involves know-how, craft, and skill.

Knowledge involves knowing how to follow procedures.

Knowledge involves knowing why, not simply when, things happen (causality).

KNOWLEDGE HAS A LOCATION

Knowledge is a cognitive event involving mental models and maps of individuals.

There is both a social and an individual basis of knowledge.

Knowledge is "sticky" (hard to move), situated (enmeshed in a firm's culture), and contextual (works only in certain situations).

KNOWLEDGE IS SITUATIONAL

Knowledge is conditional: Knowing when to apply a procedure is just as important as knowing the procedure (conditional).

Knowledge is related to context: You must know how to use a certain tool and under what circumstances.

making. This process of change is called **organizational learning**. Arguably, organizations that can sense and respond to their environments rapidly will survive longer than organizations that have poor learning mechanisms.

THE KNOWLEDGE MANAGEMENT VALUE CHAIN

Knowledge management refers to the set of business processes developed in an organization to create, store, transfer, and apply knowledge. Knowledge management increases the ability of the organization to learn from its environment and to incorporate knowledge into its business processes. Figure 11-1 illustrates the five value-adding steps in the knowledge management value chain. Each stage in the value chain adds value to raw data and information as they are transformed into usable knowledge.

In Figure 11-1, information systems activities are separated from related management and organizational activities, with information systems activities on the top of the graphic and organizational and management activities below. One apt slogan of the knowledge management field is, "Effective knowledge management is 80 percent managerial and organizational, and 20 percent technology."

In Chapter 1, we define *organizational and management capital* as the set of business processes, culture, and behavior required to obtain value from investments in information systems. In the case of knowledge management, as with other information systems investments, supportive values, structures, and behavior patterns must be built to maximize the return on investment in knowledge management projects. In Figure 11-1, the management and organizational activities in the lower half of the diagram represent the investment in organizational capital required to obtain substantial returns on the information technology (IT) investments and systems shown in the top half of the diagram.

FIGURE 11-1   THE KNOWLEDGE MANAGEMENT VALUE CHAIN

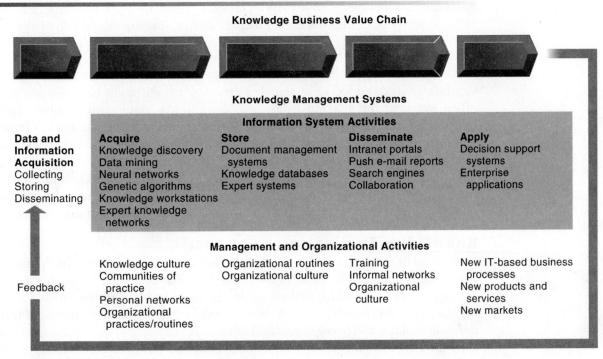

Knowledge management today involves both information systems activities and a host of enabling management and organizational activities.

Knowledge Acquisition

Organizations acquire knowledge in a number of ways, depending on the type of knowledge they seek. The first knowledge management systems sought to build corporate repositories of documents, reports, presentations, and best practices. These efforts have been extended to include unstructured documents (such as e-mail). In other cases, organizations acquire knowledge by developing online expert networks so that employees can "find the expert" in the company who has the knowledge in his or her head.

In still other cases, firms must create new knowledge by discovering patterns in corporate data or by using knowledge workstations where engineers can discover new knowledge. These various efforts are described throughout this chapter. A coherent and organized knowledge system also requires systematic data from the firm's transaction processing systems that track sales, payments, inventory, customers, and other vital data, as well as data from external sources such as news feeds, industry reports, legal opinions, scientific research, and government statistics.

Knowledge Storage

Once they are discovered, documents, patterns, and expert rules must be stored so they can be retrieved and used by employees. Knowledge storage generally involves the creation of a database. Document management systems that digitize, index, and tag documents according to a coherent framework are large databases adept at storing collections of documents. Expert systems also help corporations preserve the knowledge that is acquired by incorporating that knowledge into organizational processes and culture. Each of these is discussed later in this chapter and in the following chapter.

Management must support the development of planned knowledge storage systems, encourage the development of corporate-wide schemas for indexing documents, and reward employees for taking the time to update and store documents properly. For instance, it would reward the sales force for submitting names of prospects to a shared corporate database of prospects where all sales personnel can identify each prospect and review the stored knowledge.

Knowledge Dissemination

Portals, e-mail, instant messaging, wikis, social networks, and search engines technology have added to an existing array of collaboration technologies and office systems for sharing calendars, documents, data, and graphics (see Chapter 7). Contemporary technology seems to have created a deluge of information and knowledge. How can managers and employees discover, in a sea of information and knowledge, that which is really important for their decisions and their work? Here, training programs, informal networks, and shared management experience communicated through a supportive culture help managers focus their attention on the important knowledge and information.

Knowledge Application

Regardless of what type of knowledge management system is involved, knowledge that is not shared and applied to the practical problems facing firms and managers does not add business value. To provide a return on investment, organizational knowledge must become a systematic part of management decision making and become situated in decision-support systems (described in Chapter 12). Ultimately, new knowledge must be built into a firm's business processes and key application systems, including enterprise applications for managing key internal business processes and relationships with customers and suppliers. Management supports this process by creating—based on new knowledge—new business practices, new products and services, and new markets for the firm.

Building Organizational and Management Capital: Collaboration, Communities of Practice, and Office Environments

In addition to the activities we have just described, managers can help by developing new organizational roles and responsibilities for the acquisition of knowledge, including the creation of chief knowledge officer executive positions, dedicated staff positions (knowledge managers), and communities of practice. **Communities of practice (COPs)** are informal social networks of professionals and employees within and outside the firm who have similar work-related activities and interests. The activities of these communities include self- and group education, conferences, online newsletters, and day-to-day sharing of experiences and techniques to solve specific work problems. Many organizations, such as IBM, the U.S. Federal Highway Administration, and the World Bank have encouraged the development of thousands of online communities of practice. These communities of practice depend greatly on software environments that enable collaboration and communication.

COPs can make it easier for people to reuse knowledge by pointing community members to useful documents, creating document repositories, and filtering information for newcomers. COPs members act as facilitators, encouraging contributions and discussion. COPs can also reduce the learning curve for new employees by providing contacts with subject matter experts and access to a community's established methods and tools. Finally, COPs can act as a spawning ground for new ideas, techniques, and decision-making behavior.

TYPES OF KNOWLEDGE MANAGEMENT SYSTEMS

There are essentially three major types of knowledge management systems: enterprise-wide knowledge management systems, knowledge work systems, and intelligent techniques. Figure 11-2 shows the knowledge management system applications for each of these major categories.

Enterprise-wide knowledge management systems are general-purpose firmwide efforts to collect, store, distribute, and apply digital content and knowledge. These systems include capabilities for searching for information, storing both structured and unstructured data, and locating employee expertise within the firm. They also include supporting technologies such as portals, search engines, collaboration tools (e-mail, instant messaging, wikis, blogs, and social bookmarking), and learning management systems.

The development of powerful networked workstations and software for assisting engineers and scientists in the discovery of new knowledge has led to the creation of knowledge work systems such as computer-aided design (CAD), visualization, simulation, and virtual reality systems. **Knowledge work systems (KWS)** are specialized systems built for engineers, scientists, and other knowledge workers charged with discovering and creating new knowledge for a company. We discuss knowledge work applications in detail in Section 11.3.

Knowledge management also includes a diverse group of **intelligent techniques**, such as data mining, expert systems, neural networks, fuzzy logic, genetic algorithms, and intelligent agents. These techniques have different

FIGURE 11-2   MAJOR TYPES OF KNOWLEDGE MANAGEMENT SYSTEMS

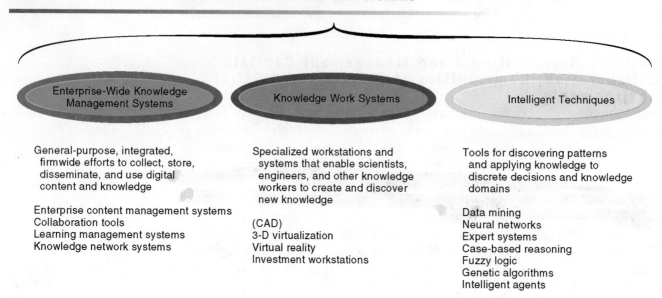

There are three major categories of knowledge management systems, and each can be broken down further into more specialized types of knowledge management systems.

objectives, from a focus on discovering knowledge (data mining and neural networks), to distilling knowledge in the form of rules for a computer program (expert systems and fuzzy logic), to discovering optimal solutions for problems (genetic algorithms). Section 11.4 provides more detail about these intelligent techniques.

11.2   ENTERPRISE-WIDE KNOWLEDGE MANAGEMENT SYSTEMS

Firms must deal with at least three kinds of knowledge. Some knowledge exists within the firm in the form of structured text documents (reports and presentations). Decision makers also need knowledge that is semistructured, such as e-mail, voice mail, chat room exchanges, videos, digital pictures, brochures, or bulletin board postings. In still other cases, there is no formal or digital information of any kind, and the knowledge resides in the heads of employees. Much of this knowledge is tacit knowledge that is rarely written down. Enterprise-wide knowledge management systems deal with all three types of knowledge.

ENTERPRISE CONTENT MANAGEMENT SYSTEMS

Businesses today need to organize and manage both structured and semistructured knowledge assets. **Structured knowledge** is explicit knowledge that exists in formal documents, as well as in formal rules that organizations derive by observing experts and their decision-making behaviors. But, according to experts, at least 80 percent of an organization's business content is semistructured or unstructured—information in folders, messages, memos, proposals, e-mails, graphics, electronic slide presentations, and even videos created in different formats and stored in many locations.

Enterprise content management systems help organizations manage both types of information. They have capabilities for knowledge capture, storage, retrieval, distribution, and preservation to help firms improve their business processes and decisions. Such systems include corporate repositories of documents, reports, presentations, and best practices, as well as capabilities for collecting and organizing semistructured knowledge such as e-mail (see Figure 11-3). Major enterprise content management systems also enable users to access external sources of information, such as news feeds and research, and to communicate via e-mail, chat/instant messaging, discussion groups, and videoconferencing. Open Text Corporation, EMC (Documentum), IBM, and Oracle Corporation are leading vendors of enterprise content management software.

Barrick Gold, the world's leading gold producer, uses Open Text LiveLink Enterprise Content Management tools to manage the massive amounts of information required for building mines. The system organizes and stores both structured and unstructured content, including CAD drawings, contracts, engineering data, and production reports. If an operational team needs to refer back to the original document, that document is in a single digital repository as opposed to being scattered over multiple systems. Barrick's electronic content management system reduces the amount of time required to search for documents, shortening project schedules, improving the quality of decisions, and minimizing rework (Open Text, 2010).

A key problem in managing knowledge is the creation of an appropriate classification scheme, or **taxonomy**, to organize information into meaningful categories so that it can be easily accessed. Once the categories for classifying knowledge have been created, each knowledge object needs to be "tagged," or classified, so that it can be easily retrieved. Enterprise content management systems have capabilities for tagging, interfacing with corporate databases where the documents are stored, and creating an enterprise portal environment for employees to use when searching for corporate knowledge.

FIGURE 11-3   AN ENTERPRISE CONTENT MANAGEMENT SYSTEM

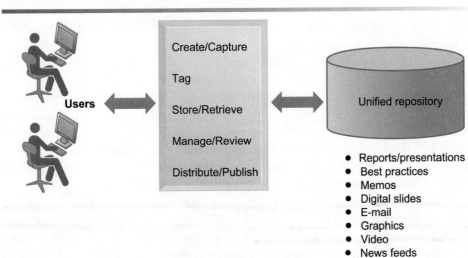

An enterprise content management system has capabilities for classifying, organizing, and managing structured and semistructured knowledge and making it available throughout the enterprise.

Firms in publishing, advertising, broadcasting, and entertainment have special needs for storing and managing unstructured digital data such as photographs, graphic images, video, and audio content. For example, Coca-Cola must keep track of all the images of the Coca-Cola brand that have been created in the past at all of the company's worldwide offices, to prevent both redundant work and variation from a standard brand image. **Digital asset management systems** help companies classify, store, and distribute these digital objects.

KNOWLEDGE NETWORK SYSTEMS

Knowledge network systems, also known as *expertise location and management systems*, address the problem that arises when the appropriate knowledge is not in the form of a digital document but instead resides in the memory of expert individuals in the firm. Knowledge network systems provide an online directory of corporate experts in well-defined knowledge domains and use communication technologies to make it easy for employees to find the appropriate expert in a company. Some knowledge network systems go further by systematizing the solutions developed by experts and then storing the solutions in a knowledge database as a best practices or frequently asked questions (FAQ) repository (see Figure 11-4). AskMe provides stand-alone knowledge network software, and some knowledge networking capabilities can be found in the leading collaboration software suites.

COLLABORATION TOOLS AND LEARNING MANAGEMENT SYSTEMS

The major enterprise content management systems include powerful portal and collaboration technologies. Enterprise knowledge portals can provide access to external sources of information, such as news feeds and research, as well as to internal knowledge resources along with capabilities for e-mail, chat/instant messaging, discussion groups, and videoconferencing.

Companies are starting to use consumer Web technologies such as blogs, wikis, and social bookmarking for internal use to foster collaboration and information exchange between individuals and teams. Blogs and wikis help capture, consolidate, and centralize this knowledge for the firm. Collaboration tools from commercial software vendors, such as Microsoft SharePoint and Lotus Connections, also offer these capabilities along with secure online collaborative workspaces.

Wikis, which we introduced in Chapters 2 and 7, are inexpensive and easy to implement. Wikis provide a central repository for all types of corporate data that can be displayed in a Web browser, including electronic pages of documents, spreadsheets, and electronic slides, and can embed e-mail and instant messages. Although users are able to modify wiki content contributed by others, wikis have capabilities for tracking these changes and tools for reverting to earlier versions. A wiki is most appropriate for information that is revised frequently but must remain available perpetually as it changes.

Social bookmarking makes it easier to search for and share information by allowing users to save their bookmarks to Web pages on a public Web site and tag these bookmarks with keywords. These tags can be used to organize and search for the documents. Lists of tags can be shared with other people to help them find information of interest. The user-created taxonomies created for

FIGURE 11-4   AN ENTERPRISE KNOWLEDGE NETWORK SYSTEM

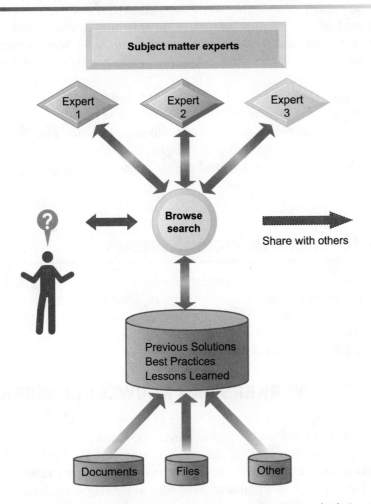

A knowledge network maintains a database of firm experts, as well as accepted solutions to known problems, and then facilitates the communication between employees looking for knowledge and experts who have that knowledge. Solutions created in this communication are then added to a database of solutions in the form of FAQs, best practices, or other documents.

shared bookmarks are called **folksonomies**. Delicious and Digg are two popular social bookmarking sites.

Suppose, for example, that you're on a corporate team researching wind power. If you did a Web search and found relevant Web pages on wind power, you'd click on a bookmarking button on a social bookmarking site and create a tag identifying each Web document you found to link it to wind power. By clicking on the "tags" button at the social networking site, you'd be able to see a list of all the tags you created and select the documents you need.

Companies need ways to keep track of and manage employee learning and to integrate it more fully into their knowledge management and other corporate systems. A **learning management system (LMS)** provides tools for the management, delivery, tracking, and assessment of various types of employee learning and training.

Contemporary LMS support multiple modes of learning, including CD-ROM, downloadable videos, Web-based classes, live instruction in classes or online, and group learning in online forums and chat sessions. The LMS consolidates mixed-media training, automates the selection and administration of courses, assembles and delivers learning content, and measures learning effectiveness.

For example, the Whirlpool Corporation uses CERTPOINT's learning management system to manage the registration, scheduling, reporting, and content for its training programs for 3,500 salespeople. The system helps Whirlpool tailor course content to the right audience, track the people who took courses and their scores, and compile metrics on employee performance.

11.3 KNOWLEDGE WORK SYSTEMS

The enterprise-wide knowledge systems we have just described provide a wide range of capabilities that can be used by many if not all the workers and groups in an organization. Firms also have specialized systems for knowledge workers to help them create new knowledge and to ensure that this knowledge is properly integrated into the business.

KNOWLEDGE WORKERS AND KNOWLEDGE WORK

Knowledge workers, which we introduced in Chapter 1, include researchers, designers, architects, scientists, and engineers who primarily create knowledge and information for the organization. Knowledge workers usually have high levels of education and memberships in professional organizations and are often asked to exercise independent judgment as a routine aspect of their work. For example, knowledge workers create new products or find ways of improving existing ones. Knowledge workers perform three key roles that are critical to the organization and to the managers who work within the organization:

- Keeping the organization current in knowledge as it develops in the external world—in technology, science, social thought, and the arts
- Serving as internal consultants regarding the areas of their knowledge, the changes taking place, and opportunities
- Acting as change agents, evaluating, initiating, and promoting change projects

REQUIREMENTS OF KNOWLEDGE WORK SYSTEMS

Most knowledge workers rely on office systems, such as word processors, voice mail, e-mail, videoconferencing, and scheduling systems, which are designed to increase worker productivity in the office. However, knowledge workers also require highly specialized knowledge work systems with powerful graphics, analytical tools, and communications and document management capabilities.

These systems require sufficient computing power to handle the sophisticated graphics or complex calculations necessary for such knowledge workers as scientific researchers, product designers, and financial analysts. Because knowledge workers are so focused on knowledge in the external world, these systems also must give the worker quick and easy access to external databases. They typically feature user-friendly interfaces that enable users to perform needed tasks without having to spend a great deal of time learning how to use the system. Knowledge workers are highly paid—wasting a knowledge worker's time is simply too expensive. Figure 11-5 summarizes the requirements of knowledge work systems.

Knowledge workstations often are designed and optimized for the specific tasks to be performed; so, for example, a design engineer requires a different workstation setup than a financial analyst. Design engineers need graphics with enough power to handle three-dimensional (3-D) CAD systems. However, financial analysts are more interested in access to a myriad number of external databases and large databases for efficiently storing and accessing massive amounts of financial data.

EXAMPLES OF KNOWLEDGE WORK SYSTEMS

Major knowledge work applications include CAD systems, virtual reality systems for simulation and modeling, and financial workstations. **Computer-aided design (CAD)** automates the creation and revision of designs, using computers and sophisticated graphics software. Using a more traditional physical design methodology, each design modification requires a mold to be made and a prototype to be tested physically. That process must be repeated many times, which is a very expensive and time-consuming process. Using a CAD workstation, the designer need only make a physical prototype toward the end of the design process because the design can be easily tested and changed on the computer. The ability of CAD software to provide design specifications for the tooling and manufacturing processes also saves a great deal of time and money while producing a manufacturing process with far fewer problems.

Troy Lee Designs, which makes sports helmets, recently invested in CAD design software that could create the helmets in 3-D. The technology defined the shapes better than traditional methods, which involved sketching an idea

FIGURE 11-5   REQUIREMENTS OF KNOWLEDGE WORK SYSTEMS

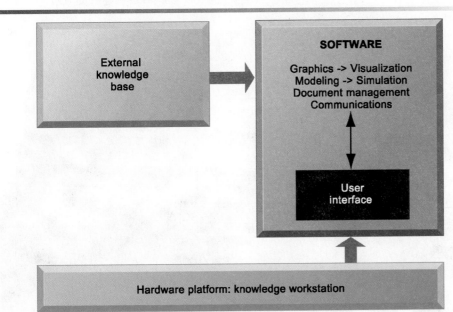

Knowledge work systems require strong links to external knowledge bases in addition to specialized hardware and software.

on paper, hand-molding a clay model, and shipping the model to Asian factories to create a plastic prototype. Production is now about six months faster and about 35 percent cheaper, with Asian factories about to produce an exact replica after receiving the digital design via e-mail (Maltby, 2010).

Virtual reality systems have visualization, rendering, and simulation capabilities that go far beyond those of conventional CAD systems. They use interactive graphics software to create computer-generated simulations that are so close to reality that users almost believe they are participating in a real-world situation. In many virtual reality systems, the user dons special clothing, headgear, and equipment, depending on the application. The clothing contains sensors that record the user's movements and immediately transmit that information back to the computer. For instance, to walk through a virtual reality simulation of a house, you would need garb that monitors the movement of your feet, hands, and head. You also would need goggles containing video screens and sometimes audio attachments and feeling gloves so that you can be immersed in the computer feedback.

A virtual reality system helps mechanics in Boeing Co.'s 25-day training course for its 787 Dreamliner learn to fix all kinds of problems, from broken lights in the cabin to major glitches with flight controls. Using both laptop and desktop computers inside a classroom with huge wall-mounted diagrams, Boeing airline mechanics train on a system that displays an interactive Boeing 787 cockpit, as well as a 3-D exterior of the plane. The mechanics "walk" around the jet by clicking a mouse, open virtual maintenance access panels, and go inside the plane to repair and replace parts (Sanders, 2010).

Augmented reality (AR) is a related technology for enhancing visualization. AR provides a live direct or indirect view of a physical real-world environment whose elements are *augmented* by virtual computer-generated imagery. The user is grounded in the real physical world, and the virtual images are merged with the real view to create the augmented display. The digital technology provides additional information to enhance the perception of reality and

CAD systems improve the quality and precision of product design by performing much of the design and testing work on the computer.

INTERACTIVE SESSION:TECHNOLOGY

AUGMENTED REALITY: REALITY GETS BETTER

Many of us are familiar with the concept of virtual reality, either from films like *Avatar* and *The Matrix*, or from science fiction novels and video games. Virtual reality is a computer-generated, interactive, three-dimensional environment in which people become immersed. But in the past few years, a new spin on virtual reality known as augmented reality has emerged as a major focus of many companies' marketing efforts. More than just science fiction, augmented reality is an exciting new way of creating richer, more interactive experiences with users and future customers.

Augmented reality differs from traditional virtual reality because users of augmented reality (also called AR) tools maintain a presence in the real world. In virtual reality, users are completely immersed in a computer-generated environment, and often use head-mounted displays that facilitate the immersion and eliminate any interference from the real world. Augmented reality mixes real-life images with graphics or other effects and can use any of three major display techniques—head-mounted displays, just as with virtual reality, spatial displays, which display graphical information on physical objects, and handheld displays.

Almost everyone has already encountered some form of AR technology. Sports fans are familiar with the yellow first-down markers shown on televised football games, or the special markings denoting the location and direction of hockey pucks in hockey games. These are examples of augmented reality. Other common usages of AR include medical procedures like image-guided surgery, where data acquired from computerized tomography (CT) and magnetic resonance imaging (MRI) scans or from ultrasound imaging are superimposed on the patient in the operating room. Other industries where AR has caught on include military training, engineering design, robotics, and consumer design.

As companies get more comfortable with augmented reality, marketers are developing creative new ways to use the technology. Print media companies see AR as a way to generate excitement about their products in an entirely new way. Esquire magazine used AR extensively in its December 2009 issue, adding several stickers with designs that, when held up to a Web camera, triggered interactive video segments featuring cover subject Robert Downey Jr. Turning the magazine in different directions yielded different images. A fashion spread describing dressing in layers showed actor Jeremy Renner adding more layers as the seasons changed. The orientation of the magazine as held up to a Web camera determined the season.

Lexus placed an advertisement in the magazine that displayed "radar waves" bouncing off of nearby objects on the page. Again, adjusting the angle of the magazine affected the content of the ad. Lexus Vice President of Marketing David Nordstrom stated that AR was attractive to him because "our job as marketers is to be able to communicate to people in interesting ways that are relevant to them and also entertaining." User response to the magazine was positive, suggesting that AR accomplished this goal. Other companies that have pursued AR as a way to attract and entertain their customers include Papa John's, which added AR tags to their pizza boxes. These tags display images of the company's founder driving a car when triggered using a Web camera. That company's president believes AR is "a great way to get customers involved in a promotion in a more interactive way than just reading or seeing an ad."

Mobile phone application developers are also excited about the growing demand for AR technologies. Most mobile phones have camera, global positioning system (GPS), Internet, and compass functionalities, which make smartphones ideal candidates for handheld AR displays. One of the major new markets for AR is in real estate, where applications that help users access real estate listings and information on the go have already taken off. An Amsterdam-based start-up, application developer Layar, has created an app for French real estate agency MeilleursAgents.com where users can point their phones at any building in Paris and within seconds the phone displays the property's value per square meter and a small photo of the property, along with a live image of the building streamed through the phone's camera.

Over 30 similar applications have been developed in other countries, including American real estate company ZipRealty, whose HomeScan application has met with early success. While the technology is still new and will take some time to develop, users can already stand in front of some houses for sale and point their phones at the property to display details superimposed on their screen. If the house is too far away, users can switch to the phone's interactive map and locate the house and other nearby houses for sale. ZipRealty is so encouraged by the early response to HomeScan that it plans to add data on restaurants, coffee shops, and other neighborhood features to the app. Another well-known application, Wikitude, allows users to view user-contributed Web-based information about their surroundings using their mobile phones.

Skeptics believe that the technology is more of a gimmick than a useful tool, but Layar's application has been downloaded over 1,000 times per week since its launch. Being able to access information on properties is more than just a gimmick—it is a legitimately useful tool to help buyers on the go.

Marketers are finding that users increasingly want their phones to have all of the functionality of desktop computers, and more AR mash-ups have been released that display information on tourist sites, chart subway stops, and restaurants, and allow interior designers to superimpose new furniture schemes onto a room so that potential customers can more easily choose what they like best. Analysts believe that AR is here to stay, predicting that the mobile AR market will grow to $732 million by 2014.

Sources: R. Scott MacIntosh, "Portable Real Estate Listings-with a Difference," *The New York Times*, March 25, 2010; Alex Viega, "Augmented Reality for Real Estate Search," Associated Press, April 16, 2010; "Augmented Reality - 5 More Examples of This 3D Virtual Experience," http://www.nickburcher.com/2009/05/augmented-reality-5-more-examples-of.html, May 30, 2009; Shira Ovide, "Esquire Flirts with Digital Reality," *The Wall Street Journal*, October 29, 2009.

CASE STUDY QUESTIONS

1. What is the difference between virtual reality and augmented reality?

2. Why is augmented reality so appealing to marketers?

3. What makes augmented reality useful for real estate shopping applications?

4. Suggest some other knowledge work applications for augmented reality

MIS IN ACTION

Find example videos of augmented reality in action (use Nick Burcher's blog if you're stuck: http://www.nickburcher.com/2009/05/augmented-reality-5-more-examples-of.html), and use them to answer the following questions:

1. Why is the example shown in the video an instance of AR?

2. Do you think it is an effective marketing tool or application? Why or why not?

3. Can you think of other products or services that would be well suited to AR?

make the surrounding real world of the user more interactive and meaningful. The Interactive Session on Technology provides more detail about AR and its applications.

Virtual reality applications developed for the Web use a standard called **Virtual Reality Modeling Language (VRML)**. VRML is a set of specifications for interactive, 3-D modeling on the World Wide Web that can organize multiple media types, including animation, images, and audio to put users in a simulated real-world environment. VRML is platform independent, operates over a desktop computer, and requires little bandwidth.

DuPont, the Wilmington, Delaware, chemical company, created a VRML application called HyperPlant, which enables users to access 3-D data over the Internet using Web browser software. Engineers can go through 3-D models as if they were physically walking through a plant, viewing objects at eye level. This level of detail reduces the number of mistakes they make during construction of oil rigs, oil plants, and other structures.

The financial industry is using specialized **investment workstations** to leverage the knowledge and time of its brokers, traders, and portfolio managers. Firms such as Merrill Lynch and UBS Financial Services have installed investment workstations that integrate a wide range of data from both internal and external sources, including contact management data, real-time and historical market data, and research reports. Previously, financial professionals had to spend considerable time accessing data from separate systems and piecing together the information they needed. By providing one-stop information faster and with fewer errors, the workstations streamline the entire investment process from stock selection to updating client records. Table 11-2 summarizes the major types of knowledge work systems.

11.4 INTELLIGENT TECHNIQUES

Artificial intelligence and database technology provide a number of intelligent techniques that organizations can use to capture individual and collective knowledge and to extend their knowledge base. Expert systems, case-based reasoning, and fuzzy logic are used for capturing tacit knowledge. Neural networks and data mining are used for **knowledge discovery**. They can discover underlying patterns, categories, and behaviors in large data sets that could not be discovered by managers alone or simply through experience. Genetic algorithms are used for generating solutions to problems that are too large and complex for human beings to analyze on their own. Intelligent agents can automate routine tasks to help firms search for and filter information for use in electronic commerce, supply chain management, and other activities.

Data mining, which we introduced in Chapter 6, helps organizations capture undiscovered knowledge residing in large databases, providing managers with new insight for improving business performance. It has become an important tool for management decision making, and we provide a detailed discussion of data mining for management decision support in Chapter 12.

The other intelligent techniques discussed in this section are based on **artificial intelligence (AI)** technology, which consists of computer-based systems (both hardware and software) that attempt to emulate human behavior. Such systems would be able to learn languages, accomplish physical tasks, use a perceptual apparatus, and emulate human expertise and decision making. Although AI applications do not exhibit the breadth, complexity, originality, and generality of human intelligence, they play an important role in contemporary knowledge management.

TABLE 11-2 EXAMPLES OF KNOWLEDGE WORK SYSTEMS

KNOWLEDGE WORK SYSTEM	FUNCTION IN ORGANIZATION
CAD/CAM (computer-aided manufacturing)	Provides engineers, designers, and factory managers with precise control over industrial design and manufacturing
Virtual reality systems	Provide drug designers, architects, engineers, and medical workers with precise, photorealistic simulations of objects
Investment workstations	High-end PCs used in the financial sector to analyze trading situations instantaneously and facilitate portfolio management

CAPTURING KNOWLEDGE: EXPERT SYSTEMS

Expert systems are an intelligent technique for capturing tacit knowledge in a very specific and limited domain of human expertise. These systems capture the knowledge of skilled employees in the form of a set of rules in a software system that can be used by others in the organization. The set of rules in the expert system adds to the memory, or stored learning, of the firm.

Expert systems lack the breadth of knowledge and the understanding of fundamental principles of a human expert. They typically perform very limited tasks that can be performed by professionals in a few minutes or hours, such as diagnosing a malfunctioning machine or determining whether to grant credit for a loan. Problems that cannot be solved by human experts in the same short period of time are far too difficult for an expert system. However, by capturing human expertise in limited areas, expert systems can provide benefits, helping organizations make high-quality decisions with fewer people. Today, expert systems are widely used in business in discrete, highly structured decision-making situations.

How Expert Systems Work

Human knowledge must be modeled or represented in a way that a computer can process. Expert systems model human knowledge as a set of rules that collectively are called the **knowledge base**. Expert systems have from 200 to many thousands of these rules, depending on the complexity of the problem. These rules are much more interconnected and nested than in a traditional software program (see Figure 11-6).

The strategy used to search through the knowledge base is called the **inference engine**. Two strategies are commonly used: forward chaining and backward chaining (see Figure 11-7).

In **forward chaining,** the inference engine begins with the information entered by the user and searches the rule base to arrive at a conclusion. The strategy is to fire, or carry out, the action of the rule when a condition is true. In Figure 11-7, beginning on the left, if the user enters a client's name with income greater than $100,000, the engine will fire all rules in sequence from left to right. If the user then enters information indicating that the same client owns real estate, another pass of the rule base will occur and more rules will fire. Processing continues until no more rules can be fired.

In **backward chaining,** the strategy for searching the rule base starts with a hypothesis and proceeds by asking the user questions about selected facts until the hypothesis is either confirmed or disproved. In our example, in Figure 11-7, ask the question, "Should we add this person to the prospect database?" Begin on the right of the diagram and work toward the left. You can see that the person should be added to the database if a sales representative is sent, term insurance is granted, or a financial adviser visits the client.

Examples of Successful Expert Systems

Expert systems provide businesses with an array of benefits including improved decisions, reduced errors, reduced costs, reduced training time, and higher levels of quality and service. Con-Way Transportation built an expert system called Line-haul to automate and optimize planning of overnight shipment routes for its nationwide freight-trucking business. The expert system captures the business rules that dispatchers follow when assigning drivers, trucks, and trailers to transport 50,000 shipments of heavy freight each night across 25 states and Canada and when plotting their routes. Line-haul runs on a

FIGURE 11-6   **RULES IN AN EXPERT SYSTEM**

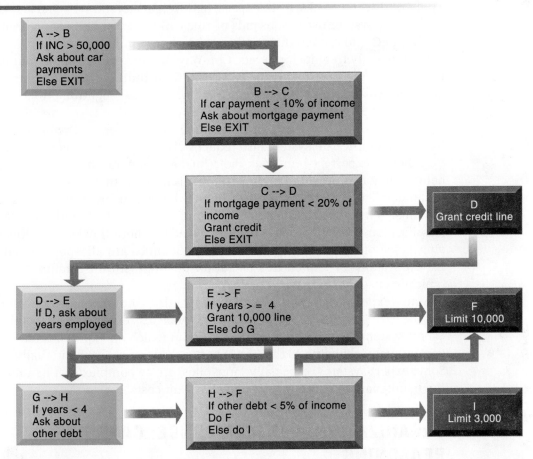

An expert system contains a number of rules to be followed. The rules are interconnected; the number of outcomes is known in advance and is limited; there are multiple paths to the same outcome; and the system can consider multiple rules at a single time. The rules illustrated are for simple credit-granting expert systems.

FIGURE 11-7   **INFERENCE ENGINES IN EXPERT SYSTEMS**

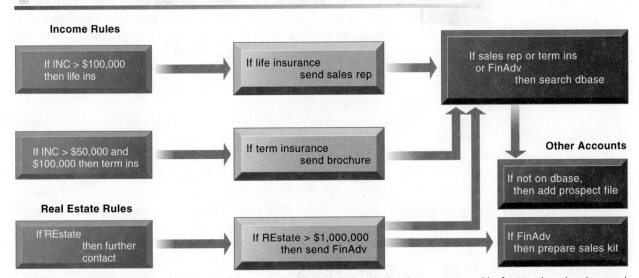

An inference engine works by searching through the rules and "firing" those rules that are triggered by facts gathered and entered by the user. Basically, a collection of rules is similar to a series of nested IF statements in a traditional software program; however, the magnitude of the statements and degree of nesting are much greater in an expert system.

Sun computer platform and uses data on daily customer shipment requests, available drivers, trucks, trailer space, and weight stored in an Oracle database. The expert system uses thousands of rules and 100,000 lines of program code written in C++ to crunch the numbers and create optimum routing plans for 95 percent of daily freight shipments. Con-Way dispatchers tweak the routing plan provided by the expert system and relay final routing specifications to field personnel responsible for packing the trailers for their nighttime runs. Con-Way recouped its $3 million investment in the system within two years by reducing the number of drivers, packing more freight per trailer, and reducing damage from rehandling. The system also reduces dispatchers' arduous nightly tasks.

Although expert systems lack the robust and general intelligence of human beings, they can provide benefits to organizations if their limitations are well understood. Only certain classes of problems can be solved using expert systems. Virtually all successful expert systems deal with problems of classification in limited domains of knowledge where there are relatively few alternative outcomes and these possible outcomes are all known in advance. Expert systems are much less useful for dealing with unstructured problems typically encountered by managers.

Many expert systems require large, lengthy, and expensive development efforts. Hiring or training more experts may be less expensive than building an expert system. Typically, the environment in which an expert system operates is continually changing so that the expert system must also continually change. Some expert systems, especially large ones, are so complex that in a few years the maintenance costs equal the development costs.

ORGANIZATIONAL INTELLIGENCE: CASE-BASED REASONING

Expert systems primarily capture the tacit knowledge of individual experts, but organizations also have collective knowledge and expertise that they have built up over the years. This organizational knowledge can be captured and stored using case-based reasoning. In **case-based reasoning (CBR)**, descriptions of past experiences of human specialists, represented as cases, are stored in a database for later retrieval when the user encounters a new case with similar parameters. The system searches for stored cases with problem characteristics similar to the new one, finds the closest fit, and applies the solutions of the old case to the new case. Successful solutions are tagged to the new case and both are stored together with the other cases in the knowledge base. Unsuccessful solutions also are appended to the case database along with explanations as to why the solutions did not work (see Figure 11-8).

Expert systems work by applying a set of IF-THEN-ELSE rules extracted from human experts. Case-based reasoning, in contrast, represents knowledge as a series of cases, and this knowledge base is continuously expanded and refined by users. You'll find case-based reasoning in diagnostic systems in medicine or customer support where users can retrieve past cases whose characteristics are similar to the new case. The system suggests a solution or diagnosis based on the best-matching retrieved case.

FUZZY LOGIC SYSTEMS

Most people do not think in terms of traditional IF-THEN rules or precise numbers. Humans tend to categorize things imprecisely using rules for making

FIGURE 11-8   HOW CASE-BASED REASONING WORKS

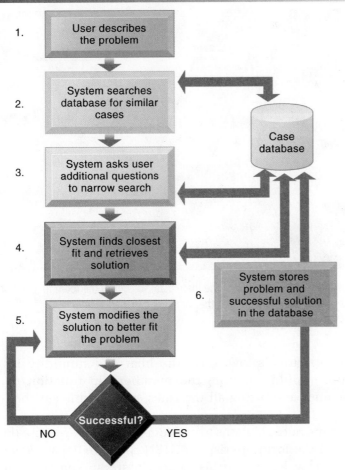

1. User describes the problem
2. System searches database for similar cases
3. System asks user additional questions to narrow search
4. System finds closest fit and retrieves solution
5. System modifies the solution to better fit the problem
6. System stores problem and successful solution in the database

Case database

Successful?   NO   YES

Case-based reasoning represents knowledge as a database of past cases and their solutions. The system uses a six-step process to generate solutions to new problems encountered by the user.

decisions that may have many shades of meaning. For example, a man or a woman can be *strong* or *intelligent*. A company can be *large, medium,* or *small* in size. Temperature can be *hot, cold, cool,* or *warm*. These categories represent a range of values.

Fuzzy logic is a rule-based technology that can represent such imprecision by creating rules that use approximate or subjective values. It can describe a particular phenomenon or process linguistically and then represent that description in a small number of flexible rules. Organizations can use fuzzy logic to create software systems that capture tacit knowledge where there is linguistic ambiguity.

Let's look at the way fuzzy logic would represent various temperatures in a computer application to control room temperature automatically. The terms (known as *membership functions*) are imprecisely defined so that, for example, in Figure 11-9, cool is between 45 degrees and 70 degrees, although the temperature is most clearly cool between about 60 degrees and 67 degrees. Note that *cool* is overlapped by *cold* or *norm*. To control the room environment using this logic, the programmer would develop similarly imprecise definitions for humidity and other factors, such as outdoor wind and temperature. The rules might include one that says: "If the temperature is *cool* or *cold* and the humidity is low while the outdoor wind is high and the

FIGURE 11-9   FUZZY LOGIC FOR TEMPERATURE CONTROL

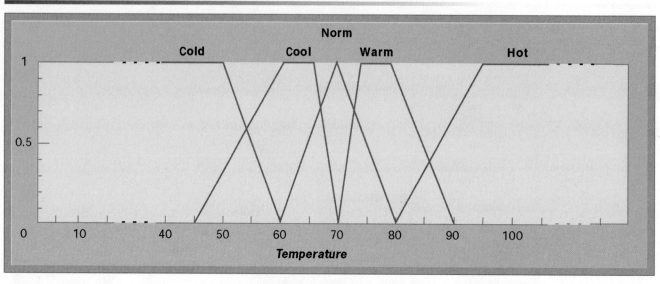

The membership functions for the input called temperature are in the logic of the thermostat to control the room temperature. Membership functions help translate linguistic expressions such as *warm* into numbers that the computer can manipulate.

outdoor temperature is low, raise the heat and humidity in the room." The computer would combine the membership function readings in a weighted manner and, using all the rules, raise and lower the temperature and humidity.

Fuzzy logic provides solutions to problems requiring expertise that is difficult to represent in the form of crisp IF-THEN rules. In Japan, Sendai's subway system uses fuzzy logic controls to accelerate so smoothly that standing passengers need not hold on. Mitsubishi Heavy Industries in Tokyo has been able to reduce the power consumption of its air conditioners by 20 percent by implementing control programs in fuzzy logic. The autofocus device in cameras is only possible because of fuzzy logic. In these instances, fuzzy logic allows incremental changes in inputs to produce smooth changes in outputs instead of discontinuous ones, making it useful for consumer electronics and engineering applications.

Management also has found fuzzy logic useful for decision making and organizational control. A Wall Street firm created a system that selects companies for potential acquisition, using the language stock traders understand. A fuzzy logic system has been developed to detect possible fraud in medical claims submitted by health care providers anywhere in the United States.

NEURAL NETWORKS

Neural networks are used for solving complex, poorly understood problems for which large amounts of data have been collected. They find patterns and relationships in massive amounts of data that would be too complicated and difficult for a human being to analyze. Neural networks discover this knowledge by using hardware and software that parallel the processing patterns of the biological or human brain. Neural networks "learn" patterns from large quantities of data by sifting through data, searching for relationships, building models, and correcting over and over again the model's own mistakes.

A neural network has a large number of sensing and processing nodes that continuously interact with each other. Figure 11-10 represents one type of neural network comprising an input layer, an output layer, and a hidden processing layer. Humans "train" the network by feeding it a set of training data for which the inputs produce a known set of outputs or conclusions. This helps the computer learn the correct solution by example. As the computer is fed more data, each case is compared with the known outcome. If it differs, a correction is calculated and applied to the nodes in the hidden processing layer. These steps are repeated until a condition, such as corrections being less than a certain amount, is reached. The neural network in Figure 11-10 has learned how to identify a fraudulent credit card purchase. Also, self-organizing neural networks can be trained by exposing them to large amounts of data and allowing them to discover the patterns and relationships in the data.

Whereas expert systems seek to emulate or model a human expert's way of solving problems, neural network builders claim that they do not program solutions and do not aim to solve specific problems. Instead, neural network designers seek to put intelligence into the hardware in the form of a generalized capability to learn. In contrast, the expert system is highly specific to a given problem and cannot be retrained easily.

Neural network applications in medicine, science, and business address problems in pattern classification, prediction, financial analysis, and control and optimization. In medicine, neural network applications are used for screening patients for coronary artery disease, for diagnosing patients with epilepsy and Alzheimer's disease, and for performing pattern recognition of pathology images. The financial industry uses neural networks to discern patterns in vast pools of data that might help predict the performance of equities, corporate bond ratings, or corporate bankruptcies. Visa International uses a neural network to help detect credit card fraud by monitoring all Visa transactions for sudden changes in the buying patterns of cardholders.

There are many puzzling aspects of neural networks. Unlike expert systems, which typically provide explanations for their solutions, neural

FIGURE 11-10   HOW A NEURAL NETWORK WORKS

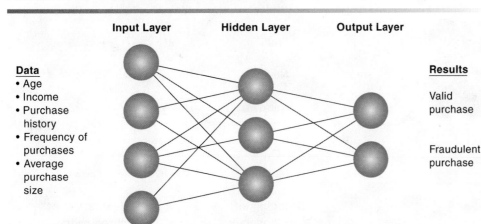

A neural network uses rules it "learns" from patterns in data to construct a hidden layer of logic. The hidden layer then processes inputs, classifying them based on the experience of the model. In this example, the neural network has been trained to distinguish between valid and fraudulent credit card purchases.

networks cannot always explain why they arrived at a particular solution. Moreover, they cannot always guarantee a completely certain solution, arrive at the same solution again with the same input data, or always guarantee the best solution. They are very sensitive and may not perform well if their training covers too little or too much data. In most current applications, neural networks are best used as aids to human decision makers instead of substitutes for them.

The Interactive Session on Organizations describes computerized stock trading applications based on a related AI technology called **machine learning.** Machine learning focuses on algorithms and statistical methods that allow computers to "learn" by extracting rules and patterns from massive data sets and make predictions about the future. Both neural networks and machine learning techniques are used in data mining. As the Interactive Session points out, the use of machine learning in the financial industry for securities trading decisions has had mixed results.

GENETIC ALGORITHMS

Genetic algorithms are useful for finding the optimal solution for a specific problem by examining a very large number of possible solutions for that problem. They are based on techniques inspired by evolutionary biology, such as inheritance, mutation, selection, and crossover (recombination).

A genetic algorithm works by representing information as a string of 0s and 1s. The genetic algorithm searches a population of randomly generated strings of binary digits to identify the right string representing the best possible solution for the problem. As solutions alter and combine, the worst ones are discarded and the better ones survive to go on to produce even better solutions.

In Figure 11-11, each string corresponds to one of the variables in the problem. One applies a test for fitness, ranking the strings in the population

FIGURE 11-11 THE COMPONENTS OF A GENETIC ALGORITHM

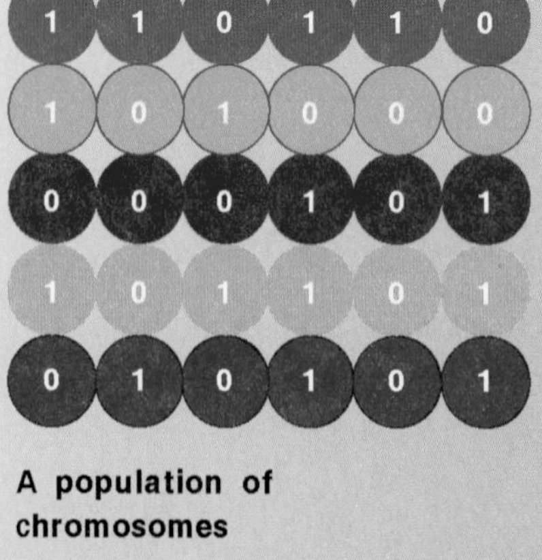

		Length	Width	Weight	Fitness
1 1 0 1 1 0	1	Long	Wide	Light	55
1 0 1 0 0 0	2	Short	Narrow	Heavy	49
0 0 0 1 0 1	3	Long	Narrow	Heavy	36
1 0 1 1 0 1	4	Short	Medium	Light	61
0 1 0 1 0 1	5	Long	Medium	Very light	74
A population of chromosomes			**Decoding of chromosomes**		**Evaluation of chromosomes**

This example illustrates an initial population of "chromosomes," each representing a different solution. The genetic algorithm uses an iterative process to refine the initial solutions so that the better ones, those with the higher fitness, are more likely to emerge as the best solution.

INTERACTIVE SESSION: ORGANIZATIONS

THE FLASH CRASH: MACHINES GONE WILD?

On May 6, 2010, the U.S. stock markets were already down and trending even lower. Concerns about European debt, primarily the possibility of Greece defaulting, added to existing investor uncertainties about the markets and the economy at that time. But at 2:42 PM, in a flash, the equity market took a plunge so fast and so deep that it could not have been motivated by investor uncertainty alone.

Before the plunge, the market was already down 300 points on the day. In less than five minutes after 2:42, the Dow Jones Industrial Average plummeted more than 600 points, representing a loss of $1 trillion in market value. At its lowest point, the Dow was down a whopping 998.50 points to 9869.62, a 9.2 percent drop from the day's opening. This represented the biggest intraday decline in Dow history. Fortunately, this loss was temporary, vanishing nearly as quickly as it appeared. By 3:07 PM, the market had already regained nearly all of the points it had lost, and eventually closed down just 347.80 points that day at 10,520.32. The hefty loss was still its worst Dow percentage decline in over a year, but it certainly could have been worse.

How could this "flash crash" have happened? It now appears that the abrupt selling activities of a single mutual fund company touched off a chain reaction. A confluence of forces was unleashed by the structural and organizational features of the electronic trading systems that execute the majority of trades on the Dow and the rest of the world's major stock exchanges. Electronic trading systems offer considerable advantages over human brokers, including speed, reduced cost, and more liquid markets. High-frequency traders (HFTs) have taken over many of the responsibilities once filled by stock exchange specialists and market makers whose job was to match buyers and sellers efficiently.

Many trading systems today, such as those used by HFTs, are automated, using algorithms to place their nearly instant trades. A number of the HFT trading firms and hedge funds now use machine learning to help their computer systems trade in and out of stocks efficiently. Machine learning programs are able to crunch vast amounts of data in short periods, "learn" what works, and adjust their stock trading strategies on the fly, based on shifting dynamics in the market and broader

economy. This method is far beyond human capability: As Michael Kearns, computer science professor at University of Pennsylvania and expert in AI investing, stated, "No human could do this. Your head would blow off." It would appear, however, in situations like the flash crash, where a computer algorithm is insufficient to handle the complexity of the event in progress, electronic trading systems have the potential to make a bad situation much worse.

At 2:32 P.M. on May 6, Waddell & Reed Financial of Overland Park, Kansas started to sell $4.1 billion of futures contracts using a computer selling algorithm that dumped 75,000 contracts onto the market over the next 20 minutes. Normally, a sale of that size would take as much as five hours, but on that day, it was executed in 20 minutes. The algorithm instructed computers to execute the trade without regard to price or time, and thus continued to sell as prices sharply dropped.

After Waddell & Reed started to sell, many of the futures contracts were bought by HFTs. As the HFTs realized prices were continuing to fall, they began to sell what they had just bought very aggressively, which caused the mutual fund's algorithm in turn to accelerate its selling. The HFT computers traded contracts back and forth, creating a "hot potato" effect. The selling pressure was then transferred from the futures market to the stock market. Frightened buyers pulled to the sidelines. The markets were overwhelmed by sell orders with no legitimate buyers available to meet those orders.

The only buy orders available at all originated from automated systems, which were submitting orders known as "stub quotes." Stub quotes are offers to buy stocks at prices so low that they are unlikely to ever be the sole buyers of that stock available; during the unique conditions of the flash crash, they were. When the only offer to buy available is a penny-priced stub quote, a market order, by its terms, will execute against the stub quote. In this respect, automated trading systems will follow their coded logic regardless of outcome, while human involvement would likely have prevented these orders from executing at absurd prices.

In the midst of the crisis, the New York Stock Exchange activated circuit breakers, measures intended to slow trading on stocks that have lost a tenth or more of their value in a short time, and routed all of their trading traffic to human brokers in an effort to stop the downward spiral. (NYSE is the only major exchange with the ability to execute trades both via computers and via human brokers.) But because of the enormous volume of orders, and because other fully electronic exchanges lacked similar circuit breakers, it may have had the reverse effect in the short term. While computerized systems simply continued to push the market lower, humans were unable to react fast enough to the situation.

Regulators are considering several different approaches to preventing future flash crashes, but it may be that there is no satisfactory solution. The Securities and Exchange Commission (SEC) may attempt to standardize circuit breakers across all financial markets, limit high-frequency trading, overhaul the stub quoting system, or stipulate that all buy and sell orders be limit orders, which places upper and lower limits on the prices at which stocks can be bought and sold. But it may be that events like the flash crash are what author and hedge fund adviser Nassim Taleb called "Black Swans" in his book of the same name—unpredictable and uncontrollable events under which we only have "the illusion of control."

After 'Black Monday' in 1987, the last crash of similar size, it was thought that computer trading prevented sudden drops in the market, but the flash crash indicates that electronic trading simply allows them to occur over a shorter time period, and may even magnify these sudden market moves in either direction because they can happen faster with less chance of intervention. But, as the flash crash has proven, if we rely solely on these automated methods of electronic trading, we'll still need to worry about machines gone wild.

Sources: Graham Bowley, "Lone $4.1 Billion Sale Led to 'Flash Crash' in May," *The New York Times*, October 1, 2010; Aaron Lucchetti, "Exchanges Point Fingers Over Human Hands," *The Wall Street Journal*, May 9, 2010; Scott Patterson, "Letting the Machines Decide," *The Wall Street Journal*, July 13, 2010; Scott Patterson and Tom Lauricella, "Did a Big Bet Help Trigger 'Black Swan' Stock Swoon?" *The Wall Street Journal*, May 10, 2010; Edward Wyatt, "Regulators Vow to Find Way to Stop Rapid Dives," *The New York Times*, May 10, 2010; Kara Scannell and Fawn Johnson, "Schapiro: Web of Rules Aided Fall," *The Wall Street Journal*, May 12, 2010; Larry Harris, "How to Prevent Another Trading Panic," *The Wall Street Journal*, May 12, 2010; Scott Patterson, "How the 'Flash Crash' Echoed Black Monday," *The Wall Street Journal*, May 17, 2010.

CASE STUDY QUESTIONS

1. Describe the conditions that preceded the flash crash.
2. What are some of the benefits of electronic trading?
3. What features of electronic trading and automated trading programs contributed to the crash?
4. Could this crash have been prevented? Why or why not?

MIS IN ACTION

Use the Web to search for information on the Black Monday crash of 1987. Then answer the following questions:

1. What features of the Black Monday crash are different from the 2010 flash crash?
2. How are the two crashes similar?
3. What measures did regulators take to ensure that no future crashes would happen again?
4. How has the advent of electronic trading affected those regulatory measures?

according to their level of desirability as possible solutions. After the initial population is evaluated for fitness, the algorithm then produces the next generation of strings, consisting of strings that survived the fitness test plus offspring strings produced from mating pairs of strings, and tests their fitness. The process continues until a solution is reached.

Genetic algorithms are used to solve problems that are very dynamic and complex, involving hundreds or thousands of variables or formulas. The problem must be one where the range of possible solutions can be represented genetically and criteria can be established for evaluating fitness. Genetic algorithms expedite the solution because they are able to evaluate many solution alternatives quickly to find the best one. For example, General Electric engineers used genetic algorithms to help optimize the design for jet turbine aircraft engines, where each design change required changes in up to 100 variables. The supply chain management software from i2 Technologies uses genetic algorithms to optimize production-scheduling models incorporating hundreds of thousands of details about customer orders, material and resource availability, manufacturing and distribution capability, and delivery dates.

HYBRID AI SYSTEMS

Genetic algorithms, fuzzy logic, neural networks, and expert systems can be integrated into a single application to take advantage of the best features of these technologies. Such systems are called **hybrid AI systems**. Hybrid applications in business are growing. In Japan, Hitachi, Mitsubishi, Ricoh, Sanyo, and others are starting to incorporate hybrid AI in products such as home appliances, factory machinery, and office equipment. Matsushita has developed a "neurofuzzy" washing machine that combines fuzzy logic with neural networks. Nikko Securities has been working on a neurofuzzy system to forecast convertible-bond ratings.

INTELLIGENT AGENTS

Intelligent agent technology helps businesses navigate through large amounts of data to locate and act on information that is considered important. **Intelligent agents** are software programs that work in the background without direct human intervention to carry out specific, repetitive, and predictable tasks for an individual user, business process, or software application. The agent uses a limited built-in or learned knowledge base to accomplish tasks or make decisions on the user's behalf, such as deleting junk e-mail, scheduling appointments, or traveling over interconnected networks to find the cheapest airfare to California.

There are many intelligent agent applications today in operating systems, application software, e-mail systems, mobile computing software, and network tools. For example, the wizards found in Microsoft Office software tools have built-in capabilities to show users how to accomplish various tasks, such as formatting documents or creating graphs, and to anticipate when users need assistance.

Of special interest to business are intelligent agents for cruising networks, including the Internet, in search of information. Chapter 7 describes how shopping bots can help consumers find products they want and assist them in comparing prices and other features.

Many complex phenomena can be modeled as systems of autonomous agents that follow relatively simple rules for interaction. **Agent-based modeling** applications have been developed to model the behavior of consumers, stock markets, and supply chains and to predict the spread of epidemics (Samuelson and Macal, 2006).

Procter & Gamble (P&G) used agent-based modeling to improve coordination among different members of its supply chain in response to changing business conditions (see Figure 11-12). It modeled a complex supply chain as a group of semiautonomous "agents" representing individual supply chain components, such as trucks, production facilities, distributors, and retail stores. The

FIGURE 11-12 INTELLIGENT AGENTS IN P&G'S SUPPLY CHAIN NETWORK

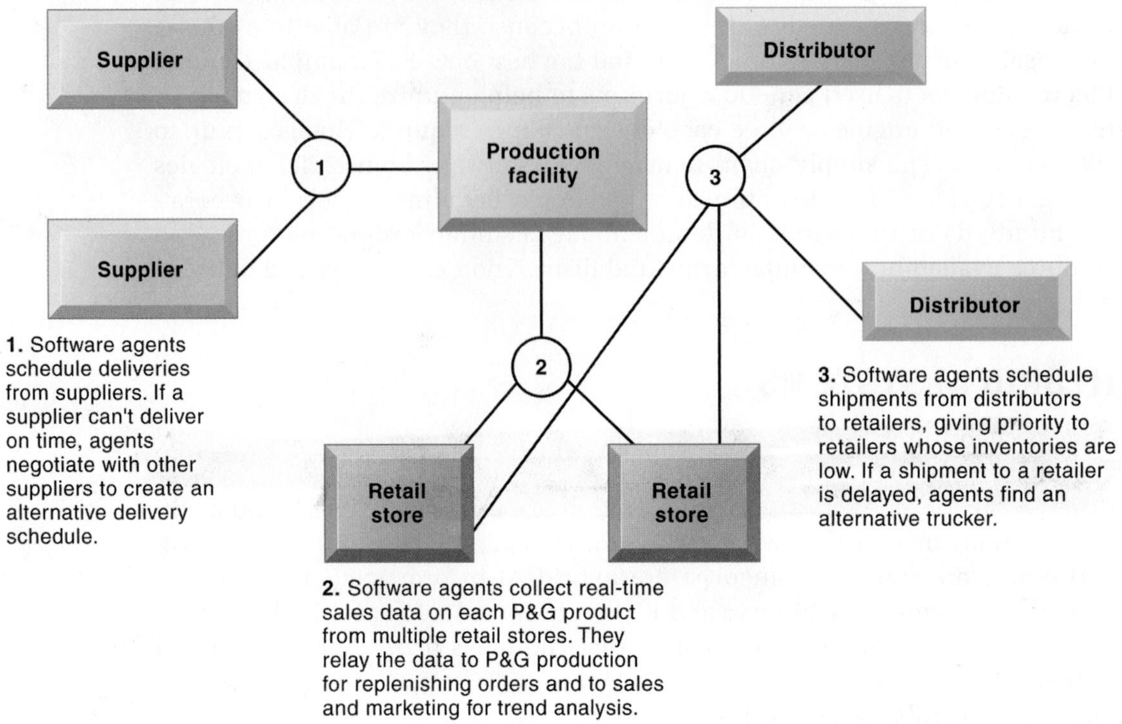
Supplier
Supplier
Production facility
Distributor
Distributor
1
2
3
Retail store
Retail store
1. Software agents schedule deliveries from suppliers. If a supplier can't deliver on time, agents negotiate with other suppliers to create an alternative delivery schedule.
2. Software agents collect real-time sales data on each P&G product from multiple retail stores. They relay the data to P&G production for replenishing orders and to sales and marketing for trend analysis.
3. Software agents schedule shipments from distributors to retailers, giving priority to retailers whose inventories are low. If a shipment to a retailer is delayed, agents find an alternative trucker.

Intelligent agents are helping P&G shorten the replenishment cycles for products such as a box of Tide.

behavior of each agent is programmed to follow rules that mimic actual behavior, such as "order an item when it is out of stock." Simulations using the agents enable the company to perform what-if analyses on inventory levels, in-store stockouts, and transportation costs.

Using intelligent agent models, P&G discovered that trucks should often be dispatched before being fully loaded. Although transportation costs would be higher using partially loaded trucks, the simulation showed that retail store stockouts would occur less often, thus reducing the amount of lost sales, which would more than make up for the higher distribution costs. Agent-based modeling has saved P&G $300 million annually on an investment of less than 1 percent of that amount.

11.5 HANDS-ON MIS PROJECTS

The projects in this section give you hands-on experience designing a knowledge portal, applying collaboration tools to solve a customer retention problem, using an expert system or spreadsheet tools to create a simple expert system, and using intelligent agents to research products for sale on the Web.

Management Decision Problems

1. U.S. Pharma Corporation is headquartered in New Jersey but has research sites in Germany, France, the United Kingdom, Switzerland, and Australia. Research and development of new pharmaceuticals is the key to ongoing profits, and U.S. Pharma researches and tests thousands of possible drugs. The company's researchers need to share information with others within and outside the company, including the U.S. Food and Drug Administration, the World Health Organization, and the International Federation of Pharmaceutical Manufacturers & Associations. Also critical is access to health information sites, such as the U.S. National Library of Medicine, and to industry conferences and professional journals. Design a knowledge portal for U.S. Pharma's researchers. Include in your design specifications relevant internal systems and databases, external sources of information, and internal and external communication and collaboration tools. Design a home page for your portal.

2. Sprint Nextel has the highest rate of customer churn (the number of customers who discontinue a service) in the cell phone industry, amounting to 2.45 percent. Over the past two years, Sprint has lost 7 million subscribers. Management wants to know why so many customers are leaving Sprint and what can be done to woo them back. Are customers deserting because of poor customer service, uneven network coverage, or the cost of Sprint cell phone plans? How can the company use tools for online collaboration and communication to help find the answer? What management decisions could be made using information from these sources?

Improving Decision Making: Building a Simple Expert System for Retirement Planning

Software skills: Spreadsheet formulas and IF function or expert system tool
Business skills: Benefits eligibility determination

Expert systems typically use a large number of rules. This project has been simplified to reduce the number of rules, but it will give you experience working with a series of rules to develop an application.

When employees at your company retire, they are given cash bonuses. These cash bonuses are based on the length of employment and the retiree's age. To receive a bonus, an employee must be at least 50 years of age and have worked for the company for five years. The following table summarizes the criteria for determining bonuses.

LENGTH OF EMPLOYMENT	BONUS
<5 years	No bonus
5–10 years	20 percent of current annual salary
11–15 years	30 percent of current annual salary
16–20 years	40 percent of current annual salary
20–25 years	50 percent of current annual salary
26 or more years	100 percent of current annual salary

Using the information provided, build a simple expert system. Find a demonstration copy of an expert system software tool on the Web that you can download. Alternatively, use your spreadsheet software to build the expert system. (If you are using spreadsheet software, we suggest using the IF function so you can see how rules are created.)

Improving Decision Making: Using Intelligent Agents for Comparison Shopping

Software skills: Web browser and shopping bot software
Business skills: Product evaluation and selection

This project will give you experience using shopping bots to search online for products, find product information, and find the best prices and vendors.

You have decided to purchase a new digital camera. Select a digital camera you might want to purchase, such as the Canon PowerShot S95 or the Olympus Stylus 7040. To purchase the camera as inexpensively as possible, try several of the shopping bot sites, which do the price comparisons for you. Visit My Simon (www.mysimon.com), BizRate.com (www.bizrate.com), and Google Product Search. Compare these shopping sites in terms of their ease of use, number of offerings, speed in obtaining information, thoroughness of information offered about the product and seller, and price selection. Which site or sites would you use and why? Which camera would you select and why? How helpful were these sites for making your decision?

LEARNING TRACK MODULE

The following Learning Track provides content relevant to topics covered in this chapter:

1. Challenges of Knowledge Management Systems

Review Summary

1. *What is the role of knowledge management and knowledge management programs in business?*

    Knowledge management is a set of processes to create, store, transfer, and apply knowledge in the organization. Much of a firm's value depends on its ability to create and manage knowledge. Knowledge management promotes organizational learning by increasing the ability of the organization to learn from its environment and to incorporate knowledge into its business processes. There are three major types of knowledge management systems: enterprise-wide knowledge management systems, knowledge work systems, and intelligent techniques.

2. *What types of systems are used for enterprise-wide knowledge management and how do they provide value for businesses?*

    Enterprise-wide knowledge management systems are firmwide efforts to collect, store, distribute, and apply digital content and knowledge. Enterprise content management systems provide databases and tools for organizing and storing structured documents and tools for organizing and storing semistructured knowledge, such as e-mail or rich media. Knowledge network systems provide directories and tools for locating firm employees with special expertise who are important sources of tacit knowledge. Often these systems include group collaboration tools (including wikis and social bookmarking), portals to simplify information access, search tools, and tools for classifying information based on a taxonomy that is appropriate for the organization. Enterprise-wide knowledge management systems can provide considerable value if they are well designed and enable employees to locate, share, and use knowledge more efficiently.

3. *What are the major types of knowledge work systems and how do they provide value for firms?*

    Knowledge work systems (KWS) support the creation of new knowledge and its integration into the organization. KWS require easy access to an external knowledge base; powerful computer hardware that can support software with intensive graphics, analysis, document management, and communications capabilities; and a user-friendly interface. Computer-aided design (CAD) systems, augmented reality applications, and virtual reality systems, which create interactive simulations that behave like the real world, require graphics and powerful modeling capabilities. KWS for financial professionals provide access to external databases and the ability to analyze massive amounts of financial data very quickly.

4. *What are the business benefits of using intelligent techniques for knowledge management?*

    Artificial intelligence lacks the flexibility, breadth, and generality of human intelligence, but it can be used to capture, codify, and extend organizational knowledge. Expert systems capture tacit knowledge from a limited domain of human expertise and express that knowledge in the form of rules. Expert systems are most useful for problems of classification or diagnosis. Case-based reasoning represents organizational knowledge as a database of cases that can be continually expanded and refined.

    Fuzzy logic is a software technology for expressing knowledge in the form of rules that use approximate or subjective values. Fuzzy logic has been used for controlling physical devices and is starting to be used for limited decision-making applications.

    Neural networks consist of hardware and software that attempt to mimic the thought processes of the human brain. Neural networks are notable for their ability to learn without programming and to recognize patterns that cannot be easily described by humans. They are being used in science, medicine, and business to discriminate patterns in massive amounts of data.

    Genetic algorithms develop solutions to particular problems using genetically based processes such as fitness, crossover, and mutation. Genetic algorithms are beginning to be applied to problems involving optimization, product design, and monitoring industrial systems where many alternatives or variables must be evaluated to generate an optimal solution.

    Intelligent agents are software programs with built-in or learned knowledge bases that carry out specific, repetitive, and predictable tasks for an individual user, business process, or software application. Intelligent agents can be programmed to navigate through large amounts of data to locate useful information and in some cases act on that information on behalf of the user.

Key Terms

Agent-based modeling, 441
Augmented reality (AR), 428
Artificial intelligence (AI), 431
Backward chaining, 432
Case-based reasoning (CBR), 434
Communities of practice (COPs), 421
Computer-aided design (CAD), 427
Data, 417
Digital asset management systems, 424
Enterprise content management systems, 423
Enterprise-wide knowledge management systems, 421
Expert systems, 432
Explicit knowledge, 417
Folksonomies, 425
Forward chaining, 432
Fuzzy logic, 435
Genetic algorithms, 438
Hybrid AI systems, 441
Inference engine, 432
Intelligent agents, 441

Intelligent techniques, 422
Investment workstations, 431
Knowledge, 417
Knowledge base, 432
Knowledge discovery, 431
Knowledge management, 419
Knowledge network systems, 424
Knowledge work systems (KWS), 421
Learning Management System (LMS), 425
Machine learning, 438
Neural networks, 436
Organizational learning, 419
Social bookmarking, 424
Structured knowledge, 422
Tacit knowledge, 417
Taxonomy, 423
Virtual Reality Modeling Language (VRML), 430
Virtual reality systems, 428
Wisdom, 417

Review Questions

1. What is the role of knowledge management and knowledge management programs in business?

   - Define knowledge management and explain its value to businesses.

   - Describe the important dimensions of knowledge.

   - Distinguish between data, knowledge, and wisdom and between tacit knowledge and explicit knowledge.

   - Describe the stages in the knowledge management value chain.

2. What types of systems are used for enterprise-wide knowledge management and how do they provide value for businesses?

   - Define and describe the various types of enterprise-wide knowledge management systems and explain how they provide value for businesses.

   - Describe the role of the following in facilitating knowledge management: portals, wikis, social bookmarking, and learning management systems.

3. What are the major types of knowledge work systems and how do they provide value for firms?

   - Define knowledge work systems and describe the generic requirements of knowledge work systems.

   - Describe how the following systems support knowledge work: CAD, virtual reality, augmented reality, and investment workstations.

4. What are the business benefits of using intelligent techniques for knowledge management?

   - Define an expert system, describe how it works, and explain its value to business.

   - Define case-based reasoning and explain how it differs from an expert system.

   - Define a neural network, and describe how it works and how it benefits businesses.

   - Define and describe fuzzy logic, genetic algorithms, and intelligent agents. Explain how each works and the kinds of problems for which each is suited.

Discussion Questions

1. Knowledge management is a business process, not a technology. Discuss.

2. Describe various ways that knowledge management systems could help firms with sales and marketing or with manufacturing and production.

3. Your company wants to do more witih knowledge management. Describe the steps it should take to develop a knowledge management program and select knowledge management applications.

Video Cases

Video Cases and Instructional Videos illustrating some of the concepts in this chapter are available. Contact your instructor to access these videos.

Collaboration and Teamwork: Rating Enterprise Content Management Systems

With a group of classmates, select two enterprise content management products, such as those from Open Text, IBM, EMC, or Oracle. Compare their features and capabilities. To prepare your analysis, use articles from computer magazines and the Web sites for the enterprise content management software vendors. If possible, use Google Sites to post links to Web pages, team communication announcements, and work assignments; to brainstorm; and to work collaboratively on project documents. Try to use Google Docs to develop a presentation of your findings for the class.

San Francisco Public Utilities Commission Preserves Expertise with Better Knowledge Management
CASE STUDY

A major challenge facing many companies and organizations is the imminent retirement of baby boomers. For certain organizations, this challenge is more daunting than usual, not only because of a larger spike in employee retirements, but also because of the business process change that must accompany significant shifts in any workforce. The San Francisco Public Utilities Commission (SFPUC) was one such organization.

SFPUC is a department of the city and county of San Francisco that provides water, wastewater, and municipal power services to the city. SFPUC has four major divisions: Regional Water, Local Water, Power, and Wastewater (collection, treatment, and disposal of water). The organization has over 2,000 employees and serves 2.4 million customers in San Francisco and the Bay Area. It is the third largest municipal utility in California.

SFPUC's Power division provides electricity to the city and county of San Francisco, including power used to operate electric streetcars and buses; the Regional and Local Water departments supply some of the purest drinking water in the world to San Francisco and neighboring Santa Clara and San Mateo counties; and the Wastewater division handles flushed and drained water to significantly reduce pollution in the San Francisco Bay and Pacific Ocean. The mission of this organization is to provide San Francisco and its Bay Area customers with reliable, high-quality, affordable water and wastewater treatment while efficiently and responsibly managing human, physical, and natural resources.

SFPUC expected that a significant portion of its employees—about 20 percent—would retire in 2009. To make matters worse, the majority of these positions were technical, which meant that the training of new employees would be more complicated, and maintaining knowledge of the retiring workers would be critical to all areas of SFPUC's business processes.

To deal with this trend, companies and organizations like SFPUC must rearrange their operations so that the generational swap doesn't adversely affect their operational capability in the upcoming decades. In particular, the organization needed a

way to capture the knowledge of its retiring employees of "baby boom" age in an efficient and cost-effective manner, and then communicate this knowledge successfully to the next generation of employees. The two major challenges SFPUC faced were successfully capturing, managing, and transferring this knowledge, and maintaining reliability and accountability despite a large influx of new workers.

SFPUC met these challenges by implementing a business process management (BPM) and workflow solution from Interfacing Technologies Corporation to drive change efforts across the organization. The system, called Enterprise Process Center, or EPC, manages knowledge retention and establishes new ways of collaborating, sharing information, and defining roles and responsibilities. SFPUC saw the retirement of its baby boomers as an opportunity to implement a structure that would alleviate similar problems in the future. With EPC, SFPUC would be able to maintain continuity from older to newer employees more easily. SFPUC was impressed that the system would span all four of its major divisions, helping to standardize common processes across multiple departments, and that it would be easy to use and train employees.

EPC sought to identify common processes, called "work crossovers," by mapping business processes across each department. EPC is unique among BPM software providers in its visual representation of these processes. Using flow charts accessible via a Web portal to clearly depict the functions performed by each department, SFPUC was able to identify redundant and inefficient tasks performed by multiple departments. This visually oriented solution for optimizing business processes catered to both technology- savvy new employees and older baby boomers.

Prior to the BPM overhaul, SFPUC employees had little incentive to share business process information. New environmental regulations were difficult to communicate. Certain inspection processes were conducted on an irregular basis, sometimes as infrequently as every 5 to 15 years. The knowledge required to execute these processes was especially valuable, because newer employees would have no way of completing these tasks without proper documentation and process knowledge. SFPUC needed ways to easily find knowledge about processes that were performed daily, as well as every 15 years, and what's more, that knowledge had to be up to date so that employees didn't stumble across obsolete information.

EPC solved that problem by creating work order flows for all tasks performed within the organization, defining the employee roles and responsibilities for each. For example, the work order flow for SFPUC's wastewater enterprise displayed each step in the process visually, with links to manuals describing how to complete the task and the documents required to complete it. EPC also identified obsolete processes that were well suited to automation or totally superfluous. Automating and eliminating outdated tasks alleviated some of SFPUC's budget and workload concerns, allowing the organization to divert extra resources to training and human resources.

SFPUC management had anticipated that eliminating outdated tasks would have the added effect of keeping employees happy, which would help SFPUC's performance by delaying retirement of older employees and increasing the likelihood that newer hires stayed at the company. EPC allowed employees to provide feedback on various tasks, helping to identify tasks that were most widely disliked. For example, the process for reimbursement of travel expenses was described as lengthy, highly labor-intensive, and valueless to the citizens of San Francisco. To be reimbursed for travel expenses, employees had to print a form, complete it by hand, attach travel receipts, and walk the documents over to their supervisors, who then had to manually review and approve each item and remit expenses for three additional levels of approval. Only then could the division controller issue the reimbursement.

To address this need for sharing information and making documents available across the organization, SFPUC started by using a wiki, but the documents lacked different levels of relevance. Critical information pertaining to everyday tasks took the same amount of time to find as information pertaining to an inspection performed every 15 years. EPC allowed users to assign levels of relevance to tasks and identify critical information so that critical information displays when employees search for certain items. For example, SFPUC employees must comply with various regulatory permits for water and air quality standards. Lack

of awareness of these standards often leads to unintentional violations. The BPM tool helped users assign risks to various tasks so that when employees queried information, the relevant regulations displayed along with the requested documents.

Identifying the experts on particular subjects for mission-critical processes is often challenging when compiling information on business processes across the enterprise. SFPUC anticipated this, using EPC to break down large-scale process knowledge into more manageable pieces, which allowed more users to contribute information. Users were reluctant to buy in to the BPM implementation at first, but management characterized the upgrade in a way that invited employees to share their thoughts about their least favorite processes and contribute their knowledge.

The final product of the knowledge management overhaul took the form of a "centralized electronic knowledge base," which graphically displays critical steps of each task and uses videos to gather information and show work being done. New employees quickly became confident that they could perform certain tasks because of these videos. The overall results of the project were overwhelmingly positive. EPC helped SFPUC take its baby boomers' individual data and knowledge and turn them into usable and actionable information that was easily shared throughout the firm. SFPUC stayed much further under budget than other comparable governmental organizations.

SFPUC's new knowledge processes made many activities more paperless, reducing printing costs, time to distribute documents, and space required for document retention. Going paperless also supported the organization's mission to become more environmentally responsible. The addition of video technology to the process maps helped employees see how they could reduce energy consumption practices and electrical costs. By automating and redesigning the unwieldy travel reimbursement process described earlier, SFPUC reduced the time to process employee reimbursement requests by as much as 50 percent.

Sources: "San Francisco Tackles Baby Boom Retirement Effect and is Selected as a Finalist for the Global Awards for Excellence in BPM-Workflow," *International Business Times,* January 12, 2010; Interfacing Technologies, Canada, "San Francisco Public Utilities Commission USA;" Catherine Curtis, "SFPUC Delivers Workforce Development Presentation at WEFTEC Conference," *Wastewater Enterprise,* October 2009; "SFPUC Water Enterprise Environmental Stewardship Policy," San Francisco Public Utilities Commission, June 27, 2006.

CASE STUDY QUESTIONS

1. What are the business goals of SFPUC? How is knowledge management related to those goals?

2. What were some of the challenges faced by SFPUC? What management, organization, and technology factors were responsible for those challenges?

3. Describe how implementing EPC improved knowledge management and operational effectiveness at SFPUC.

4. How effective was EPC as a solution for SFPUC?

Chapter 12

Enhancing Decision Making

WHAT TO SELL? WHAT PRICE TO CHARGE? ASK THE DATA

W hat's the best way to get a discount on your morning coffee at Starbucks? Well, if you live in Manhattan, you could get up an hour early and take the subway downtown to Brooklyn. A single expresso is 10 cents cheaper than in your neighborhood, as are a caffee latte and slice of lemon pound cake. But a muffin runs 10 cents more uptown in Marble Hill, and a tall Pike's Place Roast costs $1.70 no matter where you live.

Starbucks is one of many retailers using sophisticated software to analyze, store by store and item by item, how demand responds to changes in price. What customers are willing to pay for certain items depends very much on the neighborhood or even the region of the country where they live. Shoppers in certain locations are willing to pay more.

The Duane Reade drugstore chain, recently purchased by Walgreens, is also adept at adjusting prices. Software analyzing sales patterns found that parents of newborn babies are not as price-sensitive as those with toddlers, so the company was able to raise prices on diapers for newborn infants without losing sales. The chain's information systems also showed how to adjust pricing based on location. Shoppers at the Duane Reade store near 86th Street and Lexington Avenue pay 20 cents more for a box of Kleenex and 50 cents more for a bottle of Pepto-Bismol than customers in Harlem.

Business analytics software such as that used by Duane Reade typically analyzes patterns in sales data to create a "pricing profile." A store near a big commuting hub might discount convenience items to present a low-cost image while one in a family neighborhood with many young children might discount baby items to get more people through the door.

Analyzing large troves of digital sales and customer information from both online and physical stores also helps retailers decide what to sell as well. Fashion Web site HauteLook confirmed that Southerners buy more white, green, and pink than people from other regions, while ShopItToMe learned that the average woman spends less on fashion in Dallas than in Washington D.C. and that women are thinner on both coasts than in the U.S. heartland and wear more petite clothing and shoe sizes meant for smaller women.

How much of a difference does this knowledge make? Lots. 1-800-Flowers, which sells flowers and gift baskets online, has used analytics software from SAS Inc. to tweak its online storefront and marketing activities. The software helped the company quickly record and analyze buyer profiles to help improve targeting of its product, determine what "specials" to offer, and plan sales and marketing strategies based on an understanding of real customer needs. The company is able to quickly change prices and offerings on its Web site—often every hour. In the first half of 2010, 1-800-Flowers used more finely targeted Web pages and e-mail promotions to improve the conversion rate of Web site browsers to buyers by 20 percent.

Sources: Anne Kadet, "Price-Point Politics," *The Wall Street Journal*, July 24, 2010; Steve Lohr, "A Data Explosion Remakes Retailing," and Christina Binkley, "Fashion Nation: What Retailers Know about Us," *The Wall Street Journal*, July 28, 2010.

The experiences of Starbucks, Duane Reade, and 1-800-Flowers are powerful illustrations of how information systems improve decision making. Managers at these retail chains were unable to make good decisions about what prices to charge to improve profitability and what items to sell in stores to maximize sales at different locations and different time periods. They had access to customer purchase data, but they were unable to analyze millions of pieces of data on their own. Bad decisions about how much to charge and how to stock stores lowered sales revenue and prevented these companies from responding quickly to customer needs.

The chapter-opening diagram calls attention to important points raised by this case and this chapter. Starbucks, Duane Reade, and 1-800-Flowers started using business intelligence software, which is able to find patterns and trends in massive quantities of data. Information from these business intelligence systems helps managers at these companies make better decisions about pricing, shelf stocking, and product offerings. They are able to see where they can charge a higher price or where they must lower prices to maximize sales revenue, as well as what items to stock and when to change their merchandise mix. Better decision making using business intelligence has made all of these companies more profitable.

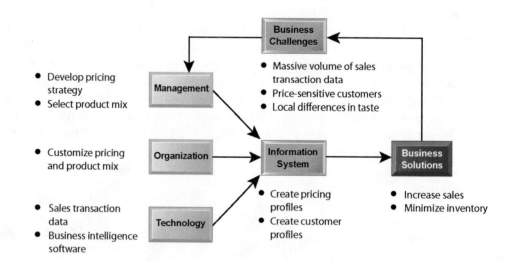

12.1 DECISION MAKING AND INFORMATION SYSTEMS

Decision making in businesses used to be limited to management. Today, lower-level employees are responsible for some of these decisions, as information systems make information available to lower levels of the business. But what do we mean by better decision making? How does decision making take place in businesses and other organizations? Let's take a closer look.

BUSINESS VALUE OF IMPROVED DECISION MAKING

What does it mean to the business to make better decisions? What is the monetary value of improved decision making? Table 12-1 attempts to measure the monetary value of improved decision making for a small U.S. manufacturing firm with $280 million in annual revenue and 140 employees. The firm has identified a number of key decisions where new system investments might improve the quality of decision making. The table provides selected estimates of annual value (in the form of cost savings or increased revenue) from improved decision making in selected areas of the business.

We can see from Table 12-1 that decisions are made at all levels of the firm and that some of these decisions are common, routine, and numerous. Although the value of improving any single decision may be small, improving hundreds of thousands of "small" decisions adds up to a large annual value for the business.

TYPES OF DECISIONS

Chapters 1 and 2 showed that there are different levels in an organization. Each of these levels has different information requirements for decision support and responsibility for different types of decisions (see Figure 12-1). Decisions are classified as structured, semistructured, and unstructured.

TABLE 12-1 BUSINESS VALUE OF ENHANCED DECISION MAKING

EXAMPLE DECISION	DECISION MAKER	NUMBER OF ANNUAL DECISIONS	ESTIMATED VALUE TO FIRM OF A SINGLE IMPROVED DECISION	ANNUAL VAUE
Allocate support to most valuable customers	Accounts manager	12	$ 100,000	$1,200,000
Predict call center daily demand	Call center management	4	150,000	600,000
Decide parts inventory levels daily	Inventory manager	365	5,000	1,825,000
Identify competitive bids from major suppliers	Senior management	1	2,000,000	2,000,000
Schedule production to fill orders	Manufacturing manager	150	10,000	1,500,000
Allocate labor to complete a job	Production floor manager	100	4,000	400,000

FIGURE 12-1   INFORMATION REQUIREMENTS OF KEY DECISION-MAKING GROUPS
                IN A FIRM

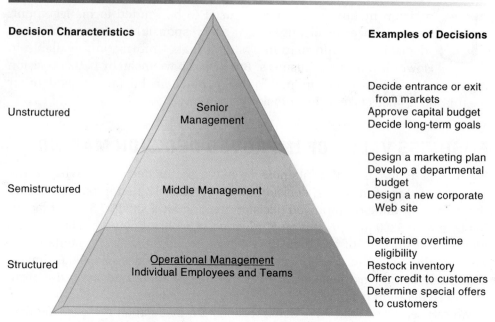

Senior managers, middle managers, operational managers, and employees have different types of decisions and information requirements.

Unstructured decisions are those in which the decision maker must provide judgment, evaluation, and insight to solve the problem. Each of these decisions is novel, important, and nonroutine, and there is no well-understood or agreed-on procedure for making them.

Structured decisions, by contrast, are repetitive and routine, and they involve a definite procedure for handling them so that they do not have to be treated each time as if they were new. Many decisions have elements of both types of decisions and are **semistructured**, where only part of the problem has a clear-cut answer provided by an accepted procedure. In general, structured decisions are more prevalent at lower organizational levels, whereas unstructured problems are more common at higher levels of the firm.

Senior executives face many unstructured decision situations, such as establishing the firm's five- or ten-year goals or deciding new markets to enter. Answering the question "Should we enter a new market?" would require access to news, government reports, and industry views as well as high-level summaries of firm performance. However, the answer would also require senior managers to use their own best judgment and poll other managers for their opinions.

Middle management faces more structured decision scenarios but their decisions may include unstructured components. A typical middle-level management decision might be "Why is the reported order fulfillment report showing a decline over the past six months at a distribution center in Minneapolis?" This middle manager will obtain a report from the firm's enterprise system or distribution management system on order activity and operational efficiency at the Minneapolis distribution center. This is the structured part of the decision. But before arriving at an answer, this middle manager will have to interview employees and gather more unstructured information from external sources about local economic conditions or sales trends.

Operational management and rank-and-file employees tend to make more structured decisions. For example, a supervisor on an assembly line has to decide whether an hourly paid worker is entitled to overtime pay. If the employee worked more than eight hours on a particular day, the supervisor would routinely grant overtime pay for any time beyond eight hours that was clocked on that day.

A sales account representative often has to make decisions about extending credit to customers by consulting the firm's customer database that contains credit information. If the customer met the firm's prespecified criteria for granting credit, the account representative would grant that customer credit to make a purchase. In both instances, the decisions are highly structured and are routinely made thousands of times each day in most large firms. The answer has been preprogrammed into the firm's payroll and accounts receivable systems.

THE DECISION-MAKING PROCESS

Making a decision is a multistep process. Simon (1960) described four different stages in decision making: intelligence, design, choice, and implementation (see Figure 12-2).

FIGURE 12-2　STAGES IN DECISION MAKING

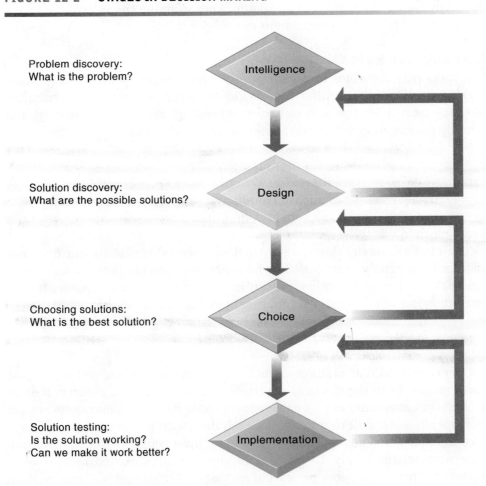

The decision-making process is broken down into four stages.

Intelligence consists of discovering, identifying, and understanding the problems occurring in the organization—why a problem exists, where, and what effects it is having on the firm.

Design involves identifying and exploring various solutions to the problem.

Choice consists of choosing among solution alternatives.

Implementation involves making the chosen alternative work and continuing to monitor how well the solution is working.

What happens if the solution you have chosen doesn't work? Figure 12-2 shows that you can return to an earlier stage in the decision-making process and repeat it if necessary. For instance, in the face of declining sales, a sales management team may decide to pay the sales force a higher commission for making more sales to spur on the sales effort. If this does not produce sales increases, managers would need to investigate whether the problem stems from poor product design, inadequate customer support, or a host of other causes that call for a different solution.

MANAGERS AND DECISION MAKING IN THE REAL WORLD

The premise of this book and this chapter is that systems to support decision making produce better decision making by managers and employees, above average returns on investment for the firm, and ultimately higher profitability. However, information systems cannot improve all the different kinds of decisions taking place in an organization. Let's examine the role of managers and decision making in organizations to see why this is so.

Managerial Roles

Managers play key roles in organizations. Their responsibilities range from making decisions, to writing reports, to attending meetings, to arranging birthday parties. We are able to better understand managerial functions and roles by examining classical and contemporary models of managerial behavior.

The **classical model of management**, which describes what managers do, was largely unquestioned for the more than 70 years since the 1920s. Henri Fayol and other early writers first described the five classical functions of managers as planning, organizing, coordinating, deciding, and controlling. This description of management activities dominated management thought for a long time, and it is still popular today.

The classical model describes formal managerial functions but does not address what exactly managers do when they plan, decide things, and control the work of others. For this, we must turn to the work of contemporary behavioral scientists who have studied managers in daily action. **Behavioral models** state that the actual behavior of managers appears to be less systematic, more informal, less reflective, more reactive, and less well organized than the classical model would have us believe.

Observers find that managerial behavior actually has five attributes that differ greatly from the classical description. First, managers perform a great deal of work at an unrelenting pace—studies have found that managers engage in more than 600 different activities each day, with no break in their pace. Second, managerial activities are fragmented; most activities last for less than nine minutes, and only 10 percent of the activities exceed one hour in duration. Third, managers prefer current, specific, and ad hoc information (printed information often will be too old). Fourth, they prefer oral forms of communication to written forms because oral media provide greater flexibility, require less effort, and bring a faster response. Fifth, managers give high priority to maintaining a diverse and complex web of contacts that acts as an informal information system and helps them execute their personal agendas and short- and long-term goals.

Analyzing managers' day-to-day behavior, Mintzberg found that it could be classified into 10 managerial roles. **Managerial roles** are expectations of the activities that managers should perform in an organization. Mintzberg found that these managerial roles fell into three categories: interpersonal, informational, and decisional.

Interpersonal Roles. Managers act as figureheads for the organization when they represent their companies to the outside world and perform symbolic duties, such as giving out employee awards, in their **interpersonal role**. Managers act as leaders, attempting to motivate, counsel, and support subordinates. Managers also act as liaisons between various organizational levels; within each of these levels, they serve as liaisons among the members of the management team. Managers provide time and favors, which they expect to be returned.

Informational Roles. In their **informational role**, managers act as the nerve centers of their organizations, receiving the most concrete, up-to-date information and redistributing it to those who need to be aware of it. Managers are therefore information disseminators and spokespersons for their organizations.

Decisional Roles. Managers make decisions. In their **decisional role**, they act as entrepreneurs by initiating new kinds of activities; they handle disturbances arising in the organization; they allocate resources to staff members who need them; and they negotiate conflicts and mediate between conflicting groups.

Table 12-2, based on Mintzberg's role classifications, is one look at where systems can and cannot help managers. The table shows that information systems are now capable of supporting most, but not all, areas of management life.

TABLE 12-2 MANAGERIAL ROLES AND SUPPORTING INFORMATION SYSTEMS

ROLE	BEHAVIOR	SUPPORT SYSTEMS
Interpersonal Roles		
Figurehead ----------------------➤		Telepresence systems
Leader ----------- Interpersonal----➤		Telepresence, social networks, Twitter
Liaison ----------------------➤		Smartphones, social networks
Informational Roles		
Nerve center -------------------➤		Management information systems, ESS
Disseminator ------- Information-----➤		E-mail, social networks
Spokesperson ------processing------➤		Webinars, telepresence
Decisional Roles		
Entrepreneur -------Decision--------➤		None exist
Disturbance handler --making--------➤		None exist
Resource allocator ----------------➤		Business intelligence, DSS systems
Negotiator ----------------------➤		None exist

Sources: Kenneth C. Laudon and Jane P. Laudon; and Mintzberg, 1971.

Real-World Decision Making

We now see that information systems are not helpful for all managerial roles. And in those managerial roles where information systems might improve decisions, investments in information technology do not always produce positive results. There are three main reasons: information quality, management filters, and organizational culture (see Chapter 3).

Information Quality. High-quality decisions require high-quality information. Table 12-3 describes information quality dimensions that affect the quality of decisions.

If the output of information systems does not meet these quality criteria, decision-making will suffer. Chapter 6 has shown that corporate databases and files have varying levels of inaccuracy and incompleteness, which in turn will degrade the quality of decision making.

Management Filters. Even with timely, accurate information, some managers make bad decisions. Managers (like all human beings) absorb information through a series of filters to make sense of the world around them. Managers have selective attention, focus on certain kinds of problems and solutions, and have a variety of biases that reject information that does not conform to their prior conceptions.

For instance, Wall Street firms such as Bear Stearns and Lehman Brothers imploded in 2008 because they underestimated the risk of their investments in complex mortgage securities, many of which were based on subprime loans that were more likely to default. The computer models they and other financial institutions used to manage risk were based on overly optimistic assumptions and overly simplistic data about what might go wrong. Management wanted to make sure that their firms' capital was not all tied up as a cushion against defaults from risky investments, preventing them from investing it to generate profits. So the designers of these risk management systems were encouraged to measure risks in a way that minimzed their importance. Some trading desks also oversimplified the information maintained about the mortgage securities to make them appear as simple bonds with higher ratings than were warranted by their underlying components (Hansell, 2008).

Organizational Inertia and Politics. Organizations are bureaucracies with limited capabilities and competencies for acting decisively. When environments change and businesses need to adopt new business models to survive,

TABLE 12-3 INFORMATION QUALITY DIMENSIONS

QUALITY DIMENSION	DESCRIPTION
Accuracy	Do the data represent reality?
Integrity	Are the structure of data and relationships among the entities and attributes consistent?
Consistency	Are data elements consistently defined?
Completeness	Are all the necessary data present?
Validity	Do data values fall within defined ranges?
Timeliness	Area data available when needed?
Accessibility	Are the data accessible, comprehensible, and usable?

strong forces within organizations resist making decisions calling for major change. Decisions taken by a firm often represent a balancing of the firm's various interest groups rather than the best solution to the problem.

Studies of business restructuring find that firms tend to ignore poor performance until threatened by outside takeovers, and they systematically blame poor performance on external forces beyond their control such as economic conditions (the economy), foreign competition, and rising prices, rather than blaming senior or middle management for poor business judgment (John, Lang, Netter, et al., 1992).

HIGH-VELOCITY AUTOMATED DECISION MAKING

Today, many decisions made by organizations are not made by managers, or any humans. For instance, when you enter a query into Google's search engine, Google has to decide which URLs to display in about half a second on average (500 milliseconds). Google indexes over 50 billion Web pages, although it does not search the entire index for every query it receives. The same is true of other search engines. The New York Stock Exchange is spending over $450 million in 2010–2011 to build a trading platform that can executes incoming orders in less that 50 milliseconds. High frequency traders at electronic stock exchanges execute their trades in under 30 milliseconds.

The class of decisions that are highly structured and automated is growing rapidly. What makes this kind of automated high-speed decision making possible are computer algorithms that precisely define the steps to be followed to produce a decision, very large databases, very high-speed processors, and software optimized to the task. In these situations, humans (including managers) are eliminated from the decision chain because they are too slow.

This also means organizations in these areas are making decisions faster than what managers can monitor or control. Inability to control automated decisions was a major factor in the "Flash Crash" experienced by U.S. stock markets on May 6, 2010, when the Dow Jones Industrial Average fell over 600 points in a matter of minutes before rebounding later that day. The stock market was overwhelmed by a huge wave of sell orders triggered primarily by high-speed computerized trading programs within a few seconds, causing shares of some companies like Proctor & Gamble to sell for pennies.

How does the Simon framework of intelligence-design-choice-implementation work in high-velocity decision environments? Essentially, the intelligence, design, choice, and implementation parts of the decision-making process are captured by the software's algorithms. The humans who wrote the software have already identified the problem, designed a method for finding a solution, defined a range of acceptable solutions, and implemented the solution. Obviously, with humans out of the loop, great care needs to be taken to ensure the proper operation of these systems lest they do significant harm to organizations and humans. And even then additional safeguards are wise to observe the behavior of these systems, regulate their performance, and if necessary, turn them off.

12.2 BUSINESS INTELLIGENCE IN THE ENTERPRISE

Chapter 2 introduced you to the different types of systems used for supporting management decision making. At the foundation of all of these decision support systems are business intelligence and business analytics infrastructure that supplies the data and the analytic tools for supporting decision making. In this section, we want to answer the following questions:

- What are business intelligence (BI) and business analytics (BA)
- Who makes business intelligence and business analytics hardware and software?
- Who are the users of business intelligence?
- What kinds of analytical tools come with a BI/BA suite?
- How do managers use these tools?
- What are some examples of firms who have used these tools?
- What management strategies are used for developing BI/BA capabilities?

WHAT IS BUSINESS INTELLIGENCE?

When we think of humans as intelligent beings we often refer to their ability to take in data from their environment, understand the meaning and significance of the information, and then act appropriately. Can the same be said of business firms? The answer appears to be a qualified "yes." All organizations, including business firms, do indeed take in information from their environments, attempt to understand the meaning of the information, and then attempt to act on the information. Just like human beings, some business firms do this well, and others poorly.

"Business intelligence" is a term used by hardware and software vendors and information technology consultants to describe the infrastructure for warehousing, integrating, reporting, and analyzing data that comes from the business environment. The foundation infrastructure collects, stores, cleans, and makes relevant information available to managers. Think databases, data warehouses, and data marts described in Chapter 6. "Business analytics" is also a vendor-defined term that focuses more on tools and techniques for analyzing and understanding data. Think online analytical processing (OLAP), statistics, models, and data mining, which we also introduced in Chapter 6.

So, stripped to its essentials, business intelligence and analytics are about integrating all the information streams produced by a firm into a single, coherent enterprise-wide set of data, and then, using modeling, statistical analysis tools (like normal distributions, correlation and regression analysis, Chi square analysis, forecasting, and cluster analysis), and data mining tools (pattern discovery and machine learning), to make sense out of all these data so managers can make better decisions and better plans, or at least know quickly when their firms are failing to meet planned targets.

One company that uses business intelligence is Hallmark Cards. The company uses SAS Analytics software to improve its understanding of buying patterns that could lead to increased sales at more than 3,000 Hallmark Gold Crown stores in the United Sates. Hallmark wanted to strengthen its relationship with frequent buyers. Using data mining and predictive modeling, the company determined how to market to various consumer segments during holidays and special occasions as well as adjust promotions on the fly. Hallmark is able to determine which customer segments are most influenced by direct mail, which should be approached through e-mail, and what specific messages to send each group. Business intelligence has helped boost Hallmark sales to its loyalty program members by 5 to 10 percent.

Business Intelligence Vendors

It is important to remember that business intelligence and analytics are products defined by technology vendors and consulting firms. They consist of hardware and software suites sold primarily by large system vendors to very large Fortune 500 firms. The largest five providers of these products are SAP, Oracle, IBM, SAS Institute, and Microsoft (see Table 12-4). Microsoft's products are aimed at small to medium size firms, and they are based on desktop tools familiar to employees (such as Excel spreadsheet software), Microsoft Sharepoint collaboration tools, and Microsoft SQL Server database software. The size of the American BI and BA marketplace in 2010 is estimated to be $10.5 billion and growing at over 20% annually (Gartner, 2010). This makes business intelligence and business analytics one of the fastest-growing and largest segments in the U.S. software market.

THE BUSINESS INTELLIGENCE ENVIRONMENT

Figure 12-3 gives an overview of a business intelligence environment, highlighting the kinds of hardware, software, and management capabilities that the major vendors offer and that firms develop over time. There are six elements in this business intelligence environment:

- **Data from the business environment:** Businesses must deal with both structured and unstructured data from many different sources, including mobile devices and the Internet. The data need to be integrated and organized so that they can be analyzed and used by human decision makers.

- **Business intelligence infrastructure:** The underlying foundation of business intelligence is a powerful database system that captures all the relevant data to operate the business. The data may be stored in transactional databases or combined and integrated into an enterprise-data warehouse or series of interrelated data marts.

- **Business analytics toolset:** A set of software tools are used to analyze data and produce reports, respond to questions posed by managers, and track the progress of the business using key indicators of performance.

- **Managerial users and methods:** Business intelligence hardware and software are only as intelligent as the human beings who use them. Managers impose order on the analysis of data using a variety of managerial methods that define strategic business goals and specify how progress will be measured. These include business performance management and balanced scorecard approaches focusing on key performance indicators and industry strategic analyses focusing on changes in the general business environment, with special attention to competitors. Without strong senior management over-

TABLE 12-4 MARKET LEADERS AND SHARE FOR THE TOP BUSINESS INTELLIGENCE VENDORS

VENDOR	MARKET SHARE	BUSINESS INTELLIGENCE SOFTWARE
SAP	25%	SAP BusinessObjects EPM Solutions
SAS Institute	15%	SAS Activity Based Management; financial, human capital, profitability, and strategy management
Oracle	14%	Enterprise Performance Management System
IBM	11%	IBM Cognos
Microsoft	7%	SQL Server with PowerPivot

FIGURE 12-3   BUSINESS INTELLIGENCE AND ANALYTICS FOR DECISION SUPPORT

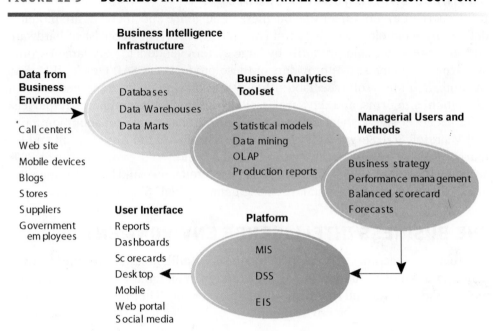

Business intelligence and analytics requires a strong database foundation, a set of analytic tools, and an involved management team that can ask intelligent questions and analyze data.

sight, business analytics can produce a great deal of information, reports, and online screens that focus on the wrong matters and divert attention from the real issues. You need to remember that, so far, only humans can ask intelligent questions.

- **Delivery platform—MIS, DSS, ESS.** The results from business intelligence and analytics are delivered to managers and employees in a variety of ways, depending on what they need to know to perform their jobs. MIS, DSS, and ESS, which we introduced in Chapter 2, deliver information and knowledge to different people and levels in the firm—operational employees, middle managers, and senior executives. In the past, these systems could not share data and operated as independent systems. Today, one suite of hardware and software tools in the form of a business intelligence and analytics package is able to integrate all this information and bring it to managers' desktop or mobile platforms.

- **User interface:** Business people are no longer tied to their desks and desktops. They often learn quicker from a visual representation of data than from a dry report with columns and rows of information. Today's business analytics software suites emphasize visual techniques such as dashboards and scorecards. They also are able to deliver reports on Blackberrys, iPhones, and other mobile handhelds as well as on the firm's Web portal. BA software is adding capabilities to post information on Twitter, Facebook, or internal social media to support decision making in an online group setting rather than in a face-to-face meeting.

BUSINESS INTELLIGENCE AND ANALYTICS CAPABILITIES

Business intelligence and analytics promise to deliver correct, nearly real-time information to decision makers, and the analytic tools help them quickly understand the information and take action. There are 5 analytic functionalities that BI systems deliver to achieve these ends:

- **Production reports:** These are predefined reports based on industry-specific requirements (see Table 12-5).
- **Parameterized reports.** Users enter several parameters as in a pivot table to filter data and isolate impacts of parameters. For instance, you might want to enter region and time of day to understand how sales of a product vary by region and time. If you were Starbucks, you might find that customers in the East buy most of their coffee in the morning, whereas in the Northwest customers buy coffee throughout the day. This finding might lead to different marketing and ad campaigns in each region. (See the discussion of pivot tables in Section 12.3).
- **Dashboards/scorecards:** These are visual tools for presenting performance data defined by users
- **Ad hoc query/search/report creation:** These allow users to create their own reports based on queries and searches
- **Drill down:** This is the ability to move from a high-level summary to a more detailed view
- **Forecasts, scenarios, models**: These include the ability to perform linear forecasting, what-if scenario analysis, and analyze data using standard statistical tools.

Who Uses Business Intelligence and Business Analytics?

In previous chapters, we have described the different information constituencies in business firms—from senior managers to middle managers, analysts, and operational employees. This also holds true for BI and BA systems (see Figure 12-4). Over 80 percent of the audience for BI consists of casual users who rely largely on production reports. Senior executives tend use BI to monitor firm activities using visual interfaces like dashboards and scorecards. Middle managers and analysts are much more likely to be immersed in the data and software, entering queries and slicing and dicing the data along different

FIGURE 12-4   BUSINESS INTELLIGENCE USERS

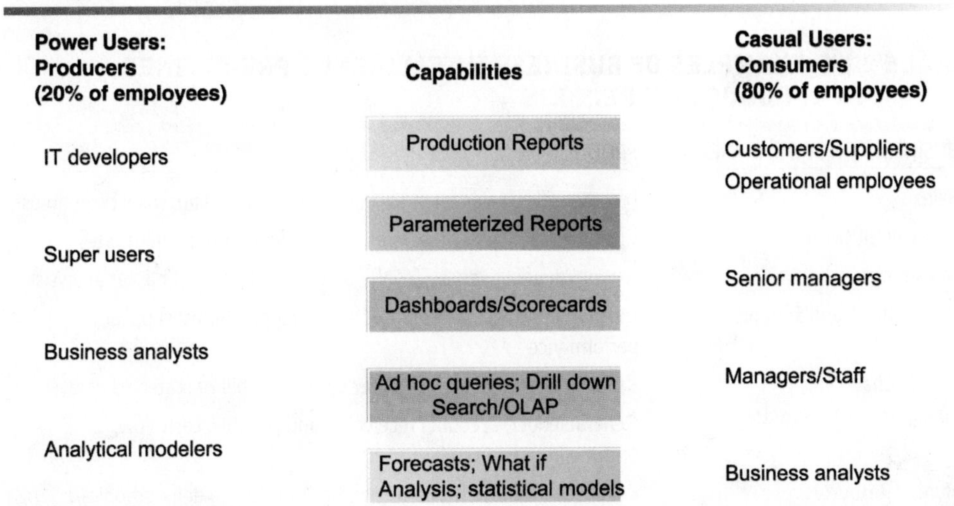

Casual users are consumers of BI output, while intense power users are the producers of reports, new analyses, models, and forecasts.

dimensions. Operational employees will, along with customers and suppliers, be looking mostly at prepackaged reports.

Examples of Business Intelligence Applications

The most widely used output of a BI suite of tools are pre-packaged production reports. Table 12-5 illustrates some common pre-defined reports from Oracle's BI suite of tools.

Predictive Analytics

Predictive analytics, which we introduced in Chapter 6, are being built into mainstream applications for everyday decision making by all types of employees, especially in finance and marketing. For example, Capital One conducts more than 30,000 experiments each year using different interest rates, incentives, direct mail packaging, and other variables to identify the best potential customers for targeting its credit card offers. These people are most likely to sign up for credit cards and to pay back Capital One for the balances they ring up in their credit card accounts. Predictive analytics have also worked especially well in the credit card industry to identify customers who are at risk for leaving.

Dealer Services, which offers inventory financing for used-car dealers, is trying to use predictive analytics to screen potential customers. Thousands of used-car dealers, who were formerly franchisees for General Motors and Chrysler, are seeking financing from companies such as Dealer Services so that they can go into business on their own. Using WebFOCUS software from Information Builders, the company is building a model that will predict the best loan prospects and eliminate up to 10 of the 15 hours required to review a financing application. The model reviews data including dealer size and type, number of locations, payment patterns, histories of bounced checks, and inventory practices and is revalidated and updated as conditions change.

FedEx is using SAS Institute's Enterprise Miner and predictive analytic tools to develop models that predict how customers will respond to price changes and new services, which customers are most at risk of switching to competitors, and how much revenue will be generated by new storefront or drop-box locations. The accuracy rate of the predictive analysis system ranges from 65 to 90 percent. FedEx is now starting to use predictive analytics in call centers to help

TABLE 12-5 EXAMPLES OF BUSINESS INTELLIGENCE PRE-DEFINED PRODUCTION REPORTS

BUSINESS FUNCTIONAL AREA	PRODUCTION REPORTS
Sales	Forecast sales; sales team performance; cross selling; sales cycle times
Service/Call Center	Customer satisfaction; service cost; resolution rates; churn rates
Marketing	Campaign effectiveness; loyalty and attrition; market basket analysis
Procurement and Support	Direct and indirect spending; off-contract purchases; supplier performance
Supply Chain	Backlog; fulfillment status; order cycle time; bill of materials analysis
Financials	General ledger; accounts receivable and payable; cash flow; profitability
Human Resources	Employee productivity; compensation; workforce demographics; retention

customer service representatives identify customers with the highest levels of dissatisfaction and take the necessary steps to make them happy.

Data Visualization and Geographic Information Systems

By presenting data in visual form, **data visualization** tools help users see patterns and relationships in large amounts of data that would be difficult to discern if the data were presented as traditional lists of text. For example, managers and employees of Day & Zimmermann, an industrial, defense, and workforce solutions provider, have detailed, real-time visibility into the company's inventory of contractors and workers through a set of dashboards populated with real-time data from a SAP ERP Human Capital Management system. The dashboards make it much easier to understand the organization's staffing levels than static paper reports. The real-time data indicate exactly what type of worker is available in what location and when a project is due to be completed. If a project is ahead of schedule, information from the dashboards helps decision makers rapidly determine when and where to reassign its workers.

Geographic information systems (GIS) help decision makers visualize problems requiring knowledge about the geographic distribution of people or other resources. Their software ties location data to points, lines, and areas on a map. Some GIS have modeling capabilities for changing the data and automatically revising business scenarios. GIS might be used to help state and local governments calculate response times to natural disasters and other emergencies or to help banks identify the best location for installing new branches or ATM terminals.

For example, Columbia, South Carolina-based First Citizens Bank uses GIS software from MapInfo to determine which markets to focus on for retaining customers and which to focus on for acquiring new customers. MapInfo also lets the bank drill down into details at the individual branch level and individualize

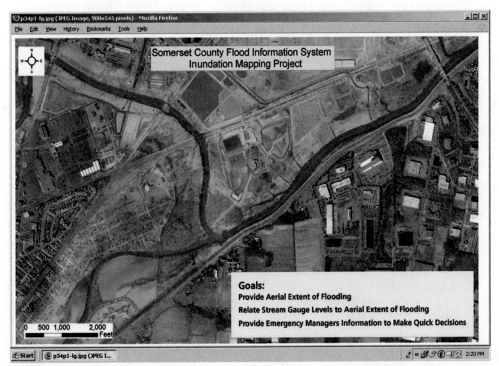

Somerset County, New Jersey, developed a GIS based on ESRI software to provide Web access to geospatial data about flood conditions. The system provides information that helps emergency responders and county residents prepare for floods and enables emergency managers to make decisions more quickly.

goals for each branch. Each branch is able to see whether the greatest revenue opportunities are from mining their database of existing customers or from finding new customers. With clearer branch segmentation and more focused service goals, the bank has moved from making cold sales calls to calls that are more service- and courtesy-oriented.

Business Intelligence in the Public Sector

Business intelligence systems are also used in the public sector. The Interactive Session on Organizations describes a school district's move to quantify and analyze student performance data to make better decisions about how to allocate resources to enhance student and teacher performance.

MANAGEMENT STRATEGIES FOR DEVELOPING BI AND BA CAPABILITIES

There are two different strategies for adopting BI and BA capabilities for the organization: one-stop integrated solutions versus multiple best-of-breed vendor solutions. The hardware firms (IBM, HP, and now Oracle, which owns Sun Microsystems) want to sell your firm integrated hardware/software solutions that tend to run only on their hardware (the totally integrated solution). It's called "one stop shopping." The software firms (SAP, SAS, and Microsoft) encourage firms to adopt the "best of breed" software and that runs on any machine they want. In this strategy, you adopt the best database and data warehouse solution, and select the best business intelligence and analytics package from whatever vendor you believe is best.

The first solution carries the risk that a single vendor provides your firm's total hardware and software solution, making your firm dependent on its pricing power. It also offers the advantage of dealing with a single vendor who can deliver on a global scale. The second solution offers greater flexibility and independence, but with the risk of potential difficulties integrating the software to the hardware platform, as well as to other software. Vendors always claim their software is "compatible" with other software, but the reality is that it can be very difficult to integrate software from different vendors. Microsoft in particular emphasizes building on its desktop interface and operating system (Windows), which are familiar to many users, and developing server applications that run on Microsoft local area networks. But data from hardware and software produced by different vendors will have to flow seamlessly into Microsoft workstations to make this strategy work. This may not be adequate for Fortune 500 firms needing a global networking solution.

Regardless of which strategy your firm adopts, all BI and BA systems lock the firm into a set of vendors and switching is very costly. Once you train thousands of employees across the world on using a particular set of tools, it is extremely difficult to switch. When you adopt these systems, you are in essence taking in a new partner.

The marketplace is very competitive and given to hyperbole. One BI vendor claims "[Our tools] bring together a portfolio of services, software, hardware and partner technologies to create business intelligence solutions. By connecting intelligence across your company, you gain a competitive advantage for creating new business opportunities." As a manager, you will have to critically evaluate such claims, understand exactly how these systems could improve your business, and determine whether the expenditures are worth the benefits.

INTERACTIVE SESSION: ORGANIZATIONS

DATA-DRIVEN SCHOOLS

As more and more reports suggest that American schoolchildren are falling behind those from other countries, improving our schools has become an increasingly urgent mission for the nation. Actually achieving that improvement is a difficult task. One approach gaining sway is more intensive use of information systems to measure educational performance at the individual and school district level and identify problem areas requiring additional resources and intervention.

The 139,000-student Montgomery County public school system in Rockville, Maryland, is at the forefront of the push for data-driven DSS in schools. Forty employees at the school district's Office of Shared Accountability generate reports on how many students take algebra in middle school or read below grade level. The district's Edline and M-Stat systems alert principals to individuals with patterns of failing so they can receive extra resources, such as after-school tutoring, study sessions, and special meetings with parents.

Earlier this decade, Montgomery County school superintendent Jerry Weast predicted that the increasing stratification between students in what he called the "green zone" (white and wealthy students) and students in the "red zone" (poor and minority students) would weigh down the school district as a whole. Having exhausted other options, administrators initiated a plan to create a data collection system for test scores, grades, and other data useful for identifying students with problems and speeding up interventions to improve their learning and educational performance.

Principals access and analyze student performance data to help make instructional decisions over the course of the year, as opposed to only when annual standardized test data arrives. This way, teachers can meet the needs of students who require additional instruction or other types of intervention before they fall behind. Test scores, grades, and other data are entered into the system in real time, and can be accessed in real time. In the past, school data were disorganized, and trends in individual student performance as well as overall student body performance were difficult to diagnose.

Kindergarten teachers are now able to monitor their students' success in reading words, noting which words each student struggles with on a handheld device like a Palm Pilot. The device calculates the accuracy with which the student reads each passage and, over time, provides information about what sorts of problems the student consistently encounters. Also, when students begin to deviate from their normal academic patterns, like getting a rash of poor grades, the system sends alerts to parents and school administrators. In many cases, this quicker response is enough to help the student reverse course before failing.

Many parents in Montgomery County have expressed concern that the new systems are an excessive and unnecessary expenditure. In the short term, President Obama's stimulus plan provides increased funding to schools over the next two years. Projects like these are likely to become more popular as it becomes clearer that a data-driven approach yields quantifiable results. But will they become the standard in American schools? The long-term sustainability of these systems is still unclear.

In Montgomery County, one of the primary goals of the implementation of data-driven systems was to close the achievement gap between white and minority students in the lower grades. Teachers and administrators would use different types of information organized by the DSS to identify gifted students earlier and challenge them with a more appropriate course load of more advanced placement (AP) classes. Data collected on each child would offer teachers insight into what methods worked best for each individual.

The results are very impressive. In Montgomery, 90 percent of kindergartners were able to read at the level required by standardized testing, with minimal differences among racial and socioeconomic groups. These numbers are up from 52 percent of African-Americans, 42 percent of Latinos, and 44 percent of low-income students just seven years ago. Also, the system has effectively identified students with abilities at an earlier age. The number of African-American students who passed at least one AP test at Montgomery has risen from 199 earlier this decade to 1,152 this year; the number of Latino students went from 218 to 1,336.

Some critics claim that the emphasis on closing the achievement gap between different student populations is shortchanging gifted students and those with disabilities. "Green zone" parents question whether their children are receiving enough attention and resources with so much emphasis being placed on the improving the red zone. Green zone districts in Montgomery County receive $13,000 per student, compared with $15,000 in the red zone. Red zone classes have only 15 students in kindergarten and 17 in the first and second grates, compared with 25 and 26 in the green zone. School administrators counter that the system not only provides appropriate help for underperforming students, but also that it provides the additional challenges that are vital to a gifted child's development.

Other evidence suggests that the gains in reducing the achievement gap earlier in childhood erode as children get older. Among eighth graders in Montgomery County, approximately 90 percent of white and Asian eighth graders tested proficient or advanced in math on state tests, compared with only half of African-Americans and Hispanics. African-American and Hispanic SAT scores were over 300 points below those of whites and Asians. Still, the data-driven implementation has been responsible for some large improvements over past statistics. Some of the red zone schools have seen the most dramatic improvement in test scores and graduation rates.

In many ways, the data-driven systems build from the wealth of standardized testing information created by the No Child Left Behind Act passed during the Bush presidency. Some parents and educators complain about the amount and frequency of standardized testing, suggesting that children should be spending more time on projects and creative tasks. But viable alternative strategies to foster improvement in struggling school districts are difficult to develop.

It's not just students that are subject to this data-driven approach. Montgomery County teachers have been enrolled in a similar program that identi-

fies struggling teachers and supplies data to help them improve. In many cases, contracts and tenure make it difficult to dismiss less-effective teachers. To try and solve this problem, teachers unions and administrators have teamed up to develop a peer review program that pairs underperforming teachers with a mentor who provides guidance and support.

After two years, teachers who fail to achieve results appear before a larger panel of teachers and principals that makes a decision regarding their potential termination or extension of another year of peer review. But teachers are rarely terminated in the program—instead, they're given tangible evidence of things they're doing well and things they can improve based on data that's been collected on their day-to-day performance, student achievement rates, and many other metrics.

Not all teachers have embraced the data-driven approach. The Montgomery Education Association, the county's main teachers' union, estimates that keeping a "running record" of student results on reading assessments and other testing adds about three to four hours to teachers' weekly workloads. According to Raymond Myrtle, principal of Highland Elementary in Silver Spring, "this is a lot of hard work. A lot of teachers don't want to do it. For those who don't like it we suggested they do something else." To date, 11 of 33 teachers at Highland have left the district or are teaching at other Montgomery schools.

Sources: www.montgomeryschoolsmd.org, accessed October 15, 2010; www.datadrivenclassroom.com, accessed October 15, 2010; John Hechinger, "Data-Driven Schools See Rising Scores," *The Wall Street Journal*, June 12, 2009; and Daniel de Vise, "Throwing a Lifeline to Struggling Teachers," *Washington Post*, June 29, 2009.

CASE STUDY QUESTIONS

1. Identify and describe the problem discussed in the case.

2. How do business intelligence systems provide a solution to this problem? What are the inputs and outputs of these systems?

3. What management, organization, and technology issues must be addressed by this solution?

4. How successful is this solution? Explain your answer.

5. Should all school districts use such a data-driven approach to education? Why or why not?

MIS IN ACTION

Explore the Web site of the Montgomery County, Maryland School District and then answer the following questions:

1. Select one of the district's elementary, middle, or high schools and describe the data available on that particular school. What kinds of decisions do these data support? How do these data help school officials improve educational performance?

2. Select one of the district's schools and then the School Survey Results. How do these surveys help decision makers improve educational quality?

12.3 BUSINESS INTELLIGENCE CONSTITUENCIES

There are many different constituencies that make up a modern business firm. Earlier in this text and in this chapter we identified three levels of management: lower supervisory (operational) management, middle management, and senior management (vice president and above, including executive or "C level" management, e.g. chief executive officer, chief financial officers, and chief operational officer.) Each of these management groups has different responsibilities and different needs for information and business intelligence, with decisions becoming less structured among higher levels of management (review Figure 12-1).

DECISION SUPPORT FOR OPERATIONAL AND MIDDLE MANAGEMENT

Operational and middle management are generally charged with monitoring the performance of key aspects of the business, ranging from the down-time of machines on a factory floor, to the daily or even hourly sales at franchise food stores, to the daily traffic at a company's Web site. Most of the decisions they make are fairly structured. Management information systems (MIS) are typically used by middle managers to support this type of decision making, and their primary output is a set of routine production reports based on data extracted and summarized from the firm's underlying transaction processing systems (TPS). Increasingly, middle managers receive these reports online on the company portal, and are able to interactively query the data to find out why events are happening. To save even more analysis time, managers turn to exception reports, which highlight only exceptional conditions, such as when the sales quotas for a specific territory fall below an anticipated level or employees have exceeded their spending limits in a dental care plan. Table 12-6 provides some examples of MIS applications.

Support for Semistructured Decisions

Some managers are "super users" and keen business analysts who want to create their own reports, and use more sophisticated analytics and models to find patterns in data, to model alternative business scenarios, or to test specific

TABLE 12-6 EXAMPLES OF MIS APPLICATIONS

COMPANY	MIS APPLICATION
California Pizza Kitchen	Inventory Express application "remembers" each restaurant's ordering patterns and compares the amount of ingredients used per menu item to predefined portion measurements established by management. The system identifies restaurants with out-of-line portions and notifies their managers so that corrective actions will be taken.
PharMark	Extranet MIS identifies patients with drug-use patterns that place them at risk for adverse outcomes.
Black & Veatch	Intranet MIS tracks construction costs for various projects across the United States.
Taco Bell	Total Automation of Company Operations (TACO) system provides information on food, labor, and period-to-date costs for each restaurant.

hypotheses. Decision support systems (DSS) are the BI delivery platform for this category of users, with the ability to support semi-structured decision making.

DSS rely more heavily on modeling than MIS, using mathematical or analytical models to perform what-if or other kinds of analysis. "What-if" analysis, working forward from known or assumed conditions, allows the user to vary certain values to test results to predict outcomes if changes occur in those values. What happens if we raise product prices by 5 percent or increase the advertising budget by $1 million? **Sensitivity analysis** models ask what-if questions repeatedly to predict a range of outcomes when one or more variables are changed multiple times (see Figure 12-5). Backward sensitivity analysis helps decision makers with goal seeking: If I want to sell 1 million product units next year, how much must I reduce the price of the product?

Chapter 6 described multidimensional data analysis and OLAP as one of the key business intelligence technologies. Spreadsheets have a similar feature for multidimensional analysis called a **pivot table**, which manager "super users" and analysts employ to identify and understand patterns in business information that may be useful for semistructured decision making.

Figure 12-6 illustrates a Microsoft Excel pivot table that examines a large list of order transactions for a company selling online management training videos and books. It shows the relationship between two dimensions: the sales region and the source of contact (Web banner ad or e-mail) for each customer order. It answers the question: does the source of the customer make a difference in addition to region? The pivot table in this figure shows that most customers come from the West and that banner advertising produces most of the customers in all the regions.

One of the Hands-on MIS projects for this chapter asks you to use a pivot table to find answers to a number of other questions using the same list of transactions for the online training company as we used in this discussion. The complete Excel file for these transactions is available in MyMISLab. We have also added a Learning Track on creating pivot tables using Excel 2010.

In the past, much of this modeling was done with spreadsheets and small stand-alone databases. Today these capabilities are incorporated into large

FIGURE 12-5 SENSITIVITY ANALYSIS

			Variable Cost per Unit			
Total fixed costs	19000					
Variable cost per unit	3					
Average sales price	17					
Contribution margin	14					
Break-even point	1357					
			Variable Cost per Unit			
Sales	1357	2	3	4	5	6
Price	14	1583	1727	1900	2111	2375
	15	1462	1583	1727	1900	2111
	16	1357	1462	1583	1727	1900
	17	1267	1357	1462	1583	1727
	18	1188	1267	1357	1462	1583

This table displays the results of a sensitivity analysis of the effect of changing the sales price of a necktie and the cost per unit on the product's break-even point. It answers the question, "What happens to the break-even point if the sales price and the cost to make each unit increase or decrease?"

FIGURE 12-6 **A PIVOT TABLE THAT EXAMINES CUSTOMER REGIONAL DISTRIBUTION AND ADVERTISING SOURCE**

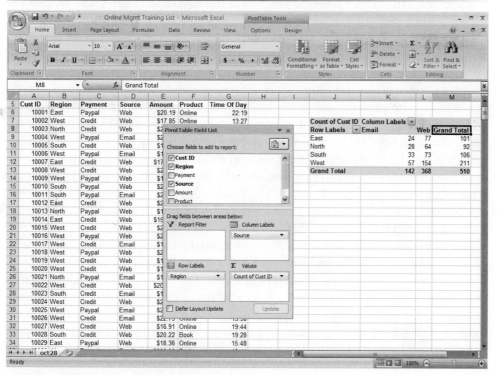

In this pivot table, we are able to examine where an online training company's customers come from in terms of region and advertising source.

enterprise BI systems where they are able to analyze data from large corporate databases. BI analytics include tools for intensive modeling, some of which we described earlier. Such capabilities help Progressive Insurance identify the best customers for its products. Using widely available insurance industry data, Progressive defines small groups of customers, or "cells," such as motorcycle riders aged 30 or above with college educations, credit scores over a certain level, and no accidents. For each "cell," Progressive performs a regression analysis to identify factors most closely correlated with the insurance losses that are typical for this group. It then sets prices for each cell, and uses simulation software to test whether this pricing arrangement will enable the company to make a profit. These analytic techniques, make it possible for Progressive to profitably insure customers in traditionally high-risk categories that other insurers would have rejected.

DECISION SUPPORT FOR SENIOR MANAGEMENT: BALANCED SCORECARD AND ENTERPRISE PERFORMANCE MANAGEMENT METHODS

The purpose of executive support systems (ESS), introduced in Chapter 2, is to help C-level executive managers focus on the really important performance information that affect the overall profitability and success of the firm. There are two parts to developing ESS. First, you will need a methodology for understanding exactly what is "the really important performance information" for a specific firm that executives need, and second, you will need to develop systems capable of delivering this information to the right people in a timely fashion.

Currently, the leading methodology for understanding the really important information needed by a firm's executives is called the **balanced scorecard method** (Kaplan and Norton, 2004; Kaplan and Norton, 1992). The balanced score card is a framework for operationalizing a firm's strategic plan by focusing on measurable outcomes on four dimensions of firm performance: financial, business process, customer, and learning and growth (Figure 12-7). Performance on each dimension is measured using **key performance indicators (KPIs)**, which are the measures proposed by senior management for understanding how well the firm is performing along any given dimension. For instance, one key indicator of how well an online retail firm is meeting its customer performance objectives is the average length of time required to deliver a package to a consumer. If your firm is a bank, one KPI of business process performance is the length of time required to perform a basic function like creating a new customer account.

The balanced scorecard framework is thought to be "balanced" because it causes managers to focus on more than just financial performance. In this view, financial performance is past history—the result of past actions—and managers should focus on the things they are able to influence today, such as business process efficiency, customer satisfaction, and employee training. Once a scorecard is developed by consultants and senior executives, the next step is automating a flow of information to executives and other managers for each of the key performance indicators. There are literally hundreds of consulting and

FIGURE 12-7   THE BALANCED SCORECARD FRAMEWORK

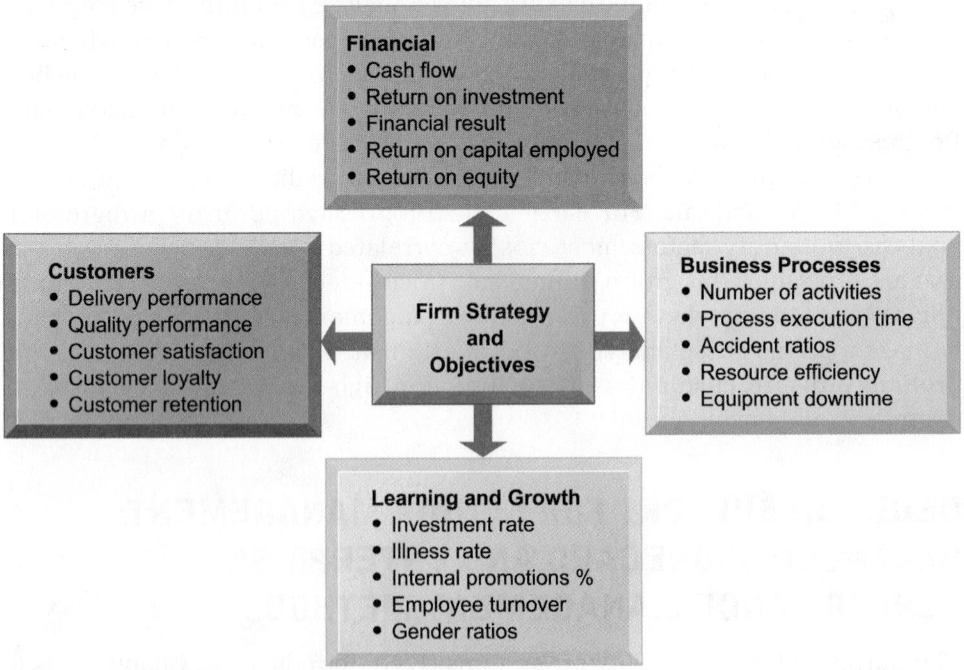

In the balanced scorecard framework, the firm's strategic objectives are operationalized along four dimensions: financial, business process, customer, and learning and growth. Each dimension is measured using several KPIs.

software firms that offer these capabilities, which are described below. Once these systems are implemented, they are often referred to as ESS.

Another closely related popular management methodology is **business performance management (BPM)**. Originally defined by an industry group in 2004 (led by the same companies that sell enterprise and database systems like Oracle, SAP, and IBM), BPM attempts to systematically translate a firm's strategies (e.g., differentiation, low-cost producer, market share growth, and scope of operation) into operational targets. Once the strategies and targets are identified, a set of KPIs are developed that measure progress towards the targets. The firm's performance is then measured with information drawn from the firm's enterprise database systems. BPM uses the same ideas as balanced scorecard but with a stronger strategy flavor (BPM Working Group, 2004).

Corporate data for contemporary ESS are supplied by the firm's existing enterprise applications (enterprise resource planning, supply chain management, and customer relationship management). ESS also provide access to news services, financial market databases, economic information, and whatever other external data senior executives require. ESS also have significant **drill-down** capabilities if managers need more detailed views of data.

Well-designed ESS enhance management effectiveness by helping senior executives monitor organizational performance, track activities of competitors, recognize changing market conditions, and identify problems and opportunities. Immediate access to data increases executives' ability to monitor activities of lower units reporting to them. That very monitoring ability enables decision making to be decentralized and to take place at lower operating levels, increasing management's span of control.

Contemporary business intelligence and analytics technology has enabled a whole new style and culture of management called "information driven management" or "management by facts." Here, information is captured at the factory floor (or sales floor) level, and immediately entered into enterprise systems and databases, and then to corporate headquarters executive dashboards for analysis—not in a matter of months, days, weeks, but in hours and seconds. It's real time management. You can see real-time management at work in hundreds of corporations in 2010, and many more are building this new decision support environment. Valero provides a good example in the Interactive Session on Management.

GROUP DECISION-SUPPORT SYSTEMS (GDSS)

The DSS we have just described focus primarily on individual decision making. However, so much work is accomplished in groups within firms that a special category of systems called **group decision-support systems (GDSS)** has been developed to support group and organizational decision making.

A GDSS is an interactive computer-based system for facilitating the solution of unstructured problems by a set of decision makers working together as a group in the same location or in different locations. Collaboration systems and Web-based tools for videoconferencing and electronic meetings described earlier in this text support some group decision processes, but their focus is primarily on communication. GDSS, however, provide tools and technologies geared explicitly toward group decision making.

GDSS-guided meetings take place in conference rooms with special hardware and software tools to facilitate group decision making. The hardware includes computer and networking equipment, overhead projectors, and display screens. Special electronic meeting software collects, documents, ranks, edits,

INTERACTIVE SESSION: MANAGEMENT

PILOTING VALERO WITH REAL-TIME MANAGEMENT

If you haven't heard of Valero, don't worry. It's largely unknown to the public although investors recognize it as one of the largest oil refiners in the United States. Valero Energy is a top-fifty Fortune 500 company headquartered in San Antonio, Texas, with annual revenues of $70 billion. Valero owns 16 refineries in the United States, Canada, and Aruba that produce gasoline, distillates, jet fuel, asphalt, petrochemicals, and other refined products. The company also owns 10 ethanol plants located in the Midwest with a combined ethanol production capacity of about 1.1 billion gallons per year.

In 2008, Valero's chief operating officer (COO) called for the development of a Refining Dashboard that would display real-time data related to plant and equipment reliability, inventory management, safety, and energy consumption. Using a series of monitors on the walls of the headquarters operations center room, with a huge central monitor screen showing a live display of the company's Refining Dashboard, the COO and other plant managers can review the performance of the firm's 16 major refineries in the United States and Canada.

The COO and his team review the performance of each refinery in terms of how each plant is performing compared to the production plan of the firm. For any deviation from plan, up or down, the plant manager is expected to provide the group an explanation, and a description of corrective actions. The headquarters group can drill down from executive level to refinery level and individual system-operator level displays of performance.

Valero's Refining Dashboard is available on the Web to plant managers in remote locations. The data are refreshed every five minutes. The dashboard taps directly into the firm's SAP Manufacturing Integration and Intelligence application where each plant's history of production and current production data is stored. Valero's management estimates that the dashboards are saving $230 million per year at the 16 refineries where they are in use.

Valero's Refining Dashboard has been so successful that the firm is developing separate dashboards that show detailed statistics on power consumption for each unit of the firm, and each plant. Using the shared data, managers will be able to share best practices with one another, and make changes in equipment to reduce energy consumption while maintaining production targets. The dashboard system has the unintended consequence of helping managers learn more about how their company actually operates, and how to improve it.

But how much do Valero's executive dashboards really make a difference? One of the dangers of real time management is not measuring the right things. Dashboards that display information unrelated to the firm's strategic goals might be largely irrelevant, although pretty to look at. Valero's goals and measures of performance were inspired by Solomon benchmark performance studies used in the oil and gas industry. How helpful were they?

Valero's stock price fell from a high of $80 in June 2008, to about $20 in November 2010. As it turns out, Valero's profits are not strongly related to small changes in its refining efficiency. Instead, its profitability is largely determined by the spread between the price of refined products and the price of crude oil, referred to as the "refined product margin." The global economic slowdown beginning in 2008 and extending through 2010 weakened demand for refined petroleum products, which put pressure on refined product margins throughout 2009 and 2010. This reduced demand, combined with increased inventory levels, caused a significant decline in diesel and jet fuel profit margins.

The price of crude and aggregate petroleum demand are largely beyond the control of Valero management. The cost of refining crude varies within a very narrow range over time, and there are no technological breakthroughs expected in refining technology. Although Valero's dashboard focuses on one of the things management can control within a narrow range (namely refining costs), the dashboard does not display a number of strategic factors beyond its control, which nevertheless powerfully impact company performance. Bottom line: a powerful dashboard system does not turn an unprofitable operation into a profitable one.

Another limitation of real-time management is that it is most appropriate for process industries such as oil refining where the process is relatively unchanging, well known and understood, and central to the revenues of a firm. Dashboard systems say nothing about innovation in products, marketing, sales, or any other area of the firm where innovation is important. Apple Corporation did not invent the

Apple iPhone using a performance dashboard, although it might have such a dashboard today to monitor iPhone manufacturing and sales. Managers have to be sensitive to, and reflect upon, all the factors that shape the success of their business even if they are not reflected in the firm's dashboards.

Sources: Chris Kahn, "Valero Energy Posts 3Q Profit, Reverses Loss," *Business Week*, October 26, 2010; Valero Energy Corporation, Form 10K Annual Report for the fiscal year ended December 31, 2009, filed with the Securities and Exchange Commission, February 28, 2010; and Doug Henderson, "Execs Want Focus on Goals, Not Just Metrics," *InformationWeek*, Novemvber 13, 2009.

CASE STUDY QUESTIONS

1. What management, organization, and technology issues had to be addressed when developing Valero's dashboard?

2. What measures of performance do the dashboards display? Give examples of several management decisions that would benefit from the information provided by Valero's dashboards.

3. What kinds of information systems are required by Valero to maintain and operate its refining dashboard?

4. How effective are Valero's dashboards in helping management pilot the company? Explain your answer.

5. Should Valero develop a dashboard to measure the many factors in its environment that it does not control? Why or why not?

MIS IN ACTION

1. Visit Valero.com and click on its annual report in the Investor Relations section. On page 2 of the annual report you will find Valero's corporate vision statement. Read its corporate vision statement of strategic objectives (especially vision statement #2). Based on the firm's vision, what other corporate dashboards might be appropriate for senior management?

2. Read the annual report and develop a list of factors mentioned in the report that explain the company's poor performance over the last two years. Devise a method for measuring these profitability factors, and then using electronic presentation software create a corporate profitability dashboard for senior managers.

and stores the ideas offered in a decision-making meeting. The more elaborate GDSS use a professional facilitator and support staff. The facilitator selects the software tools and helps organize and run the meeting.

A sophisticated GDSS provides each attendee with a dedicated desktop computer under that person's individual control. No one will be able to see what individuals do on their computers until those participants are ready to share information. Their input is transmitted over a network to a central server that stores information generated by the meeting and makes it available to all on the meeting network. Data can also be projected on a large screen in the meeting room.

GDSS make it possible to increase meeting size while at the same time increasing productivity because individuals contribute simultaneously rather than one at a time. A GDSS promotes a collaborative atmosphere by guaranteeing contributors' anonymity so that attendees focus on evaluating the ideas themselves without fear of personally being criticized or of having their ideas rejected based on the contributor. GDSS software tools follow structured methods for organizing and evaluating ideas and for preserving the results of meetings, enabling nonattendees to locate needed information after the meeting. GDSS effectiveness depends on the nature of the problem and the group and on how well a meeting is planned and conducted.

12.4 HANDS-ON MIS PROJECTS

The projects in this section give you hands-on experience analyzing opportunities for DSS, using a spreadsheet pivot table to analyze sales data, and using online retirement planning tools for financial planning.

Management Decision Problems

1. Applebee's is the largest casual dining chain in the world, with 1,970 locations throughout the United States and nearly 20 other countries worldwide. The menu features beef, chicken, and pork items, as well as burgers, pasta, and seafood. The Applebee's CEO wants to make the restaurant more profitable by developing menus that are tastier and contain more items that customers want and are willing to pay for despite rising costs for gasoline and agricultural products. How might information systems help management implement this strategy? What pieces of data would Applebee's need to collect? What kinds of reports would be useful to help management make decisions on how to improve menus and profitability?

2. During the 1990s, the Canadian Pacific Railway used a tonnage-based operating model in which freight trains ran only when there was sufficient traffic to justify the expense. This model focused on minimizing the total number of freight trains in service and maximizing the size of each train. However, it did not necessarily use crews, locomotives, and equipment efficiently, and it resulted in inconsistent transit times and delivery schedules. Canadian Pacific and other railroads were losing business to trucking firms, which offered more flexible deliveries that could be scheduled at the times most convenient for customers. How could a DSS help Canadian Pacific and other railroads compete with trucking firms more effectively?

Improving Decision Making: Using Pivot Tables to Analyze Sales Data

Software skills: Pivot tables
Business skills: Analyzing sales data

This project gives you an opportunity to learn how to use Excel's PivotTable functionality to analyze a database or data list.

Use the data list for Online Management Training Inc. (OMT) described earlier in the chapter. This is a list of the sales transactions at OMT for one day. You can find this spreadsheet file MyMISLab.

Use Excel's PivotTable to help you answer the following questions:

- Where are the average purchases higher? The answer might tell managers where to focus marketing and sales resources, or pitch different messages to different regions.

- What form of payment is the most common? The answer might be used to emphasize in advertising the most preferred means of payment.

- Are there any times of day when purchases are most common? Do people buy products while at work (likely during the day) or at home (likely in the evening)?

- What's the relationship between region, type of product purchased, and average sales price?

Improving Decision Making: Using a Web-Based DSS for Retirement Planning

Software skills: Internet-based software
Business skills: Financial planning

This project will help develop your skills in using Web-based DSS for financial planning.

The Web sites for CNN Money and MSN Money Magazine feature Web-based DSS for financial planning and decision making. Select either site to plan for retirement. Use your chosen site to determine how much you need to save to have enough income for your retirement. Assume that you are 50 years old and plan to retire in 16 years. You have one dependant and $100,000 in savings. Your current annual income is $85,000. Your goal is to be able to generate an annual retirement income of $60,000, including Social Security benefit payments.

• To calculate your estimated Social Security benefit, search for and use the Quick Calculator at the Social Security Administration Web site.

• Use the Web site you have selected to determine how much money you need to save to help you achieve your retirement goal.

• Critique the site—its ease of use, its clarity, the value of any conclusions reached, and the extent to which the site helps investors understand their financial needs and the financial markets.

LEARNING TRACK MODULE

The following Learning Track provides content relevant to topics covered in this chapter:

1. Building and Using Pivot Tables

Review Summary

1. *What are the different types of decisions and how does the decision-making process work?*

   The different levels in an organization (strategic, management, operational) have different decision-making requirements. Decisions can be structured, semistructured, or unstructured, with structured decisions clustering at the operational level of the organization and unstructured decisions at the strategic level. Decision making can be performed by individuals or groups and includes employees as well as operational, middle, and senior managers. There are four stages in decision making: intelligence, design, choice, and implementation. Systems to support decision making do not always produce better manager and employee decisions that improve firm performance because of problems with information quality, management filters, and organizational culture.

2. *How do information systems support the activities of managers and management decision making?*

   Early classical models of managerial activities stress the functions of planning, organizing, coordinating, deciding, and controlling. Contemporary research looking at the actual behavior of managers has found that managers' real activities are highly fragmented, variegated, and brief in duration and that managers shy away from making grand, sweeping policy decisions.

Information technology provides new tools for managers to carry out both their traditional and newer roles, enabling them to monitor, plan, and forecast with more precision and speed than ever before and to respond more rapidly to the changing business environment. Information systems have been most helpful to managers by providing support for their roles in disseminating information, providing liaisons between organizational levels, and allocating resources. However, information systems are less successful at supporting unstructured decisions. Where information systems are useful, information quality, management filters, and organizational culture can degrade decision-making.

3. *How do business intelligence and business analytics support decision making?*

Business intelligence and analytics promise to deliver correct, nearly real-time information to decision makers, and the analytic tools help them quickly understand the information and take action. A business intelligence environment consists of data from the business environment, the BI infrastructure, a BA toolset, managerial users and methods, a BI delivery platform (MIS, DSS, or ESS), and the user interface. There are six analytic functionalities that BI systems deliver to achieve these ends: pre-defined production reports, parameterized reports, dashboards and scorecards, ad hoc queries and searches, the ability to drill down to detailed views of data, and the ability to model scenarios and create forecasts.

4. *How do different decision-making constituencies in an organization use business intelligence?*

Operational and middle management are generally charged with monitoring the performance of their firm. Most of the decisions they make are fairly structured. Management information systems (MIS) producing routine production reports are typically used to support this type of decision making. For making unstructured decisions, middle managers and analysts will use decision-support systems (DSS) with powerful analytics and modeling tools, including spreadsheets and pivot tables. Senior executives making unstructured decisions use dashboards and visual interfaces displaying key performance information affecting the overall profitability, success, and strategy of the firm. The balanced scorecard and business performance management are two methodologies used in designing executive support systems (ESS).

5. *What is the role of information systems in helping people working in a group make decisions more efficiently?*

Group decision-support systems (GDSS) help people working together in a group arrive at decisions more efficiently. GDSS feature special conference room facilities where participants contribute their ideas using networked computers and software tools for organizing ideas, gathering information, making and setting priorities, and documenting meeting sessions.

Key Terms

Balanced scorecard method, 474	Implementation, 458
Behavioral models, 458	Informational role, 459
Business performance management (BPM), 475	Intelligence, 458
Choice, 458	Interpersonal role, 459
Classical model of management, 458	Key performance indicators (KPIs), 474
Data visualization, 467	Managerial roles, 454
Decisional role, 459	Pivot table, 472
Design, 458	Sensitivity analysis, 472
Drill down, 475	Semistructured decisions, 456
Geographic information systems (GIS), 467	Structured decisions, 456
Group decision-support systems (GDSS), 475	Unstructured decisions, 456

Review Questions

1. What are the different types of decisions and how does the decision-making process work?

- List and describe the different levels of decision making and decision-making constituencies in organizations. Explain how their decision-making requirements differ.
- Distinguish between an unstructured, semistructured, and structured decision.
- List and describe the stages in decision making.

2. How do information systems support the activities of managers and management decision making?

- Compare the descriptions of managerial behavior in the classical and behavioral models.
- Identify the specific managerial roles that can be supported by information systems.

3. How do business intelligence and business analytics support decision making?

- Define and describe business intelligence and business analytics.
- List and describe the elements of a business intelligence environment.

- List and describe the analytic functionalities provided by BI systems.
- Compare two different management strategies for developing BI and BA capabilities.

4. How do different decision-making constituencies in an organization use business intelligence?

- List each of the major decision-making constituencies in an organization and describe the types of decisions each makes.
- Describe how MIS, DSS, or ESS provide decision support for each of these groups.
- Define and describe the balanced scorecard method and business performance management.

5. What is the role of information systems in helping people working in a group make decisions more efficiently?

- Define a group decision-support system (GDSS) and explain how it differs from a DSS.
- Explain how a GDSS works and how it provides value for a business.

Discussion Questions

1. As a manager or user of information systems, what would you need to know to participate in the design and use of a DSS or an ESS? Why?

2. If businesses used DSS, GDSS, and ESS more widely, would managers and employees make better decisions? Why or why not?

3. How much can business intelligence and business analytics help companies refine their business strategy? Explain your answer.

Video Cases

Video Cases and Instructional Videos illustrating some of the concepts in this chapter are available. Contact your instructor to access these videos.

Collaboration and Teamwork: Designing a University GDSS

With three or four of your classmates, identify several groups in your university that might benefit from a GDSS. Design a GDSS for one of those groups, describing its hardware, software, and people elements. If possible, use Google Sites to post links to Web pages, team communication announcements, and work assignments; to brainstorm; and to work collaboratively on project documents. Try to use Google Docs to develop a presentation of your findings for the class.

Does CompStat Reduce Crime?
CASE STUDY

CompStat (short for COMPuter STATistics or COMParative STATistics) originated in the New York City Police Department (NYPD) in 1994 when William Bratton was police commissioner. CompStat is a comprehensive, city-wide database that records all reported crimes or complaints, arrests, and summonses issued in each of the city's 76 precincts. City officials had previously believed that crime could not be prevented by better information and analytical tools but instead by using more foot patrols in neighborhoods along with the concept of "community policing" in which efforts were made to strengthen the involvement of community groups. In contrast, Bratton and Rudy Giuliani, then the mayor of New York City, believed that police could be more effective in reducing crime if operational decisions took place at the precinct level and if decision makers had better information. Precinct commanders were in a better position than police headquarters to understand the specific needs of the communities they served and to direct the work of the 200 to 400 police officers they managed. CompStat gave precinct commanders more authority and responsibility, but also more accountability.

At weekly meetings, representatives from each of the NYPD's precincts, service areas, and transit districts are put on the "hot seat" at police headquarters and required to provide a statistical summary of the week's crime complaint, arrest, and summons activity, as well as significant cases, crime patterns, and police activities. Commanders must explain what has been done to reduce crime in the districts under their command, and if crime has gone up, they must explain why. Commanders are held directly accountable for reducing crime in their area of command. In the past, they were evaluated primarily on the basis of their administrative skills, such as staying within budget and deploying resources efficiently.

The data these commanders provide, including specific times and locations of crimes and enforcement activities, are forwarded to the NYPD's CompStat Unit where they are loaded into a city-wide database. The system analyzes the data and produces a weekly CompStat report on crime complaint and arrest activity at the precinct, patrol borough, and city wide levels. The data are summa-rized by week, prior 30 days, and year-to-date for comparison with the previous year's activity and for establishing trends. The CompStat Unit also issues weekly commander profile reports to measure the performance of precinct commanders.

The weekly commander profile reports include information on the commander's date of appointment, years in rank, education and specialized training, most recent performance evaluation rating, the units that person previously commanded, the amount of overtime generated by police under that commander, absence rates, community demographics, and civilian complaints.

Using MapInfo geographic information system (GIS) software, the CompStat data can be displayed on maps showing crime and arrest locations, crime "hot spots," and other relevant information. Comparative charts, tables, and graphs can also be projected simultaneously. These visual presentations help precinct commanders and members of the NYPD's executive staff to quickly identify patterns and trends. Depending on the intelligence gleaned from the system, police chiefs and captains develop a targeted strategy for fighting crime, such as dispatching more foot patrols to high-crime neighborhoods, or issuing warnings to the public when a particular model of vehicle is susceptible to theft.

During Bratton's 27-month tenure, serious crime in New York dropped by 25 percent and homicides went down by 44 percent. Crime in New York City has dropped by 69 percent in the last 12 years. Skeptics do not believe that CompStat was responsible for these results. They point to the decline in the number of young, poor men, an improved economy, programs that reduced welfare rolls while giving poor people access to better housing, increasing the size of the NYC police force, and giving precinct commanders more decision-making responsibility and accountability.

Nevertheless, Bratton, convinced that CompStat was the catalyst for New York's drop in crime, implemented the system in Los Angeles to further prove its worth. Since the introduction of CompStat, combined violent and property crimes in Los Angeles dropped for six consecutive years. Yet the ratio of police officers to residents is only half that of New York and Chicago. CompStat has also been adopted in Philadelphia, Austin, San Francisco, Baltimore, and Vancouver, British Columbia.

Skeptics point out that crime has fallen in all urban areas in the United States since 1990 regardless of whether the cities used CompStat. In fact, a critical study of CompStat by the Police Foundation found that CompStat encouraged police to be only reactive rather than pro-active in fighting crime. Sending police to where crime has become a problem is, in other words, too late. CompStat encouraged what the Police Foundation called "whack-a-mole" theory of policing, similar to the game played in amusement parks. Rather than change police departments into nimble crime fighters, the Foundation found that a database had been attached to traditional organizations, which themselves remained unchanged.

Because of the emphasis placed on reducing crime and because of the newfound importance of crime statistics to officers' careers, CompStat has created pressure on some precinct commanders to manipulate crime statistics to produce favorable results. Officers must continue to improve their crime statistics, despite shrinking budgets and dwindling numbers of officers. A study conducted in 2009 via a questionnaire given to 1,200 retired police captains and more senior officers concluded that nearly a third of respondents were aware of unethical manipulation of crime data.

More than 100 survey respondents said that intense pressure to produce annual crime reductions led some supervisors and precinct commanders to manipulate crime statistics. For example, officers were known to check catalogs, eBay, and other sites for items similar to those reported stolen, looking for lower prices they could use to reduce the values of the stolen goods for record-keeping purposes. Grand larceny, a felony, is considered to be theft of goods valued at $1,000 or more, whereas theft of goods valued at less than $1,000 is only a misdemeanor. Using this method, precincts could reduce the number of felony thefts, considered an "index crime" and tracked by CompStat. Surveys and anecdotal evidence also indicated a lack of receptiveness on the part of police in some areas, possibly motivated by a

desire to reduce the number of crime incidents reported.

Some survey respondents stated that precinct commanders or aides dispatched to crime scenes sometimes tried to persuade victims not to file complaints or urged them to change their accounts of what happened in ways that could downgrade offenses to lesser crimes.

Previous studies of CompStat encountered an unwillingness by the NYPD to disclose their data reporting methods. A professor performing a study that ultimately praised CompStat's influence on crime in New York City was given full access to NYPD crime data, but the NYPD did not cooperate with the Commission to Combat Police Corruption (CCPC), an independent board that monitors police corruption. The commission sought subpoena power to demand the NYPD turn over its data and data collection procedures to uncover potential wrong doing by the police. Unfortunately, the commission was denied access to this data after strong police department opposition.

On the other hand, versions of CompStat have been adopted by hundreds of other police departments across the United States, and the CompStat approach has been credited with improving police work in many cities. In New York City itself, much of the public believes that crime is down, and that the city has become a safer and more pleasant place to live.

Sources: William K. Rashbaum, "Retired Officers Raise Questions on Crime Data," *The New York Times,* February 6, 2010; A.G. Sulzberger and Karen Zraick, "Forget Police Data, New Yorkers Rely on Own Eyes, *The New York Times*, February 7, 2010; Luis Garicano, "How Does Information Technology Help Police Reduce Crime?" TNIT Newsletter 3 (December, 2009); and New York City Police Department, "COMPSTAT Process," www.nyc.gov/html/nypd/html, accessed October 9, 2006.

CASE STUDY QUESTIONS

1. What management, organization, and technology factors make CompStat effective?

2. Can police departments effectively combat crime without the CompStat system? Is community policing incompatible with CompStat? Explain your answer.

3. Why would officers misreport certain data to CompStat? What should be done about the misreporting of data? How can it be detected?

Building and Managing Systems

Part Four focuses on building and managing systems in organizations. This part answers questions such as: What activities are required to build a new information system? What alternative approaches are available for building system solutions? How should information systems projects be managed to ensure that new systems provide genuine business benefits and work successfully in the organization? What issues must be addressed when building and managing global systems?

Chapter 13

Building Information Systems

CIMB GROUP REDESIGNS ITS ACCOUNT OPENING PROCESS

CIMB Group, headquartered in Kuala Lumpur, is Malaysia's second largest financial services provider and the third largest company on the Malaysian stock exchange. It offers a full range of financial products and services, including consumer banking, corporate and investment banking, insurance, and asset management, and its retail banking network of over 1,100 branches is the largest in Southeast Asia.

What's wrong with this? Not much, except management wants to do even better. The company launched a five-year information technology transformation initiative in January 2008 to align its information technology investments more closely with its resources. It used the ARIS business process management (BPM) tool from IDS Scheer to identify 25 different areas for improving technology, people, and processes. The ARIS software helped identify gaps and inefficiencies in existing processes.

The process of opening an account at a retail branch was singled out as needing improvement. Improving this process was given high priority because it provided customers with their first impression of CIMB Group's service and customer experience.

The old account-opening process was cumbersome and time-consuming, requiring filling out four separate data entry screens for customer information, account details, name and address details, and details concerning the automated teller machine (ATM) card. New technology created opportunities for a short-cut. Malaysia has a compulsory identity card for its citizens and permanent residents known as the Government Multipurpose Card, or MyKad. It is the world's first smart identity card, incorporating a microchip with identification data (such as name, address, gender, and religion) and capabilities for user authentication, government services, electronic payments, education, loyalty programs, mobile applications, and other conveniences.

CIMB Group's systems-building team modified the front end of the customer account system to reduce the number of data entry screens and to accept customer data obtained from scanning a MyKad card. By automatically extracting most of the identification data required to open an account from a MyKad card, CIMB only needs to use a single data entry screen to set up a new account. CIMB Group was thus able to streamline the account-opening process, reducing the time required to open a bank account by 56 percent. The experience became more personal and engaging for both the bank officer and the customer. Productivity has increased, lowering CIMB Group's cost by 8 to 9 percent annually.

Sources: Avanti Kumar, "Reaching for the Skies," MIS Asia, April 16, 2010; www.ids-scheer.com, accessed October 5, 2010; and www.cimb.com, accessed October 5, 2010.

The experience of CIMB Group illustrates some of the steps required to design and build new information systems. Building the new system entailed analyzing the organization's problems with existing information systems, assessing people's information requirements, selecting appropriate technology, and redesigning business processes and jobs. Management had to monitor the system-building effort and evaluate its benefits and costs. The new information system represented a process of planned organizational change.

The chapter-opening diagram calls attention to important points raised by this case and this chapter. CIMB Group's process for setting up a new account was excessively manual and inefficient, dragging down business operations and raising costs. It also detracted from the brand image the company wanted to project as a company with high-quality customer service. Management had an opportunity to use information technology and the information stored in MyKad smart cards to streamline and redesign this process.

CIMB's system-building team evaluated alternative system solutions. It selected a solution that replaces the process of entering customer and account data manually into a series of four data entry screens with one that obtains most of the same data by swiping a MyKad smart card. This solution did not replace CIMB's existing banking system entirely, but enhanced it with a more efficient and streamlined user interface. These enhancements improved business operations by reducing the amount of time to open a new account with CIMB Group and making the customer experience more pleasant.

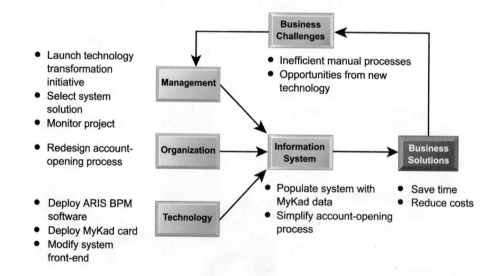

13.1 SYSTEMS AS PLANNED ORGANIZATIONAL CHANGE

Building a new information system is one kind of planned organizational change. The introduction of a new information system involves much more than new hardware and software. It also includes changes in jobs, skills, management, and organization. When we design a new information system, we are redesigning the organization. System builders must understand how a system will affect specific business processes and the organization as a whole.

SYSTEMS DEVELOPMENT AND ORGANIZATIONAL CHANGE

Information technology can promote various degrees of organizational change, ranging from incremental to far-reaching. Figure 13-1 shows four kinds of structural organizational change that are enabled by information technology: (1) automation, (2) rationalization, (3) business process redesign, and (4) paradigm shifts. Each carries different risks and rewards.

The most common form of IT-enabled organizational change is **automation**. The first applications of information technology involved assisting employees with performing their tasks more efficiently and effectively. Calculating paychecks and payroll registers, giving bank tellers instant access to customer deposit records, and developing a nationwide reservation network for airline ticket agents are all examples of early automation.

FIGURE 13-1   ORGANIZATIONAL CHANGE CARRIES RISKS AND REWARDS

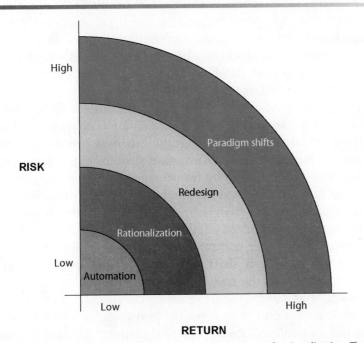

The most common forms of organizational change are automation and rationalization. These relatively slow-moving and slow-changing strategies present modest returns but little risk. Faster and more comprehensive change—such as redesign and paradigm shifts—carries high rewards but offers substantial chances of failure.

A deeper form of organizational change—one that follows quickly from early automation—is **rationalization of procedures**. Automation frequently reveals new bottlenecks in production and makes the existing arrangement of procedures and structures painfully cumbersome. Rationalization of procedures is the streamlining of standard operating procedures. For example, CIMB Bank's system for handling retail banking accounts is effective not only because it uses computer technology but also because the company simplified the business process for opening a customer account. CIMB streamlined its workflow to take advantage of the system's new user interface and software for importing personal data from MyKad.

Rationalization of procedures is often found in programs for making a series of continuous quality improvements in products, services, and operations, such as total quality management (TQM) and six sigma. **Total quality management (TQM)** makes achieving quality an end in itself and the responsibility of all people and functions within an organization. TQM derives from concepts developed by American quality experts such as W. Edwards Deming and Joseph Juran, but it was popularized by the Japanese. **Six sigma** is a specific measure of quality, representing 3.4 defects per million opportunities. Most companies cannot achieve this level of quality, but use six sigma as a goal for driving ongoing quality improvement programs.

A more powerful type of organizational change is **business process redesign**, in which business processes are analyzed, simplified, and redesigned. Business process redesign reorganizes workflows, combining steps to cut waste and eliminate repetitive, paper-intensive tasks. (Sometimes the new design eliminates jobs as well.) It is much more ambitious than rationalization of procedures, requiring a new vision of how the process is to be organized.

A widely cited example of business process redesign is Ford Motor Company's invoiceless processing, which reduced headcount in Ford's North American Accounts Payable organization of 500 people by 75 percent. Accounts payable clerks used to spend most of their time resolving discrepancies between purchase orders, receiving documents, and invoices. Ford redesigned its accounts payable process so that the purchasing department enters a purchase order into an online database that can be checked by the receiving department when the ordered items arrive. If the received goods match the purchase order, the system automatically generates a check for accounts payable to send to the vendor. There is no need for vendors to send invoices.

Rationalizing procedures and redesigning business processes are limited to specific parts of a business. New information systems can ultimately affect the design of the entire organization by transforming how the organization carries out its business or even the nature of the business. For instance, the long-haul trucking and transportation firm Schneider National used new information systems to change its business model. Schneider created a new business managing logistics for other companies. This more radical form of business change is called a **paradigm shift**. A paradigm shift involves rethinking the nature of the business and the nature of the organization.

Paradigm shifts and reengineering often fail because extensive organizational change is so difficult to orchestrate (see Chapter 14). Why, then, do so many corporations contemplate such radical change? Because the rewards are equally high (see Figure 13-1). In many instances, firms seeking paradigm shifts and pursuing reengineering strategies achieve stunning, order-of-magnitude increases in their returns on investment (or productivity). Some of these success stories, and some failure stories, are included throughout this book.

BUSINESS PROCESS REDESIGN

Like CIMB Group, described in the chapter-opening case, many businesses today are trying to use information technology to improve their business processes. Some of these systems entail incremental process change, but others require more far-reaching redesign of business processes. To deal with these changes, organizations are turning to business process management. **Business process management** provides a variety of tools and methodologies to analyze existing processes, design new processes, and optimize those processes. BPM is never concluded because process improvement requires continual change. Companies practicing business process management go through the following steps:

1. **Identify processes for change:** One of the most important strategic decisions that a firm can make is not deciding how to use computers to improve business processes, but understanding what business processes need improvement. When systems are used to strengthen the wrong business model or business processes, the business can become more efficient at doing what it should not do. As a result, the firm becomes vulnerable to competitors who may have discovered the right business model. Considerable time and cost may also be spent improving business processes that have little impact on overall firm performance and revenue. Managers need to determine what business processes are the most important and how improving these processes will help business performance.

2. **Analyze existing processes:** Existing business processes should be modeled and documented, noting inputs, outputs, resources, and the sequence of activities. The process design team identifies redundant steps, paper-intensive tasks, bottlenecks, and other inefficiencies.

Figure 13-2 illustrates the "as-is" process for purchasing a book from a physical bookstore. Consider what happens when a customer visits a physical book-

FIGURE 13-2   AS-IS BUSINESS PROCESS FOR PURCHASING A BOOK FROM A PHYSICAL BOOKSTORE

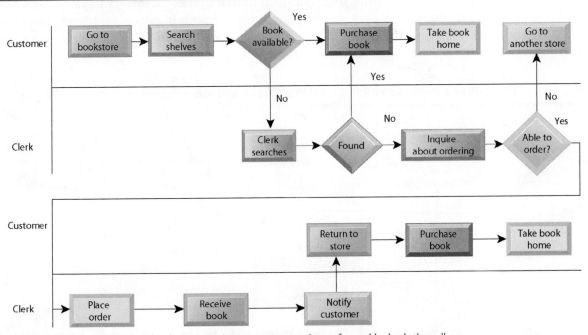

Purchasing a book from a physical bookstore requires many steps to be performed by both the seller and the customer.

store and searches its shelves for a book. If he or she finds the book, that person takes it to the checkout counter and pays for it via credit card, cash, or check. If the customer is unable to locate the book, he or she must ask a bookstore clerk to search the shelves or check the bookstore's inventory records to see if it is in stock. If the clerk finds the book, the customer purchases it and leaves. If the book is not available locally, the clerk inquires about ordering it for the customer, from the bookstore's warehouse or from the book's distributor or publisher. Once the ordered book arrives at the bookstore, a bookstore employee telephones the customer with this information. The customer would have to go to the bookstore again to pick up the book and pay for it. If the bookstore is unable to order the book for the customer, the customer would have to try another bookstore. You can see that this process has many steps and might require the customer to make multiple trips to the bookstore.

3. **Design the new process:** Once the existing process is mapped and measured in terms of time and cost, the process design team will try to improve the process by designing a new one. A new streamlined "to-be" process will be documented and modeled for comparison with the old process.

Figure 13-3 illustrates how the book-purchasing process can be redesigned by taking advantage of the Internet. The customer accesses an online bookstore over the Internet from his or her computer. He or she searches the bookstore's online catalog for the book he or she wants. If the book is available, the customer orders the book online, supplying credit card and shipping address information, and the book is delivered to the customer's home. If the online bookstore does not carry the book, the customer selects another online bookstore and searches for the book again. This process has far fewer steps than that for purchasing the book in a physical bookstore, requires much less effort on the part of the customer, and requires less sales staff for customer service. The new process is therefore much more efficient and time-saving.

The new process design needs to be justified by showing how much it reduces time and cost or enhances customer service and value. Management first measures the time and cost of the existing process as a baseline. In our example, the time required for purchasing a book from a physical bookstore might range from 15 minutes (if the customer immediately finds what he or she wants) to 30 minutes if the book is in stock but has to be located by sales staff.

FIGURE 13-3   REDESIGNED PROCESS FOR PURCHASING A BOOK ONLINE

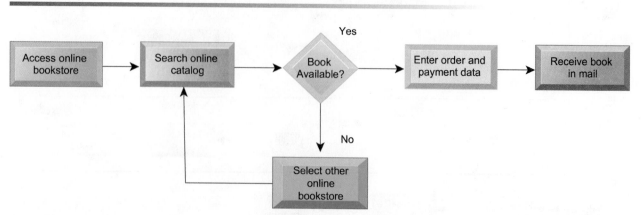

Using Internet technology makes it possible to redesign the process for purchasing a book so that it requires fewer steps and consumes fewer resources.

If the book has to be ordered from another source, the process might take one or two weeks and another trip to the bookstore for the customer. If the customer lives far away from the bookstore, the time to travel to the bookstore would have to be factored in. The bookstore will have to pay the costs for maintaining a physical store and keeping the book in stock, for sales staff on site, and for shipment costs if the book has to be obtained from another location.

The new process for purchasing a book online might only take several minutes, although the customer might have to wait several days or a week to receive the book in the mail and will have to pay a shipping charge. But the customer saves time and money by not having to travel to the bookstore or make additional visits to pick up the book. Booksellers' costs are lower because they do not have to pay for a physical store location or for local inventory.

4. **Implement the new process:** Once the new process has been thoroughly modeled and analyzed, it must be translated into a new set of procedures and work rules. New information systems or enhancements to existing systems may have to be implemented to support the redesigned process. The new process and supporting systems are rolled out into the business organization. As the business starts using this process, problems are uncovered and addressed. Employees working with the process may recommend improvements.

5. **Continuous measurement:** Once a process has been implemented and optimized, it needs to be continually measured. Why? Processes may deteriorate over time as employees fall back on old methods, or they may lose their effectiveness if the business experiences other changes.

Although many business process improvements are incremental and ongoing, there are occasions when more radical change must take place. Our example of a physical bookstore redesigning the book-purchasing process so that it can be carried out online is an example of this type of radical, far-reaching change. When properly implemented, business process redesign produces dramatic gains in productivity and efficiency, and may even change the way the business is run. In some instances, it drives a "paradigm shift" that transforms the nature of the business itself.

This actually happened in book retailing when Amazon challenged traditional physical bookstores with its online retail model. By radically rethinking the way a book can be purchased and sold, Amazon and other online bookstores have achieved remarkable efficiencies, cost reductions, and a whole new way of doing business.

BPM poses challenges. Executives report that the largest single barrier to successful business process change is organizational culture. Employees do not like unfamiliar routines and often try to resist change. This is especially true of projects where organizational changes are very ambitious and far-reaching. Managing change is neither simple nor intuitive, and companies committed to extensive process improvement need a good change management strategy (see Chapter 14).

Tools for Business Process Management

Over 100 software firms provide tools for various aspects of BPM, including IBM, Oracle, and TIBCO. These tools help businesses identify and document processes requiring improvement, create models of improved processes, capture and enforce business rules for performing processes, and integrate existing systems to support new or redesigned processes. BPM software tools also provide analytics for verifying that process performance has been improved and for measuring the impact of process changes on key business performance indicators.

Some BPM tools document and monitor business processes to help firms identify inefficiencies, using software to connect with each of the systems a company uses for a particular process to identify trouble spots. Canadian mutual fund company AIC used Sajus BPM monitoring software to check inconsistencies in its process for updating accounts after each client transaction. Sajus specializes in "goal-based" process management, which focuses on finding the causes of organizational problems through process monitoring before applying tools to address those problems.

Another category of tools automate some parts of a business process and enforce business rules so that employees perform that process more consistently and efficiently.

For example, American National Insurance Company (ANCO), which offers life insurance, medical insurance, property casualty insurance, and investment services, used Pegasystems BPM workflow software to streamline customer service processes across four business groups. The software built rules to guide customer service representatives through a single view of a customer's information that was maintained in multiple systems. By eliminating the need to juggle multiple applications simultaneously to handle customer and agent requests, the improved process increased customer service representative workload capacity by 192 percent.

A third category of tools helps businesses integrate their existing systems to support process improvements. They automatically manage processes across the business, extract data from various sources and databases, and generate transactions in multiple related systems. For example, the Star Alliance of 15 airlines, including United and Lufthansa, used BPM to create common processes shared by all of its members by integrating their existing systems. One project created a new service for frequent fliers on member airlines by consolidating 90 separate business processes across nine airlines and 27 legacy systems. The BPM software documented how each airline processed frequent flier information to help airline managers model a new business process that showed how to share data among the various systems.

The Interactive Session on Organizations describes how several companies used similar tools for their business process management programs. As you read this case, think about the kinds of changes the companies using these BPM tools were able to make in the way they ran their businesses.

13.2 OVERVIEW OF SYSTEMS DEVELOPMENT

New information systems are an outgrowth of a process of organizational problem solving. A new information system is built as a solution to some type of problem or set of problems the organization perceives it is facing. The problem may be one in which managers and employees realize that the organization is not performing as well as expected, or that the organization should take advantage of new opportunities to perform more successfully.

The activities that go into producing an information system solution to an organizational problem or opportunity are called **systems development**. Systems development is a structured kind of problem solved with distinct activities. These activities consist of systems analysis, systems design, programming, testing, conversion, and production and maintenance.

INTERACTIVE SESSION: ORGANIZATIONS

CAN BUSINESS PROCESS MANAGEMENT MAKE A DIFFERENCE?

If you're a large successful company, business process management might be just what you're looking for. AmerisourceBergen and Diebold Inc. are two examples. AmerisourceBergen is one of the world's largest pharmaceutical services companies and a member of the Fortune 25, with $70 billion in revenue in 2009. It provides drug distribution and related services designed to reduce costs and improve patient outcomes, servicing both pharmaceutical manufacturers and healthcare providers.

Because it is so large, AmerisourceBergen has numerous and complicated relationships with manufacturers, pharmacies, and hospitals. Frequently changing business conditions cause contract prices to fluctuate. When they do, both the distributor and manufacturer need to analyze these changes and make sure they comply with their business rules and federal regulations. Managing these contract and pricing details associated with each of these relationships had been very time-consuming and paper-intensive, relying heavily on e-mail, telephone, fax, and postal mail. Many of these processes were redundant.

AmerisourceBergen's management believed the company had many old and inefficient business processes. After an extensive BPM vendor analysis, the company selected Metastorm BPM software. Metastorm BPM provides a complete set of tools for analyzing, managing, and redesigning business processes. Business professionals, managers, and information systems specialists are able to create rich graphical models of business processes as well as new user interfaces and business rules. Metastorm has an engine for deploying redesigned processes along with capabilities for integrating the processes it manages with external systems.

For its first BPM project, AmerisourceBergen decided to automate and implement an online collaborative contract and chargeback process, which is responsible for a $10 billion annual cash flow. This process drives the establishment of pricing and terms with each of the company's manufacturers and also controls compliance with pricing terms and the payment of rebates from the manufacturer if the company is forced to sell at a lower price to compete. Any disputes or inaccurate pricing data create costly delays in obtaining the refunds the company is owed.

Metasource BPM makes it possible for all contract changes to be recorded into the system and validated against internal business rules, and also enables AmerisourceBergen to link with its trading partners for collaborative BPM. All contract information is housed in a single repository, making it much easier to investigate chargebacks and communicate contract and pricing information with trading partners and among internal departments.

The BPM project was successful and resulted in lower headcount, fewer disputes, more accurate pricing information, and a high return on investment. This early success encouraged the company to expand BPM to other areas of the business and use it to support a broader business transformation program. AmerisourceBergen used Metastorm BPM to create six new specialized processes for managing and automating high-volume, highly specialized supplier credits which interface with its SAP enterprise system.

To meet federal and industry-specific regulations, AmerisourceBergen must carefully track and match all direct, indirect, and third-party credits with the appropriate product inflows and outflows. The company used Metastorm BPM to create specialized processes that interface with SAP, including the ability to receive, track, reconcile, and expedite all credit variances, such as discrepancies in invoices and purchase orders. After the SAP system identifies the variances, it passes credits to Metastorm BPM for exception handling, resolution, and reconciliation with master credit data. Reconciled credits are then returned to the SAP system. More than 1.2 million credit/debit adjustment documents and paper-based credits are seamlessly passed between Metastorm and SAP this way.

To date, AmerisourceBergen has automated nearly 300 processes, benefiting from more efficient and accurate record tracking, faster turnaround times, greater management into key performance indicators, and an online audit trail of all activities. AmerisourceBergen's BPM projects had such positive outcomes that the company won a Global Excellence in BPM and Workflow award in 2009.

Diebold, Inc. is another recent convert to business process management. Diebold is a global leader in integrated self-service delivery and security systems and services, with 17,000 associates across 90 countries. The company makes, installs, and services ATMs, vaults, currency-processing systems, and other security equipment used in financial, retail, and government markets. Diebold hoped to use business process management to understand and improve its order fulfillment process. The company selected Progress Savvion's BusinessManager BPM solution for this task.

BusinessManager provides a platform for defining an organization's business processes and deploying those processes as Web-accessible applications. The platform gives managers real-time visibility to monitor, analyze, control, and improve the execution of those processes and can integrate these processes with existing operational systems. BusinessManager receives and organizes data from multiple sources to provide a more complete view of the Diebold order process. Diebold managers are able to track orders in

real time at any step in the process and also predict future performance based on past data. Since the tool enables managers to learn how long each step of the process usually takes, they can forecast where orders ought to be and compare that with where the system says the orders actually are. BusinessManager can detect whether production of an item is complete and where specific items are located.

Pleased with these capabilities, Diebold immediately used BusinessManager for other processes, such as issue resolution. The system aggregates input from various sources, such as workers in the field and in factories. Diebold is now able to quickly identify issues raised by employees and customers and determine how long it takes to resolve them.

Sources: Judith Lamont, "BPM, Enterprisewide and Beyond," *KMWorld*, February 1, 2010; "Customer Success Story: AmerisourceBergen," www.metastorm.com, accessed November 4, 2010; and www.progress.com, accessed November 4, 2010.

CASE STUDY QUESTIONS

1. Why are large companies such as AmerisourceBergen and Diebold good candidates for business process management?

2. What were the business benefits for each company from redesigning and managing their business processes?

3. How did BPM change the way these companies ran their businesses?

4. What might be some of the problems with extending BPM software across a large number of business processes?

5. What companies stand to gain the most by implementing BPM?

MIS IN ACTION

Search online for "BPM software provider" or "enterprise-wide BPM" and visit the Web site of a major BPM vendor not mentioned in this case. Then answer the following questions:

1. What types of companies have benefited from this software?

2. What are some of the important functionalities of the BPM products offered?

3. Would this company have been a better fit than Savvion or Metastorm for Diebold or AmerisourceBergen, respectively? Why or why not?

Figure 13-4 illustrates the systems development process. The systems development activities depicted usually take place in sequential order. But some of the activities may need to be repeated or some may take place simultaneously, depending on the approach to system building that is being employed (see Section 13.4).

SYSTEMS ANALYSIS

Systems analysis is the analysis of a problem that a firm tries to solve with an information system. It consists of defining the problem, identifying its causes,

FIGURE 13-4   THE SYSTEMS DEVELOPMENT PROCESS

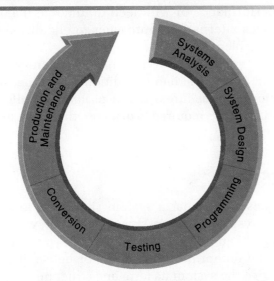

Building a system can be broken down into six core activities.

specifying the solution, and identifying the information requirements that must be met by a system solution.

The systems analyst creates a road map of the existing organization and systems, identifying the primary owners and users of data along with existing hardware and software. The systems analyst then details the problems of existing systems. By examining documents, work papers, and procedures; observing system operations; and interviewing key users of the systems, the analyst can identify the problem areas and objectives a solution would achieve. Often the solution requires building a new information system or improving an existing one.

The systems analysis also includes a **feasibility study** to determine whether that solution is feasible, or achievable, from a financial, technical, and organizational standpoint. The feasibility study determines whether the proposed system is expected to be a good investment, whether the technology needed for the system is available and can be handled by the firm's information systems specialists, and whether the organization can handle the changes introduced by the system.

Normally, the systems analysis process identifies several alternative solutions that the organization can pursue and assess the feasibility of each. A written systems proposal report describes the costs and benefits, and the advantages and disadvantages, of each alternative. It is up to management to determine which mix of costs, benefits, technical features, and organizational impacts represents the most desirable alternative.

Establishing Information Requirements

Perhaps the most challenging task of the systems analyst is to define the specific information requirements that must be met by the chosen system solution. At the most basic level, the **information requirements** of a new system involve identifying who needs what information, where, when, and how. Requirements analysis carefully defines the objectives of the new or modified system and develops a detailed description of the functions that the new system must perform. Faulty requirements analysis is a leading cause of systems failure and high systems development costs (see Chapter 14). A system designed around the wrong set of requirements will either have to be discarded because of poor performance or will need to undergo major modifications. Section 13.3 describes alternative approaches to eliciting requirements that help minimize this problem.

Some problems do not require an information system solution but instead need an adjustment in management, additional training, or refinement of existing organizational procedures. If the problem is information related, systems analysis still may be required to diagnose the problem and arrive at the proper solution.

SYSTEMS DESIGN

Systems analysis describes what a system should do to meet information requirements, and **systems design** shows how the system will fulfill this objective. The design of an information system is the overall plan or model for that system. Like the blueprint of a building or house, it consists of all the specifications that give the system its form and structure.

The systems designer details the system specifications that will deliver the functions identified during systems analysis. These specifications should address all of the managerial, organizational, and technological components of the system solution. Table 13-1 lists the types of specifications that would be produced during systems design.

Like houses or buildings, information systems may have many possible designs. Each design represents a unique blend of all technical and organizational components. What makes one design superior to others is the ease and efficiency with which it fulfills user requirements within a specific set of technical, organizational, financial, and time constraints.

TABLE 13-1 DESIGN SPECIFICATIONS

OUTPUT	PROCESSING	DOCUMENTATION
Medium	Computations	Operations documentation
Content	Program modules	Systems documentation
Timing	Required reports	User documentation
	Timing of outputs	
INPUT		CONVERSION
Origins	MANUAL PROCEDURES	Transfer files
Flow	What activities	Initiate new procedures
Data entry	Who performs them	Select testing method
	When	Cut over to new system
USER INTERFACE	How	
Simplicity	Where	TRAINING
Efficiency		Select training techniques
Logic	CONTROLS	Develop training modules
Feedback	Input controls (characters, limit,	Identify training facilities
Errors	reasonableness)	
	Processing controls (consistency, record counts)	ORGANIZATIONAL CHANGES
DATABASE DESIGN	Output controls (totals, samples of output)	Task redesign
Logical data model	Procedural controls (passwords, special forms)	Job design
Volume and speed requirements		Process design
File organization and design	SECURITY	Organization structure design
Record specifications	Access controls	Reporting relationships
	Catastrophe plans	
	Audit trails	

The Role of End Users

User information requirements drive the entire system-building effort. Users must have sufficient control over the design process to ensure that the system reflects their business priorities and information needs, not the biases of the technical staff. Working on design increases users' understanding and acceptance of the system. As we describe in Chapter 14, insufficient user involvement in the design effort is a major cause of system failure. However, some systems require more user participation in design than others, and Section 13.3 shows how alternative systems development methods address the user participation issue.

COMPLETING THE SYSTEMS DEVELOPMENT PROCESS

The remaining steps in the systems development process translate the solution specifications established during systems analysis and design into a fully operational information system. These concluding steps consist of programming, testing, conversion, production, and maintenance.

Programming

During the **programming** stage, system specifications that were prepared during the design stage are translated into software program code. Today, many organizations no longer do their own programming for new systems. Instead, they purchase the software that meets the requirements for a new system from external sources such as software packages from a commercial software vendor, software services from an application service provider, or outsourcing firms that develop custom application software for their clients (see Section 13.3).

Testing

Exhaustive and thorough **testing** must be conducted to ascertain whether the system produces the right results. Testing answers the question, "Will the system produce the desired results under known conditions?" As Chapter 5 noted, some companies are starting to use cloud computing services for this work.

The amount of time needed to answer this question has been traditionally underrated in systems project planning (see Chapter 14). Testing is time-consuming: Test data must be carefully prepared, results reviewed, and corrections made in the system. In some instances, parts of the system may have to be redesigned. The risks resulting from glossing over this step are enormous.

Testing an information system can be broken down into three types of activities: unit testing, system testing, and acceptance testing. **Unit testing**, or program testing, consists of testing each program separately in the system. It is widely believed that the purpose of such testing is to guarantee that programs are error-free, but this goal is realistically impossible. Testing should be viewed instead as a means of locating errors in programs, focusing on finding all the ways to make a program fail. Once they are pinpointed, problems can be corrected.

System testing tests the functioning of the information system as a whole. It tries to determine whether discrete modules will function together as planned and whether discrepancies exist between the way the system actually works and the way it was conceived. Among the areas examined are performance time, capacity for file storage and handling peak loads, recovery and restart capabilities, and manual procedures.

Acceptance testing provides the final certification that the system is ready to be used in a production setting. Systems tests are evaluated by users and reviewed by management. When all parties are satisfied that the new system meets their standards, the system is formally accepted for installation.

The systems development team works with users to devise a systematic test plan. The **test plan** includes all of the preparations for the series of tests we have just described.

Figure 13-5 shows an example of a test plan. The general condition being tested is a record change. The documentation consists of a series of test plan screens maintained on a database (perhaps a PC database) that is ideally suited to this kind of application.

Conversion is the process of changing from the old system to the new system. Four main conversion strategies can be employed: the parallel strategy, the direct cutover strategy, the pilot study strategy, and the phased approach strategy.

In a **parallel strategy,** both the old system and its potential replacement are run together for a time until everyone is assured that the new one functions correctly. This is the safest conversion approach because, in the event of errors or processing disruptions, the old system can still be used as a backup. However, this approach is very expensive, and additional staff or resources may be required to run the extra system.

The **direct cutover strategy** replaces the old system entirely with the new system on an appointed day. It is a very risky approach that can potentially be more costly than running two systems in parallel if serious problems with the new system are found. There is no other system to fall back on. Dislocations, disruptions, and the cost of corrections may be enormous.

The **pilot study strategy** introduces the new system to only a limited area of the organization, such as a single department or operating unit. When this pilot version is complete and working smoothly, it is installed throughout the rest of the organization, either simultaneously or in stages.

The **phased approach strategy** introduces the new system in stages, either by functions or by organizational units. If, for example, the system is

FIGURE 13-5   A SAMPLE TEST PLAN TO TEST A RECORD CHANGE

Procedure	Address and Maintenance "Record Change Series"		Test Series 2		
Prepared By:		Date:	Version:		
Test Ref.	Condition Tested	Special Requirements	Expected Results	Output On	Next Screen
2.0	Change records				
2.1	Change existing record	Key field	Not allowed		
2.2	Change nonexistent record	Other fields	"Invalid key" message		
2.3	Change deleted record	Deleted record must be available	"Deleted" message		
2.4	Make second record	Change 2.1 above	OK if valid	Transaction file	V45
2.5	Insert record		OK if valid	Transaction file	V45
2.6	Abort during change	Abort 2.5	No change	Transaction file	V45

When developing a test plan, it is imperative to include the various conditions to be tested, the requirements for each condition tested, and the expected results. Test plans require input from both end users and information systems specialists.

introduced by function, a new payroll system might begin with hourly workers who are paid weekly, followed six months later by adding salaried employees (who are paid monthly) to the system. If the system is introduced by organizational unit, corporate headquarters might be converted first, followed by outlying operating units four months later.

Moving from an old system to a new one requires that end users be trained to use the new system. Detailed **documentation** showing how the system works from both a technical and end-user standpoint is finalized during conversion time for use in training and everyday operations. Lack of proper training and documentation contributes to system failure, so this portion of the systems development process is very important.

Production and Maintenance

After the new system is installed and conversion is complete, the system is said to be in **production**. During this stage, the system will be reviewed by both users and technical specialists to determine how well it has met its original objectives and to decide whether any revisions or modifications are in order. In some instances, a formal **postimplementation audit** document is prepared. After the system has been fine-tuned, it must be maintained while it is in production to correct errors, meet requirements, or improve processing efficiency. Changes in hardware, software, documentation, or procedures to a production system to correct errors, meet new requirements, or improve processing efficiency are termed **maintenance**.

Approximately 20 percent of the time devoted to maintenance is used for debugging or correcting emergency production problems. Another 20 percent is concerned with changes in data, files, reports, hardware, or system software. But 60 percent of all maintenance work consists of making user enhancements, improving documentation, and recoding system components for greater processing efficiency. The amount of work in the third category of maintenance problems could be reduced significantly through better systems analysis and design practices. Table 13-2 summarizes the systems development activities.

TABLE 13-2 SYSTEMS DEVELOPMENT

CORE ACTIVITY	DESCRIPTION
Systems analysis	Identify problem(s) Specify solutions Establish information requirements
Systems design	Create design specifications
Programming	Translate design specifications into program code
Testing	Perform unit testing Perform systems testing Perform testing
Conversion	Plan conversion Prepare documentation Train users and technical staff
Production and maintenance	Operate the system Evaluate the system Modify the system

MODELING AND DESIGNING SYSTEMS: STRUCTURED AND OBJECT-ORIENTED METHODOLOGIES

There are alternative methodologies for modeling and designing systems. Structured methodologies and object-oriented development are the most prominent.

Structured Methodologies

Structured methodologies have been used to document, analyze, and design information systems since the 1970s. **Structured** refers to the fact that the techniques are step by step, with each step building on the previous one. Structured methodologies are top-down, progressing from the highest, most abstract level to the lowest level of detail—from the general to the specific.

Structured development methods are process-oriented, focusing primarily on modeling the processes, or actions that capture, store, manipulate, and distribute data as the data flow through a system. These methods separate data from processes. A separate programming procedure must be written every time someone wants to take an action on a particular piece of data. The procedures act on data that the program passes to them.

The primary tool for representing a system's component processes and the flow of data between them is the **data flow diagram (DFD)**. The data flow diagram offers a logical graphic model of information flow, partitioning a system into modules that show manageable levels of detail. It rigorously specifies the processes or transformations that occur within each module and the interfaces that exist between them.

Figure 13-6 shows a simple data flow diagram for a mail-in university course registration system. The rounded boxes represent processes, which portray the

FIGURE 13-6 **DATA FLOW DIAGRAM FOR MAIL-IN UNIVERSITY REGISTRATION SYSTEM**

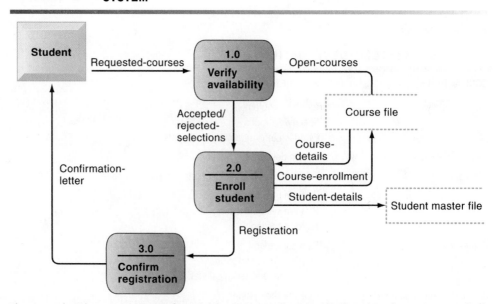

The system has three processes: Verify availability (1.0), Enroll student (2.0), and Confirm registration (3.0). The name and content of each of the data flows appear adjacent to each arrow. There is one external entity in this system: the student. There are two data stores: the student master file and the course file.

transformation of data. The square box represents an external entity, which is an originator or receiver of information located outside the boundaries of the system being modeled. The open rectangles represent data stores, which are either manual or automated inventories of data. The arrows represent data flows, which show the movement between processes, external entities, and data stores. They contain packets of data with the name or content of each data flow listed beside the arrow.

This data flow diagram shows that students submit registration forms with their name, identification number, and the numbers of the courses they wish to take. In process 1.0, the system verifies that each course selected is still open by referencing the university's course file. The file distinguishes courses that are open from those that have been canceled or filled. Process 1.0 then determines which of the student's selections can be accepted or rejected. Process 2.0 enrolls the student in the courses for which he or she has been accepted. It updates the university's course file with the student's name and identification number and recalculates the class size. If maximum enrollment has been reached, the course number is flagged as closed. Process 2.0 also updates the university's student master file with information about new students or changes in address. Process 3.0 then sends each student applicant a confirmation-of-registration letter listing the courses for which he or she is registered and noting the course selections that could not be fulfilled.

The diagrams can be used to depict higher-level processes as well as lower-level details. Through leveled data flow diagrams, a complex process can be broken down into successive levels of detail. An entire system can be divided into subsystems with a high-level data flow diagram. Each subsystem, in turn, can be divided into additional subsystems with second-level data flow diagrams, and the lower-level subsystems can be broken down again until the lowest level of detail has been reached.

Another tool for structured analysis is a data dictionary, which contains information about individual pieces of data and data groupings within a system (see Chapter 6). The data dictionary defines the contents of data flows and data stores so that systems builders understand exactly what pieces of data they contain. **Process specifications** describe the transformation occurring within the lowest level of the data flow diagrams. They express the logic for each process.

In structured methodology, software design is modeled using hierarchical structure charts. The **structure chart** is a top-down chart, showing each level of design, its relationship to other levels, and its place in the overall design structure. The design first considers the main function of a program or system, then breaks this function into subfunctions, and decomposes each subfunction until the lowest level of detail has been reached. Figure 13-7 shows a high-level structure chart for a payroll system. If a design has too many levels to fit onto one structure chart, it can be broken down further on more detailed structure charts. A structure chart may document one program, one system (a set of programs), or part of one program.

Object-Oriented Development

Structured methods are useful for modeling processes, but do not handle the modeling of data well. They also treat data and processes as logically separate entities, whereas in the real world such separation seems unnatural. Different modeling conventions are used for analysis (the data flow diagram) and for design (the structure chart).

FIGURE 13-7   **HIGH-LEVEL STRUCTURE CHART FOR A PAYROLL SYSTEM**

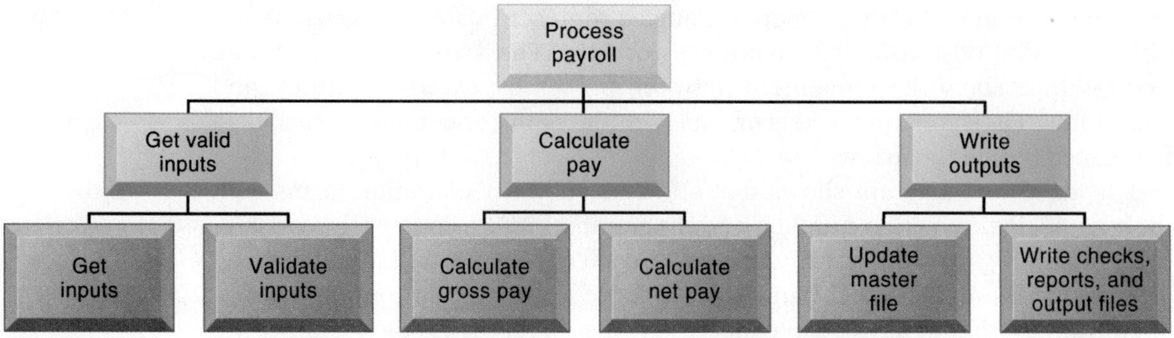

This structure chart shows the highest or most abstract level of design for a payroll system, providing an overview of the entire system.

Object-oriented development addresses these issues. Object-oriented development uses the **object** as the basic unit of systems analysis and design. An object combines data and the specific processes that operate on those data. Data encapsulated in an object can be accessed and modified only by the operations, or methods, associated with that object. Instead of passing data to procedures, programs send a message for an object to perform an operation that is already embedded in it. The system is modeled as a collection of objects and the relationships among them. Because processing logic resides within objects rather that in separate software programs, objects must collaborate with each other to make the system work.

Object-oriented modeling is based on the concepts of *class* and *inheritance*. Objects belonging to a certain class, or general categories of similar objects, have the features of that class. Classes of objects in turn can inherit all the structure and behaviors of a more general class and then add variables and behaviors unique to each object. New classes of objects are created by choosing an existing class and specifying how the new class differs from the existing class, instead of starting from scratch each time.

We can see how class and inheritance work in Figure 13-8, which illustrates the relationships among classes concerning employees and how they are paid. Employee is the common ancestor, or superclass, for the other three classes. Salaried, Hourly, and Temporary are subclasses of Employee. The class name is in the top compartment, the attributes for each class are in the middle portion of each box, and the list of operations is in the bottom portion of each box. The features that are shared by all employees (id, name, address, date hired, position, and pay) are stored in the Employee superclass, whereas each subclass stores features that are specific to that particular type of employee. Specific to hourly employees, for example, are their hourly rates and overtime rates. A solid line from the subclass to the superclass is a generalization path showing that the subclasses Salaried, Hourly, and Temporary have common features that can be generalized into the superclass Employee.

Object-oriented development is more iterative and incremental than traditional structured development. During analysis, systems builders document the functional requirements of the system, specifying its most important properties and what the proposed system must do. Interactions between the system and its users are analyzed to identify objects, which include both data and processes. The object-oriented design phase describes how the objects will

FIGURE 13-8 CLASS AND INHERITANCE

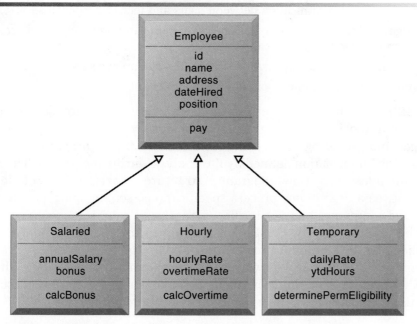

This figure illustrates how classes inherit the common features of their superclass.

behave and how they will interact with one other. Similar objects are grouped together to form a class, and classes are grouped into hierarchies in which a subclass inherits the attributes and methods from its superclass.

The information system is implemented by translating the design into program code, reusing classes that are already available in a library of reusable software objects and adding new ones created during the object-oriented design phase. Implementation may also involve the creation of an object-oriented database. The resulting system must be thoroughly tested and evaluated.

Because objects are reusable, object-oriented development could potentially reduce the time and cost of writing software because organizations can reuse software objects that have already been created as building blocks for other applications. New systems can be created by using some existing objects, changing others, and adding a few new objects. Object-oriented frameworks have been developed to provide reusable, semicomplete applications that the organization can further customize into finished applications.

Computer-Aided Software Engineering

Computer-aided software engineering (CASE)—sometimes called *computer-aided systems engineering*—provides software tools to automate the methodologies we have just described to reduce the amount of repetitive work the developer needs to do. CASE tools also facilitate the creation of clear documentation and the coordination of team development efforts. Team members can share their work easily by accessing each other's files to review or modify what has been done. Modest productivity benefits can also be achieved if the tools are used properly.

CASE tools provide automated graphics facilities for producing charts and diagrams, screen and report generators, data dictionaries, extensive reporting facilities, analysis and checking tools, code generators, and documentation generators. In general, CASE tools try to increase productivity and quality by :

- Enforcing a standard development methodology and design discipline
- Improving communication between users and technical specialists
- Organizing and correlating design components and providing rapid access to them using a design repository
- Automating tedious and error-prone portions of analysis and design
- Automating code generation and testing and control rollout

CASE tools contain features for validating design diagrams and specifications. CASE tools thus support iterative design by automating revisions and changes and providing prototyping facilities. A CASE information repository stores all the information defined by the analysts during the project. The repository includes data flow diagrams, structure charts, entity-relationship diagrams, data definitions, process specifications, screen and report formats, notes and comments, and test results.

To be used effectively, CASE tools require organizational discipline. Every member of a development project must adhere to a common set of naming conventions and standards as well as to a development methodology. The best CASE tools enforce common methods and standards, which may discourage their use in situations where organizational discipline is lacking.

13.3   ALTERNATIVE SYSTEMS-BUILDING APPROACHES

Systems differ in terms of their size and technological complexity and in terms of the organizational problems they are meant to solve. A number of systems-building approaches have been developed to deal with these differences. This section describes these alternative methods: the traditional systems life cycle, prototyping, application software packages, end-user development, and outsourcing.

TRADITIONAL SYSTEMS LIFE CYCLE

The **systems life cycle** is the oldest method for building information systems. The life cycle methodology is a phased approach to building a system, dividing systems development into formal stages. Systems development specialists have different opinions on how to partition the systems-building stages, but they roughly correspond to the stages of systems development that we have just described.

The systems life cycle methodology maintains a very formal division of labor between end users and information systems specialists. Technical specialists, such as system analysts and programmers, are responsible for much of the systems analysis, design, and implementation work; end users are limited to providing information requirements and reviewing the technical staff's work. The life cycle also emphasizes formal specifications and paperwork, so many documents are generated during the course of a systems project.

The systems life cycle is still used for building large complex systems that require a rigorous and formal requirements analysis, predefined specifications, and tight controls over the system-building process. However, the systems life cycle approach can be costly, time-consuming, and inflexible. Although systems builders can go back and forth among stages in the life cycle, the systems life cycle is predominantly a "waterfall" approach in which tasks in one stage are completed before work for the next stage begins. Activities can be repeated, but volumes of new documents must be generated and steps retraced if requirements and specifications need to be revised. This encourages freezing of specifications relatively early in the development process. The life cycle approach is also not suitable for many small desktop systems, which tend to be less structured and more individualized.

PROTOTYPING

Prototyping consists of building an experimental system rapidly and inexpensively for end users to evaluate. By interacting with the prototype, users can get a better idea of their information requirements. The prototype endorsed by the users can be used as a template to create the final system.

The **prototype** is a working version of an information system or part of the system, but it is meant to be only a preliminary model. Once operational, the prototype will be further refined until it conforms precisely to users' requirements. Once the design has been finalized, the prototype can be converted to a polished production system.

The process of building a preliminary design, trying it out, refining it, and trying again has been called an **iterative** process of systems development because the steps required to build a system can be repeated over and over again. Prototyping is more explicitly iterative than the conventional life cycle, and it actively promotes system design changes. It has been said that prototyping replaces unplanned rework with planned iteration, with each version more accurately reflecting users' requirements.

Steps in Prototyping

Figure 13-9 shows a four-step model of the prototyping process, which consists of the following:

Step 1: Identify the user's basic requirements. The system designer (usually an information systems specialist) works with the user only long enough to capture the user's basic information needs.

Step 2: Develop an initial prototype. The system designer creates a working prototype quickly, using tools for rapidly generating software.

Step 3: Use the prototype. The user is encouraged to work with the system to determine how well the prototype meets his or her needs and to make suggestions for improving the prototype.

Step 4: Revise and enhance the prototype. The system builder notes all changes the user requests and refines the prototype accordingly. After the prototype has been revised, the cycle returns to Step 3. Steps 3 and 4 are repeated until the user is satisfied.

When no more iterations are required, the approved prototype then becomes an operational prototype that furnishes the final specifications for the application. Sometimes the prototype is adopted as the production version of the system.

Advantages and Disadvantages of Prototyping

Prototyping is most useful when there is some uncertainty about requirements or design solutions and often used for designing an information system's **end-user interface** (the part of the system with which end users interact, such as online display and data entry screens, reports, or Web pages). Because prototyping encourages intense end-user involvement throughout the systems

FIGURE 13-9   THE PROTOTYPING PROCESS

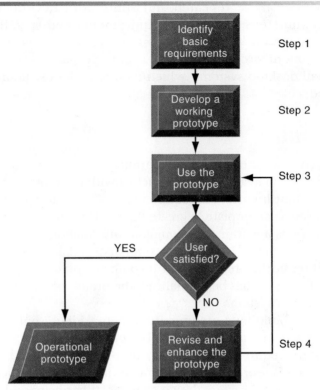

The process of developing a prototype can be broken down into four steps. Because a prototype can be developed quickly and inexpensively, systems builders can go through several iterations, repeating steps 3 and 4, to refine and enhance the prototype before arriving at the final operational one.

development life cycle, it is more likely to produce systems that fulfill user requirements.

However, rapid prototyping can gloss over essential steps in systems development. If the completed prototype works reasonably well, management may not see the need for reprogramming, redesign, or full documentation and testing to build a polished production system. Some of these hastily constructed systems may not easily accommodate large quantities of data or a large number of users in a production environment.

END-USER DEVELOPMENT

Some types of information systems can be developed by end users with little or no formal assistance from technical specialists. This phenomenon is called **end-user development**. A series of software tools categorized as fourth-generation languages makes this possible. **Fourth-generation languages** are software tools that enable end users to create reports or develop software applications with minimal or no technical assistance. Some of these fourth-generation tools also enhance professional programmers' productivity.

Fourth-generation languages tend to be nonprocedural, or less procedural, than conventional programming languages. Procedural languages require specification of the sequence of steps, or procedures, that tell the computer what to do and how to do it. Nonprocedural languages need only specify what has to be accomplished rather than provide details about how to carry out the task.

Table 13-3 shows that there are seven categories of fourth-generation languages: PC software tools, query languages, report generators, graphics languages, application generators, application software packages, and very high-level programming languages. The table shows the tools ordered in terms of ease of use by nonprogramming end users. End users are most likely to work with PC software tools and query languages. **Query languages** are software tools that provide immediate online answers to requests for information that are not predefined, such as "Who are the highest-performing sales representatives?" Query languages are often tied to data management software and to database management systems (see Chapter 6).

On the whole, end-user-developed systems can be completed more rapidly than those developed through the conventional systems life cycle. Allowing users to specify their own business needs improves requirements gathering and often leads to a higher level of user involvement and satisfaction with the system. However, fourth-generation tools still cannot replace conventional tools for some business applications because they cannot easily handle the processing of large numbers of transactions or applications with extensive procedural logic and updating requirements.

TABLE 13-3 CATEGORIES OF FOURTH-GENERATION LANGUAGES

FOURTH-GENERATION TOOL	DESCRIPTION	EXAMPLE	
PC software tools	General-purpose application software packages for PCs.	WordPerfect Microsoft Access	**Oriented toward end users**
Query language	Languages for retrieving data stored in databases or files. Capable of supporting requests for information that are not predefined.	SQL	
Report generator	Extract data from files or databases to create customized reports in a wide range of formats not routinely produced by an information system. Generally provide more control over the way data are formatted, organized, and displayed than query languages.	Crystal Reports	
Graphics language	Retrieve data from files or databases and display them in graphic format. Some graphics software can perform arithmetic or logical operations on data as well.	SAS/GRAPH Systat	
Application generator	Contain preprogrammed modules that can generate entire applications, including Web sites, greatly speeding development. A user can specify what needs to be done, and the application generator will create the appropriate program code for input, validation, update, processing, and reporting.	WebFOCUS QuickBase	
Application software package	Software programs sold or leased by commercial vendors that eliminate the need for custom-written, in-house software.	Oracle PeopleSoft HCM mySAP ERP	
Very high-level programming language	Generate program code with fewer instructions than conventional languages, such as COBOL or FORTRAN. Designed primarily as productivity tools for professional programmers.	APL Nomad2	**Oriented toward IS professionals**

End-user computing also poses organizational risks because it occurs outside of traditional mechanisms for information systems management and control. When systems are created rapidly, without a formal development methodology, testing and documentation may be inadequate. Control over data can be lost in systems outside the traditional information systems department. To help organizations maximize the benefits of end-user applications development, management should control the development of end-user applications by requiring cost justification of end-user information system projects and by establishing hardware, software, and quality standards for user-developed applications.

APPLICATION SOFTWARE PACKAGES AND OUTSOURCING

Chapter 5 points out that much of today's software is not developed in-house but is purchased from external sources. Firms can rent the software from a software service provider, they can purchase a software package from a commercial vendor, or they can have a custom application developed by an outside outsourcing firm.

Application Software Packages

During the past several decades, many systems have been built on an application software package foundation. Many applications are common to all business organizations—for example, payroll, accounts receivable, general ledger, or inventory control. For such universal functions with standard processes that do not change a great deal over time, a generalized system will fulfill the requirements of many organizations.

If a software package can fulfill most of an organization's requirements, the company does not have to write its own software. The company can save time and money by using the prewritten, predesigned, pretested software programs from the package. Package vendors supply much of the ongoing maintenance and support for the system, including enhancements to keep the system in line with ongoing technical and business developments.

If an organization has unique requirements that the package does not address, many packages include capabilities for customization. **Customization** features allow a software package to be modified to meet an organization's unique requirements without destroying the integrity of the packaged software. If a great deal of customization is required, additional programming and customization work may become so expensive and time-consuming that they negate many of the advantages of software packages.

When a system is developed using an application software package, systems analysis will include a package evaluation effort. The most important evaluation criteria are the functions provided by the package, flexibility, user friendliness, hardware and software resources, database requirements, installation and maintenance efforts, documentation, vendor quality, and cost. The package evaluation process often is based on a **Request for Proposal (RFP)**, which is a detailed list of questions submitted to packaged-software vendors.

When a software package is selected, the organization no longer has total control over the system design process. Instead of tailoring the system design specifications directly to user requirements, the design effort will consist of trying to mold user requirements to conform to the features of the package. If the organization's requirements conflict with the way the package works and the package cannot be customized, the organization will have to adapt to the package and change its procedures.

The Interactive Session on Technology describes the experience of Zimbra, a software company that selected a software package solution for its new marketing automation system. As you read this case, be aware of the management, technology, and organization issues that had to be addressed when the company selected a new software package solution.

Outsourcing

If a firm does not want to use its internal resources to build or operate information systems, it can outsource the work to an external organization that specializes in providing these services. Cloud computing and SaaS providers, which we described in Chapter 5, are one form of outsourcing. Subscribing companies would use the software and computer hardware provided by the service as the technical platform for their systems. In another form of outsourcing, a company could hire an external vendor to design and create the software for its system, but that company would operate the system on its own computers. The outsourcing vendor might be domestic or in another country.

Domestic outsourcing is driven primarily by the fact that outsourcing firms possess skills, resources, and assets that their clients do not have. Installing a new supply chain management system in a very large company might require hiring an additional 30 to 50 people with specific expertise in supply chain management software, licensed from a vendor. Rather than hire permanent new employees, most of whom would need extensive training in the software package, and then release them after the new system is built, it makes more sense, and is often less expensive, to outsource this work for a 12-month period.

In the case of **offshore outsourcing**, the decision tends to be much more cost-driven. A skilled programmer in India or Russia earns about USD $9,000 per year, compared to $65,000 per year for a comparable programmer in the United States. The Internet and low-cost communications technology have drastically reduced the expense and difficulty of coordinating the work of global teams in faraway locations. In addition to cost savings, many offshore outsourcing firms offer world-class technology assets and skills. Wage inflation outside the United States has recently eroded some of these advantages, and some jobs have moved back to the United States.

Nevertheless, there is a very strong chance that at some point in your career, you'll be working with offshore outsourcers or global teams. Your firm is most likely to benefit from outsourcing if it takes the time to evaluate all the risks and to make sure outsourcing is appropriate for its particular needs. Any company that outsources its applications must thoroughly understand the project, including its requirements, method of implementation, anticipated benefits, cost components, and metrics for measuring performance.

Many firms underestimate costs for identifying and evaluating vendors of information technology services, for transitioning to a new vendor, for improving internal software development methods to match those of outsourcing vendors, and for monitoring vendors to make sure they are fulfilling their contractual obligations. Companies will need to allocate resources for documenting requirements, sending out RFPs, travel expenses, negotiating contracts, and project management. Experts claim it takes from three months to a full year to fully transfer work to an offshore partner and make sure the vendor thoroughly understands your business.

INTERACTIVE SESSION: TECHNOLOGY

ZIMBRA ZOOMS AHEAD WITH ONEVIEW

Zimbra is a software company whose flagship product is its Zimbra Collaboration Suite (ZCS), an open source messaging and communications software package, that relies heavily on Ajax to provide a variety of business functions. Purchased by Yahoo in 2007, the company now has accumulated 50 million paid mailboxes. In addition to e-mail, ZCS combines contact lists, a shared calendar, instant messaging, hosted documents, search, and VoIP into one package, and can be used from any mobile Web browser.

As an open source software company, Zimbra uses viral marketing models, word-of-mouth marketing, and open standards to grow its business. Customers are as free to criticize Zimbra and ZCS as they are to praise the company and its flagship offering. For the most part, this strategy has proven very successful for the company thus far.

Zimbra makes sales via its Web site and offers both free and commercial versions. Zimbra's business model hinges on driving large numbers of visitors to its Web site, allowing them to try the most basic version of the software for free, and then persuading them to purchase one of its more full-featured commercial versions. Zimbra has over 200,000 visitors to its Web site each week.

Zimbra's sales process begins when one of these 200,000 weekly visitors downloads a 60-day trial version. Salespeople try to identify which people using the trial version are most likely to upgrade to one of its commercial versions and then contact them via e-mail and telephone to try to close the sale. To make this work, Zimbra's sales team needs to be able to weed out the interested buyers from its huge volume of Web visitors. As Greg Armanini, Zimbra's director of marketing, pointed out, the sales team will be overwhelmed with a large number of unqualified leads unless sales and marketing automation tools are able to focus sales reps only on the leads that will generate revenue.

Zimbra uses its Web site to track visitor activity and tie it to sales lead information in its Salesforce.com customer relationship management (CRM) system. Identifying sales prospects that visit the Web site regularly and alerting sales reps when those prospects are visiting the site helps the sales team select which prospects to contact by telephone and when to call them.

Zimbra initially used marketing automation software from Eloqua, which had a large number of features but was too complicated for both marketing and sales staff to use. For example, the Eloqua system required salespeople to code conditional logic for any data field containing data they wanted to collect. Though doable, this task was a poor use of Zimbra's sales staff time. Eloqua only worked with the Microsoft Internet Explorer Web browser, while two-thirds of Zimbra's sales department used Mozilla Firefox. And Eloqua was expensive. Zimbra could only afford its entry-level package, which provided access to only five salespeople and one marketing person.

Zimbra did not need many of Eloqua's features, but it did need a more streamlined solution that focused on the core areas of its marketing strategy: lead generation, e-mail marketing, and Web analytics. The new marketing automation system had to be easy to install and maintain. Many available options required several well-trained administrators, and Zimbra could not afford to allocate even one employee for this purpose.

After examining several software products, Zimbra choose OneView, an on-demand marketing automation solution from LoopFuse, a Georgia-based software company that specializes in sales and marketing automation. OneView was more highly targeted than the Eloqua software. Not only that, but much of OneView consists of automated processes that allowed Zimbra to quickly implement the solution and to maintain it without dedicating someone to the task full-time. The core functions of OneView include Web site visitor tracking, automated marketing program communication, customer activity alerts, and CRM integration.

Zimbra was also pleased with LoopFuse's convenient pricing options, including its "unlimited seating" and "pay-per-use" options, which allowed Zimbra to pay for only the services it needed for as many users as it required. Because of these options, Zimbra was able to deploy LoopFuse across almost its entire 30-person sales force.

Other benefits of OneView include easy integration with Salesforce.com, Zimbra's preferred CRM solution, simplified reporting processes, and the ability to manage a larger number of leads thanks to having more salespeople and time to devote to demand generation. OneView works with multiple Web browsers, including Firefox. The old solution offered so many ways to handle and organize data that generating data reports could take a long time. With OneView's simplified reporting processes, the sales staff can generate reports in a fraction of the time.

Has OneView improved Zimbra's bottom line? OneView reduced the amount of time Zimbra spent using and maintaining its marketing system by 50 percent. Zimbra reports that since changing vendors, it has witnessed a jump in its close rate on qualified sales leads from 10 to 15 percent, a huge increase. The answer appears to be a resounding "yes."

Sources: Jessica Tsai, "Less is More," *Customer Relationship Management*, August 2009, www.destinationCRM.com; and "LoopFuse OneView helps Zimbra Raise Sales and Marketing Efficiency by 50 Percent," www.loopfuse.com, May 2009.

CASE STUDY QUESTIONS

1. Describe the steps in Zimbra's sales process. How well did its old marketing automation system support that process? What problems did it create? What was the business impact of these problems?

2. List and describe Zimbra's requirements for a new marketing software package. If you were preparing the RFP for Zimbra's new system, what questions would you ask?

3. How did the new marketing system change the way Zimbra ran its business? How successful was it?

MIS IN ACTION

Visit the LoopFuse Web site and then answer the following questions:

1. List and describe each of the major features of LoopFuse OneView.

2. Select two of these features and describe how they would help Zimbra's sales team.

Outsourcing offshore incurs additional costs for coping with cultural differences that drain productivity and dealing with human resources issues, such as terminating or relocating domestic employees. All of these hidden costs undercut some of the anticipated benefits from outsourcing. Firms should be especially cautious when using an outsourcer to develop or to operate applications that give it some type of competitive advantage.

Figure 13-10 shows best- and worst-case scenarios for the total cost of an offshore outsourcing project. It shows how much hidden costs affect the total project cost. The best case reflects the lowest estimates for additional costs, and the worst case reflects the highest estimates for these costs. As you can see, hidden costs increase the total cost of an offshore outsourcing project by an extra 15 to 57 percent. Even with these extra costs, many firms will benefit from offshore outsourcing if they manage the work well. Under the worst-case scenario, a firm would still save about 15 percent.

13.4 APPLICATION DEVELOPMENT FOR THE DIGITAL FIRM

In the digital firm environment, organizations need to be able to add, change, and retire their technology capabilities very rapidly to respond to new opportu-

FIGURE 13-10   **TOTAL COST OF OFFSHORE OUTSOURCING**

TOTAL COST OF OFFSHORE OUTSOURCING				
Cost of outsourcing contract			$10, 000, 000	
Hidden Costs	Best Case	Additional Cost ($)	Worst Case	Additional Cost ($)
1. Vendor selection	0%	20,000	2%	200,000
2. Transition costs	2%	200,000	3%	300,000
3. Layoffs & retention	3%	300,000	5%	500,000
4. Lost productivity/cultural issues	3%	300,000	27%	2,700,000
5. Improving development processes	1%	100,000	10%	1,000,000
6. Managing the contract	6%	600,000	10%	1,000,000
Total additional costs		1,520,000		5,700,000
	Outstanding Contract ($)	Additional Cost ($)	Total Cost ($)	Additional Cost
Total cost of outsourcing (TCO) best case	10,000,000	1,520,000	11,520,000	15.2%
Total cost of outsourcing (TCO) worst case	10,000,000	5,700,000	15,700,000	57.0%

If a firm spends $10 million on offshore outsourcing contracts, that company will actually spend 15.2 percent in extra costs even under the best-case scenario. In the worst-case scenario, where there is a dramatic drop in productivity along with exceptionally high transition and layoff costs, a firm can expect to pay up to 57 percent in extra costs on top of the $10 million outlay for an offshore contract.

nities. Companies are starting to use shorter, more informal development processes that provide fast solutions. In addition to using software packages and external service providers, businesses are relying more heavily on fast-cycle techniques such as rapid application development, joint application design, agile development, and reusable standardized software components that can be assembled into a complete set of services for e-commerce and e-business.

RAPID APPLICATION DEVELOPMENT (RAD)

Object-oriented software tools, reusable software, prototyping, and fourth-generation language tools are helping systems builders create working systems much more rapidly than they could using traditional systems-building methods and software tools. The term **rapid application development (RAD)** is used to describe this process of creating workable systems in a very short period of time. RAD can include the use of visual programming and other tools for building graphical user interfaces, iterative prototyping of key system elements, the automation of program code generation, and close teamwork among end users and information systems specialists. Simple systems often can be assembled from prebuilt components. The process does not have to be sequential, and key parts of development can occur simultaneously.

Sometimes a technique called **joint application design (JAD)** is used to accelerate the generation of information requirements and to develop the initial systems design. JAD brings end users and information systems specialists together in an interactive session to discuss the system's design. Properly prepared and facilitated, JAD sessions can significantly speed up the design phase and involve users at an intense level.

Agile development focuses on rapid delivery of working software by breaking a large project into a series of small subprojects that are completed in short periods of time using iteration and continuous feedback. Each mini-project is worked on by a team as if it were a complete project, including planning, requirements analysis, design, coding, testing, and documentation. Improvement or addition of new functionality takes place within the next iteration as developers clarify requirements. This helps to minimize the overall risk, and allows the project to adapt to changes more quickly. Agile methods emphasize face-to-face communication over written documents, encouraging people to collaborate and make decisions quickly and effectively.

COMPONENT-BASED DEVELOPMENT AND WEB SERVICES

We have already described some of the benefits of object-oriented development for building systems that can respond to rapidly changing business environments, including Web applications. To further expedite software creation, groups of objects have been assembled to provide software components for common functions such as a graphical user interface or online ordering capability that can be combined to create large-scale business applications. This approach to software development is called **component-based development**, and it enables a system to be built by assembling and integrating existing software components. Increasingly, these software components are coming from cloud services. Businesses are using component-based development to create their e-commerce applications by combining commercially available components for shopping carts, user authentication, search engines, and catalogs with pieces of software for their own unique business requirements.

Web Services and Service-Oriented Computing

Chapter 5 introduced *Web services* as loosely coupled, reusable software components delivered using Extensible Markup Language (XML) and other open protocols and standards that enable one application to communicate with another with no custom programming required to share data and services. In addition to supporting internal and external integration of systems, Web services can be used as tools for building new information system applications or enhancing existing systems. Because these software services use a universal set of standards, they promise to be less expensive and less difficult to weave together than proprietary components.

Web services can perform certain functions on their own, and they can also engage other Web services to complete more complex transactions, such as checking credit, procurement, or ordering products. By creating software components that can communicate and share data regardless of the operating system, programming language, or client device, Web services can provide significant cost savings in systems building while opening up new opportunities for collaboration with other companies.

13.5 | HANDS-ON MIS PROJECTS

The projects in this section give you hands-on experience analyzing business process problems, designing and building a customer system for auto sales, and redesigning business processes for a company that wants to purchase goods over the Web.

Management Decision Problems

1. A customer purchasing a Sears Roebuck appliance, such as a washing machine, can also purchase a three-year service contract for an additional fee. The contract provides free repair service and parts for the specified appliance using an authorized Sears service provider. When a person with a Sears service contract needs an appliance repaired, such as a washing machine, he or she calls the Sears Repairs & Parts department to schedule an appointment. The department makes the appointment and gives the caller the date and approximate time of the appointment. The repair technician arrives during the designated time frame and diagnoses the problem. If the problem is caused by a faulty part, the technician either replaces the part if he is carrying the part with him or orders the replacement part from Sears. If the part is not in stock at Sears, Sears orders the part and gives the customer an approximate time when the part will arrive. The part is shipped directly to the customer. After the part has arrived, the customer must call Sears to schedule a second appointment for a repair technician to replace the ordered part. This process is very lengthy.

   It may take two weeks for the first repair visit to occur, another two weeks to receive the ordered part, and another week for the second repair visit to occur in which the ordered part is installed.

   - Diagram the existing process.
   - What is the impact of the existing process on Sears' operational efficiency and customer relationships?
   - What changes could be made to make this process more efficient? How could information systems support these changes? Diagram the new improved process.

2. Management at your agricultural chemicals corporation has been dissatisfied with production planning. Production plans are created using best guesses of demand for each product, which are based on how much of each product has been ordered in the past. If a customer places an unexpected order or requests a change to an existing order after it has been placed, there is no way to adjust production plans. The company may have to tell customers it can't fill their orders, or it may run up extra costs maintaining additional inventory to prevent stock-outs.

   At the end of each month, orders are totaled and manually keyed into the company's production planning system. Data from the past month's production and inventory systems are manually entered into the firm's order management system. Analysts from the sales department and from the production department analyze the data from their respective systems to determine what the sales targets and production targets should be for the next month. These estimates are usually different. The analysts then get together at a high-level planning meeting to revise the production and sales targets to take into account senior management's goals for market share, revenues, and profits. The outcome of the meeting is a finalized production master schedule.

   The entire production planning process takes 17 business days to complete. Nine of these days are required to enter and validate the data. The remaining days are spent developing and reconciling the production and sales targets and finalizing the production master schedule.

- Draw a diagram of the existing production planning process.
- Analyze the problems this process creates for the company.
- How could an enterprise system solve these problems? In what ways could it lower costs? Diagram what the production planning process might look like if the company implemented enterprise software.

Improving Decision Making: Using Database Software to Design a Customer System for Auto Sales

Software skills: Database design, querying, reporting, and forms
Business skills: Sales lead and customer analysis

This project requires you to perform a systems analysis and then design a system solution using database software.

Ace Auto Dealers specializes in selling new vehicles from Subaru. The company advertises in local newspapers and is also listed as an authorized dealer on the Subaru Web site and other major auto-buyer Web sites. The company benefits from a good local word-of-mouth reputation and name recognition and is a leading source of information for Subaru vehicles in the Portland, Oregon, area.

When a prospective customer enters the showroom, he or she is greeted by an Ace sales representative. The sales representative manually fills out a form with information such as the prospective customer's name, address, telephone number, date of visit, and make and model of the vehicle in which the customer is interested. The representative also asks where the prospect heard about Ace—whether it was from a newspaper ad, the Web, or word of mouth—and this information is noted on the form also. If the customer decides to purchase an auto, the dealer fills out a bill of sale form.

Ace does not believe it has enough information about its customers. It cannot easily determine which prospects have made auto purchases, nor can it identify which customer touch points have produced the greatest number of sales leads or actual sales so it can focus advertising and marketing more on the channels that generate the most revenue. Are purchasers discovering Ace from newspaper ads, from word of mouth, or from the Web?

Prepare a systems analysis report detailing Ace's problem and a system solution that can be implemented using PC database management software. The company has a PC with Internet access and the full suite of Microsoft Office desktop productivity tools. Then use database software to develop a simple system solution. Your systems analysis report should include the following:

- Description of the problem and its organizational and business impact
- Proposed solution, solution objectives, and solution feasibility
- Costs and benefits of the solution you have selected
- Information requirements to be addressed by the solution
- Management, organization, and technology issues to be addressed by the solution, including changes in business processes

On the basis of the requirements you have identified, design the database and populate it with at least 10 records per table. Consider whether you can use or modify Ace's existing customer database in your design. You can find this database on the myMISlab. Print out the database design. Then use the system you have created to generate queries and reports that would be most useful to management. Create several prototype data input forms for the system and review them with your instructor. Then revise the prototypes.

Achieving Operational Excellence: Redesigning Business Processes for Web Procurement

Software skills: Web browser software
Business skills: Procurement

This project requires you to rethink how a business should be redesigned when it moves to the Web.

You are in charge of purchasing for your firm and would like to use the Grainger (www.grainger.com) B2B e-commerce site for this purpose. Find out how to place an order for painting supplies by exploring the Catalog, Order Form, and Repair Parts Order capabilities of this site. Do not register at the site. Describe all the steps your firm would need to take to use this system to place orders online for 30 gallons of paint thinner. Include a diagram of what you think your firm's business process for purchasing should be and the pieces of information required by this process.

In a traditional purchase process, whoever is responsible for making the purchase fills out a requisition form and submits it for approval based on the company's business rules. When the requisition is approved, a purchase order with a unique purchase order identification number is sent to the supplier. The purchaser might want to browse supplier catalogs to compare prices and features before placing the order. The purchaser might also want to determine whether the items to be purchased are available. If the purchasing firm were an approved customer, that company would be granted credit to make the purchase and would be billed for the total cost of the items purchased and shipped after the order was shipped. Alternatively, the purchasing company might have to pay for the order in advance or pay for the order using a credit card. Multiple payment options might be possible. How might this process have to change to make purchases electronically from the Grainger site?

LEARNING TRACK MODULES

The following Learning Tracks provide content relevant to topics covered in this chapter:

1. Unified Modeling Language (UML)
2. A Primer on Business Process Design and Documentation
3. A Primer on Business Process Management

Review Summary

1. *How does building new systems produce organizational change?*

    Building a new information system is a form of planned organizational change. Four kinds of technology-enabled change are (a) automation, (b) rationalization of procedures, (c) business process redesign, and (d) paradigm shift, with far-reaching changes carrying the greatest risks and rewards. Many organizations are using business process management to redesign work flows and business processes in the hope of achieving dramatic productivity break-throughs. Business process management is also useful for promoting, total quality management (TQM), six sigma, and other initiatives for incremental process improvement.

2. *What are the core activities in the systems development process?*

    The core activities in systems development are systems analysis, systems design, programming, testing, conversion, production, and maintenance. Systems analysis is the study and analysis of problems of existing systems and the identification of requirements for their solutions. Systems design provides the specifications for an information system solution, showing how its technical and organizational components fit together.

3. *What are the principal methodologies for modeling and designing systems?*

    The two principal methodologies for modeling and designing information systems are structured methodologies and object-oriented development. Structured methodologies focus on modeling processes and data separately. The data flow diagram is the principal tool for structured analysis, and the structure chart is the principal tool for representing structured software design. Object-oriented development models a system as a collection of objects that combine processes and data. Object-oriented modeling is based on the concepts of class and inheritance.

4. *What are the alternative methods for building information systems?*

    The oldest method for building systems is the systems life cycle, which requires that information systems be developed in formal stages. The stages must proceed sequentially and have defined outputs; each requires formal approval before the next stage can commence. The systems life cycle is useful for large projects that need formal specifications and tight management control over each stage of systems building, but it is very rigid and costly.

    Prototyping consists of building an experimental system rapidly and inexpensively for end users to interact with and evaluate. Prototyping encourages end-user involvement in systems development and iteration of design until specifications are captured accurately. The rapid creation of prototypes can result in systems that have not been completely tested or documented or that are technically inadequate for a production environment.

    Using a software package reduces the amount of design, programming, testing, installation, and maintenance work required to build a system. Application software packages are helpful if a firm does not have the internal information systems staff or financial resources to custom develop a system. To meet an organization's unique requirements, packages may require extensive modifications that can substantially raise development costs.

    End-user development is the development of information systems by end users, either alone or with minimal assistance from information systems specialists. End-user-developed systems can be created rapidly and informally using fourth-generation software tools. However, end-user development may create information systems that do not necessarily meet quality assurance standards and that are not easily controlled by traditional means.

Outsourcing consists of using an external vendor to build (or operate) a firm's information systems instead of the organization's internal information systems staff. Outsourcing can save application development costs or enable firms to develop applications without an internal information systems staff. However, firms risk losing control over their information systems and becoming too dependent on external vendors. Outsourcing also entails "hidden" costs, especially when the work is sent offshore.

5.   *What are new approaches for system building in the digital firm era?*

Companies are turning to rapid application design, joint application design (JAD), agile development, and reusable software components to accelerate the systems development process. Rapid application development (RAD) uses object-oriented software, visual programming, prototyping, and fourth-generation tools for very rapid creation of systems. Agile development breaks a large project into a series of small subprojects that are completed in short periods of time using iteration and continuous feedback. Component-based development expedites application development by grouping objects into suites of software components that can be combined to create large-scale business applications. Web services provide a common set of standards that enable organizations to link their systems regardless of their technology platform through standard plug- and-play architecture

Key Terms

Acceptance testing, 499
Agile development, 514
Automation, 489
Business process management, 491
Business process redesign, 490
Component-based development, 515
Computer-aided software
  engineering (CASE), 505
Conversion, 500
Customization, 510
Data flow diagram (DFD), 502
Direct cutover strategy, 500
Documentation, 501
End-user development, 508
End-user interface, 507
Feasibility study, 497
Fourth-generation languages, 508
Information requirements, 497
Iterative, 507
Joint application design (JAD), 514
Maintenance, 501
Object, 504
Object-oriented development, 504
Offshore outsourcing, 511
Paradigm shift, 490

Parallel strategy, 500
Phased approach strategy, 500
Pilot study strategy, 500
Postimplementation audit, 501
Process specifications, 503
Production, 501
Programming, 499
Prototype, 507
Prototyping, 507
Query languages, 509
Rapid application development (RAD), 514
Rationalization of procedures, 490
Request for Proposal (RFP), 510
Six sigma, 490
Structure chart, 503
Structured, 502
Systems analysis, 496
Systems design, 498
Systems development, 494
Systems life cycle, 506
System testing, 499
Test plan, 500
Testing, 499
Total quality management (TQM), 490
Unit testing, 499

Review Questions

1. How does building new systems produce organizational change?

• Describe each of the four kinds of organizational change that can be promoted with information technology.

• Define business process management and describe the steps required to carry it out.

• Explain how information systems support process changes that promote quality in an organization.

2. What are the core activities in the systems development process?

   - Distinguish between systems analysis and systems design. Describe the activities for each.

   - Define information requirements and explain why they are difficult to determine correctly.

   - Explain why the testing stage of systems development is so important. Name and describe the three stages of testing for an information system.

   - Describe the role of programming, conversion, production, and maintenance in systems development.

3. What are the principal methodologies for modeling and designing systems?

   - Compare object-oriented and traditional structured approaches for modeling and designing systems.

4. What are the alternative methods for building information systems?

   - Define the traditional systems life cycle. Describe each of its steps and its advantages and disadvantages for systems building.

   - Define information system prototyping. Describe its benefits and limitations. List and describe the steps in the prototyping process.

   - Define an application software package. Explain the advantages and disadvantages of developing information systems based on software packages.

   - Define end-user development and describe its advantages and disadvantages. Name some policies and procedures for managing end-user development.

   - Describe the advantages and disadvantages of using outsourcing for building information systems.

5. What are new approaches for system building in the digital firm era?

   - Define rapid application development (RAD) and agile development and explain how they can speed up system-building.

   - Explain how component-based development and Web services help firms build and enhance their information systems.

Discussion Questions

1. Why is selecting a systems development approach an important business decision? Who should participate in the selection process?

2. Some have said that the best way to reduce systems development costs is to use application software packages or fourth-generation tools. Do you agree? Why or why not?

3. Why is is so important to understand how a business process works when trying to develop a new information system?

Video Cases

Video Cases and Instructional Videos illustrating some of the concepts in this chapter are available. Contact your instructor to access these videos.

Collaboration and Teamwork: Preparing Web Site Design Specifications

With three or four of your classmates, select a system described in this text that uses the Web. Review the Web site for the system you select. Use what you have learned from the Web site and the description in this book to prepare a report describing some of the design specifications for the system you select.

If possible, use Google Sites to post links to Web pages, team communication announcements, and work assignments; to brainstorm; and to work collaboratively on project documents. Try to use Google Docs to develop a presentation of your findings for the class.

Are Electronic Medical Records a Cure for Health Care?
CASE STUDY

Creating more efficient health care systems in the United States has been a pressing medical, social, and political issue for decades. Despite the fact that 15 percent of Americans are uninsured and another 20 percent on top of that are underinsured, or unable to pay for necessary health care, the U.S. spends more money per person on health care than any other country in the world. In 2009, the United States spent $2.5 trillion on health care, which was 17.6 percent of its gross domestic product (GDP). Approximately 12 percent of that figure was spent on administrative costs, most of which involve the upkeep of medical records.

The astronomical health care spending figures in the United States are inflated by inefficiency, errors, and fraud. The good news is that information technology may present an opportunity for health care providers to save money and provide better care. Health care providers have begun to create electronic medical record (EMR) systems at the urging of the government in an effort to eliminate much of the inefficiency inherent in paper-based recordkeeping. Many insurance companies are also lending their support to the development of EMR systems.

An electronic medical record system contains all of a person's vital medical data, including personal information, a full medical history, test results, diagnoses, treatments, prescription medications, and the effect of those treatments. A physician would be able to immediately and directly access needed information from the EMR without having to pore through paper files. If the record holder went to the hospital, the records and results of any tests performed at that point would be immediately available online.

Many experts believe that electronic records will reduce medical errors and improve care, create less paperwork, and provide quicker service, all of which will lead to dramatic savings in the future: an estimated $77.8 billion per year. The government's short-term goal is for all health care providers in the United States to have functional EMR systems in place that meet a set of basic functional criteria by the year 2015. Its long-term goal is to have a fully functional nationwide electronic medical recordkeeping network.

Evidence of EMR systems in use today suggests that these benefits are possible for doctors and hospitals, but the challenges of setting up individual systems, let alone a nationwide system, are daunting. Even with stimulus money, many smaller practices are finding it difficult to afford the costs and time commitment for upgrading their recordkeeping systems. In 2010, 80 percent of physicians and 90 percent of hospitals in the United States are still using paper medical records.

It's also unclear whether the systems being developed and implemented in 2010 will be compatible with one another in 2015 and beyond, jeopardizing the goal of a national system where all health care providers can share information. And there are many other smaller obstacles that health care providers, health IT developers, and insurance companies need to overcome for electronic health records to catch on nationally, including patients' privacy concerns, data quality issues, and resistance from health care workers.

The government plans to pay out the stimulus money provided by the American Recovery and Reinvestment Act to health care providers in two ways. First, $2 billion will be provided up front to hospitals and physicians to help set up electronic records. Another $17 billion will also be available as a reward for providers that successfully implement electronic records by 2015. The stimulus specifies that to qualify for these rewards, providers must demonstrate "meaningful use" of electronic health record systems. The bill defines this as the successful implementation of certified e-record products, the ability to write at least 40 percent of their total prescriptions electronically, and the ability to exchange and report data to government health agencies. Individual practices can receive up to $64,000 for successful implementations, and hospitals can make as much as $11.5 million.

But in addition to the reward of stimulus money, the government will also assess penalties on practices that fail to comply with the new electronic recordkeeping standards. Providers that cannot meet the standards by 2015 will have their Medicare and Medicaid reimbursements reduced by 1 percent per year until 2018, with further, more stringent penalties coming beyond that time if a sufficiently low number of providers are using electronic health records.

Electronic medical recordkeeping systems typically cost around $30,000 to $50,000 per doctor. While the stimulus money should eventually be enough to cover that cost, only a small amount of it is available up front. For many providers, especially medical practices with fewer than four doctors and hospitals with fewer than 50 beds, this creates a significant problem. The expenditure of overhauling recordkeeping systems represents a significant increase in the short-term budgets and workloads of smaller health care providers. Smaller providers are also less likely to have started digitizing their records compared to their larger counterparts.

Many smaller practices and hospitals have balked at the transition to EMR systems for these reasons, but the evidence of these systems in action suggests that the move may be well worth the effort. The most prominent example of electronic medical records in use today is the Veterans Affairs (VA) system of doctors and hospitals. The VA system switched to digital records years ago, and far exceeds the private sector and Medicare in quality of preventive services and chronic care. The 1,400 VA facilities use VistA, record-sharing software developed by the government that allows doctors and nurses to share patient histories. A typical VistA record lists a patient's health problems, weight and blood pressure since beginning treatment at the VA, images of the patient's X-rays, lab results, and other test results, lists of medications, and reminders about upcoming appointments.

But VistA is more than a database; it also has many features that improve quality of care. For example, nurses scan tags for patients and medications to ensure that the correct dosages of medicines are going to the correct patients. This feature reduces medication errors, which is one of the most common and costly types of medical errors, and speeds up treatment as well. The system also generates automatic warnings based on specified criteria. It can notify providers if a patient's blood pressure goes over a certain level or if a patient is overdue for a regularly scheduled procedure like a flu shot or a cancer screening. Devices that measure patients' vital signs can automatically transmit their results into the VistA system, which automatically updates doctors at the first sign of trouble.

The results suggest that electronic records offer significant advantages to hospitals and patients alike. The 40,000 patients in the VA's in-home monitoring program reduced their hospital admissions by 25 percent and the length of their hospital stays by 20 percent. More patients receive necessary periodic treatments under VistA (from 27 percent to 83 percent for flu vaccines and from 34 percent to 84 percent for colon cancer screenings).

Patients also report that the process of being treated at the VA is effortless compared to paper-based providers. That's because instant processing of claims and payments are among the benefits of EMR systems. Insurance companies traditionally pay claims about two weeks after receiving them, despite quickly processing them soon after they are received. Additionally, today's paper-based health care providers must assign the appropriate diagnostic codes and procedure codes to claims. Because there are thousands of these codes, the process is even slower, and most providers employ someone solely to perform this task. Electronic systems hold the promise of immediate processing, or real-time claims adjudication, just like when you pay using a credit card—claim data would be sent immediately, and diagnostic and procedure code information are automatically entered.

VistA is far from the only option for doctors and hospitals starting the process of updating their records. Many health IT companies are eagerly awaiting the coming spike in demand for their EMR products and have developed a variety of different health record structures. Humana, Aetna, and other health insurance companies are helping to defray the cost of setting up EMR systems for some doctors and hospitals. Humana has teamed up with health IT company Athenahealth to subsidize EMR systems for approximately 100 primary care practices within Humana's network. Humana pays most of the bill and offers further rewards to practices that meet governmental performance standards. Aetna and IBM, on the other hand, have launched a cloud-based system that pools patient records and is licensed to doctors both inside and outside of Aetna.

There are two problems with the plethora of options available to health care providers. First, there are likely to be many issues with the sharing of medical data between different systems. While the majority of EMR systems are likely to satisfy the specified criteria of reporting data electronically to governmental agencies, they may not be able to report the same data to one another, a key requirement for a nation-wide system. Many fledgling systems are designed using VistA as a guide, but many are not. Even if medical data are easily shared, it's another problem altogether for doctors to actually locate the information they need quickly and easily.

Many EMR systems have no capacity to drill down for more specific data, forcing doctors to wade through large repositories of information to find the one piece of data they need. EMR vendors are developing search engine technology intended for use in medical records. Only after EMR systems become more widespread will the extent of the problems with data sharing and accessibility become clearer.

The second problem is that there is a potential conflict of interest for the insurance companies involved in the creation of health record systems. Insurers are often accused of seeking ways to avoid providing care for sick people. While most insurers are adamant that only doctors and patients will be able to access data in these systems, many prospective patients are skeptical. In 2009, a poll conducted for National Public Radio found that 59 percent of respondents said they doubted the confidentiality of online medical records; even if the systems are secure, the perception of poor privacy

could affect the success of the system and quality of care provided. One in eight Americans have skipped doctor visits or regular tests, asked a doctor to change a test result, or paid privately for a test, motivated mostly by privacy concerns. A poorly designed EMR network would amplify these concerns.

Sources: Katherine Gammon, "Connecting Electronic Medical Records," *Technology Review*, August 9, 2010; Avery Johnson, "Doctors Get Dose of Technology From Insurers," *The Wall Street Journal*, August 8, 2010; David Talbot, "The Doctor Will Record Your Data Now," *Technology Review*, July 23, 2010; Tony Fisher and Joyce Norris-Montanari, "The Current State of Data in Health Care," Information-Management.com, June 15, 2010; Laura Landro, "Breaking Down the Barriers," *The Wall Street Journal*, April 13, 2010; Jacob Goldstein, "Can Technology Cure Health Care?", *The Wall Street Journal*, April 13, 2010; Deborah C. Peel, "Your Medical Records Aren't Secure," *The Wall Street Journal*, March 23, 2010; Jane E. Brody, "Medical Paper Trail Takes Electronic Turn," *The New York Times*, February 23, 2010; and Laura Landro, "An Affordable Fix for Modernizing Medical Records," *The Wall Street Journal*, April 30, 2009.

CASE STUDY QUESTIONS

1. What management, organization, and technology factors are responsible for the difficulties in building electronic medical record systems? Explain your answer.

2. What stages of system building will be the most difficult when creating electronic medical record systems? Explain your answer.

3. What is the business and social impact of not digitizing medical records (to individual physicians, hospitals, insurers, patients)?

4. What are the business and social benefits of digitizing medical recordkeeping?

5. Name two important information requirements for physicians, two for patients, and two for hospitals that should be addressed by electronic medical records systems.

6. Diagram the "as-is" and "to-be" processes for prescribing a medication for a patient before and after an EMR system is implemented.

Chapter 14

Managing Projects

LEARNING OBJECTIVES

After reading this chapter, you will be able to answer the following questions:

1. What are the objectives of project management and why is it so essential in developing information systems?

2. What methods can be used for selecting and evaluating information systems projects and aligning them with the firm's business goals?

3. How can firms assess the business value of information systems projects?

4. What are the principal risk factors in information systems projects?

5. What strategies are useful for managing project risk and system implementation?

Interactive Sessions:

DST Systems Scores with Scrum and Application Life Cycle Management

Motorola Turns to Project Portfolio Management

CHAPTER OUTLINE

14.1 THE IMPORTANCE OF PROJECT MANAGEMENT
Runaway Projects and System Failure
Project Management Objectives

14.2 SELECTING PROJECTS
Management Structure for Information Systems Projects
Linking Systems Projects to the Business Plan
Critical Success Factors
Portfolio Analysis
Scoring Models

14.3 ESTABLISHING THE BUSINESS VALUE OF INFORMATION SYSTEMS
Information System Costs and Benefits
Real Options Pricing Models
Limitations of Financial Models

14.4 MANAGING PROJECT RISK
Dimensions of Project Risk
Change Management and the Concept of Implementation
Controlling Risk Factors
Designing for the Organization
Project Management Software Tools

14.5 HANDS-ON MIS PROJECTS
Management Decision Problems
Improving Decision Making: Using Spreadsheet Software for Capital Budgeting for a New CAD System
Improving Decision Making: Using Web Tools for Buying and Financing a Home

LEARNING TRACK MODULES
Capital Budgeting Methods for Information System Investments
Information Technology Investments and Productivity
Enterprise Analysis (Business Systems Planning)

COCA-COLA: "OPENING HAPPINESS" WITH A NEW PROJECT MANAGEMENT SYSTEM

The Coca-Cola Company is the world's leading owner and marketer of nonalcoholic beverage brands and the world's largest manufacturer, distributor, and marketer of concentrates and syrups used to produce nonalcoholic beverages. Coke sells its concentrates to independent bottlers in 200 countries. Arguably, Coke is the most valuable brand in the world. In fact, Coke owns 12 brands that sell more than $1 billion a year. It's corporate branding handle in 2010 is "Open happiness." Coke revenues were $30.9 billion in 2009.

Coca-Cola Bottling Co. Consolidated ("Coke Bottling") is the second largest Coca-Cola bottler in the United States. The company operates in 11 states in the Southeast and has revenues of $1.5 billion. The company has hundreds of projects under management at any point in time. Coke Bottling had been using an older project management software tool to coordinate these projects, but by 2010 it lacked many of the features that good project managers want. Not all projects in the company used the system, and information about these projects was spread across many legacy systems. The software could not track cost elements, such as labor and material cost, in one repository. Senior management wanted cost details and capital requirements for projects that could not be delivered. Project teams would typically need to go back and ask senior management for more money because projects routinely exceeded their budgets. Time and money were wasted gathering data from several locations and performing ad hoc analyses on spreadsheets. The software was unable to report on compliance of projects with various federal laws, including Sarbanes-Oxley.

Management wanted a new project management tool that could track all projects in the firm, utilize existing SAP databases and reporting tools, and integrate with its Microsoft Server environment. Coke Bottling chose the Microsoft Office Enterprise Project Management (EPM) Solution, which includes Microsoft Office Project Portfolio Server 2007, Microsoft Office Project Server 2007, and Microsoft Office Project Professional 2007. The hope was to simplify the firm's software footprint to consist primarily of SAP and Microsoft products and thereby reduce maintenance costs.

The EPM is integrated with Windows SharePoint Services so that users can update project information, manage documents, and track risks and issues using common SharePoint sites, known as project workspaces. For training employees and help implementing the system, Coke Bottling hired Project Solutions Group, a consulting and training firm in Marlborough, Massachusetts.

A number of benefits have flowed from this choice of project management software. For the first time, the company has a centralized repository of the cash flow and capital requirements of projects. This helps reduce its financing costs. With the EPM solution, managers can request the amount of capital they need with a high degree of accuracy from the start of a project. The firm can manage its human resources and schedules more effectively because it now knows who is working on which projects. Based on the

number of hours people spend on tasks, resource managers can see whether they have the right spread of resources, and they can take decisive and informed action by seeing where people are spending their time.

The firm also implemented a Project Gate methodology that consists of five gates: qualify need, define, design, build/test, and deploy/measure. In the past, managers just used checklists to manage projects, and there was no consistency across projects or managers. The gate methodology ensures all projects go through the same management process. To ensure enterprise-wide implementation of its new EPM solution, Coke Bottling created a new project management office to bring consistency and structure to all the firm's projects.

Sources: Microsoft Corporation, "Microsoft Case Studies. Coca-Cola Bottling Co. Improves Project Cost Reporting," August 2009, www.microsoft.com/casestudies, accessed November 5, 2010; Microsoft Corporation, "Microsoft Enterprise Project Management (EPM) Solution." www.microsoft.com/project., accessed November 5, 2010; and The Coca-Cola Company, Form 10K for the Fiscal Year Ending December 31, 2009, filed with the Securities and Exchange Commission, February 26, 2010.

One of the principal challenges posed by information systems is ensuring they deliver genuine business benefits. Many information systems projects don't succeed because organizations incorrectly assess their business value or because firms fail to manage the organizational change surrounding the introduction of new technology.

Coke Bottling's management knew this when it implemented its enterprise project management system. The new system involved an enterprise-wide change in management and organizational behavior, in addition to careful introduction of an entire set of software tools. Coke Bottling succeeded in this project because it took a balanced view of the management, organizational, and technical changes needed.

The chapter-opening diagram calls attention to important points raised by this case and this chapter. Coke Bottling manages several hundred projects each year. The existing software was unable to account for costs, predict financial needs, comply with federal regulations and due diligence requests, and allocate resources efficiently. This increased the likelihood of project failure, and raised the costs of company operations. The company was able to improve project inventory management by implementing an enterprise-wide project management software tool that was tightly integrated with its existing enterprise database environment and its desktop software. Management was wise enough to also change the organization by creating a Project Management Office, and develop a new set of management practices to ensure the software performed up to expectations.

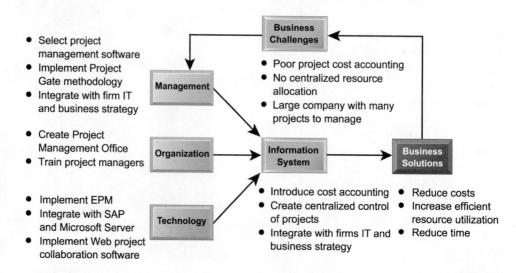

14.1 THE IMPORTANCE OF PROJECT MANAGEMENT

There is a very high failure rate among information systems projects. In nearly every organization, information systems projects take much more time and money to implement than originally anticipated or the completed system does not work properly. When an information system does not meet expectations or costs too much to develop, companies may not realize any benefit from their information system investment, and the system may not be able to solve the problems for which it was intended. The development of a new system must be carefully managed and orchestrated, and the way a project is executed is likely to be the most important factor influencing its outcome. That's why it's essential to have some knowledge about managing information systems projects and the reasons why they succeed or fail.

RUNAWAY PROJECTS AND SYSTEM FAILURE

How badly are projects managed? On average, private sector projects are underestimated by one-half in terms of budget and time required to deliver the complete system promised in the system plan. Many projects are delivered with missing functionality (promised for delivery in later versions). The Standish Group consultancy, which monitors IT project success rates, found that only 29 percent of all technology investments were completed on time, on budget, and with all features and functions originally specified (Levinson, 2006). A 2007 Tata Consultancy Services study of IT effectiveness reported similar findings (Blair, 2010). Between 30 and 40 percent of all software projects are "runaway" projects that far exceed the original schedule and budget projections and fail to perform as originally specified.

As illustrated in Figure 14-1, a systems development project without proper management will most likely suffer these consequences:

- Costs that vastly exceed budgets
- Unexpected time slippage
- Technical performance that is less than expected
- Failure to obtain anticipated benefits

The systems produced by failed information projects are often not used in the way they were intended, or they are not used at all. Users often have to develop parallel manual systems to make these systems work.

The actual design of the system may fail to capture essential business requirements or improve organizational performance. Information may not be provided quickly enough to be helpful, it may be in a format that is impossible to digest and use, or it may represent the wrong pieces of data.

FIGURE 14-1   CONSEQUENCES OF POOR PROJECT MANAGEMENT

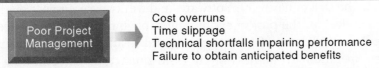

Without proper management, a systems development project takes longer to complete and most often exceeds the allocated budget. The resulting information system most likely is technically inferior and may not be able to demonstrate any benefits to the organization.

The way in which nontechnical business users must interact with the system may be excessively complicated and discouraging. A system may be designed with a poor user interface. The user interface is the part of the system with which end users interact. For example, an online input form or data entry screen may be so poorly arranged that no one wants to submit data or request information. System outputs may be displayed in a format that is too difficult to comprehend.

Web sites may discourage visitors from exploring further if the Web pages are cluttered and poorly arranged, if users cannot easily find the information they are seeking, or if it takes too long to access and display the Web page on the user's computer.

Additionally, the data in the system may have a high level of inaccuracy or inconsistency. The information in certain fields may be erroneous or ambiguous, or it may not be organized properly for business purposes. Information required for a specific business function may be inaccessible because the data are incomplete.

PROJECT MANAGEMENT OBJECTIVES

A **project** is a planned series of related activities for achieving a specific business objective. Information systems projects include the development of new information systems, enhancement of existing systems, or upgrade or replacement of the firm's information technology (IT) infrastructure.

Project management refers to the application of knowledge, skills, tools, and techniques to achieve specific targets within specified budget and time constraints. Project management activities include planning the work, assessing risk, estimating resources required to accomplish the work, organizing the work, acquiring human and material resources, assigning tasks, directing activities, controlling project execution, reporting progress, and analyzing the results. As in other areas of business, project management for information systems must deal with five major variables: scope, time, cost, quality, and risk.

Scope defines what work is or is not included in a project. For example, the scope of project for a new order processing system might be to include new modules for inputting orders and transmitting them to production and accounting but not any changes to related accounts receivable, manufacturing, distribution, or inventory control systems. Project management defines all the work required to complete a project successfully, and should ensure that the scope of a project does not expand beyond what was originally intended.

Time is the amount of time required to complete the project. Project management typically establishes the amount of time required to complete major components of a project. Each of these components is further broken down into activities and tasks. Project management tries to determine the time required to complete each task and establish a schedule for completing the work.

Cost is based on the time to complete a project multiplied by the cost of human resources required to complete the project. Information systems project costs also include the cost of hardware, software, and work space. Project management develops a budget for the project and monitors ongoing project expenses.

Quality is an indicator of how well the end result of a project satisfies the objectives specified by management. The quality of information systems projects usually boils down to improved organizational performance and decision making. Quality also considers the accuracy and timeliness of information produced by the new system and ease of use.

Risk refers to potential problems that would threaten the success of a project. These potential problems might prevent a project from achieving its objectives by increasing time and cost, lowering the quality of project outputs, or preventing the project from being completed altogether. Section 14.3 describes the most important risk factors for information systems.

14.2 SELECTING PROJECTS

Companies typically are presented with many different projects for solving problems and improving performance. There are far more ideas for systems projects than there are resources. Firms will need to select from this group the projects that promise the greatest benefit to the business. Obviously, the firm's overall business strategy should drive project selection.

MANAGEMENT STRUCTURE FOR INFORMATION SYSTEMS PROJECTS

Figure 14-2 shows the elements of a management structure for information systems projects in a large corporation. It helps ensure that the most important projects are given priority.

At the apex of this structure is the corporate strategic planning group and the information system steering committee. The corporate strategic planning group is responsible for developing the firm's strategic plan, which may require the development of new systems.

The information systems steering committee is the senior management group with responsibility for systems development and operation. It is composed of

FIGURE 14-2   MANAGEMENT CONTROL OF SYSTEMS PROJECTS

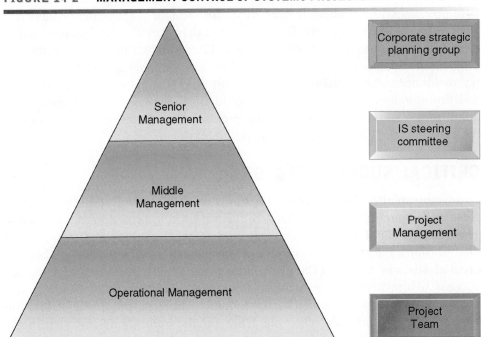

Each level of management in the hierarchy is responsible for specific aspects of systems projects, and this structure helps give priority to the most important systems projects for the organization.

department heads from both end-user and information systems areas. The steering committee reviews and approves plans for systems in all divisions, seeks to coordinate and integrate systems, and occasionally becomes involved in selecting specific information systems projects.

The project team is supervised by a project management group composed of information systems managers and end-user managers responsible for overseeing several specific information systems projects. The project team is directly responsible for the individual systems project. It consists of systems analysts, specialists from the relevant end-user business areas, application programmers, and perhaps database specialists. The mix of skills and the size of the project team depend on the specific nature of the system solution.

LINKING SYSTEMS PROJECTS TO THE BUSINESS PLAN

In order to identify the information systems projects that will deliver the most business value, organizations need to develop an **information systems plan** that supports their overall business plan and in which strategic systems are incorporated into top-level planning. The plan serves as a road map indicating the direction of systems development (the purpose of the plan), the rationale, the current systems/situation, new developments to consider, the management strategy, the implementation plan, and the budget (see Table 14-1).

The plan contains a statement of corporate goals and specifies how information technology will support the attainment of those goals. The report shows how general goals will be achieved by specific systems projects. It identifies specific target dates and milestones that can be used later to evaluate the plan's progress in terms of how many objectives were actually attained in the time frame specified in the plan. The plan indicates the key management decisions concerning hardware acquisition; telecommunications; centralization/decentralization of authority, data, and hardware; and required organizational change. Organizational changes are also usually described, including management and employee training requirements, recruiting efforts, changes in business processes, and changes in authority, structure, or management practice.

In order to plan effectively, firms will need to inventory and document all of their information system applications and IT infrastructure components. For projects in which benefits involve improved decision making, managers should try to identify the decision improvements that would provide the greatest additional value to the firm. They should then develop a set of metrics to quantify the value of more timely and precise information on the outcome of the decision (see Chapter 12 for more detail on this topic).

CRITICAL SUCCESS FACTORS

To develop an effective information systems plan, the organization must have a clear understanding of both its long- and short-term information requirements. The strategic analysis, or critical success factors, approach argues that an organization's information requirements are determined by a small number of **critical success factors (CSFs)** of managers. If these goals can be attained, success of the firm or organization is assured (Rockart, 1979; Rockart and Treacy, 1982). CSFs are shaped by the industry, the firm, the manager, and the broader environment. For example, CSFs for the automobile industry might include styling, quality, and cost to meet the goals of increasing market share and raising profits. New information systems should focus on providing information that helps the firm meet these goals.

TABLE 14-1 INFORMATION SYSTEMS PLAN

1. Purpose of the Plan
   Overview of plan contents
   Current business organization and future organization
   Key business processes
   Management strategy

2. Strategic Business Plan Rationale
   Current situation
   Current business organization
   Changing environments
   Major goals of the business plan
   Firm's strategic plan

3. Current Systems
   Major systems supporting business functions and processes
   Current infrastructure capabilities
      Hardware
      Software
      Database
      Telecommunications and Internet
   Difficulties meeting business requirements
   Anticipated future demands

4. New Developments
   New system projects
      Project descriptions
      Business rationale
      Applications' role in strategy
   New infrastructure capabilities required
      Hardware
      Software
      Database
      Telecommunications and Internet

5. Management Strategy
   Acquisition plans
   Milestones and timing
   Organizational realignment
   Internal reorganization
   Management controls
   Major training initiatives
   Personnel strategy

6. Implementation Plan
   Anticipated difficulties in implementation
   Progress reports

7. Budget Requirements
   Requirements
   Potential savings
   Financing
   Acquisition cycle

The principal method used in CSF analysis is personal interviews—three or four—with a number of top managers identifying their goals and the resulting CSFs. These personal CSFs are aggregated to develop a picture of the firm's CSFs. Then systems are built to deliver information on these CSFs. (For the method of developing CSFs in an organization, see Figure 14-3.)

Only top managers are interviewed, and the questions focus on a small number of CSFs rather than requiring a broad inquiry into what information is used in the organization. It is especially suitable for top management and for the development of decision-support systems (DSS) and executive support systems (ESS). The CSF method focuses organizational attention on how information should be handled.

The method's primary weakness is that there is no particularly rigorous way in which individual CSFs can be aggregated into a clear company pattern. In addition, interviewees (and interviewers) often become confused when distinguishing between *individual* and *organizational* CSFs. These types of CSFs are not necessarily the same. What may be considered critical to a manager may not be important for the organization as a whole. This method is clearly biased toward top managers, although it could be extended to elicit ideas for promising new systems from lower-level members of the organization (Peffers and Gengler, 2003).

PORTFOLIO ANALYSIS

Once strategic analyses have determined the overall direction of systems development, **portfolio analysis** can be used to evaluate alternative system projects. Portfolio analysis inventories all of the organization's information systems projects and assets, including infrastructure, outsourcing contracts,

FIGURE 14-3   USING CSFs TO DEVELOP SYSTEMS

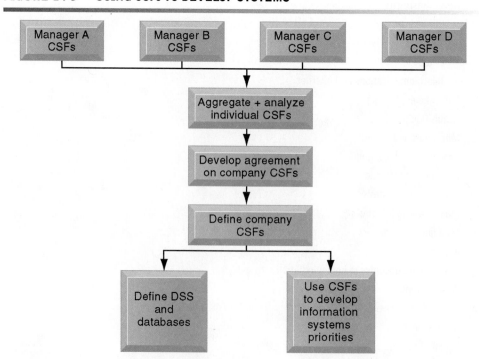

The CSF approach relies on interviews with key managers to identify their CSFs. Individual CSFs are aggregated to develop CSFs for the entire firm. Systems can then be built to deliver information on these CSFs.

and licenses. This portfolio of information systems investments can be described as having a certain profile of risk and benefit to the firm (see Figure 14-4) similar to a financial portfolio.

Each information systems project carries its own set of risks and benefits. (Section 14-4 describes the factors that increase the risks of systems projects.) Firms would try to improve the return on their portfolios of IT assets by balancing the risk and return from their systems investments. Although there is no ideal profile for all firms, information-intensive industries (e.g., finance) should have a few high-risk, high-benefit projects to ensure that they stay current with technology. Firms in non-information-intensive industries should focus on high-benefit, low-risk projects.

Most desirable, of course, are systems with high benefit and low risk. These promise early returns and low risks. Second, high-benefit, high-risk systems should be examined; low-benefit, high-risk systems should be totally avoided; and low-benefit, low-risk systems should be reexamined for the possibility of rebuilding and replacing them with more desirable systems having higher benefits. By using portfolio analysis, management can determine the optimal mix of investment risk and reward for their firms, balancing riskier high-reward projects with safer lower-reward ones. Firms where portfolio analysis is aligned with business strategy have been found to have a superior return on their IT assets, better alignment of IT investments with business objectives, and better organization-wide coordination of IT investments (Jeffrey and Leliveld, 2004).

SCORING MODELS

A **scoring model** is useful for selecting projects where many criteria must be considered. It assigns weights to various features of a system and then calculates the weighted totals. Using Table 14-2, the firm must decide among two alternative enterprise resource planning (ERP) systems. The first column lists the criteria that decision makers will use to evaluate the systems. These criteria are usually the result of lengthy discussions among the decision-making group. Often the most important outcome of a scoring model is not the score but agreement on the criteria used to judge a system.

Table 14-2 shows that this particular company attaches the most importance to capabilities for sales order processing, inventory management, and warehousing. The second column in Table 14-2 lists the weights that decision

FIGURE 14-4   A SYSTEM PORTFOLIO

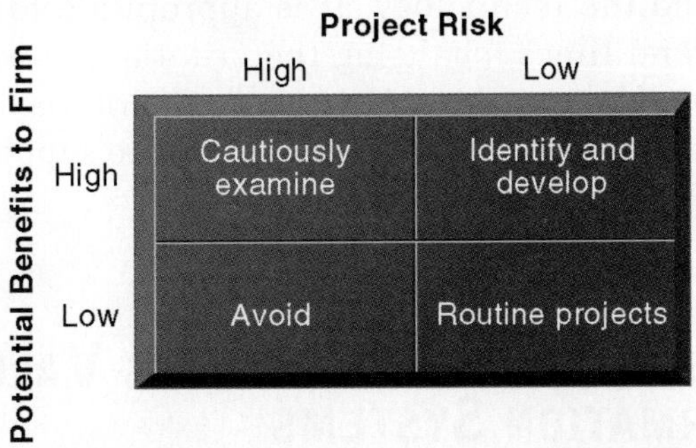

Companies should examine their portfolio of projects in terms of potential benefits and likely risks. Certain kinds of projects should be avoided altogether and others developed rapidly. There is no ideal mix. Companies in different industries have different profiles.

TABLE 14-2 EXAMPLE OF A SCORING MODEL FOR AN ERP SYSTEM

CRITERIA	WEIGHT	ERP SYSTEM A %	ERP SYSTEM A SCORE	ERP SYSTEM B %	ERP SYSTEM B SCORE
1.0 Order Processing					
1.1 Online order entry	4	67	268	73	292
1.2 Online pricing	4	81	324	87	348
1.3 Inventory check	4	72	288	81	324
1.4 Customer credit check	3	66	198	59	177
1.5 Invoicing	4	73	292	82	328
Total Order Processing			1,370		1,469
2.0 Inventory Management					
2.1 Production forecasting	3	72	216	76	228
2.2 Production planning	4	79	316	81	324
2.3 Inventory control	4	68	272	80	320
2.4 Reports	3	71	213	69	207
Total Inventory Management			1,017		1,079
3.0 Warehousing					
3.1 Receiving	2	71	142	75	150
3.2 Picking/packing	3	77	231	82	246
3.3 Shipping	4	92	368	89	356
Total Warehousing			741		752
Grand Total			3,128		3,300

makers attached to the decision criteria. Columns 3 and 5 show the percentage of requirements for each function that each alternative ERP system can provide. Each vendor's score can be calculated by multiplying the percentage of requirements met for each function by the weight attached to that function. ERP System B has the highest total score.

As with all "objective" techniques, there are many qualitative judgments involved in using the scoring model. This model requires experts who understand the issues and the technology. It is appropriate to cycle through the scoring model several times, changing the criteria and weights, to see how sensitive the outcome is to reasonable changes in criteria. Scoring models are used most commonly to confirm, to rationalize, and to support decisions, rather than as the final arbiters of system selection.

14.3 ESTABLISHING THE BUSINESS VALUE OF INFORMATION SYSTEMS

Even if a system project supports a firm's strategic goals and meets user information requirements, it needs to be a good investment for the firm. The value of systems from a financial perspective essentially revolves around the issue of return on invested capital. Does a particular information system investment produce sufficient returns to justify its costs?

INFORMATION SYSTEM COSTS AND BENEFITS

Table 14-3 lists some of the more common costs and benefits of systems. **Tangible benefits** can be quantified and assigned a monetary value. **Intangible benefits**, such as more efficient customer service or enhanced decision making, cannot be immediately quantified but may lead to quantifiable gains in the long run. Transaction and clerical systems that displace labor and save space always produce more measurable, tangible benefits than management information systems, decision-support systems, and computer-supported collaborative work systems (see Chapters 2 and 11).

TABLE 14-3 COSTS AND BENEFITS OF INFORMATION SYSTEMS

COSTS

Hardware
Telecommunications
Software
Services
Personnel

TANGIBLE BENEFITS (COST SAVINGS)

Increased productivity

Lower operational costs

Reduced workforce

Lower computer expenses

Lower outside vendor costs

Lower clerical and professional costs

Reduced rate of growth in expenses

Reduced facility costs

INTANGIBLE BENEFITS

Improved asset utilization
Improved resource control
Improved organizational planning
Increased organizational flexibility
More timely information
More information
Increased organizational learning
Legal requirements attained
Enhanced employee goodwill
Increased job satisfaction
Improved decision making
Improved operations
Higher client satisfaction
Better corporate image

Chapter 5 introduced the concept of total cost of ownership (TCO), which is designed to identify and measure the components of information technology expenditures beyond the initial cost of purchasing and installing hardware and software. However, TCO analysis provides only part of the information needed to evaluate an information technology investment because it does not typically deal with benefits, cost categories such as complexity costs, and "soft" and strategic factors discussed later in this section.

Capital Budgeting for Information Systems

To determine the benefits of a particular project, you'll need to calculate all of its costs and all of its benefits. Obviously, a project where costs exceed benefits should be rejected. But even if the benefits outweigh the costs, additional financial analysis is required to determine whether the project represents a good return on the firm's invested capital. **Capital budgeting** models are one of several techniques used to measure the value of investing in long-term capital investment projects.

Capital budgeting methods rely on measures of cash flows into and out of the firm; capital projects generate those cash flows. The investment cost for information systems projects is an immediate cash outflow caused by expenditures for hardware, software, and labor. In subsequent years, the investment may cause additional cash outflows that will be balanced by cash inflows resulting from the investment. Cash inflows take the form of increased sales of more products (for reasons such as new products, higher quality, or increasing market share) or reduced costs in production and operations. The difference between cash outflows and cash inflows is used for calculating the financial worth of an investment. Once the cash flows have been established, several alternative methods are available for comparing different projects and deciding about the investment.

The principal capital budgeting models for evaluating IT projects are: the payback method, the accounting rate of return on investment (ROI), net present value, and the internal rate of return (IRR). You can find out more about how these capital budgeting models are used to justify information system investments in the Learning Tracks for this chapter.

REAL OPTIONS PRICING MODELS

Some information systems projects are highly uncertain, especially investments in IT infrastructure. Their future revenue streams are unclear and their up-front costs are high. Suppose, for instance, that a firm is considering a $20 million investment to upgrade its IT infrastructure—its hardware, software, data management tools, and networking technology. If this upgraded infrastructure were available, the organization would have the technology capabilities to respond more easily to future problems and opportunities. Although the costs of this investment can be calculated, not all of the benefits of making this investment can be established in advance. But if the firm waits a few years until the revenue potential becomes more obvious, it might be too late to make the infrastructure investment. In such cases, managers might benefit from using real options pricing models to evaluate information technology investments.

Real options pricing models (ROPMs) use the concept of options valuation borrowed from the financial industry. An *option* is essentially the right, but not the obligation, to act at some future date. A typical *call option*, for instance, is a financial option in which a person buys the right (but not the obligation) to purchase an underlying asset (usually a stock) at a fixed price (strike price) on or before a given date.

For instance, let's assume that on October 15, 2010, you could purchase a call option for $14.25 that would give you the right to buy a share of P&G common stock for $50 per share on a certain date. Options expire over time, and this call option has a maturity date in December. If the price of P&G stock does not rise above $50 per share by the end of December, you would not exercise the option, and the value of the option would fall to zero on the strike date. If, however, the price of P&G common stock rose to, say, $100 per share, you could purchase the stock for the strike price of $50 and retain the profit of $50 per share minus the cost of the option. (Because the option is sold as a 100-share contract, the cost of the contract would be 100 × $14.25 before commissions, or $1,425, and you would be purchasing and obtaining a profit from 100 shares of Procter & Gamble.) The stock option enables the owner to benefit from the upside potential of an opportunity while limiting the downside risk.

ROPMs value information systems projects similar to stock options, where an initial expenditure on technology creates the right, but not the obligation, to obtain the benefits associated with further development and deployment of the technology as long as management has the freedom to cancel, defer, restart, or expand the project. ROPMs give managers the flexibility to stage their IT investment or test the waters with small pilot projects or prototypes to gain more knowledge about the risks of a project before investing in the entire implementation. The disadvantages of this model are primarily in estimating all the key variables affecting option value, including anticipated cash flows from the underlying asset and changes in the cost of implementation. Models for determining option value of information technology platforms are being developed (Fichman, 2004; McGrath and MacMillan, 2000).

LIMITATIONS OF FINANCIAL MODELS

The traditional focus on the financial and technical aspects of an information system tends to overlook the social and organizational dimensions of information systems that may affect the true costs and benefits of the investment. Many companies' information systems investment decisions do not adequately consider costs from organizational disruptions created by a new system, such as the cost to train end users, the impact that users' learning curves for a new system have on productivity, or the time managers need to spend overseeing new system-related changes. Benefits, such as more timely decisions from a new system or enhanced employee learning and expertise, may also be overlooked in a traditional financial analysis (Ryan, Harrison, and Schkade, 2002).

14.4 MANAGING PROJECT RISK

We have already introduced the topic of information system risks and risk assessment in Chapter 8. In this chapter, we describe the specific risks to information systems projects and show what can be done to manage them effectively.

DIMENSIONS OF PROJECT RISK

Systems differ dramatically in their size, scope, level of complexity, and organizational and technical components. Some systems development projects are more likely to create the problems we have described earlier or to suffer delays because they carry a much higher level of risk than others. The level of project risk is influenced by project size, project structure, and the level of technical expertise of the information systems staff and project team.

- *Project size.* The larger the project—as indicated by the dollars spent, the size of the implementation staff, the time allocated for implementation, and the number of organizational units affected—the greater the risk. Very large-scale systems projects have a failure rate that is 50 to 75 percent higher than that for other projects because such projects are complex and difficult to control. The organizational complexity of the system—how many units and groups use it and how much it influences business processes—contribute to the complexity of large-scale systems projects just as much as technical characteristics, such as the number of lines of program code, length of project, and budget (Xia and Lee, 2004; Concours Group, 2000; Laudon, 1989). In addition, there are few reliable techniques for estimating the time and cost to develop large-scale information systems.

- *Project structure.* Some projects are more highly structured than others. Their requirements are clear and straightforward so outputs and processes can be easily defined. Users know exactly what they want and what the system should do; there is almost no possibility of the users changing their minds. Such projects run a much lower risk than those with relatively undefined, fluid, and constantly changing requirements; with outputs that cannot be fixed easily because they are subject to users' changing ideas; or with users who cannot agree on what they want.

- *Experience with technology.* The project risk rises if the project team and the information system staff lack the required technical expertise. If the team is unfamiliar with the hardware, system software, application software, or database management system proposed for the project, it is highly likely that the project will experience technical problems or take more time to complete because of the need to master new skills.

Although the difficulty of the technology is one risk factor in information systems projects, the other factors are primarily organizational, dealing with the complexity of information requirements, the scope of the project, and how many parts of the organization will be affected by a new information system.

CHANGE MANAGEMENT AND THE CONCEPT OF IMPLEMENTATION

The introduction or alteration of an information system has a powerful behavioral and organizational impact. Changes in the way that information is defined, accessed, and used to manage the organization's resources often lead to new distributions of authority and power. This internal organizational change breeds resistance and opposition and can lead to the demise of an otherwise good system.

A very large percentage of information systems projects stumble because the process of organizational change surrounding system building was not properly addressed. Successful system building requires careful **change management**.

The Concept of Implementation

To manage the organizational change surrounding the introduction of a new information system effectively, you must examine the process of implementation. **Implementation** refers to all organizational activities working toward the adoption, management, and routinization of an innovation, such as a new information system. In the implementation process, the systems analyst is a **change agent**. The analyst not only develops technical solutions but also redefines the configurations, interactions, job activities, and power relationships of various organizational groups. The analyst is the catalyst for the entire change process and is responsible for ensuring that all parties involved accept the changes created by a new system. The change agent communicates with users, mediates between competing interest groups, and ensures that the organizational adjustment to such changes is complete.

The Role of End Users

System implementation generally benefits from high levels of user involvement and management support. User participation in the design and operation of information systems has several positive results. First, if users are heavily involved in systems design, they have more opportunities to mold the system according to their priorities and business requirements, and more opportunities to control the outcome. Second, they are more likely to react positively to the completed system because they have been active participants in the change process. Incorporating user knowledge and expertise leads to better solutions.

The relationship between users and information systems specialists has traditionally been a problem area for information systems implementation efforts. Users and information systems specialists tend to have different backgrounds, interests, and priorities. This is referred to as the **user-designer communications gap**. These differences lead to divergent organizational loyalties, approaches to problem solving, and vocabularies.

Information systems specialists, for example, often have a highly technical, or machine, orientation to problem solving. They look for elegant and sophisticated technical solutions in which hardware and software efficiency is optimized at the expense of ease of use or organizational effectiveness. Users prefer systems that are oriented toward solving business problems or facilitating organizational tasks. Often the orientations of both groups are so at odds that they appear to speak in different tongues.

These differences are illustrated in Table 14-4, which depicts the typical concerns of end users and technical specialists (information systems designers) regarding the development of a new information system. Communication problems between end users and designers are a major reason why user requirements are not properly incorporated into information systems and why users are driven out of the implementation process.

Systems development projects run a very high risk of failure when there is a pronounced gap between users and technical specialists and when these groups continue to pursue different goals. Under such conditions, users are often

TABLE 14-4 THE USER-DESIGNER COMMUNICATIONS GAP

USER CONCERNS	DESIGNER CONCERNS
Will the system deliver the information I need for my work?	How much disk storage space will the master file consume?
How quickly can I access the data?	How many lines of program code will it take to perform this function?
How easily can I retrieve the data?	How can we cut down on CPU time when we run the system?
How much clerical support will I need to enter data into the system?	What is the most efficient way of storing these data?
How will the operation of the system fit into my daily business schedule?	What database management system should we use?

driven away from the project. Because they cannot comprehend what the technicians are saying, users conclude that the entire project is best left in the hands of the information specialists alone.

Management Support and Commitment

If an information systems project has the backing and commitment of management at various levels, it is more likely to be perceived positively by both users and the technical information services staff. Both groups will believe that their participation in the development process will receive higher-level attention and priority. They will be recognized and rewarded for the time and effort they devote to implementation. Management backing also ensures that a systems project receives sufficient funding and resources to be successful. Furthermore, to be enforced effectively, all the changes in work habits and procedures and any organizational realignments associated with a new system depend on management backing. If a manager considers a new system a priority, the system will more likely be treated that way by his or her subordinates.

Change Management Challenges for Business Process Reengineering, Enterprise Applications, and Mergers and Acquisitions

Given the challenges of innovation and implementation, it is not surprising to find a very high failure rate among enterprise application and business process reengineering (BPR) projects, which typically require extensive organizational change and which may require replacing old technologies and legacy systems that are deeply rooted in many interrelated business processes. A number of studies have indicated that 70 percent of all business process reengineering projects fail to deliver promised benefits. Likewise, a high percentage of enterprise applications fail to be fully implemented or to meet the goals of their users even after three years of work.

Many enterprise application and reengineering projects have been undermined by poor implementation and change management practices that failed to address employees' concerns about change. Dealing with fear and anxiety throughout the organization, overcoming resistance by key managers, changing job functions, career paths, and recruitment practices have posed greater threats to reengineering than the difficulties companies faced visualizing and designing breakthrough changes to business processes. All of the enterprise applications require tighter coordination among different functional groups as well as extensive business process change (see Chapter 9).

Projects related to mergers and acquisitions have a similar failure rate. Mergers and acquisitions are deeply affected by the organizational characteristics of the merging companies as well as by their IT infrastructures. Combining the information systems of two different companies usually requires considerable organizational change and complex systems projects to manage. If the integration is not properly managed, firms can emerge with a tangled hodgepodge of inherited legacy systems built by aggregating the systems of one firm after another. Without a successful systems integration, the benefits anticipated from the merger cannot be realized, or, worse, the merged entity cannot execute its business processes effectively.

CONTROLLING RISK FACTORS

Various project management, requirements gathering, and planning methodologies have been developed for specific categories of implementation problems. Strategies have also been devised for ensuring that users play appropriate roles throughout the implementation period and for managing the organizational change process. Not all aspects of the implementation process can be easily controlled or planned. However, anticipating potential implementation problems and applying appropriate corrective strategies can increase the chances for system success.

The first step in managing project risk involves identifying the nature and level of risk confronting the project (Schmidt et al., 2001). Implementers can then handle each project with the tools and risk-management approaches geared to its level of risk (Iversen, Mathiassen, and Nielsen, 2004; Barki, Rivard, and Talbot, 2001; McFarlan, 1981).

Managing Technical Complexity

Projects with challenging and complex technology for users to master benefit from **internal integration tools**. The success of such projects depends on how well their technical complexity can be managed. Project leaders need both heavy technical and administrative experience. They must be able to anticipate problems and develop smooth working relationships among a predominantly technical team. The team should be under the leadership of a manager with a strong technical and project management background, and team members should be highly experienced. Team meetings should take place frequently. Essential technical skills or expertise not available internally should be secured from outside the organization.

Formal Planning and Control Tools

Large projects benefit from appropriate use of **formal planning tools** and **formal control tools** for documenting and monitoring project plans. The two most commonly used methods for documenting project plans are Gantt charts and PERT charts. A **Gantt chart** lists project activities and their corresponding start and completion dates. The Gantt chart visually represents the timing and duration of different tasks in a development project as well as their human resource requirements (see Figure 14-5). It shows each task as a horizontal bar whose length is proportional to the time required to complete it.

Although Gantt charts show when project activities begin and end, they don't depict task dependencies, how one task is affected if another is behind schedule, or how tasks should be ordered. That is where **PERT charts** are useful. PERT stands for Program Evaluation and Review Technique, a methodology developed by the U.S. Navy during the 1950s to manage the Polaris submarine missile program. A PERT chart graphically depicts project tasks and their interrelationships. The PERT chart lists the specific activities that make up a project and the activities that must be completed before a specific activity can start, as illustrated in Figure 14-6.

The PERT chart portrays a project as a network diagram consisting of numbered nodes (either circles or rectangles) representing project tasks. Each node is numbered and shows the task, its duration, the starting date, and the completion date. The direction of the arrows on the lines indicates the sequence of tasks and shows which activities must be completed before the commencement of another activity. In Figure 14-6, the tasks in nodes 2, 3, and 4 are not dependent on each other and can be undertaken simultaneously, but each is dependent on completion of the first task. PERT charts for complex projects can be difficult to interpret, and project managers often use both techniques.

These project management techniques can help managers identify bottlenecks and determine the impact that problems will have on project comple-

FIGURE 14-5 A GANTT CHART

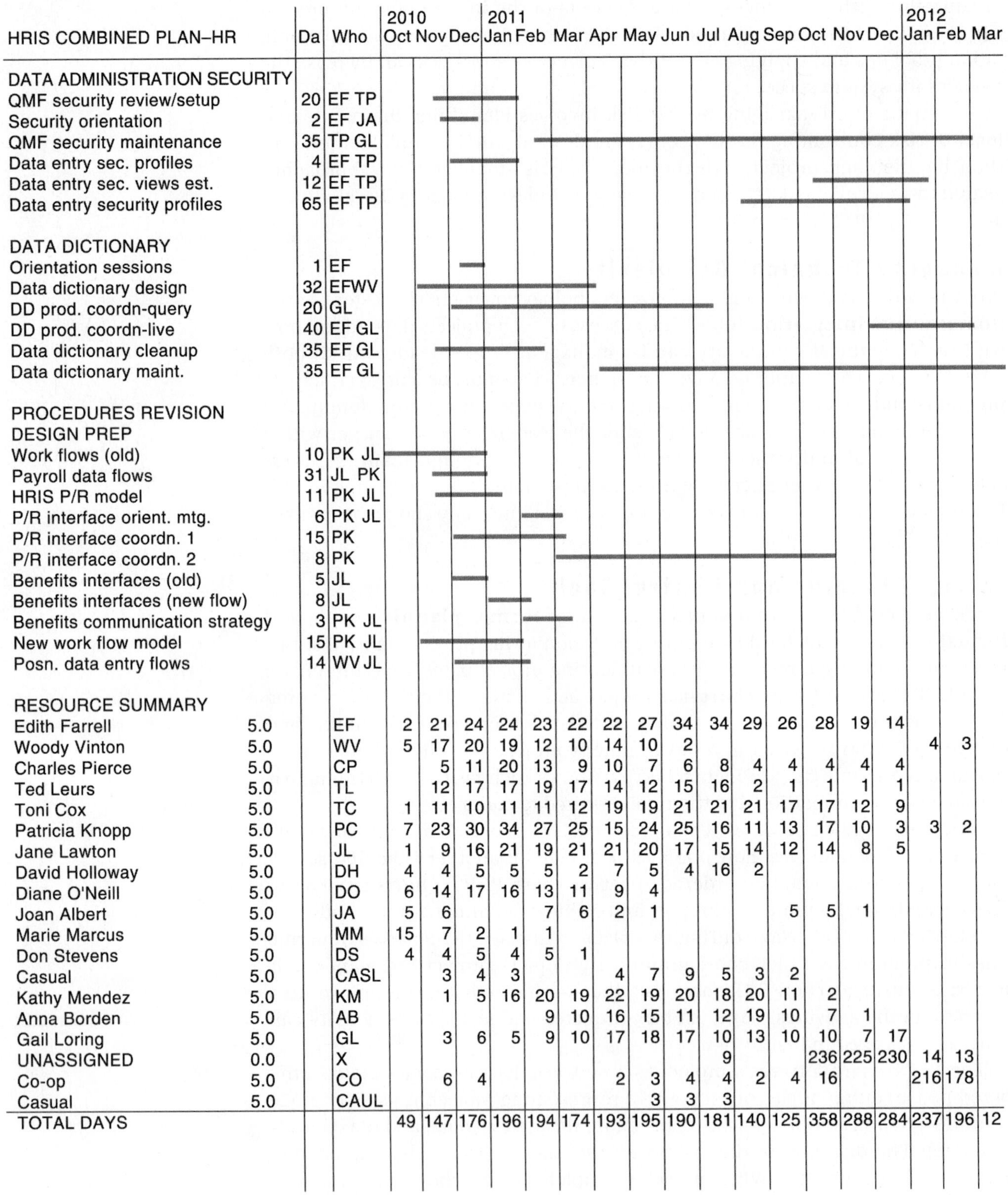
HRIS COMBINED PLAN–HR
Da
Who
2010 Oct Nov Dec
2011 Jan Feb Mar Apr May Jun Jul Aug Sep Oct Nov Dec
2012 Jan Feb Mar
DATA ADMINISTRATION SECURITY
QMF security review/setup 20 EF TP
Security orientation 2 EF JA
QMF security maintenance 35 TP GL
Data entry sec. profiles 4 EF TP
Data entry sec. views est. 12 EF TP
Data entry security profiles 65 EF TP
DATA DICTIONARY
Orientation sessions 1 EF
Data dictionary design 32 EFWV
DD prod. coordn-query 20 GL
DD prod. coordn-live 40 EF GL
Data dictionary cleanup 35 EF GL
Data dictionary maint. 35 EF GL
PROCEDURES REVISION
DESIGN PREP
Work flows (old) 10 PK JL
Payroll data flows 31 JL PK
HRIS P/R model 11 PK JL
P/R interface orient. mtg. 6 PK JL
P/R interface coordn. 1 15 PK
P/R interface coordn. 2 8 PK
Benefits interfaces (old) 5 JL
Benefits interfaces (new flow) 8 JL
Benefits communication strategy 3 PK JL
New work flow model 15 PK JL
Posn. data entry flows 14 WV JL

RESOURCE SUMMARY		Who	Oct	Nov	Dec	Jan	Feb	Mar	Apr	May	Jun	Jul	Aug	Sep	Oct	Nov	Dec	Jan	Feb	Mar
Edith Farrell	5.0	EF	2	21	24	24	23	22	22	27	34	34	29	26	28	19	14			
Woody Vinton	5.0	WV	5	17	20	19	12	10	14	10	2							4	3	
Charles Pierce	5.0	CP		5	11	20	13	9	10	7	6	8	4	4	4	4	4			
Ted Leurs	5.0	TL		12	17	17	19	17	14	12	15	16	2	1	1	1	1			
Toni Cox	5.0	TC	1	11	10	11	11	12	19	19	21	21	21	17	17	12	9			
Patricia Knopp	5.0	PC	7	23	30	34	27	25	15	24	25	16	11	13	17	10	3	3	2	
Jane Lawton	5.0	JL	1	9	16	21	19	21	21	20	17	15	14	12	14	8	5			
David Holloway	5.0	DH	4	4	5	5	5	2	7	5	4	16	2							
Diane O'Neill	5.0	DO	6	14	17	16	13	11	9	4										
Joan Albert	5.0	JA	5	6			7	6	2	1				5	5	1				
Marie Marcus	5.0	MM	15	7	2	1	1													
Don Stevens	5.0	DS	4	4	5	4	5	1												
Casual	5.0	CASL		3	4	3			4	7	9	5	3	2						
Kathy Mendez	5.0	KM		1	5	16	20	19	22	19	20	18	20	11	2					
Anna Borden	5.0	AB					9	10	16	15	11	12	19	10	7	1				
Gail Loring	5.0	GL		3	6	5	9	10	17	18	17	10	13	10	10	7	17			
UNASSIGNED	0.0	X										9			236	225	230	14	13	
Co-op	5.0	CO		6	4				2	3	4	4	2	4	16			216	178	
Casual	5.0	CAUL								3	3	3								
TOTAL DAYS			49	147	176	196	194	174	193	195	190	181	140	125	358	288	284	237	196	12

The Gantt chart in this figure shows the task, person-days, and initials of each responsible person, as well as the start and finish dates for each task. The resource summary provides a good manager with the total person-days for each month and for each person working on the project to manage the project successfully. The project described here is a data administration project.

FIGURE 14-6   A PERT CHART

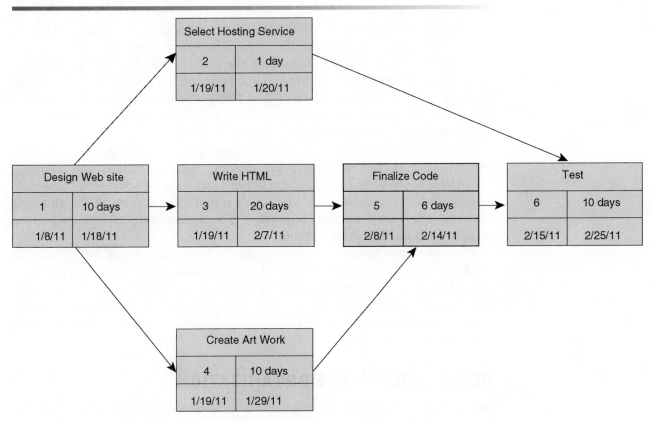

This is a simplified PERT chart for creating a small Web site. It shows the ordering of project tasks and the relationship of a task with preceding and succeeding tasks.

tion times. They can also help systems developers partition projects into smaller, more manageable segments with defined, measurable business results. Standard control techniques can successfully chart the progress of the project against budgets and target dates, so deviations from the plan can be spotted.

Increasing User Involvement and Overcoming User Resistance

Projects with relatively little structure and many undefined requirements must involve users fully at all stages. Users must be mobilized to support one of many possible design options and to remain committed to a single design. **External integration tools** consist of ways to link the work of the implementation team to users at all organizational levels. For instance, users can become active members of the project team, take on leadership roles, and take charge of installation and training. The implementation team can demonstrate its responsiveness to users, promptly answering questions, incorporating user feedback, and showing their willingness to help (Gefen and Ridings, 2002).

Participation in implementation activities may not be enough to overcome the problem of user resistance to organizational change. Different users may be affected by the system in different ways. Whereas some users may welcome a new system because it brings changes they perceive as beneficial to them, others may resist these changes because they believe the shifts are detrimental to their interests.

If the use of a system is voluntary, users may choose to avoid it; if use is mandatory, resistance will take the form of increased error rates, disruptions, turnover, and even sabotage. Therefore, the implementation strategy must not only encourage user participation and involvement, but it must also address the issue of counterimplementation (Keen, 1981). **Counterimplementation** is a deliberate strategy to thwart the implementation of an information system or an innovation in an organization.

Strategies to overcome user resistance include user participation (to elicit commitment as well as to improve design), user education and training, management edicts and policies, and better incentives for users who cooperate. The new system can be made more user friendly by improving the end-user interface. Users will be more cooperative if organizational problems are solved prior to introducing the new system.

The Interactive Session on Organizations illustrates some of these issues at work. Software firm DST Systems had trouble managing its projects because it had a high level of technical complexity and needed more powerful tools for planning and control. DST also needed buy-in from end users. As you read this case, try to determine how DST's selection of software development methods addressed these problems.

DESIGNING FOR THE ORGANIZATION

Because the purpose of a new system is to improve the organization's performance, information systems projects must explicitly address the ways in which the organization will change when the new system is installed, including installation of intranets, extranets, and Web applications. In addition to procedural changes, transformations in job functions, organizational structure, power relationships, and the work environment should be carefully planned.

Areas where users interface with the system require special attention, with sensitivity to ergonomics issues. **Ergonomics** refers to the interaction of people and machines in the work environment. It considers the design of jobs, health issues, and the end-user interface of information systems. Table 14-5 lists the organizational dimensions that must be addressed when planning and implementing information systems.

Although systems analysis and design activities are supposed to include an organizational impact analysis, this area has traditionally been neglected. An **organizational impact analysis** explains how a proposed system will

TABLE 14-5 ORGANIZATIONAL FACTORS IN SYSTEMS PLANNING AND IMPLEMENTATION

Employee participation and involvement
Job design
Standards and performance monitoring
Ergonomics (including equipment, user interfaces, and the work environment)
Employee grievance resolution procedures
Health and safety
Government regulatory compliance

INTERACTIVE SESSION: ORGANIZATIONS

DST SYSTEMS SCORES WITH SCRUM AND APPLICATION LIFE CYCLE MANAGEMENT

Companies like DST Systems have recognized the value in Scrum development to their bottom lines, but making the transition from traditional developmental methods to Scrum development can be challenging. DST Systems is a software development company whose flagship product, Automated Work Distributor (AWD), increases back-office efficiency and helps offices become paperless. DST was founded in 1969 and its headquarters are in Kansas City, Missouri. The company has approximately 10,000 employees, 1,200 of whom are software developers.

This development group had used a mixture of tools, processes, and source code control systems without any unified repository for code or a standardized developer tool set. Different groups within the organization used very different tools for software development, like Serena PVCS, Eclipse, or other source code software packages. Processes were often manual and time-consuming. Managers were unable to easily determine how resources were being allocated, which of their employees were working on certain projects, and the status of specific assets.

All of this meant that DST struggled to update its most important product, AWD, in a timely fashion. Its typical development schedule was to release a new version once every two years, but competitors were releasing versions faster. DST knew that it needed a better method than the traditional "waterfall" method for designing, coding, testing, and integrating its products. In the waterfall model of software development, progression flows sequentially from one step to the next like a waterfall, with each step unable to start until the previous step has been completed. While DST had used this method with great success previously, DST began searching for viable alternatives.

The development group started exploring Scrum, a framework for agile software development in which projects progress via a series of iterations called sprints. Scrum projects make progress in a series of sprints, which are timeboxed iterations no more than a month long. At the start of a sprint, team members commit to delivering some number of features that were listed on a project's product backlog. These features are supposed to be completed by the end of the sprint—coded, tested, and integrated into the evolving product or system. At the end of the sprint, a sprint review allows the team to demonstrate the new functionality to the product owner and other interested stakeholders who provide feedback that could influence the next sprint.

Scrum relies on self-organizing, cross-functional teams supported by a ScrumMaster and a product owner. The ScrumMaster acts as a coach for the team, while the product owner represents the business, customers, or users in guiding the team toward building the right product.

DST tried Scrum with its existing software development tools and experienced strong results. The company accelerated its software development cycle from 24 to 6 months and developer productivity increased 20 percent, but Scrum didn't work as well as DST had hoped with its existing tools. Processes broke down and the lack of standardization among the tools and processes used by DST prevented Scrum from providing its maximum benefit to the company. DST needed an application life cycle management (ALM) product that would unify its software development environment.

DST set up a project evaluation team to identify the right development environment for them. Key factors included cost-effectiveness, ease of adoption, and feature-effectiveness. DST wanted the ability to use the new software without significant training and software they could quickly adopt without jeopardizing AWD's development cycle. After considering several ALM products and running test projects with each one, DST settled on CollabNet's offering, TeamForge, for its ALM platform.

CollabNet specializes in software designed to work well with agile software development methods such as Scrum. Its core product is TeamForge, an integrated suite of Web-based development and collaboration tools for agile software development that centralizes management of users, projects, processes, and assets. DST also adopted CollabNet's Subversion product to help with the management and control of changes to project documents, programs, and other information stored as computer files. DST's adoption of CollabNet's products was fast, requiring only 10 weeks, and DST developers now do all of their work within this ALM platform. TeamForge was not forced on developers, but the ALM platform was so appealing compared to DST's previous environment that developers adopted the product virally.

Jerry Tubbs, the systems development manager at DST Systems, says that DST was successful in its attempts to revamp its software group because of a few factors. First, it looked for simplicity rather than complicated, do-everything offerings. Simpler wasn't just better for DST—it was also less expensive than some of the alternatives. DST also involved developers in the decision-making process to ensure

that changes would be greeted enthusiastically. Last, by allowing developers to adopt ALM software on their own, DST avoided the resentment associated with mandating unwelcome change. DST's move from waterfall to Scrum development was a success because the company selected the right development framework as well as the right software to make that change a reality and skillfully managed the change process.

Sources: Jerry Tubbs, "Team Building Goes Viral," *Information Week*, February 22, 2010; www.collab.net, accessed August 2010; Mountain Goat Software, "Introduction to Scrum - An Agile Process," www.mountaingoatsoftware.com/topics/scrum, accessed August 2010.

CASE STUDY QUESTIONS

1. What were some of the problems with DST Systems' old software development environment?

2. How did Scrum development help solve some of those problems?

3. What other adjustments did DST make to be able to use Scrum more effectively in its software projects? What management, organization, and technology issues had to be addressed?

MIS IN ACTION

Search the Internet for videos or Web sites explaining Scrum or agile development. Then answer the following questions:

1. Describe some of the benefits and drawbacks of Scrum development.

2. How does Scrum differ from other software development methodologies?

3. What are the potential benefits to companies using Scrum development?

affect organizational structure, attitudes, decision making, and operations. To integrate information systems successfully with the organization, thorough and fully documented organizational impact assessments must be given more attention in the development effort.

Sociotechnical Design

One way of addressing human and organizational issues is to incorporate **sociotechnical design** practices into information systems projects. Designers set forth separate sets of technical and social design solutions. The social design plans explore different workgroup structures, allocation of tasks, and the design of individual jobs. The proposed technical solutions are compared with the proposed social solutions. The solution that best meets both social and technical objectives is selected for the final design. The resulting sociotechnical design is expected to produce an information system that blends technical efficiency with sensitivity to organizational and human needs, leading to higher job satisfaction and productivity.

PROJECT MANAGEMENT SOFTWARE TOOLS

Commercial software tools that automate many aspects of project management facilitate the project management process. Project management software typically features capabilities for defining and ordering tasks, assigning resources to tasks, establishing starting and ending dates to tasks, tracking progress, and facilitating modifications to tasks and resources. Many automate the creation of Gantt and PERT charts.

Some of these tools are large sophisticated programs for managing very large projects, dispersed work groups, and enterprise functions. These high-end tools can manage very large numbers of tasks and activities and complex relationships.

Microsoft Office Project 2010 has become the most widely used project management software today. It is PC-based, with capabilities for producing PERT and Gantt charts and for supporting critical path analysis, resource allocation, project tracking, and status reporting. Project also tracks the way changes in one aspect of a project affect others. Project Professional 2010 provides collaborative project management capabilities when used with Microsoft Office Project Server 2010. Project Server stores project data in a central SQL Server database, enabling authorized users to access and update the data over the Internet. Project Server 2010 is tightly integrated with the Microsoft Windows SharePoint Services collaborative workspace platform. These features help large enterprises manage projects in many different locations. Products such as EasyProjects .NET and Vertabase are also useful for firms that want Web-based project management tools.

Going forward, delivery of project management software as a software service (SaaS) will make this technology accessible to more organizations, especially smaller ones. Open source versions of project management software such as Project Workbench and OpenProj will further reduce the total cost of ownership and attract new users. Thanks to the popularity of social media such as Facebook and Twitter, project management software is also likely to become more flexible, collaborative, and user-friendly.

While project management software helps organizations track individual projects, the resources allocated to them, and their costs, project portfolio management software helps organizations manage portfolios of projects and dependencies among them. The Interactive Session on Management describes how Hewlett-Packard's project portfolio management software helped Motorola Inc. coordinate projects and determine the right mix of projects and resources to accomplish its strategic goals.

INTERACTIVE SESSION: MANAGEMENT

MOTOROLA TURNS TO PROJECT PORTFOLIO MANAGEMENT

Motorola Inc. is a large multinational technology company based in Schaumburg, Illinois, specializing in broadband communications infrastructure, enterprise mobility, public safety solutions, high-definition video, mobile devices, and a wide variety of other mobile technologies. Motorola earned $22 billion in revenue in 2009, with 53,000 employees worldwide. Motorola has grown organically through mergers and acquisitions, and consequently has thousands of systems performing various functions throughout the business. Motorola knew that if it could better manage its systems and its projects, it could drastically lower its operating costs. In today's weakened economic climate, saving money and increasing efficiency have become more important than ever.

Motorola is organized into three major segments. The Mobile Devices segment of the business designs, manufactures, sells, and services wireless handsets, including smartphones. Motorola expects to face increasingly intense competition in this segment from a growing number of challengers hoping to cash in on the smartphone craze. Motorola's Home and Networks segment develops infrastructure and equipment used by cable television operators, wireless service providers, and other communications providers, and its Enterprise Mobility Solutions segment develops and markets voice and data communications products, wireless broadband systems, and a host of applications and devices to a variety of enterprise customers.

Weak economic conditions had driven Motorola's numbers down across all major segments of the business. The company used the downturn to review its business in depth to locate areas where it could become more efficient. Motorola first analyzed each of its business functions in terms of its importance and value to the business. Then, it analyzed the complexity and cost of that function. For example, engineering at Motorola is very important to the company's success, and differentiates it from its competitors. Engineering is also one of Motorola's most complicated and costly business functions.

Motorola repeated this analysis for all of its business functions, and then determined which areas required adjustment. Processes that were not as critical to the company's success, but were still highly complex and costly became candidates to be scaled down. Processes that were critical to the company but poorly funded were candidates for better support. After performing this exercise, Motorola hoped to automate many of the management tasks that it had classified as less complex, but the sheer size of the company made automation challenging.

Motorola has 1,800 information systems and 1,500 information systems employees who are responsible for 1,000 projects per year. The company also outsources much of its IT work to outside contractors, further increasing the number of regular users of its systems. Managing that many workers is difficult and often leads to inefficiency. Many of the company's employees were working on similar projects or compiling the same data sets, unaware that other groups within the company were doing the same work. Motorola hoped to identify and eliminate these groups, also known as "redundant silos" of activity within the company, both to cut costs and increase productivity. Management also hoped to prioritize resource usage so that projects that were most valuable to the company received the resources they needed to be successful first.

Motorola's managers hoped to achieve their goals of automating processes and lowering operating costs by adopting HP's Project and Portfolio Management Center software, or HP PPM. This software helps managers compare proposals, projects, and operational activities against budgets and resource capacity levels. All of the information Motorola gathered from its process analysis is located in a central location with HP PPM, which also serves as the centralized source of other critical information such as the amount of investment dollars used by a process and the priorities of business requests coming through Motorola's systems. HP PPM allows Motorola's IT employees and managers quick and easy access to any and all data pertaining to the company's business processes.

HP PPM allows Motorola to govern its entire IT portfolio using a broad array of tools, including objective prioritization; multiple levels of input, review, and approval; and. real-time visibility into all areas of the business. HP PPM users have up-to-the-minute data on resources, budgets, forecasts, costs, programs, projects, and overall IT demand. HP PPM can be accessed by Motorola employees on the premises or as software as a service (SaaS). Motorola used the on-site version, but converted to SaaS with no effect on usability. Motorola employees rave that HP has been responsive and reliable with its service and customer support. Using SaaS reduces Motorola's support costs by approximately 50 percent.

HP PPM uses a series of graphical displays and highly targeted data to effectively capture real-time IT program and project status. It also features what-if scenario planning that automatically creates an optimal mix of projects, proposals, and assets. This means that users can use HP PPM to perform a similar analysis of business processes that Motorola initially made by hand to begin its IT overhaul and to generate recommendations based on that analysis. Users can also use the what-if scenario planning

tools to predict the value and usefulness of new projects.

The results have been just what Motorola had hoped for. In two years, the company reduced its cost structure by 40 percent, and on larger projects using HP PPM, Motorola has achieved an average of 150 percent ROI. Motorola's IT support costs decreased by 25 percent. The redundant silos of workers performing the same tasks were all but eliminated, removing 25 percent of the company's "wasted work." Motorola also hopes to use HP PPM for resource management and application support.

Sources: HP, "Motorola: Excellence in Cost Optimization" (2010) and "HP Project and Portfolio Management (PPM) Portfolio Management Module Data Sheet, "www.hp.com, accessed November 9, 2010; Dana Gardner, "Motorola Shows Dramatic Savings in IT Operations Costs with 'ERP for IT' Tools," *ZD Net*, June 18, 2010; "Motorola Inc. Form 10-K", for the fiscal year ended Dec 31, 2009, accessed via www.sec.gov.

CASE STUDY QUESTIONS

1. What are some of the challenges Motorola faces as a business? Why is project management so critical at this company?

2. What features of HP PPM were most useful to Motorola?

3. What management, organization, and technology factors had to be addressed before Motorola could implement and successfully use HP PPM?

4. Evaluate the business impact of adopting HP PPM at Motorola.

MIS IN ACTION

Use a search engine to search for "IT portfolio management software" or "IT project management software" and find a competing offering to HP PPM. Then answer the following questions:

1. What makes this solution different from HP PPM?

2. What types of companies is this solution best geared towards?

3. Find a case study of this solution in action. Did the company described in the case realize similar benefits to Motorola?

14.5 HANDS-ON MIS PROJECTS

The projects in this section give you hands-on experience evaluating information systems projects, using spreadsheet software to perform capital budgeting analyses for new information systems investments, and using Web tools to analyze the financing for a new home.

Management Decision Problems

1. In 2001, McDonald's Restaurants undertook a project called Innovate to create an intranet connecting headquarters with its 30,000 restaurants in 120 countries to provide detailed operational information in real time. The new system would, for instance, inform a manager at the company's Oak Brook, Illinois, headquarters immediately if sales were slowing at a franchise in London, or if the grill temperature in a Rochester, Minnesota, restaurant wasn't hot enough. The idea was to create a global ERP application touching the workings of every McDonald's restaurant. Some of these restaurants were in countries that lacked network infrastructures. After spending over $1 billion over several years, including $170 million on consultants and initial implementation period planning, McDonalds terminated the project. What should management have known or done at the outset to prevent this outcome?

2. Caterpillar is the world's leading maker of earthmoving machinery and supplier of agricultural equipment. Caterpillar wants to end its support for its Dealer Business System (DBS), which it licenses to its dealers to help them run their businesses. The software in this system is becoming out of date, and senior management wants to transfer support for the hosted version of the software to Accenture Consultants so it can concentrate on its core business. Caterpillar never required its dealers to use DBS, but the system had become a de facto standard for doing business with the company. The majority of the 50 Cat dealers in North America use some version of DBS, as do about half of the 200 or so Cat dealers in the rest of the world. Before Caterpillar turns the product over to Accenture, what factors and issues should it consider? What questions should it ask? What questions should its dealers ask?

Improving Decision Making: Using Spreadsheet Software for Capital Budgeting for a New CAD System

Software skills: Spreadsheet formulas and functions
Business skills: Capital budgeting

This project provides you with an opportunity to use spreadsheet software to use the capital budgeting models discussed in this chapter to analyze the return on an investment for a new CAD system.

Your company would like to invest in a new CAD system that requires purchasing hardware, software, and networking technology, as well as expenditures for installation, training, and support. MyMISlab contains tables showing each cost component for the new system as well as annual maintenance costs over a five-year period. It also features a Learning Track on capital budgeting models. You believe the new system will produce annual savings by reducing the amount of labor required to generate designs and design specifications, thus increasing your firm's annual cash flow.

- Using the data provided in these tables, create a worksheet that calculates the costs and benefits of the investment over a five-year period and analyzes the investment using the capital budgeting models presented in this chapter's Learning Track.

- Is this investment worthwhile? Why or why not?

Improving Decision Making: Using Web Tools for Buying and Financing a Home

Software skills: Internet-based software
Business skills: Financial planning

This project will develop your skills using Web-based software for searching for a home and calculating mortgage financing for that home.

You have found a new job in Denver, Colorado, and would like to purchase a home in that area. Ideally, you would like to find a single-family house with at least three bedrooms and one bathroom that costs between $150,000 and $225,000 and finance it with a 30-year fixed-rate mortgage. You can afford a down payment that is 20 percent of the value of the house. Before you purchase a house, you would like to find out what homes are available in your price range, find a mortgage, and determine the amount of your monthly payment. You would also like to see how much of your mortgage payment represents principal and how much represents interest. Use the Yahoo! Real Estate Web site to help you with the following tasks:

- Locate homes in your price range in Denver, Colorado. Find out as much information as you can about the houses, including the real estate listing agent, condition of the house, number of rooms, and school district.

- Find a mortgage for 80 percent of the list price of the home. Compare rates from at least three sites (use search engines to find sites other than Yahoo!).

- After selecting a mortgage, calculate your closing costs.

- Calculate the monthly payment for the mortgage you select.

- Calculate how much of your monthly mortgage payment represents principal and how much represents interest, assuming you do not plan to make any extra payments on the mortgage.

When you are finished, evaluate the whole process. For example, assess the ease of use of the site and your ability to find information about houses and mortgages, the accuracy of the information you found, the breadth of choice of homes and mortgages, and how helpful the whole process would have been for you if you were actually in the situation described in this project.

LEARNING TRACK MODULES

The following Learning Tracks provide content relevant to topics covered in this chapter:

1. Capital Budgeting Methods for Information System Investments
2. Information Technology Investments and Productivity
3. Enterprise Analysis (Business Systems Planning)

Review Summary

1. *What are the objectives of project management and why is it so essential in developing information systems?*

   Good project management is essential for ensuring that systems are delivered on time, on budget, and provide genuine business benefits. Project management activities include planning the work, assessing the risk, estimating and acquiring resources required to accomplish the work, organizing the work, directing execution, and analyzing the results. Project management must deal with five major variables: scope, time, cost, quality, and risk.

2. *What methods can be used for selecting and evaluating information systems projects and aligning them with the firm's business goals?*

   Organizations need an information systems plan that describes how information technology supports the attainment of their business goals and documents all their system applications and IT infrastructure components. Large corporations will have a management structure to ensure the most important systems projects receive priority. Critical success factors, portfolio analysis, and scoring models can be used to identify and evaluate alternative information systems projects.

3. *How can firms assess the business value of information systems projects?*

   To determine whether an information systems project is a good investment, one must calculate its costs and benefits. Tangible benefits are quantifiable, and intangible benefits that cannot be immediately quantified may provide quantifiable benefits in the future. Benefits that exceed costs should be analyzed using capital budgeting methods to make sure a project represents a good return on the firm's invested capital. Real options pricing models, which apply the same techniques for valuing financial options to systems investments, can be useful when considering highly uncertain IT investments.

4. *What are the principal risk factors in information systems projects?*

   The level of risk in a systems development project is determined by (1) project size, (2) project structure, and (3) experience with technology. IS projects are more likely to fail when there is insufficient or improper user participation in the systems development process, lack of management support, and poor management of the implementation process. There is a very high failure rate among projects involving business process reengineering, enterprise applications, and mergers and acquisitions because they require extensive organizational change.

5. *What strategies are useful for managing project risk and system implementation?*

   Implementation refers to the entire process of organizational change surrounding the introduction of a new information system. User support and involvement and management support and control of the implementation process are essential, as are mechanisms for dealing with the level of risk in each new systems project. Project risk factors can be brought under some control by a contingency approach to project management. The risk level of each project determines the appropriate mix of external integration tools, internal integration tools, formal planning tools, and formal control tools to be applied.

Key Terms

Capital budgeting, 538
Change agent, 541
Change management, 540
Counterimplementation, 546
Critical success factors (CSFs), 533
Ergonomics, 546
External integration tools, 545
Formal control tools, 543
Formal planning tools, 543
Gantt chart, 543
Implementation, 540
Information systems plan, 532

Intangible benefits, 537
Internal integration tools, 543
Organizational impact analysis, 546
PERT chart, 543
Portfolio analysis, 534
Project, 530
Project management, 530
Real options pricing models (ROPMs), 538
Scope, 530
Scoring model, 535
Sociotechnical design, 548
Tangible benefits, 537
User-designer communications gap, 541
User interface, 530

Review Questions

1. What are the objectives of project management and why is it so essential in developing information systems?

- Describe information system problems resulting from poor project management.
- Define project management. List and describe the project management activities and variables addressed by project management.

2. What methods can be used for selecting and evaluating information systems projects and aligning them with the firm's business goals?

- Name and describe the groups responsible for the management of information systems projects.
- Describe the purpose of an information systems plan and list the major categories in the plan.
- Explain how critical success factors, portfolio analysis, and scoring models can be used to select information systems projects.

3. How can firms assess the business value of information systems projects?

- List and describe the major costs and benefits of information systems.
- Distinguish between tangible and intangible benefits.

- Explain how real options pricing models can help manages evaluate information technology investments.

4. What are the principal risk factors in information systems projects?

- Identify and describe each of the principal risk factors in information systems projects.
- Explain why builders of new information systems need to address implementation and change management.
- Explain why eliciting support of management and end users is so essential for successful implementation of information systems projects.
- Explain why there is such a high failure rate for implementations involving enterprise applications, business process reengineering, and mergers and acquisitions.

5. What strategies are useful for managing project risk and system implementation?

- Identify and describe the strategies for controlling project risk.
- Identify the organizational considerations that should be addressed by project planning and implementation.
- Explain how project management software tools contribute to successful project management.

Discussion Questions

1. How much does project management impact the success of a new information system?

2. It has been said that most systems fail because systems builders ignore organizational behavior problems. Why might this be so?

3. What is the role of end users in information systems project management?

Video Cases

Video Cases and Instructional Videos illustrating some of the concepts in this chapter are available. Contact your instructor to access these videos.

Collaboration and Teamwork: Identifying Implementation Problems

Form a group with two or three other students. Write a description of the implementation problems you might expect to encounter in one of the systems described in the Interactive Sessions or chapter-ending cases in this text. Write an analysis of the steps you would take to solve or prevent these problems. If possible, use Google Sites to post links to Web pages, team communication announcements, and work assignments; to brainstorm; and to work collaboratively on project documents. Try to use Google Docs to develop a presentation of your findings for the class.

JetBlue and WestJet: A Tale of Two IS Projects
CASE STUDY

In recent years, the airline industry has seen several low-cost, high-efficiency carriers rise to prominence using a recipe of extremely competitive fares and outstanding customer service. Two examples of this business model in action are JetBlue and WestJet. Both companies were founded within the past two decades and have quickly grown into industry powerhouses. But when these companies need to make sweeping IT upgrades, their relationships with customers and their brands can be tarnished if things go awry. In 2009, both airlines upgraded their airline reservation systems, and one of the two learned this lesson the hard way.

JetBlue was incorporated in 1998 and founded in 1999 by David Neeleman. The company is headquartered in Queens, New York. Its goal is to provide low-cost travel along with unique amenities like TV in every seat, and its development of state-of-the-art IT throughout the business was a critical factor in achieving that goal. JetBlue met with early success, and the airline was one of the few that remained profitable in the wake of the 9/11 attacks. JetBlue continued to grow at a rapid pace, remaining profitable throughout, until 2005, when the company lost money in a quarter for the first time since going public. Undaunted, the airline quickly returned to profitability in the next year after implementing its "Return to Profitability" plan, and consistently ranks at the top of customer satisfaction surveys and rankings for U.S. airlines.

Headquartered in Calgary, Canada, WestJet was founded by a group of airline industry veterans in 1996, including Neeleman, who left to start JetBlue shortly thereafter. The company began with approximately 40 employees and three aircraft. Today, the company has 7,700 employees and operates 380 flights per day. Earlier in this decade, WestJet underwent rapid expansion spurred by its early success and began adding more Canadian destinations and then U.S. cities to its flight schedule. By 2010, WestJet held nearly 40 percent of the Canadian airline market, with Air Canada dropping to 55 percent.

JetBlue is slightly bigger, with 151 aircraft in use compared to WestJet's 88, but both have used the same low-cost, good-service formula to achieve profitability in the notoriously treacherous airline marketplace. The rapid growth of each airline rendered their existing information systems obsolete, including their airline reservation systems.

Upgrading a reservation system carries special risks. From a customer perspective, only one of two things can happen: Either the airline successfully completes its overhaul and the customer notices no difference in the ability to book flights, or the implementation is botched, angering customers and damaging the airline's brand.

The time had come for both JetBlue and WestJet to upgrade their reservation systems. Each carrier had started out using a system designed for smaller start-up airlines, and both needed more processing power to deal with a far greater volume of customers. They also needed features like the ability to link prices and seat inventories to other airlines with whom they cooperated.

Both JetBlue and WestJet contracted with Sabre Holdings, one of the most widely used airline IT providers, to upgrade their airline reservation systems. The difference between WestJet and JetBlue's implementation of Sabre's SabreSonic CSS reservation system illustrates the dangers inherent in any large-scale IT overhaul. It also serves as yet another reminder of how successfully planning for and implementing new technology is just as valuable as the technology itself.

SabreSonic CSS performs a broad array of services for any airline. It sells seats, collects payments, allows customers to shop for flights on the airline's Web site, and provides an interface for communication with reservation agents. Customers can use it to access airport kiosks, select specific seats, check their bags, board, rebook, and receive refunds for flight cancellations. All of the data generated by these transactions are stored centrally within the system. JetBlue selected SabreSonic CSS over its legacy system developed by Sabre rival Navitaire, and WestJet was upgrading from an older Sabre reservation system of its own.

The first of the two airlines to implement SabreSonic CSS was WestJet. When WestJet went live with the new system in October 2009, customers struggled to place reservations, and the WestJet Web site crashed repeatedly. WestJet's call centers were also overwhelmed, and customers experienced slowdowns at airports. For a company that built its business on the strength of good customer service, this was a nightmare. How did WestJet allow this to happen?

[handwritten: What went wrong?]

[handwritten left margin: Customer expressed their dissatisfaction (against WestJet?)]

The critical issue was the transfer of WestJet's 840,000 files containing data on transactions for past WestJet customers who had already purchased flights, from WestJet's old reservation system servers in Calgary to Sabre servers in Oklahoma. The migration required WestJet agents to go through complex steps to process the data. WestJet had not anticipated the transfer time required to move the files and failed to reduce its passenger loads on flights operating immediately after the changeover. Hundreds of thousands of bookings for future flights that were made before the changeover were inaccessible during the file transfer and for a period of time thereafter, because Sabre had to adjust the flights using the new system.

[handwritten right margin: Problem(s) WestJet stalled their growth plans]

This delay provoked a deluge of customer dissatisfaction, a rarity for WestJet. In addition to the increase in customer complaint calls, customers also took to the Internet to express their displeasure. Angry flyers expressed outrage on Facebook and flooded WestJet's site, causing the repeated crashes. WestJet quickly offered an apology to customers on its site once it went back up, explaining why the errors had occurred. WestJet employees had trained with the new system for a combined 150,000 hours prior to the upgrade, but WestJet spokesman Robert Palmer explained that the company "encounter(ed) some problems in the live environment that simply did not appear in the test environment," foremost among them the issues surrounding the massive file transfer.

WestJet's latest earnings reports show that the company weathered the storm successfully and remained profitable, but the incident forced the airline to scale back its growth plans. WestJet has put its frequent flyer program and co-branded credit card, the RBC WestJet MasterCard, on hold, in addition to code-sharing plans with other airlines including Southwest, KLM, and British Airways. These plans would allow one airline to sell flights under its own name on aircraft operated by other airlines. For the time being, WestJet is hoping to return to growth before pursuing these measures.

In contrast, JetBlue had the advantage of seeing WestJet begin its implementation months before, so it was able to avoid many of the pitfalls that WestJet endured. For example, they built a backup Web site to prepare for the worst-case scenario. The company also hired 500 temporary call center workers to manage potential spikes in customer service calls. (WestJet also ended up hiring temporary offshore call center workers, but only after the problem had gotten out of hand.) JetBlue made sure to switch its files over to Sabre's servers on a Friday night, because Saturday flight traffic is typically very low. JetBlue also sold smaller numbers of seats on the flights that did take off that day.

JetBlue experienced a few glitches—call wait times increased, and not all airport kiosks and ticket printers came online right away. In addition, JetBlue needed to add some booking functions. But compared to what WestJet endured, the company was extremely well prepared to handle these problems. JetBlue ended up using its backup site several times.

However, JetBlue had also experienced its own customer service debacles in the past. In February 2007, JetBlue tried to operate flights during a blizzard when all other major airlines had already canceled their flights. This turned out to be a poor decision, as the weather conditions prevented the flights from taking off and passengers were stranded for as long as 10 hours. JetBlue had to continue canceling flights for days afterwards, reaching a total of 1,100 flights canceled and a loss of $30 million. JetBlue management realized in the wake of the crisis that the airline's IT infrastructure, although sufficient to deal with normal day-to-day conditions, was not robust enough to handle a crisis of this magnitude. This experience, coupled with the observation of WestJet's struggles when implementing its new system, motivated JetBlue's cautious approach to its own IT implementation.

Sources: Susan Carey, "Two Paths to Software Upgrade," *The Wall Street Journal*, April 13, 2010; Aaron Karp, "WestJet Offers 'Heartfelt Apologies' on Res System Snafus; Posts C$31 Million Profit,"*Air Transport World*, November 5, 2009; Ellen Roseman, "WestJet Reservation Change Frustrates," thestar.com, December 2, 2009; Calgary Herald, "WestJet Reservation-System Problems Affecting Sales," Kelowna.com; "JetBlue Selects SabreSonic CSS for Revenue and Operational Systems," Shepard.com, February 17, 2009; "Jilted by JetBlue for Sabre," Tnooz.com, February 5, 2010.

CASE STUDY QUESTIONS

1. How important is the reservation system at airlines such as WestJet and JetBlue? How does it impact operational activities and decision making?

2. Evaluate the key risk factors of the projects to upgrade the reservation systems of WestJet and JetBlue.

3. Classify and describe the problems each airline faced in implementing its new reservation system. What management, organization, and technology factors caused those problems?

4. Describe the steps you would have taken to control the risk in these projects.

Chapter 15

Managing Global Systems

LEARNING OBJECTIVES

After reading this chapter, you will be able to answer the following questions:

1. What major factors are driving the internationalization of business?

2. What are the alternative strategies for developing global businesses?

3. How can information systems support different global business strategies?

4. What are the challenges posed by global information systems and management solutions for these challenges?

5. What are the issues and technical alternatives to be considered when developing international information systems?

CHAPTER OUTLINE

Interactive Sessions:

"Fonterra: Managing the World's Milk Trade"

How Cell Phones Support Economic Development

Chapter 15 is located online at www.pearsonhighered.com/laudon.

References

CHAPTER 1

Belson, Ken. "Technology Lets High-End Hotels Anticipate Guests' Whims." *The New York Times* (November 16, 2005).

Brynjolfsson, Erik and Lorin M. Hitt. "Beyond Computation: Information Technology, Organizational Transformation, and Business Performance." Journal of Economic Perspectives 14, no. 4 (2000).

Brynjolfsson, Erik. "VII Pillars of IT Productivity." *Optimize* (May 2005).

Bureau of Economic Analysis. National Income and Product Accounts, 2009. Table 5.3.5. Private Fixed Investment by Type (A) (Q)

Carr, Nicholas. "IT Doesn't Matter." *Harvard Business Review* (May 2003).

Dataxis Intelligence Press Release (February 11, 2010).

Davern, Michael J. and Robert J. Kauffman. "Discovering Potential and Realizing Value form Information Technology Investments." *Journal of Management Information Systems* 16, no. 4 (Spring 2000).

Dedrick, Jason, Vijay Gurbaxani, and Kenneth L. Kraemer. "Information Technology and Economic Performance: A Critical Review of the Empirical Evidence." Center for Research on Information Technology and Organizations, University of California, Irvine (December 2001).

EMarketer. "Always-On Devices and Networks." (January 2010).

Friedman, Thomas. *The World is Flat.* New York: Farrar, Straus, and Giroux (2006).

Garretson, Rob. "IT Still Matters." *CIO Insight* 81 (May 2007).

Hughes, Alan and Michael S. Scott Morton. "The Transforming Power of Complementary Assets." *MIT Sloan Management Review* 47. No. 4 (Summer 2006).

Ives, Blake, Joseph S. Valacich, Richard T. Watson, and Robert W. Zmud. "What Every Business Student Needs to Know about Information Systems." *CAIS* 9, Article 30 (December 2002).

Lamb, Roberta, Steve Sawyer, and Rob Kling. "A Social Informatics Perspective of Socio-Technical Networks." http://lamb.cba.hawaii.edu/pubs (2004).

Laudon, Kenneth C. *Computers and Bureaucratic Reform.* New York: Wiley (1974).

Lev, Baruch. "Intangibles: Management, Measurement, and Reporting." The Brookings Institution Press (2001).

Marchand, Donald A. "Exgtracting the Business Value of IT: IT Is Usage, Not Just Deployment that Counts!" The Copco Institute Journal of Financial Transformation (2004).

Nevo, Saggi and Michael R. Wade. "The Formation and Value of IT-Enabled Resources: Antecedents and Consequences of Synergistic Relationships."MIS Quarterly 34, No. 1 (March 2010).

Pew Internet and American Life Project. "Internet Activities." (2008). wwwpewinternet.org, accessed 9/20/08.

Pew Internet & American Life Project. "Daily Internet Activity," September 2010.

Quinn, Francis J. "eBusiness Evangelist; An Interview with Erik Brynjolfsson." *Supply Chain Management Review* (May/June 2006).

Ross, Jeanne W., and Peter Weill. "Six IT Decisions Your IT People Shouldn't Make." *Harvard Business Review* (November 2002).

Teece, David. *Economic Performance and Theory of the Firm: The Selected Papers of David Teece.* London: Edward Elgar Publishing (1998).

Tuomi, Ilkka. "Data Is More Than Knowledge. *Journal of Management Information Systems* 16, no. 3 (Winter 1999–2000).

U.S. Bureau of Labor Statistics. Occupational Outlook Handbook, 2009-2010 Edition. Washington D.C.: Bureau of Labor Statistics (2010).

U.S. Department of Commerce, Bureau of the Census. Statistical Abstract of the United States, 2009. Washington D.C. (2010).

Verisign Inc., The Domain Name Industry Brief, Volume 7, September 2010.

Weill, Peter and Jeanne Ross. IT Savvy: What Top Executives Must Know to Go from Pain to Gain. Boston: Harvard Business School Press (2009).

Weill, Peter, Jeanne Ross, and David Robertson. "Digitizing Down to the Core." *Optimize Magazine* (September 2006).

CHAPTER 2

"The Radicati Group Releases 'Instant Messaging Market, 2008-2012.'" Reuters.com (January 7, 2009).

Alter, Allan. "Unlocking the Power of Teams." CIO Insight (March 2008).

Aral, Sinan; Erik Brynjolfsson; and Marshall Van Alstyne, "Productivity Effects of Information Diffusion in Networks," MIT Center for Digital Business (July 2007).

Berlind, David. "Google Apps Refresh Sets up Death Match with Microsoft," Information Week, April 12, 2010;

Bernoff, Josh and Charlene Li."Harnessing the Power of Social Applications." MIT Sloan Management Review (Spring 2008).

Boulton, Clint. "Google Wave Used for Disparate Collaboration Cases." eWeek (February 22, 2010).

Broadbent, Marianne and Ellen Kitzis. The New CIO Leader. Boston, MA: Harvard Business Press (2004).

Cash, James I. Jr., Michael J. Earl, and Robert Morison. "Teaming Up to Crack Innovation and Enterprise Integration." Harvard Business Review (November 2008).

Chen, Daniel Q., David S. Preston, and Weidong Xia. "Antecedents and Effects of CIO Supply-Side and Demand-Side Leadership: A Staged Maturity Model."Journal of Management Information Systems 27, No. 1 (Summer 2010).

Easley, Robert F., Sarv Devaraj, and J. Michael Crant."Relating Collaborative Technology Use to Teamwork Quality and Performance: An Empirical Analysis." Journal of Management Information Systems 19, no. 4 (Spring 2003)

Frost & White. "Meetings Around the World II: Charting the Course of Advanced Collaboration." (October 14, 2009).

IBM. "EuroChem Finds the Right Chemistry for Collaboration with IBM." (June 30, 2009).

Johnson, Bradfor, James Manyika, and Lareina Yee. "The Next Revolution in Interactions," McKinsey Quarterly No. 4 (2005).

Johnston, Russell, and Michael J. Vitale. "Creating Competitive Advantage with Interorganizational Information Systems." MIS Quarterly 12, no. 2 (June 1988).

Lardi-Nadarajan, Kamales. "Doing Business in Virtual Worlds." CIO Insight (March 2008).

Malone, Thomas M., Kevin Crowston, Jintae Lee, and Brian Pentland. "Tools for Inventing Organizations: Toward a Handbook of Organizational Processes." Management Science 45, no. 3 (March 1999).

McAfee, Andrew P. "Shattering the Myths About Enterprise 2.0." Harvard Business Review (November 2009).

Microsoft Corporation. "Sony Electronics Improves Collaboration, Information Access, and Productivity." (May 12, 2010).

Nolan, Richard, and F. Warren McFarland. "Information Technology and the Board of Directors." Harvard Business Review (October 1, 2005).

Oracle Corporation. "Alcoa Implements Oracle Solution 20% below Projected Cost, Eliminates 43 Legacy Systems." www.oracle.com, accessed August 21, 2005.

Perez, Dan. "The New Data-Driven U.S. Government." SIPA News (January 2010).

Picarelle, Lisa. Planes, Trains, and Automobiles." Customer Relationship Management (February 2004).

Poltrock, Steven and Mark Handel. "Models of Collaboration as the Foundation for Collaboration Technologies. Journal of Management Information Systems 27, No. 1 (Summer 2010).

Premier Global Services. "Evaluating Shift to Online Communication Tools."(2010).

Raghupathi, W. "RP". "Corporate Governance of IT: A Framework for Development." Communications of the ACM 50, No. 8 (August 2007).

Rapoza, Jim. "SharePoint Moves into the Modern Web Age." eWeek (January 18, 2010).

Reinig, Bruce A. "Toward an Understanding of Satisfaction with the Process and Outcomes of Teamwork." Journal of Management Information Systems 19, no. 4 (Spring 2003)

SAP. "Alcan Packaging Implements mySAP SCM to Increase Shareholder Value." www.mysap.com, accessed August 20, 2005.

Siebdrat, Frank, Martin Hoegl, and Holger Ernst. "How to Manage Virtual Teams." MIT Sloan Management Review 50, No. 4 (Summer 2009).

Soat, John. "Tomorrow's CIO." Information Week (June 16, 2008).

Vance, Ashlee. "Microsoft's SharePoint Thrives in the Recession." The New York Times (August 7, 2009).

Weill, Peter and Jeanne W. Ross. IT Governance. Boston: Harvard Business School Press (2004).

Weill, Peter, and Jeanne Ross. "A Matrixed Approach to Designing IT Governance." MIT Sloan Management Review 46, no. 2 (Winter 2005).

CHAPTER 3

Attewell, Paul, and James Rule. "Computing and Organizations: What We Know and What We Don't Know." Communications of the ACM 27, no. 12 (December 1984).

Bhatt, Ganesh D., and Varun Grover. "Types of Information Technology Capabilities and Their Role in Competitive Advantage." Journal of Management Information Systems 22, no.2 (Fall 2005).

Bresnahan, Timohy F., Erik Brynjolfsson, and Lorin M. Hitt, "Information Technology, Workplace Organization, and the Demand for Skilled Labor." Quarterly Journal of Economics 117 (February 2002).

Bughin, Jacques, Michael Chui, and Brad Johnson. "The Next Step in Open Innovation." The McKinsey Quarterly (June 2008).

Cash, J. I., and Benn R. Konsynski. "IS Redraws Competitive Boundaries." Harvard Business Review (March-April 1985).

Champy, James. Outsmart: How to Do What Your Competitors Can't. Upper Saddle River, NJ: FT Press (2008).

Chen, Daniel Q., Martin Mocker, David S. Preston, and Alexander Teubner. "Information Systems Strategy: Reconceptualization, Measurement, and Implications." MIS Quarterly 34, no. 2 (June 2010).

Christensen, Clayton M. The Innovator's Dilemma : The Revolutionary Book That Will Change the Way You Do Business, New York: HarperCollins (2003)

Christensen, Clayton. "The Past and Future of Competitive Advantage." Sloan Management Review 42, no. 2 (Winter 2001).

Clemons, Eric K. "Evaluation of Strategic Investments in Information Technology." Communications of the ACM (January 1991).

Clemons, Eric. "The Power of Patterns and Pattern Recognition When Developing Information-Based Strategy. Journal of Management Information Systems 27, No. 1 (Summer 2010).

Coase, Ronald H. "The Nature of the Firm."(1937) in Putterman, Louis and Randall Kroszner. The Economic Nature of the Firm: A Reader, Cambridge University Press, 1995.

Drucker, Peter. "The Coming of the New Organization." Harvard Business Review (January-February 1988).

Eisenhardt, Kathleen M. "Has Strategy Changed?" Sloan Management Review 43, no.2 (Winter 2002).

Feeny, David E., and Blake Ives. "In Search of Sustainability: Reaping Long-Term Advantage from Investments in Information Technology." Journal of Management Information Systems (Summer 1990).

Ferguson, Glover, Sanjay Mathur, and Baiju Shah. "Evolving from Information to Insight." MIT Sloan Management Review 46, no. 2 (Winter 2005).

Fine, Charles H., Roger Vardan, Robert Pethick, and Jamal E-Hout. "Rapid-Response Capability in Value-Chain Design." Sloan Management Review 43, no.2 (Winter 2002).

Freeman, John, Glenn R. Carroll, and Michael T. Hannan. "The Liability of Newness: Age Dependence in Organizational Death Rates." American Sociological Review 48 (1983).

Gallaugher, John M. and Yu-Ming Wang. "Understanding Network Effects in Software Markets: Evidence from Web Server Pricing." MIS Quarterly 26, no. 4 (December 2002).

Garretson, Rob."IS IT Still Strategic?" CIO Insight (May 2007).

Gilbert, Clark and Joseph L. Bower, "Disruptive Change." Harvard Business Review (May 2002),

Gurbaxani, V., and S. Whang, "The Impact of Information Systems on Organizations and Markets." Communications of the ACM 34, no. 1 (Jan. 1991).

Hitt, Lorin M. "Information Technology and Firm Boundaries: Evidence from Panel Data." Information Systems Research 10, no. 2 (June 1999).

Hitt, Lorin M., and Erik Brynjolfsson. "Information Technology and Internal Firm Organization: An Exploratory Analysis." Journal of Management Information Systems 14, no. 2 (Fall 1997).

Huber, George. "Organizational Learning: The Contributing Processes and Literature." Organization Science, 2 (1991), pp. 88-115.

-----. "The Nature and Design of Post-Industrial Organizations." Management Science 30, no. 8 (August 1984).

Iansiti, Marco, and Roy Levien. "Strategy as Ecology." Harvard Business Review (March 2004).

Ives, Blake and Gabriele Piccoli. "Custom Made Apparel and Individualized Service at Lands' End." Communications of the AIS 11 (2003).

Iyer, Bala and Thomas H. Davenport. "Reverse Engineering Google's Innovation Machine." Harvard Business Review (April 2008).

Jensen, M. C., and W. H. Meckling. "Specific and General Knowledge and Organizational Science." In Contract

Economics, edited by L. Wetin and J. Wijkander. Oxford: Basil Blackwell (1992).

Jensen, Michael C., and William H. Meckling. "Theory of the Firm: Managerial Behavior, Agency Costs, and Ownership Structure." Journal of Financial Economics 3 (1976).

Kauffman, Robert J. and Yu-Ming Wang. "The Network Externalities Hypothesis and Competitive Network Growth." Journal of Organizational Computing and Electronic Commerce 12, no. 1 (2002).

Kettinger, William J., Varun Grover, Subashish Guhan, and Albert H. Segors. "Strategic Information Systems Revisited: A Study in Sustainability and Performance." MIS Quarterly 18, no. 1 (March 1994).

King, J. L., V. Gurbaxani, K. L. Kraemer, F. W. McFarlan, K. S. Raman, and C. S. Yap. "Institutional Factors in Information Technology Innovation." Information Systems Research 5, no. 2 (June 1994).

Kling, Rob. "Social Analyses of Computing: Theoretical Perspectives in Recent Empirical Research." Computing Survey 12, no. 1 (March 1980).

Kolb, D. A., and A. L. Frohman. "An Organization Development Approach to Consulting." Sloan Management Review 12, no. 1 (Fall 1970).

Koulopoulos, Thomas, and James Champy. "Building Digital Value Chains." Optimize (September 2005).

Kraemer, Kenneth, John King, Debora Dunkle, and Joe Lane. Managing Information Systems. Los Angeles: Jossey-Bass (1989).

Lamb, Roberta and Rob Kling. "Reconceptualizing Users as Social Actors in Information Systems Research." MIS Quarterly 27, no. 2 (June 2003).

Laudon, Kenneth C. "A General Model of the Relationship Between Information Technology and Organizations." Center for Research on Information Systems, New York University. Working paper, National Science Foundation (1989).

------. "Environmental and Institutional Models of Systems Development." Communications of the ACM 28, no. 7 (July 1985).

------. Dossier Society: Value Choices in the Design of National Information Systems. New York: Columbia University Press (1986).

Laudon, Kenneth C. and Kenneth L. Marr, "Information Technology and Occupational Structure." (April 1995).

Lawrence, Paul, and Jay Lorsch. Organization and Environment. Cambridge, MA: Harvard University Press (1969).

Leavitt, Harold J. "Applying Organizational Change in Industry: Structural, Technological, and Humanistic Approaches." In Handbook of Organizations, edited by James G. March. Chicago: Rand McNally (1965).

Leavitt, Harold J., and Thomas L. Whisler. "Management in the 1980s." Harvard Business Review (November-December 1958).

Luftman, Jerry. Competing in the Information Age: Align in the Sand,. Oxford University Press , USA; 2 edition (August 6, 2003).

Malone, Thomas W., JoAnne Yates, and Robert I. Benjamin. "Electronic Markets and Electronic Hierarchies." Communications of the ACM (June 1987).

March, James G., and Herbert A. Simon. Organizations. New York: Wiley (1958).

Markus, M. L. "Power, Politics, and MIS Implementation." Communications of the ACM 26, no. 6 (June 1983).

McAfee, Andrew and Erik Brynjolfsson. "Investing in the IT That Makes a Competitive Difference." Harvard Business Review (July/August 2008).

McFarlan, F. Warren. "Information Technology Changes the Way You Compete." Harvard Business Review (May-June 1984).

Mendelson, Haim, and Ravindra R. Pillai. "Clock Speed and Informational Response: Evidence from the Information Technology Industry." Information Systems Research 9, no. 4 (December 1998).

Mintzberg, Henry. "Managerial Work: Analysis from Observation." Management Science 18 (October 1971).

------. The Structuring of Organizations. Englewood Cliffs, NJ: Prentice Hall (1979).

Murray, Alan. "The End of Management." The Wall Street Journal (August 21, 2010).

Orlikowski, Wanda J., and Daniel Robey. "Information Technology and the Structuring of Organizations." Information Systems Research 2, no. 2 (June 1991).

Piccoli, Gabriele, and Blake Ives. "Review: IT-Dependent Strategic Initiatives and Sustained Competitive Advantage: A Review and Synthesis of the Literature." MIS Quarterly 29, no. 4 (December 2005).

Pindyck, Robert S., and Daniel L. Rubinfeld. Microeconomics, Seventh Ed. Upper Saddle River, NJ: Prentice Hall (2009).

Porter, Michael E. "The Five Competitive Forces that Shape Strategy." Harvard Business Review (January 2008).___

Porter, Michael E. and Scott Stern. "Location Matters." Sloan Management Review 42, no. 4 (Summer 2001).

Porter, Michael. Competitive Advantage. New York: Free Press (1985).

------. Competitive Strategy. New York: Free Press (1980).

------ "Strategy and the Internet." Harvard Business Review (March 2001).

Prahalad, C.K.and M.S.Krishnan. The New Age of Innovation. Driving Cocreated Value Through Global Networks. New York: McGraw Hill (2008).

Robey, Daniel and Marie-Claude Boudreau. "Accounting for the Contradictory Organizational Consequences of Information Technology: Theoretical Directions and Methodological Implications." Information Systems Research 10, no. 42 (June1999).

Shapiro, Carl, and Hal R. Varian. Information Rules. Boston, MA: Harvard Business School Press (1999).

Shpilberg, David, Steve Berez, Rudy Puryear, and Sachin Shah. "Avoiding the Alignment Trap in Information Technology." MIT Sloan Management Review 49, no. 1 (Fall 2007).

Starbuck, William H. "Organizations as Action Generators." American Sociological Review 48 (1983).

Tushman, Michael L., and Philip Anderson. "Technological Discontinuities and Organizational Environments." Administrative Science Quarterly 31 (September 1986).

Wallace, Amy. "Putting Customers In Charge of Design." The New York Times (May 14, 2010).

Watson, Brian P. "Is Strategic Alignment Still a Priority?" CIO Insight (October 2007).

Weber, Max. The Theory of Social and Economic Organization. Translated by Talcott Parsons. New York: Free Press (1947).

Williamson, Oliver E. The Economic Institutions of Capitalism. New York: Free Press, (1985).

CHAPTER 4

Angst, Corey M. and Ritu Agarwal. "Adoption of Electronic Health Records in the Presence of Privacy Concerns: The Elaboration Likelihood Model and Individual Persuasion." MIS Quarterly 33, No. 2 (June 2009).

Baumstein, Avi. "New Tools Close Holes in Cam-Spam." Information Week (February 23, 2009).

Beck, Melinda. "Becoming a Squinter Nation." The Wall Street Journal (August 17, 2010).

Bhattacharjee, Sudip, Ram D. Gopal, and G. Lawrence Sanders. "Digital Music and Online Sharing: Software Piracy 2.0?" Communications of the ACM 46, no.7 (July 2003).

Bowen, Jonathan. "The Ethics of Safety-Critical Systems."Communications of the ACM 43, no. 3 (April 2000).

Brown Bag Software vs. Symantec Corp. 960 F2D 1465 (Ninth Circuit, 1992).

Carr, Nicholas. "Tracking Is an Assault on Liberty, with Real Dangers." The Wall Street Journal (August 7, 2010).

Chellappa, Ramnath K. and Shivendu Shivendu. "An Economic Model of Privacy: A Property Rights Approach to Regulatory Choices for Online Personalization." Journal of Management Information Systems 24, no. 3 (Winter 2008).

Clifford, Stephanie. "Web Coupons Know Lots About You, and They Tell." The New York Times (April 16, 2010).

Culnan, Mary J. and Cynthia Clark Williams. "How Ethics Can Enhance Organizational Privacy." MIS Quarterly 33, No. 4 (December 2009).

Farmer, Dan and Charles C. Mann. "Surveillance Nation." Part I Technology Review (April 2003) and Part II (Technology Review (May 2003)..

Goodman, Joshua, Gordon V. Cormack, and David Heckerman. "Spam and the Ongoing Battle for the Inbox." Communications of the ACM 50, No. 2 (February 2007).

Grimes, Galen A. "Compliance with the CAN-SPAM Act of 2003." Communications of the ACM 50, No. 2 (February 2007).

Harper, Jim. "It's Modern Trade: Web Users Get as Much as They Give." The Wall Street Journal (August 7, 2010).

Hsieh, J.J. Po-An, Arun Rai, and Mark Keil. "Understanding Digital Inequality: Comparing Continued Use Behavioral Models of the Socio-Economically Advantaged and Disadvantaged." MIS Quarterly 32, no. 1 (March 2008).

Jackson, Linda A., Alexander von Eye, Gretchen Barbatsis, Frank Biocca, Hiram E. Fitzgerald, and Yong Zhao. "The Impact of Internet Use on the Other Side of the Digital Divide." Communications of the ACM 47, no. 7 (July 2004).

Jackson, Thomas W., Ray Dawson, and Darren Wilson. "Understanding Email Interaction Increases Organizational Productivity." Communications of the ACM 46, no. 8 (August 2003).

Laudon, Kenneth C. and Carol Guercio Traver. E-Commerce: Business, Technology, Society 7th Edition. Upper Saddle River, NJ: Prentice-Hall (2011).

Laudon, Kenneth C. Dossier Society: Value Choices in the Design of National Information Systems. New York: Columbia University Press (1986b).

Lohr, Steve. "How Privacy Vanishes Online." The New York Times (March 16, 2010.)

Matt Richtel, "Hooked on Gadgets,and Paying a Mental Price," The New York Times, June 7, 2010.

Nord, G. Daryl, Tipton F. McCubbins, and Jeretta Horn Nord. "E-Monitoring in the Workplace: Privacy, Legislation, and Surveillance Software. Communications of the ACM 49, No. 8 (August 2006).

Okerson, Ann. "Who Owns Digital Works?" Scientific American (July 1996).

Payton, Fay Cobb."Rethinking the Digital Divide." Communications of the ACM 46, no. 6 (June 2003)

Rapoza, Jim. "Web Bug Alert." EWeek (June 15, 2009).

Rifkin, Jeremy. "Watch Out for Trickle-Down Technology." The New York Times (March 16, 1993).

Rigdon, Joan E. "Frequent Glitches in New Software Bug Users." The Wall Street Journal (January 18, 1995).

Risen, James and Eric Lichtblau. "E-Mail Surveillance Renews Concerns in Congress." The New York Times (June 17, 2009).

Schwartz, Matthew J. "Sophos: U.S. Leads List of Spam Originators." Information Week (July 15, 2010).

Singer, Natasha. "Shoppers Who Can't Have Secrets." The New York Times (April 30, 2010).

Smith, H. Jeff, and John Hasnas. "Ethics and Information Systems: The Corporate Domain." MIS Quarterly 23, no. 1 (March 1999).

Smith, H. Jeff. "The Shareholders vs. Stakeholders Debate." MIS Sloan Management Review 44, no. 4 (Summer 2003).

Steel, Emily and Jessica E. Vascellaro. "Facebook, MySpace Confront Privacy Loophole." The Wall Street Journal (May 21, 2010).

Steel, Emily and Julia Angwin,"On the Web's Cutting Edge, Anonymity in Name Only." The Wall Street Journal (August 4, 2010).

Steel, Emily. "Marketers Watch Friends Interact Online." The Wall Street Journal (April 15, 2010)
_____"Exploring Ways to Build a Better Consumer Profile," The Wall Street Journal (March 15, 2010)

Story, Louise. "To Aim Ads, Web Is Keeping Closer Eye on You." The New York Times (March 10, 2008).

Stout, Hilary. "Antisocial Networking?" The New York Times (May 2, 2010).

United States Department of Health, Education, and Welfare. Records, Computers, and the Rights of Citizens. Cambridge: MIT Press (1973).

Urbaczewski, Andrew and Leonard M. Jessup. "Does Electronic Monitoring of Employee Internet Usage Work?" Communications of the ACM 45, no. 1 (January 2002).

Valentino-Devries, Jennifer and Emily Steel. "'Cookies' Cause Bitter Backlash." The Wall Street Journal (September 19, 2010).

Vascellaro, Jessica E. "Google Agonizes on Privacy as Ad World Vaults Ahead." The Wall Street Journal (August 10, 2010).

Xu, Heng, Hock-Hai Teo, Bernard C.Y. Tan, and Ritu Agarwal. "The Role of Push-Pull Technology in Privacy Calculus: The Case of Location-Based Services." Journal of Management Information Systems 26, No. 3 (Winter 2010).

CHAPTER 5

"IBM Launches New Autonomic Offerings for Self-Managing IT Systems." IBM Media Relations (June 30, 2005).

Amazon Web Services."AWS Case Study: Envoy Media Group." (January 2010).

Babcock, Charles. "Linux No Longer the Cool New Kid on the Block. Now What?" Information Week (April 14, 2008).

Carr, Nicholas. The Big Switch. New York: Norton (2008).

Cheng, Roger. "'Cloud Computing': What Exactly Is It Anyway?" The Wall Street Journal (February 8, 2010).

Chickowski, Ericka."How Good Are Your Service-Level Agreements?" Baseline (January 2008).

Choi, Jae, Derek L. Nazareth, and Hemant K. Jain. "Implementing Service-Oriented Architecture in Organizations." Journal of Management Information Systems 26, No. 4 (Spring 2010).

Cole, Arthur. "Mainframes They Are A'Changin," ITBusinessEdge.com, January 20, 2010.

David, Julie Smith, David Schuff, and Robert St. Louis. "Managing Your IT Total Cost of Ownership." Communications of the ACM 45, no. 1 (January 2002).

Dempsey, Bert J.. Debra Weiss, Paul Jones, and Jane Greenberg. "What Is an Open Source Software Developer?" Communications of the ACM 45, no. 1 (January 2001).

Dubney, Abhijit and Dilip Wagle. "Delivering Software as a Service." The McKinsey Quarterly (June 2007).

Fitzgerald, Brian. "The Transformation of Open Source Software." MIS Quarterly 30, No. 3 (September 2006).

Fox, Armando, and David Patterson. "Self-Repairing Computers." Scientific American (May 2003.).

Ganek, A. G., and T. A. Corbi. "The Dawning of the Autonomic Computing Era." IBM Systems Journal 42, no 1, (2003).

Gerlach, James, Bruce Neumann, Edwin Moldauer, Martha Argo, and Daniel Frisby." Determining the Cost of IT Services."Communications of the ACM 45, no. 9 (September 2002).

Glader, Paul and Don Clark. "IBM Launches Next Generation of Chips and Servers." The Wall Street Journal (February 7, 2010).

Gomes, Lee and Taylor Buley. "Is the PC Dead?" Forbes (December 28, 2010).

Google Inc., "Cloud Computing: Latest Buzzword or a Glimpse of the Future?" CBS Interactive (2009).

Duvall, Mel. "Study Predicts Big Shift to Cloud By 2020." Information Management (June 10, 2010).

Hagel III, John and John Seeley Brown. "Your Next IT Strategy." Harvard Business Review (October, 2001).

Hay, Timothy."IBM Survey Says Mobile Apps Will Dominate Enterprise." The Wall Street Journal (October 7, 2010).

Helft, Miguel. "Google Offers Peek at Operating System, a Potential Challenger to Windows." The New York Times (November 20, 2009).

IBM. "IBM Launches New Autonomic Offerings for Self-Managing IT Systems." IBM Media Relations (June 30, 2005).

IBM. "Seeding the Clouds: Key Infrastructure Elements for Cloud Computing." (February 2009).

IBM. "The Benefits of Cloud Computing." (2009).

Kauffman, Robert J. and Julianna Tsai. "The Unified Procurement Strategy for Enterprise Software: A Test of the 'Move to the Middle' Hypothesis." Journal of Management Information Systems 26, No. 2 (Fall 2009).

King, John. "Centralized vs. Decentralized Computing: Organizational Considerations and Management Options." Computing Surveys (October 1984).

Kurzweil, Ray. "Exponential Growth an Illusion?: Response to Ilkka Tuomi." KurzweilAI.net, September 23, 2003

Leong, Lydia and Ted Chamberlin. "Magic Quadrant for Web Hosting and Hosted Cloud System Infrastructure Services (On Demand)." Gartner Inc. (July 2, 2009).

Lohr, Steve. "Bundling Hardware and Software to Do Big Jobs." The New York Times (February 8, 2010).

Markoff, John. "After the Transistor, a Leap into the Microcosm." The New York Times, August 31, 2009

McAfee, Andrew. "Will Web Services Really Transform Collaboration?" MIT Sloan Management Review 46, no. 2 (Winter 2005).

McCafferty, Dennis. "Cloudy Skies: Public Vs. Private Option Still Up in the Air." Baseline (March/April 2010).

Mell, Peter and Tim Grance. "The NIST Definition of Cloud Computing" Version 15. NIST (October 17, 2009).

Metrics 2.0. "Worldwide PC Shipments to Reach 334 Million in 2010." Metrics2.com, accessed October 2, 2010.

Moore, Gordon. "Cramming More Components Onto Integrated Circuits," Electronics 38, Number 8 (April 19, 1965).

Mueller, Benjamin, Goetz Viering, Christine Legner, and Gerold Riempp. "Understanding the Economic Potential of Service-Oriented Architecture." Journal of Management Information Systems 26, No. 4 (Spring 2010).

Oltsik, Jon. "What's Needed For Cloud Computing." Enterprise Strategy Group (June 2010).

Patel, Samir, and Suneel Saigal. "When Computers Learn to Talk: A Web Services Primer." McKinsey Quarterly no. 1 (2002).

Rogow, Rruce."Tracking Core Assets." Optimize Magazine (April 2006).

Schuff, David and Robert St. Louis. "Centralization vs. Decentralization of Application Software." Communications of the ACM 44, no. 6 (June 2001).

Seltzer, Larry. "OS of the Future: Built for Security." eWeek (April 19,2010).

Stango, Victor. "The Economics of Standards Wars." Review of Network Economics 3, Issue 1 (March 2004).

Stone, Brad and Ashlee Vance. "Companies Slowly Join Cloud Computing," The New York Times (April 18, 2010).

Susarla, Anjana, Anitesh Barua, and Andrew B. Whinston. "A Transaction Cost Perspective of the 'Software as a Service' Business Model. " Journal of Management Information Systems 26, No. 2 (Fall 2009).

Tuomi, Ilkka. "The Lives and Death of Moore's Law." FirstMonday, Col 7, No. 11

(November 2002). www.firstmonday.org.

Vance, Ashlee. "Open Source as a Model for Business Is Elusive." The New York Times (November 29, 2009).

Vance, Ashlee. "Microsoft Office 2010 Starts Ascension to the Cloud." The New York Times (July 13, 2009).

Walsh, Lawrence. "Outsourcing: A Means of Business Enablement." Baseline (May 2008).

Weill, Peter, and Marianne Broadbent. Leveraging the New Infrastructure. Cambridge, MA: Harvard Business School Press (1998).

Weill, Peter, Mani Subramani and Marianne Broadbent. "Building IT Infrastructure for Strategic Agility." Sloan Management Review 44, no. 1 (Fall 2002).

Weitzel, Tim. Economics of Standards in Information Networks. Springer (2004).

CHAPTER 6

Cappiello, Cinzia, Chiara Francalanci, and Barbara Pernici. "Time-Related Factors of Data Quality in Multichannel Information Systems." Journal of Management Information Systems 20, no. 3 (Winter 2004).

Chen, Andrew N. K., Paulo B. Goes, and James R. Marsden. "A Query-Driven Approach to the Design and Management of Flexible Database Systems." Journal of Management Information Systems 19, no. 3 (Winter 2002-2003).

Clifford, James, Albert Croker, and Alex Tuzhilin. "On Data Representation and Use in a Temporal Relational DBMS." Information Systems Research 7, no. 3 (September 1996).

Eckerson, Wayne W. "Data Quality and the Bottom Line." The Data Warehousing Institute (2002).

Fayyad, Usama, Ramasamy Ramakrishnan, and Ramakrisnan Srikant. "Evolving Data Mining into Solutions for Insights." Communications of the ACM 45, no.8 (August 2002).

Foshay, Neil, Avinandan Mukherjee and Andrew Taylor. "Does Data Warehouse End-User Metadata Add Value? Communications of the ACM 50, no. 11 (November 2007).

Gartner Inc. "'Dirty Data' is a Business Problem, not an IT Problem, Says Gartner." Sydney, Australia (March 2, 2007).

Henschen, Doug. "Text Mining for Customer Insight." Information Week (November 30, 2009).

Henschen, Doug. "Big and Fast." Information Week (August 9, 2010).

Henschen, Doug. "The Data Warehouse Revised." Information Week (May 26, 2008).

Henschen, Doug."Wendy's Taps Text Analytics to Mine Customer Feedback," Information Week (March 23, 2010).

Hoffer, Jeffrey A., Mary Prescott, and Heikki Toppi. Modern Database Management, 10th ed. Upper Saddle River, NJ: Prentice-Hall (2011).

Jinesh Radadia. "Breaking the Bad Data Bottlenecks." Information Management (May/June 2010).

Kim, Yong Jin, Rajiv Kishore, and G. Lawrence Sanders. "From DQ to EQ: Understanding Data Quality in the Context of E-Business System"Communications of the ACM 48, no. 10 (October 2005).

Klau, Rick. "Data Quality and CRM." Line56.com, accessed March 4, 2003.

Kroenke, David M. and David Auer. Database Processing 11e. Upper Saddle River, NJ: Prentice-Hall (2010).

Lee, Yang W., and Diane M. Strong. "Knowing-Why about Data Processes and Data Quality." Journal of Management Information Systems 20, no. 3 (Winter 2004).

Loveman, Gary. "Diamonds in the Datamine." Harvard Business Review (May 2003).

McKnight, William. "Seven Sources of Poor Data Quality." Information Management (April 2009).

Pierce, Elizabeth M. "Assessing Data Quality with Control Matrices." Communications of the ACM 47, no. 2 (February 2004).

Redman, Thomas. Data Driven: Profiting from Your Most Important Business Asset. Boston: Harvard Business Press (2008).

Stodder, David. "Customer Insights." Information Week (February 1, 2010)

CHAPTER 7

Borland, John. "A Smarter Web." Technology Review (March/April 2007).

Bustillo, Maguel. "Wal-Mart Radio Tags to Track Clothing." The Wall Street Journal (July 23, 2010).

Cheng, Roger. "Verizon Readies 4G Launch." The Wall Street Journal (October 7, 2010).

Dekleva, Sasha, J.P. Shim, Upkar Varshney, and Geoffrey Knoerzer. "Evolution and Emerging Issues in Mobile Wireless Networks." Communications of the ACM 50, No. 6 (June 2007).

EMarketer. "US Ad Spending." (June 2010)

Fish, Lynn A. and Wayne C. Forrest. "A Worldwide Look at RFID." Supply Chain Management Review (April 1, 2007).

Furchgott, Roy. "That's a Nice Phone, But How's the Network?" The New York Times (June 30, 2010).

Helft, Miguel. "Google Makes a Case That It Isn't So Big." The New York Times (June 29, 2009)

Housel, Tom, and Eric Skopec. Global Telecommunication Revolution: The Business Perspective. New York: McGraw-Hill (2001).

ICANN. "ICANN Policy Update." 10, No. 9 (September 2010).

Kocas, Cenk. "Evolution of Prices in Electronic Markets under Diffusion of Price-Comparison Shopping." Journal of Management Information Systems 19, no. 3 (Winter 2002-2003).

Nicopolitidis, Petros, Georgios Papademitriou, Mohammad S. Obaidat, and Adreas S. Pomportsis. "The Economics of Wireless Networks." Communications of the ACM 47, no. 4 (April 2004).

Panko, Raymond. Business Data Networks and Telecommunications 8e. Upper Saddle River, NJ: Prentice-Hall (2011).

Papazoglou, Mike P. "Agent-Oriented Technology in Support of E-Business." Communications of the ACM 44, no. 4 (April 2001).

Phillips, Lisa E. "US Internet Users, 2010."eMarketer (April 2010).

Pottie, G. J., and W.J Kaiser. "Wireless Integrated Network Sensors." Communications of the ACM 43, no. 5 (May 2000).

St. Clair, Scott and Keefe Bailey. "Prognosis: Opportunity." Information Week (March 8, 2010).

"The Internet of Things." McKinsey Quarterly (March 2010).

Varshney, Upkar, Andy Snow, Matt McGivern, and Christi Howard. "Voice Over IP." Communications of the ACM 45, no. 1 (January 2002).

Wingfield, Nick and Amir Efrati. "Google Rekindles the Browser War." The Wall Street Journal (July 7, 2010).

Xiao, Bo and Izak Benbasat. "E-Commerce Product Recommendation Agents: Use, Characteristics, and Impact." MIS Quarterly 31, no. 1 (March 2007).

CHAPTER 8

Ante, Spencer E. "Dark Side Arises for Phone Apps." The Wall Street Journal (June 6, 2010).

------. "Get Smart: Targeting Phone Security Flaws." The Wall Street Journal (June 15, 2010).

Bernstein, Corinne. "The Cost of Data Breaches." Baseline (April 2009)

Bray, Chad. "Global Cyber Scheme Hits Bank Accounts." The Wall Street Journal (October 1, 2010.)

Brenner, Susan W. "U.S. Cybercrime Law: Defining Offenses." Information Systems Frontiers 6, no. 2 (June 2004).

Cavusoglu, Huseyin, Birendra Mishra, and Srinivasan Raghunathan. "A Model for Evaluating IT Security Investments." Communications of the ACM 47, no. 7 (July 2004).

Chickowski, Ericka. "Is Your Information Really Safe?" Baseline (April 2009).

Choe Sang-Hun, "Cyberattacks Hit U.S. and South Korean Web Sites," The New York Times, July 9, 2009.

Computer Security Institute. "2009 CSI Computer Crime and Security Survey" (2009).

Consumer Reports. "State of the Net 2010." (June 2010)

Coopes, Amy. "Australian 17-Year-Old Takes Blame for Twitter Chaos." AFP (September 22, 2010).

D'Arcy, John and Anat Hovav. "Deterring Internal Information Systems Use." Communications of the ACM 50, no. 10 (October 2007).

Danchev, Dancho."Malware Watch: Rogue Facebook Apps, Fake Amazon Orders, and Bogus Adobe Updates." ZD Net (May 19, 2010).

Dash, Eric. "Online Woes Plague Chase for 2nd Day." The New York Times (September 15, 2010).

Ely, Adam. "Browser as Attack Vector." Information Week (August 9, 2010).

Feretic, Eileen. "Security Lapses More Costly."baselinemag.com, January 26, 2010

Foley, John. "P2P Peril." Information Week (March 17, 2008).

Fratto, Mike. "What's Your Appetite for Risk?" Information Week (June 22, 2009).

Giordano, Scott M. "Electronic Evidence and the Law." Information Systems Frontiers 6, No. 2 (June 2004).

Galbreth, Michael R. and Mikhael Shor. "The Impact of Malicious Agents on the Enterprise Software Industry." MIS Quarterly 34, no. 3 (September 2010).

Gorman, Siobhan. "Broad New Hacking Attack Detected." The Wall Street Journal (February 18, 2010).

Housley, Russ, and William Arbaugh. "Security Problems in 802.11b Networks." Communications of the ACM 46, no. 5 (May 2003).

IBM. "Secure By Design: Building Identity-Based Security into Today's Information Systems." (March 2010).

Ives, Blake, Kenneth R. Walsh, and Helmut Schneider. "The Domino Effect of Password Reuse." Communications of the ACM 47, no.4 (April 2004).

Jagatic Tom, Nathaniel Johnson, Markus Jakobsson, and Filippo Menczer. "Social Phishing." Communications of the ACM 50, no. 10 (October 2007).

Javelin Strategy & Research. "2010 Identity Fraud Survey Report." (2010).

Markoff, John. "Vast Spy System Loots Computers in 103 Countries." The New York Times (March 29, 2009).

McDougall, Paul. "High Cost of Data Loss." Information Week (March 20,2006).

McGraw, Gary. "Real-World Software Security." Information Week (August 9, 2010).

Meckbach, Greg. "MasterCard's Robust Data Centre: Priceless." Computerworld Canada (March 26, 2008).

Mercuri, Rebeca T. "Analyzing Security Costs." Communications of the ACM 46, no. 6 (June 2003).

Mills, Elinor. "Facebook Disables Rogue Data-Stealing, Spamming Apps." CNET News (August 20, 2009).

Mitchell, Dan. "It's Here: It's There; It's Spyware." The New York Times (May 20, 2006).

Null, Christopher. "WPA Cracked in 1 Minute." Yahoo! Tech (August 27, 2009).

Panko, Raymond R. Corporate Computer and Network Security 2e. Upper Saddle River, NJ: Pearson Prentice Hall (2010).

"Plea in Case of Stolen Code from Goldman," Reuters, February 17, 2010

Prince, Brian. "The Growing E-Mail Security Challenge." eWeek (April 21, 2008).

Pug, Ivan P.L. and Qiu-Hong Wang. "Information Security: Facilitating User Precautions Vis a Vis Enforcement Against Attackers. "Journal of Management Information Systems 26, No. 2 (Fall 2009).

Roche, Edward M., and George Van Nostrand. Information Systems, Computer Crime and Criminal Justice. New York: Barraclough Ltd. (2004).

Ryan Naraine, "ActiveX Under Siege," eWeek (February 11, 2008).

Sample, Char and Diana Kelley. "Cloud Computing Security: Infrastructure Issues." Security Curve June 23, 2009).

Sarrel, Matthew. "The Biggest Security Threats Right Now." eWeek (May 6, 2010).

Schwerha, Joseph J., IV. "Cybercrime: Legal Standards Governing the Collection of Digital Evidence." Information Systems Frontiers 6, no. 2 (June 2004).

Sophos. "Security Threat Report: Midyear 2010." (2010).

Spears. Janine L. and Henri Barki. "User Participation in Information Systems Security Risk Management." MIS Quarterly 34, No. 3 (September 2010).

Steel, Emily. "Web Ad Sales Open Door to Viruses." The Wall Street Journal (June 15, 2009).

Symantec. "Symantec Global Internet Security Threat Report: Trends for 2009." (April 2010).

Vance, Ashlee. "If Your Password is 123456, Just Make It HackMe." The New York Times (January 20, 2010).

Volonino, Linda and Stephen R. Robinson. Principles and Practice of Information Security. Upper Saddle River, NJ: Pearson Prentice Hall (2004).

Wang, Huaiqing, and Chen Wang. "Taxonomy of Security Considerations and Software Quality." Communications of the ACM 46, no. 6 (June 2003).

Warkentin, Merrill, Xin Luo, and Gary F. Templeton. "A Framework for Spyware Assessement." Communications of the ACM 48, no. 8 (August 2005).

Westerman, George. IT Risk: Turning Business Threats into Competitive Advantage. Harvard Business School Publishing (2007)

Wright, Ryan T. and Kent Marrett."The Influence of Experiental and Dispositional Factors in Phishing: An Empirical Investigation of the Deceived." Journal of Management Information Systems 27, No. 1 (Summer 2010).

CHAPTER 9

Aeppel, Timothy."'Bullwhip' Hits Firms as Growth Snaps Back." The Wall Street Journal (January 27, 2010).

Barrett, Joe. "Whirlpool Cleans Up Its Delivery Act." The Wall Street Journal (September 24, 2009).

Chickowski, Ericka. "5 ERP Disasters Explained."www.Baselinemag.com, accessed October 8, 2009.

D'Avanzo, Robert, Hans von Lewinski, and Luk N. Van Wassenhove. "The Link between Supply Chain and Financial Performance." Supply Chain Management Review (November 1, 2003).

Davenport, Thomas H. Mission Critical: Realizing the Promise of Enterprise Systems. Boston: Harvard Business School Press (2000).

Ferrer, Jaume, Johan Karlberg, and Jamie Hintlian."Integration: The Key to Global Success." Supply Chain Management Review (March 1, 2007).

Fleisch, Elgar, Hubert Oesterle, and Stephen Powell. "Rapid Implementation of Enterprise Resource Planning Systems." Journal of Organizational Computing and Electronic Commerce 14, no. 2 (2004).

Garber, Randy and Suman Sarkar. "Want a More Flexible Supply Chain?" Supply Chain Management Review (January 1, 2007).

Goodhue, Dale L., Barbara H. Wixom, and Hugh J. Watson. "Realizing Business Benefits through CRM: Hitting the Right Target in the Right Way." MIS Quarterly Executive 1, no. 2 (June 2002).

Greenbaum, Joshiua. "Is ERP Dead? Or Has It Just Gone Underground?" SAP NetWeaver Magazine 3 (2007).

Guinipero, Larry, Robert B. Handfield, and Douglas L. Johansen. "Beyond Buying." The Wall Street Journal (March 10, 2008).

Handfield, Robert B. and Ernest L. Nichols. Supply Chain Redesign: Transforming Supply Chains into Integrated Value Systems. Financial Times Press (2002).

Henschen, Doug. "Salesforce's Facebook Envy Goes Mobile." Information Week (September 13, 2010).

_____ "SAP Tries One Platform for Mobile Devices," Information Week (March 8, 2010).

Hitt, Lorin, D. J. Wu, and Xiaoge Zhou. "Investment in Enterprise Resource Planning: Business Impact and Productivity Measures." Journal of Management Information Systems 19, no. 1 (Summer 2002).

Johnson, Maryfran."What's Happening with ERP Today." CIO (January 27, 2010).

Kalakota, Ravi, and Marcia Robinson. E-Business 2.0. Boston: Addison-Wesley (2001).

_____. Services Blueprint: Roadmap for Execution. Boston: Addison-Wesley (2003).

Kanakamedala, Kishore, Glenn Ramsdell, and Vats Srivatsan. "Getting Supply Chain Software Right." McKinsey Quarterly No. 1 (2003).

Klein, Richard and Arun Rai. "Interfirm Strategic Information Flows in Logistics Supply Chain Relationships. MIS Quarterly 33, No. 4 (December 2009).

Kopczak, Laura Rock, and M. Eric Johnson. "The Supply-Chain Management Effect." MIT Sloan Management Review 44, no. 3 (Spring 2003).

Laudon, Kenneth C. "The Promise and Potential of Enterprise Systems and Industrial Networks." Working paper, The Concours Group. Copyright Kenneth C. Laudon (1999).

Lee, Hau, L., V. Padmanabhan, and Seugin Whang. "The Bullwhip Effect in Supply Chains." Sloan Management Review (Spring 1997).

Lee, Hau. "The Triple-A Supply Chain." Harvard Business Review (October 2004).

Li, Xinxin and Lorin M. Hitt. "Price Effects in Online Product Reviews: An Analytical Model and Empirical Analysis." MIS Quarterly 34, No. 4 (December 2010).

Liang, Huigang, Nilesh Sharaf, Quing Hu, and Yajiong Xue. "Assimilation of Enterprise Systems: The Effect of

Institutional Pressures and the Mediating Role of Top Management." MIS Quarterly 31, no. 1 (March 2007).

Malhotra, Arvind, Sanjay Gosain, and Omar A. El Sawy. "Absorptive Capacity Configurations in Supply Chains: Gearing for Partner-Enabled Market Knowledge Creation." MIS Quarterly 29, no. 1 (March 2005).

Oracle Corporation. "Alcoa Implements Oracle Solution 20% below Projected Cost, Eliminates 43 Legacy Systems." www.oracle.com, accessed August 21, 2005.

Rai, Arun, Ravi Patnayakuni, and Nainika Seth. "Firm Performance Impacts of Digitally Enabled Supply Chain Integration Capabilities." MIS Quarterly 30 No. 2 (June 2006).

Ranganathan, C. and Carol V. Brown. "ERP Iinvestments and the Market Value of Firms: Toward an Understanding of Influential ERP Project Variables." Information Systems Research 17, No. 2 (June 2006).

Robey, Daniel, Jeanne W. Ross, and Marie-Claude Boudreau. "Learning to Implement Enterprise Systems: An Exploratory Study of the Dialectics of Change." Journal of Management Information Systems 19, no. 1 (Summer 2002).

Schwartz, Ephraim. "Does ERP Matter-Industry Stalwarts Speak Out." InfoWorld (April 10, 2007).

Scott, Judy E., and Iris Vessey. "Managing Risks in Enterprise Systems Implementations." Communications of the ACM 45, no. 4 (April 2002).

Seldon, Peter B., Cheryl Calvert, and Song Yang. "A Multi-Project Model of Key Factors Affecting Organizational Benefits from Enterprise Systems." MIS Quarterly 34, no. 2 (June 2010).

Strong, Diane M. and Olga Volkoff. "Understanding Organization-Enterprise System Fit: A Path to Theorizing the Information Technology Artifact." MIS Quarterly 34, No.4 (December 2010).

Wailgum, Thomas. "Why ERP Is Still So Hard." CIO (September 9. 2009).

Wailgum, Thomas. "The Future of ERP." Parts I and II. CIO (November 20, 2009).

Wing, George. "Unlocking the Value of ERP." Baseline (January/February 2010).

CHAPTER 10

Adomavicius, Gediminas and Alexander Tuzhilin. "Personalization Technologies: A Process-Oriented Perspective." Communications of the ACM 48, no. 10 (October 2005).

Bakos, Yannis. "The Emerging Role of Electronic Marketplaces and the Internet." Communications of the ACM 41, no. 8 (August 1998).

Bluefly, Inc. Form 10K Report for Fiscal Year Ended December 31, 2009, filed with the Securities and Exchange Commission (February 19, 2010).

Bluefly, Inc. "Form 10K Report for Fiscal Year Ended December 31, 2009." Filed with the Securities and Exchange Commission (February 19, 2010).

Bo, Xiao and Izak Benbasat. "E-Commerce Product Recommendation Agents: Use, Characteristics, and Impact." MIS Quarterly 31, no. 1 (March 2007).

Brynjolfsson, Erik, Yu Hu, and Michael D. Smith. "Consumer Surpus in the Digital Economy: Estimating the Value of Increased Product Variety at Online Booksellers." Management Science 49, no. 11 (November 2003).

Cain Miller, Claire and Jenna Wortham. "Technology Aside, Most People Still Decline to be Located." The New York Times (August 29, 2010).

Cain Miller, Claire. "Take a Step Closer for an Invitation to Shop." The New York Times (February 23, 2010).

Clemons, Eric K. "Business Models for Monetizing Internet Applications and Web Sites: Experience, Theory, and Predictions. "Journal of Management Information Systems 26, No. 2 (Fall 2009).

Clifford, Stephanie. "Aisle by Aisle, an App that Pushes Bargains." The New York Times (August 17, 2010).

Clifford, Stephanie. "Web Coupons Know Lots About You, and They Tell." The New York Times (April 16, 2010).

ComScore. "Facebook Takes Lead In Time Spent. comScore Media Metrix. Press Release. (September 9, 2010).

ComScore. "Facebook Takes Lead In Time Spent." comScore Media Metrix Press Release (September 9, 2010).

Dou, Wenyu, Kai H. Lim, Chenting Su, Nan Zhou, and Nan Cui. "Brand Positioning Strategy Using Search Engine Marketing." MIS Quarterly 34, no.2 (June 2010).

EMarketer, "US Mobile Subscribers." eMarketer Chart (March 2010d)

EMarketer. "Paid Music Content." eMarketer Report (January 2010c).

EMarketer. "US Internet Users, 2010." eMarketer Report (April 2010b).

EMarketer. "US Portal Advertising Revenues." Report. April 2010e

EMarketer. "US Retail E-commerce Forecast." eMarketer Report (March 2010a).

Evans, Philip and Thomas S. Wurster. Blown to Bits: How the New Economics of Information Transforms Strategy. Boston, MA: Harvard Business School Press (2000).

Fuller, Johann, Hans Muhlbacher, Kurt Matzler, and Gregor Jawecki. "Customer Empowerment Through Internet-Based Co-Creation." Journal of Management Information Systems 26, No. 3 (Winter 2010).

Hallerman, David. "US Advertising Spending: The New Reality." EMarketer (April 2009a).

Helft, Miguel. "Data, Not Design, Is King in the Age of Google." The New York Times (May 10, 2009).

Howe, Heff. Crowdsourcing: Why the Power of the Crowd Is Driving the Future of Business. New York: Random House (2008).

Internetworldstats.com. "Internet Usage Statistics: The Big Picture." Internetworldstats.com, (June 2010).

Jiang, Zhengrui and Sumit Sarkar. "Speed Matters: The Role of Free Software Offer in Software Diffusion." Journal of Management Information Systems 26, No. 3 (Winter 2010).

Kaplan, Steven and Mohanbir Sawhney. "E-Hubs: the New B2B Marketplaces." Harvard Business Review (May-June 2000).

Kauffman, Robert J. and Bin Wang. "New Buyers' Arrival Under Dynamic Pricing Market Microstructure: The Case of Group-Buying Discounts on the Internet, Journal of Management Information Systems 18, no. 2 (Fall 2001).

Laseter, Timothy M., Elliott Rabinovich, Kenneth K. Boyer, and M. Johnny Rungtusanatham. "Critical Issues in Internet Retailing." MIT Sloan Management Review 48, no. 3 (Spring 2007).

Laudon, Kenneth C. and Carol Guercio Traver. E-Commerce: Business, Technology, Society, 7th edition. Upper Saddle River, NJ: Prentice-Hall (2011).

Leimeister, Jan Marco, Michael Huber, Ulrich Bretschneider, and Helmut Krcmar." Leveraging Crowdsourcing: Activation-Supporting Components for IT-Based Ideas Competition." Journal of Management Information Systems 26, No. 1 (Summer 2009).

Magretta, Joan. "Why Business Models Matter." Harvard Business Review (May 2002).

Markoff, John. "The Cellphone, Navigating Our Lives." The New York Times (February 17, 2009).

Mattioli, Dana. "Retailers Answer Call of Smartphones." The Wall Street Journal (June 11, 2010).

Michael D. Smith and Rahul Telang. "Competing with Free: The Impact of Movie Broadcasts on DVD Sales and Internet Piracy." MIS Quarterly 33, No. 2 (June 2009).

Miller, Claire Cain. "Google Campaign to Build Up Its Display Ads." The New York Times (September 21, 2010).

Pavlou, Paul A., Huigang Liang, and Yajiong Xue. "Understanding and Mitigating Uncertainty in Online Exchange Relationships: A Principal-Agent Perspective." MIS Quarterly 31, no. 1 (March 2007).

Pew Internet & American Life Project. "US Daily Internet Activities." (2010).

Pew Internet & American Life Project. "US Daily Internet Activities." Pew Internet & American Life Project, 2010.

Philips, Jeremy. "To Rake It In, Give It Away." The Wall Street Journal (July 8, 2009).

Rayport, Jeffrey. "Demand-Side Innovation: Where IT Meets Marketing." Optimize Magazine (February 2007).

Resnick, Paul and Hal Varian. "Recommender Systems." Communications of the ACM (March 2007).

Rosenbloom, Stephanie. "Cellphones Let Shoppers Point, Click, and Purchase." The New York Times (February 26, 2010).

Schoder, Detlef and Alex Talalavesky. "The Price Isn't Right." MIT Sloan Management Review (August 22, 2010).

Schultze, Ulrike and Wanda J. Orlikowski. "A Practice Perspective on Technology-Mediated Network Relations: The Use of Internet-Based Self-Serve Technologies." Information Systems Research 15, no. 1 (March 2004).

Smith, Michael D., Joseph Bailey and Erik Brynjolfsson."Understanding Digital Markets: Review and Assessment" in Erik Brynjolfsson and Brian Kahin, ed. Understanding the Digital Economy. Cambridge, MA: MIT Press (1999).

Steel, Emily, "Exploring Ways to Build a Better Consumer Profile." The Wall Street Journal (March 15, 2010).

Steel, Emily."Marketers Watch Friends Interact Online." The Wall Street Journal (April 15, 2010).

Story, Louise. "To Aim Ads, Web Is Keeping Closer Eye on You. " The New York Times (March 10, 2008).

Stross, Randall. "Just Browsing? A Web Store May Follow You Out the Door." The New York Times (May 17, 2009).

Surowiecki, James. The Wisdom of Crowds: Why the Many Are Smarter Than the Few and How Collective Wisdom Shapes Business, Economies, Societies and Nations. Boston: Little, Brown (2004).

Tsai, Jessica. "Social Media: The Five-Year Forecast." destinationcrm.com (April 27, 2009).

United States Census Bureau. "E-Stats Report. Measuring the Electronic Economy." (May 27, 2010).

United States Census Bureau. "E-Stats Report. Measuring the Electronic Economy." (May 27, 2010). www.census.gov.

Worthen, Ben. "Branching Out: Mobile Banking Finds New Users." The Wall Street Journal (February 3, 2009).

CHAPTER 11

Alavi, Maryam and Dorothy Leidner. "Knowledge Management and Knowledge Management Systems: Conceptual Foundations and Research Issues," MIS Quarterly 25, No. 1ARTON (March 2001).

Alavi, Maryam, Timothy R. Kayworth, and Dorothy E. Leidner. "An Empirical Investigation of the Influence of Organizational Culture on Knowledge Management Practices." Journal of Management Information Systems 22, No.3 (Winter 2006).

Allen, Bradley P. "CASE-Based Reasoning: Business Applications." Communications of the ACM 37, no. 3 (March 1994).

Anthes, Gary H. "Agents Change."Computerworld (January 27, 2003).

Awad, Elias and Hassan M Ghaziri. Knowledge Management. Upper Saddle River, NJ: Prentice-Hall (2004).

Bargeron, David, Jonathan Grudin, Anoop Gupta, Elizabeth Sanocki, Francis Li, and Scott Le Tiernan."Asynchronous Collaboration Around Multimedia Applied to On-Demand Education." Journal of Management Information Systems 18, No. 4 (Spring 2002).

Barker, Virginia E., and Dennis E. O'Connor. "Expert Systems for Configuration at Digital: XCON and Beyond." Communications of the ACM (March 1989).

Becerra-Fernandez, Irma, Avelino Gonzalez, and Rajiv Sabherwal. Knowledge Management. Upper Saddle River, NJ: Prentice-Hall (2004).

Bieer, Michael, Douglas Englebart Richard Furuta, Starr Roxanne Hiltz, John Noll, Jennifer Preece, Edward A. Stohr, Murray Turoff, and Bartel Van de Walle." Toward Virtual Community Knowledge Evolution." Journal of Management Information Systems 18, No. 4 (Spring 2002).

Birkinshaw, Julian and Tony Sheehan. "Managing the Knowledge Life Cycle." MIT Sloan Management Review 44, no. 1 (Fall 2002).

Booth, Corey and Shashi Buluswar. "The Return of Artificial Intelligence," The McKinsey Quarterly No. 2 (2002).

Burtka, Michael. "Generic Algorithms." The Stern Information Systems Review 1, no. 1 (Spring 1993).

Churchland, Paul M., and Patricia Smith Churchland. "Could a Machine Think?" Scientific American (January 1990).

Cross, Rob and Lloyd Baird."Technology is Not Enough: Improving Performance by Building Organizational Memory." Sloan Management Review 41, no. 3 (Spring 2000).

Cross, Rob, Nitin Nohria, and Andrew Parker. "Six Myths about Informal Networks-and How to Overcome Them," Sloan Management Review 43, no. 3 (Spring 2002)

Davenport, Thomas H., and Lawrence Prusak. Working Knowledge: How Organizations Manage What They Know. Boston, MA: Harvard Business School Press (1997).

Davenport, Thomas H., David W. DeLong, and Michael C. Beers. "Successful Knowledge Management Projects." Sloan Management Review 39, no. 2 (Winter 1998).

Davenport, Thomas H., Laurence Prusak, and Bruce Strong. "Putting Ideas to Work." The Wall Street Journal (March 10, 2008).

Davenport, Thomas H., Robert J. Thomas and Susan Cantrell. "The Mysterious Art and Science of Knowledge-Worker Performance." MIT Sloan Management Review 44, no. 1 (Fall 2002).

Davis, Gordon B. "Anytime/ Anyplace Computing and the Future of Knowledge Work." Communications of the ACM 42, no.12 (December 2002).

Desouza, Kevin C. "Facilitating Tacit Knowledge Exchange." Communications of the ACM 46, no. 6 (June 2003).

Dhar, Vasant, and Roger Stein. Intelligent Decision Support Methods: The Science of Knowledge Work. Upper Saddle River, NJ: Prentice Hall (1997).

Du, Timon C., Eldon Y. Li, and An-pin Chang. "Mobile Agents in Distributed Network Management." Communications of the ACM 46, no.7 (July 2003).

Earl, Michael J., and Ian A. Scott. "What Is a Chief Knowledge Officer?" Sloan Management Review 40, no. 2 (Winter 1999).

Earl, Michael. "Knowledge Management Strategies: Toward a Taxonomy." Journal of Management Information Systems 18, no. 1 (Summer 2001).

El Najdawi, M. K., and Anthony C. Stylianou. "Expert Support Systems: Integrating AI Technologies." Communications of the ACM 36, no. 12 (December 1993).

Flash, Cynthia. "Who is the CKO?" Knowledge Management (May 2001).

Gelernter, David. "The Metamorphosis of Information Management." Scientific American (August 1989).

Gordon, Steven R. and Monideepa Tarafdar. "The IT Audit that Boosts Innovation." MIT Sloan Management Review (June 26, 2010).

Gregor, Shirley and Izak Benbasat. "Explanations from Intelligent Systems: Theoretical Foundations and Implications for Practice." MIS Quarterly 23, no. 4 (December 1999).

Griffith, Terri L., John E. Sawyer, and Margaret A Neale. "Virtualness and Knowledge in Teams: Managing the Love Triangle of Organizations, Individuals, and Information Technology." MIS Quarterly 27, no. 2 (June 2003).

Grover, Varun and Thomas H. Davenport. "General Perspectives on Knowledge Management: Fostering a Research Agenda." Journal of Management Information Systems 18, no. 1 (Summer 2001).

Gu, Feng and Baruch Lev. "Intangible Assets. Measurements, Drivers, Usefulness." http://pages.stern.nyu.edu/~blev/.

Holland, John H. "Genetic Algorithms." Scientific American (July 1992).

Housel Tom and Arthur A. Bell. Measuring and Managing Knowledge. New York: McGraw-Hill (2001).

Jarvenpaa, Sirkka L. and D. Sandy Staples. "Exploring Perceptions of Organizational Ownership of Information and Expertise."Journal of Management Information Systems 18, no. 1 (Summer 2001).

Jones, Quentin, Gilad Ravid, and Sheizaf Rafaeli. "Information Overload and the Message Dynamics of Online Interaction Spaces: A Theoretical Model and Empirical Exploration." Information Systems Research 15, no. 2 (June 2004).

Kankanhalli, Atreyi, Frasiska Tanudidjaja, Juliana Sutanto, and Bernard C.Y Tan."The Role of IT in Successful Knowledge Management Initiatives." Communications of the ACM 46, no. 9 (September 2003).

King, William R., Peter V. Marks, Jr. and Scott McCoy. "The Most Important Issues in Knowledge Management." Communications of the ACM 45, no.9 (September 2002).

Kuo, R.J., K. Chang, and S.Y.Chien."Integration and Self-Organizing Feature Maps and Genetic-Algorithm-Based Clustering Method for Market Segmentation." Journal of Organizational Computing and Electronic Commerce 14, no. 1 (2004).

Lamont, Judith."Communities of Practice Leverage Knowledge." KMWorld (July/August 2006).

Lamont, Judith."Open Source ECM Platforms Bring Mobility to Market," KMWorld March 2010.

Leonard-Barton, Dorothy and Walter Swap. "Deep Smarts.' Harvard Business Review (September 1, 2004).

Leonard-Barton, Dorothy, and John J. Sviokla. "Putting Expert Systems to Work." Harvard Business Review (March-April 1988).

Lev, Baruch, and Theodore Sougiannis. "Penetrating the Book-to-Market Black Box: The R&D Effect," Journal of Business Finance and Accounting (April/May 1999).

Lev, Baruch. "Sharpening the Intangibles Edge." Harvard Business Review (June 1, 2004).

Maes, Patti. "Agents that Reduce Work and Information Overload." Communications of the ACM 38, no. 7 (July 1994).

Maglio, Paul P. and Christopher S. Campbell. "Attentive Agents." Communications of the ACM 46, no. 3 (March 2003).

Malone, Thomas W., Robert Laubacher, and Chrysanthos Dellarocas. "The Collective Intelligence Genome." MIT Sloan Management Review 51, No. 3 (Spring 2010).

Maltby, Emily. "Affordable 3-D Arrives." The Wall Street Journal (July 29, 2010).

Markus, M. Lynne, Ann Majchrzak, and Less Gasser." A Design Theory for Systems that Support Emergent Knowledge Processes." MIS Quarterly 26, no. 3 (September 2002).

Markus, M. Lynne. "Toward a Theory of Knowledge Reuse: Types of Knowledge Reuse Situations and Factors in Reuse Success."

Journal of Management Information Systems 18, no. 1 (Summer 2001).

Maryam Alavi and Dorothy E. Leidner. "Knowledge Management and Knowledge Management Systems." MIS Quarterly 25, no. 1 (March 2001).

McCarthy, John. "Generality in Artificial Intelligence." Communications of the ACM (December 1987).

Moravec, Hans. "Robots, After All." Communications of the ACM 46, no. 10 (October 2003).

Nidumolu, Sarma R. Mani Subramani and Alan Aldrich. " Situated Learning and the Situated Knowledge Web: Exploring the Ground Beneath Knowledge Management." Journal of Management Information Systems 18, no. 1 (Summer 2001).

Open Text Corporation. "Barrick Gold Turns to Open Text to Help Streamline Information Flow." (2010).

Orlikowski, Wanda J. "Knowing in Practice: Enacting a Collective Capability in Distributed Organizing." Organization Science 13, no. 3 (May-June 2002).

Patterson, Scott. "Letting the Machines Decide." The Wall Street Journal (July 13, 2010).

Ranft, Annette L. and Michael D. Lord. "Acquiring New Technologies and Capabilities: A Grounded Model of Acquisition Implementation." Organization Science 13, no. 4 (July-August 2002).

Sadeh, Norman, David W. Hildum, and Dag Kjenstad."Agent-Based E-Supply Chain Decision Support." Journal of Organizational Computing and Electronic Commerce 13, no. 3 & 4 (2003)

Samuelson, Douglas A. and Charles M. Macal. "Agent-Based Simulation." OR/MS Today (August 2006).

Sanders, Peter. "Boeing 787 Training Takes Virtual Path." The Wall Street Journal (September 2, 2010).

Schultze, Ulrike and Dorothy Leidner."Studying Knowledge Management in Information Systems Research: Discourses and Theoretical Assumptions." MIS Quarterly 26, no. 3 (September 2002).

Spangler, Scott, Jeffrey T. Kreulen, and Justin Lessler. "Generating and Browsing Multiple Taxonomies over a Document Collection." Journal of Management Information Systems 19, no. 4 (Spring 2003)

Spender, J. C. "Organizational Knowledge, Learning and Memory: Three Concepts In Search of a Theory." Journal of Organizational Change Management 9, 1996.

Starbuck, William H. "Learning by Knowledge-Intensive Firms." Journal of Management Studies 29, no. 6 (November 1992).

Tiwana, Amrit. "Affinity to Infinity in Peer-to-Peer Knowledge Platforms." Communications of the ACM 46, no. 5 (May 2003).

Trippi, Robert, and Efraim Turban. "The Impact of Parallel and Neural Computing on Managerial Decision Making." Journal of Management Information Systems 6, no. 3 (Winter 1989-1990).

Voekler, Michael. "Staying a Step Ahead of Fraud." Intelligent Enterprise (September 2006).

Walczak, Stephen."An Emprical Analysis of Data Requirements for Financial Forecasting with Neural Networks." Journal of Management Information Systems 17, no. 4 (Spring 2001).

Wang, Huaiqing, John Mylopoulos, and Stephen Liao. "Intelligent Agents and Financial Risk Monitoring Systems." Communications of the ACM 45, no. 3 (March 2002).

Yimam-Seid, Dawit and Alfred Kobsa." Expert-Finding Systems for Organizations: Problem and Domain Analysis and the DEMOIR Approach." Journal of Organizational Computing and Electronic Commerce 13, no. 1 (2003)

Zack, Michael H "Rethinking the Knowledge-Based Organization." MIS Sloan Management Review 44, no. 4 (Summer 2003).

Zadeh, Lotfi A. "Fuzzy Logic, Neural Networks, and Soft Computing." Communications of the ACM 37, no. 3 (March 1994).

Zadeh, Lotfi A. "The Calculus of Fuzzy If/Then Rules." AI Expert (March 1992).

CHAPTER 12

Anson, Rob and Bjorn Erik Munkvold. "Beyond Face-to-Face: A Field Study of Electronic Meetings in Different Time and Place Modes." Journal of Organizational Computing and Electronic Commerce 14, no. 2 (2004).

Barkhi, Reza. "The Effects of Decision Guidance and Problem Modeling on Group Decision-Making. "Journal of Management Information Systems 18, no. 3 (Winter 2001-2002).

Bazerman, Max H. and Dolly Chugh. "Decisions Without Blinders." Harvard Business Review (January 2006).

BPM Working Group. "Performance Management Industry Leaders Form BPM Standards Group." Working Group to Establish Common Definition for BPM and Deliver Business Performance Management Framework Press Release. Stamford, CT (March 25, 2004)

Clark, Thomas D., Jr., Mary C. Jones, and Curtis P. Armstrong. "The Dynamic Structure of Management Support Systems: Theory Development, Research Focus, and Direction." MIS Quarterly 31, no. 3 (September 2007).

Davenport, Thomas H. and Jeanne G. Harris. Competing on Analytics: The New Science of Winning. Boston: Harvard Business School Press (2007).

Davenport, Thomas H., Jeanne Harris, and Jeremy Shapiro. "Competing on Talent Analytics." Harvard Business Review (October 2010).

Davenport, Thomas H., Jeanne G. Harris, and Robert Morison. Analytics at Work: Smarter Decisions, Better Results. Boston: Harvard Business Press (2010).

Dennis, Alan R., Jay E. Aronson, William G. Henriger, and Edward D. Walker III. "Structuring Time and Task in Electronic Brainstorming." MIS Quarterly 23, no. 1 (March 1999).

Dennis, Alan R., Joey F. George, Len M. Jessup, Jay F. Nunamaker, and Douglas R. Vogel. "Information Technology to Support Electronic Meetings." MIS Quarterly 12, no. 4 (December 1988).

DeSanctis, Geraldine, and R. Brent Gallupe. "A Foundation for the Study of Group Decision Support Systems." Management Science 33, no. 5 (May 1987).

El Sherif, Hisham, and Omar A. El Sawy. "Issue-Based Decision Support Systems for the Egyptian Cabinet." MIS Quarterly 12, no. 4 (December 1988).

Gallupe, R. Brent, Geraldine DeSanctis, and Gary W. Dickson. "Computer-Based Support for Group Problem-Finding: An Experimental Investigation." MIS Quarterly 12, no. 2 (June 1988).

Gartner, Inc. "Gartner, Inc. Business intelligence market grows 22%." Press Release, June 15, 2010.

Gorry, G. Anthony, and Michael S. Scott Morton. "A Framework for Management Information Systems." Sloan Management Review 13, no. 1 (Fall 1971).

Greengard, Samuel. "Business Intelligence and Business Analytics, Big Time." Baseline (February 2010).

Henschen, Doug. "Next-Gen BI Is Here." Information Week (August 31, 2009).

Hoover, J. Nicholas." Search, Mobility BI Keys to Hotel Chain's Growth." Information Week (September 13, 2010).

Jensen, Matthew, Paul Benjamin Lowry, Judee K. Burgoon, and Jay Nunamaker. "Technology Dominance in Complex Decisionmaking." Journal of Management Information Systems 27, No. 1 (Summer 2010).

Kaplan, Robert S. and David P. Norton. "The Balanced Scorecard: Measures that Drive Performance", Harvard Business Review (Jan - Feb 1992).

Kaplan, Robert S. and David P. Norton. Strategy Maps: Converting Intangible Assets into Tangible Outcomes. Boston: Harvard Business School Press (2004).

LaValle, Steve, Michael S. Hopkins, Eric Lesser, Rebecca Shockley, and Nina Kruschwitz. "Analytics: The New Path to Value." MIT Sloan Management Review and IBM Institute for Business Value (Fall 2010).

Leidner, Dorothy E., and Joyce Elam. "The Impact of Executive Information Systems on Organizational Design, Intelligence, and Decision Making." Organization Science 6, no. 6 (November-December 1995).

Lilien, Gary L., Arvind Rangaswamy, Gerrit H. Van Bruggen, and Katrin Starke. "DSS Effectiveness in Marketing Resource Allocation Decisions: Reality vs. Perception. "Information Systems Research 15, no. 3 (September 2004).

Rockart, John F., and David W. DeLong. Executive Support Systems: The Emergence of Top Management Computer Use. Homewood, IL: Dow-Jones Irwin (1988).

Rockart, John F., and Michael E. Treacy. "The CEO Goes On-Line." Harvard Business Review (January-February 1982).

Scanlon, Robert J. "A New Route to Performance Management." Baseline Magazine. (January/February 2009).

Schwabe, Gerhard. "Providing for Organizational Memory in Computer-Supported Meetings." Journal of Organizational Computing and Electronic Commerce 9, no. 2 and 3 (1999).

Simon, H. A. The New Science of Management Decision. New York: Harper & Row (1960).

Turban, Efraim, Ramesh Sharda, and Dursun Delen. Decision Support and Business Intelligence Systems, 9th ed. Upper Saddle River, NJ: Prentice Hall (2011).

Turban, Efraim, Ramesh Sharda, Dursun Delen, and David King. Business Intelligence, 2nd ed. Upper Saddle River, NJ: Prentice Hall (2011).

Yoo, Youngjin and Maryam Alavi. "Media and Group Cohesion: Relative Influences on Social Presence, Task Participation, and Group Consensus." MIS Quarterly 25, no. 3 (September 2001).

CHAPTER 13

Albert, Terri C., Paulo B. Goes, and Alok Gupta. "GIST: A Model for Design and Management of Content and Interactivity of Customer-Centric Web Sites." MIS Quarterly 28, no. 2 (June 2004).

Alter, Allan E. "I.T. Outsourcing: Expect the Unexpected." CIO Insight (March 7, 2007).

Armstrong, Deborah J. and Bill C. Hardgrove. "Understanding Mindshift Learning: The Transition to Object-Oriented Development." MIS Quarterly 31, no. 3 (September 2007).

Aron, Ravi, Eric K.Clemons, and Sashi Reddi. "Just Right Outsourcing: Understanding and Managing Risk." Journal of Management Information Systems 22, no. 1 (Summer 2005).

Ashrafi, Noushin and Hessam Ashrafi. Object-Oriented Systems Analysis and Design. Upper Saddle River, NY: Prentice-Hall (2009).

Avison, David E. and Guy Fitzgerald. "Where Now for Development Methodologies?" Communications of the ACM 41, no. 1 (January 2003).

Babcock, Charles. "Platform as a Service: What Vendors Offer." Information Week (October 3, 2009).

Baily, Martin N. and Diana Farrell. "Exploding the Myths of Offshoring." The McKinsey Quarterly (July 2004).

Barthelemy, Jerome. "The Hidden Costs of IT Outsourcing." Sloan Management Review (Spring 2001).

Broadbent, Marianne, Peter Weill, and Don St. Clair. "The Implications of Information Technology Infrastructure for Business Process Redesign." MIS Quarterly 23, no. 2 (June 1999).

Brown, Susan A., Norman L. Chervany, and Bryan A. Reinicke. "What Matters When Introducing New Technology." Communications of the ACM 50, No. 9 (September 2007).

Cha, Hoon S., David E. Pingry, and Matt E. Thatcher. "A Learning Model of Information Technology Outsourcing: Normative Implications. "Journal of Management Information Systems 26, No. 2 (Fall 2009).

Champy, James A.X-Engineering the Corporation: Reinventing Your Business in the Digital Age. New York: Warner Books (2002).

Curbera, Francisco, Rania Khalaf, Nirmal Mukhi, Stefan Tai, and Sanjiva Weerawarana."The Next Step in Web Services." Communications of the ACM 46, no 10 (October 2003).

Davidson, Elisabeth J. "Technology Frames and Framing: A Socio-Cognitive Investigation of Requirements Determination. MIS Quarterly 26, no. 4 (December 2002).

DeMarco, Tom. Structured Analysis and System Specification. New York: Yourdon Press (1978).

Den Hengst, Marielle and Gert-Jan DeVreede. "Collaborative Business Engineering: A Decade of Lessons from the Field." Journal of Management Information Systems 20, no. 4 (Spring 2004).

Dibbern, Jess, Jessica Winkler, and Armin Heinzl. "Explaining Variations in Client Extra Costs between Software Projects Offshored to India." MIS Quarterly 32, no. 2 (June 2008).

El Sawy, Omar A.Redesigning Enterprise Processes for E-Business. McGraw-Hill (2001).

Erickson, Jonathan. "Dr Dobb's Report: Agile Development." Information Week (April 27, 2009).

Feeny, David, Mary Lacity, and Leslie P. Willcocks. "Taking the Measure of Outsourcing Providers." MIT Sloan Management Review 46, No. 3 (Spring 2005).

Fischer, G., E. Giaccardi. Y.Ye, A.G. Sutcliffe, and N. Mehandjiev. " Meta-Design: A Manifesto for End-User Development." Communications of the ACM 47, no. 9 (September 2004).

Gefen, David and Catherine M. Ridings. "Implementation Team Responsiveness and User Evaluation of Customer Relationship Management: A Quasi-Experimental Design Study of Social Exchange Theory." Journal of Management Information Systems 19, no. 1 (Summer 2002).

Gefen, David and Erran Carmel. "Is the World Really Flat? A Look at Offshoring in an Online Programming Marketplace." MIS Quarterly 32, no. 2 (June 2008).

Gemino, Andrew and Yair Wand." Evaluating Modeling Techniques Based on Models of Learning." Communications of the ACM 46, no. 10 (October 2003).

George, Joey, Dinesh Batra, Joseph S. Valacich, and Jeffrey A. Hoffer. Object Oriented System Analysis and Design, 2nd ed. Upper Saddle River, NJ: Prentice Hall (2007).

Goo, Jahyun, Rajive Kishore, H. R. Rao, and Kichan Nam. The Role of Service Level Agreements in Relational Management of Information Technology Outsourcing: An Empirical Study." MIS Quarterly 33, No. 1 (March 2009).

Grunbacher, Paul, Michael Halling, Stefan Biffl, Hasan Kitapci, and Barry W. Boehm. "Integrating Collaborative Processes and Quality Assurance Techniques: Experiences from Requirements Negotiation." Journal of Management Information Systems 20, no. 4 (Spring 2004).

Hahn, Eugene D., Jonathan P. Doh, and Kraiwinee Bunyaratavej. "The Evolution of Risk in Information Systems Offshoring: The Impact of Home Country Risk, Firm Learning, and Competitive Dynamics. MIS Quarterly 33, No. 3 (September 2009).

Hammer, Michael, and James Champy. Reengineering the Corporation. New York: HarperCollins (1993).

Hammer, Michael."Process Management and the Future of Six Sigma." "Sloan Management Review 43, no.2 (Winter 2002)

Hickey, Ann M., and Alan M. Davis. "A Unified Model of Requirements Elicitation." Journal of Management Information Systems 20, no. 4 (Spring 2004).

Hirscheim, Rudy and Mary Lacity."The Myths and Realities of Information Technology Insourcing." Communications of the ACM 43, no. 2 (February 2000).

Hoffer, Jeffrey, Joey George, and Joseph Valacich. Modern Systems Analysis and Design, 5th ed. Upper Saddle River, NJ: Prentice Hall (2008).

Hopkins, Jon. "Component Primer." Communications of the ACM 43, no. 10 (October 2000).

Irwin, Gretchen. "The Role of Similarity in the Reuse of Object-Oriented Analysis Models." Journal of Management Information Systems 19, no. 2 (Fall 2002).

Ivari, Juhani, Rudy Hirscheim, and Heinz K. Klein. "A Dynamic Framework for Classifying Information Systems Development Methodologies and Approaches." Journal of Management Information Systems 17, no. 3 (Winter 2000-2001).

Iyer, Bala, Jim Freedman, Mark Gaynor, and George Wyner. "Web Services: Enabling Dynamic Business Networks." Communications of the Association for Information Systems 11 (2003).

Johnson, Richard A. The Ups and Downs of Object-Oriented Systems Development." Communications of the ACM 43, no.10 (October 2000).

Kendall, Kenneth E., and Julie E. Kendall. Systems Analysis and Design, 9th ed. Upper Saddle River, NJ: Prentice Hall (2011).

Kettinger, William J., and Choong C. Lee. "Understanding the IS-User Divide in IT Innovation." Communications of the ACM 45, no.2 (February 2002).

Kindler, Noah B., Vasantha Krishnakanthan, and Ranjit Tinaikar. "Applying Lean to Application Development and Maintenance." The McKinsey Quarterly (May 2007).

Koh, Christine, Song Ang, and Detmar W. Straub. "IT Outsourcing Success: A Psychological Contract Perspective." Information Systems Research 15 no. 4 (December 2004).

Krishna, S., Sundeep Sahay, and Geoff Walsham. "Managing Cross-Cultural Issues in Global Software Outsourcing." Communications of the ACM 47, No. 4 (April 2004).

Lee, Gwanhoo and Weidong Xia. "Toward Agile: An Integrated Analysis of Quantitative and Qualitative Field Data." MIS Quarterly 34, no. 1 (March 2010).

Lee, Jae Nam, Shaila M. Miranda, and Yong-Mi Kim."IT Outsourcing Strategies: Universalistic, Contingency, and Configurational Explanations of Success." Information Systems Research 15, no. 2 (June 2004).

Levina, Natalia, and Jeanne W. Ross. "From the Vendor's Perspective: Exploring the Value Proposition in Information Technology Outsourcing." MIS Quarterly 27, no. 3 (September 2003).

Limayem, Moez, Mohamed Khalifa, and Wynne W. Chin. "Case Tools Usage and Impact on System Development Performance." Journal of Organizational Computing and Electronic Commerce 14, no. 3 (2004).

Majchrzak, Ann, Cynthia M. Beath, and Ricardo A. Lim. "Managing Client Dialogues during Information Systems Design to Facilitate Client Learning." MIS Quarterly 29, no. 4 (December 2005).

Mani, Deepa, Anitesh Barua, and Andrew Whinston. "An Empirical Analysis of the Impact of Information Capabilities Design on Business Process Outsourcing Performance." MIS Quarterly 34, no. 1 (March 2010).

Nelson, H. James, Deborah J. Armstrong, and Kay M. Nelson. Patterns of Transition: The Shift from Traditional to Object-Oriented Development." Journal of Management Information Systems 25, No. 4 (Spring 2009).

Nidumolu, Sarma R. and Mani Subramani."The Matrix of Control: Combining Process and Structure Approaches to Managing Software Development." Journal of Management Information Systems 20, no. 4 (Winter 2004).

O'Donnell, Anthony. "BPM: Insuring Business Success." Optimize Magazine (April 2007).

Overby, Stephanie, "The Hidden Costs of Offshore Outsourcing," CIO Magazine (Sept.1, 2003).

Palmer, Jonathan W. "Web Site Usability, Design and Performance Metrics." Information Systems Research 13, no.3 (September 2002).

Phillips, James and Dan Foody. "Building a Foundation for Web Services." EAI Journal (March 2002).

Pitts, Mitzi G. and Glenn J. Browne."Stopping Behavior of Systems Analysts During Information Requirements Elicitation." Journal of Management Information Systems 21, no. 1 (Summer 2004).

Prahalad, C. K. and M.S.. Krishnan."Synchronizing Strategy and Information Technology." Sloan Management Review 43, no. 4 (Summer 2002).

Ravichandran, T. and Marcus A. Rothenberger."Software Reuse Strategies and Component Markets." Communications of the ACM 46, no. 8 (August 2003).

Silva, Leiser and Rudy Hirschheim. "Fighting Against Windmills: Strategic Information Systems and Organizational Deep Structures." MIS Quarterly 31, no. 2 (June 2007).

Sircar, Sumit, Sridhar P. Nerur, and Radhakanta Mahapatra. "Revolution or Evolution? A Comparison of Object-Oriented and Structured Systems Development Methods." MIS Quarterly 25, no. 4 (December 2001).

Smith, Howard and Peter Fingar. Business Process Management: The Third Wave Tampa, Florida: Meghan-Kiffer Press (2002).

Swanson, E. Burton and Enrique Dans. "System Life Expectancy and the Maintenance Effort: Exploring their Equilibration." MIS Quarterly 24, no. 2 (June 2000).

Turetken, Ozgur, David Schuff, Ramesh Sharda, and Terence T. Ow. "Supporting Systems Analysis and Design through Fisheye Views." Communications of the ACM 47, no. 9 (September 2004).

Van Den Heuvel, Willem-Jan and Zakaria Maamar. "Moving Toward a Framework to Compose Intelligent Web Services." Communications of the ACM 46, no. 10 (October 2003).

Vitharana, Padmal."Risks and Challenges of Component-Based Software Development." Communications of the ACM 46, no. 8 (August 2003).

Watad, Mahmoud M. and Frank J. DiSanzo. "Case Study: The Synergism of Telecommuting and Office Automation." Sloan Management Review 41, no. 2 (Winter 2000).

Wulf, Volker, and Matthias Jarke. "The Economics of End-User Development." Communications of the ACM 47, no. 9 (September 2004).

Yourdon, Edward, and L. L. Constantine. Structured Design. New York: Yourdon Press (1978).

CHAPTER 14

Aladwani, Adel M. "An Integrated Performance Model of Information Systems Projects." Journal of Management Information Systems 19, no.1 (Summer 2002).

Alleman, James."Real Options Real Opportunities." Optimize Magazine (January 2002).

Andres, Howard P. and Robert W. Zmud. "A Contingency Approach to Software Project Coordination." Journal of Management Information Systems 18, no. 3 (Winter 2001-2002).

Armstrong, Curtis P. and V. Sambamurthy."Information Technology Assimilation in Firms: The Influence of Senior Leadership and IT Infrastructures." Information Systems Research 10, no. 4 (December 1999).

Banker, Rajiv. "Value Implications of Relative Investments in Information Technology." Department of Information Systems and Center for Digital Economy Research, University of Texas at Dallas, January 23, 2001.

Barki, Henri and Jon Hartwick. "Interpersonal Conflict and Its Management in Information Systems Development. " MIS Quarterly 25, no.2 (June 2001).

Barki, Henri, Suzanne Rivard, and Jean Talbot. "An Integrative Contingency Model of Software Project Risk Management." Journal of Management Information Systems 17, no. 4 (Spring 2001).

Beath, Cynthia Mathis, and Wanda J. Orlikowski. "The Contradictory Structure of Systems Development Methodologies: Deconstructing the IS-User Relationship in Information Engineering." Information Systems Research 5, no. 4 (December 1994).

Benaroch, Michel and Robert J. Kauffman. "Justifying Electronic Banking Network Expansion Using Real Options Analysis." MIS Quarterly 24, no. 2 (June 2000).

Benaroch, Michel, Sandeep Shah, and Mark Jeffrey. "On the Valuation of Multistage Information Technology Investments Embedding Nested Real Options." Journal of Management Information Systems 23, No. 1 (Summer 2006).

Benaroch, Michel. "Managing Information Technology Investment Risk: A Real Options Perspective." Journal of Management Information Systems 19, no. 2 (Fall 2002).

Bhattacherjee, Anol and G. Premkumar. "Understanding Changes In Belief and Attitude Toward Information Technology Usage: A Theoretical Model and Longitudinal Test." MIS Quarterly 28, no. 2 (June 2004).

Blair, Leslie. "Reconciling IT Spend with C-Suite Expectations." Baseline (March/April 2010).

Bostrom, R. P., and J. S. Heinen. "MIS Problems and Failures: A Socio-Technical Perspective. Part I: The Causes." MIS Quarterly 1 (September 1977); "Part II: The Application of Socio-Technical Theory." MIS Quarterly 1 (December 1977).

Brewer, Jeffrey and Kevin Dittman. Methods of IT Project Management. Upper Saddle River, NJ: Prentice-Hall (2010).

Brooks, Frederick P. "The Mythical Man-Month." Datamation (December 1974).

Brynjolfsson, Erik, and Lorin M. Hitt. "Information Technology and Organizational Design: Evidence from Micro Data." (January 1998).

Bullen, Christine, and John F. Rockart. "A Primer on Critical Success Factors." Cambridge, MA: Center for Information Systems Research, Sloan School of Management (1981).

Chatterjee, Debabroto, Rajdeep Grewal, and V. Sabamurthy. "Shaping Up for E-Commerce: Institutional Enablers of the Organizational Assimilation of Web Technologies." MIS Quarterly 26, no. 2 (June 2002).

Chickowski, Ericka. "Projects Gone Wrong." Baseline (May 15, 2009).

Clement, Andrew, and Peter Van den Besselaar. "A Retrospective Look at PD Projects." Communications of the ACM 36, no. 4 (June 1993).

Concours Group. "Delivering Large-Scale System Projects." (2000).

Cooper, Randolph B. "Information Technology Development Creativity: A Case Study of Attempted Radical Change." MIS Quarterly 24, no. 2 (June 2000).

Datz, Todd. "Portfolio Management: How to Do It Right," CIO Magazine (May 1, 2003).

De Meyer, Arnoud, Christoph H. Loch and Michael T. Pich." Managing Project Uncertainty: From Variation to Chaos." Sloan Management Review 43, no.2 (Winter 2002).

Delone, William H. and Ephraim R. McLean. "The Delone and McLean Model of Information Systems Success: A Ten-Year Update. Journal of Management Information Systems 19, no. 4 (Spring 2003).

Doll, William J., Xiaodung Deng, T. S. Raghunathan, Gholamreza Torkzadeh, and Weidong Xia. "The Meaning and Measurement of User Satisfaction: A Multigroup Invariance Analysis of End-User Computing Satisfaction Instrument." Journal of Management Information Systems 21, no. 1 (Summer 2004).

Feldman, Jonathan. "Get Your Projects In Line." Information Week (March 8, 2010).

Fichman, Robert G. "Real Options and IT Platforms Adoption: Implications for Theory and Practice." Information Systems Research 15, no. 2 (June 2004).

Fuller, Mark, Joe Valacich, and Joey George. Information Systems Project Management: A Process and Team Approach. Upper Saddle River, NJ: Prentice Hall (2008).

Geng, Xianjun, Lihui Lin, and Andrew B. Whinston."Effects of Organizational Learning and Knowledge Transfer on Investment Decisions Under Uncertainty." "Journal of Management Information Systems 26, No. 2 (Fall 2009).

Goff, Stacy A. "The Future of IT Project Management Software." CIO (January 6, 2010).

Gordon, Steven R. and Monideepa Tarafdar. "The IT Audit that Boosts Innovation." MIT Sloan Management Review 51, No. 4 (Summer 2010).

Hitt, Lorin, D.J. Wu, and Xiaoge Zhou. "Investment in Enterprise Resource Planning: Business Impact and Productivity Measures." Journal of Management Information Systems 19, no. 1 (Summer 2002).

Housel, Thomas J., Omar El Sawy, Jianfang Zhong, and Waymond Rodgers. "Measuring the Return on e-Business Initiatives at the Process Level: The Knowledge Value-Added Approach." ICIS (2001).

Iversen, Jakob H., Lars Mathiassen, and Peter Axel Nielsen."Managing Risk in Software Process Improvement: An Action Research Approach." MIS Quarterly 28, no. 3 (September 2004).

Jeffrey, Mark, and Ingmar Leliveld. "Best Practices in IT Portfolio Management." MIT Sloan Management Review 45, no. 3 (Spring 2004).

Jiang, James J., Gary Klein, Debbie Tesch, and Hong-Gee Chen. "Closing the User and Provider Service Quality Gap," Communications of the ACM 46, no.2 (February 2003).

Jun He and William R. King. "The Role of User Participation In Information Systems Development: Implications from a Meta-Analysis." Journal of Management Information Systems 25, no. 1 (Summer 2008).

Keen, Peter W. "Information Systems and Organizational Change." Communications of the ACM 24 (January 1981).

Keil, Mark and Daniel Robey. "Blowing the Whistle on Troubled Software Projects." Communications of the ACM 44, no. 4 (April 2001).

Keil, Mark and Ramiro Montealegre. "Cutting Your Losses: Extricating Your Organization When a Big Project Goes Awry." Sloan Management Review 41, no. 3 (Spring 2000).

Keil, Mark, Joan Mann, and Arun Rai. "Why Software Projects Escalate: An Empirical Analysis and Test of Four Theoretical Models." MIS Quarterly 24, no. 4 (December 2000).

Keil, Mark, Paul E. Cule, Kalle Lyytinen, and Roy C. Schmidt. "A Framework for Identifying Software Project Risks." Communications of the ACM 41, 11 (November 1998).

Kettinger, William J. and Choong C. Lee. "Understanding the IS-User Divide in IT Innovation." Communications of the ACM 45, no.2 (February 2002).

Kim, Hee Woo and Atreyi Kankanhalli. "Investigating User Resistance to Information Systems Implementation: A Status Quo Bias Perspective." MIS Quarterly 33, No. 3 (September 2009).

Klein, Gary, James J. Jiang, and Debbie B. Tesch. "Wanted: Project Teams with a Blend of IS Professional Orientations." Communications of the ACM 45, no. 6 (June 2002).

Kolb, D. A., and A. L. Frohman. "An Organization Development Approach to Consulting." Sloan Management Review 12 (Fall 1970).

Lapointe, Liette, and Suzanne Rivard. "A Multilevel Model of Resistance to Information Technology Implementation." MIS Quarterly 29, no. 3 (September 2005).

Laudon, Kenneth C. "CIOs Beware: Very Large Scale Systems." Center for Research on Information Systems, New York University Stern School of Business, working paper (1989).

Liang, Huigang, Nilesh Sharaf, Qing Hu, and Yajiong Xue. "Assimilation of Enterprise Systems: The Effect of Institutional Pressures and the Mediating Role of Top Management." MIS Quarterly 31, no 1 (March 2007).

Lipin, Steven and Nikhil Deogun. "Big Mergers of 90s Prove Disappointing to Shareholders." The Wall Street Journal (October 30, 2000).

Mahmood, Mo Adam, Laura Hall, and Daniel Leonard Swanberg, "Factors Affecting Information Technology Usage: A Meta-Analysis of the Empirical Literature." Journal of Organizational Computing and Electronic Commerce 11, no. 2 (November 2, 2001)

Markus, M. Lynne, and Robert I. Benjamin. "Change Agentry-The Next IS Frontier." MIS Quarterly 20, no. 4 (December 1996).

Markus, M. Lynne, and Robert I. Benjamin. "The Magic Bullet Theory of IT-Enabled Transformation." Sloan Management Review (Winter 1997).

McFarlan, F. Warren. "Portfolio Approach to Information Systems." Harvard Business Review (September-October 1981).

McGrath, Rita Gunther and Ian C.McMillan. "Assessing Technology Projects Using Real Options Reasoning." Industrial Research Institute (2000)

Mumford, Enid, and Mary Weir. Computer Systems in Work Design: The ETHICS Method. New York: John Wiley (1979).

Murray, Diane and Al Kagan. "Reinventing Program Management." CIO Insight (2nd Quarter 2010).

Nidumolu, Sarma R. and Mani Subramani. "The Matrix of Control: Combining Process and Structure Approaches to Management Software Development." Journal of Management Information Systems 20, no. 3 (Winter 2004).

Palmer, Jonathan W. "Web Site Usability, Design and Performance Metrics." Information Systems Research 13, no.3 (September 2002).

Peffers, Ken and Timo Saarinen. "Measuring the Business Value of IT Investments: Inferences from a Study of Senior Bank Executives." Journal of Organizational Computing and Electronic Commerce 12, no. 1 (2002).

Quan, Jin "Jim", Quing Hu, and Paul J. Hart."Information Technology Investments and Firms' Performance-A Duopoly Perspective." Journal of Management Information Systems 20, no. 3 (Winter 2004).

Rai, Arun, Sandra S. Lang, and Robert B. Welker. "Assessing the Validity of IS Success Models: An Empirical Test and Theoretical Analysis." Information Systems Research 13, no. 1 (March 2002).

Rapoza, Jim. "Next-Gen Project Management." eWeek (March 3, 2008).

Robey, Daniel, Jeanne W. Ross, and Marie-Claude Boudreau. "Learning to Implement Enterprise Systems: An Exploratory Study of the Dialectics of Change." Journal of Management Information Systems 19, no. 1 (Summer 2002).

Ross, Jeanne W. and Cynthia M. Beath." Beyond the Business Case: New Approaches to IT Investment." Sloan Management Review 43, no.2 (Winter 2002).

Ryan, Sherry D. and David A. Harrison. "Considering Social Subsystem Costs and Benefits in Information Technology Investment Decisions: A View from the Field on Anticipated Payoffs." Journal of Management Information Systems 16, no. 4 (Spring 2000).

Ryan, Sherry D., David A. Harrison, and Lawrence L Schkade." Information Technology Investment Decisions: When Do Cost and Benefits in the Social Subsystem Matter?" Journal of Management Information Systems 19, no. 2 (Fall 2002).

Sakthivel, S. "Managing Risk in Offshore Systems Development." Communications of the ACM 50, No. 4 (April 2007).

Sambamurthy, V., Anandhi Bharadwaj, and Varun Grover. "Shaping Agility Through Digital Options: Reconceptualizing the Role of Information Technology in Contemporary Firms." MIS Quarterly 27, no. 2 (June 2003).

Santhanam, Radhika and Edward Hartono. "Issues in Linking Information Technology Capability to Firm Performance." MIS Quarterly 27, no. 1 (March 2003).

Sauer, Chris and Leslie P. Willcocks, "The Evolution of the Organizational Architect." Sloan Management Review 43, no. 3 (Spring 2002).

Sauer, Chris, Andrew Gemino, and Blaize Horner Reich. "The Impact of Size and Volatility on IT Project Performance. "Communications of the ACM 50, no. 11 (November 2007).

Schmidt, Roy, Kalle Lyytinen, Mark Keil, and Paul Cule. "Identifying Software Project Risks: An International Delphi Study." Journal of Management Information Systems 17, no. 4 (Spring 2001)

Schneiderman, Ben." Universal Usability." Communications of the ACM 43, no. 5 (May 2000).

Schwalbe, Kathy. Information Technology Project Management, 6/e. Course Technology (2010).

Shank, Michael E., Andrew C. Boynton, and Robert W. Zmud. "Critical Success Factor Analysis as a Methodology for MIS Planning." MIS Quarterly (June 1985).

Sharma, Rajeev and Philip Yetton. "The Contingent Effects of Training, Technical Complexity, and Task Interdependence on Successful Information Systems Implementation." MIS Quarterly 31, no. 2 (June 2007).

Siewiorek, Daniel P. "New Frontiers of Application Design." Communications of the ACM 45, no.12 (December 2002.)

Smith, H. Jeff, Mark Keil, and Gordon Depledge. "Keeping Mum as the Project Goes Under." Journal of Management Information Systems 18, no. 2 (Fall 2001)

Speier, Cheri and Michael. G. Morris. "The Influence of Query Interface Design on Decision-Making Performance." MIS Quarterly 27, no. 3 (September 2003).

Straub, Detmar W., Arun Rai and Richard Klein. "Measuring Firm Performance at the Network Level: A Nomology of the Business Impact of Digital Supply Networks." Journal of Management Information Systems 21, no 1 (Summer 2004).

Swanson, E. Burton. Information System Implementation. Homewood, IL: Richard D. Irwin (1988).

Tallon, Paul P, Kenneth L. Kraemer, and Vijay Gurbaxani. "Executives' Perceptions of the Business Value of Information Technology: A Process-Oriented Approach." Journal of Management Information Systems 16, no. 4 (Spring 2000).

Taudes, Alfred, Markus Feurstein, and Andreas Mild. "Options Analysis of Software Platform Decisions: A Case Study." MIS Quarterly 24, no. 2 (June 2000).

Thatcher, Matt E. and Jim R. Oliver. "The Impact of Technology Investments on a Firm's Production Efficiency, Product Quality, and Productivity." Journal of Management Information Systems 18, no. 2 (Fall 2001).

Tiwana, Amrit, and Mark Keil. "Control in Internal and Outsourced Software Projects." Journal of Management Information Systems 26, No. 3 (Winter 2010).

Tornatsky, Louis G., J. D. Eveland, M. G. Boylan, W. A. Hetzner, E. C. Johnson, D. Roitman, and J. Schneider. The Process of Technological Innovation: Reviewing the Literature. Washington, DC: National Science Foundation (1983).

Venkatesh, Viswanath, Michael G. Morris, Gordon B Davis, and Fred D. Davis." User Acceptance of Information Technology: Toward a Unified View." MIS Quarterly 27, No. 3 (September 2003).

Wallace, Linda and Mark Keil. "Software Project Risks and Their Effect on Outcomes." Communications of the ACM 47, no. 4 (April 2004).

Wang, Eric T.G., Gary Klein, and James J. Jiang. "ERP Misfit: Country of Origin and Organizational Factors." Journal of Management Information Systems 23, No. 1 (Summer 2006).

Xia, Weidong and Gwanhoo Lee. "Grasping the Complexity of IS Development Projects." Communications of the ACM 47, no. 5 (May 2004).

Xia, Weidong, and Gwanhoo Lee. "Complexity of Information Systems Development Projects." Journal of Management Information Systems 22, no. 1 (Summer 2005).

Xue, Yajion, Huigang Liang, and William R. Boulton. "Information Technology Governance in Information Technology Investment Decision Processes: The Impact of Investment Characteristics, External Environment, and Internal Context." MIS Quarterly 32, no. 1 (March 2008).

Yin, Robert K. "Life Histories of Innovations: How New Practices Become Routinized." Public Administration Review (January-February 1981).

Zhu, Kevin and Kenneth L. Kraemer."E-Commerce Metrics for Net-Enhanced Organizations: Assessing the Value of e-Commerce to Firm Performance in the Manufacturing Sector." Information Systems Research 13, no.3 (September 2002).

Zhu, Kevin, Kenneth L. Kraemer, Sean Xu, and Jason Dedrick. "Information Technology Payoff in E-Business Environments: An International Perspective on Value Creation of E-business in the Financial Services Industry." Journal of Management Information Systems 21, no. 1 (Summer 2004).

Zhu, Kevin. "The Complementarity of Information Technology Infrastructure and E-Commerce Capability: A Resource-Based Assessment of Their Business Value." Journal of Management Information Systems 21, no. 1 (Summer 2004).

CHAPTER 15

Barboza, David. "Supply Chain for IPhone Highlights Costs in China." The New York Times (July 5, 2010).

Biehl, Markus. "Success Factors For Implementing Global Information Systems." Communications of the ACM 50, No. 1 (January 2007).

Bisson, Peter, Elizabeth Stephenson, and S. Patrick Viguerie. "Global Forces: An Introduction." McKinsey Quarterly (June 2010).

Cox, Butler. Globalization: The IT Challenge. Sunnyvale, CA: Amdahl Executive Institute (1991).

Davis. Bob. "Rise of Nationalism Frays Global Ties." The Wall Street Journal (April 28, 2008).

Davison, Robert. "Cultural Complications of ERP." Communications of the ACM 45, no. 7 (July 2002).

Deans, Candace P., and Michael J. Kane. International Dimensions of Information Systems and Technology. Boston, MA: PWS-Kent (1992).

Ein-Dor, Philip, Seymour E. Goodman, and Peter Wolcott." From Via Maris to Electronic Highway: The Internet in Canaan." Communications of the ACM 43, no. 7 (July 2000).

Farhoomand, Ali, Virpi Kristiina Tuunainen, and Lester W. Yee. "Barrier to Global Electronic Commerce: A Cross-Country

Study of Hong Kong and Finland." Journal of Organizational Computing and Electronic Commerce 10, no. 1 (2000).

Ghislanzoni, Giancarlo, Risto Penttinen, an David Turnbull. "The Multilocal Challenge: Managing Cross-Border Functions." The McKinsey Quarterly (March 2008).

Giridharadas, Anand. "Where a Cellphone Is Still Cutting Edge," The New York Times (April 11, 2010).

Ives, Blake, and Sirkka Jarvenpaa. "Applications of Global Information Technology: Key Issues for Management." MIS Quarterly 15, no. 1 (March 1991).

Ives, Blake, S. L. Jarvenpaa, R. O. Mason, "Global business drivers: Aligning Information Technology to Global Business Strategy," IBM Systems Journal Vol 32, No 1, 1993.

King, William R. and Vikram Sethi. "An Empirical Analysis of the Organization of Transnational Information Systems." Journal of Management Information Systems 15, no. 4 (Spring 1999).

Kirsch, Laurie J. "Deploying Common Systems Globally: The Dynamic of Control." Information Systems Research 15, no. 4 (December 2004).

Lai, Vincent S. and Wingyan Chung. "Managing International Data Communication." Communications of the ACM 45, no.3 (March 2002).

Liang, Huigang, Yajiong Xue, William R. Boulton, and Terry Anthony Byrd. "Why Western Vendors Don't Dominate China's ERP Market." Communications of the ACM 47, no. 7 (July 2004).

Mann, Catherine L. "What Global Sourcing Means for U.S. I.T. Workers and for the U.S. Economy." Communications of the ACM 47, no. 7 (July 2004)

Martinsons, Maris G. "ERP In China: One Package Two Profiles," Communications of the ACM 47, no. 7 (July 2004).

Petrazzini, Ben, and Mugo Kibati. "The Internet in Developing Countries." Communications of the ACM 42, no. 6 (June 1999).

Quelch, John A., and Lisa R. Klein. "The Internet and International Marketing." Sloan Management Review (Spring 1996).

Roche, Edward M. Managing Information Technology in Multinational Corporations. New York: Macmillan (1992).

Shore, Barry. "Enterprise Integration Across the Globally Dispersed Service Organization. Communications of the ACM 49, NO. 6 (June 2006).

Soh, Christina, Sia Siew Kien, and Joanne Tay-Yap. "Cultural Fits and Misfits: Is ERP a Universal Solution? "Communications of the ACM 43, no. 3 (April 2000).

Steel, Emily and Amol Sharma. "U.S. Web Sites Draw Traffic from Abroad but Few Ads." The Wall Street Journal (July 10, 2008).

Tan, Zixiang, William Foster, and Seymour Goodman. "China's State-Coordinated Internet Infrastructure." Communications of the ACM 42, no. 6 (June 1999).

Tractinsky, Noam, and Sirkka L. Jarvenpaa. "Information Systems Design Decisions in a Global Versus Domestic Context." MIS Quarterly 19, no. 4 (December 1995).

Watson, Richard T., Gigi G. Kelly, Robert D. Galliers, and James C. Brancheau. "Key Issues in Information Systems Management: An International Perspective." Journal of Management Information Systems 13, no. 4 (Spring 1997)

Glossary

3G networks Cellular networks based on packet-switched technology with speeds ranging from 144 Kbps for mobile users to over 2 Mbps for stationary users, enabling users to transmit video, graphics, and other rich media, in addition to voice.

4G networks The next evolution in wireless communication is entirely packet switched and capable of providing between 1 Mbps and 1 Gbps speeds; up to ten times faster than 3G networks. Not widely deployed in 2010.

acceptable use policy (AUP) Defines acceptable uses of the firm's information resources and computing equipment, including desktop and laptop computers, wireless devices, telephones, and the Internet, and specifies consequences for noncompliance.

acceptance testing Provides the final certification that the system is ready to be used in a production setting.

accountability The mechanisms for assessing responsibility for decisions made and actions taken.

accumulated balance digital payment systems Systems enabling users to make micropayments and purchases on the Web, accumulating a debit balance on their credit card or telephone bills.

affiliate revenue model an e-commerce revenue model in which Web sites are paid as "affiliates" for sending their visitors to other sites in return for a referral fee.

agent-based modeling Modeling complex phenomena as systems of autonomous agents that follow relatively simple rules for interaction.

agency theory Economic theory that views the firm as a nexus of contracts among self-interested individuals who must be supervised and managed.

agile development Rapid delivery of working software by breaking a large project into a series of small sub-projects that are completed in short periods of time using iteration and continuous feedback.

Ajax Development technique for creating interactive Web applications capable of updating the user interface without reloading the entire browser page.

analog signal A continuous waveform that passes through a communications medium; used for voice communications.

analytical CRM Customer relationship management applications dealing with the analysis of customer data to provide information for improving business performance.

Android A mobile operating system developed by Android, Inc. (purchased by Google) and later the Open Handset Alliance as a flexible, upgradeable mobile device platform.

antivirus software Software designed to detect, and often eliminate, computer viruses from an information system.

application controls: Specific controls unique to each computerized application that ensure that only authorized data are completely and accurately processed by that application.

application server Software that handles all application operations between browser-based computers and a company's back-end business applications or databases.

application software package A set of prewritten, precoded application software programs that are commercially available for sale or lease.

application software Programs written for a specific application to perform functions specified by end users.

apps Small pieces of software that run on the Internet, on your computer, or on your cell phone and are generally delivered over the Internet.

artificial intelligence (AI) The effort to develop computer-based systems that can behave like humans, with the ability to learn languages, accomplish physical tasks, use a perceptual apparatus, and emulate human expertise and decision making.

attribute A piece of information describing a particular entity.

augmented reality A technology for enhancing visualization. Provides a live direct or indirect view of a physical real-world environment whose elements are augmented by virtual computer-generated imagery.

authentication The ability of each party in a transaction to ascertain the identity of the other party.

authorization management systems Systems for allowing each user access only to those portions of a system or the Web that person is permitted to enter, based on information established by a set of access rules.

authorization policies Determine differing levels of access to information assets for different levels of users in an organization.

automation Using the computer to speed up the performance of existing tasks.

autonomic computing Effort to develop systems that can manage themselves without user intervention.

backward chaining A strategy for searching the rule base in an expert system that acts like a problem solver by beginning with a hypothesis and seeking out more information until the hypothesis is either proved or disproved.

balanced scorecard method Framework for operationalizing a firms strategic plan by focusing on measurable financial, business process, customer, and learning and growth outcomes of firm performance.

bandwidth The capacity of a communications channel as measured by the difference between the highest and lowest frequencies that can be transmitted by that channel.

banner ad A graphic display on a Web page used for advertising. The banner is linked to the advertiser's Web site so that a person clicking on it will be transported to the advertiser's Web site.

batch processing A method of collecting and processing data in which transactions are accumulated and stored until a specified time when it is convenient or necessary to process them as a group.

baud A change in signal from positive to negative or vice versa that is used as a measure of transmission speed.

behavioral models Descriptions of management based on behavioral scientists' observations of what managers actually do in their jobs.

behavioral targeting Tracking the click-streams (history of clicking behavior) of individuals across multiple Web sites for the purpose of understanding their interests and intentions, and exposing them to advertisements which are uniquely suited to their interests.

benchmarking Setting strict standards for products, services, or activities and measuring organizational performance against those standards.

best practices The most successful solutions or problem-solving methods that have been developed by a specific organization or industry.

biometric authentication Technology for authenticating system users that compares a person's unique characteristics such as fingerprints, face, or retinal image, against a stored set profile of these characteristics.

bit A binary digit representing the smallest unit of data in a computer system. It can only have one of two states, representing 0 or 1.

blade server Entire computer that fits on a single, thin card (or blade) and that is plugged into a single chassis to save space, power and complexity.

blog Popular term for Weblog, designating an informal yet structured Web site where individuals can publish stories, opinions, and links to other Web sites of interest.

blogosphere Totality of blog-related Web sites.

Bluetooth Standard for wireless personal area networks that can transmit up to 722 Kbps within a 10-meter area.

botnet A group of computers that have been infected with bot malware without users' knowledge, enabling a hacker to use the amassed resources of the computers to launch distributed denial-of-service attacks, phishing campaigns or spam.

broadband High-speed transmission technology. Also designates a single communications medium that can transmit multiple channels of data simultaneously.

bugs Software program code defects.

bullwhip effect Distortion of information about the demand for a product as it passes from one entity to the next across the supply chain.

bus topology Network topology linking a number of computers by a single circuit with all messages broadcast to the entire network.

business continuity planning Planning that focuses on how the company can restore business operations after a disaster strikes.

business driver A force in the environment to which businesses must respond and that influences the direction of business.

business ecosystem Loosely coupled but interdependent networks of suppliers, distributors, outsourcing firms, transportation service firms, and technology manufacturers

business functions Specialized tasks performed in a business organization, including manufacturing and production, sales and marketing, finance and accounting, and human resources.

business intelligence Applications and technologies to help users make better business decisions.

business model An abstraction of what an enterprise is and how the enterprise delivers a product or service, showing how the enterprise creates wealth.

business performance management Attempts to systematically translate a firm's strategies (e.g., differentiation, low-cost producer, market share growth, and scope of operation) into operational targets.

business process management Business process management (BPM) is an approach to business which aims to continuously improve and manage business processes.

business process redesign Type of organizational change in which business processes are analyzed, simplified, and redesigned.

business processes The unique ways in which organizations coordinate and organize work activities, information, and knowledge to produce a product or service.

business-to-business (B2B) electronic commerce Electronic sales of goods and services among businesses.

business-to-consumer (B2C) electronic commerce Electronic retailing of products and services directly to individual consumers.

byte A string of bits, usually eight, used to store one number or character in a computer system.

cable Internet connections Internet connections that use digital cable lines to deliver high-speed Internet access to homes and businesses.

call center An organizational department responsible for handling customer service issues by telephone and other channels.

capacity planning The process of predicting when a computer hardware system becomes saturated to ensure that adequate computing resources are available for work of different priorities and that the firm has enough computing power for its current and future needs.

capital budgeting The process of analyzing and selecting various proposals for capital expenditures.

carpal tunnel syndrome (CTS) Type of RSI in which pressure on the median nerve through the wrist's bony carpal tunnel structure produces pain.

case-based reasoning (CBR) Artificial intelligence technology that represents knowledge as a database of cases and solutions.

cell phone A device that transmits voice or data, using radio waves to communicate with radio antennas placed within adjacent geographic areas called cells.

centralized processing Processing that is accomplished by one large central computer.

change agent In the context of implementation, the individual acting as the catalyst during the change process to ensure successful organizational adaptation to a new system or innovation.

change management Managing the impact of organizational change associated with an innovation, such as a new information system.

channel conflict Competition between two or more different distribution chains used to sell the products or services of the same company.

channel The link by which data or voice are transmitted between sending and receiving devices in a network.

chat Live, interactive conversations over a public network.

chief information officer (CIO) Senior manager in charge of the information systems function in the firm.

chief knowledge officer (CKO) Senior executive in charge of the organization's knowledge management program.

chief privacy officer (CPO) Responsible for ensuring the company complies with existing data privacy laws.

chief security officer (CSO) Heads a formal security function for the organization and is responsible for enforcing the firm's security policy.

choice Simon's third stage of decision making, when the individual selects among the various solution alternatives.

Chrome OS Google's lightweight computer operating system for users who do most of their computing on the Internet; runs on computers ranging from netbooks to desktop computers.

churn rate Measurement of the number of customers who stop using or purchasing products or services from a company. Used as an indicator of the growth or decline of a firm's customer base.

classical model of management Traditional description of management that focused on its formal functions of planning, organizing, coordinating, deciding, and controlling.

click fraud Fraudulently clicking on an online ad in pay per click advertising to generate an improper charge per click.

clicks-and-mortar Business model where the Web site is an extension of a traditional bricks-and-mortar business.

clickstream tracking Tracking data about customer activities at Web sites and storing them in a log.

client The user point-of-entry for the required function in client/server computing. Normally a desktop computer, workstation, or laptop computer.

client/server computing A model for computing that splits processing between clients and servers on a network, assigning functions to the machine most able to perform the function.

cloud computing Web-based applications that are stored on remote servers and accessed via the "cloud" of the Internet using a standard Web browser.

coaxial cable A transmission medium consisting of thickly insulated copper wire; can transmit large volumes of data quickly.

Code Division Multiple Access (CDMA) Major cellular transmission standard in the United States that transmits over several frequencies, occupies the entire spectrum, and randomly assigns users to a range of frequencies over time.

collaboration Working with others to achieve shared and explicit goals.

collaborative commerce The use of digital technologies to enable multiple organizations to collaboratively design, develop, build and manage products through their life cycles.

collaborative filtering Tracking users' movements on a Web site, comparing the information gleaned about a user's behavior against data about other customers with similar interests to predict what the user would like to see next.

collaborative planning, forecasting, and replenishment (CPFR) Firms collaborating with their suppliers and buyers to formulate demand forecasts, develop production plans, and coordinate shipping, warehousing, and stocking activities.

co-location a kind of Web site hosting in which firm purchase or rent a physical server computer at a hosting company's location in order to operate a Web site.

community provider a Web site business model that creates a digital online environment where people with similar interests can transact (buy and sell goods); share interests, photos, videos; communicate with like-minded people; receive interest-related information; and even play out fantasies by adopting online personalities called avatars.

competitive forces model Model used to describe the interaction of external influences, specifically threats and opportunities, that affect an organization's strategy and ability to compete.

complementary assets Additional assets required to derive value from a primary investment.

component-based development Building large software systems by combining pre-existing software components.

computer Physical device that takes data as an input, transforms the data by executing stored instructions, and outputs information to a number of devices.

computer abuse The commission of acts involving a computer that may not be illegal but are considered unethical.

computer crime The commission of illegal acts through the use of a computer or against a computer system.

computer forensics The scientific collection, examination, authentication, preservation, and analysis of data held on or retrieved from computer storage media in such a way that the information can be used as evidence in a court of law.

computer hardware Physical equipment used for input, processing, and output activities in an information system.

computer literacy Knowledge about information technology, focusing on understanding of how computer-based technologies work.

computer software Detailed, preprogrammed instructions that control and coordinate the work of computer hardware components in an information system.

computer virus Rogue software program that attaches itself to other software programs or data files in order to be executed, often causing hardware and software malfunctions.

computer vision syndrome (CVS) Eyestrain condition related to computer display screen use; symptoms include headaches, blurred vision, and dry and irritated eyes.

computer-aided design (CAD) Information system that automates the creation and revision of designs using sophisticated graphics software.

computer-aided software engineering (CASE) Automation of step-by-step methodologies for software and systems development to reduce the amounts of repetitive work the developer needs to do.

computer-based information systems (CBIS) Information systems that rely on computer hardware and software for processing and disseminating information.

connectivity The ability of computers and computer-based devices to communicate with each other and share information in a meaningful way without human intervention..

consumer-to-consumer (C2C) electronic commerce Consumers selling goods and services electronically to other consumers.

controls All of the methods, policies, and procedures that ensure protection of the organization's assets, accuracy and reliability of its records, and operational adherence to management standards.

conversion The process of changing from the old system to the new system.

cookies Tiny file deposited on a computer hard drive when an individual visits certain Web sites. Used to identify the visitor and track visits to the Web site.

cooptation Bringing the opposition into the process of designing and implementing a solution without giving up control of the direction and nature of the change.

copyright A statutory grant that protects creators of intellectual property against copying by others for any purpose for a minimum of 70 years.

core competency Activity at which a firm excels as a world-class leader.

core systems Systems that support functions that are absolutely critical to the organization.

cost transparency the ability of consumers to discover the actual costs merchants pay for products.

counterimplementation A deliberate strategy to thwart the implementation of an information system or an innovation in an organization.

critical success factors (CSFs) A small number of easily identifiable operational goals shaped by the industry, the firm, the manager, and the broader environment that are believed to assure the success of an organization. Used to determine the information requirements of an organization.

cross-selling Marketing complementary products to customers.

crowdsourcing Using large Internet audiences for advice, market feedback, new ideas and solutions to business problems. Related to the 'wisdom of crowds' theory.

culture The set of fundamental assumptions about what products the organization should produce, how and where it should produce them, and for whom they should be produced.

customer lifetime value (CLTV) Difference between revenues produced by a specific customer and the expenses for acquiring and servicing that customer minus the cost of promotional marketing over the lifetime of the customer relationship, expressed in today's dollars.

customer relationship management (CRM) Business and technology discipline that uses information systems to coordinate all of the business processes surrounding the firm's interactions with its customers in sales, marketing, and service.

customer relationship management systems Information systems that track all the ways in which a company interacts with its customers and analyze these interactions to optimize revenue, profitability, customer satisfaction, and customer retention.

customization The modification of a software package to meet an organization's unique requirements without destroying the package software's integrity.

cybervandalism Intentional disruption, defacement, or destruction of a Web site or corporate information system.

data Streams of raw facts representing events occurring in organizations or the physical environment before they have been organized and arranged into a form that people can understand and use.

data administration A special organizational function for managing the organization's data resources, concerned with information policy, data planning, maintenance of data dictionaries, and data quality standards.

data cleansing Activities for detecting and correcting data in a database or file that are incorrect, incomplete, improperly formatted, or redundant. Also known as data scrubbing.

data definition DBMS capability that specifies the structure and content of the database.

data dictionary An automated or manual tool for storing and organizing information about the data maintained in a database.

data-driven DSS A system that supports decision making by allowing users to extract and analyze useful information that was previously buried in large databases.

data element A field.

data flow diagram (DFD) Primary tool for structured analysis that graphically illustrates a system's component process and the flow of data between them.

data governance Policies and processes for managing the availability, usability, integrity, and security of the firm's data.

data inconsistency The presence of different values for same attribute when the same data are stored in multiple locations.

data management software Software used for creating and manipulating lists, creating files and databases to store data, and combining information for reports.

data manipulation language A language associated with a database management system that end users and programmers use to manipulate data in the database.

data mart A small data warehouse containing only a portion of the organization's data for a specified function or population of users.

data mining Analysis of large pools of data to find patterns and rules that can be used to guide decision making and predict future behavior.

data quality audit A survey and/or sample of files to determine accuracy and completeness of data in an information system.

data redundancy The presence of duplicate data in multiple data files.

data visualization Technology for helping users see patterns and relationships in large amounts of data by presenting the data in graphical form.

data warehouse A database, with reporting and query tools, that stores current and historical data extracted from various operational systems and consolidated for management reporting and analysis.

data workers People such as secretaries or bookkeepers who process the organization's paperwork.

database A group of related files.

database (rigorous definition) A collection of data organized to service many applications at the same time by storing and managing data so that they appear to be in one location.

database administration Refers to the more technical and operational aspects of managing data, including physical database design and maintenance.

database management system (DBMS) Special software to create and maintain a database and enable individual business applications to extract the data they need without having to create separate files or data definitions in their computer programs.

database server A computer in a client/server environment that is responsible for running a DBMS to process SQL statements and perform database management tasks.

dataconferencing Teleconferencing in which two or more users are able to edit and modify data files simultaneously.

decisional roles Mintzberg's classification for managerial roles where managers initiate activities, handle disturbances, allocate resources, and negotiate conflicts.

decision-support systems (DSS) Information systems at the organization's management level that combine data and sophisticated analytical models or data analysis tools to support semistructured and unstructured decision making.

dedicated lines Telephone lines that are continuously available for transmission by a lessee. Typically conditioned to transmit data at high speeds for high-volume applications.

deep packet inspection (DPI) Technology for managing network traffic by examining data packets, sorting out low-priority data from higher priority business-critical data, and sending packets in order of priority.

demand planning Determining how much product a business needs to make to satisfy all its customers' demands.

denial of service (DoS) attack Flooding a network server or Web server with false communications or requests for services in order to crash the network.

Descartes' rule of change A principle that states that if an action cannot be taken repeatedly, then it is not right to be taken at any time.

design Simon's second stage of decision making, when the individual conceives of possible alternative solutions to a problem.

digital asset management systems Classify, store, and distribute digital objects such as photographs, graphic images, video, and audio content.

digital certificate An attachment to an electronic message to verify the identity of the sender and to provide the receiver with the means to encode a reply.

digital checking Systems that extend the functionality of existing checking accounts so they can be used for online shopping payments.

digital credit card payment system Secure services for credit card payments on the Internet that protect information transmitted among users, merchant sites, and processing banks.

digital dashboard Displays all of a firm's key performance indicators as graphs and charts on a single screen to provide one-page overview of all the critical measurements necessary to make key executive decisions.

digital divide Large disparities in access to computers and the Internet among different social groups and different locations.

digital firm Organization where nearly all significant business processes and relationships with customers, suppliers, and employees are digitally enabled, and key corporate assets are managed through digital means.

digital goods Goods that can be delivered over a digital network.

digital market A marketplace that is created by computer and communication technologies that link many buyers and sellers.

Digital Millennium Copyright Act (DMCA) Adjusts copyright laws to the Internet Age by making it illegal to make, distribute, or use devices that circumvent technology-based protections of copy-righted materials.

digital signal A discrete waveform that transmits data coded into two discrete states as 1-bits and 0-bits, which are represented as on-off electrical pulses; used for data communications.

digital subscriber line (DSL) A group of technologies providing high-capacity transmission over existing copper telephone lines.

digital wallet Software that stores credit card, electronic cash, owner identification, and address information and provides this data automatically during electronic commerce purchase transactions.

direct cutover A risky conversion approach where the new system completely replaces the old one on an appointed day.

disaster recovery planning Planning for the restoration of computing and communications services after they have been disrupted.

disintermediation The removal of organizations or business process layers responsible for certain intermediary steps in a value chain.

disruptive technologies Technologies with disruptive impact on industries and businesses, rendering existing products, services and business models obsolete.

distance learning Education or training delivered over a distance to individuals in one or more locations.

distributed database A database that is stored in more than one physical location. Parts or copies of the database are physically stored in one location, and other parts or copies are stored and maintained in other locations.

distributed denial-of-service (DDoS) attack Numerous computers inundating and overwhelming a network from numerous launch points.

distributed processing The distribution of computer processing work among multiple computers linked by a communications network.

documentation Descriptions of how an information system works from either a technical or end-user standpoint.

domain name English-like name that corresponds to the unique 32-bit numeric Internet Protocol (IP) address for each computer connected to the Internet

Domain Name System (DNS) A hierarchical system of servers maintaining a database enabling the conversion of domain names to their numeric IP addresses.

domestic exporter Form of business organization characterized by heavy centralization of corporate activities in the home county of origin.

downsizing The process of transferring applications from large computers to smaller ones.

downtime Period of time in which an information system is not operational.

drill down The ability to move from summary data to lower and lower levels of detail.

due process A process in which laws are well-known and understood and there is an ability to appeal to higher authorities to ensure that laws are applied correctly.

dynamic pricing Pricing of items based on real-time interactions between buyers and sellers that determine what a item is worth at any particular moment.

e-government Use of the Internet and related technologies to digitally enable government and public sector agencies' relationships with citizens, businesses, and other arms of government.

e-learning Instruction delivered through purely digital technology, such as CD-ROMs, the Internet, or private networks.

efficient customer response system System that directly links consumer behavior back to distribution, production, and supply chains.

electronic billing and payment presentation system Systems used for paying routine monthly bills that allow users to view their bills electronically and pay them through electronic funds transfers from banks or credit card accounts.

electronic business (e-business) The use of the Internet and digital technology to execute all the business processes in the enterprise. Includes e-commerce as well as processes for the internal management of the firm and for coordination with suppliers and other business partners.

electronic commerce The process of buying and selling goods and services electronically involving transactions using the Internet, networks, and other digital technologies.

electronic data interchange (EDI) The direct computer-to-computer exchange between two organizations of standard business transactions, such as orders, shipment instructions, or payments.

electronic payment system The use of digital technologies, such as credit cards, smart cards and Internet-based payment systems, to pay for products and services electronically.

e-mail The computer-to-computer exchange of messages.

employee relationship management (ERM) Software dealing with employee issues that are closely related to CRM, such as setting objectives, employee performance management, performance-based compensation, and employee training.

encryption The coding and scrambling of messages to prevent their being read or accessed without authorization.

end-user development The development of information systems by end users with little or no formal assistance from technical specialists.

end-user interface The part of an information system through which the end user interacts with the system, such as on-line screens and commands.

end users Representatives of departments outside the information systems group for whom applications are developed.

enterprise applications Systems that can coordinate activities, decisions, and knowledge across many different functions, levels, and business units in a firm. Include enterprise systems, supply chain management systems, and knowledge management systems.

enterprise content management systems Help organizations manage structured and semistructured knowledge, providing corporate repositories of documents, reports, presentations, and best practices and capabilities for collecting and organizing e-mail and graphic objects.

enterprise portal Web interface providing a single entry point for accessing organizational information and services, including information from various enterprise applications and in-house legacy systems so that information appears to be coming from a single source.

enterprise software Set of integrated modules for applications such as sales and distribution, financial accounting, investment management, materials management, production planning, plant maintenance, and human resources that allow data to be used by multiple functions and business processes.

enterprise systems Integrated enterprise-wide information systems that coordinate key internal processes of the firm.

enterprise-wide knowledge management systems General-purpose, firmwide systems that collect, store, distribute, and apply digital content and knowledge.

entity A person, place, thing, or event about which information must be kept.

entity-relationship diagram A methodology for documenting databases illustrating the relationship between various entities in the database.

ergonomics The interaction of people and machines in the work environment, including the design of jobs, health issues, and the end-user interface of information systems.

e-tailer Online retail stores from the giant Amazon to tiny local stores that have Web sites where retail goods are sold.

ethical "no free lunch" rule Assumption that all tangible and intangible objects are owned by someone else, unless there is a specific declaration otherwise, and that the creator wants compensation for this work.

ethics Principles of right and wrong that can be used by individuals acting as free moral agents to make choices to guide their behavior.

evil twins Wireless networks that pretend to be legitimate to entice participants to log on and reveal passwords or credit card numbers.

exchange Third-party Net marketplace that is primarily transaction oriented and that connects many buyers and suppliers for spot purchasing.

executive support systems (ESS) Information systems at the organization's strategic level designed to address unstructured decision making through advanced graphics and communications.

expert system Knowledge-intensive computer program that captures the expertise of a human in limited domains of knowledge.

explicit knowledge Knowledge that has been documented.

external integration tools Project management technique that links the work of the implementation team to that of users at all organizational levels.

extranet Private intranet that is accessible to authorized outsiders.

Fair Information Practices (FIP) A set of principles originally set forth in 1973 that governs the collection and use of information about individuals and forms the basis of most U.S. and European privacy laws.

fault-tolerant computer systems Systems that contain extra hardware, software, and power supply components that can back a system up and keep it running to prevent system failure.

feasibility study As part of the systems analysis process, the way to determine whether the solution is achievable, given the organization's resources and constraints.

feedback Output that is returned to the appropriate members of the organization to help them evaluate or correct input.

fiber-optic cable A fast, light, and durable transmission medium consisting of thin strands of clear glass fiber bound into cables. Data are transmitted as light pulses.

field A grouping of characters into a word, a group of words, or a complete number, such as a person's name or age.

file transfer protocol (FTP) Tool for retrieving and transferring files from a remote computer.

file A group of records of the same type.

firewall Hardware and software placed between an organization's internal network and an external network to prevent outsiders from invading private networks.

focused differentiation Competitive strategy for developing new market niches for specialized products or services where a business can compete in the target area better than its competitors.

folksonomies User-created taxonomies for classifying and sharing information.

foreign key Field in a database table that enables users find related information in another database table.

formal control tools Project management technique that helps monitor the progress toward completion of a task and fulfillment of goals.

formal planning tools Project management technique that structures and sequences tasks, budgeting time, money, and technical resources required to complete the tasks.

forward chaining A strategy for searching the rule base in an expert system that begins with the information entered by the user and searches the rule base to arrive at a conclusion.

fourth-generation language A programming language that can be employed directly by end users or less-skilled programmers to develop computer applications more rapidly than conventional programming languages.

franchiser Form of business organization in which a product is created, designed, financed, and initially produced in the home country, but for product-specific reasons relies heavily on foreign personnel for further production, marketing, and human resources.

free/fremium revenue model an e-commerce revenue model in which a firm offers basic services or content for free, while charging a premium for advanced or high value features.

fuzzy logic Rule-based AI that tolerates imprecision by using nonspecific terms called membership functions to solve problems.

Gantt chart Visually representats the timing, duration, and resource requirements of project tasks.

general controls Overall control environment governing the design, security, and use of computer programs and the security of data files in general throughout the organization's information technology infrastructure.

genetic algorithms Problem-solving methods that promote the evolution of solutions to specified problems using the model of living organisms adapting to their environment.

geographic information system (GIS) System with software that can analyze and display data using digitized maps to enhance planning and decision-making.

global culture The development of common expectations, shared artifacts, and social norms among different cultures and peoples

global positioning system (GPS) Worldwide satellite navigational system.

graphical user interface (GUI) The part of an operating system users interact with that uses graphic icons and the computer mouse to issue commands and make selections.

Gramm-Leach-Bliley Act Requires financial institutions to ensure the security and confidentiality of customer data.

green computing Refers to practices and technologies for designing, manufacturing, using, and disposing of computers, servers, and associated devices such as monitors, printers, storage devices, and networking and communications systems to minimize impact on the environment.

grid computing Applying the resources of many computers in a network to a single problem.

group decision-support system (GDSS) An interactive computer-based system to facilitate the solution to unstructured problems by a set of decision makers working together as a group.

hacker A person who gains unauthorized access to a computer network for profit, criminal mischief, or personal pleasure.

hertz Measure of frequency of electrical impulses per second, with 1 Hertz equivalent to 1 cycle per second.

high-availability computing Tools and technologies ,including backup hardware resources, to enable a system to recover quickly from a crash.

HIPAA Law outlining rules for medical security, privacy, and the management of health care records.

hotspot A specific geographic location in which an access point provides public Wi-Fi network service.

hubs Very simple devices that connect network components, sending a packet of data to all other connected devices.

hybrid AI systems Integration of multiple AI technologies into a single application to take advantage of the best features of these technologies.

hypertext markup language (HTML) Page description language for creating Web pages and other hypermedia documents.

hypertext transfer protocol (HTTP) The communications standard used to transfer pages on the Web. Defines how messages are formatted and transmitted.

identity management Business processes and software tools for identifying the valid users of a system and controlling their access to system resources.

identity theft Theft of key pieces of personal information, such as credit card or Social Security numbers, in order to obtain merchandise and services in the name of the victim or to obtain false credentials.

Immanuel Kant's Categorical Imperative A principle that states that if an action is not right for everyone to take it is not right for anyone.

implementation Simon's final stage of decision-making, when the individual puts the decision into effect and reports on the progress of the solution.

industry structure The nature of participants in an industry and their relative bargaining power. Derives from the competitive forces and establishes the general business environment in an industry and the overall profitability of doing business in that environment.

inference engine The strategy used to search through the rule base in an expert system; can be forward or backward chaining.

information Data that have been shaped into a form that is meaningful and useful to human beings.

information asymmetry Situation where the relative bargaining power of two parties in a transaction is determined by one party in the transaction possessing more information essential to the transaction than the other party.

information density The total amount and quality of information available to all market participants, consumers, and merchants.

information policy Formal rules governing the maintenance, distribution, and use of information in an organization.

information requirements A detailed statement of the information needs that a new system must satisfy; identifies who needs what information, and when, where, and how the information is needed.

information rights The rights that individuals and organizations have with respect to information that pertains to themselves.

information system Interrelated components working together to collect, process, store, and disseminate information to support decision making, coordination, control, analysis, and visualization in an organization.

information systems department The formal organizational unit that is responsible for the information systems function in the organization.

information systems literacy Broad-based understanding of information systems that includes behavioral knowledge about organizations and individuals using information systems as well as technical knowledge about computers.

information systems managers Leaders of the various specialists in the information systems department.

information systems plan A road map indicating the direction of systems development: the rationale, the current situation, the management strategy, the implementation plan, and the budget.

information technology (IT) All the hardware and software technologies a firm needs to achieve its business objectives.

information technology (IT) infrastructure Computer hardware, software, data, storage technology, and networks providing a portfolio of shared IT resources for the organization.

informational roles Mintzberg's classification for managerial roles where managers act as the nerve centers of their organizations, receiving and disseminating critical information.

informed consent Consent given with knowledge of all the facts needed to make a rational decision.

input The capture or collection of raw data from within the organization or from its external environment for processing in an information system.

instant messaging Chat service that allows participants to create their own private chat channels so that a person can be alerted whenever someone on his or her private list is on-line to initiate a chat session with that particular individual.

intangible benefits Benefits that are not easily quantified; they include more efficient customer service or enhanced decision making.

intellectual property Intangible property created by individuals or corporations that is subject to protections under trade secret, copyright, and patent law.

intelligence The first of Simon's four stages of decision making, when the individual collects information to identify problems occurring in the organization.

intelligent agent Software program that uses a built-in or learned knowledge base to carry out specific, repetitive, and predictable tasks for an individual user, business process, or software application.

internal integration tools Project management technique that ensures that the implementation team operates as a cohesive unit.

international information systems architecture The basic information systems required by organizations to coordinate worldwide trade and other activities.

Internet Global network of networks using universal standards to connect millions of different networks.

Internet Protocol (IP) address Four-part numeric address indicating a unique computer location on the Internet.

Internet Service Provider (ISP) A commercial organization with a permanent connection to the Internet that sells temporary connections to subscribers.

Internet telephony Technologies that use the Internet Protocol's packet-switched connections for voice service.

Internet2 Research network with new protocols and transmission speeds that provides an infrastructure for supporting high-bandwidth Internet applications.

interorganizational systems Information systems that automate the flow of information across organizational boundaries and link a company to its customers, distributors, or suppliers.

interpersonal roles Mintzberg's classification for managerial roles where managers act as figureheads and leaders for the organization.

intranet An internal network based on Internet and World Wide Web technology and standards.

intrusion detection system Tools to monitor the most vulnerable points in a network to detect and deter unauthorized intruders.

investment workstation Powerful desktop computer for financial specialists, which is optimized to access and manipulate massive amounts of financial data.

IT governance Strategy and policies for using information technology within an organization, specifying the decision rights and accountabilities to ensure that information technology supports the organization's strategies and objectives.

iterative A process of repeating over and over again the steps to build a system.

Java Programming language that can deliver only the software functionality needed for a particular task, such as a small applet downloaded from a network; can run on any computer and operating system.

Joint Application Design (JAD) Process to accelerate the generation of information requirements by having end users and information systems specialists work together in intensive interactive design sessions.

just-in-time Scheduling system for minimizing inventory by having components arrive exactly at the moment they are needed and finished goods shipped as soon as they leave the assembly line.

key field A field in a record that uniquely identifies instances of that record so that it can be retrieved, updated, or sorted.

key performance indicators Measures proposed by senior management for understanding how well the firm is performing along specified dimensions.

keylogger Spyware that records every keystroke made on a computer to steal personal information or passwords or to launch Internet attacks.

knowledge Concepts, experience, and insight that provide a framework for creating, evaluating, and using information.

knowledge—and information-intense products Products that require a great deal of learning and knowledge to produce.

knowledge base Model of human knowledge that is used by expert systems.

knowledge discovery Identification of novel and valuable patterns in large databases.

knowledge management The set of processes developed in an organization to create, gather, store, maintain, and disseminate the firm's knowledge.

knowledge management systems Systems that support the creation, capture, storage, and dissemination of firm expertise and knowledge.

knowledge network system Online directory for locating corporate experts in well-defined knowledge domains.

knowledge workers People such as engineers or architects who design products or services and create knowledge for the organization.

learning management system (LMS) Tools for the management, delivery, tracking, and assessment of various types of employee learning.

legacy system A system that has been in existence for a long time and that continues to be used to avoid the high cost of replacing or redesigning it.

legitimacy The extent to which one's authority is accepted on grounds of competence, vision, or other qualities. Making judgments and taking actions on the basis of narrow or personal characteristics.

liability The existence of laws that permit individuals to recover the damages done to them by other actors, systems, or organizations.

Linux Reliable and compactly designed operating system that is an offshoot of UNIX and that can run on many different hardware platforms and is available free or at very low cost. Used as alternative to UNIX and Windows NT.

local area network (LAN) A telecommunications network that requires its own dedicated channels and that encompasses a limited distance, usually one building or several buildings in close proximity.

long tail marketing Refers to the ability of firms to profitably market goods to very small online audiences, largely because of the lower costs of reaching very small market segments (people who fall into the long tail ends of a Bell curve).

mainframe Largest category of computer, used for major business processing.

maintenance Changes in hardware, software, documentation, or procedures to a production system to correct errors, meet new requirements, or improve processing efficiency.

malware Malicious software programs such as computer viruses, worms, and Trojan horses.

managed security service provider (MSSP) Company that provides security management services for subscribing clients.

management information systems (MIS) The study of information systems focusing on their use in business and management..

management-level systems Information systems that support the monitoring, controlling, decision-making, and administrative activities of middle managers.

managerial roles Expectations of the activities that managers should perform in an organization.

man-month The traditional unit of measurement used by systems designers to estimate the length of time to complete a project. Refers to the amount of work a person can be expected to complete in a month.

market creator An e-commerce business model in which firms provide a digital online environment where buyers and sellers can meet, search for products, and engage in transactions.

marketspace A marketplace extended beyond traditional boundaries and removed from a temporal and geographic location.

mashups Composite software applications that depend on high-speed networks, universal communication standards, and open-source code.

mass customization The capacity to offer individually tailored products or services using mass production resources..

megahertz A measure of cycle speed, or the pacing of events in a computer; one megahertz equals one million cycles per second.

menu costs Merchants' costs of changing prices.

metric A standard measurement of performance.

metropolitan area network (MAN) Network that spans a metropolitan area, usually a city and its major suburbs. Its geographic scope falls between a WAN and a LAN.

microbrowser Web browser software with a small file size that can work with low-memory constraints, tiny screens of handheld wireless devices, and low bandwidth of wireless networks.

micropayment Payment for a very small sum of money, often less than $10.

microprocessor Very large scale integrated circuit technology that integrates the computer's memory, logic, and control on a single chip.

microwave A high-volume, long-distance, point-to-point transmission in which high-frequency radio signals are transmitted through the atmosphere from one terrestrial transmission station to another.

middle management People in the middle of the organizational hierarchy who are responsible for carrying out the plans and goals of senior management.

midrange computer Middle-size computer that is capable of supporting the computing needs of smaller organizations or of managing networks of other computers.

minicomputer Middle-range computer used in systems for universities, factories, or research laboratories.

MIS audit Identifies all the controls that govern individual information systems and assesses their effectiveness.

mobile commerce (m-commerce) The use of wireless devices, such as cell phones or handheld digital information appliances, to conduct both business-to-consumer and business-to-business e-commerce transactions over the Internet.

mobile wallets (m-wallets) Store m-commerce shoppers' personal information and credit card numbers to expedite the purchase process.

moblog Specialized blog featuring photos with captions posted from mobile phones.

model An abstract representation that illustrates the components or relationships of a phenomenon.

model-driven DSS Primarily stand-alone system that uses some type of model to perform "what-if" and other kinds of analyses.

modem A device for translating a computer's digital signals into analog form for transmission over ordinary telephone lines, or for translating analog signals back into digital form for reception by a computer.

module A logical unit of a program that performs one or several functions.

Moore's Law Assertion that the number of components on a chip doubles each year

MP3 (MPEG3) Compression standard that can compress audio files for transfer over the Internet with virtually no loss in quality.

multicore processor Integrated circuit to which two or more processors have been attached for enhanced performance, reduced power consumption and more efficient simultaneous processing of multiple tasks.

multimedia The integration of two or more types of media such as text, graphics, sound, voice, full-motion video, or animation into a computer-based application.

multinational Form of business organization that concentrates financial management and control out of a central home base while decentralizing

multiplexing Ability of a single communications channel to carry data transmissions from multiple sources simultaneously.

multitiered (N-tier) client/server architecture Client/server network which the work of the entire network is balanced over several different levels of servers.

nanotechnology Technology that builds structures and processes based on the manipulation of individual atoms and molecules.

natural language Nonprocedural language that enables users to communicate with the computer using conversational commands resembling human speech.

net marketplace A single digital marketplace based on Internet technology linking many buyers to many sellers.

netbook Small low-cost, lightweight subnotebook optimized for wireless communication and Internet access.

network The linking of two or more computers to share data or resources, such as a printer.

network economics Model of strategic systems at the industry level based on the concept of a network where adding another participant entails zero marginal costs but can create much larger marginal gains.

network interface card (NIC) Expansion card inserted into a computer to enable it to connect to a network.

network operating system (NOS) Special software that routes and manages communications on the network and coordinates network resources.

networking and telecommunications technology Physical devices and software that link various computer hardware components and transfer data from one physical location to another.

neural network Hardware or software that attempts to emulate the processing patterns of the biological brain.

nonobvious relationship awareness (NORA) Technology that can find obscure hidden connections between people or other entities by analyzing information from many different sources to correlate relationships.

normalization The process of creating small stable data structures from complex groups of data when designing a relational database.

object Software building block that combines data and the procedures acting on the data.

object-oriented DBMS An approach to data management that stores both data and the procedures acting on the data as objects that can be automatically retrieved and shared; the objects can contain multimedia.

object-oriented development Approach to systems development that uses the object as the basic unit of systems analysis and design. The system is modeled as a collection o objects and the relationship between them.

object-oriented programming An approach to software development that combines data and procedures into a single object.

object-relational DBMS A database management system that combines the capabilities of a relational DBMS for storing traditional information and the capabilities of an object-oriented DBMS for storing graphics and multimedia.

Office 2010 The latest version of Microsoft desktop software suite with capabilities for supporting collaborative work on the Web or incorporating information from the Web into documents.

offshore outsourcing Outsourcing systems development work or maintenance of existing systems to external vendors in another country.

on-demand computing Firms off-loading peak demand for computing power to remote, large-scale data processing centers, investing just enough to handle average processing loads and paying for only as much additional computing power as the market demands. Also called utility computing.

on-line analytical processing (OLAP) Capability for manipulating and analyzing large volumes of data from multiple perspectives.

online processing A method of collecting and processing data in which transactions are entered directly into the computer system and processed immediately.

online transaction processing Transaction processing mode in which transactions entered on-line are immediately processed by the computer.

open-source software Software that provides free access to its program code, allowing users to modify the program code to make improvements or fix errors.

operating system Software that manages the resources and activities of the computer.

operational CRM Customer-facing applications, such as sales force automation, call center and customer service support, and marketing automation.

operational management People who monitor the day-to-day activities of the organization.

operational-level systems Information systems that monitor the elementary activities and transactions of the organization.

opt-in Model of informed consent permitting prohibiting an organization from collecting any personal information unless the individual specifically takes action to approve information collection and use.

opt-out Model of informed consent permitting the collection of personal information until the consumer specifically requests that the data not be collected.

organization (behavioral definition) A collection of rights, privileges, obligations, and responsibilities that are delicately balanced over a period of time through conflict and conflict resolution.

organization (technical definition) A stable, formal, social structure that takes resources from the environment and processes them to produce outputs.

organizational and management capital Investments in organization and management such as new business processes, management behavior, organizational culture, or training.

organizational impact analysis Study of the way a proposed system will affect organizational structure, attitudes, decision making, and operations.

organizational learning Creation of new standard operating procedures and business processes that reflect organizations' experience.

output The distribution of processed information to the people who will use it or to the activities for which it will be used.

outsourcing The practice of contracting computer center operations, telecommunications networks, or applications development to external vendors.

P3P Industry standard designed to give users more control over personal information gathered on Web sites they visit. Stands for Platform for Privacy Preferences Project.

packet switching Technology that breaks messages into small, fixed bundles of data and routes them in the most economical way through any available communications channel..

paradigm shift Radical reconceptualization of the nature of the business and the nature of the organization.

parallel strategy A safe and conservative conversion approach where both the old system and its potential replacement are run together for a time until everyone is assured that the new one functions correctly.

particularism Making judgments and taking action on the basis of narrow or personal characteristics, in all its forms (religious, nationalistic, ethnic, regionalism, geopolitical position).

partner relationship management (PRM) Automation of the firm's relationships with its selling partners using customer data and analytical tools to improve coordination and customer sales.

patch Small pieces of software to repair the software flaws without disturbing the proper operation of the software.

patent A legal document that grants the owner an exclusive monopoly on the ideas behind an invention for 17 years; designed to ensure that inventors of new machines or methods are rewarded for their labor while making widespread use of their inventions.

peer-to-peer Network architecture that gives equal power to all computers on the network; used primarily in small networks.

personal area network (PAN) Computer network used for communication among digital devices (including telephones and PDAs) that are close to one person.

personal digital assistant (PDA) Small, pen-based, handheld computer with built-in wireless telecommunications capable of entirely digital communications transmission.

personalization Ability of merchants to target marketing messages to specific individuals by adjusting the message for a person's name, interests, and past purchases.

PERT chart Network diagram depicting project tasks and their interrelationships.

pharming Phishing technique that redirects users to a bogus Web page, even when an individual enters the correct Web page address.

phased approach Introduces the new system in stages either by functions or by organizational units.

phishing Form of spoofing involving setting up fake Web sites or sending e-mail messages that resemble those of legitimate businesses that ask users for confidential personal data.

pilot study A strategy to introduce the new system to a limited area of the organization until it is proven to be fully functional; only then can the conversion to the new system across the entire organization take place.

pivot table Spreadsheet tool for reorganizing and summarizing two or more dimensions of data in a tabular format.

podcasting Publishing audio broadcasts via the Internet so that subscribing users can download audio files onto their personal computers or portable music players.

pop-up ad Ad that opens automatically and does not disappear until the user clicks on it.

portal Web interface for presenting integrated personalized content from a variety of sources. Also refers to a Web site service that provides an initial point of entry to the Web.

portfolio analysis An analysis of the portfolio of potential applications within a firm to determine the risks and benefits, and to select among alternatives for information systems.

post-implementation audit Formal review process conducted after a system has been placed in production to determine how well the system has met its original objectives.

prediction markets An analysis of the portfolio of potential applications within a firm to determine the risks and benefits, and to select among alternatives for information systems.

predictive analytics The use of data mining techniques, historical data, and assumptions about future conditions to predict outcomes of events, such as the probability a customer will respond to an offer or purchase a specific product.

price discrimination Selling the same goods, or nearly the same goods, to different targeted groups at different prices.

price transparency The ease with which consumers can find out the variety of prices in a market.

primary activities Activities most directly related to the production and distribution of a firm's products or services.

primary key Unique identifier for all the information in any row of a database table.

privacy The claim of individuals to be left alone, free from surveillance or interference from other individuals, organizations, or the state.

private cloud A proprietary network or a data center that ties together servers, storage, networks, data, and applications as a set of virtualized services that are shared by users inside a company.

private exchange Another term for a private industrial network.

private industrial networks Web-enabled networks linking systems of multiple firms in an industry for the coordination of trans-organizational business processes.

process specifications Describe the logic of the processes occurring within the lowest levels of a data flow diagram.

processing The conversion, manipulation, and analysis of raw input into a form that is more meaningful to humans.

procurement Sourcing goods and materials, negotiating with suppliers, paying for goods, and making delivery arrangements.

product differentiation Competitive strategy for creating brand loyalty by developing new and unique products and services that are not easily duplicated by competitors.

production The stage after the new system is installed and the conversion is complete; during this time the system is reviewed by users and technical specialists to determine how well it has met its original goals.

production or service workers People who actually produce the products or services of the organization.

profiling The use of computers to combine data from multiple sources and create electronic dossiers of detailed information on individuals.

program-data dependence The close relationship between data stored in files and the software programs that update and maintain those files. Any change in data organization or format requires a change in all the programs associated with those files.

programmers Highly trained technical specialists who write computer software instructions.

programming The process of translating the system specifications prepared during the design stage into program code.

project Planned series of related activities for achieving a specific business objective.

project management Application of knowledge, tools, and techniques to achieve specific targets within a specified budget and time period.

protocol A set of rules and procedures that govern transmission between the components in a network.

prototype The preliminary working version of an information system for demonstration and evaluation purposes.

prototyping The process of building an experimental system quickly and inexpensively for demonstration and evaluation so that users can better determine information requirements.

public cloud A cloud maintained by an external service provider, accessed through the Internet, and available to the general public.

public key encryption Uses two keys: one shared (or public) and one private.

public key infrastructure(PKI) System for creating public and private keys using a certificate authority (CA) and digital certificates for authentication.

pull-based model Supply chain driven by actual customer orders or purchases so that members of the supply chain produce and deliver only what customers have ordered.

pure-play Business models based purely on the Internet.

push-based model Supply chain driven by production master schedules based on forecasts or best guesses of demand for products, and products are "pushed" to customers.

query language Software tool that provides immediate online answers to requests for information that are not predefined.

radio-frequency identification (RFID) Technology using tiny tags with embedded microchips containing data about an item and its location to transmit short-distance radio signals to special RFID readers that then pass the data on to a computer for processing.

Rapid Application Development (RAD) Process for developing systems in a very short time period by using prototyping, fourth-generation tools, and close teamwork among users and systems specialists.

rational model Model of human behavior based on the belief that people, organizations, and nations engage in basically consistent, value-maximizing calculations.

rationalization of procedures The streamlining of standard operating procedures, eliminating obvious bottlenecks, so that automation makes operating procedures more efficient.

real options pricing models Models for evaluating information technology investments with uncertain returns by using techniques for valuing financial options.

record A group of related fields.

recovery-oriented computing Computer systems designed to recover rapidly when mishaps occur.

referential integrity Rules to ensure that relationships between coupled database tables remain consistent.

relational DBMS A type of logical database model that treats data as if they were stored in two-dimensional tables. It can relate data stored in one table to data in another as long as the two tables share a common data element.

Repetitive Stress Injury (RSI) Occupational disease that occurs when muscle groups are forced through repetitive actions with high-impact loads or thousands of repetitions with low-impact loads.

Request for Proposal (RFP) A detailed list of questions submitted to vendors of software or other services to determine how well the vendor's product can meet the organization's specific requirements.

resource allocation The determination of how costs, time, and personnel are assigned to different phases of a systems development project.

responsibility Accepting the potential costs, duties, and obligations for the decisions one makes.

revenue model A description of how a firm will earn revenue, generate profits, and produce a return on investment.

richness Measurement of the depth and detail of information that a business can supply to the customer as well as information the business collects about the customer.

ring topology A network topology in which all computers are linked by a closed loop in a manner that passes data in one direction from one computer to another.

risk assessment Determining the potential frequency of the occurrence of a problem and the potential damage if the problem were to occur. Used to determine the cost/benefit of a control.

Risk Aversion Principle Principle that one should take the action that produces the least harm or incurs the least cost.

router Specialized communications processor that forwards packets of data from one network to another network.

routines Precise rules, procedures and practices that have been developed to cope with expected situations.

RSS Technology using aggregator software to pull content from Web sites and feed it automatically to subscribers' computers.

SaaS (Software as a Service) Services for delivering and providing access to software remotely as a Web-based service.

safe harbor Private self-regulating policy and enforcement mechanism that meets the objectives of government regulations but does not involve government regulation or enforcement.

Sarbanes-Oxley Act Law passed in 2002 that imposes responsibility on companies and their management to protect investors by safeguarding the accuracy and integrity of financial information that is used internally and released externally.

scalability The ability of a computer, product, or system to expand to serve a larger number of users without breaking down.

scope Defines what work is and is not included in a project.

scoring model A quick method for deciding among alternative systems based on a system of ratings for selected objectives.

search costs The time and money spent locating a suitable product and determining the best price for that product.

search engine A tool for locating specific sites or information on the Internet.

search engine marketing Use of search engines to deliver in their results sponsored links, for which advertisers have paid.

search engine optimization (SEO) the process of changing a Web site's content, layout, and format in order to increase the ranking of the site on popular search engines, and to generate more site visitors.

Secure Hypertext Transfer Protocol (S-HTTP) Protocol used for encrypting data flowing over the Internet; limited to individual messages.

Secure Sockets Layer (SSL) Enables client and server computers to manage encryption and decryption activities as they communicate with each other during a secure Web session.

security Policies, procedures, and technical measures used to prevent unauthorized access, alteration, theft, or physical damage to information systems.

security policy Statements ranking information risks, identifying acceptable security goals, and identifying the mechanisms for achieving these goals.

Semantic Web Ways of making the Web more "intelligent," with machine-facilitated understanding of information so that searches can be more intuitive, effective, and executed using intelligent software agents.

semistructured decisions Decisions in which only part of the problem has a clear-cut answer provided by an accepted procedure.

senior management People occupying the topmost hierarchy in an organization who are responsible for making long-range decisions.

sensitivity analysis Models that ask "what-if" questions repeatedly to determine the impact of changes in one or more factors on the outcomes.

server Computer specifically optimized to provide software and other resources to other computers over a network.

server farm Large group of servers maintained by a commercial vendor and made available to subscribers for electronic commerce and other activities requiring heavy use of servers.

service level agreement (SLA) Formal contract between customers and their service providers that defines the specific responsibilities of the service provider and the level of service expected by the customer.

service platform Integration of multiple applications from multiple business functions, business units, or business partners to deliver a seamless experience for the customer, employee, manager, or business partner.

service-oriented architecture Software architecture of a firm built on a collection of software programs that communicate with each other to perform assigned tasks to create a working software application

shopping bot Software with varying levels of built-in intelligence to help electronic commerce shoppers locate and evaluate products or service they might wish to purchase.

Simple Object Access Protocol (SOAP) Set of rules that allows Web services applications to pass data and instructions to one another.

six sigma A specific measure of quality, representing 3.4 defects per million opportunities; used to designate a set of methodologies and techniques for improving quality and reducing costs.

smart card A credit-card-size plastic card that stores digital information and that can be used for electronic payments in place of cash.

smartphone Wireless phone with voice, text, and Internet capabilities.

sniffer Type of eavesdropping program that monitors information traveling over a network.

social bookmarking Capability for users to save their bookmarks to Web pages on a public Web site and tag these bookmarks with keywords to organize documents and share information with others.

social engineering Tricking people into revealing their passwords by pretending to rrbe legitimate users or members of a company in need of information.

social networking sites Online community for expanding users' business or social contacts by making connections through their mutual business or personal connections.

social shopping Use of Web sites featuring user-created Web pages to share knowledge about items of interest to other shoppers.

sociotechnical design Design to produce information systems that blend technical efficiency with sensitivity to organizational and human needs.

sociotechnical view Seeing systems as composed of both technical and social elements.

software localization Process of converting software to operate in a second language.

software package A prewritten, precoded, commercially available set of programs that eliminates the need to write software programs for certain functions.

spam Unsolicited commercial e-mail.

spamming Form of abuse in which thousands and even hundreds of thousands of unsolicited e-mail and electronic messages are sent out, creating a nuisance for both businesses and individual users.

spyware Technology that aids in gathering information about a person or organization without their knowledge.

SQL injection attack Attacks against a Web site that take advantage of vulnerabilities in poorly coded SQL (a standard and common database software application) applications in order to introduce malicious program code into a company's systems and networks.

star topology A network topology in which all computers and other devices are connected to a central host computer. All communications between network devices must pass through the host computer.

storage area network (SAN) A high-speed network dedicated to storage that connects different kinds of storage devices, such as tape libraries and disk arrays so they can be shared by multiple servers.

storage technology Physical media and software governing the storage and organization of data for use in an information system.

stored value payment systems Systems enabling consumers to make instant on-line payments to merchants and other individuals based on value stored in a digital account.

strategic information systems Computer systems at any level of the organization that change goals, operations, products, services, or environmental relationships to help the organization gain a competitive advantage.

strategic transitions A movement from one level of sociotechnical system to another. Often required when adopting strategic systems that demand changes in the social and technical elements of an organization.

streaming A publishing method for music and video files that flows a continuous stream of content to a user's device without being stored locally on the device.

structure chart System documentation showing each level of design, the relationship among the levels, and the overall place in the design structure; can document one program, one system, or part of one program.

structured Refers to the fact that techniques are carefully drawn up, step by step, with each step building on a previous one.

structured decisions Decisions that are repetitive, routine, and have a definite procedure for handling them.

structured knowledge Knowledge in the form of structured documents and reports.

Structured Query Language (SQL) The standard data manipulation language for relational database management systems.

supply chain Network of organizations and business processes for procuring materials, transforming raw materials into intermediate and finished products, and distributing the finished products to customers.

supply chain execution systems Systems to manage the flow of products through distribution centers and warehouses to ensure that products are delivered to the right locations in the most efficient manner.

supply chain management Integration of supplier, distributor, and customer logistics requirements into one cohesive process.

supply chain management systems Information systems that automate the flow of information between a firm and its suppliers in order to optimize the planning, sourcing, manufacturing, and delivery of products and services.

supply chain planning systems Systems that enable a firm to generate demand forecasts for a product and to develop sourcing and manufacturing plans for that product.

support activities Activities that make the delivery of a firm's primary activities possible. Consist of the organization's infrastructure, human resources, technology, and procurement.

switch Device to connect network components that has more intelligence than a hub and can filter and forward data to a specified destination.

switching costs The expense a customer or company incurs in lost time and expenditure of resources when changing from one supplier or system to a competing supplier or system.

syndicator Business aggregating content or applications from multiple sources, packaging them for distribution, and reselling them to third-party Web sites.

system testing Tests the functioning of the information system as a whole in order to determine if discrete modules will function together as planned.

systems analysis The analysis of a problem that the organization will try to solve with an information system.

systems analysts Specialists who translate business problems and requirements into information requirements and systems, acting as liaison between the information systems department and the rest of the organization.

systems design Details how a system will meet the information requirements as determined by the systems analysis.

systems development The activities that go into producing an information systems solution to an organizational problem or opportunity.

systems life cycle A traditional methodology for developing an information system that partitions the systems development process into formal stages that must be completed sequentially with a very formal division of labor between end users and information systems specialists.

T lines High-speed data lines leased from communications providers, such as T-1 lines (with a transmission capacity of 1.544 Mbps).

tacit knowledge Expertise and experience of organizational members that has not been formally documented.

tangible benefits Benefits that can be quantified and assigned a monetary value; they include lower operational costs and increased cash flows.

taxonomy Method of classifying things according to a predetermined system.

teams Teams are formal groups whose members collaborate to achieve specific goals.

teamware Group collaboration software that is customized for teamwork.

technology standards Specifications that establish the compatibility of products and the ability to communicate in a network.

technostress Stress induced by computer use; symptoms include aggravation, hostility toward humans, impatience, and enervation.

telecommunications system A collection of compatible hardware and software arranged to communicate information from one location to another.

teleconferencing The ability to confer with a group of people simultaneously using the telephone or electronic-mail group communication software.

telepresence Telepresence is a technology that allows a person to give the appearance of being present at a location other than his or her true physical location.

Telnet Network tool that allows someone to log on to one computer system while doing work on another.

test plan Prepared by the development team in conjunction with the users; it includes all of the preparations for the series of tests to be performed on the system.

testing The exhaustive and thorough process that determines whether the system produces the desired results under known conditions.

text mining Discovery of patterns and relationships from large sets of unstructured data.

token Physical device similar to an identification card that is designed to prove the identity of a single user.

touch point Method of firm interaction with a customer, such as telephone, e-mail, customer service desk, conventional mail, or point-of-purchase.

topology The way in which the components of a network are connected.

Total Cost of Ownership (TCO) Designates the total cost of owning technology resources, including initial purchase costs, the cost of hardware and software upgrades, maintenance, technical support, and training.

Total Quality Management (TQM) A concept that makes quality control a responsibility to be shared by all people in an organization.

trade secret Any intellectual work or product used for a business purpose that can be classified as belonging to that business, provided it is not based on information in the public domain.

transaction costs Costs incurred when a firm buys on the marketplace what it cannot make itself.

transaction cost theory Economic theory stating that firms grow larger because they can conduct marketplace transactions internally more cheaply than they can with external firms in the marketplace.

transaction fee revenue model An online e-commerce revenue model where the firm receives a fee for enabling or executing transactions.

transaction processing systems (TPS) Computerized systems that perform and record the daily routine transactions necessary to conduct the business; they serve the organization's operational level.

transborder data flow The movement of information across international boundaries in any form.

Transmission Control Protocol/Internet Protocol (TCP/IP) Dominant model for achieving connectivity among different networks. Provides a universally agree-on method for breaking up digital messages into packets, routing them to the proper addresses, and then reassembling them into coherent messages.

transnational Truly global form of business organization with no national headquarters; value-added activities are managed from a global perspective without reference to national borders, optimizing sources of supply and demand and local competitive advantage.

Trojan horse A software program that appears legitimate but contains a second hidden function that may cause damage.

tuple A row or record in a relational database.

twisted wire A transmission medium consisting of pairs of twisted copper wires; used to transmit analog phone conversations but can be used for data transmission.

Unified communications Integrates disparate channels for voice communications, data communications, instant messaging, e-mail, and electronic conferencing into a single experience where users can seamlessly switch back and forth between different communication modes.

Unified Modeling Language (UML) Industry standard methodology for analysis and design of an object-oriented software system.

unified threat management (UTM) Comprehensive security management tool that combines multiple security tools, including firewalls, virtual private networks, intrusion detection systems, and Web content filtering and anti-spam software.

uniform resource locator (URL) The address of a specific resource on the Internet.

unit testing The process of testing each program separately in the system. Sometimes called program testing.

UNIX Operating system for all types of computers, which is machine independent and supports multiuser processing, multitasking, and networking. Used in high-end workstations and servers.

unstructured decisions Nonroutine decisions in which the decision maker must provide judgment, evaluation, and insights into the problem definition; there is no agreed-upon procedure for making such decisions.

Usenet Forums in which people share information and ideas on a defined topic through large electronic bulletin boards where anyone can post messages on the topic for others to see and to which others can respond.

user interface The part of the information system through which the end user interacts with the system; type of hardware and the series of on-screen commands and responses required for a user to work with the system.

user-designer communications gap The difference in backgrounds, interests, and priorities that impede communication and problem solving among end users and information systems specialists.

Utilitarian Principle Principle that assumes one can put values in rank order and understand the consequences of various courses of action.

utility computing Model of computing in which companies pay only for the information technology resources they actually use during a specified time period. Also called on-demand computing or usage-based pricing.

value chain model Model that highlights the primary or support activities that add a margin of value to a firm's products or services where information systems can best be applied to achieve a competitive advantage.

value web Customer-driven network of independent firms who use information technology to coordinate their value chains to collectively produce a product or service for a market.

Value-Added Network (VAN) Private, multipath, data-only, third-party-managed network that multiple organizations use on a subscription basis.

videoconferencing Teleconferencing in which participants see each other over video screens.

virtual company Organization using networks to link people, assets and ideas to create and distribute products and services without being limited to traditional organizational boundaries or physical location.

Virtual Private Network (VPN) A secure connection between two points across the Internet to transmit corporate data. Provides a low-cost alternative to a private network.

Virtual Reality Modeling Language (VRML) A set of specifications for interactive three-dimensional modeling on the World Wide Web.

virtual reality systems Interactive graphics software and hardware that create computer-generated simulations that provide sensations that emulate real-world activities.

virtual world Computer-based simulated environment intended for its users to inhabit and interact via graphical representations called avatars.

virtualization Presenting a set of computing resources so that they can all be accessed in ways that are not restricted by physical configuration or geographic location.

Voice over IP (VoIP) Facilities for managing the delivery of voice information using the Internet Protocol (IP).

war driving Technique in which eavesdroppers drive by buildings or park outside and try to intercept wireless network traffic.

Web 2.0 Second-generation, interactive Internet-based services that enable people to collaborate, share information, and create new services online, including mashups, blogs, RSS, and wikis.

Web 3.0 Future vision of the Web where all digital information is woven together with intelligent search capabilities.

Web beacons Tiny objects invisibly embedded in e-mail messages and Web pages that are designed to monitor the behavior of the user visiting a Web site or sending e-mail.

Web browser An easy-to-use software tool for accessing the World Wide Web and the Internet.

Web bugs Tiny graphic files embedded in e-mail messages and Web pages that are designed to monitor online Internet user behavior.

Web hosting service Company with large Web server computers to maintain the Web sites of fee-paying subscribers.

Web mining Discovery and analysis of useful patterns and information from the World Wide Web.

Web server Software that manages requests for Web pages on the computer where they are stored and that delivers the page to the user's computer.

Web services Set of universal standards using Internet technology for integrating different applications from different sources without time-consuming custom coding. Used for linking systems of different organizations or for linking disparate systems within the same organization.

Web site All of the World Wide Web pages maintained by an organization or an individual.

Wi-Fi Standards for Wireless Fidelity and refers to the 802.11 family of wireless networking standards.

Wide Area Network (WAN) Telecommunications network that spans a large geographical distance. May consist of a variety of cable, satellite, and microwave technologies.

wiki Collaborative Web site where visitors can add, delete, or modify content, including the work of previous authors.

WiMax Popular term for IEEE Standard 802.16 for wireless networking over a range of up to 31 miles with a data transfer rate of up to 75 Mbps. Stands for Worldwide Interoperability for Microwave Access.

Windows Microsoft family of operating systems for both network servers and client computers. The most recent version is Windows Vista.

Windows 7 The successor to Microsoft Windows Vista operating system released in 2009.

Wintel PC Any computer that uses Intel microprocessors (or compatible processors) and a Windows operating system.

wireless portals Portals with content and services optimized for mobile devices to steer users to the information they are most likely to need.

wireless sensor networks (WSNs) Networks of interconnected wireless devices with built-in processing, storage, and radio frequency sensors and antennas that are embedded into the physical environment to provide measurements of many points over large spaces.

wisdom The collective and individual experience of applying knowledge to the solution of problems.

wisdom of crowds The belief that large numbers of people can make better decisions about a wide range of topics or products than a single person or even a small committee of experts (first proposed in a book by James Surowiecki).

WML (Wireless Markup Language) Markup language for Wireless Web sites; based on XML and optimized for tiny displays.

workflow management The process of streamlining business procedures so that documents can be moved easily and efficiently from one location to another.

World Wide Web A system with universally accepted standards for storing, retrieving, formatting, and displaying information in a networked environment.

worms Independent software programs that propagate themselves to disrupt the operation of computer networks or destroy data and other programs.

XML (eXtensible Markup Language) General-purpose language that describes the structure of a document and supports links to multiple documents, allowing data to be manipulated by the computer. Used for both Web and non-Web applications.

Photo and Screen Shot Credits

Index

Name Index

A

Abdulmutallab, Umar Farouk, 242
Aleynikov, Sergey, 300
Anderson, John, 241

B

Benioff, Marc, 203, 204
Berman, Craig, 411
Berson, Anthony M., 157
Bertarelli, Ignacio, 41
Bewkes, Jeff, 119
Bezos, Jeff, 411
Bratton, William, 482
Brin, Sergey, 270

C

Casey, Jim, 22
Childs, Timothy, 9
Columbus, Christopher, 8

D

Deming, W. Edwards, 490
DeWalt, David, 305
DiSisto, Rob, 205

E

Ellison, Larry, 41, 42

F

Fathi, David, 241
Fayol, Henri, 458
Filo, David, 270
Fine, Glenn A., 242
Flax, Daniel, 321
Fletcher, Clyde, 575
Ford, Henry, 57
Friedman, Thomas, 8

G

Gates, Bill, 57
Gates, Robert, 330
Gonzalez, Alberto, 301
Gosling, James, 188

H

Hawiger, Marcel, 37
Heltsley, Laurie, 76
Hicks, Michael, 241
Horan, Tim, 119

I

Iqbal, Asif, 241

J

Jensen, Robert, 581
Jeppsen, Bryan, 227
Jerome-Parks, Scott, 157, 158

Jn-Charles, Alexandra, 157, 158
Jobs, Steve, 57, 103, 178, 288
Juran, Joseph, 490

K

Kalach, Nina, 157
Kant, Immanuel, 130
Kearns, Michael, 439
Kelly, Chris, 390
Kennedy, Ted, 241
Kilar, Jason, 118
King, Stephen, 411

L

Lafley, A.G., 75
Leavitt, 93
Levy, Scott, 9

M

McConnell, Mike, 330
McDonald, Robert, 75
McPherson, Barry, 305
Merzenich, Michael, 152
Mettler, Fred A., Jr., 159
Mintzberg, 88, 459
Monaghan, Tom, 52
Moore, Gordon, 170
Mullin, Mike, 321
Myrtle, Raymond, 470

N

Neeleman, David, 556
Nordstrom, David, 429

O

Obama, Barack, 469
Oliphant, Jack, 37
Ophir, Eyal, 152
Orwell, George, 144
Oxley, Michael, 306

P

Page, Larry, 270
Peterson, Bud, 36–37
Peterson, Val, 36–37
Porter, Michael, 95, 115

R

Reed, Alan, 178
Ricci, Ron, 3
Ryan, Claude, 22

S

Sarbanes, Paul, 306
Schueller, Joe, 75
Shahzad, Faisal, 242
Skaare, Jerry, 358, 359

T

Taleb, Nassim, 440

Tijerina, Louis, 147
Tubbs, Jerry, 548

V

Vazquez, Raul, 411

W

Weast, Jerry, 469
Woo, Edward, 119

Y

Yang, Jerry, 270

Z

Zimmerman, Richard, 345, 346
Zuckerberg, Mark, 390

Organizations Index

A

A&E, 118
ABC Inc., 118
Accenture, 62, 92, 181
Acxiom, 121
Adobe, 180, 404
Advanced Micro Design, 8
Aetna, 523
Agora Games, 321
AIC, 494
Air France, 65
Airborne Express, 22
Ajax Boiler, 266
Alcatel-Lucent, 180, 400, 401
Alcoa, 339–340
Allot Communications, 319
Alta Vista, 87
Amazon, 12, 45, 94, 98, 102, 103, 106, 111, 135,
    170, 183, 184, 204, 262, 301, 321, 374, 383,
    384, 386, 388, 389, 401, 404, 410, 411
AMD, 176
America Online (AOL), 118, 137, 382
America West, 113
American Airlines, 111
American Express, 13, 100
American Library Association, 262
American National Insurance Co. (ANCO), 494
AmerisourceBergen, 495
Android, Inc., 288
Ann Taylor, 109
Apple, 7, 9, 13, 14, 57, 79, 87, 94, 97, 103–104,
    110, 111, 140, 168, 178, 185, 204, 287–289,
    295, 388
Applebee's, 478
Armani Exchange, 401
Art Technology Group, 405
AstraTech, 181
AstraZeneca, 123
AT&T, 9, 79–80, 87, 110, 118, 180, 262, 276, 401

Subject Index

REVIEWERS AND CONSULTANTS

CONSULTANTS

AUSTRALIA
Robert MacGregor, *University of Wollongong*
Alan Underwood, *Queensland University of Technology*

CANADA
Wynne W. Chin, *University of Calgary*
Len Fertuck, *University of Toronto*
Robert C. Goldstein, *University of British Columbia*
Rebecca Grant, *University of Victoria*
Kevin Leonard, *Wilfrid Laurier University*
Anne B. Pidduck, *University of Waterloo*

GERMANY
Lutz M. Kolbe, *University of Göttingen*
Detlef Schoder, *University of Cologne*

GREECE
Anastasios V. Katos, *University of Macedonia*

HONG KONG
Enoch Tse, *Hong Kong Baptist University*

INDIA
Sanjiv D. Vaidya, *Indian Institute of Management, Calcutta*

ISRAEL
Phillip Ein-Dor, *Tel-Aviv University*
Peretz Shoval, *Ben Gurion University*

MEXICO
Noe Urzua Bustamante, *Universidad Tecnológica de México*

NETHERLANDS
E.O. de Brock, *University of Groningen*
Theo Thiadens, *University of Twente*
Charles Van Der Mast, *Delft University of Technology*

PUERTO RICO,
Commonwealth of the United States
Brunilda Marrero, *University of Puerto Rico*

SOUTH AFRICA
Daniel Botha, *University of Stellenbosch*

SWEDEN
Mats Daniels, *Uppsala University*

SWITZERLAND
Andrew C. Boynton, *International Institute for Management Development*
Walter Brenner, *University of St. Gallen*
Donald A. Marchand, *International Institute for Management Development*

UNITED KINGDOM

ENGLAND
G.R. Hidderley, *University of Central England, Birmingham*
Christopher Kimble, *University of York*
Jonathan Liebenau, *London School of Economics and Political Science*
Kecheng Liu, *Staffordshire University*

SCOTLAND
William N. Dyer, *Falkirk College of Technology*

UNITED STATES OF AMERICA
Tom Abraham, *Kean University*
Evans Adams, *Fort Lewis College*
Kamal Nayan Agarwal, *Howard University*
Roy Alvarez, *Cornell University*
Chandra S. Amaravadi, *Western Illinois University*
Beverly Amer, *Northern Arizona University*
John Anderson, *Northeastern State University*
Arben Asllani, *Faytetteville State University*
David Bahn, *Metropolitan State University of Minnesota*
Rahul C. Basole, *Georgia Institute of Technology*
Jon W. Beard, *University of Richmond*
Patrick Becka, *Indiana University Southeast*
Michel Benaroch, *Syracuse University*
Cynthia Bennett, *University of Arkansas at Pine Bluff*
David Bradbard, *Winthrop University*
Nancy Brome, *Southern NH University*
Kimberly Cass, *University of Redlands*
Jason Chen, *Gonzaga University*
P. C. Chu, *Ohio State University, Columbus*
Kungwen Chu, *Purdue University, Calumet*
Richard Clemens, *West Virginia Wesleyan College*
Lynn Collen, *St. Cloud State University*
Daniel Connolly, *University of Denver*
Jakov Crnkovic, *SUNY Albany*
Albert Cruz, *National University*
John Dalphin, *SUNY Potsdam*
Marica Deeb, *Waynesburg College*
William DeLone, *American University*
Vasant Dhar, *New York University*
Cindy Drexel, *Western State College of Colorado*
Warren W. Fisher, *Stephen F. Austin State University*
William B. Fredenberger, *Valdosta State University*
Bob Fulkerth, *Golden Gate University*